MW00612718

An Introduction to Linear Control Systems

CONTROL AND SYSTEMS THEORY

A Series of Monographs and Textbooks

Volume 1: Discrete Techniques of Parameter Estimation: The Equation Error Formulation, JERRY M. MENDEL

Volume 2: Mathematical Description of Linear Systems, WILSON J. RUGH

Volume 3: The Qualitative Theory of Optimal Processes, R. GABASOV *and* F. KIRILLOVA

Volume 4: Self-Organizing Control of Stochastic Systems, GEORGE N. SARIDIS

Volume 5: An Introduction to Linear Control Systems, THOMAS E. FORTMANN *and* KONRAD L. HITZ

OTHER VOLUMES IN PREPARATION

An Introduction to Linear Control Systems

THOMAS E. FORTMANN
Bolt Beranek and Newman Inc.
Cambridge, Massachusetts

KONRAD L. HITZ
Mechanical Engineering Department
University of Newcastle
Newcastle, New South Wales
Australia

MARCEL DEKKER, INC. New York and Basel

Library of Congress Cataloging in Publication Data

Fortmann, Thomas E.
An introduction to linear control systems

(Control and systems theory ; v. 5)
Bibliography: p.
Includes index.
1. Automatic control. 2. Control theory.
I. Hitz, Konrad L., joint author. II. Title.
TJ213.F62 629.8′32 76-29499
ISBN 0-8247-6512-5

MARCEL DEKKER, INC.

270 Madison Avenue, New York, New York 10016

Current Printing (last digit):

10 9 8 7 6 5 4 3

PRINTED IN THE UNITED STATES OF AMERICA

To my parents, Mary and Dan Fortmann

—T. E. F.

In memory of my father, Conrad Hitz

—K. L. H.

CONTENTS

Part IV: Related Topics

PREFACE

This book is intended for a two-semester introductory course in control systems at the advanced undergraduate or graduate level. It assumes as a prerequisite some familiarity with differential equations, linear algebra, and Laplace transforms. With appropriate omissions it is also suitable for shorter courses, and it is sufficiently self-contained to be used for self-study. Students in all branches of engineering and in mathematics should feel equally comfortable with it. At the University of Newcastle, for example, it forms the basis for a two-semester junior-year engineering sequence in which students may choose to take one or both semesters. It is also being used in draft form at M.I.T. for a senior/graduate level course.

One author (K. L. H.) combines several years' experience in the design and implementation of classical control systems with current research interests involving modern control and optimization theory. The other (T. E. F.) has a more theoretical background spanning modern control, optimization, estimation, and computer systems. This text grew out of two jointly held convictions about the teaching of control.

First, the linear theory and many of the basic analysis and synthesis techniques underlying both modern and classical control systems can and should be integrated and taught jointly. Moreover, much of what has been called "advanced" modern state-space analysis is no more mathematically advanced than the traditional transform approaches and can be safely included in an introductory course (a rigorous proof of Nyquist's theorem, for example, requires mathematics every bit as sophisticated as does a derivation of the Jordan canonical form). Accordingly, the different points of view — time-domain and transform methods, state-space and input-output models — should be developed side by side and linked together as closely as possible.

Second, an introduction to control systems should provide the student with a solid understanding of the fundamental mathematical principles, a clear perception of the physical problems that motivate the theory, and a familiarity with the most important analysis and design techniques, whether his ultimate objectives lie in practical hardware design, theoretical research, or somewhere in between. In addition, it should point out extensions, advanced topics, and research areas that lie beyond the introductory material.

The book's organization and content reflect these convictions. Part I deals principally with the mathematical description of physical systems. After a brief history of control systems and an example to motivate the use of feedback in Chapter 1, differential equations in both state-space and input-output form are introduced in Chapter 2. The modeling of physical devices and systems with such equations, including linearization of nonlinear models, is explored using various examples. These and other examples are developed further in order to illustrate additional material in subsequent chapters.

Part II concerns the solution of linear, time-invariant, differential equations and those properties of solutions which are relevant to control systems. Time-domain and Laplace transform solutions are derived in Chapters 3 and 4, respectively, and their equivalence is carefully pointed out. Chapters 5 and 6 are complementary: one deals with stability and other aspects of transient solutions, the other with steady-state and frequency response. Solution of differential equations by means of analog and digital computation is covered in Chapter 7.

Part III explores feedback control in some depth. Chapter 8 covers single-loop feedback control: steady-state accuracy, root-locus analysis, and Nyquist's stability criterion. Chapter 9 discusses performance specifications for single-loop systems and then introduces compensation as a means of improving performance. Multivariable systems and their structural properties are the subject of Chapter 10, including controllability and observability, canonical forms, realizations, and the significance of pole-zero cancellations. Chapter 11 covers multi-loop control: pole placement using state feedback, steady-state accuracy, and decoupling. Observers are introduced in Chapter 12, including use of the state estimate in a feedback control law and a short digression to discuss the Kalman-Bucy filter as a particular example of an observer.

Part IV contains relatively brief excursions into linear optimal control, time-varying systems, and discrete-time systems (Chapters 13-15), followed by a survey of more advanced topics and research areas in Chapter 16. The bibliography is extensive and includes articles appearing through mid-1977.

For students who need a bit of review or practice in linear algebra or Laplace transform theory, Appendices A and B provide concise but complete summaries of the concepts and results in these areas which are relevant to the main text.

Problems are provided at the end of each chapter, grouped by section; these range from mundane to challenging, and a few are intentionally vague to encourage free thinking. Many of these request completion or extension of material in the text. We recommend reading all of the problems even if solutions are not written out formally. Solutions have been found at one time or another for most of the difficult ones (a solutions manual will be published if demand is sufficient), and only one is known to be a potential thesis topic.

A great many people have influenced and assisted in the preparation of this book. We are especially grateful to all of the following:

- Brian Anderson and Shelly Baron, whose encouragement and support were indispensable.
- Lorraine Pogonoski, who typed and corrected draft after draft with great skill and patience; Pauline Levreault, whose splendid final copy speaks for itself; Peter McLauchlan, who drew many of the figures; and Lorraine King, Ann Sarajian, and Brenda Aighes, who typed various sections of the drafts.
- Tim Johnson, Ian Rhodes, Mike Athans, Vishu Viswanathan, Dick Pew, Nils Sandell, Al Willsky, Rob Evans, Dave Kleinman, Ted Davison, and Jeff Berliner, who provided many helpful comments and criticisms.
- Ramal Muraldharan and Oliver Selfridge, who took on the thankless task of proofreading the final copy.
- Carla Fortmann and Janice Hitz, who provided love and good spirits when they were needed most.

Thomas E. Fortmann
Lexington, Massachusetts

Konrad L. Hitz
Newcastle, Australia

ORGANIZATION AND NOTATION

This book contains 16 chapters and 2 appendixes, each of which is subdivided into sections such as 2.3 and 9.2. Theorems, examples, definitions, and the like are numbered consecutively as they appear within each section. Equations are numbered in the same way, as are figures. Problems appear at the ends of the chapters, numbered separately for each section. References to items within a section use a single number, for example, Theorem 5, Equation (8), Fig. 3, Problem 6. References to items in a different section include its number, e.g., Definition 8.4.3, Equation (4.3-12), Fig. 2.4-3, Problem A.3-9. The ends of theorems, examples, etc. are denoted by $\Delta\Delta\Delta$. References to the bibliography at the end of the text appear in the form [312] or Ref. 86; those below 200 refer to books, those above 200 to papers.

Standard notation is used for proportional $\propto$, approximately equal $\approx$, identically equal $\equiv$, definition $\triangleq$, much less than $\ll$, much greater than $\gg$, implication $\Rightarrow$, and equivalence $\Leftrightarrow$. Vectors $\underline{y}$ and matrices $\underline{\underline{C}}$ are indicated by underscoring; other vector/matrix notation may be found in Appendix A.

Standard set notation is also used. If A and B are sets, then $b \in B$ means that b is a member of the set B, and $A \subseteq B$ means that A is contained in B. If, for example, B is the set all real numbers, then $\{b \in B: b \geq 0\}$ denotes the set of all nonnegative real numbers. The intersection and union of A and B are denoted $A \cap B$ and $A \cup B$, respectively.

If z is a complex number, its conjugate, magnitude, and angle are denoted $\bar{z}$, $|z|$, and $\angle z$, respectively. Derivatives of a function x(t) with respect to time t are denoted as follows,

$$\dot{x} \triangleq \frac{dx}{dt} \qquad \ddot{x} \triangleq \frac{d^2x}{dt^2} \qquad \cdots \qquad x^{(n)} \triangleq \frac{d^nx}{dt^n}$$

When the context is clear, we will assume that the statement $x(t) = y(t)$ includes the implicit proviso, "for all t in the region of interest," and that x(t) refers to a whole function $x(\cdot)$ rather than its value at a single point. The unit step function 1(t) is defined as

$$1(t) \triangleq \begin{cases} 0 & (t \leq 0) \\ 1 & (t > 0) \end{cases}$$

Part 1

DESCRIPTION OF DYNAMIC SYSTEMS

Chapter 1

INTRODUCTION

1.0 INTRODUCTION

A cornerstone in the development of modern technology has been man's capacity to construct devices which automatically control or regulate the operation of an enormous range of machines and processes. This has allowed him to operate plants which would otherwise require an intolerable amount of monotonous effort, and it has allowed him to build and use machines whose control is quite beyond his own physical capacity. Consider transportation: a horse-drawn coach could be quite adequately handled by the coachman, using simple and direct controls. Operation of a modern car is still within the physical capacity of most people, although power-assisted steering and braking has become very common. A large aircraft, on the other hand, simply cannot be flown without the assistance of various devices which automatically regulate the operation of the engines, and which translate the pilot's manual actions into movements of huge control surfaces requiring very large forces.

The design of devices which monitor and regulate the operation of machines and processes has become known as *control engineering*. Like most engineering disciplines, it began as an art, practiced by gifted craftsman-engineers with an unusual amount of common sense and ingenuity. However, the early inventions soon led to problems which were difficult to explain with intuition alone and which motivated the first theoretical studies of automatic control systems. Since then, the history of control engineering has been a fascinating example of the interaction between theory and practice in the development of an engineering science: practical problems suggested theoretical analyses which in turn permitted the formulation of systematic rules of design and extended the range of control problems which could be attacked by practical engineers.

It will be useful to briefly tour the highlights in this historical development. We hope to give the reader some idea of the current state of the science of control, and to make it easier for him to put the content and purpose of this book into perspective. For a more extensive historical discussion see Refs. 123 and 227a.

1.1. HISTORY AND MOTIVATION

The key concept in automatic control devices is that of feedback: deviations of a machine or process from its desired state of operation are made to activate a device which in turn forces the machine or process back to its desired state. Feedback systems abound in nature: the regulation of body temperature by metabolic heating and evaporative cooling, the automatic control of blood sugar level by the opposing hormones insulin and glucagon, and the control of body movements by optical feedback are examples from human physiology. However, to recognize the abstract idea of feedback long before such natural examples were understood, and to use it to invent new devices, has been one of man's outstanding intellectual achievements. It is by no means a recent one; feedback systems for regulating the speed of clock mechanisms, water wheels, and windmills date back at least to Christiaan Huygens in the 17th century. They took on major importance, however, during the industrial revolution; the centrifugal steam engine governor invented by James Watt is perhaps the most famous early example.

Confronted with the problem of maintaining a constant engine speed when there were unpredictable changes in the load driven by the engine, Watt hit on the idea of using a set of flyballs, or rotating weights suspended from levers, driven by the engine and connected to the steam valve in such a way that an increase in engine speed raised the flyballs and thereby reduced the opening of the steam valve [see Fig. 1(a)]. Schematically, Watt's governor system can be represented by the "block diagram" shown in Fig. 1(b), in which the blocks represent components in the system and the lines with arrowheads represent variables.

It was soon discovered that if the governor used a direct mechanical link between the flyball sleeve and the steam valve, perfect regulation of engine speed was impossible: a sustained deviation of the flyballs from their usual height was required to achieve the change in steam flow corresponding to the change in load. This led to the invention of the "resetting" governor, and with it to the first major problem in automatic control. Under some circumstances the governed engine would spontaneously break into violent oscillations of speed: the feedback mechanism produced an increase in steam flow just when a reduction was called for.

In a classic paper published in 1868 [25], J. Clerk Maxwell explained this instability and formulated rules for governor design which would avoid it. He showed that according to elementary principles of mechanics the governor-engine-load system could be described by an ordinary linear differential equation, and that engine instability corresponded to the occurrence of growing exponential functions in the unforced solution of this equation. Maxwell considered how the occurrence of such growing exponentials might be deduced directly from the coefficients of the equation, i.e., from the known characteristics of the "plant," without first solving the equation, and was successful for differential equations of low order. He was able to state simple algebraic criteria which the plant parameters had to satisfy to guarantee stability. The extension of this idea to more complex systems, described by high-order differential equations, was given later (1895) by two mathematicians,

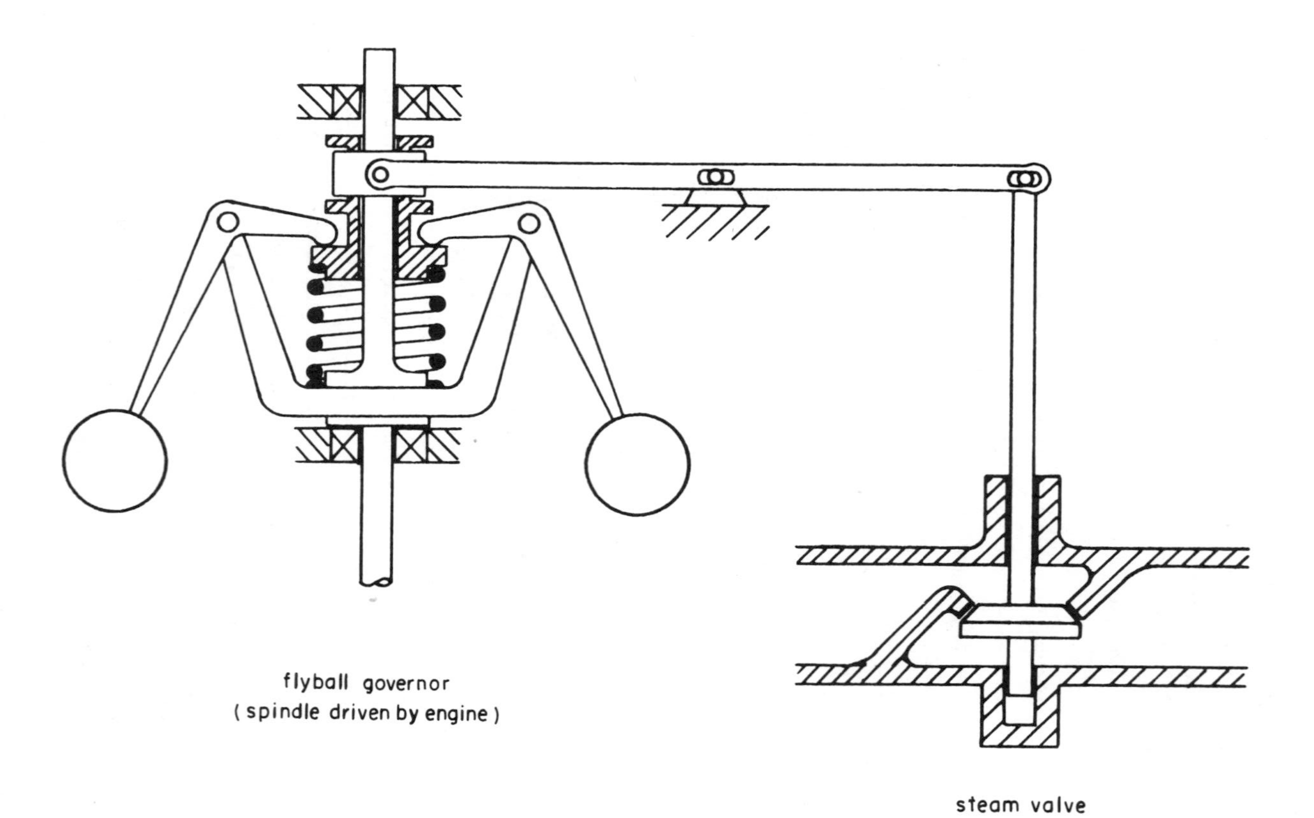

Fig. 1(a). Watt's steam engine governor system.

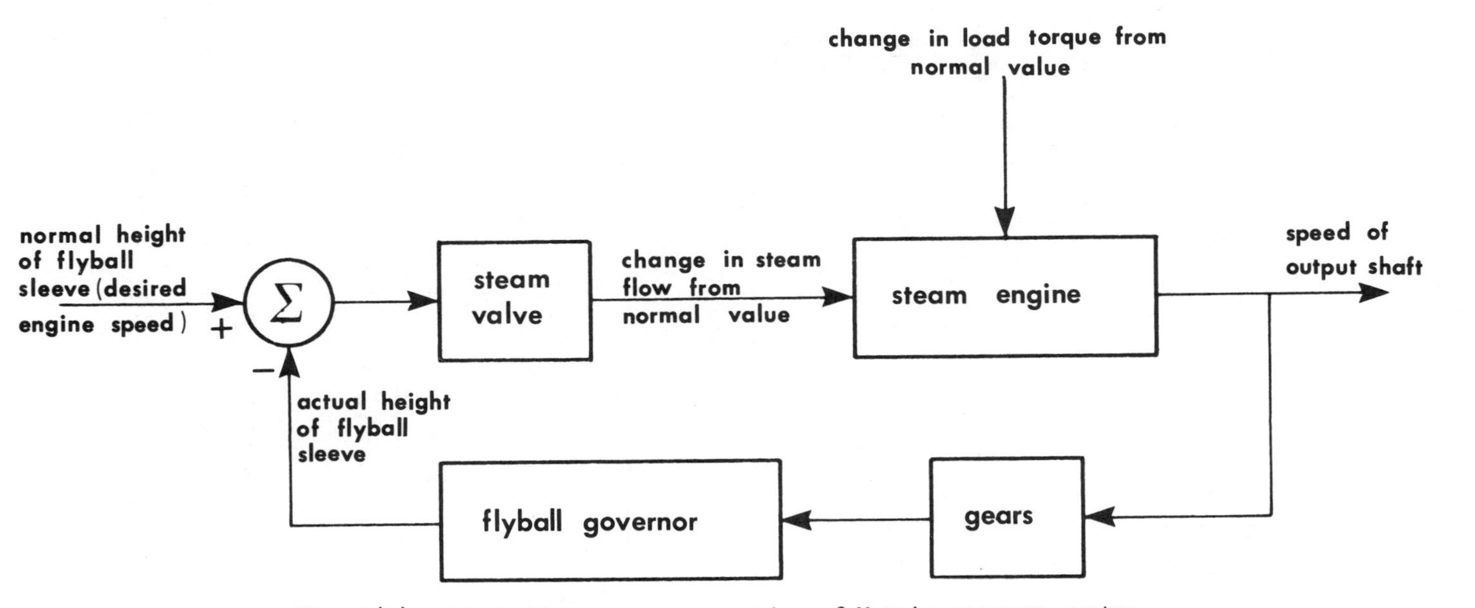

Fig. 1(b), Block-diagram representation of Watt's governor system.

Routh and Hurwitz, who independently discovered two forms of a remarkable algebraic criterion for the stability of a system described by an ordinary linear differential equation of arbitrary order [25].

This result was essentially sufficient for the needs of control engineers until the 1940s. The automatic control systems built in that era were regulating systems: the speed of an engine or of a hydraulic turbine had to be maintained accurately at a constant value. If an occasional disturbance such as a change in load occurred, the speed had to return to its former value, but the precise way in which it behaved during its return was only of secondary importance. Static accuracy and avoidance of instability were the major design criteria, while speed and smoothness of the "transient" response were considered less important.

This emphasis changed rapidly during the second world war. Advances in weapons such as ships' guns and antiaircraft batteries created an urgent need for a whole range of automatic control systems (servomechanisms), whose ability to accurately follow or compensate for rapidly varying signals was of paramount importance. It was then discovered that much of the theory needed for the design of such systems had already been developed in the quite unrelated field of communications engineering.

This theory had its origins in the practical problems of building long-distance telephone networks in the United States. In a transcontinental telephone line using an economically feasible amount of copper conductors, the number of repeater amplifiers required to make up transmission losses was so large that even the small amounts of noise and distortion introduced by each amplifier were sufficient to completely corrupt the signal in its passage through the line. A very high quality amplifier was therefore required, and this led (in 1927) to the invention of the negative feedback amplifier by H. S. Black. In an "ordinary" amplifier, the signal voltage passes through the amplifier circuits once, emerging slightly distorted and with an amplification, say, of α_c. Black discovered that if the amplifier is made to amplify not just the incoming signal but the difference

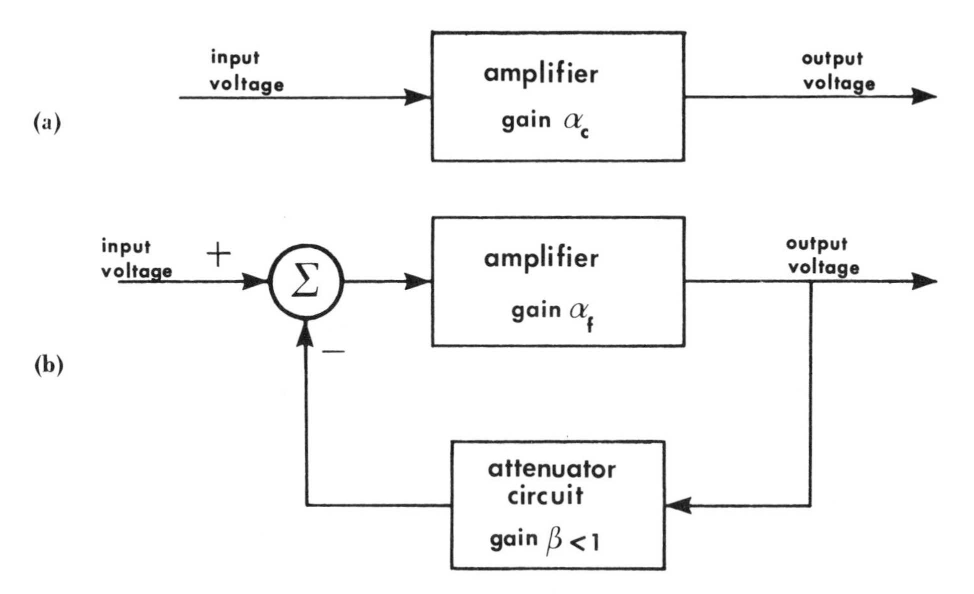

Fig. 2. (a) Conventional amplifier and
(b) negative feedback amplifier.

between the signal and a fraction β of its own output (see Fig. 2) with a much higher amplification α_f than before ($\alpha_f >> \alpha_c$), then the same overall signal amplification can be achieved, but the corruption of the signal is drastically reduced. Black in effect discovered a fundamentally important property of feedback systems: the overall behavior of a properly designed negative feedback system is quite insensitive to the precise values of individual system components. The fact that the gain of the amplifier varied slightly with the magnitude of the input voltage, producing a distorted output, became much less significant if the amplifier was part of a feedback structure.

Just as in the case of Watt's governed engine, it was soon discovered that a feedback amplifier could in some circumstances become unstable and behave as an oscillator instead. Therefore, before the feedback idea could be thoroughly exploited, it was necessary to have design criteria which guaranteed stability. The rules of Routh and Hurwitz were of little practical use: typical amplifiers required for their adequate description a differential equation of such a high order that the computational effort involved in its use was prohibitive. Instead, communications engineers had become used to describing their circuits in terms of "frequency response," that is, the attenuation and phase shift of the circuit output when excited by sinusoidal voltages. This was readily measured in the laboratory, and conveniently handled in graphic form. The required stability criterion for Black's amplifier was formulated in 1932 by H. Nyquist [25], who discovered a very simple property which the frequency response of the amplifier and attenuator had to have to guarantee the stability of the feedback structure of Fig. 2(b). Moreover, Nyquist's criterion, unlike that of Routh and Hurwitz, gave a simple quantitative way of measuring just how close to instability a stable feedback system is, and this stability margin turned out to be closely related, at least qualitatively, to the transient response of the feedback system. Thus, the frequency response theory developed by communications engineers from Nyquist's key result provided just the analytical tools needed by control engineers for the design of military control systems having a high quality transient performance as well as static accuracy.

This work was made public after the war. An important contribution to the theory was made in 1948 by W. R. Evans. His "root-locus" method provided a simple and effective way of deducing, from the differential equation model of the system, how the set of exponential functions occurring in the transient response of a feedback system can be changed by varying one of the parameters of the system. This formed the basis for design methods complementary to, and in some cases simpler and more direct than, the frequency response methods derived from Nyquist's theorem. With this contribution, the first stage in the development of control engineering was essentially complete; the theory based on Nyquist's criterion and Evans' root-locus method is now commonly known as *classical control theory*.

It rapidly led to an almost explosive growth in the application of feedback control systems. Much effort went into refining the details of theory and design methods, numerous textbooks were published (among the classics are Refs. 41, 124, 137, and 168), formal academic courses in control engineering were introduced, and instrument manufacturers began marketing a wide range of standard components for servomechanisms and industrial process control systems. The widespread success of control engineering in the postwar years raised hopes that its principles could be readily generalized to encompass more complex situations. The term *cybernetics* was coined at that time by Norbert Wiener [178], an eminent mathematician who speculated that the theoretical knowledge of feedback systems then available would, in short order, lead to major advances in our understanding of such highly complex systems as biological control mechanisms and nervous systems, and yield methods for more effective control of the many complex economic and social processes in an industrial society.

Such speculations were premature. In fact, the classical theory turned out to have three quite serious limitations, which prevented its immediate extension to more complex control problems, and which motivated the major theoretical developments that have since taken place.

First, it was restricted to linear time-invariant systems, i.e., systems whose components could be adequately described by ordinary linear differential equations with constant coefficients. Even such simple nonlinear devices as the "on-off" thermostat controls on household cookers and heating and air conditioning plants fall outside this category. A great deal of research effort has gone into overcoming this limitation. Generalized versions of Nyquist's stability criterion have been found for certain classes of nonlinear and time-varying feedback systems, and a variety of optimization techniques have been adapted to such problems.

The second limitation of the classical theory was its restriction to what are now called "scalar" or single-loop feedback systems. In such systems there is just one variable, called the output, which is controlled by manipulating a single input variable. Watt's governed steam engine is a typical example, in that engine speed is controlled by adjusting the steam valve opening. In many important engineering systems, however, one must deal with multiple inputs and outputs which are dynamically coupled. As an example, consider the multistand rolling mill for finishing steel strip, shown in Fig. 3. The strip passes through several pairs of rolls in series, where each roll stand is driven independently and has an independent mechanism for adjusting the roll gap. For such a mill to operate properly, it is necessary not only to control the final strip speed and thickness but also to maintain an appropriate balance in the reductions of thickness achieved at the different stands, and to control the strip tension between them. This is done by adjusting the roll speed and gap at each stand. Thus, for a mill of four stands, there are eight outputs to be controlled simultaneously by manipulating up to eight different inputs. The difficulty is, of course, that a change in any one of the eight inputs will affect not just one but several (or even all) of the outputs. Thus, before one can begin to design feedback controls for such a multivariable system, it is necessary to know precisely which of the outputs will be affected by each of the available inputs and in what way changes to the various inputs oppose or assist each other.

This kind of complexity could not be satisfactorily attacked with the methods provided by the classical feedback theory, even if the systems were linear. These methods all used an "input-output"

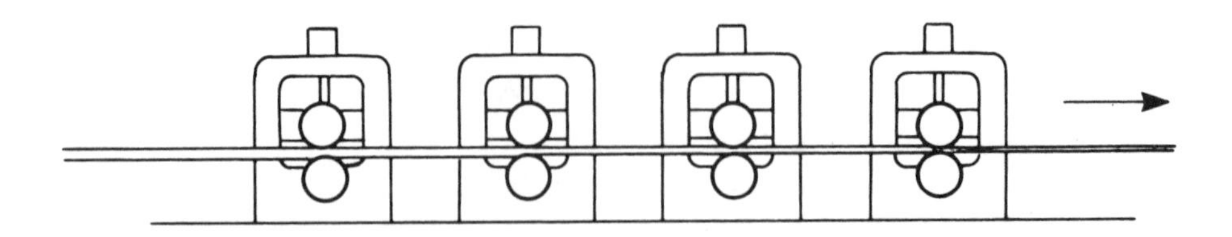

Fig. 3. Multistand rolling mill.

description of the system which essentially ignored its internal behavior and could not effectively deal with more than one input-output pair at a time. To treat nonlinear, time-varying, and multivariable systems it was necessary, at least initially, to augment the input-output approach with a theory in which the internal behavior of the system is considered explicitly. This *modern control theory,* which developed rapidly during the 1960s, is founded in the mathematical disciplines of linear algebra and, to some degree, functional analysis. Many of the resulting analysis and design procedures involve very extensive and time-consuming calculations and manipulations, and it was no coincidence that these new developments in control theory occurred just as large general-purpose digital computers were becoming readily available. More recently, there has been a resurgence of interest in extensions of the input-output theory to handle multivariable (but still linear and time-invariant) systems.

The third limitation of classical feedback theory was, at least to begin with, a problem more for theoreticians than for practical engineers. Broadly speaking, the classical theory provided techniques of analysis rather than synthesis. Given a particular system, one can check whether a given set of performance specifications is satisfied; if not, the theory suggests how the system should be changed to give better performance. Design is usually by trial and error: starting with the simplest configuration which experience suggests might be suitable, a series of systems is generated and analyzed, and the process is stopped with the first design found to be just good enough. Although this procedure is typical of much of engineering design, it is unsatisfying in an intellectual sense. It is natural to ask whether there might not be a best possible design for a given application.

As an illustration, consider the operation of a bloom mill in a steel works, where thick cast ingots of steel are reduced to long slender bars prior to being rolled into their final shapes in finishing mills. In the bloom mill, the rectangular steel bar is passed back and forth through the gap between a single set of rolls. The rolls are stopped and reversed and the gap is reduced after each pass by a control system in which desired values of speed and gap are remotely set by the mill operator. In the classical approach to the design of these control systems, one might specify that roll speed and gap should reach their new values without a final steady-state error, and that the change should be rapid and smooth, i.e., without perceptible oscillations. On the other hand, it is quite natural (and economically significant) to ask for control systems which will change these variables in the least possible time, subject to limitations on power available from the drive motors.

Such problems have led to the development of what is now called *optimal control theory*. It is a branch as much of applied mathematics as of control engineering: methods for designing optimal control systems require sophisticated mathematical tools, many of which were invented only during the development of the subject. With one important exception which will be discussed in Chapter 13, optimal control systems are typically nonlinear and they can usually be implemented only with digital computers, in contrast to the quite simple hardware required for linear feedback systems obtained using classical methods. The first widespread practical applications of optimal control ideas, at least in non-military areas, occurred in the various space missions conducted in the 1960s. The success of these missions depended in no small way on the solution of such optimal control problems as transferring a space vehicle into a new trajectory or correcting deviations from a desired path with minimum fuel cost. The cost of solving such problems (by means of extensive digital computation in ground control centers) was insignificant compared with the benefits obtained. By contrast, industrial applications of optimal control systems are still comparatively rare, but with the rapid development of powerful small digital computers they will undoubtedly become increasingly important.

These recent developments in control theory, arising from the most serious limitations of the classical theory, are by no means complete; each area abounds with unsolved problems, and much research is still going on. Other problems have also become of importance to control scientists. One of these is *system identification,* i.e., where one is confronted with a "black box," a system whose internal structure is so complex or obscure that a tractable mathematical model of it cannot be derived a priori from known laws of science. All that is available are observations of inputs and outputs of the black box, and the problem is to derive a suitable mathematical model from these observations. Such problems occur frequently in process plants such as chemical plants, paper mills, and the metal refining industries, and nearly all biological and social systems are of this nature. For instance, no one has yet formulated laws of economics which allow us to model the dynamic behavior of an economy with the same level of confidence that we have in an electrical circuit model derived from the laws of physics.

This brief survey should convince the reader that control engineering has become a large field covering a wide range of engineering activities. At one end of the spectrum are practical engineers concerned with the installation and maintenance of industrial automatic control systems using standard devices. At the other end are applied mathematicians working on hitherto intractable problems in optimal control, system identification, and a variety of other areas, many of which are discussed in the final chapter.

Obviously, no single book can hope to give an adequate introduction to all branches of the subject. In this text, we shall discuss the theory of *linear* systems. One reason for this choice is that a large variety of real systems can in fact be adequately described by linear models. Another is that they form the only class of systems for which a powerful general theory is available; by contrast, most nonlinear systems must be treated individually and few generalizations about their behavior are possible. Moreover, many of the more effective tools for tackling nonlinear systems are based on methods developed originally for linear systems. Thus, a sound background in linear systems is an essential prerequisite to the study of more advanced topics.

We conclude this introduction by considering a typical linear control system.

1. *EXAMPLE (Rotating antenna).* Assume that we wish to design a system for remotely controlling the azimuth angle of the rotating antenna shown in Fig. 4. The antenna might be part of a radar tracking installation, where the radar beam must be aligned as closely as possible with the line of sight to a target; it might be part of a satellite communications system in which the antenna dish is to be kept aimed at a satellite; or it might be the dish of an optical or radio telescope, in which case the antenna axis is to remain fixed while the control system compensates for the rotation of the earth. To achieve these objectives, of course,

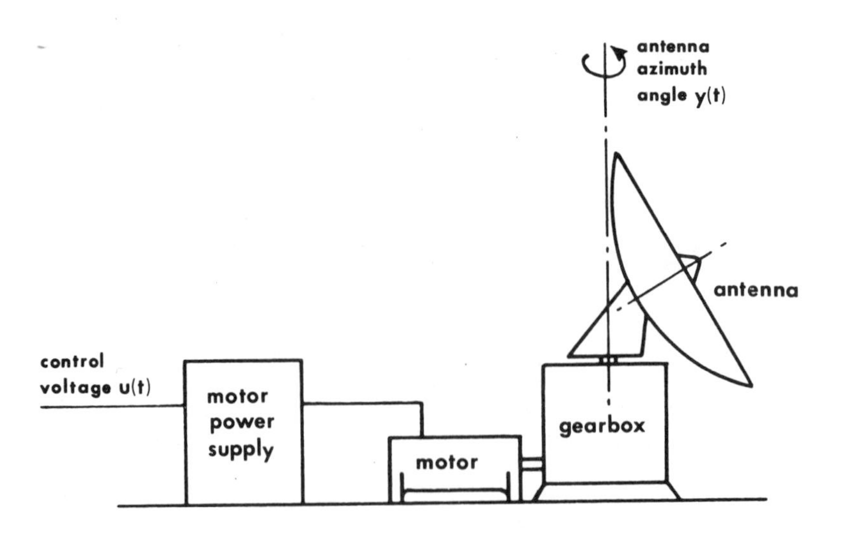

FIG. 4. Antenna system.

it is also necessary to control the antenna elevation, but we assume that this is done by a separate control system.

The voltage applied to the motor determines antenna position, and our task is to design a system which will produce whatever voltage is necessary to move the antenna to any specified azimuth angle. Moreover, we assume that the antenna is located in the open air and therefore expect that wind forces will affect the antenna position; the system is to automatically produce whatever change in u is necessary to make the antenna as insensitive as possible to such wind loads, represented by the variable d(t).

The antenna drive system is assumed to develop a torque as long as a nonzero voltage u is applied; the direction of the torque will correspond to the sign of the applied voltage. Clearly, therefore, a particular azimuth angle cannot be specified by giving u a constant value; the voltage must depend not only on the desired azimuth angle y_r but also on the actual antenna position y. The simplest possibility for a control system is to use the *feedback control law*:

$$u(t) = y_r(t) - y(t) \tag{1}$$

Then the motor will develop no torque, and thus leave the antenna at rest, whenever it is pointing in the desired direction. The problem now is to determine whether the control strategy embodied in Eq. (1) will actually work. More precisely, we must determine how the whole system will behave in various situations.

The only satisfactory way of doing this, short of building the system and testing it, is to construct a model of it in the form of a set of equations known to describe the behavior of the various components. This description must be valid at all instants of time since we are interested not only in finding out whether the antenna ultimately points in the correct direction when y_r and d have constant values, but also in the antenna response to sudden changes: whether it will accurately follow a rapidly changing command (important in a radar tracking system, for example) and how long it will take to settle down after a momentary wind gust. For this reason, we shall write all the variables as functions of time.

A useful first step in deriving the system equations is to draw a more detailed schematic diagram (Fig. 5). For the sake of simplicity, we shall assume the antenna drive to be a field-controlled d.c. motor; such motors have a separate armature supply with constant current. Figure 5 also illustrates one method of implementing Eq. (1): a potentiometer is attached to the antenna or one of the gearbox shafts so that the wiper rotates with the antenna. The wiper voltage will then be proportional to the actual antenna position, and we assume a proportionality constant of 1 for simplicity. An amplifier of (adjustable) gain k_a then amplifies the difference between the voltage $y_r(t)$, representing the desired antenna position, and the potentiometer voltage:

$$u(t) = k_a[y_r(t) - y(t)] \tag{2}$$

We shall next derive a relationship between the applied voltage u(t), the torque applied to the antenna by wind loading, d(t), and the antenna position.

Consider the motor. For d.c. motors, the torque developed is almost directly proportional to the product of the magnetic flux produced by the current in the stator windings, and the current carried by the armature conductors. In our case, the armature current is held constant by a special armature supply. If we assume that the iron core for the stator winding is not saturated at any time, and that the magnetization curve for the stator field is linear and shows negligible hysteresis, then the stator magnetic flux is very nearly proportional to the current in the stator field windings, i(t). Hence if m(t) is the torque developed by the motor, we can write

$$m(t) = k_m i(t) \tag{3}$$

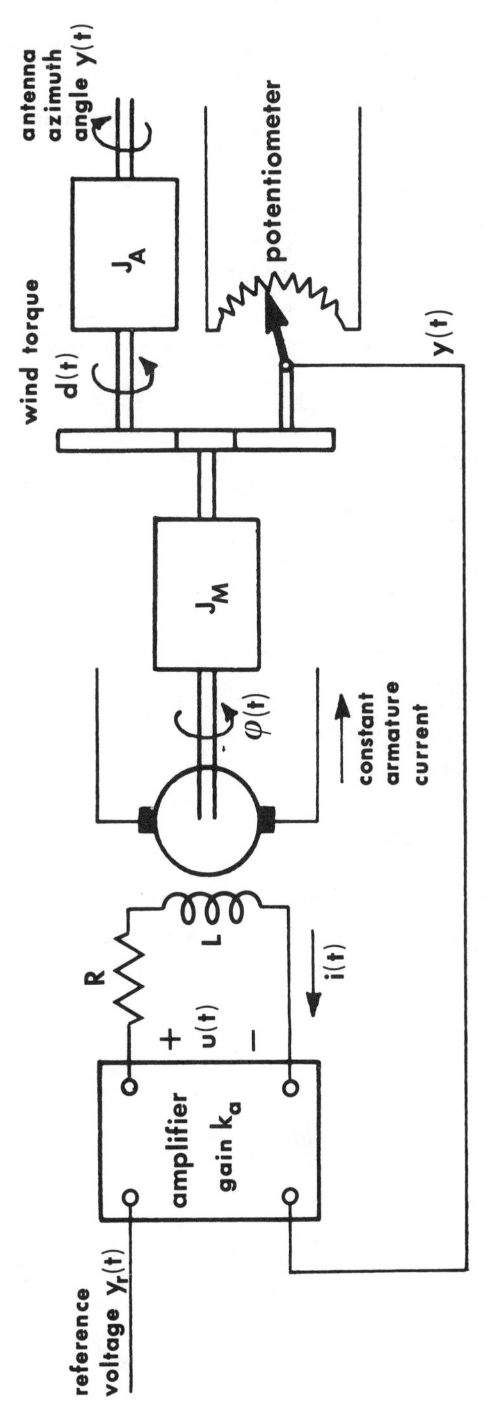

FIG. 5. Circuit diagram of antenna system.

where k_m is some constant for the particular motor used. Now the field current i(t) depends only on the voltage applied to the field circuit; summing voltage drops in this circuit gives

$$u(t) = R\,i(t) + L\,\dot{i}(t) \tag{4}$$

where R and L are the resistance and inductance, respectively, of the stator field circuit. Let us assume, to simplify matters, that the voltage drop across the inductance is always negligibly small compared with the voltage drop across the resistance. Then (4) simplifies to

$$i(t) = \frac{1}{R}\,u(t) \tag{5}$$

Eliminating u(t) and i(t) from (2), (3) and (5), we have

$$m(t) = \frac{k_a k_m}{R}\,[y_r(t) - y(t)] \tag{6}$$

Now consider the rotating components of the system. Let $m_1(t)$ be the torque applied to the motor shaft by the gearbox. Then Newton's law of motion for rotating masses, applied to the armature shaft, yields

$$m(t) - m_1(t) = J_M\,\ddot{\phi}(t) \tag{7}$$

where $\phi(t)$ is the angular position of the armature and J_M is its mass moment of inertia. If n is the gear ratio between the motor and antenna shafts, and backlash in the gears can be neglected,

$$\phi(t) = n\,y(t)$$

or

$$\ddot{\phi}(t) = n\,\ddot{y}(t)$$

and therefore,

$$m(t) - m_1(t) = n\,J_M\,\ddot{y}(t) \tag{8}$$

If the gearbox applies a torque $m_1(t)$ to the armature and friction is neglected, then by the law of action and reaction, a torque $n\,m_1(t)$ is applied to the antenna shaft. Hence a sum of torques on the antenna gives

$$n\,m_1(t) - d(t) = J_A\,\ddot{y}(t) \tag{9}$$

where J_A is the mass moment of inertia of the gearbox and antenna.

Elimination of $m_1(t)$ and $m(t)$ from (6), (8), and (9) yields the final model for the antenna system, a linear second-order differential equation with constant coefficients:

$$\ddot{y}(t) + \omega^2 y(t) = \omega^2 y_r(t) - \beta\,d(t) \tag{10}$$

where

$$\omega^2 \triangleq \frac{k_a k_m}{RJ}$$

$$\beta \triangleq \frac{1}{nJ}$$

$$J \triangleq nJ_M + \frac{J_A}{n}$$

It is worth recapitulating the assumptions made in deriving Eq. (10):

(a) The motor was assumed to have negligible hysteresis and a linear magnetization curve.

(b) Friction and backlash in the gearbox and antenna bearings were neglected.

(c) The voltage drop across the stator inductance, $L\,\dot{i}(t)$, was considered negligible. (This means that our model will certainly not be accurate if the rates of change of system variables are high, i.e., if the system is expected to respond very rapidly to any inputs.)

Keeping these in mind, we may now ask how the antenna will respond to various inputs. In particular, suppose the antenna is at rest, $y = 0$, and pointing in the desired direction, say $y = y_r = 0$, with no wind loading when it is decided to move to a new position Y. If the command input is given very rapidly, it can be represented by the *step function*

$$y_r(t) = Y1(t) = \begin{cases} 0 & (t \leq 0) \\ Y & (t > 0) \end{cases} \tag{11}$$

The antenna response to this command can be computed by solving (10) for $d(t) \equiv 0$, $y(0) = \dot{y}(0) = 0$, and $y_r(t)$ given by (11). As we shall see in Chapters 3 and 4, the solution (for $t > 0$) is

$$y(t) = Y(1 - \cos \omega t) \tag{12}$$

Quite clearly, the antenna control law we have assumed with (2) is useless: the antenna will never settle down to a new steady value following a step change in desired position, but will oscillate with constant amplitude about the desired value Y. A similar conclusion can be drawn if the antenna is initially at rest and a sudden wind gust arrives and applies a constant torque: it will keep oscillating about a nonzero mean with constant amplitude. The antenna system behaves just like an ordinary harmonic oscillator, the simplest example of which is a mass suspended from a linear frictionless spring. In that case, it is well known that the introduction of a dashpot or viscous damper will attenuate the oscillations: the dashpot applies a force proportional to the velocity of the mass and in the opposite direction, thus giving the system a measure of "anticipation." By analogy, we should expect such damping of the antenna oscillations if the motor voltage can be made to depend on the antenna velocity as well as the antenna position. Such a strategy can in this case be quite readily implemented by installing a tachogenerator, a d.c. generator whose output voltage is directly proportional to its rotor speed. We therefore replace (2) by

$$u(t) = k_a[y_r(t) - y(t) - k_v\dot{y}(t)] \tag{13}$$

where k_v is the gain constant of the tachogenerator. Combining this with the other system equations yields the antenna equation

$$\ddot{y}(t) + \omega^2 k_v \dot{y}(t) + \omega^2 y(t) = \omega^2 y_r(t) - \beta d(t) \tag{14}$$

where ω^2 and β were defined above. It has become standard to define $\zeta \triangleq k_v\omega/2$ and write this as

$$\ddot{y}(t) + 2\zeta\omega\dot{y}(t) + \omega^2 y(t) = \omega^2 y_r(t) - \beta d(t) \tag{15}$$

For the same inputs and initial conditions as above, a closed-form solution of (15) can be obtained (for $t > 0$) using standard methods:

$$y(t) = Y\left(1 - \frac{1}{\gamma} e^{-\zeta\omega t} \sin(\gamma\omega t - \phi)\right) \qquad (\zeta < 1) \tag{16}$$

where $\gamma \triangleq \sqrt{1 - \zeta^2}$ and $\phi \triangleq \tan^{-1}(-\gamma/\zeta)$;

$$y(t) = Y\left(1 + \frac{e^{-\alpha_1\omega t}}{2\delta\alpha_1} - \frac{e^{-\alpha_2\omega t}}{2\delta\alpha_2}\right) \qquad (\zeta > 1) \tag{17}$$

where $\delta \triangleq \sqrt{\zeta^2 - 1}$ and $\alpha_{1,2} \triangleq \zeta \pm \delta$; and

$$y(t) = Y(1 - e^{-\omega t} - \omega t\, e^{-\omega t}) \qquad (\zeta = 1) \tag{18}$$

In all three of these equations, we note that

$$\lim_{t\to\infty} y(t) = Y$$

so that the antenna will in fact come to rest at its desired new position Y. Secondly, the parameter ω occurs only as a multiplier of t. Thus it determines only the "time scale," or the speed of response of the antenna, while the shape of the response depends on the parameter ζ. Graphs of the response for some typical values of ζ are shown in Fig. 6.

We conclude that the new control strategy of feeding back a signal proportional to antenna velocity as well as antenna position is successful, at least in that the antenna can now be pointed in any desired direction. By choosing a large value of ω, i.e., by making the amplifier gain k_a large [see Eq. (10)], we obtain a *fast*

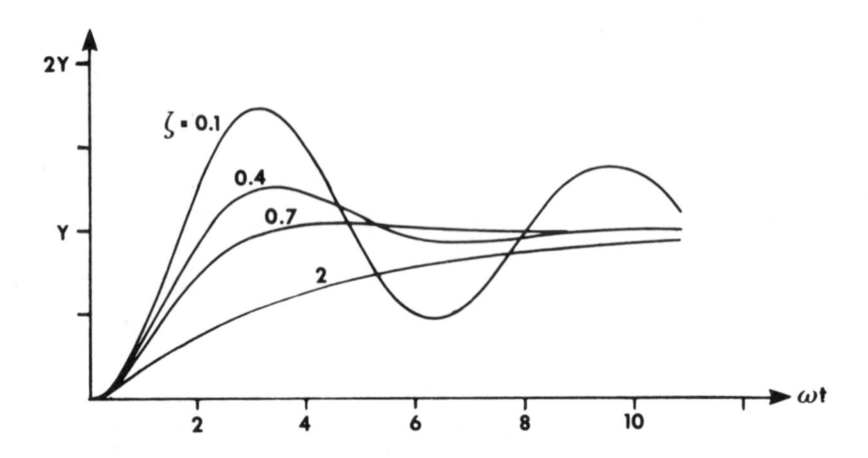

Fig. 6. Step response of damped antenna system.

transition from 0 to Y, and by adjusting ζ, i.e, by selecting the appropriate value of tachogenerator gain k_v for the given ω, we obtain a *smooth* response (see Fig. 6).

Now consider the behavior of the modified antenna system when a sudden wind gust applies a step change in torque to the antenna. As before, we assume that $y(0) = \dot{y}(0) = 0$, and that no change in desired antenna position occurs, $y_r(t) \equiv 0$. The wind loading will be assumed to be

$$d(t) = D1(t) \tag{19}$$

Under these conditions, the solution of Eq. (15) differs from Eqs. (16)-(18) only in the constant factor. If $\zeta < 1$, for example, the solution is

$$y(t) = -\frac{D\beta}{\omega^2}\left[1 - \frac{1}{\gamma} e^{-\zeta\omega t} \sin(\gamma\omega t - \phi)\right] \tag{20}$$

Since

$$\lim_{t\to\infty} y(t) = -\frac{D\beta}{\omega^2},$$

the antenna will not return to its desired position, $y_r = 0$, when a constant torque D due to wind loads occurs. The steady-state error can be made small by choosing a large value of ω (that is, of k_a) but it cannot be made zero. This is just the phenomenon that was observed by Watt with his steam engine governor. Thus, if the design specifications for the antenna control system call for a zero steady-state error under conditions of constant wind torque, our system is inadequate and needs to be modified further, as we shall see in Chapter 8.

So far, we have made the simplifying assumption that the voltage drop $L\dot{i}(t)$ across the stator field inductance in (4) was negligible. We leave it as an exercise for the reader to reinstate this term and show that the antenna model (15) must then be replaced by

$$\dddot{y}(t) + \frac{R}{L}\ddot{y}(t) + ck_v\dot{y}(t) + cy(t)$$

$$= c\,y_r(t) - \frac{\beta R}{L}d(t) - \beta\dot{d}(t) \tag{21}$$

where $c \triangleq k_a k_m/JL$.

We shall not attempt to determine explicit solutions of (21). Indeed, this can no longer be done unless numerical values are assigned to the various parameters. Thus it is impossible, by producing a closed solution in terms of undetermined coefficients, to investigate how the behavior of the system depends on these coefficients. An obvious alternative is the brute-force approach of varying each of the "free" parameters k_a and k_v over a range of values and solving (21) for each set. Clearly this involves an enormous amount of computation, and at the end of it, we might well be forced to the conclusion that no set of parameter values exists for which the system has a wholly satisfactory behavior. In fact, the antenna system described by (21) turns out to suffer from the same defect as that described by (15): a zero steady-state error cannot be achieved for constant wind torques. Moreover, all the computation needed to reach this conclusion would not indicate how the system or, more accurately, the control strategy embodied in (13), ought to be changed to give better performance.

This is an example of the fundamental problem that control theory attempts to solve. The design of our antenna control system consists essentially of two parts, the first of which is to determine a suitable control law or the basic structure of the feedback system. In the example, this means a choice of the way in which u(t) is to depend on the desired and actual antenna positions, as specified by (2) or (13). The next step is to determine numerical values of the free parameters in the control law (k_a and k_v in the example) for which the system exhibits "good" behavior. In all but the very simplest cases, the dependence of system behavior on various adjustable parameters cannot be analyzed by means of a closed-form solution of the equations.

ΔΔΔ

Given a mathematical model of the *plant* (that is, the system which is to be controlled), a set of performance specifications, and a particular control law, it is therefore desirable to have short-cut methods which will

(a) Enable us to make at least a qualitative estimate of the behavior of the control system for a variety of typical inputs directly from the parameters of the system equations and without requiring their explicit solution

(b) Indicate how the control law should be changed if the analysis in (a) indicates that satisfactory performance cannot be achieved.

For systems described by linear differential equations with constant coefficients (and this means linear plants as well as linear control laws) such methods are available; the major part of this book is devoted to their study.

In Part I (this chapter and the next), we show by means of a number of examples how to derive linear models of dynamic systems. Solution of the resulting differential equations must be considered before we can discuss control of such systems, and in Part II we systematically exploit the fact that linear constant systems all share two fundamental properties. The first is that their response to a momentary disturbance such as an impulse is a linear combination of exponential functions. This should not be surprising since a homogeneous differential equation, such as (14) with $y_r(t)$ and d(t) zero, states that the output is a linear combination of its own derivatives, and the only functions having this property are the exponential ones. Just which of them occur in the response of any given system can be determined by algebraic manipulations of the coefficients in the system equations. Thus a great deal of qualitative information about the system response can be obtained directly from its parameters without solving the equations. In particular, the overriding question of stability can be answered in this way.

The other fundamental property of linear constant systems, at least of stable ones, is that their response to an exponential input tends to an exponential function of the same form as the input. For example, a sinusoidal input will provoke a sinusoidal response at the same frequency as the input, only attenuated and shifted in phase. A knowledge of this attenuation and phase shift for all input frequencies constitutes a complete description of the system, and much about the response to any other input can be directly deduced from it, again without an explicit solution of the system equations.

The first two parts form the essential background for Part III, which is devoted to the theory of control. The first chapters in this part introduce the classical theory of controlling scalar or single-input, single-output systems, such as the antenna system considered above. The later chapters introduce the fundamental ideas required for the control of multivariable systems, i.e., systems with several inputs and several outputs.

Part IV contains a brief discussion of optimal linear systems and extensions of the earlier theory to the time-varying and discrete-time cases, and concludes with a brief survey of current research in control and systems theory.

The appendixes review material on linear algebra and Laplace transforms which is used throughout the book. The reader is assumed to have some background in these areas, however a brief perusal of both appendixes is recommended before proceeding to the remaining chapters.

Although the book is primarily concerned with theoretical matters, one must not lose sight of the practical problems which have motivated these abstractions; examples and exercises are provided to illustrate applications of the theory in a variety of physical situations. We encourage the reader to heed the words of Richard Bellman:

> "Theory without application is like the smile of the Cheshire cat; application without theory is blindman's buff."

Chapter 2

FORMULATION OF MATHEMATICAL MODELS[†]

2.0. INTRODUCTION

In this chapter we shall discuss the description of physical phenomena using mathematical models, and observe some of the forms in which these models may appear.

Devising adequate mathematical descriptions of complex physical systems is in general a highly subjective and often iterative task. It requires a large measure of intuition, i.e., an ability to determine which physical variables and relationships are negligible and which are crucial to the accuracy of the model. In many ways it is as much an art as a science.

The reader should bear in mind that the standard characterizations of electrical and mechanical circuits as interconnections of linear, lumped, time-invariant elements are highly idealized. Nevertheless, this sort of analysis can yield a great deal of useful information, provided one does not lose sight of its limitations.

This book will deal primarily with just such idealized systems: they will usually be assumed *linear, time-invariant,* and *finite-dimensional.* Linear analysis, in particular, is applicable to a

[†]An understanding of Appendix A through Sec. A.3 is assumed from this point on.

wide class of problems, as illustrated in Sec. 2.4 by the linearization of nonlinear systems about known operating points.

Time-varying systems will be introduced in Chapter 14, where it will be seen that many of the results for time-invariant systems extend readily. The case of a *discrete-time system,* which arises, for instance, when the variables of a continuous-time system are periodically sampled, is discussed in Chapter 15. Numerous other extensions and applications may be found in the final chapter.

One class of systems has been omitted entirely, namely, those in which the variables of interest take on a finite number of discrete values. Included in this class are *logical* (or *Boolean*) *systems*, whose variables take on only the values 0 and 1, and *finite-state machines*. Most methods for dealing with such systems are very different from those presented here; the interested reader is referred to any of the current texts on switching circuits, logical design, and automata theory.

2.1. STATE-VARIABLE FORM OF SYSTEM EQUATIONS

The concept of an abstract dynamic system is quite complex and not entirely standard [98,185,278]. For the purposes of this book a more restricted definition will suffice, and we will apply this term to any process whose behavior as a function of time is governed by a *finite set of ordinary differential equations*.

Such equations arise because the physical laws governing a system or process usually relate certain variables to rates of change of others: Newton's second law relates forces or torques to rates of change of velocities; voltages induced in coils are proportional to rates of change of current; flow rates through restrictions in fluid conduits are determined by pressure differences across them; rates at which chemical reactions take place frequently depend on the concentrations of reagents in a vessel.

The *state* of a dynamic system at any particular time may be (loosely) defined as the collection of information which is both necessary and sufficient to determine the future behavior of the system, assuming that all future external stimuli are also known. The *state space* consists of all those values which the state may take on.

The term *input* refers to all stimuli which are applied to the system from external sources, and the *output* comprises that portion of the system's state which can be directly determined by external measurements.

1. EXAMPLE. The dynamic system shown in Fig. 1 consists of a simple linear circuit with a signal generator supplying the *input voltage* u(t) and the ammeter measuring the *output current* y(t).

As any student of elementary circuit theory knows, the behavior of this circuit can be determined from time t on if three quantities are known: the initial inductor current i(t), the initial capacitor voltage v(t), and the input voltage function $u(\tau)$, $\tau \geq t$. Thus the *state* of this system at time t consists of the *state variables* i(t) and v(t), which may be collected into a 2-dimensional *state vector*

$$\underline{x}(t) = \begin{bmatrix} x_1(t) \\ y_2(t) \end{bmatrix} = \begin{bmatrix} i(t) \\ v(t) \end{bmatrix} \tag{1}$$

We may derive the differential equations governing this system's behavior by summing voltages around the left-hand loop to get

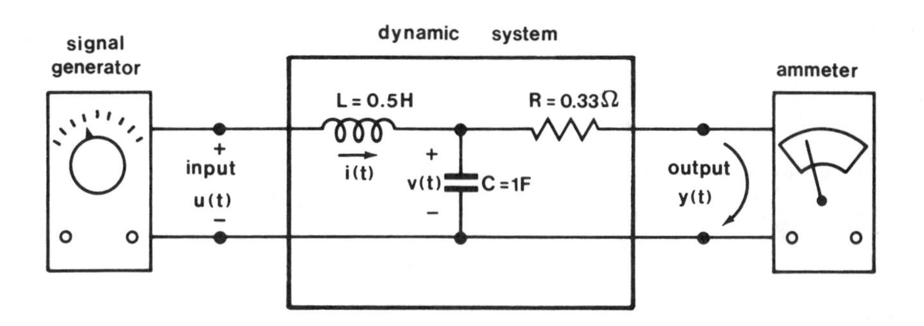

Fig. 1. Electrical circuit of Example 1.

$$\frac{1}{2}\frac{di(t)}{dt} = -v(t) + u(t) \tag{2}$$

and summing currents at the upper node to get

$$\frac{dv(t)}{dt} = i(t) - 3v(t) \tag{3}$$

The output current is

$$y(t) = 3v(t) \tag{4}$$

In terms of the state vector $\underline{x}(t)$, these equations may be written as

$$\frac{d}{dt}\begin{bmatrix} x_1(t) \\ x_2(t) \end{bmatrix} = \begin{bmatrix} -2x_2(t) + 2u(t) \\ x_1(t) - 3x_2(t) \end{bmatrix} = \begin{bmatrix} 0 & -2 \\ 1 & -3 \end{bmatrix}\begin{bmatrix} x_1(t) \\ x_2(t) \end{bmatrix} + \begin{bmatrix} 2 \\ 0 \end{bmatrix} u(t)$$

$$y(t) = 3x_2(t) = \begin{bmatrix} 0 & 3 \end{bmatrix}\begin{bmatrix} x_1(t) \\ x_2(t) \end{bmatrix} \tag{5}$$

This *first-order vector differential equation* is generally expressed as

$$\dot{\underline{x}}(t) = \underline{A}\underline{x}(t) + \underline{b}u(t)$$
$$y(t) = \underline{c}'\underline{x}(t) \tag{6}$$

where the dot denotes differentiation and

$$\underline{A} \triangleq \begin{bmatrix} 0 & -2 \\ 1 & -3 \end{bmatrix} \qquad \underline{b} \triangleq \begin{bmatrix} 2 \\ 0 \end{bmatrix} \qquad \underline{c}' \triangleq \begin{bmatrix} 0 & 3 \end{bmatrix} \tag{7}$$

Alternatively, the system's behavior may be described by an equation involving only the input and output. Differentiating (3) and substituting (2) and (4), we get a *second-order scalar differential equation,*

$$\ddot{y}(t) + 3\dot{y}(t) + 2y(t) = 6u(t) \tag{8}$$

ΔΔΔ

2. *EXAMPLE*. Consider the simple mechanical system shown in Fig. 2. Applied to the vehicle are an *input force* u(t) and a friction force proportional to the vehicle's velocity. The *output* y(t) of this system is the vehicle's position, measured from a fixed origin.

As any student of elementary dynamics knows, the behavior of this vehicle can be determined from time t on if three quantities are known: the vehicle's initial position y(t) and initial velocity v(t), and the input forcing function $u(\tau)$, $\tau \geq t$. Thus the state of this system at time t consists of the state variables y(t) and v(t), which may be collected into a 2-dimensional state vector

$$\underline{x}(t) = \begin{bmatrix} x_1(t) \\ x_2(t) \end{bmatrix} = \begin{bmatrix} y(t) \\ v(t) \end{bmatrix}$$

The differential equations governing this system's behavior may be derived by noting that velocity is the derivative of position,

$$\dot{y}(t) = v(t) \tag{9}$$

and that the sum of the forces applied is equal to the mass times the acceleration,

$$\frac{1}{2}\ddot{y}(t) = \frac{1}{2}\dot{v}(t) = u(t) - 3v(t) \tag{10}$$

In terms of the state vector $\underline{x}(t)$, these equations may be written

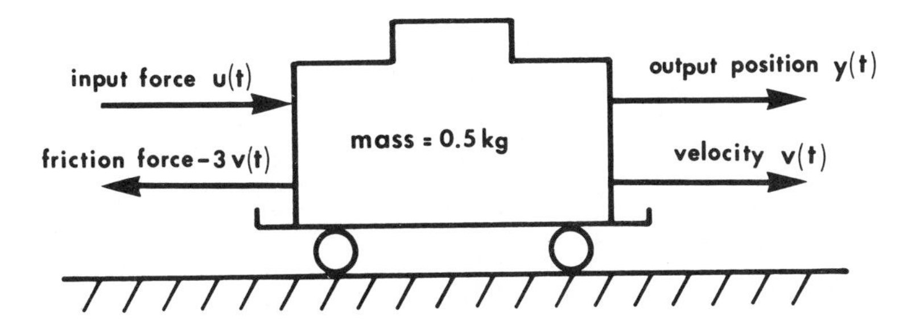

Fig. 2. Mechanical system of Example 2.

as

$$\frac{d}{dt}\begin{bmatrix} x_1(t) \\ x_2(t) \end{bmatrix} = \begin{bmatrix} x_2(t) \\ -6x_2(t) + 2u(t) \end{bmatrix} = \begin{bmatrix} 0 & 1 \\ 0 & -6 \end{bmatrix}\begin{bmatrix} x_1(t) \\ x_2(t) \end{bmatrix} + \begin{bmatrix} 0 \\ 2 \end{bmatrix} u(t) \tag{11}$$

$$y(t) = \quad x_1(t) \quad = \begin{bmatrix} 1 & 0 \end{bmatrix}\begin{bmatrix} x_1(t) \\ x_2(t) \end{bmatrix}$$

This first-order vector differential equation is generally expressed as

$$\begin{aligned} \dot{\underline{x}}(t) &= \underline{A}\underline{x}(t) + \underline{b}u(t) \\ y(t) &= \underline{c}'\underline{x}(t) \end{aligned} \tag{12}$$

where

$$\underline{A} \triangleq \begin{bmatrix} 0 & 1 \\ 0 & -6 \end{bmatrix} \qquad \underline{b} \triangleq \begin{bmatrix} 0 \\ 2 \end{bmatrix} \qquad \underline{c}' \triangleq \begin{bmatrix} 1 & 0 \end{bmatrix} \tag{13}$$

Alternatively, the system's behavior may be described by an equation involving only the input and output. Differentiating (9) and substituting (10) and (9), we get a second-order scalar differential equation,

$$\ddot{y}(t) + 6\dot{y}(t) = 2u(t) \tag{14}$$

ΔΔΔ

In general, the state, input, and output of a dynamic system may all be vector quantities, denoted $\underline{x}(t)$, $\underline{u}(t)$, and $\underline{y}(t)$, having dimensions n, m, and p, respectively. The system itself is often represented schematically by a *block diagram,* as shown in Fig. 3.

A general *nonlinear, time-varying*[†] *dynamic system* may be described by a differential equation and an output equation in the

[†]This system is said to be *time-varying* because the time t appears explicitly in $\underline{f}$ and $\underline{h}$. A system of the form $\dot{\underline{x}} = \underline{f}(\underline{x},\underline{u})$, $\underline{y} = \underline{h}(\underline{x},\underline{u})$ is said to be *time-invariant* (but still nonlinear).

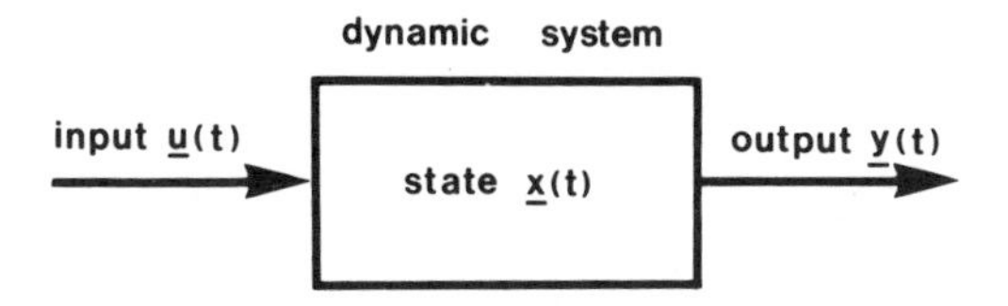

FIG. 3. Block diagram representation of dynamic system.

following *state-variable format:*

$$\begin{aligned} \dot{\underline{x}}(t) &= \underline{f}(\underline{x}(t),\underline{u}(t),t) \\ \underline{y}(t) &= \underline{h}(\underline{x}(t),\underline{u}(t),t) \end{aligned} \tag{15}$$

where $\underline{f}$ and $\underline{h}$ are *vector-valued functions* of the state $\underline{x}$ and the input $\underline{u}$ (see Examples 2.4.2 and 2.4.3).

In the case of a *linear, time-varying dynamic system,* the equations become

$$\begin{aligned} \dot{\underline{x}}(t) &= \underline{A}(t)\underline{x}(t) + \underline{B}(t)\underline{u}(t) \\ \underline{y}(t) &= \underline{C}(t)\underline{x}(t) + \underline{D}(t)\underline{u}(t) \end{aligned} \tag{16}$$

where $\underline{A}(t)$, $\underline{B}(t)$, $\underline{C}(t)$, and $\underline{D}(t)$ are time-dependent matrices and the right-hand sides are *linear* functions of the state and input.

Finally a *linear, time-invariant system* is described by

$$\begin{aligned} \dot{\underline{x}}(t) &= \underline{A}\underline{x}(t) + \underline{B}\underline{u}(t) \\ \underline{y}(t) &= \underline{C}\underline{x}(t) + \underline{D}\underline{u}(t) \end{aligned} \tag{17}$$

where $\underline{A}$, $\underline{B}$, $\underline{C}$, and $\underline{D}$ are constant matrices. Time invariance simply means that the system's parameters (mass, inductance, resistance, etc.) do not change as time goes on.

It is this latter class of systems which we shall consider throughout most of the book, and the second output term will usually be discarded, leaving[†]

$$\dot{\underline{x}}(t) = \underline{A}\underline{x}(t) + \underline{B}\underline{u}(t)$$
$$\underline{y}(t) = \underline{C}\underline{x}(t) \tag{18}$$

The matrices $\underline{A}$, $\underline{B}$, and $\underline{C}$ will be known as the *system matrix, input matrix,* and *output matrix,* respectively. Because the input and output are vector quantities, this is said to be a *multivariable* (or *multiple-input, multiple-output*) system. The subclass of *scalar* (or *single-input, single-output*) systems has the form

$$\dot{\underline{x}}(t) = \underline{A}\underline{x}(t) + \underline{b}u(t)$$
$$y(t) = \underline{c}'\underline{x}(t) \tag{19}$$

and includes Examples 1 and 2 above.

In some instances the output of a system is taken to be the entire state ($\underline{y} = \underline{x}$), in which case the output equation is omitted.

All differential equations, of course, require *boundary conditions* to be specified before a solution can be uniquely determined. For systems described in state-variable format, the initial state $\underline{x}(t_0)$ or $\underline{x}(0)$ will serve this purpose.

2.2. INPUT-OUTPUT FORM OF SYSTEM EQUATIONS

In this section we restrict our attention to linear, time-invariant *scalar* systems with a single input u(t) and a single output y(t). As was pointed out in Examples 1 and 2 of Sec. 2.1, an alternative form of the differential equation governing the system's behavior in this case is[††]

[†] The term $\underline{D}u(t)$ in the output is simply proportional to the input and hence independent of the system dynamics. Little generality is lost in ignoring it, provided we remember that it can be reappended to the equations if necessary. A system of the form (17) is said to be *proper* and one of the form (18) is *strictly proper* [39,185].

[††] In some cases a term $q_n u^{(n)}(t)$ might be included on the right-hand side, but it is omitted for the same reason that the term $\underline{D}u(t)$ was dropped from (2.1-17). See Problem 2.

$$y^{(n)}(t) + p_{n-1}y^{(n-1)}(t) + \dots p_1\dot{y}(t) + p_0y(t)$$
$$= q_{n-1}u^{(n-1)}(t) + \dots q_1\dot{u}(t) + q_0u(t) \qquad (1)$$

an n-th order scalar differential equation involving only the input u(t) and the output y(t). Time invariance means that the coefficients p_i and q_i are all constant.

This is the form in which system behavior is usually described in the conventional (or "classical") approach to control theory. Although it is adequate for some aspects of the analysis of scalar systems, it becomes very cumbersome in the case of multivariable systems, which may be described using coupled sets of equations of the form (1).

The following theorem verifies the interchangeability of the input-output format of (1) and the state-variable format of (2.1-18).

1. THEOREM. Consider a system described by an n-th order scalar differential equation of the form (1). A completely equivalent, state-variable description is the first-order vector differential equation†

$$\dot{\underline{x}}(t) = \begin{bmatrix} 0 & 1 & 0 & \dots & 0 & 0 \\ 0 & 0 & 1 & \dots & 0 & 0 \\ \dots & \dots & \dots & \dots & \dots & \dots \\ 0 & 0 & 0 & \dots & 1 & 0 \\ 0 & 0 & 0 & \dots & 0 & 1 \\ -p_0 & -p_1 & -p_2 & \dots & -p_{n-2} & -p_{n-1} \end{bmatrix} \underline{x}(t) + \begin{bmatrix} 0 \\ 0 \\ \vdots \\ 0 \\ 0 \\ 1 \end{bmatrix} u(t) \qquad (2)$$

and the output equation

$$y(t) = \begin{bmatrix} q_0 & q_1 & q_2 & \dots & q_{n-2} & q_{n-1} \end{bmatrix} \underline{x}(t) \qquad (3)$$

---

†In Sec. 10.4 this will be called the *controllable canonical form* of the state equations.

Proof: A general proof (using Laplace transforms) is requested in Problem 4.3-4. We shall prove the theorem only for the particular case where

$$q_1 = q_2 = \cdots = q_{n-1} = 0 \tag{4}$$

and the differential equation (1) becomes

$$y^{(n)}(t) + p_{n-1}y^{(n-1)}(t) + \cdots + p_1\dot{y}(t) + p_0 y(t) = q_0 u(t) \tag{5}$$

Substitution of (4) into (3) yields

$$y(t) = \begin{bmatrix} q_0 & 0 & 0 & \cdots & 0 & 0 \end{bmatrix} \underline{x}(t) = q_0 x_1(t) \tag{6}$$

Differentiating this equation and using the first row of (2), we have

$$\dot{y}(t) = q_0\dot{x}_1(t) = q_0 x_2(t) \tag{7}$$

Continuing in this manner and using successive rows of (2), we have

$$\begin{aligned} \ddot{y}(t) &= q_0\dot{x}_2(t) = q_0 x_3(t) \\ &\cdots\cdots\cdots\cdots\cdots \\ y^{(n-1)}(t) &= q_0\dot{x}_{n-1}(t) = q_0 x_n(t) \end{aligned} \tag{8}$$

and we may identify the state vector as

$$\underline{x}(t) = q_0^{-1} \begin{bmatrix} y(t) \\ \dot{y}(t) \\ \vdots \\ y^{(n-1)}(t) \end{bmatrix} \tag{9}$$

The last row of (2) is

$$\dot{x}_n(t) = -p_0 x_1(t) - p_1 x_2(t) - \cdots - p_{n-1} x_n(t) + u(t) \tag{10}$$

and substitution of (7) and (8) verifies that this is equivalent to (1).

ΔΔΔ

The reader should recall from the study of elementary differential equations that the information required to solve (5) from any time t_0 onward consists of the initial conditions $y(t_0)$, $\dot{y}(t_0)$, ..., $y^{(n-1)}(t_0)$ and the forcing function $u(t)$, $t \geq t_0$. Thus the definition (9) of the state is a natural one.

Theorem 1 provides an easy method for putting an n-th order differential equation of the form (1) into state-variable format: simply substitute the values of the coefficients p_i and q_i into (2) and (3). An alternative state-variable form is the "transpose" of (2) and (3), as set out in Problem 2.

One may also establish a converse result, that any scalar system in the state-variable form (2.1.19) can be described equivalently[†] in the input-output form (1): this will follow from Lemma 6 of Section 4.3.

2. *EXAMPLE*. Recall that a state vector in Example 1 of Sec. 2.1 was chosen on the basis of a physical argument to be

$$\underline{x}(t) = \begin{bmatrix} i(t) \\ v(t) \end{bmatrix} \tag{11}$$

where i and v represent an inductor current and a capacitor voltage, and the system equations were found to be

$$\dot{\underline{x}}(t) = \begin{bmatrix} 0 & -2 \\ 1 & -3 \end{bmatrix} \underline{x}(t) + \begin{bmatrix} 2 \\ 0 \end{bmatrix} u(t) \tag{12}$$

$$y(t) = \begin{bmatrix} 0 & 3 \end{bmatrix} \underline{x}(t)$$

However, it was also found that the input and output in this example are related by

---

[†]Actually, the equivalence in this case only applies to input-output behavior since, in general, a system may have *uncontrollable* and/or *unobservable* states which do not affect its input-output characteristics. This will become clear in Chapter 10.

$$\ddot{y}(t) + 3\dot{y}(t) + 2y(t) = 6u(t) \tag{13}$$

Since this equation has the form (5), we may choose a state vector as in (9),

$$\hat{\underline{x}}(t) = \begin{bmatrix} y(t)/6 \\ \dot{y}(t)/6 \end{bmatrix} \tag{14}$$

so that the state equations (2) and (3) are

$$\begin{aligned} \dot{\hat{\underline{x}}}(t) &= \begin{bmatrix} 0 & 1 \\ -2 & -3 \end{bmatrix} \hat{\underline{x}}(t) + \begin{bmatrix} 0 \\ 1 \end{bmatrix} u(t) \\ y(t) &= \begin{bmatrix} 6 & 0 \end{bmatrix} \hat{\underline{x}}(t) \end{aligned} \tag{15}$$

We thus have two alternative formulations of the problem, in terms of the state vectors $\underline{x}$ and $\hat{\underline{x}}$. Since the system is the same in both cases, these formulations must be completely equivalent. This equivalence will be made explicit in Example 2 of Sec. 2.5.

ΔΔΔ

2.3. CONSTRUCTION OF LINEAR SYSTEM MODELS

We have already stated that our main concern will be the study of systems for which linear, time-invariant differential equations provide a useful model. Before going on to consider what mathematical analysis allows us to do with such models, it is appropriate to illustrate their formulation with a number of examples. It is hoped that these will not only give the reader some familiarity with the process of model-building but also satisfy him that linear differential equations are, in fact, reasonable models for a wide range of real systems. One such model has already been derived in Example 1.1.1.

1. *EXAMPLE (Armature-controlled motor).* Consider the armature-controlled d.c. motor shown schematically in Fig. 1. Such motors are fairly common components in systems of interest to control engineers. The torque m(t) developed by the motor and hence its angular velocity $\omega(t)$ (or in some applications, its angular position) are controlled by varying the voltage v(t) applied to the armature winding. The stator (field) winding is connected to a constant voltage source, so that the current and hence the magnetic flux ϕ_f in the field winding are approximately constant. As in all d.c. motors, the torque developed depends on the strength of the magnetic field due to the stator windings and the current carried by the armature conductors. More precisely, it is approximately proportional to the product of the field density and the current, some nonlinearity being introduced by the distortion of the stator field due to the flux generated by the armature current. This so-called "armature reaction" is kept small by careful motor design. We may therefore write

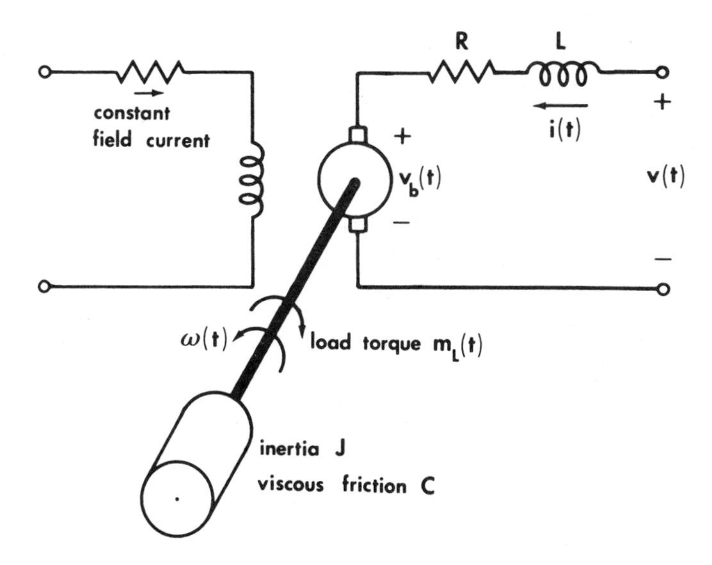

FIG. 1. Schematic diagram of armature-controlled d.c. motor.

$$m(t) = ki(t) \tag{1}$$

where i(t) is the armature current and k a constant depending on the particular motor.

Now consider the armature circuit. The armature conductors rotate in the magnetic field of the stator and so have a voltage $v_b(t)$ induced in them proportional to the stator field strength and the speed of rotation:

$$v_b(t) = k_c\phi_f\omega(t) \equiv k\omega(t) \tag{2}$$

Note that $k_c\phi_f$ must equal k since the electrical power converted to mechanical power is $v_b(t)i(t)$, and this equals $m(t)\omega(t)$. The voltage $v_b(t)$ opposes the externally applied armature voltage; so a sum of voltages in the armature circuit gives

$$L\frac{di(t)}{dt} = -Ri(t) + v(t) - v_b(t) \tag{3}$$

and combination of Eqs. (2) and (3) yields

$$\dot{i}(t) = -\frac{R}{L}i(t) - \frac{k}{L}\omega(t) + \frac{1}{L}v(t) \tag{4}$$

which is a linear differential equation relating the armature current to speed and applied voltage.

To complete our model of the motor, we need a relation between the motor torque m(t) and the angular velocity ω(t), which is provided by Newton's second law of dynamics as applied to rotating masses. Denoting the mass moment of inertia of the armature as J, and assuming that there is an externally applied load torque $m_L(t)$ and a viscous friction torque $C\omega(t)$ opposing the motion, we write

$$J\frac{d\omega(t)}{dt} = -C\omega(t) + m(t) - m_L(t) \tag{5}$$

Finally, substitution of Eq. (1) yields

$$\dot{\omega}(t) = -\frac{C}{J}\omega(t) + \frac{k}{J}i(t) - \frac{1}{J}m_L(t) \tag{6}$$

Equations (4) and (6) comprise our model for the motor. If we now choose some convenient numerical values for the various parameters,

$$\begin{aligned} R &= 1\ \Omega & C &= 2\ \text{N}\cdot\text{m}\cdot\text{sec} \\ L &= \frac{1}{2}\ \text{H} & k &= 3\ \frac{\text{N}\cdot\text{m}}{\text{A}} \\ J &= \frac{1}{2}\ \text{kg}\cdot\text{m}^2 \end{aligned} \tag{7}$$

these equations become

$$\dot{i}(t) = -2i(t) - 6\omega(t) + 2v(t) \tag{8}$$

$$\dot{\omega}(t) = 6i(t) - 4\omega(t) - 2m_L(t) \tag{9}$$

The armature voltage and external load torque are inputs to the system. If they are zero, then clearly a knowledge of $i(t)$ and $\omega(t)$ at some value of t, say $t = 0$, is sufficient to allow the equations to be solved for all future values of time. Thus, by our definition of state, the variables $i(t)$ and $\omega(t)$ form a suitable *state vector*

$$\underline{x}(t) \triangleq \begin{bmatrix} i(t) \\ \omega(t) \end{bmatrix} \tag{10}$$

Defining a vector of inputs

$$\underline{u}(t) \triangleq \begin{bmatrix} v(t) \\ m_L(t) \end{bmatrix} \tag{11}$$

and considering the speed $\omega(t)$ to be the single output of the system, we may write Eqs. (8) and (9) in state-variable format:

$$\begin{aligned} \dot{\underline{x}}(t) &= \underline{\underline{A}}\underline{x}(t) + \underline{\underline{B}}\underline{u}(t) = \begin{bmatrix} -2 & -6 \\ 6 & -4 \end{bmatrix} \underline{x}(t) + \begin{bmatrix} 2 & 0 \\ 0 & -2 \end{bmatrix} \underline{u}(t) \\ y(t) &= \underline{c}'\underline{x}(t) \qquad\qquad\quad = \begin{bmatrix} 0 & 1 \end{bmatrix} \underline{x}(t) \end{aligned} \tag{12}$$

As an alternative to this form, (8) and (9) may be combined into a single scalar differential equation relating the output ω to the inputs v and m_L. To do this, differentiate (9):

$$\ddot{\omega}(t) = 6\dot{i}(t) - 4\dot{\omega}(t) - 2\dot{m}_L(t) \tag{13}$$

Next, eliminate i between (8) and (9) leaving

$$\dot{i}(t) = \frac{1}{3}\left[-\dot{\omega}(t) - 22\omega(t) + 6v(t) - 2m_L(t)\right] \tag{14}$$

and substitute this into (13) to get

$$\ddot{\omega}(t) + 6\dot{\omega}(t) + 44\omega(t) = 12v(t) - 4m_L(t) - 2\dot{m}_L(t) \tag{15}$$

which is the required equation in input-output format.

The above analysis has assumed motor speed ω to be the variable of interest. In some applications, of course, the angular position of the motor shaft will also be important, in which case the model may be augmented to include it (see Problem 1).

ΔΔΔ

2. *EXAMPLE (Broom-balancer).* The dynamic system shown in Fig. 2 consists of a cart with an inverted pendulum attached, and is

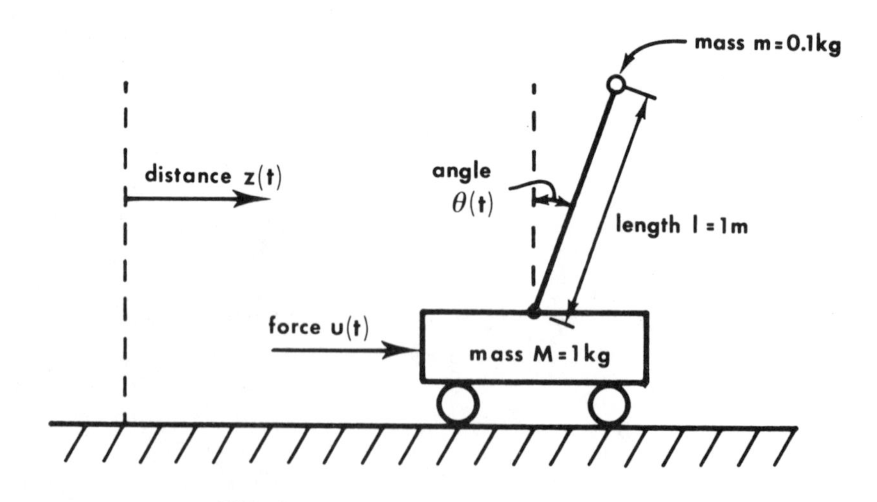

FIG. 2. Broom-balancing system.

commonly known as a *broom-balancer* because the problem of applying a force u(t) to keep the pendulum upright is very similar to that of balancing a broomstick on one's hand. It has a more practical interpretation in terms of launching a rocket, which must be balanced on its own thrust vector by rotating the engine.

We shall simplify this example by constraining the cart and pendulum to move in only one plane and neglecting such factors as the mass of the rod, motor dynamics, friction, and gusts of wind, but the essence of the problem remains. Note in particular that the system is fundamentally *unstable* in the sense that the broom will inevitably fall if no controlling force is applied.

The dynamic behavior of this system is completely described by the position and velocity of the cart and the angular position and velocity of the pendulum; so we may define the state vector to be

$$\underline{x}(t) = \left[z(t) \;\; \dot{z}(t) \;\; \theta(t) \;\; \dot{\theta}(t) \right] \tag{16}$$

To determine the differential equations governing this system, we apply Newton's second law to the horizontal motion of the cart, and to the horizontal and vertical motions of the pendulum mass. If the pendulum rod is rigid and has negligible mass, then it will have a negligible moment of inertia and the force T exerted by the cart on the pendulum at the pivot must always act along the rod. Newton's law then yields

$$u - T \sin \theta = M\ddot{z} \tag{17a}$$

$$T \sin \theta = m \frac{d^2}{dt^2}(z + \ell \sin \theta) = m(\ddot{z} + \ell \ddot{\theta} \cos \theta - \ell \dot{\theta}^2 \sin \theta) \tag{17b}$$

$$mg - T \cos \theta = m \frac{d^2}{dt^2}(\ell - \ell \cos \theta) = m \ell (\ddot{\theta} \sin \theta - \dot{\theta}^2 \cos \theta) \tag{17c}$$

where g is the gravitational acceleration. Eliminating T gives

$$(M + m)\ddot{z} + m \ell (\ddot{\theta} \cos \theta - \dot{\theta}^2 \sin \theta) = u \tag{18}$$

$$mz \cos \theta + m \ell \ddot{\theta} = mg \sin \theta \tag{19}$$

These differential equations are quite nonlinear, since they contain various products and trigonometric functions of the state variables, and they are virtually impossible to solve except with numerical methods. Consequently, we need a further simplification. Since the object of controlling the system is to keep the pendulum upright, it seems reasonable to assume that $\Theta(t)$ and $\dot{\Theta}(t)$ will remain close to zero. With this assumption, we can *linearize* the equations by retaining only those terms which are linear in Θ and $\dot{\Theta}$ and neglecting higher order terms such as Θ^2, $\dot{\Theta}^2$, and $\Theta\dot{\Theta}$ on the grounds that they will be insignificantly small.

Applying this process to the trigonometric expansions, we have

$$\begin{aligned} \sin\Theta &= \Theta - \frac{\Theta^3}{3!} + \frac{\Theta^5}{5!} - \cdots \approx \Theta \\ \cos\Theta &= 1 - \frac{\Theta^2}{2!} + \frac{\Theta^4}{4!} - \cdots \approx 1 \end{aligned} \tag{20}$$

Substituting these into Eqs. (18) and (19) and dropping the higher order terms, we get a set of *approximate* differential equations for the system,

$$(M + m)\ddot{z}(t) + m\ell\ddot{\Theta}(t) \approx u(t) \tag{21}$$

$$m\ddot{z}(t) + m\ell\ddot{\Theta}(t) \approx mg\Theta(t) \tag{22}$$

The validity of these equations depends, of course, on the validity of the assumption that $\Theta(t) \approx \dot{\Theta}(t) \approx 0$. As long as a control force is applied which maintains this condition, our mathematical model will be accurate; otherwise, it will break down.

Equations (21) and (22) may be solved simultaneously for†

$$\ddot{z} = -\frac{mg}{M}\Theta + \frac{1}{M}u \tag{23}$$

$$\ddot{\Theta} = \frac{(M + m)g}{M\ell}\Theta - \frac{1}{M\ell}u \tag{24}$$

---

†We shall drop the $\approx$ notation for the sake of convenience.

If we also include the identities

$$\frac{d}{dt} z = \dot{z} \qquad \frac{d}{dt} \Theta = \dot{\Theta} \tag{25}$$

and designate z as the system output, then these relations may be expressed in state-variable format as

$$\frac{d}{dt}\begin{bmatrix} z \\ \dot{z} \\ \Theta \\ \dot{\Theta} \end{bmatrix} = \begin{bmatrix} 0 & 1 & 0 & 0 \\ 0 & 0 & -\frac{mg}{M} & 0 \\ 0 & 0 & 0 & 1 \\ 0 & 0 & \frac{(M+m)g}{M\ell} & 0 \end{bmatrix} \begin{bmatrix} z \\ \dot{z} \\ \Theta \\ \dot{\Theta} \end{bmatrix} + \begin{bmatrix} 0 \\ \frac{1}{M} \\ 0 \\ -\frac{1}{M\ell} \end{bmatrix} u$$

$$z = \begin{bmatrix} 1 & 0 & 0 & 0 \end{bmatrix} \begin{bmatrix} z \\ \dot{z} \\ \Theta \\ \dot{\Theta} \end{bmatrix} \tag{26}$$

Suppose that the system parameters are

$$M = 1 \text{ kg} \qquad m = 0.1 \text{ kg} \qquad \ell = 1 \text{ m} \tag{27}$$

and recall that the acceleration of gravity is $g = 9.81 \text{ m/sec}^2$. Then the entries in the system matrix are

$$\begin{aligned} \frac{mg}{M} &= 0.981 \approx 1 \\ \frac{(M+m)g}{M\ell} &= 10.79 \approx 11 \\ \frac{1}{M} &= \frac{1}{M\ell} = 1 \end{aligned} \tag{28}$$

and (26) becomes

$$\dot{\underline{x}} = \begin{bmatrix} 0 & 1 & 0 & 0 \\ 0 & 0 & -1 & 0 \\ 0 & 0 & 0 & 1 \\ 0 & 0 & 11 & 0 \end{bmatrix} \underline{x} + \begin{bmatrix} 0 \\ 1 \\ 0 \\ -1 \end{bmatrix} u \tag{29}$$

$$z = \begin{bmatrix} 1 & 0 & 0 & 0 \end{bmatrix} \underline{x}$$

This is a linearized set of differential equations in state-variable format which governs the behavior of the broom-balancer as long as the angle Θ remains small. We shall examine various properties of this system, as well as methods of controlling it, in subsequent chapters.

Note that the system dynamics can be described by combining (23) and (24) into a fourth-order scalar differential equation,

$$\ddot{\ddot{z}} - \frac{(M+m)g}{M\ell}\ddot{z} = \frac{1}{M}\ddot{u} - \frac{g}{M\ell}u \tag{30}$$

However, this form is much less convenient than (29) for studying the system, since the variable to be controlled, Θ, does not appear explicitly in the equation.

ΔΔΔ

3. EXAMPLE (Bubonic plague). In this example we suppose that a city is afflicted with bubonic plague, and construct a mathematical model in order to predict the course of the epidemic and to analyze various strategies for dealing with it.

The population of the city can be divided into four groups at any time t (measured in days),

$x_1(t)$ persons are *susceptible* to the disease
$x_2(t)$ persons are currently *infected* with it
$x_3(t)$ persons have been *immunized* against it
$x_4(t)$ persons have already *died* from it

and we define the state vector† for our model to be

$$\underline{x}(t) = [x_1(t) \quad x_2(t) \quad x_3(t) \quad x_4(t)]' \tag{31}$$

In order to analyze the dynamic behavior of the model, we assume that each day

A fraction α of susceptibles are immunized

A fraction β of susceptibles are infected

A fraction γ of infectives die

A fraction δ of infectives are cured and immunized

We also assume that new susceptibles and new infectives are added to the population at input rates of $u_1(t)$ and $u_2(t)$ per day, respectively.

With these assumptions we can specify the derivatives of the state variables,

$$\begin{aligned} \dot{x}_1 &= -\alpha x_1 - \beta x_1 + u_1 \\ \dot{x}_2 &= \beta x_1 - \gamma x_2 - \delta x_2 + u_2 \\ \dot{x}_3 &= \alpha x_1 + \delta x_2 \\ \dot{x}_4 &= \gamma x_2 \end{aligned} \tag{32}$$

Collecting these differential equations into the standard state-variable format and defining the outputs (somewhat arbitrarily) to be the number immunized and the number dead, we have

$$\begin{aligned} \dot{\underline{x}} &= \begin{bmatrix} -(\alpha+\beta) & 0 & 0 & 0 \\ \beta & -(\gamma+\delta) & 0 & 0 \\ \alpha & \delta & 0 & 0 \\ 0 & \gamma & 0 & 0 \end{bmatrix} \underline{x} + \begin{bmatrix} 1 & 0 \\ 0 & 1 \\ 0 & 0 \\ 0 & 0 \end{bmatrix} \underline{u} \\ \underline{y} &= \begin{bmatrix} 0 & 0 & 1 & 0 \\ 0 & 0 & 0 & 1 \end{bmatrix} \underline{x} \end{aligned} \tag{33}$$

†Strictly speaking, the numbers x_1, x_2, x_3 and x_4 should all be integers, but we shall assume that the population is sufficiently large for this restriction to be ignored.

Later, in Problem 4.2-7, the behavior of $\underline{x}(t)$ will be analyzed for some specific values of α, β, γ, and δ. Strategies for dealing with the epidemic, such as the importation of medical teams to alter α and δ, will also be examined.

One might argue that a more realistic model would assume the rate of infection to be related not only to the number of susceptibles but also to the number who are already infected and hence spreading the disease. For example, the term βx_1 in the first two differential equations could be replaced by $\beta x_1 x_2$, resulting in a *bilinear* set of differential equations. The interested reader may wish to consider this and other refinements and to investigate linearization of the resulting nonlinear models (see Problem 2).

ΔΔΔ

2.4. LINEARIZATION OF NONLINEAR MODELS

From the examples discussed in the previous sections, it should be clear that linear differential equations are accurate models over wide ranges of the system variables only in the case of simple electrical and mechanical networks, and then only if the dynamic response of the systems is sufficiently slow so that energy transfers by electromagnetic radiation or by transmission of shock waves along components are negligible. In most of the examples presented so far, the equations describing the systems were in fact quite nonlinear. Linear relationships could be obtained only by restricting the system variables to sufficiently small deviations from a normal "operating point" or equilibrium value. If such assumptions cannot validly be made, one has no choice but to deal with the nonlinear differential equations directly, using ad hoc techniques, numerical algorithms, and other methods which are beyond the scope of this book.

However, the use of linear models which are accurate only for relatively small ranges of the variables, known as *small-signal analysis*, has a great deal of practical value. One reason is that many automatic control systems are in fact designed to maintain themselves as close to a desired equilibrium as possible; the governor system of a power station turbine is an example of this, as is the autopilot system of an aircraft. Another reason is that even in highly nonlinear systems, small perturbations from a known solution of the nonlinear equations of the system usually can be described by a linear model.

The main purpose of this section will be to present a somewhat more systematic method of deriving such small-signal linear models from nonlinear system equations. We begin with two examples.

1. EXAMPLE (Pneumatic controller). In many industrial applications of automatic control, pneumatic equipment is used for measuring, transmitting and processing signals. The flapper amplifier shown schematically in Fig. 1 is a key component in the pneumatic controllers commonly used in such systems. It is essentially a sensitive amplifier which transforms the input u(t), the position of the flapper, into a change in air pressure y(t) in the output line. We assume

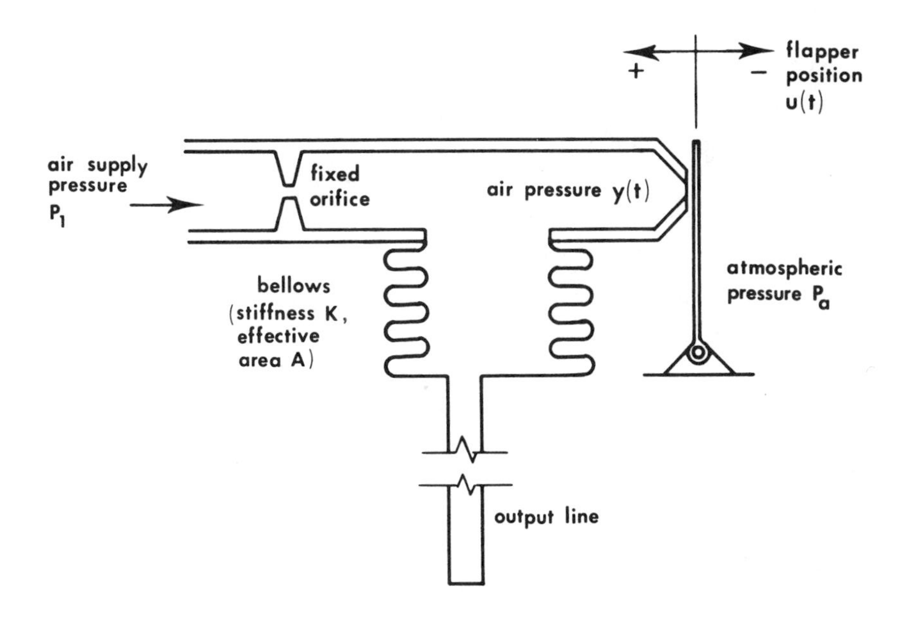

FIG. 1. Pneumatic flapper amplifier.

that the output line is closed at its other end so that no air can escape through it from the bellows chamber. This means that for each fixed position of the flapper, an equilibrium state must be reached in which the mass flow rate of air into the bellows chamber through the fixed orifice equals the mass flow rate of air out of the chamber through the orifice formed by the nozzle and flapper; in the equilibrium state, the mass of air stored inside the chamber cannot change.

The equations describing the mass flow rate of a compressible fluid through orifices or restrictions in a conduit are quite complex and very nonlinear. However, we do not need them in their precise form for this example. If we make the quite reasonable assumptions that the air supply line delivers air at constant pressure P_1 and constant temperature, that the air velocities at all places other than in the restrictions are negligible, that the flow through the restrictions is adiabatic (i.e., there is no heat loss to the metal walls of the orifices), and that the atmospheric pressure P_a remains constant, then it turns out that the mass flow rates through the restrictions depend only on the pressure drop across them, as in incompressible fluid flow. Thus for any *fixed* flapper position u^*, the chamber pressure must assume the constant value y^* necessary to give equal mass flow rates through the orifices. Figure 2 shows a typical graph relating equilibrium values of flapper position u^* and bellows pressure y^*. Let $m_1(t)$ and $m_2(t)$ be the flow rates through the fixed and variable restrictions, respectively. They depend on the pressure drops across the restrictions and in the latter case also on the orifice area and hence on $u(t)$. We indicate this by writing

$$m_1(t) = f_1(P_1 - y(t)) = f(y(t)) \tag{1}$$

$$m_2(t) = g_1(y(t) - P_a, u(t)) = g(y(t),u(t)) \tag{2}$$

Suppose now that the input is perturbed from its equilibrium value:

$$u(t) = u^* + \varepsilon\hat{u}(t) \tag{3}$$

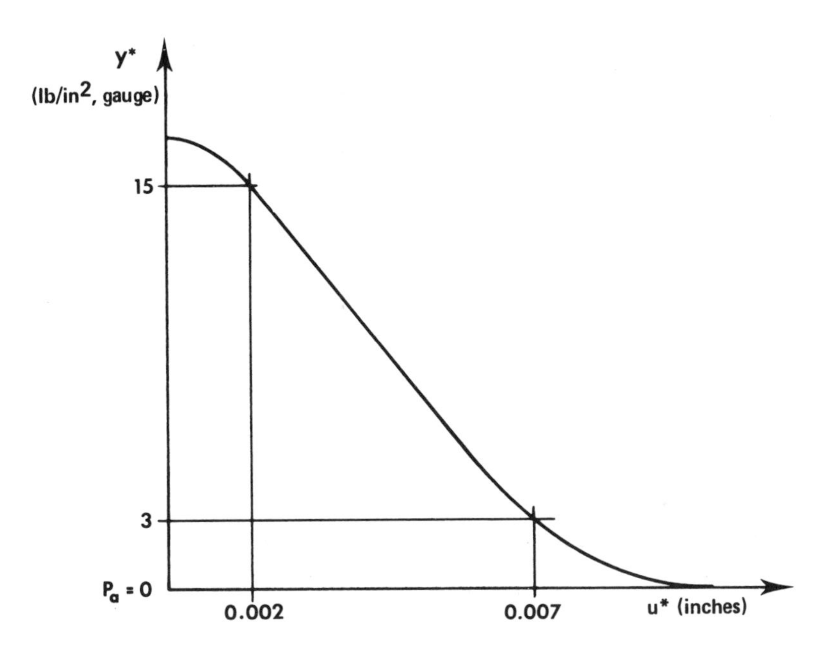

FIG. 2. The steady state characteristic of a flapper amplifier.

where ε is a small positive number. As a result, all other variables will also be perturbed from their equilibrium values:

$$\begin{aligned} y(t) &= y^* + \varepsilon\hat{y}(t) \\ m_1(t) &= m^* + \varepsilon\hat{m}_1(t) \\ m_2(t) &= m^* + \varepsilon\hat{m}_2(t) \end{aligned} \tag{4}$$

Substituting these into (1) and (2), and expanding f and g in Taylor series about the equilibrium, we obtain

$$m^* + \varepsilon\hat{m}_1(t) = f(y^*) + \varepsilon\frac{\partial f(y^*)}{\partial y}\hat{y}(t) + o(\varepsilon^2) \tag{5}$$

$$m^* + \varepsilon\hat{m}_2(t) = g(y^*,u^*) + \varepsilon\frac{\partial g(y^*,u^*)}{\partial y}\hat{y}(t) + \varepsilon\frac{\partial g(y^*,u^*)}{\partial u}\hat{u}(t) + o(\varepsilon^2) \tag{6}$$

where $o(\varepsilon^2)$ indicates terms proportional to ε^2, ε^3, ... If ε is sufficiently small, these may be neglected. Moreover, since $f(y^*) = g(y^*,u^*) = m^*$ at equilibrium, (5) and (6) simplify to

$$\varepsilon\hat{m}_1(t) = \varepsilon\frac{\partial f(y^*)}{\partial y}\hat{y}(t)$$

$$\varepsilon\hat{m}_2(t) = \varepsilon\frac{\partial g(y^*,u^*)}{\partial y}\hat{y}(t) + \varepsilon\frac{\partial g(y^*,u^*)}{\partial u}\hat{u}(t)$$

and division by ε yields

$$\hat{m}_1(t) = -c_1\hat{y}(t) \tag{7}$$

$$\hat{m}_2(t) = c_2\hat{y}(t) - c_3\hat{u}(t) \tag{8}$$

where the positive constants c_i are obtained by evaluating the partial derivatives. It is worth pointing out that the process of replacing Eqs. (1) and (2) by (7) and (8) amounts to translating the origin of the coordinates to the equilibrium point, and replacing the curves (1) and (2) by their tangents at the new origin.

Now let $x(t)$ denote the total mass of air stored in the bellows chamber, and $\varepsilon\hat{x}(t)$ the change in x from its equilibrium value. The continuity principle then yields

$$\dot{x}(t) = m_1(t) - m_2(t) \tag{9}$$

from which it follows immediately that

$$\dot{\hat{x}}(t) = \hat{m}_1(t) - \hat{m}_2(t) \tag{10}$$

Equations (7), (8), and (10) are not yet sufficient to obtain a relation between the input perturbation $\hat{u}(t)$ and the output change $\hat{y}(t)$. To complete the model, we assume that the air inside the bellows chamber obeys the perfect gas equation

$$y(t) = RT\frac{x(t)}{v(t)} \tag{11}$$

where $v(t)$ is the total chamber volume, T is the stagnation temperature of the air (assumed constant), and R is the gas content. If we again consider only small perturbations from equilibrium, then a Taylor series expansion allows the nonlinear gas equation to be replaced by

$$\hat{y}(t) = c_4\hat{x}(t) - c_5\hat{v}(t) \tag{12}$$

where the c_i are again positive constants.

Finally, we sum forces on the bellows:

$$\frac{K\hat{v}(t)}{A} = A\hat{y}(t) \tag{13}$$

The above equations may be combined into

$$\dot{\hat{x}}(t) = -c_6(c_1 + c_2)\hat{x}(t) + c_3\hat{u}(t) \tag{14}$$

$$\hat{y}(t) = c_6\hat{x}(t) \tag{15}$$

where we have set $c_6 = c_4 / (1 + c_5A^2/K)$. This is a first-order system with state $x(t)$. In input-output format, it becomes

$$\dot{\hat{y}}(t) = -c_6(c_1 + c_2)\hat{y}(t) + c_6c_3\hat{u}(t) \tag{16}$$

This linear, time-invariant differential equation describing the behavior of the perturbations remains valid as long as $\varepsilon\hat{u}$, $\varepsilon\hat{x}$, and $\varepsilon\hat{y}$ remain sufficiently small.

ΔΔΔ

2. *EXAMPLE (Tunnel diode).* Consider the nonlinear circuit shown in Fig. 3(a), where the tunnel diode has the nonlinear voltage-current relationship shown in Fig. 3(b). In other words, the current $i(t)$ through the diode at any given time t is a function only of the voltage $v(t)$,

$$i(t) = p(v(t)) \tag{17}$$

Since the only energy storage device in the circuit (i.e., the only device with "memory") is the blocking capacitor C, we may take its voltage x as the state of the system. Summing the currents at the upper node, we obtain the differential equation

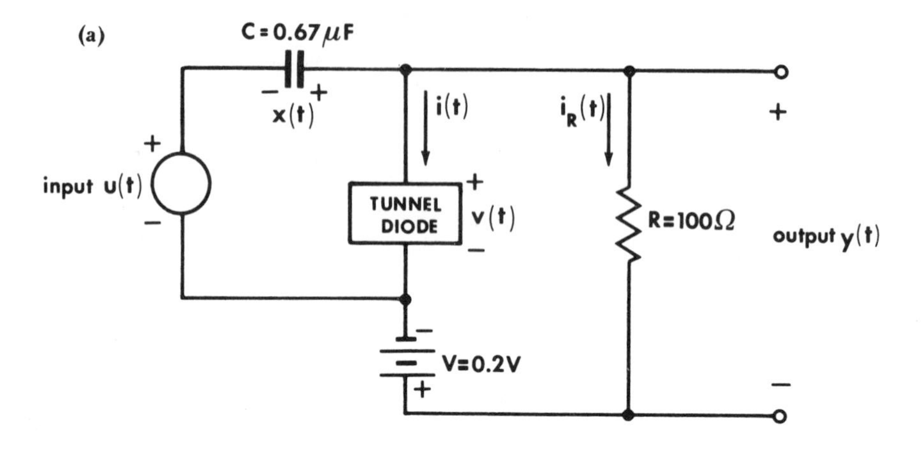

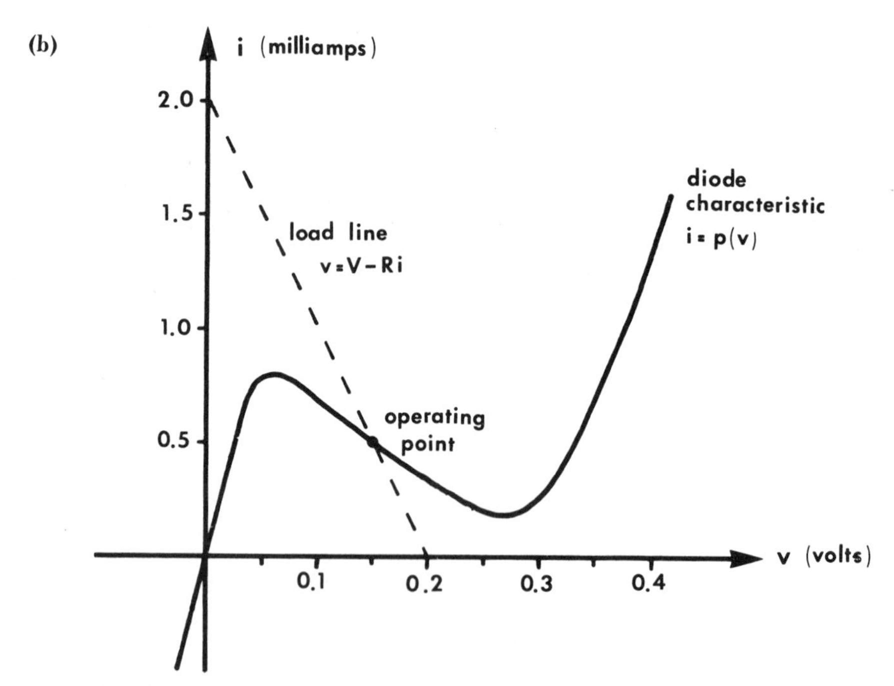

FIG. 3. Tunnel diode of Example 2: (a) amplifier circuit and (b) characteristic curve.

$$C\dot{x} = -i - i_R = -p(v) - i_R \tag{18}$$

and the output equation is

$$y = R\,i_R \tag{19}$$

Summing voltages around the loops, we have

$$v = u + x \tag{20}$$

$$R\,i_R = v - V = u + x - V \tag{21}$$

which may be substituted into (18) and (19) to yield

$$\dot{x} = -\frac{1}{C}p(u+x) - \frac{u+x-V}{RC} \triangleq f(x,u) \tag{22}$$

$$y = u + x - V \triangleq h(x,u) \tag{23}$$

The system is thus described by a nonlinear differential equation (22) which cannot be solved in general, except by numerical means.

Suppose for a moment that the input u is zero. Then it is easy to verify that Eqs. (22) and (23) have a constant solution[†]

$$\begin{aligned} u^*(t) &= 0\text{ V} \\ x^*(t) &= 0.15\text{ V} \\ \dot{x}^*(t) &= f(x^*,u^*) = 0\text{ V} \\ y^*(t) &= h(x^*,u^*) = -0.05\text{ V} \end{aligned} \tag{24}$$

Now suppose that the input is perturbed slightly from its nominal value of zero, that is,

$$u(t) = u^*(t) + \varepsilon\hat{u}(t) = \varepsilon\hat{u}(t) \tag{25}$$

where $0 < \varepsilon \ll 1$. Then x and y will also be perturbed, so that

$$\begin{aligned} x(t) &= x^*(t) + \varepsilon\hat{x}(t) = 0.15 + \varepsilon\hat{x}(t) \\ y(t) &= y^*(t) + \varepsilon\hat{y}(t) = -0.05 + \varepsilon\hat{y}(t) \end{aligned} \tag{26}$$

[†]Since $x = v$, the solution must satisfy $p(x) = (V-x)/R$, and is found graphically in Fig. 3(b) as the point of intersection of the *load line* and the diode's characteristic curve.

Substituting these variables into Eqs. (22) and (23), and expanding the right-hand sides in Taylor series, we have

$$\dot{x}^* + \varepsilon\dot{\hat{x}} = f(x^*,u^*) + \frac{\partial f(x^*,u^*)}{\partial x}\varepsilon\hat{x} + \frac{\partial f(x^*,u^*)}{\partial u}\varepsilon\hat{u} + o(\varepsilon^2)$$
$$y^* + \varepsilon\hat{y} = h(x^*,u^*) + \frac{\partial h(x^*,u^*)}{\partial x}\varepsilon\hat{x} + \frac{\partial h(x^*,u^*)}{\partial u}\varepsilon\hat{u} + o(\varepsilon^2) \qquad (27)$$

As in the previous example, we discard the higher-order terms in ε, subtract out (24), and divide by ε to get

$$\dot{\hat{x}} = \frac{\partial f(x^*,u^*)}{\partial x}\hat{x} + \frac{\partial f(x^*,u^*)}{\partial u}\hat{u}$$
$$\hat{y} = \frac{\partial h(x^*,u^*)}{\partial x}\hat{x} + \frac{\partial h(x^*,u^*)}{\partial x}\hat{u} \qquad (28)$$

The partial derivatives are found by differentiating (22) and (23) and noting that the slope of the graph at the operating point is $-(300\ \Omega)^{-1}$:

$$\dot{\hat{x}}(t) = -(10^4)\hat{x}(t) - (10^4)\hat{u}(t)$$
$$\hat{y}(t) = \hat{x}(t) + \hat{u}(t) \qquad (29)$$

Thus we have a set of linear, time-invariant equations describing the perturbations, which remain valid as long as $\varepsilon\hat{u}$, $\varepsilon\hat{x}$, and $\varepsilon\hat{y}$ remain sufficiently small.

ΔΔΔ

The analysis used in Examples 1 and 2 can be applied to the general case of a nonlinear system described by

$$\dot{\underline{x}}(t) = \underline{f}(\underline{x}(t),\underline{u}(t))$$
$$\underline{y}(t) = \underline{h}(\underline{x}(t),\underline{u}(t)) \qquad (30)$$

We assume that the partial derivatives

$$\frac{\partial f_j(\underline{x}(t),\underline{u}(t))}{\partial x_i} \qquad \frac{\partial f_j(\underline{x}(t),\underline{u}(t))}{\partial u_k} \qquad (i,j = 1,2,\ldots,n)$$

$$\frac{\partial h_\ell(\underline{x}(t),\underline{u}(t))}{\partial x_i} \qquad \frac{\partial h_\ell(\underline{x}(t),\underline{u}(t))}{\partial u_k} \qquad \begin{matrix}(k = 1,2,\ldots,m)\\(\ell = 1,2,\ldots,p)\end{matrix} \tag{31}$$

exist and are continuous for all $\underline{x}(t)$ and $\underline{u}(t)$.

Now suppose that the time functions $\underline{x}^*(t)$, $\underline{u}^*(t)$, and $\underline{y}^*(t)$ constitute a nominal solution of the nonlinear equations in (30), that is,

$$\begin{aligned} \dot{\underline{x}}^*(t) &= \underline{f}(\underline{x}^*(t),\underline{u}^*(t)) \\ \underline{y}^*(t) &= \underline{h}(\underline{x}^*(t),\underline{u}^*(t)) \end{aligned} \tag{32}$$

We shall consider *perturbations* about this solution of the form

$$\begin{aligned} \underline{x}(t) &= \underline{x}^*(t) + \varepsilon\hat{\underline{x}}(t) \\ \underline{u}(t) &= \underline{u}^*(t) + \varepsilon\hat{\underline{u}}(t) \\ \underline{y}(t) &= \underline{y}^*(t) + \varepsilon\hat{\underline{y}}(t) \end{aligned} \tag{33}$$

where $0 < \varepsilon << 1$, and determine what equations $\hat{\underline{x}}$, $\hat{\underline{u}}$, and $\hat{\underline{y}}$ should satisfy in order that $\underline{x}$, $\underline{u}$, $\underline{y}$ is still a solution of (30).

The rows of the first equation in (30) are

$$\begin{aligned} \dot{x}_j &= f_j(\underline{x},\underline{u}) \\ &= f_j(x_1,x_2,\ldots,x_n\,;\,u_1,u_2,\ldots,u_m) \qquad (j = 1,2,\ldots,n) \end{aligned} \tag{34}$$

Substituting (33) and expanding the right-hand side in a Taylor series, we find

$$\begin{aligned} \dot{x}_j^* + \varepsilon\dot{\hat{x}}_j = f_j(\underline{x}^*,\underline{u}^*) &+ \left.\frac{\partial f_j}{\partial x_1}\right|_* \varepsilon\hat{x}_1 + \cdots + \left.\frac{\partial f_j}{\partial x_n}\right|_* \varepsilon\hat{x}_n \\ &+ \left.\frac{\partial f_j}{\partial u_1}\right|_* \varepsilon\hat{u}_1 + \cdots + \left.\frac{\partial f_j}{\partial u_m}\right|_* \varepsilon\hat{u}_m + o_j(\varepsilon^2) \\ &\qquad (j = 1,2,\ldots,n) \end{aligned} \tag{35}$$

where we have adopted the notation

$$\left.\frac{\partial f_j}{\partial x_i}\right|_* \triangleq \frac{\partial f_j(\underline{x}^*(t),\underline{u}^*(t))}{\partial x_i} \tag{36}$$

and $o_j(\varepsilon^2)$ indicates terms proportional to ε^2, ε^3, ... If we now define the *partial derivative of a vector with respect to a vector* to be a matrix

$$\frac{\partial \underline{f}(\underline{x},\underline{u})}{\partial \underline{x}} \triangleq \begin{bmatrix} \frac{\partial f_1(\underline{x},\underline{u})}{\partial x_1} & \cdots & \frac{\partial f_1(\underline{x},\underline{u})}{\partial x_n} \\ \cdots & \cdots & \cdots \\ \frac{\partial f_n(\underline{x},\underline{u})}{\partial x_1} & \cdots & \frac{\partial f_n(\underline{x},\underline{u})}{\partial x_n} \end{bmatrix} \tag{37}$$

the n equations in (35) may be collected into†

$$\dot{\underline{x}}^* + \varepsilon\dot{\hat{\underline{x}}} = \underline{f}(\underline{x}^*,\underline{u}^*) + \left.\frac{\partial \underline{f}}{\partial \underline{x}}\right|_* \varepsilon\hat{\underline{x}} + \left.\frac{\partial \underline{f}}{\partial \underline{u}}\right|_* \varepsilon\hat{\underline{u}} + \underline{o}(\varepsilon^2) \tag{38}$$

Similarly, the output equation may be expanded into†

$$\underline{y}^* + \varepsilon\hat{\underline{y}} = \underline{h}(\underline{x}^*,\underline{u}^*) + \left.\frac{\partial \underline{h}}{\partial \underline{x}}\right|_* \varepsilon\hat{\underline{x}} + \left.\frac{\partial \underline{h}}{\partial \underline{u}}\right|_* \varepsilon\hat{\underline{u}} + \underline{o}(\varepsilon^2) \tag{39}$$

Now we may use the fact that $\underline{x}^*, \underline{u}^*$, and $\underline{y}^*$ are a solution of the original equations, subtract (32) from Eqs. (38) and (39), and divide by ε to get

$$\begin{aligned} \dot{\hat{\underline{x}}} &= \left.\frac{\partial \underline{f}}{\partial \underline{x}}\right|_* \hat{\underline{x}} + \left.\frac{\partial \underline{f}}{\partial \underline{u}}\right|_* \hat{\underline{u}} + \underline{o}(\varepsilon) \\ \hat{\underline{y}} &= \left.\frac{\partial \underline{h}}{\partial \underline{x}}\right|_* \hat{\underline{x}} + \left.\frac{\partial \underline{h}}{\partial \underline{u}}\right|_* \hat{\underline{u}} + \underline{o}(\varepsilon) \end{aligned} \tag{40}$$

---

†The reader should verify this to his own satisfaction (See Problem 1).

Recall that $\underline{o}(\varepsilon)$ represents terms proportional to ε, ε^2, ... We *assume* that for sufficiently small ε, the equations in (40) remain approximately valid when these terms are neglected, that is,

$$\begin{aligned} \dot{\hat{\underline{x}}}(t) &= \underline{A}(t)\hat{\underline{x}}(t) + \underline{B}(t)\hat{\underline{u}}(t) \\ \hat{\underline{y}}(t) &= \underline{C}(t)\hat{\underline{x}}(t) + \underline{D}(t)\hat{\underline{u}}(t) \end{aligned} \tag{41}$$

where

$$\begin{aligned} \underline{A}(t) &\triangleq \left.\frac{\partial \underline{f}}{\partial \underline{x}}\right|_* = \frac{\partial \underline{f}(\underline{x}^*(t),\underline{u}^*(t))}{\partial \underline{x}} \\ \underline{B}(t) &\triangleq \left.\frac{\partial \underline{f}}{\partial \underline{u}}\right|_* = \frac{\partial \underline{f}(\underline{x}^*(t),\underline{u}^*(t))}{\partial \underline{u}} \\ \underline{C}(t) &\triangleq \left.\frac{\partial \underline{h}}{\partial \underline{x}}\right|_* = \frac{\partial \underline{h}(\underline{x}^*(t),\underline{u}^*(t))}{\partial \underline{x}} \\ \underline{D}(t) &\triangleq \left.\frac{\partial \underline{h}}{\partial \underline{u}}\right|_* = \frac{\partial \underline{h}(\underline{x}^*(t),\underline{u}^*(t))}{\partial \underline{u}} \end{aligned} \tag{42}$$

Thus we have derived a set of linear (and, in general, time-varying) equations describing the behavior of the perturbations $\hat{\underline{x}}(t)$, $\hat{\underline{u}}(t)$, and $\hat{\underline{y}}(t)$ about the solution $\underline{x}^*(t)$, $\underline{u}^*(t)$, $\underline{y}^*(t)$ of the original equations in (30). In many cases (e.g., Examples 1 and 2) the matrices in (42) will turn out to be constants, so that the perturbations are described by a set of time-invariant equations,

$$\begin{aligned} \dot{\hat{\underline{x}}}(t) &= \underline{\underline{A}}\hat{\underline{x}}(t) + \underline{\underline{B}}\hat{\underline{u}}(t) \\ \hat{\underline{y}}(t) &= \underline{\underline{C}}\hat{\underline{x}}(t) + \underline{\underline{D}}\hat{\underline{u}}(t) \end{aligned} \tag{43}$$

It is important to bear in mind the fundamental assumption here, that the perturbations remain *sufficiently small* to validate this linear analysis. The question of how small this should be is difficult to answer with any generality, and one is usually reduced to verifying the validity of the assumption in various specific cases.

3. EXAMPLE (Satellite) [32]. A satellite in earth orbit is shown in Fig. 4. Its mass is m, its position is specified at any given time t by the polar coordinates $r(t)$, $\theta(t)$, and $\phi(t)$, and the variables $u_r(t)$, $u_\theta(t)$, and $u_\phi(t)$ are thrusts (forces) which may be applied in each of the three orthogonal directions by small rocket engines or gas jets.

If we assume the earth to be stationary, the state of this system is the position and velocity of the satellite, that is, the variables r, θ, and ϕ and their time derivatives. The inputs are the thrusts u_r, u_θ, and u_ϕ, and we shall arbitrarily define the outputs to be the position variables, r, θ, and ϕ.

To derive the equations of motion of the system, we express the *kinetic energy* as

$$K = \frac{mv^2}{2} = \frac{m}{2}\left(\dot{r}^2 + (r\dot{\phi})^2 + (r\dot{\theta}\cos\phi)^2\right) \tag{44}$$

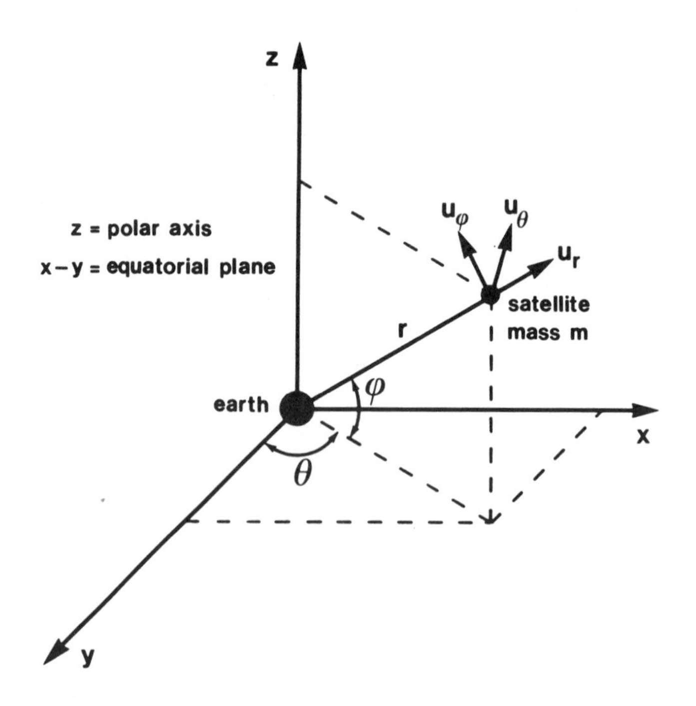

Fig. 4. Satellite in earth orbit.

where v is the magnitude of the satellite's velocity, and the *potential energy* as

$$P = -\frac{km}{r} \tag{45}$$

where k is a known physical constant ($4 \times 10^{14}\ \mathrm{N \cdot m^2 / kg}$). The *Lagrangian function* is defined as the difference of these,

$$L = K - P \tag{46}$$

and the dynamic behavior of the system is specified by *Lagrange's equations*,

$$\begin{aligned} \frac{d}{dt}\left(\frac{\partial L}{\partial \dot{r}}\right) - \frac{\partial L}{\partial r} &= u_r \\ \frac{d}{dt}\left(\frac{\partial L}{\partial \dot{\theta}}\right) - \frac{\partial L}{\partial \theta} &= u_\theta \\ \frac{d}{dt}\left(\frac{\partial L}{\partial \dot{\phi}}\right) - \frac{\partial L}{\partial \phi} &= u_\phi \end{aligned} \tag{47}$$

Taking derivatives as indicated, we have

$$\begin{aligned} m\left(\ddot{r} - r\dot{\theta}^2\cos^2\phi - r\dot{\phi}^2 + \frac{k}{r^2}\right) &= u_r \\ m\left(\ddot{\theta}r^2\cos^2\phi + 2r\dot{r}\dot{\theta}\cos^2\phi - 2r^2\dot{\theta}\dot{\phi}\cos\phi\sin\phi\right) &= (r\cos\phi)\,u_\theta \\ m\left(\ddot{\phi}r^2 + r^2\dot{\theta}^2\cos\phi\sin\phi + 2r\dot{r}\dot{\phi}\right) &= ru_\phi \end{aligned} \tag{48}$$

In terms of the state, input, and output vectors

$$\underline{x}(t) = \begin{bmatrix} r(t) \\ \dot{r}(t) \\ \theta(t) \\ \dot{\theta}(t) \\ \phi(t) \\ \dot{\phi}(t) \end{bmatrix} \qquad \underline{u}(t) = \begin{bmatrix} u_r(t) \\ u_\theta(t) \\ u_\phi(t) \end{bmatrix} \qquad \underline{y}(t) = \begin{bmatrix} r(t) \\ \theta(t) \\ \phi(t) \end{bmatrix} \tag{49}$$

these may be written as

$$\dot{\underline{x}} = \underline{f}(\underline{x},\underline{u}) = \begin{bmatrix} \dot{r} \\ r\dot{\theta}^2 \cos^2\phi + r\dot{\phi}^2 - k/r^2 + u_r/m \\ \dot{\theta} \\ -2\dot{r}\dot{\theta}/r + 2\dot{\theta}\dot{\phi}\sin\phi / \cos\phi + u_\theta / mr\cos\phi \\ \dot{\phi} \\ -\dot{\theta}^2 \cos\phi \sin\phi - 2\dot{r}\dot{\phi} / r + u_\phi / mr \end{bmatrix} \tag{50}$$

and

$$\underline{y} = \underline{C}\underline{x} = \begin{bmatrix} 1 & 0 & 0 & 0 & 0 & 0 \\ 0 & 0 & 1 & 0 & 0 & 0 \\ 0 & 0 & 0 & 0 & 1 & 0 \end{bmatrix} \underline{x} \tag{51}$$

One solution of these equations which is of particular interest (e.g., for communications satellites) is a *circular, equatorial orbit*, specified by

$$\underline{x}^*(t) = \begin{bmatrix} r^*(t) \\ \dot{r}^*(t) \\ \theta^*(t) \\ \dot{\theta}^*(t) \\ \phi^*(t) \\ \dot{\phi}^*(t) \end{bmatrix} = \begin{bmatrix} r_0 \\ 0 \\ \omega t \\ \omega \\ 0 \\ 0 \end{bmatrix} \qquad \underline{u}^*(t) = \underline{0} \tag{52}$$

where the radius r_0 and angular velocity ω are related by

$$r_0^3 \omega^2 = k \tag{53}$$

The satellite will remain in this uncontrolled (free) trajectory as long as there are no disturbances. If the satellite varies from this trajectory, a *control problem* arises: What inputs should be applied in order to return the satellite to its circular orbit? In this case we may consider the perturbed variables

$$\underline{x}(t) = \underline{x}^*(t) + \varepsilon\hat{\underline{x}}(t) = \begin{bmatrix} r_0 + \varepsilon\hat{x}_1(t) \\ \varepsilon\hat{x}_2(t) \\ \omega t + \varepsilon\hat{x}_3(t) \\ \omega + \varepsilon\hat{x}_4(t) \\ \varepsilon\hat{x}_5(t) \\ \varepsilon\hat{x}_6(t) \end{bmatrix} \qquad (54)$$

$$\underline{u}(t) = \underline{u}^*(t) + \varepsilon\hat{\underline{u}}(t) = \begin{bmatrix} \varepsilon\hat{u}_1(t) \\ \varepsilon\hat{u}_2(t) \\ \varepsilon\hat{u}_3(t) \end{bmatrix} \qquad (55)$$

$$\underline{y}(t) = \underline{y}^*(t) + \varepsilon\hat{\underline{y}}(t) = \begin{bmatrix} r_0 + \varepsilon\hat{x}_1(t) \\ \omega t + \varepsilon\hat{x}_3(t) \\ \varepsilon\hat{x}_5(t) \end{bmatrix} \qquad (56)$$

If the perturbations are sufficiently small ($\varepsilon \ll 1$), they may be described by a *linearized* system of equations as in (41) and (42),

$$\begin{aligned} \dot{\hat{\underline{x}}}(t) &= \underline{\underline{A}}\hat{\underline{x}}(t) + \underline{\underline{B}}\hat{\underline{u}}(t) \\ \hat{\underline{y}}(t) &= \underline{\underline{C}}\hat{\underline{x}}(t) \end{aligned} \qquad (57)$$

where

$$\underline{A} \triangleq \left.\frac{\partial \underline{f}}{\partial \underline{x}}\right|_{*} = \begin{bmatrix} 0 & 1 & 0 & 0 & 0 & 0 \\ 3\omega^2 & 0 & 0 & 2\omega r_0 & 0 & 0 \\ 0 & 0 & 0 & 1 & 0 & 0 \\ 0 & -\frac{2\omega}{r_0} & 0 & 0 & 0 & 0 \\ 0 & 0 & 0 & 0 & 0 & 1 \\ 0 & 0 & 0 & 0 & -\omega^2 & 0 \end{bmatrix} \tag{58}$$

$$\underline{B} \triangleq \left.\frac{\partial \underline{f}}{\partial \underline{u}}\right|_{*} = \begin{bmatrix} 0 & 0 & 0 \\ \frac{1}{m} & 0 & 0 \\ 0 & 0 & 0 \\ 0 & \frac{1}{mr_0} & 0 \\ 0 & 0 & 0 \\ 0 & 0 & \frac{1}{mr_0} \end{bmatrix} \tag{59}$$

and the output equation (51) is already linear.

For the sake of simplicity, we shall normalize our units so that

$$m = r_0 = 1 \tag{60}$$

in which case (57) becomes

$$\dot{\hat{\underline{x}}}(t) = \left[\begin{array}{cccc:cc} 0 & 1 & 0 & 0 & 0 & 0 \\ 3\omega^2 & 0 & 0 & 2\omega & 0 & 0 \\ 0 & 0 & 0 & 1 & 0 & 0 \\ 0 & -2\omega & 0 & 0 & 0 & 0 \\ \hdashline 0 & 0 & 0 & 0 & 0 & 1 \\ 0 & 0 & 0 & 0 & -\omega^2 & 0 \end{array}\right] \hat{\underline{x}}(t) + \left[\begin{array}{cc:c} 0 & 0 & 0 \\ 1 & 0 & 0 \\ 0 & 0 & 0 \\ 0 & 1 & 0 \\ \hdashline 0 & 0 & 0 \\ 0 & 0 & 1 \end{array}\right] \hat{\underline{u}}(t)$$

$$\hat{\underline{y}}(t) = \left[\begin{array}{cccc:cc} 1 & 0 & 0 & 0 & 0 & 0 \\ 0 & 0 & 1 & 0 & 0 & 0 \\ \hdashline 0 & 0 & 0 & 0 & 1 & 0 \end{array}\right] \hat{\underline{x}}(t) \tag{61}$$

As long as the deviation from a circular orbit remains small, these equations may be used to analyze the problem of controlling the satellite.

The dotted lines in (61) divide the matrices into blocks and indicate that this system is *uncoupled*, that is, the variables x_5, x_6, u_3, and y_3 are completely independent of the others. This means that the sixth-order system (61) may be broken into a fourth-order system describing the satellite's motion in the equatorial (r,θ) plane,

$$\begin{bmatrix} \dot{\hat{x}}_1 \\ \dot{\hat{x}}_2 \\ \dot{\hat{x}}_3 \\ \dot{\hat{x}}_4 \end{bmatrix} = \begin{bmatrix} 0 & 1 & 0 & 0 \\ 3\omega^2 & 0 & 0 & 2\omega \\ 0 & 0 & 0 & 1 \\ 0 & -2\omega & 0 & 0 \end{bmatrix} \begin{bmatrix} \hat{x}_1 \\ \hat{x}_2 \\ \hat{x}_3 \\ \hat{x}_4 \end{bmatrix} + \begin{bmatrix} 0 & 0 \\ 1 & 0 \\ 0 & 0 \\ 0 & 1 \end{bmatrix} \begin{bmatrix} \hat{u}_1 \\ \hat{u}_2 \end{bmatrix}$$

$$\begin{bmatrix} \hat{y}_1 \\ \hat{y}_2 \end{bmatrix} = \begin{bmatrix} 1 & 0 & 0 & 0 \\ 0 & 0 & 1 & 0 \end{bmatrix} \begin{bmatrix} \hat{x}_1 \\ \hat{x}_2 \\ \hat{x}_3 \\ \hat{x}_4 \end{bmatrix} \tag{62}$$

and a second-order system describing its (azimuthal) motion orthogonal to this plane (ϕ),

$$\begin{bmatrix} \dot{\hat{x}}_5 \\ \dot{\hat{x}}_6 \end{bmatrix} = \begin{bmatrix} 0 & 1 \\ -\omega^2 & 0 \end{bmatrix} \begin{bmatrix} \hat{x}_5 \\ \hat{x}_6 \end{bmatrix} + \begin{bmatrix} 0 \\ 1 \end{bmatrix} \hat{u}_3$$

$$y_3 = \begin{bmatrix} 1 & 0 \end{bmatrix} \begin{bmatrix} \hat{x}_5 \\ \hat{x}_6 \end{bmatrix} \tag{65}$$

We shall examine various other features of this example in subsequent chapters. The reader may wish to verify that for a non-circular orbit the linearized equations are time varying and periodic.

ΔΔΔ

2.5. STATE TRANSFORMATIONS

In Sec. 2.1 we introduced the concept of the state $\underline{x}(t)$ of a dynamic system and the standard state-variable format (2.1-18) for the differential equations governing a linear, time-invariant system,

$$\begin{aligned} \dot{\underline{x}}(t) &= \underline{A}\underline{x}(t) + \underline{B}\underline{u}(t) \\ \underline{y}(t) &= \underline{C}\underline{x}(t) \end{aligned} \tag{1}$$

It became clear in Example 2 of Sec. 2.2 that the choice of a state and the corresponding state equations are not unique. Indeed, suppose $\underline{Q}$ is any nonsingular $n \times n$ matrix and we define

$$\hat{\underline{x}}(t) = \underline{Q}\underline{x}(t) \tag{2}$$

Then $\hat{\underline{x}}(t)$ also qualifies as a state vector for the system; the two vectors contain the same information since $\underline{x}(t)$ can be recovered by

$$\underline{x}(t) = \underline{Q}^{-1}\hat{\underline{x}}(t) \tag{3}$$

1. LEMMA. Suppose (1) describes a linear system in terms of the state vector $\underline{x}(t)$. Then in terms of the alternate state vector $\hat{\underline{x}}(t)$ in Eq. (2), the same system is described by

$$\begin{aligned} \dot{\hat{\underline{x}}}(t) &= \hat{\underline{A}}\hat{\underline{x}}(t) + \hat{\underline{B}}\underline{u}(t) \\ \underline{y}(t) &= \hat{\underline{C}}\hat{\underline{x}}(t) \end{aligned} \tag{4}$$

where

$$\hat{\underline{A}} \triangleq \underline{Q}\,\underline{A}\,\underline{Q}^{-1} \qquad \hat{\underline{B}} \triangleq \underline{Q}\,\underline{B} \qquad \hat{\underline{C}} \triangleq \underline{C}\,\underline{Q}^{-1} \tag{5}$$

Proof: This follows easily if we differentiate (2) and use (1)-(3).

$$\dot{\hat{\underline{x}}}(t) = \underline{Q}\dot{\underline{x}}(t) = \underline{Q}\underline{A}\underline{x}(t) + \underline{Q}\underline{B}\underline{u}(t) = \underline{Q}\underline{A}\underline{Q}^{-1}\hat{\underline{x}}(t) + \underline{Q}\underline{B}\underline{u}(t) \tag{6}$$

$$\underline{y}(t) = \underline{C}\underline{x}(t) = \underline{C}\underline{Q}^{-1}\hat{\underline{x}}(t) \tag{7}$$

$\Delta\Delta\Delta$

An infinite number of state vectors may be generated in this manner, one for every nonsingular matrix. The change from $\underline{x}$ to $\hat{\underline{x}}$ and from $\underline{A}$ to $\hat{\underline{A}}$ is known as a *state transformation*.† It simply provides a different "point of view" from which to study the system, since (1) and (4) are equivalent descriptions of the same process.

2. *EXAMPLE*. Recall that in Example 2.2.2, two different sets of equations were given for the same system, namely,

$$\dot{\underline{x}} = \begin{bmatrix} 0 & -2 \\ 1 & -3 \end{bmatrix} \underline{x} + \begin{bmatrix} 2 \\ 0 \end{bmatrix} u = \underline{A}\underline{x} + \underline{b}u$$

$$y = \begin{bmatrix} 0 & 3 \end{bmatrix} \underline{x} = \underline{c}'\underline{x} \tag{8}$$

and

$$\dot{\hat{\underline{x}}} = \begin{bmatrix} 0 & 1 \\ -2 & -3 \end{bmatrix} \hat{\underline{x}} + \begin{bmatrix} 0 \\ 1 \end{bmatrix} u = \hat{\underline{A}}\hat{\underline{x}} + \hat{\underline{b}}u$$

$$y = \begin{bmatrix} 6 & 0 \end{bmatrix} \hat{\underline{x}} = \hat{\underline{c}}'\hat{\underline{x}} \tag{9}$$

We shall now show that these two representations are related by a state transformation

$$\hat{\underline{x}}(t) = \underline{Q}\underline{x}(t) = \begin{bmatrix} 0 & 1/2 \\ 1/2 & -3/2 \end{bmatrix} \underline{x}(t)$$

$$\underline{x}(t) = \underline{Q}^{-1}\hat{\underline{x}}(t) = \begin{bmatrix} 6 & 2 \\ 2 & 0 \end{bmatrix} \hat{\underline{x}}(t) \tag{10}$$

---

†This is also called a *similarity* or *equivalence* transformation. It is the same as a change of basis (or coordinates) from the standard basis (A.1-12) to a new one specified by the columns of $\underline{Q}$.

by verifying that (5) holds:

$$\underline{Q}\underline{A}\underline{Q}^{-1} = \begin{bmatrix} 0 & 1/2 \\ 1/2 & -3/2 \end{bmatrix}\begin{bmatrix} 0 & -2 \\ 1 & -3 \end{bmatrix}\begin{bmatrix} 6 & 2 \\ 2 & 0 \end{bmatrix} = \begin{bmatrix} 0 & 1/2 \\ 1/2 & -3/2 \end{bmatrix}\begin{bmatrix} -4 & 0 \\ 0 & 2 \end{bmatrix}$$

$$= \begin{bmatrix} 0 & 1 \\ -2 & -3 \end{bmatrix} = \hat{\underline{A}} \tag{11}$$

$$\underline{Q}\underline{b} = \begin{bmatrix} 0 & 1/2 \\ 1/2 & -3/2 \end{bmatrix}\begin{bmatrix} 2 \\ 0 \end{bmatrix} = \begin{bmatrix} 0 \\ 1 \end{bmatrix} = \hat{\underline{b}} \tag{12}$$

$$\underline{c}'\underline{Q}^{-1} = \begin{bmatrix} 0 & 3 \end{bmatrix}\begin{bmatrix} 6 & 2 \\ 2 & 0 \end{bmatrix} = \begin{bmatrix} 6 & 0 \end{bmatrix} = \hat{\underline{c}}' \tag{13}$$

This example illustrates that two equivalent systems may be related by a state transformation. More generally, we will see in Chapter 10 that two systems whose input-output behavior is equivalent are always related by a state transformation, provided they have the same dimension and this dimension is *minimal*.

ΔΔΔ

It will become clear in the sequel that all the fundamental properties of a linear system are independent of the choice of state vector (i.e., they are *invariant* under state transformations). In particular, the eigenvalues of the system matrix are unaffected by such a transformation (Theorem A.3.2). However, such transformations may be useful in revealing the underlying structure of a system, as illustrated in the following.

3. EXAMPLE. Consider the scalar system whose block diagram is shown in Fig. 1(a),

$$\begin{aligned} \dot{\underline{x}} &= \underline{A}\underline{x} + \underline{b}u \\ y &= \underline{c}'\underline{x} \end{aligned} \tag{14}$$

(a)

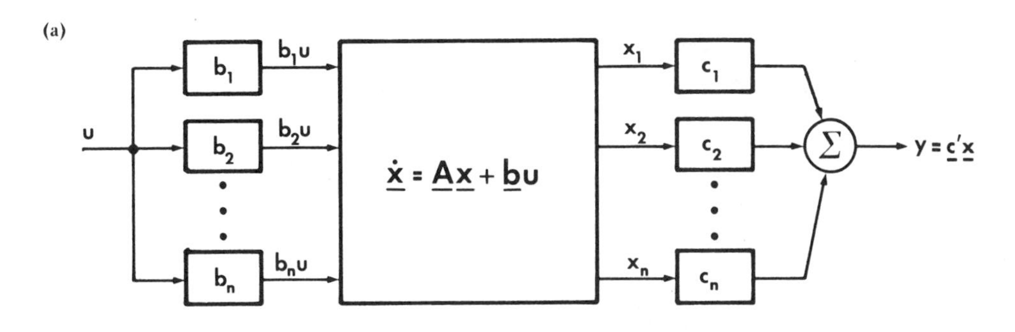

(b)

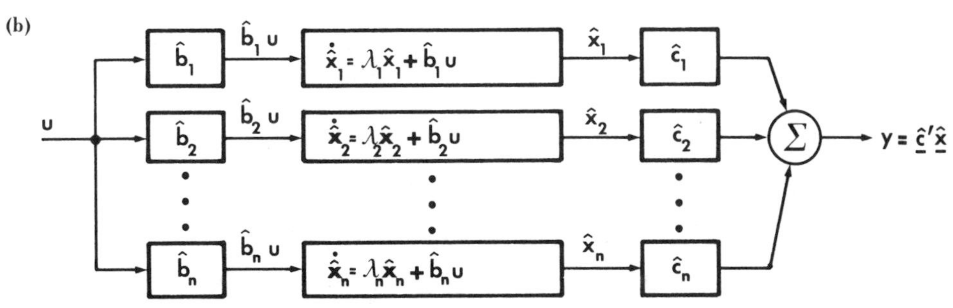

FIG. 1. System of Example 3: (a) linear system and (b) the same system uncoupled by similarity transformation.

where $\underline{A}$ has eigenvalues $\lambda_1, \lambda_2, \ldots, \lambda_n$. If its eigenvectors $\underline{v}_1$, $\underline{v}_2, \ldots, \underline{v}_n$ form a linearly independent set, the matrix

$$\underline{V} = [\underline{v}_1 \quad \underline{v}_2 \quad \cdots \quad \underline{v}_n] \tag{15}$$

will have an inverse $\underline{V}^{-1}$. We may define a state transformation

$$\hat{\underline{x}} = \underline{V}^{-1}\underline{x} \qquad \underline{x} = \underline{V}\hat{\underline{x}} \tag{16}$$

so that in terms of the new state vector $\hat{\underline{x}}$, the system equations are

$$\dot{\hat{\underline{x}}} = \underline{V}^{-1}\underline{AV}\hat{\underline{x}} + \underline{V}^{-1}\underline{b}u = \hat{\underline{A}}\hat{\underline{x}} + \hat{\underline{b}}u$$
$$y = \underline{c}'\underline{V}\hat{\underline{x}} = \hat{\underline{c}}'\hat{\underline{x}} \tag{17}$$

Using Theorem A.3.3, we have

$$\hat{\underline{A}} = \underline{V}^{-1}\underline{AV} = \underline{\Lambda} = \begin{bmatrix} \lambda_1 & 0 & \cdots & 0 \\ 0 & \lambda_2 & \cdots & 0 \\ \cdots & \cdots & \cdots & \cdots \\ 0 & 0 & \cdots & \lambda_n \end{bmatrix} \tag{18}$$

and (17) may be written as

$$\dot{\hat{\underline{x}}} = \begin{bmatrix} \lambda_1 & 0 & \cdots & 0 \\ 0 & \lambda_2 & \cdots & 0 \\ \cdots & \cdots & \cdots & \cdots \\ 0 & 0 & \cdots & \lambda_n \end{bmatrix} \hat{\underline{x}} + \begin{bmatrix} \hat{b}_1 \\ \hat{b}_2 \\ \vdots \\ \hat{b}_n \end{bmatrix} u$$
$$y = \begin{bmatrix} \hat{c}_1 & \hat{c}_2 & \cdots & \hat{c}_n \end{bmatrix} \hat{\underline{x}} \tag{19}$$

The system is now said to be in *uncoupled* or *diagonal* form. In other words, (19) consists of n independent one-dimensional subsystems,

$$\dot{\hat{x}}_i = \lambda_i \hat{x}_i + \hat{b}_i u \qquad (i = 1,2,\ldots,n)$$
$$y = \sum_{i=1}^{n} \hat{c}_i \hat{x}_i \tag{20}$$

as indicated in Fig. 1(b). This is a convenient form since it is often easier to analyze n separate one-dimensional systems than one n-dimensional system. ΔΔΔ

In later chapters we will make use of state transformations to put systems in a variety of other useful forms.

2.6. SUMMARY

This chapter has been concerned with mathematical modeling of physical systems, using ordinary differential equations. In Sec. 2.1 we introduced the concept of state and the state-variable format for the differential equations of a system. In Sec. 2.2 an alternative formulation involving only the input and output was presented, and its relation to the state-variable format was indicated.

In Sec. 2.3 we illustrated the construction of linear system models with a series of examples. It was shown that in all but the simplest electrical and mechanical networks, linear models are only an approximate description of real systems, valid if appropriate assumptions can be made. The most important such assumption is that the behavior of the system need be studied only for small changes of the variables from the steady state. This allows nonlinear relations to be replaced by linear approximations; a formal procedure for doing this was set out in Sec. 2.4.

Finally, in Sec. 2.5 we saw how the nonuniqueness of the state vector may be exploited via state transformations to study a system in various coordinate systems.

PROBLEMS

Section 2.2

1. Express each of the following scalar differential equations in state-variable format using Theorem 2.2.1:
   (a) $\dddot{y} + 2\ddot{y} + 3\dot{y} + 4y = 5\ddot{u} + 6\dot{u} + 7u$
   (b) $\ddddot{y} + 3\dddot{y} + 2\dot{y} = -\dot{u}$
   (c) $2\dddot{y} - 3y = \ddot{u} - 2u$
   (d) $\ddot{y} = 0$

2. The "transpose" of Theorem 2.2.1 states that a scalar differential equation of the form (2.2-1) is equivalent to[†]

[†] In Sec. 10.4 this will be called the *observable canonical form* of the state equations.

$$\dot{\underline{x}}(t) = \begin{bmatrix} 0 & 0 & \dots & 0 & 0 & -p_0 \\ 1 & 0 & \dots & 0 & 0 & -p_1 \\ 0 & 0 & \dots & 0 & 0 & -p_2 \\ \dots & \dots & \dots & \dots & \dots & \dots \\ 0 & 0 & \dots & 1 & 0 & -p_{n-2} \\ 0 & 0 & \dots & 0 & 1 & -p_{n-1} \end{bmatrix} \underline{x}(t) + \begin{bmatrix} q_0 \\ q_1 \\ q_2 \\ \vdots \\ q_{n-2} \\ q_{n-1} \end{bmatrix} u(t)$$

$$y(t) = \begin{bmatrix} 0 & 0 & \dots & 0 & 0 & 1 \end{bmatrix} \underline{x}(t)$$

(a) Prove this result for the case $q_1 = q_2 = \dots = q_{n-1} = 0$ by identifying the states as

$$x_n(t) = y(t)$$

$$x_{n-1}(t) = \dot{y}(t) + p_{n-1}y(t)$$

$$\dots\dots\dots\dots\dots\dots\dots$$

$$x_1(t) = y^{(n-1)}(t) + p_{n-1}\, y^{(n-2)}(t) + \dots + p_1 y(t)$$

(b) Prove the result in general and indicate how the expressions for $x_1, x_2, \dots, x_n$ must be modified.

(c) Modify this result to account for an additional term $q_n u^{(n)}(t)$ in the right-hand side of Eq. (2.2-1).

3. (a) Find an alternative state-variable formulation for Example 2.1.1 in the form of Problem 2.2-2.

(b) Find two alternative state-variable formulations for Example 2.1.2, in the forms of Theorem 2.2.1 and Problem 2.2-2, respectively.

Section 2.3

1. Alter the model of Example 2.3.1 so that the angular position $\Theta(t)$ of the motor shaft is the system output. Determine the scalar differential equation (input-output format) relating Θ to the inputs v and m_L.

2. Show how to alter the dynamic equations of the broom-balancing system (Example 2.3.2) in order to account for horizontal wind gusts which exert a force w(t) on the pendulum and kw(t) on the cart.

3. Derive a set of linearized differential equations for each of the dual-broom systems described in Problem 10.1-15.

4. Alter the bubonic plague model of Example 2.3.3 so that the infection rate contains a term proportional to the product of susceptibles and infectives (i.e., replace βx_1 in (2.3-32) with $\beta x_1 x_2$). Show how to linearize this model, stating explicitly any assumptions which you make. Can you suggest any other alterations which would improve the accuracy of the model?

5. Figure P1 shows a hydraulic power amplifier or *servomotor*, consisting of a combination of a four-way hydraulic valve and a piston. The input to the device is the displacement u(t) of the left-hand end of the floating link; the output is the displacement y(t) of the piston rod.

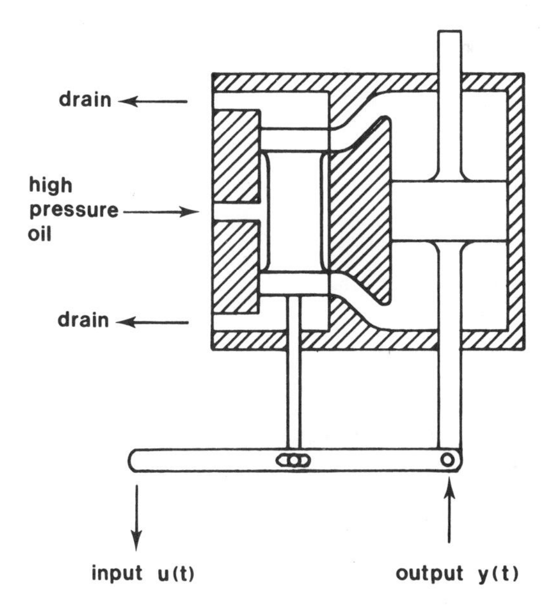

FIG. P1. Hydraulic servomotor.

(a) With the floating link connecting the valve spool and piston rod removed, show that the device acts as an *integrator* (i.e., that

$$\dot{y}(t) = az(t)$$

where a is a positive constant and z the displacement of the valve spool), provided that certain assumptions can be made. Carefully list these assumptions.

(b) If the assumptions made in (a) are valid, how are y(t) and u(t) related when the link is reconnected?

Section 2.4

1. (a) Verify that (2.4-38) follows from (2.4-35) for the case $n = m = 2$.

   (b) Derive (2.4-39) similarly, for the case $n = p = 2$.

2. Write the nonlinear differential equation

$$\ddot{y} + 3\dot{y} + 2y = uy^3$$

in state-variable format. Verify that

$$y^*(t) = e^{-t} \qquad u^*(t) = 0$$

is a solution, and linearize the equations about this solution.

3. Verify (2.4-58) and (2.4-59).

4. Linearize the equations of motion of the satellite (Example 2.4.3) about an elliptical orbit.

Section 2.5

1. (a) Deduce the state transformation (2.5-10) *directly* from Example 2.1.1 and the definition

$$\underline{x}(t) = \begin{bmatrix} y(t)/6 \\ \dot{y}(t)/6 \end{bmatrix}$$

   (b) Find state transformations relating the various state-variable representations of systems in Problem 2.2-3.

2. For some applications it is useful to transform the input and output vectors as well as the state, defining

$$\hat{\underline{x}} = \underline{Q}\underline{x} \qquad \hat{\underline{y}} = \underline{R}\underline{y} \qquad \hat{\underline{u}} = \underline{P}\underline{u}$$

where $\underline{Q}$, $\underline{R}$, and $\underline{P}$ are nonsingular square matrices. The system (2.5-1) may then be described by

$$\dot{\hat{\underline{x}}} = \hat{\underline{A}}\hat{\underline{x}} + \hat{\underline{B}}\hat{\underline{u}}$$

$$\hat{\underline{y}} = \hat{\underline{C}}\hat{\underline{x}}$$

Find expressions for $\hat{\underline{A}}$, $\hat{\underline{B}}$, and $\hat{\underline{C}}$.

3. Consider the antenna of Example 1.1.1, described by Eqs. (1.1-3), (1.1-4), (1.1-8), (1.1-9), and (1.1-13), *without* the simplifying assumption (1.1-5).

   (a) Give a description of the antenna system in state-variable form (2.5-1), where the components of $\underline{u}$ are the desired antenna position y_r and the wind load d.

   (b) Setting $d = 0$, use (1.1-21) and Theorem 2.2.1 to derive an alternative state-variable form of the system equations.

   (c) Assign convenient numerical values to the coefficients of the equations and show that the state vectors of (a) and (b) are related by a state transformation of the form (2.5-2) when $d = 0$.

Part 2

SOLUTION OF DIFFERENTIAL EQUATIONS

Chapter 3

DIRECT SOLUTION OF DIFFERENTIAL EQUATIONS

3.0. INTRODUCTION

In Chapter 2 we examined a variety of physical systems whose behavior could be described (at least to an acceptable degree of accuracy) by *ordinary, linear, time-invariant differential equations*, either in state-variable or input-output format.

Having thus arrived at a mathematical model for a physical system, our next task is one of analysis. In order to *explicitly* determine the behavior of the system (or more precisely, the model), we must find *solutions* of the differential equations.

In this chapter we restrict ourselves to equations in state-variable format and investigate the general form of such solutions.

3.1. SOLUTION OF UNFORCED EQUATIONS

We consider first the solution of an *unforced*† (or *homogeneous*) *differential equation*,

$$\dot{\underline{x}}(t) = \underline{\underline{A}}\,\underline{x}(t) \tag{1}$$

---

†In other words, no input (forcing function) is applied to the system.

The problem is incompletely defined without a set of *boundary conditions*. These conditions are provided by the *initial state*

$$\underline{x}(t_0) = \underline{x}_0 \tag{2}$$

at some initial time t_0. When dealing with time-invariant systems, it is usually convenient to let $t_0 = 0$; in the event that $t_0 \neq 0$, we need only replace t by $t - t_0$ in the solution.

Given any set of differential equations and boundary conditions, the first questions one must answer concern the *existence* and *uniqueness* of solutions. Uniqueness means that there is *at most one* solution of the differential equation passing through any given initial state. It is quite easy to establish that solutions are *always* unique in the case of linear differential equations [32] and we shall not pursue this question any further.

Existence of solutions will be established in the linear case† by deriving the form of the solution for every possible initial state. We first consider a scalar example.

1. EXAMPLE. Consider the one-dimensional system

$$\dot{x}(t) = ax(t) \tag{3}$$

$$x(0) = x_0 \tag{4}$$

and assume for a moment that the solution may be written as a power series about $t = 0$,

$$x(t) = f_0 + f_1 t + f_2 t^2 + \ldots \tag{5}$$

Substituting this into (3), we have

---

†For nonlinear equations of the form $\dot{\underline{x}} = \underline{f}(\underline{x},\underline{u},t)$, statements of such generality cannot be made. However a *sufficient* condition for both existence and uniqueness is that the differential equation satisfies a *Lipschitz condition*

$$\| \underline{f}(\underline{x}_1,\underline{u},t) - \underline{f}(\underline{x}_2,\underline{u},t) \| \leq K \| \underline{x}_1 - \underline{x}_2 \|$$

for some constant $K > 0$ and for all $\underline{x}_1$, $\underline{x}_2$, $\underline{u}$, and t [42].

$$\dot{x}(t) = f_1 + 2f_2 t + 3f_3 t^2 + \dots$$
$$= af_0 + af_1 t + af_2 t^2 + \dots \tag{6}$$

Since (6) must hold for all t, we may equate the various powers of t and find that

$$f_1 = af_0$$

$$f_2 = \frac{af_1}{2} = \frac{a^2 f_0}{2}$$

$$f_3 = \frac{af_2}{3} = \frac{a^2 f_1}{3 \times 2} = \frac{a^3 f_0}{3!} \tag{7}$$

....

$$f_k = \frac{af_{k-1}}{k} = \dots = \frac{a^k f_0}{k!}$$

....

The value of f_0 is found by equating (4) and (5), with $t = 0$:

$$f_0 = x_0 = x(0) \tag{8}$$

so that the series becomes

$$x(t) = (1 + at + \frac{a^2 t^2}{2!} + \dots + \frac{a^k t^k}{k!} + \dots)\, x_0 \tag{9}$$

It is well known that the power series in (9) converges absolutely and uniformly on any time interval to the exponential e^{at} (this validates our original assumption), so that

$$x(t) = e^{at} x_0 \tag{10}$$

Substitution in (3) and (4) verifies that this is the correct solution,

$$\dot{x}(t) = ae^{at} x_0 = ax(t)$$
$$x(0) = e^{a0} x_0 = x_0 \tag{11}$$

ΔΔΔ

The general case may be treated in precisely the same way as this example. We assume a (vector) power series solution about $t = 0$,

$$\underline{x}(t) = \underline{f}_0 + \underline{f}_1 t + \underline{f}_2 t^2 + \ldots \tag{12}$$

and substitute this into (1) to find

$$\begin{aligned} \dot{\underline{x}}(t) &= \underline{f}_1 + 2\underline{f}_2 t + 3\underline{f}_3 t^2 + \ldots \\ &= \underline{A}\underline{f}_0 + \underline{A}\underline{f}_1 t + \underline{A}\underline{f}_2 f^2 + \ldots \end{aligned} \tag{13}$$

Equating powers of t, we have

$$\begin{aligned} \underline{f}_1 &= \underline{A}\underline{f}_0 \\ \underline{f}_2 &= \frac{\underline{A}\underline{f}_1}{2} = \frac{\underline{A}^2\underline{f}_2}{2} \\ \underline{f}_3 &= \frac{\underline{A}\underline{f}_2}{3} = \frac{\underline{A}^2\underline{f}_1}{3\times 2} = \frac{\underline{A}^3\underline{f}_0}{3!} \\ &\ldots\ldots\ldots\ldots\ldots\ldots \\ \underline{f}_k &= \frac{\underline{A}\underline{f}_{k-1}}{k} = \ldots = \frac{\underline{A}^k\underline{f}_0}{k!} \\ &\ldots\ldots\ldots\ldots\ldots\ldots \end{aligned} \tag{14}$$

and using the initial state (2) in (12) yields

$$\underline{f}_0 = \underline{x}_0 = \underline{x}(0) \tag{15}$$

so that the series becomes

$$\underline{x}(t) = \left(\underline{I} + \underline{A}t + \frac{\underline{A}^2 t^2}{2!} + \ldots + \frac{\underline{A}^k t^k}{k!} + \ldots\right) \underline{x}_0 \tag{16}$$

It can be shown (see Problem 2) that this matrix power series converges absolutely and uniformly on any time interval. By analogy with the power series in (9) for the ordinary exponential function, we adopt the following nomenclature.

2. *DEFINITION*. If $\underline{A}$ is an $n \times n$ matrix, the *matrix exponential* of $\underline{A}t$ is

$$e^{\underline{A}t} = \underline{I} + \underline{A}t + \frac{\underline{A}^2 t^2}{2!} + \dots + \frac{\underline{A}^k t^k}{k!} + \dots \tag{17}$$

This is also known as a *transition matrix*[†] (or a *fundamental matrix*), and it will occasionally be written in the alternative form $\exp \underline{A}t$.

ΔΔΔ

3. *THEOREM*. The matrix exponential $e^{\underline{A}t}$ satisfies the same differential equation as the original system,

$$\frac{d}{dt} e^{\underline{A}t} = \underline{A}\, e^{\underline{A}t} = e^{\underline{A}t}\, \underline{A} \tag{18}$$

with the initial condition

$$e^{\underline{A}0} = \underline{I} \tag{19}$$

Proof: To establish (18), we differentiate the series (17) term by term[††]

$$\begin{aligned}\frac{d}{dt} e^{\underline{A}t} &= \underline{A} + \underline{A}^2 t + \frac{\underline{A}^3 t^2}{2!} + \dots \\ &= \underline{A}\left(\underline{I} + \underline{A}t + \frac{\underline{A}^2 t^2}{2!} + \dots\right) = \underline{A}\, e^{\underline{A}t} \\ &= \left(\underline{I} + \underline{A}t + \frac{\underline{A}^2 t^2}{2!} + \dots\right)\underline{A} = e^{\underline{A}t}\, \underline{A}\end{aligned} \tag{20}$$

[†] See Chapter 14.

[††] This gives the correct derivative because both series converge uniformly on any time interval. [153]

and for (19) we simply note that

$$e^{\underline{A}0} = \underline{I} + \underline{A}0 + \frac{\underline{A}^2 0^2}{2!} + \ldots = \underline{I} \tag{21}$$

ΔΔΔ

Returning to the original system, we see that the *natural* (or *unforced*, or *zero-input*) *response* is given in the following theorem.

4. *THEOREM (Natural response).* The solution of

$$\begin{aligned} \dot{\underline{x}}(t) &= \underline{A}\underline{x}(t) \\ \underline{x}(0) &= \underline{x}_0 \end{aligned} \tag{22}$$

is given by

$$\underline{x}(t) = e^{\underline{A}t}\,\underline{x}_0 \tag{23}$$

Proof: We differentiate (23) and use Theorem 3 to verify that

$$\dot{\underline{x}}(t) = \frac{d}{dt} e^{\underline{A}t}\,\underline{x}_0 = \underline{A}\, e^{\underline{A}t}\,\underline{x}_0 = \underline{A}\underline{x}(t) \tag{24}$$

and we substitute $t = 0$ to find

$$\underline{x}(0) = e^{\underline{A}0}\,\underline{x}_0 = \underline{x}_0 \tag{25}$$

ΔΔΔ

Because the system is time-invariant, the initial time is entirely arbitrary, and if $t_0 \neq 0$, we need only replace t by $t - t_0$. In this case the solution given by (23) becomes

$$\underline{x}(t) = e^{\underline{A}(t-t_0)}\,\underline{x}_0 \tag{26}$$

where

$$e^{\underline{A}(t-t_0)} = \underline{I} + \underline{A}(t - t_0) + \frac{\underline{A}^2 (t - t_0)^2}{2!} + \ldots \tag{27}$$

5. *EXAMPLE.* Recall from Example 2.4.3 that the linearized azimuthal motion of an earth satellite was described by (2.4-63),

$$\dot{\underline{x}}(t) = \begin{bmatrix} 0 & 1 \\ -\omega^2 & 0 \end{bmatrix} \underline{x}(t) + \begin{bmatrix} 0 \\ 1 \end{bmatrix} u_\phi(t) \tag{28}$$

where

$$\underline{x}(t) = \begin{bmatrix} \phi(t) \\ \dot{\phi}(t) \end{bmatrix} \tag{29}$$

and ϕ, $\dot{\phi}$, and u_ϕ are perturbations of the azimuthal angle, angular velocity and thrust about their nominal values of zero. Setting $u_\phi(t) \equiv 0$, we have

$$\dot{\underline{x}}(t) = \underline{A}\underline{x}(t) \qquad \underline{A} = \begin{bmatrix} 0 & 1 \\ -\omega^2 & 0 \end{bmatrix} \tag{30}$$

and the matrix exponential series is

$$e^{\underline{A}t} = \underline{I} + \underline{A}t + \frac{\underline{A}^2t^2}{2!} + \frac{\underline{A}^3t^3}{3!} + \cdots$$

$$= \begin{bmatrix} 1 & 0 \\ 0 & 1 \end{bmatrix} + \begin{bmatrix} 0 & t \\ -\omega^2 t & 0 \end{bmatrix} + \begin{bmatrix} \frac{-\omega^2t^2}{2!} & 0 \\ 0 & \frac{-\omega^2t^2}{2!} \end{bmatrix} + \begin{bmatrix} 0 & \frac{-\omega^2t^3}{3!} \\ \frac{\omega^4t^3}{3!} & 0 \end{bmatrix} + \cdots$$

$$= \begin{bmatrix} \alpha_{11}(t) & \alpha_{12}(t) \\ \alpha_{21}(t) & \alpha_{22}(t) \end{bmatrix} \tag{31}$$

We recognize the elements of $e^{\underline{A}t}$ as the familiar series

$$\alpha_{11}(t) = \alpha_{22}(t)$$

$$= 1 - \frac{\omega^2 t^2}{2!} + \frac{\omega^4 t^4}{4!} - \cdots \tag{32}$$

$$= \cos \omega t$$

$$\omega\alpha_{12}(t) = -\frac{\alpha_{21}(t)}{\omega}$$

$$= \omega t - \frac{\omega^3 t^3}{3!} + \frac{\omega^5 t^5}{5!} - \cdots \tag{33}$$

$$= \sin \omega t$$

so that

$$e^{\underline{A}t} = \begin{bmatrix} \cos \omega t & \omega^{-1} \sin \omega t \\ -\omega \sin \omega t & \cos \omega t \end{bmatrix} \tag{34}$$

and

$$\underline{x}(t) = e^{\underline{A}t}\underline{x}_0 = \begin{bmatrix} \cos \omega t & \omega^{-1} \sin \omega t \\ -\omega \sin \omega t & \cos \omega t \end{bmatrix} \begin{bmatrix} \phi_0 \\ \dot{\phi}_0 \end{bmatrix}$$

$$= \begin{bmatrix} \phi(t) \\ \dot{\phi}(t) \end{bmatrix} = \begin{bmatrix} \phi_0 \cos \omega t + \dot{\phi}_0 \omega^{-1} \sin \omega t \\ -\phi_0 \omega \sin \omega t + \dot{\phi}_0 \cos \omega t \end{bmatrix} \tag{35}$$

Thus we see that in the absence of disturbances, any perturbation ($\underline{x}_0 \neq 0$) of the azimuthal angle results in sinusoidal oscillations about the zero value. These will continue unattenuated unless the control $u_\phi(t)$ (which was assumed zero here) is used to drive ϕ and $\dot{\phi}$ back to zero.

ΔΔΔ

3.2. PROPERTIES OF THE MATRIX EXPONENTIAL

Equation (3.1-26) specified that the solution of the differential equation

$$\dot{\underline{x}}(t) = \underline{A}\underline{x}(t) \tag{1}$$

passing through the point $\underline{x}_0$ at time $t = t_0$ has the form

$$\underline{x}(t) = e^{\underline{A}(t-t_0)} \underline{x}_0 \tag{2}$$

The motion of the state as a function of time is known as a *state trajectory* and is shown in Fig. 1 where $\underline{x}(t)$ is represented schematically by a two-dimensional vector.

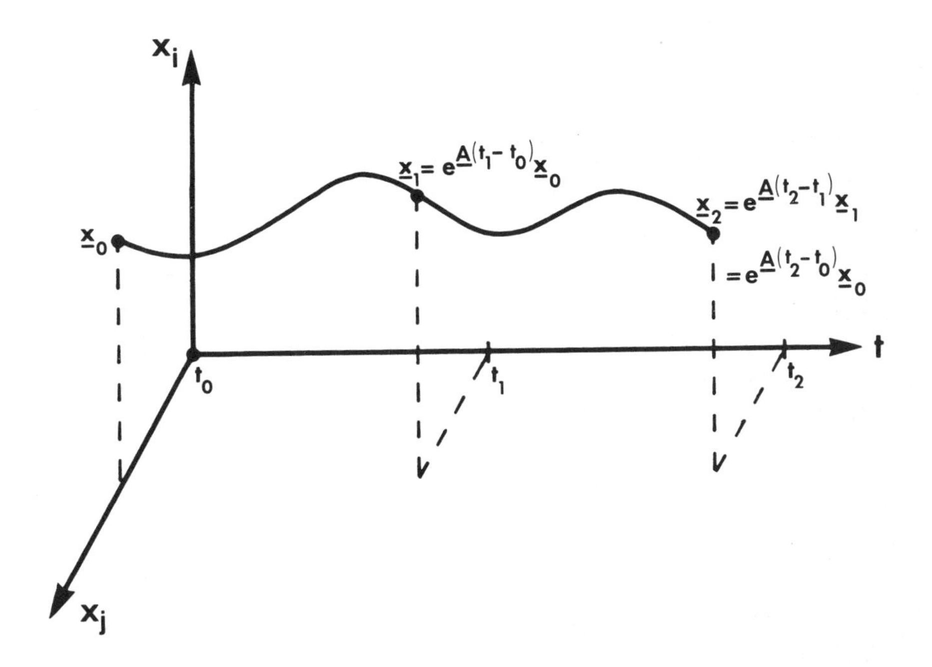

FIG. 1. State trajectory for the system $\dot{\underline{x}} = \underline{A}\underline{x}$.

At subsequent times $t_2 > t_1 > t_0$, the state trajectory will pass through the points

$$\underline{x}_1 \triangleq \underline{x}(t_1) = e^{\underline{A}(t_1 - t_0)} \underline{x}_0 \tag{3}$$

$$\underline{x}_2 \triangleq \underline{x}(t_2) = e^{\underline{A}(t_2 - t_0)} \underline{x}_0 \tag{4}$$

This illustrates the earlier reference to $e^{\underline{A}(t-t_0)}$ as a *transition matrix:* it governs the transition of the state from one point to the next along the trajectory.

However, we could also consider t_1 as an initial time, the state at t_1 being $\underline{x}_1$, and arrive at another expression for $\underline{x}_2$,

$$\underline{x}_2 = \underline{x}(t_2) = e^{\underline{A}(t_2 - t_1)} \underline{x}_1 \tag{5}$$

Substitution of (3) and (4) yields

$$\underline{x}_2 = e^{\underline{A}(t_2 - t_1)} e^{\underline{A}(t_1 - t_0)} \underline{x}_0 = e^{\underline{A}(t_2 - t_0)} \underline{x}_0 \tag{6}$$

and since this must hold for all $\underline{x}_0$, we conclude† that

$$e^{\underline{A}(t_2 - t_1)} e^{\underline{A}(t_1 - t_0)} = e^{\underline{A}(t_2 - t_0)} \tag{7}$$

This is known as the *composition rule* (or *group property)* and is stated more generally as follows:

1. *THEOREM (Composition rule).* For any $n \times n$ matrix $\underline{A}$,

$$e^{\underline{A}t} e^{\underline{A}\tau} = e^{\underline{A}(t+\tau)} \qquad \text{(for all } t, \tau) \tag{8}$$

Proof: This may be verified directly from the power series (3.1-17) defining $e^{\underline{A}t}$,

---

†See Problem A.2-6.

$$e^{\underline{A}t} e^{\underline{A}\tau} = (\underline{I} + \underline{A}t + \frac{\underline{A}^2 t^2}{2!} + \ldots)(\underline{I} + \underline{A}\tau + \frac{\underline{A}^2 \tau^2}{2!} + \ldots)$$

$$= \underline{I} + \underline{A}\tau + \frac{\underline{A}^2\tau^2}{2!} + \frac{\underline{A}^3\tau^3}{3!} + \ldots$$

$$+ \underline{A}t + \underline{A}^2 t\tau + \frac{\underline{A}^3 t\tau^2}{2!} + \ldots$$

$$+ \frac{\underline{A}^2 t^2}{2!} + \frac{\underline{A}^3 t^2 \tau}{2!} + \ldots$$

$$+ \frac{\underline{A}^3 t^3}{3!} + \ldots$$

$$= \underline{I} + \underline{A}(t + \tau) + \frac{\underline{A}^2 (t+\tau)^2}{2!} + \frac{\underline{A}^3 (t+\tau)^3}{3!} + \ldots$$

$$= e^{\underline{A}(t+\tau)} \tag{9}$$

ΔΔΔ

2. *COROLLARY.* The matrix exponential is nonsingular for all values of t, and

$$[e^{\underline{A}t}]^{-1} = e^{-\underline{A}t} \tag{10}$$

Proof: Set $\tau = -t$ in Theorem 1.

ΔΔΔ

The implication of Corollary 2 is that (1) can just as easily be solved backwards in time. In other words, given $\underline{x}(t_1)$, we can always determine $\underline{x}(t)$ for any previous time $t < t_1$. Another useful property is the following.

3. *THEOREM.* The determinant of a matrix exponential is given by

$$\det e^{\underline{A}t} = e^{\mathrm{tr}(\underline{A})t} \tag{11}$$

where $\mathrm{tr}(\underline{A})$ is the *trace* of $\underline{A}$ (the sum of its diagonal elements).

Proof: See Ref. 32.

ΔΔΔ

Theorem 1 is an extension of a familiar rule involving scalar exponentials,

$$e^{at} e^{a\tau} = e^{a(t+\tau)} \tag{12}$$

This property may also be expressed as

$$e^{at} e^{bt} = e^{(a+b)t} \tag{13}$$

but this form does *not* extend to the matrix case, that is

$$e^{\underline{A}t} e^{\underline{B}t} \neq e^{(\underline{A}+\underline{B})t} \tag{14}$$

in general. More specifically, we have the following theorem.

4. THEOREM. For any two $n \times n$ matrices $\underline{A}$ and $\underline{B}$,

$$e^{\underline{A}t} e^{\underline{B}t} = e^{(\underline{A}+\underline{B})t} \tag{15}$$

if and only if $\underline{A}$ and $\underline{B}$ commute, that is,

$$\underline{A}\underline{B} = \underline{B}\underline{A} \tag{16}$$

Proof: Using the power series (3.1-17), we have

$$\begin{aligned} e^{\underline{A}t} e^{\underline{B}t} &= \left(\underline{I} + \underline{A}t + \frac{\underline{A}^2 t^2}{2!} + \ldots\right)\left(\underline{I} + \underline{B}t + \frac{\underline{B}^2 t^2}{2!} + \ldots\right) \\ &= \underline{I} + \underline{B}t + \frac{\underline{B}^2 t^2}{2!} + \ldots \\ &\quad + \underline{A}t + \underline{A}\underline{B}t^2 + \ldots \\ &\quad + \frac{\underline{A}^2 t^2}{2!} + \ldots \\ &= \underline{I} + (\underline{A} + \underline{B})t + \frac{\underline{A}^2 + 2\underline{A}\underline{B} + \underline{B}^2}{2!} + \ldots \end{aligned} \tag{17}$$

and

$$e^{(\underline{A} + \underline{B})t} = \underline{I} + (\underline{A} + \underline{B}) + \frac{(\underline{A} + \underline{B})^2 t^2}{2!} + \cdots$$
$$= \underline{I} + (\underline{A} + \underline{B}) + \frac{(\underline{A}^2 + \underline{AB} + \underline{BA} + \underline{B}^2)t^2}{2!} + \cdots \tag{18}$$

These two series will be equal if and only if they agree term by term. The first two terms are identical, but the quadratic terms are clearly equal if and only if

$$\underline{AB} = \underline{BA} \tag{19}$$

It is easy to show (Problem 1) that this condition also implies equality in each of the subsequent terms.

ΔΔΔ

The following is a useful consequence of Theorem 4.

5. *COROLLARY*. If

$$\underline{A} = \sigma\underline{I} + \underline{B} \tag{20}$$

for some scalar σ and any matrix $\underline{B}$, then

$$e^{\underline{A}t} = e^{\sigma t} e^{\underline{B}t} \tag{21}$$

Proof: Left to the reader (Problem 2).

ΔΔΔ

We next state the form of the matrix exponential for a number of common matrices.

6. LEMMA.

$$\exp\begin{bmatrix}\lambda_1 & 0 & \cdots & 0\\ 0 & \lambda_2 & \cdots & 0\\ \cdots & \cdots & \cdots & \cdots\\ 0 & 0 & \cdots & \lambda_n\end{bmatrix} t = \begin{bmatrix}e^{\lambda_1 t} & 0 & \cdots & 0\\ 0 & e^{\lambda_2 t} & \cdots & 0\\ \cdots & \cdots & \cdots & \cdots\\ 0 & 0 & \cdots & e^{\lambda_n t}\end{bmatrix} \tag{22}$$

$$\exp\begin{bmatrix}\underline{A}_1 & \underline{0} & \cdots & \underline{0}\\ \underline{0} & \underline{A}_2 & \cdots & \underline{0}\\ \cdots & \cdots & \cdots & \cdots\\ \underline{0} & \underline{0} & \cdots & \underline{A}_m\end{bmatrix} t = \begin{bmatrix}e^{\underline{A}_1 t} & \underline{0} & \cdots & \underline{0}\\ \underline{0} & e^{\underline{A}_2 t} & \cdots & \underline{0}\\ \cdots & \cdots & \cdots & \cdots\\ \underline{0} & \underline{0} & \cdots & e^{\underline{A}_m t}\end{bmatrix} \tag{23}$$

$$\exp\begin{bmatrix}0 & 1 & 0 & \cdots & 0\\ 0 & 0 & 1 & \cdots & 0\\ 0 & 0 & 0 & \cdots & 0\\ \cdots & \cdots & \cdots & \cdots & \cdots\\ 0 & 0 & 0 & \cdots & 1\\ 0 & 0 & 0 & \cdots & 0\end{bmatrix} t = \begin{bmatrix}1 & t & \frac{t^2}{2} & \cdots & \frac{t^{n-1}}{(n-1)!}\\ 0 & 1 & t & \cdots & \frac{t^{n-2}}{(n-2)!}\\ \cdots & \cdots & \cdots & \cdots & \cdots\\ 0 & 0 & 0 & \cdots & t\\ 0 & 0 & 0 & \cdots & 1\end{bmatrix} \tag{24}$$

$$\exp\begin{bmatrix}\lambda & 1 & 0 & \cdots & 0\\ 0 & \lambda & 1 & \cdots & 0\\ 0 & 0 & \lambda & \cdots & 0\\ \cdots & \cdots & \cdots & \cdots & \cdots\\ 0 & 0 & 0 & \cdots & 1\\ 0 & 0 & 0 & \cdots & \lambda\end{bmatrix} t = e^{\lambda t}\begin{bmatrix}1 & t & \frac{t^2}{2} & \cdots & \frac{t^{n-1}}{(n-1)!}\\ 0 & 1 & t & \cdots & \frac{t^{n-2}}{(n-2)!}\\ \cdots & \cdots & \cdots & \cdots & \cdots\\ 0 & 0 & 0 & \cdots & t\\ 0 & 0 & 0 & \cdots & 1\end{bmatrix} \tag{25}$$

$$\exp\begin{bmatrix} 0 & \omega \\ -\omega & 0 \end{bmatrix} t = \begin{bmatrix} \cos \omega t & \sin \omega t \\ -\sin \omega t & \cos \omega t \end{bmatrix} \tag{26}$$

$$\exp\begin{bmatrix} \sigma & \omega \\ -\omega & \sigma \end{bmatrix} t = \begin{bmatrix} e^{\sigma t} \cos \omega t & e^{\sigma t} \sin \omega t \\ -e^{\sigma t} \sin \omega t & e^{\sigma t} \cos \omega t \end{bmatrix} \tag{27}$$

Proof: See Problems 3 and 4.

ΔΔΔ

It is clear that if a system

$$\dot{\underline{x}}(t) = \underline{A}\underline{x}(t)$$
$$\underline{x}(0) = \underline{x}_0 \tag{28}$$

has $\underline{A}$ in one of the above forms, its response

$$\underline{x}(t) = e^{\underline{A}t} \underline{x}_0 \tag{29}$$

will be straightforward to compute. Moreover, we shall now see that a state transformation can be used to reduce any system to such a form. In particular, Lemma 2.5.1 established that a transformation of the state to

$$\hat{\underline{x}}(t) = \underline{Q}\underline{x}(t) \tag{30}$$

puts the system in the form

$$\dot{\hat{\underline{x}}}(t) = \hat{\underline{A}}\hat{\underline{x}}(t) = \underline{Q}\underline{A}\underline{Q}^{-1} \hat{\underline{x}}(t)$$
$$\hat{\underline{x}}(t) = \hat{\underline{x}}_0 = \underline{Q}\underline{x}_0 \tag{31}$$

with response

$$\hat{\underline{x}}(t) = e^{\hat{\underline{A}}t} \hat{\underline{x}}_0 \tag{32}$$

This implies (see Problem 5) that

$$e^{\underline{A}t} = \underline{Q}^{-1} e^{\hat{\underline{A}}t} \underline{Q} \qquad e^{\hat{\underline{A}}t} = \underline{Q} e^{\underline{A}t} \underline{Q}^{-1} \tag{33}$$

The following make use of state transformations to provide some *canonical forms* of $e^{\hat{\underline{A}}t}$.

7. *THEOREM*. If a matrix $\underline{A}$ can be diagonalized[†] so that

$$\hat{\underline{A}} = \underline{QAQ}^{-1} = \begin{bmatrix} \lambda_1 & 0 & \dots & 0 \\ 0 & \lambda_2 & \dots & 0 \\ \dots & \dots & \dots & \dots \\ 0 & 0 & \dots & \lambda_n \end{bmatrix} \tag{34}$$

where $\lambda_1, \lambda_2, \dots, \lambda_n$ are the eigenvalues of $\underline{A}$ (and $\hat{\underline{A}}$), then its exponential is

$$e^{\underline{A}t} = \underline{Q}^{-1} e^{\hat{\underline{A}}t} \underline{Q} = \underline{Q}^{-1} \begin{bmatrix} e^{\lambda_1 t} & 0 & \dots & 0 \\ 0 & e^{\lambda_2 t} & \dots & 0 \\ \dots & \dots & \dots & \dots \\ 0 & 0 & \dots & e^{\lambda_n t} \end{bmatrix} \underline{Q} \tag{35}$$

Otherwise, it has some repeated eigenvalues and can always be put in Jordan form, as in Theorem A.3.4,

$$\hat{\underline{A}} = \underline{QAQ}^{-1} = \begin{bmatrix} \underline{J}_1 & \underline{0} & \dots & \underline{0} \\ \underline{0} & \underline{J}_2 & \dots & \underline{0} \\ \dots & \dots & \dots & \dots \\ \underline{0} & \underline{0} & \dots & \underline{J}_p \end{bmatrix} \tag{36}$$

[†]A sufficient condition for this is that the eigenvalues be distinct (see Theorem A.3.3).

where the $\mu_i \times \mu_i$ diagonal blocks each have the form

$$\underline{J}_i = \left[\begin{array}{ccccc|cccc} \lambda_1 & 1 & 0 & \cdots & 0 & & & & \\ 0 & \lambda_i & 1 & \cdots & 0 & & & & \\ \cdots & \cdots & \cdots & \cdots & \cdots & & \underline{0} & & \\ 0 & 0 & 0 & \cdots & 1 & & & & \\ 0 & 0 & 0 & \cdots & \lambda_i & & & & \\ \hline & & & & & \lambda_i & 0 & \cdots & 0 \\ & & \underline{0} & & & 0 & \lambda_i & \cdots & 0 \\ & & & & & \cdots & \cdots & \cdots & \cdots \\ & & & & & 0 & 0 & \cdots & \lambda_i \end{array}\right] \quad (i=1,2,\ldots,p) \tag{37}$$

where the upper-left block spans η_i columns and the lower-right block spans $\mu_i - \eta_i$ columns.

The matrix exponential is then

$$e^{\underline{A}t} = \underline{Q}^{-1} e^{\underline{A}t} \underline{Q} = \underline{Q}^{-1} \begin{bmatrix} e^{\underline{J}_1 t} & \underline{0} & \cdots & \underline{0} \\ \underline{0} & e^{\underline{J}_2 t} & \cdots & \underline{0} \\ \cdots & \cdots & \cdots & \cdots \\ \underline{0} & \underline{0} & \cdots & e^{\underline{J}_p t} \end{bmatrix} \underline{Q} \tag{38}$$

where the diagonal blocks are

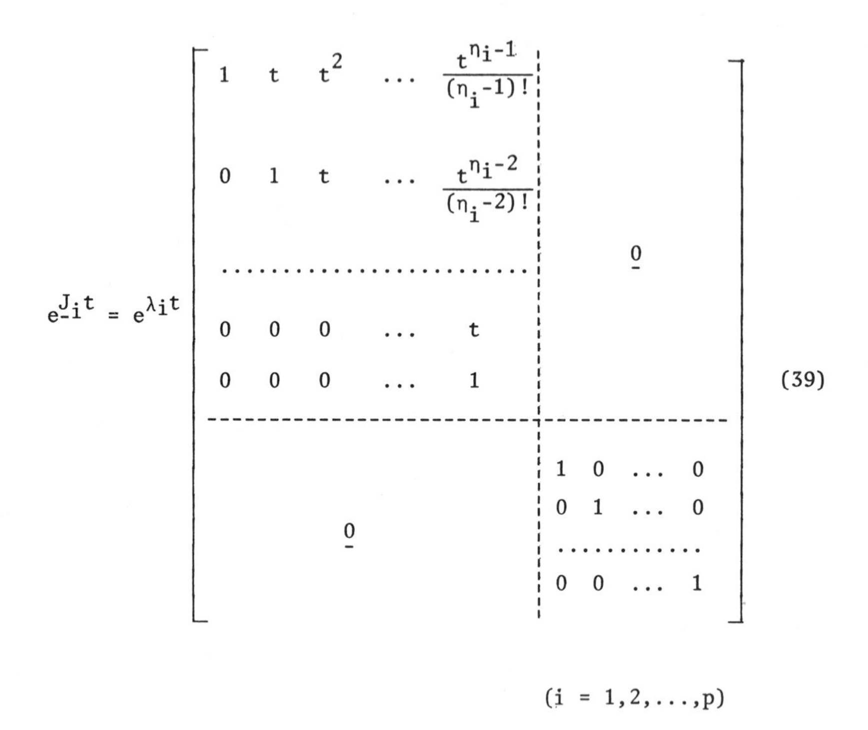

$$e^{\underline{J}_i t} = e^{\lambda_i t} \left[\begin{array}{ccccc|cccc} 1 & t & t^2 & \cdots & \frac{t^{\eta_i - 1}}{(\eta_i - 1)!} & & & & \\ 0 & 1 & t & \cdots & \frac{t^{\eta_i - 2}}{(\eta_i - 2)!} & & & & \\ \cdots & \cdots & \cdots & \cdots & \cdots & & \underline{0} & & \\ 0 & 0 & 0 & \cdots & t & & & & \\ 0 & 0 & 0 & \cdots & 1 & & & & \\ \hline & & & & & 1 & 0 & \cdots & 0 \\ & & \underline{0} & & & 0 & 1 & \cdots & 0 \\ & & & & & \cdots & \cdots & \cdots & \cdots \\ & & & & & 0 & 0 & \cdots & 1 \end{array}\right] \tag{39}$$

$(i = 1,2,\ldots,p)$

Proof: Equation (35) follows from (22), and (38) from (23). The form of the individual blocks (39) was given in (25).

ΔΔΔ

It should be noted that $\underline{Q}$ and $\underline{Q}^{-1}$ above will be complex in general. The following result allows us to avoid complex numbers entirely.

8. THEOREM. A matrix which can be diagonalized may also be put in the alternative block-diagonal form of Theorem A.3.6,

$$\hat{\underline{A}} = \underline{QAQ}^{-1} = \begin{bmatrix} \underline{\Lambda}_1 & \underline{0} & \cdots & \underline{0} & \underline{0} \\ \underline{0} & \underline{\Lambda}_3 & \cdots & \underline{0} & \underline{0} \\ \cdots & \cdots & \cdots & \cdots & \cdots \\ \underline{0} & \underline{0} & \cdots & \underline{\Lambda}_{m-1} & \underline{0} \\ \underline{0} & \underline{0} & \cdots & \underline{0} & \underline{\Lambda}_{m+1} \end{bmatrix} \tag{40}$$

where

$$\underline{\Lambda}_i = \begin{bmatrix} \sigma_i & \omega_i \\ -\omega_i & \sigma_i \end{bmatrix} \qquad (i = 1,3,5,\ldots,m-1) \tag{41}$$

are formed from the complex eigenvalues of $\underline{A}$ (and $\hat{\underline{A}}$) ,

$$\begin{aligned} \lambda_i &= \sigma_i + j\omega_i \\ \lambda_{i+1} &= \sigma_i - j\omega_i = \overline{\lambda_i} \end{aligned} \qquad (i = 1,3,5,\ldots,m-1) \tag{42}$$

and

$$\underline{\Lambda}_{m+1} = \begin{bmatrix} \lambda_{m+1} & 0 & \cdots & 0 \\ 0 & \lambda_{m+2} & \cdots & 0 \\ \cdots & \cdots & \cdots & \cdots \\ 0 & 0 & \cdots & \lambda_n \end{bmatrix} \tag{43}$$

is formed from the real eigenvalues

$$\lambda_{m+1}, \lambda_{m+2}, \ldots, \lambda_n \tag{44}$$

In this case the matrix exponential is

$$e^{\underline{A}t} = \underline{Q}^{-1} e^{\underline{A}t} \underline{Q} = \underline{Q}^{-1} \begin{bmatrix} e^{\underline{\Lambda}_1 t} & \underline{0} & \cdots & \underline{0} \\ \underline{0} & e^{\underline{\Lambda}_3 t} & \cdots & \underline{0} \\ \cdots & \cdots & \cdots & \cdots \\ \underline{0} & \underline{0} & \cdots & e^{\underline{\Lambda}_{m+1} t} \end{bmatrix} \underline{Q} \tag{45}$$

where

$$e^{\underline{\Lambda}_i t} = e^{\sigma_i t} \begin{bmatrix} \cos\omega_i t & \sin\omega_i t \\ -\sin\omega_i t & \cos\omega_i t \end{bmatrix} \qquad (i = 1,3,5,\ldots,m-1) \tag{46}$$

and

$$e^{\underline{\Lambda}_{m+1}t} = \begin{bmatrix} e^{\lambda_{m+1}t} & 0 & \cdots & 0 \\ 0 & e^{\lambda_{m+2}t} & \cdots & 0 \\ \cdots & \cdots & \cdots & \cdots \\ 0 & 0 & \cdots & e^{\lambda_n t} \end{bmatrix} \tag{47}$$

Otherwise, the matrix has repeated eigenvalues and may always be put into the block-Jordan form of Theorem A.3.7. The form of the matrix exponential in this case is left for the reader to work out (Problem 6).

Proof: This follows from Eqs. (23) and (27).

ΔΔΔ

Theorems 7 and 8 establish that the dynamic behavior of (28) is determined primarily by the eigenvalues of $\underline{A}$ since each component of the state trajectory $\underline{x}(t) = e^{\underline{A}t}\underline{x}_0$ may be expressed as a linear combination of terms such as $e^{\lambda t}$, $e^{\sigma t}\sin\omega t$, and $e^{\sigma t}\cos\omega t$, possibly multiplied by powers of t. It is standard practice to refer to these eigenvalues as the *poles* of the system (28), and to the exponential and sinusoidal terms in the state trajectory as its *modes*.

These observations lead naturally into the concept of system stability, which is the subject of Chapter 5. The system (28) will be called *stable* if every initial state $\underline{x}_0$ produces a state trajectory $\underline{x}(t)$ which eventually decays to $\underline{0}$. It should be clear from the discussion above that this condition will hold when the real parts (λ, σ, etc.) of all the poles are negative, so that the corresponding exponentials approach zero.

9. *EXAMPLE.* Consider the linear system

$$\dot{\underline{x}}(t) = \underline{A}\underline{x}(t) = \begin{bmatrix} -4 & 3 \\ -6 & 5 \end{bmatrix} \underline{x}(t) \tag{48}$$

The poles of this system (see Section A.3) are the roots of

$$\det(\lambda \underline{I} - \underline{A}) = \det \begin{bmatrix} \lambda + 4 & -3 \\ 6 & \lambda - 5 \end{bmatrix} \tag{49}$$

$$= \lambda^2 - \lambda - 2 = (\lambda + 1)(\lambda - 2)$$

that is, $\lambda_1 = -1$ and $\lambda_2 = 2$. To diagonalize $\underline{A}$ we must compute its eigenvectors $\underline{u}_1$ and $\underline{u}_2$ by solving

$$\underline{A}\underline{u}_1 = \begin{bmatrix} -4 & 3 \\ -6 & 5 \end{bmatrix} \begin{bmatrix} u_{11} \\ u_{21} \end{bmatrix} = \lambda_1 \underline{u}_1 = -1 \begin{bmatrix} u_{11} \\ u_{21} \end{bmatrix} \tag{50}$$

and

$$\underline{A}\underline{u}_2 = \begin{bmatrix} -4 & 3 \\ -6 & 5 \end{bmatrix} \begin{bmatrix} u_{12} \\ u_{22} \end{bmatrix} = \lambda_2 \underline{u}_2 = 2 \begin{bmatrix} u_{12} \\ u_{22} \end{bmatrix} \tag{51}$$

Equation (50) yields

$$\begin{aligned} -4u_{11} + 3u_{21} &= -u_{11} \\ -6u_{11} + 5u_{21} &= -u_{21} \end{aligned} \tag{52}$$

both of which reduce to

$$u_{11} = u_{21} \tag{53}$$

Choosing $u_{11} = 1$ (the magnitude of an eigenvector is arbitrary), we have $\underline{u}_1 = [1 \quad 1]'$. Similarly, both equations in (51) yield

$$u_{22} = 2u_{12} \tag{54}$$

and the choice $u_{12} = 1$ provides $\underline{u}_2 = [1 \quad 2]'$.

Collecting the eigenvectors into a matrix,

$$\underset{\sim}{Q}^{-1} = \begin{bmatrix} \underset{\sim}{u}_1 & \underset{\sim}{u}_2 \end{bmatrix} = \begin{bmatrix} 1 & 1 \\ 1 & 2 \end{bmatrix} \tag{55}$$

we may now make use of the state transformation

$$\hat{\underset{\sim}{x}}(t) = \underset{\sim}{Q}\underset{\sim}{x}(t) = \begin{bmatrix} 2 & -1 \\ -1 & 1 \end{bmatrix} \underset{\sim}{x}(t) \tag{56}$$

to put (48) in the form

$$\begin{aligned} \dot{\hat{\underset{\sim}{x}}} = \underset{\sim}{Q}\underset{\sim}{A}\underset{\sim}{Q}^{-1}\hat{\underset{\sim}{x}} &= \begin{bmatrix} 2 & -1 \\ -1 & 1 \end{bmatrix}\begin{bmatrix} -4 & 3 \\ -6 & 5 \end{bmatrix}\begin{bmatrix} 1 & 1 \\ 1 & 2 \end{bmatrix} \hat{\underset{\sim}{x}} \\ &= \begin{bmatrix} -1 & 0 \\ 0 & 2 \end{bmatrix} \hat{\underset{\sim}{x}} = \hat{\underset{\sim}{A}}\hat{\underset{\sim}{x}} \end{aligned} \tag{57}$$

The response from an initial state $\hat{\underset{\sim}{x}}_0 = \underset{\sim}{Q}\underset{\sim}{x}_0$ is thus

$$\hat{\underset{\sim}{x}}(t) = e^{\hat{\underset{\sim}{A}}t}\hat{\underset{\sim}{x}}_0 = \begin{bmatrix} e^{-t} & 0 \\ 0 & e^{2t} \end{bmatrix}\begin{bmatrix} \hat{x}_{10} \\ \hat{x}_{20} \end{bmatrix} = \begin{bmatrix} \hat{x}_{10}e^{-t} \\ \hat{x}_{20}e^{2t} \end{bmatrix} \tag{58}$$

or, in terms of the original state vector,

$$\begin{aligned} \underset{\sim}{x}(t) &= \underset{\sim}{Q}^{-1}\hat{\underset{\sim}{x}}(t) \\ &= \underset{\sim}{Q}^{-1} e^{\hat{\underset{\sim}{A}}t} \underset{\sim}{Q}\underset{\sim}{x}_0 \\ &= \begin{bmatrix} 1 & 1 \\ 1 & 2 \end{bmatrix}\begin{bmatrix} e^{-t} & 0 \\ 0 & e^{2t} \end{bmatrix}\begin{bmatrix} 2 & -1 \\ -1 & 1 \end{bmatrix}\begin{bmatrix} x_{10} \\ x_{20} \end{bmatrix} \\ &= \begin{bmatrix} (2x_{10} - x_{20})e^{-t} + (x_{20} - x_{10})e^{2t} \\ (2x_{10} - x_{20})e^{-t} + 2(x_{20} - x_{10})e^{2t} \end{bmatrix} \end{aligned} \tag{59}$$

where the modes of this system are e^{-t} and e^{2t}.

ΔΔΔ

The theorems above also lead to the following *spectral* (or *modal*) *representation* of the matrix exponential.

10. THEOREM. If $\underline{A}$ is an $n \times n$ matrix which can be diagonalized, then its exponential may be written as

$$e^{\underline{A}t} = \sum_{i=1}^{n} \underline{R}_i \, e^{\lambda_i t} \tag{60}$$

where $\lambda_1, \lambda_2, \ldots, \lambda_n$ are the eigenvalues of $\underline{A}$.

Otherwise, it has repeated eigenvalues and can be written as

$$e^{\underline{A}t} = \sum_{i=1}^{p} \sum_{j=1}^{\eta_i} \underline{R}_i^j \, t^{j-1} \, e^{\lambda_i t} \tag{61}$$

Proof: Referring to (35), we may write

$$e^{\underline{\hat{A}}t} = \sum_{i=1}^{n} \underline{E}_i e^{\lambda_i t} \tag{62}$$

where $\underline{E}_i$ is a matrix with 1 in the i-th diagonal position and 0's everywhere else. Using (33),

$$\begin{aligned} e^{\underline{A}t} &= \underline{Q}^{-1} e^{\underline{\hat{A}}t} \underline{Q} = \underline{Q}^{-1} \sum_{i=1}^{n} \underline{E}_i e^{\lambda_i t} \underline{Q} \\ &= \sum_{i=1}^{n} \underline{Q}^{-1} \underline{E}_i \underline{Q} \, e^{\lambda_i t} \end{aligned} \tag{63}$$

which establishes (60), with $\underline{R}_i = \underline{Q}^{-1} \underline{E}_i \underline{Q}$.

Equation (61) follows from (38) and (39) in a straightforward extension of the preceding argument. The details are left to the reader (see Problem 7).

ΔΔΔ

Since $\underline{Q}$ and $\underline{Q}^{-1}$ may be complex, $\underline{R}_i$ and $\underline{R}_i^j$ will contain complex numbers in general. The following result avoids this inconvenience.

11. THEOREM. If a matrix can be diagonalized, then its exponential may also be written as

$$e^{\underline{A}t} = \sum_{i=1,3,\ldots}^{m-1} e^{\sigma_i t}(\underline{C}_i \cos \omega_i t + \underline{S}_i \sin \omega_i t) + \sum_{i=m+1}^{n} \underline{R}_i e^{\lambda_i t} \tag{64}$$

where $\underline{C}_i$, $\underline{S}_i$, and $\underline{R}_i$ are real, and the eigenvalues of $\underline{A}$ are

$$\left.\begin{aligned} \lambda_i &= \sigma_i + j\omega_i \\ \lambda_{i+1} &= \sigma_i - j\omega_i = \overline{\lambda}_i \end{aligned}\right\} \quad (i = 1,3,5,\ldots,m-1)$$
$$\lambda_i = \overline{\lambda}_i \qquad (i = m+1, m+2, \ldots, n) \tag{65}$$

Otherwise, the matrix has repeated eigenvalues and a more complicated formula applies, derivation of which is left to the reader (Problem 6).

Proof: This follows from Theorem 8 in a straightforward extension of the proof of Theorem 10. Details are left to the reader (Problem 7).

ΔΔΔ

12. EXAMPLE. Applying Theorem 10 to the system of Example 9, we find that

$$\begin{aligned} \underline{R}_1 &= \underline{Q}^{-1}\underline{E}_1\underline{Q} = \begin{bmatrix} 1 & 1 \\ 1 & 2 \end{bmatrix}\begin{bmatrix} 1 & 0 \\ 0 & 0 \end{bmatrix}\begin{bmatrix} 2 & -1 \\ -1 & 1 \end{bmatrix} = \begin{bmatrix} 2 & -1 \\ 2 & -1 \end{bmatrix} \\ \underline{R}_2 &= \underline{Q}^{-1}\underline{E}_2\underline{Q} = \begin{bmatrix} 1 & 1 \\ 1 & 2 \end{bmatrix}\begin{bmatrix} 0 & 0 \\ 0 & 1 \end{bmatrix}\begin{bmatrix} 2 & -1 \\ -1 & 1 \end{bmatrix} = \begin{bmatrix} -1 & 1 \\ -2 & 2 \end{bmatrix} \end{aligned} \tag{66}$$

and the matrix exponential may be expressed as

$$e^{\underline{A}t} = \sum_{i=1}^{2} \underline{R}_i e^{\lambda_i t} = \begin{bmatrix} 2 & -1 \\ 2 & -1 \end{bmatrix} e^{-t} + \begin{bmatrix} -1 & 1 \\ -2 & 2 \end{bmatrix} e^{2t} \tag{67}$$

If the initial state in (48) is $\underline{x}_0 = [x_{10} \;\; x_{20}]'$, then the state trajectory will be

$$\underline{x}(t) = e^{\underline{A}t}\underline{x}_0 = \begin{bmatrix} 2x_{10} - x_{20} \\ 2x_{10} - x_{20} \end{bmatrix} e^{-t} + \begin{bmatrix} x_{20} - x_{10} \\ 2(x_{20} - x_{10}) \end{bmatrix} e^{2t} \tag{68}$$

which agrees with (59).

ΔΔΔ

In addition to these spectral representations, a finite sum representation which is often useful expresses $e^{\underline{A}t}$ as an (n-1)-th degree matrix polynomial.

13. THEOREM. If $\underline{A}$ is an $n \times n$ matrix, then

$$e^{\underline{A}t} = \sum_{i=0}^{n-1} \underline{A}^i \alpha_i(t) \tag{69}$$

where $\alpha_i(t)$, $i = 0,1,2,\ldots,n-1$, are scalar functions of time.

Proof: This is a consequence of the Cayley-Hamilton theorem (see Problem 8), but it is more convenient to defer its proof until the next chapter (see Corollary 4.3.3). For an alternate derivation and a proof that the α_i are linearly independent functions, see Ref. 345.

ΔΔΔ

In this section we have examined the basic structure of the matrix exponential $e^{\underline{A}t}$. This led us to conclude that solutions of unforced, linear, time-invariant differential equations always belong to a special class of functions: linear combinations of terms containing exponentials, sinusoids, and occasionally powers of t. In the next section we shall see that the solution of a forced equation splits into an unforced solution of this form plus another term due to the forcing function.

3.3. SOLUTION OF FORCED EQUATIONS

Now we shall consider the case where an input is applied to the system under consideration, so that Eq.(3.1-1) generalizes to a *forced* (or *inhomogeneous*) *differential equation*,

$$\begin{aligned} \dot{\underline{x}}(t) &= \underline{\underline{A}}\,\underline{x}(t) + \underline{v}(t) \\ \underline{x}(t_0) &= \underline{x}_0 \end{aligned} \tag{1}$$

We asserted in Sec. 3.1 that a solution of (1) must be unique, and now we shall see that a solution always exists and is equal to the sum of the unforced solution and another term involving the forcing function. We begin by extending Example 3.1.1.

1. EXAMPLE. Consider the one-dimensional system

$$\dot{x}(t) = ax(t) + v(t) \tag{2}$$

with $x(0) = x_0$, where $v(t)$ is a known (integrable) function of time. Rearranging (2) and multiplying by the integrating factor e^{-at}, we have

$$e^{-at}\,[\dot{x}(t) - ax(t)] = e^{-at}\,v(t) \tag{3}$$

Since the left-hand side is a perfect differential, this is

$$\frac{d}{dt}\,[e^{-at}\,x(t)] = e^{-at}\,v(t) \tag{4}$$

which can be integrated from 0 to t,

$$\begin{aligned} \int_0^t \frac{d}{d\tau}\,[e^{-a\tau}\,x(\tau)]\,d\tau &= e^{-at}\,x(t) - e^{-a0}\,x(0) \\ &= \int_0^t e^{-a\tau}\,v(\tau)\,d\tau \end{aligned} \tag{5}$$

Finally, this can be rearranged and multiplied by e^{at} to yield the solution of (2),

$$x(t) = e^{at}\,x_0 + \int_0^t e^{a(t-\tau)}\,v(\tau)\,d\tau \tag{6}$$

We see that this consists of the solution $e^{at}x_0$ of the unforced equation (the response to the initial condition) plus an integral involving v(t) (the response to the forcing function).

ΔΔΔ

The general case is analogous to this example.

2. *THEOREM.* The solution of

$$\dot{\underline{x}}(t) = \underline{\underline{A}}\underline{x}(t) + \underline{v}(t) \tag{7}$$

with $\underline{x}(t_0) = \underline{x}_0$ is given by

$$\underline{x}(t) = e^{\underline{A}(t-t_0)}\underline{x}_0 + \int_{t_0}^{t} e^{\underline{A}(t-\tau)} \underline{v}(\tau)\, d\tau \tag{8}$$

Proof: As in the example, we rearrange (7), multiply by an integrating factor $e^{-\underline{A}t}$, and use Theorems 3.1.3 and A.2.10 to obtain

$$e^{-\underline{A}t}[\dot{\underline{x}}(t) - \underline{\underline{A}}\underline{x}(t)] = \frac{d}{dt}[e^{-\underline{A}t}\underline{x}(t)]$$

$$= e^{-\underline{A}t}\underline{v}(t) \tag{9}$$

Integrating this, we have

$$\int_{t_0}^{t} \frac{d}{d\tau}[e^{-\underline{A}\tau}\underline{x}(\tau)]\, d\tau = e^{-\underline{A}t}\underline{x}(t) - e^{-\underline{A}t_0}\underline{x}(t_0)$$

$$= \int_{t_0}^{t} e^{-\underline{A}\tau}\underline{v}(\tau)\, d\tau \tag{10}$$

which can be rearranged and multiplied by $e^{\underline{A}t}$ to yield (8).

ΔΔΔ

The results have thus far been stated in terms of the system state $\underline{x}(t)$ and a forcing function $\underline{v}(t)$, both n-dimensional vectors. We shall now reintroduce the m-dimensional input vector $\underline{u}(t)$ and the p-dimensional output vector $\underline{y}(t)$ which appear in the formulation (2.1-18),

$$\underline{v}(t) = \underline{B}\underline{u}(t) \tag{11}$$

$$\underline{y}(t) = \underline{C}\underline{x}(t) \tag{12}$$

3. *COROLLARY (Forced response).* The solution of

$$\begin{aligned} \dot{\underline{x}}(t) &= \underline{A}\underline{x}(t) + \underline{B}\underline{u}(t) \\ \underline{y}(t) &= \underline{C}\underline{x}(t) \end{aligned} \tag{13}$$

with $\underline{x}(t_0) = \underline{x}_0$ is given by

$$\underline{y}(t) = \underline{C}\underline{x}(t) = \underline{C}e^{\underline{A}(t-t_0)}\underline{x}_0 + \int_{t_0}^{t} \underline{C}e^{\underline{A}(t-\tau)}\underline{B}\underline{u}(\tau)\, d\tau \tag{14}$$

The first term on the right-hand side, due to the initial state $\underline{x}_0$, is known as the *natural* (or *zero-input*) *response*. The second term, due to the input $\underline{u}(t)$, is called the *forced* (or *zero-state*) *response*.

Proof: Combine Eqs. (8), (11), and (12).

ΔΔΔ

Note that the output $\underline{y}(t)$ in (14) depends not just on the present input $\underline{u}(t)$, but on past inputs $\underline{u}(\tau)$ for $t_0 < \tau \leq t$ as well; thus the system is said to have *memory*. Because $\underline{y}(t)$ does not depend upon $\underline{u}(\tau)$ for $\tau > t$, the system is said to be *causal* (or *nonanticipatory*).

When $\underline{x}_0 = \underline{0}$, (14) reduces to the forced response

$$\underline{y}(t) = \int_{t_0}^{t} \underline{C}\, e^{\underline{A}(t-\tau)}\underline{B}\underline{u}(\tau)\, d\tau \tag{15}$$

from which it is clear that the input-output behavior is determined entirely by the matrix

$$\underline{H}(t - \tau) \triangleq \underline{C}\, e^{\underline{A}(t-\tau)}\underline{B} \tag{16}$$

sometimes called the *weighting pattern* of the system. Its interpretation as an *impulse response matrix* will be discussed in the next section.

Equation (15) makes explicit the fundamental characteristic of linear systems which distinguishes them from nonlinear systems and makes linear analysis such a fruitful activity. This is the celebrated *superposition property,* which amounts to the fact that in the absence of initial conditions ($\underline{x}_0 = \underline{0}$), the output $\underline{y}$ of the linear system (13) depends *linearly* upon the input $\underline{u}$.

4. *COROLLARY (Superposition property).* Let the system (13) have an initial state $\underline{x}_0 = \underline{0}$. If the input functions $\underline{u}_1(t)$ and $\underline{u}_2(t)$ evoke responses $\underline{y}_1(t)$ and $\underline{y}_2(t)$, respectively, then the forcing function

$$\underline{u}(t) = \underline{u}_1(t) + \beta\underline{u}_2(t) \tag{17}$$

where β is any scalar constant, evokes a response

$$\underline{y}(t) = \underline{y}_1(t) + \beta\underline{y}_2(t) \tag{18}$$

Proof: This follows by substitution in (15):

$$\underline{y}(t) = \int_{t_0}^{t} \underline{C}e^{\underline{A}(t-\tau)}\underline{B}[\underline{u}_1(\tau) + \beta\underline{u}_2(\tau)]\, d\tau$$

$$= \int_{t_0}^{t} \underline{C}e^{\underline{A}(t-\tau)}\underline{B}\underline{u}_1(\tau)\, d\tau + \beta\int_{t_0}^{t} \underline{C}e^{\underline{A}(t-\tau)}\underline{B}\underline{u}_2(\tau)\, d\tau \tag{19}$$

$$= \underline{y}_1(t) + \beta\underline{y}_2(t)$$

ΔΔΔ

5. *EXAMPLE.* Recall from Example 2.4.3 that the linearized azimuthal motion of an earth satellite was described by (2.4-63),

$$\begin{aligned} \dot{\underline{x}}(t) &= \begin{bmatrix} 0 & 1 \\ -\omega^2 & 0 \end{bmatrix} \underline{x}(t) + \begin{bmatrix} 0 \\ 1 \end{bmatrix} u_\phi(t) \\ y(t) &= \phi(t) = \begin{bmatrix} 1 & 0 \end{bmatrix} \underline{x}(t) \end{aligned} \tag{20}$$

where

$$\underline{x}(t) = \begin{bmatrix} \phi(t) \\ \dot{\phi}(t) \end{bmatrix} \tag{21}$$

and ϕ, $\dot{\phi}$, and u_ϕ are perturbations of the azimuthal angle, angular velocity, and thrust about their nominal values of zero.

This scalar system has the form (13), where

$$\underline{A} = \begin{bmatrix} 0 & 1 \\ -\omega^2 & 0 \end{bmatrix} \qquad \underline{b} = \begin{bmatrix} 0 \\ 1 \end{bmatrix} \qquad \underline{c}' = \begin{bmatrix} 1 & 0 \end{bmatrix} \tag{22}$$

and the exponential of $\underline{A}t$ was found in Example 3.1.5 to be

$$e^{\underline{A}t} = \begin{bmatrix} \cos \omega t & \omega^{-1} \sin \omega t \\ -\omega \cos \omega t & \cos \omega t \end{bmatrix} \tag{23}$$

From (14), the output is

$$\begin{aligned} y(t) = \phi(t) &= \underline{c}' e^{\underline{A}t} \underline{x}(0) + \int_0^t \underline{c}' e^{\underline{A}(t-\tau)} \underline{b} u_\phi(\tau)\, d\tau \\ &= \phi(0) \cos \omega t + \dot{\phi}(0)\, \omega^{-1} \sin \omega t + \int_0^t [\omega^{-1} \sin \omega(t-\tau)]\, u_\phi(\tau)\, d\tau \end{aligned} \tag{24}$$

Suppose the system is initially at rest (no azimuthal perturbations),

$$\phi(0) = \dot{\phi}(0) = 0 \tag{25}$$

and a *unit step* input is applied,

$$u_\phi(t) = 1(t) = \begin{cases} 0 & (t \le 0) \\ 1 & (t > 0) \end{cases} \tag{26}$$

as shown in Fig. 1(a). Then the response will be oscillatory

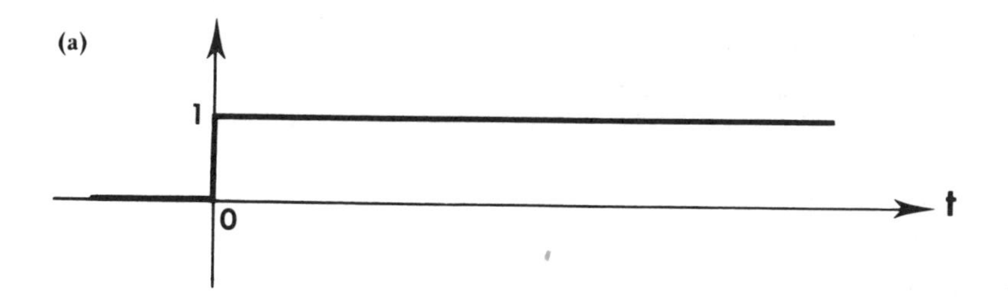

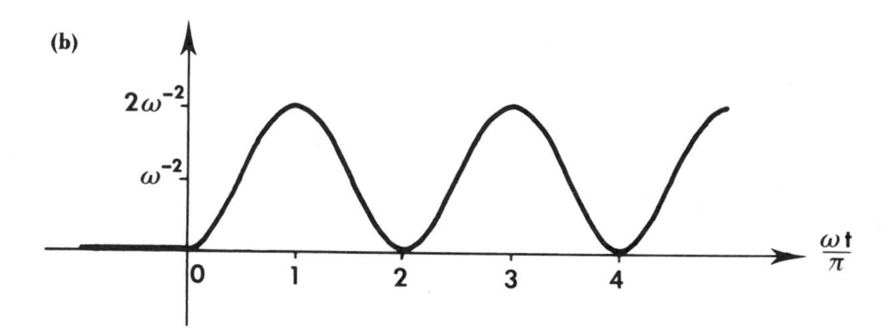

FIG. 1. Input and output functions of Example 5: (a) unit step input $u_\phi(t)$, and (b) oscillatory output $\phi(t)$.

$$
\begin{aligned}
y(t) = \phi(t) &= \int_0^t \omega^{-1} \sin \omega(t-\tau)\, d\tau \\
&= \omega^{-2} \cos \omega(t-\tau)\Big|_0^t = \omega^{-2}(1 - \cos \omega t)
\end{aligned}
\tag{27}
$$

as shown in Fig. 1(b).

ΔΔΔ

3.4. IMPULSE RESPONSE AND CONVOLUTION

For a zero initial state ($\underline{x}_0 = \underline{0}$), the output of the scalar system

$$\begin{aligned} \dot{\underline{x}}(t) &= \underline{A}\underline{x}(t) + \underline{b}u(t) \\ y(t) &= \underline{c}'\underline{x}(t) \end{aligned} \tag{1}$$

is given by (3.3-15),

$$y(t) = \int_0^t \underline{c}' e^{\underline{A}(t-\tau)} \underline{b} u(\tau)\, d\tau \tag{2}$$

where the initial time is taken as zero for convenience.

If the input u(t) is a high, narrow pulse of unit area, as shown in Fig. 1(a),

$$u(t) = \begin{cases} \varepsilon^{-1} & (0 < t \le \varepsilon) \\ 0 & (\text{otherwise}) \end{cases} \tag{3}$$

then the output (2) is

$$\begin{aligned} y(t) &= \int_0^t \underline{c}' e^{\underline{A}(t-\tau)} \underline{b} u(\tau)\, d\tau = \varepsilon^{-1} \int_0^\varepsilon \underline{c}' e^{\underline{A}(t-\tau)} \underline{b}\, d\tau \\ &= \varepsilon^{-1} \underline{c}' e^{\underline{A}t} \int_0^\varepsilon e^{-\underline{A}\tau} \underline{b}\, d\tau \end{aligned} \tag{4}$$

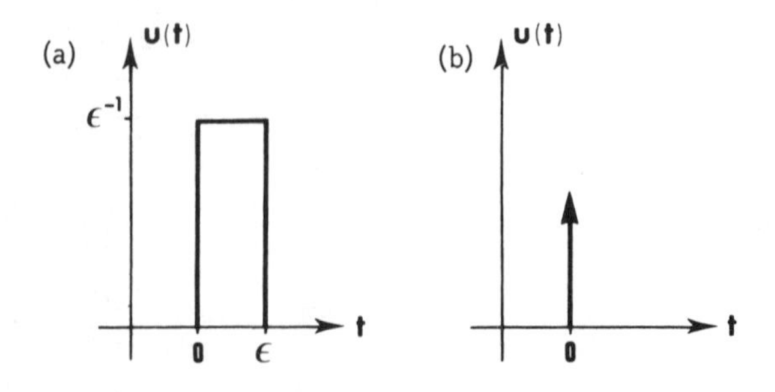

FIG. 1. Unit impulse function: (a) limiting pulse of unit area ($\varepsilon \to 0$) and (b) impulse $\delta(t)$.

The integrand is

$$e^{-\underline{A}\tau}\underline{b} = \left[\underline{I} + \underline{A}(-\tau) + \frac{\underline{A}^2(-\tau)^2}{2!} + \cdots\right]\underline{b} \tag{5}$$

where

$$|\tau| \leq \varepsilon \tag{6}$$

If ε is sufficiently small, all the terms except the first become negligible, so that

$$\lim_{\varepsilon\to 0} y(t) = \lim_{\varepsilon\to 0} \left(\varepsilon^{-1}\underline{c}'\, e^{\underline{A}t}\int_0^{\varepsilon}\underline{b}\, d\tau\right) = \underline{c}'\, e^{\underline{A}t}\,\underline{b} \tag{7}$$

1. DEFINITION. As ε becomes very small, we denote the input pulse (3) as

$$\lim_{\varepsilon\to 0} u(t) = \delta(t) \tag{8}$$

where δ is a *unit impulse*, or *Dirac delta function*, at $t = 0$, and is indicated schematically as in Fig. 1(b).

ΔΔΔ

In fact, $\delta(t)$ is not really a function and the limit (8) does not exist in the ordinary sense,[†] but we can overlook these difficulties because the limit in (7) is perfectly well defined. From an engineering point of view, it is not necessary to take the mathematical limit as $\varepsilon \to 0$ since (7) will hold with sufficient accuracy as long as ε is small in relation to the speed of response of the system being considered.

[†]This can all be made rigorous with a branch of mathematics known as *distribution theory*. The impulse δ is then defined by the following two properties:

(a) $\delta(t) = 0$ for $t \neq 0$.

(b) $\int f(t)\,\delta(t)\,dt = f(0)$ for any continuous function f.

For more details, see Appendix A of Ref. 185.

Thus we see that the output of (1) evoked by a unit impulse at the input is given by (7), which justifies the following definition.

2. *DEFINITION.* The scalar function

$$h(t) = \underline{c}' e^{\underline{A}t} \underline{b} \tag{9}$$

is called the *impulse response* of the scalar system (1). More generally, the multivariable system

$$\begin{aligned} \dot{\underline{x}}(t) &= \underline{A}\underline{x}(t) + \underline{B}\underline{u}(t) \\ \underline{y}(t) &= \underline{C}\underline{x}(t) \end{aligned} \tag{10}$$

has an *impulse response matrix*

$$\underline{H}(t) = \underline{C}e^{\underline{A}t}\underline{B} \tag{11}$$

and it is easy to see that an input vector

$$\underline{u}(t) = \underline{u}_0 \, \delta(t) \tag{12}$$

evokes an output

$$\underline{y}(t) = \underline{H}(t)\underline{u}_0 \tag{13}$$

In Theorem 5 we shall see that the impulse response is unchanged by a similarity transformation.

ΔΔΔ

It is significant that the impulse response matrix (11) is identical to the weighting pattern (3.3-16). In order to illuminate this relationship, consider a scalar system of the form (1) and a continuous input function $u(t)$ on the interval $0 < t \leq T$, as shown in Fig. 2(a). In Fig. 2(b) this function is approximated by a sequence of narrow pulses,

$$u(t) \approx \sum_{i=0}^{n-1} u_i(t) \tag{14}$$

where for $i = 0, 1, 2, \ldots, n-1$,

$$u_i(t) = \begin{cases} u(iT/n) & (iT/n < t \leq (i+1)T/n) \\ 0 & (\text{otherwise}) \end{cases} \tag{15}$$

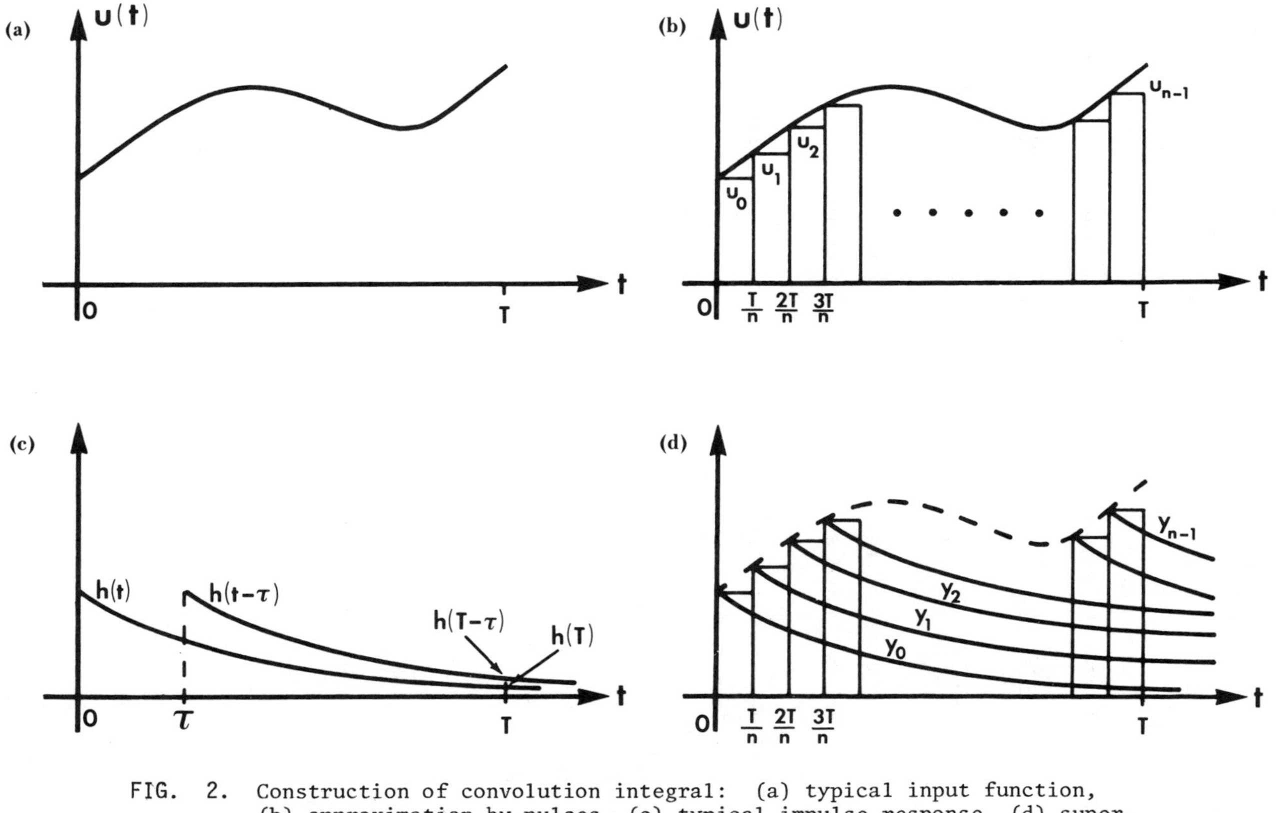

FIG. 2. Construction of convolution integral: (a) typical input function, (b) approximation by pulses, (c) typical impulse response, (d) superposition of impulse response.

For large enough n, each input pulse in (15) approximates an impulse,

$$u_i(t) \approx u(iT/n)(T/n)\delta(t - iT/n) \qquad (i = 0, 1, 2, \ldots, n-1) \qquad (16)^{\dagger}$$

The impulse response of (1) is $h(t) = \underline{c}' e^{\underline{A}t} \underline{b}$, as illustrated in Fig. 2(c), and the response to the i-th impulse $u_i(t)$ is

$$y_i(t) = h(t - iT/n)u(iT/n)(T/n) \qquad (t > iT/n) \qquad (17)$$

This is the impulse response delayed by iT/n and multiplied by the impulse magnitude, as shown in Fig. 2(d).

Now we make use of the superposition property (Corollary 3.3.4) of linear systems, which says that the response of the system at time T to the sum of input pulses in (14) is precisely the sum of the individual responses (17) which would result if each pulse were applied separately, that is,

$$y(T) = \sum_{i=0}^{n-1} y_i(T) = \sum_{i=0}^{n-1} h(T - iT/n)u(iT/n)(T/n) \qquad (18)$$

As $n \to \infty$, we let

$$\frac{iT}{n} = \tau$$

$$\frac{T}{n} = d\tau \qquad (19)$$

and see that (18) becomes an integral,

$$y(T) = \int_0^T h(T - \tau)\, u(\tau)\, d\tau \qquad (20)$$

Since this equation holds for any value of $T \geq 0$, we have proved that for $\underline{x}_0 = \underline{0}$, the input u(t) applied to (1) evokes a response

---

† The factor of T/n is necessary for the pulse approximating δ to have unit area.

$$y(t) = \int_0^t h(t - \tau)u(\tau)\, d\tau = \int_0^t \underline{c}'\, e^{\underline{A}(t-\tau)}\underline{b}u(\tau)\, d\tau \tag{21}$$

Thus we have reestablished (3.3-15) for a scalar system.[†]

The generalization to multivariable systems is straightforward and yields

$$\underline{y}(t) = \int_0^t \underline{H}(t - \tau)\, \underline{u}(\tau)\, d\tau = \int_0^t \underline{C}e^{\underline{A}(t-\tau)}\, \underline{B}\underline{u}(\tau)\, d\tau \tag{22}$$

3. *DEFINITION.* Equations (21) and (22) are known as *convolution* (or *superposition*) *integrals* and are sometimes denoted as

$$y = h * u \qquad \text{and} \qquad \underline{y} = \underline{H} * \underline{u} \tag{23}$$

respectively.

ΔΔΔ

4. *EXAMPLE.* In Example 3.3.5, the impulse response of the azimuthal motion of the satellite was found to be

$$h(t) = \underline{c}'\, e^{\underline{A}t}\, \underline{b} = \omega^{-1} \sin \omega t \tag{24}$$

If $\phi(0) = \dot{\phi}(0) = 0$ and a unit impulse of thrust is applied at time $t = 0$, the response will be

$$y(t) = \omega^{-1} \sin \omega t \tag{25}$$

If any other input function $u(t)$ is applied, the output can be found by convolution,

$$y(t) = \int_0^t [\omega^{-1} \sin \omega(t - \tau)]\, u(\tau)\, d\tau \tag{26}$$

as illustrated by (3.3-27).

ΔΔΔ

[†]The extension to $t_0 \neq 0$ is a simple one.

The impulse response $\underline{H}(t)$ of a system characterizes its input-output behavior via (22), and this behavior is independent of the particular state vector chosen to represent the system dynamics. Thus it is not surprising that $\underline{H}(t)$ is invariant under a state transformation.

5. *THEOREM*. The two systems

$$\dot{\underline{x}} = \underline{A}\underline{x} + \underline{B}\underline{u}$$
$$\underline{y} = \underline{C}\underline{x} \tag{27}$$

and

$$\dot{\hat{\underline{x}}} = \underline{Q}\underline{A}\underline{Q}^{-1}\hat{\underline{x}} + \underline{Q}\underline{B}\underline{u}$$
$$\underline{y} = \underline{C}\underline{Q}^{-1}\hat{\underline{x}} \tag{28}$$

which are related by a state transformation

$$\hat{\underline{x}} = \underline{Q}\underline{x} \qquad \underline{x} = \underline{Q}^{-1}\hat{\underline{x}} \tag{28}$$

have the same impulse response matrix

$$\underline{H}(t) = \underline{C}\, e^{\underline{A}t}\, \underline{B} \tag{30}$$

Proof: First note that $\underline{H}(t)$ is the impulse response of (27) by definition. From (3.2-33), the impulse response of (28) is

$$\underline{C}\underline{Q}^{-1} e^{\underline{Q}\underline{A}\underline{Q}^{-1}t}\, \underline{Q}\underline{B} = \underline{C}\underline{Q}^{-1}\, \underline{Q}e^{\underline{A}t}\, \underline{Q}^{-1}\, \underline{Q}\underline{B}$$
$$= \underline{C}\, e^{\underline{A}t}\, \underline{B} = \underline{H}(t) \tag{31}$$

ΔΔΔ

3.5. SUMMARY

In this chapter we have investigated the problem of solving ordinary, linear, time-invariant differential equations expressed in the state-variable format. The case of an unforced equation was considered in Sec. 3.1, and we saw in Theorem 3.1.4 that the unforced solution (or natural response) is a matrix exponential $e^{\underline{A}t}$ multiplying the initial state. Various properties of the matrix exponential were examined in Sec. 3.2, including several canonical forms and representations for $e^{\underline{A}t}$ in terms of the eigenvalues (or spectrum) of $\underline{A}$.

In Sec. 3.3 a forcing function was introduced, and the general solution was seen in Corollary 3.3.3 to consist of the natural response plus an integral term known as the forced response. This led us to the superposition property of linear systems (Corollary 3.3.4), whereby the output function is always linearly dependent upon the input function. We introduced the concept of an impulse response in Sec. 3.4, and this was used to interpret the forced response of the differential equation as a superposition (or convolution) integral.

In the next chapter we shall investigate an alternative method for solving the differential equations using Laplace transforms.

PROBLEMS

Section 3.1

1. The *adjoint equation* associated with (3.1-1) is

$$\dot{\underline{p}}(t) = -\underline{A}'\underline{p}(t)$$

Show that all solutions $\underline{p}(t)$ of this equation have the property that

$$\underline{p}'(t)\,\underline{x}(t) = \text{constant}$$

where $\underline{x}(t)$ is any solution of (3.1-1).

2. Prove that for any $\underline{A}$, the exponential power series (3.1-17) defining $e^{\underline{A}t}$ converges absolutely and uniformly on any time interval, by using the following result:

Lemma (Weierstrass M-test) [8,54,153]. Suppose

$$f_k(t) \qquad (k = 0, 1, 2, \ldots)$$

is a sequence of functions on the interval $t_0 \leq t \leq t_1$ which are bounded by

$$|f_k(t)| \leq M_k \qquad t_0 \leq t \leq t_1 \quad ; \quad k = 0,1,2,\ldots$$

If the infinite sum $\sum_{k=0}^{\infty} M_k$ converges, then $\sum_{k=0}^{\infty} f_k(t)$ converges absolutely and uniformly on the interval.

3. Determine $e^{\underline{A}t}$ for each of the following matrices:

$$\begin{bmatrix} 0 & 0 \\ 0 & 0 \end{bmatrix} \quad \begin{bmatrix} 0 & 1 \\ 0 & 0 \end{bmatrix} \quad \begin{bmatrix} 1 & 0 \\ 0 & 2 \end{bmatrix} \quad \begin{bmatrix} 2 & 1 \\ 0 & 2 \end{bmatrix}$$

4. Determine the solution of the differential equation

$$\dot{\underline{x}}(t) = \begin{bmatrix} -1 & 2 \\ -2 & -1 \end{bmatrix} \underline{x}(t)$$

$$\underline{x}(0) = \begin{bmatrix} 0 \\ 1 \end{bmatrix}$$

Section 3.2

1. Complete the proof of Theorem 3.2.4 by examining the cubic and higher order terms.

2. Prove Corollary 3.2.5 by making use of (3.2-22).

3. Prove Eqs.(22)-(24) and (26) in Lemma 3.2.6 directly from the series definition of $e^{\underline{A}t}$.

4. Prove Eqs.(25) and (27) in Lemma 3.2.6 by using Corollary 3.2.5.

5. Prove that

$$e^{(\underset{\sim}{Q}\underset{\sim}{A}\underset{\sim}{Q}^{-1})t} = \underset{\sim}{Q}\, e^{\underset{\sim}{A}t}\, \underset{\sim}{Q}^{-1}$$

in two different ways:

(a) Directly from Eq.(3.2-28) through (3.2-32).

(b) Using only the power series (3.1-17).

6. (a) Generalize Corollary 3.2.5 to matrices which are partitioned into blocks.

(b) Similarly generalize (3.2.24) and (3.2.25).

(c) Derive the matrix exponential of the block-Jordan form given in Theorem A.3.7.

(d) Derive the formula analogous to (3.2-64) for this case.

7. (a) Supply the proof of Theorem 3.2.11.

(b) Supply the second half of the proof of Theorem 3.2.10.

8. Indicate how the Cayley-Hamilton theorem (A.3.9) might be used to prove Theorem 3.2.13. What would it take to make the proof rigorous?

9. Express $e^{\underset{\sim}{A}t}$ in Example 3.1.5 in the form (3.2-64).

10. (a) Show that each matrix $\underset{\sim}{R}_i$ in the first part of Theorem 3.2.10 has rank 1 (and hence is nonzero). Show that $\underset{\sim}{R}_i$ may also be expressed as a *dyad* $\underset{\sim}{R}_i = \underset{\sim}{u}_i \underset{\sim}{v}_i'$, where $\underset{\sim}{u}_i$ is the i-th column of $\underset{\sim}{Q}^{-1}$ (i.e., the i-th eigenvector of $\underset{\sim}{A}$) and $\underset{\sim}{v}_i'$ is the i-th row of $\underset{\sim}{Q}$.

(b) In the case of repeated eigenvalues, show that $\underset{\sim}{R}_i^1 \neq 0$, $i = 1,2,\ldots,p$.

Section 3.3

1. Extend Corollary 3.3.3 to apply to the system (2.1-17), which has an output

$$\underline{y}(t) = \underline{\underline{C}}\underline{x}(t) + \underline{\underline{D}}\underline{u}(t)$$

2. Suppose the input in Example 3.3.5 is altered to

$$u_\phi(t) = \begin{cases} 0 & (t \leq 0) \\ 1 & (0 < t \leq T) \\ 0 & (t > T) \end{cases}$$

Determine the output $y(t) = \phi(t)$, for the two cases $T = \pi/\omega$ and $T = 2\pi/\omega$.

3. Determine the solution of the differential equation

$$\dot{\underline{x}}(t) = \begin{bmatrix} 0 & 1 \\ 0 & 0 \end{bmatrix} \underline{x}(t) + \begin{bmatrix} 0 \\ 1 \end{bmatrix} u(t)$$

$$y(t) = \begin{bmatrix} 1 & 0 \end{bmatrix} \underline{x}(t)$$

where the initial state is $\underline{x}(t_0) = \begin{bmatrix} 1 & 1 \end{bmatrix}'$ and the input $u(t)$ is a unit step at $t = 0$.

Chapter 4

LAPLACE TRANSFORM SOLUTION OF DIFFERENTIAL EQUATIONS†

4.0. INTRODUCTION

In Chapter 3 we determined the general form for solutions of differential equations in state-variable form. This was done directly (in the time domain) by means of the matrix exponential and convolution integral.

Appendix B provides a brief review of an indirect method for deriving the same results. This involves mapping the differential equations in the time domain via a *Laplace transform* into algebraic equations in the complex frequency domain, whose solution may be found very simply. The result is then mapped back into the time domain via an *inverse transform* to provide the solution to the original differential equation.

In this chapter the Laplace transform method will be used to rederive the principal results of Chapter 3 and to establish certain additional properties relating to the structure of linear systems.

4.1. SOLUTION OF EQUATIONS IN INPUT-OUTPUT FORM

The input-output form of differential equations for scalar systems was introduced in Sec. 2.2 as

---

†An understanding of Appendix B is assumed from this point on.

$$y^{(n)}(t) + p_{n-1}y^{(n-1)}(t) + \dots + p_1\dot{y}(t) + p_0y(t) = q_{n-1}u^{(n-1)}(t) + \dots + q_1\dot{u}(t) + q_0u(t) \tag{1}$$

where u and y are the system input and output, respectively. We take the initial time to be $t = 0$, and assume that the initial conditions $y(0)$, $\dot{y}(0)$, ..., $y^{(n-1)}(0)$ and the input $u(t)$ are known. Furthermore, we shall adopt the convention that[†]

$$u(0) = \dot{u}(0) = \dots = u^{(n-1)}(0) = 0 \tag{2}$$

in order to avoid confusion.

Using the linearity and differentiation properties of the Laplace transform (Theorems B.1.3 and B.1.10), the differential equation (1) may be transformed to

$$\begin{aligned} &[s^n\hat{y}(s) - s^{n-1}y(0) - s^{n-2}\dot{y}(0) - \dots - y^{(n-1)}(0)] \\ &\quad + p_{n-1}[s^{n-1}\hat{y}(s) - s^{n-2}y(0) - s^{n-3}\dot{y}(0) - \dots - y^{(n-2)}(0)] \\ &\quad + \dots + p_1[s\hat{y}(s) - y(0)] + p_0\hat{y}(s) \\ &= q_{n-1}s^{n-1}\hat{u}(s) + q_{n-2}s^{n-2}\hat{u}(s) + \dots + q_1s\hat{u}(s) + q_0\hat{u}(s) \end{aligned} \tag{3}$$

Rearranging terms, we find that

$$\begin{aligned} &[s^n + p_{n-1}s^{n-1} + \dots + p_1s + p_0]\hat{y}(s) \\ &= [s^{n-1}y(0) + s^{n-2}\{p_{n-1}y(0) + \dot{y}(0)\} + \dots \\ &\quad + \{p_1y(0) + p_2\dot{y}(0) + \dots + y^{(n-1)}(0)\}] \\ &\quad + [q_{n-1}s^{n-1} + q_{n-2}s^{n-2} + \dots + q_1s + q_0]\hat{u}(s) \end{aligned} \tag{4}$$

[†]In other words, the initial conditions are specified at $t = 0$ and the input begins slightly later at $t = 0^+$. If the initial time is $t_0 \neq 0$, we simply replace t by $t - t_0$.

or

$$p(s)\,\hat{y}(s) = r(s) + q(s)\,\hat{u}(s) \tag{5}$$

where the polynomials in brackets have been defined as p, r, and q, respectively. Finally, Eq. (5) may be solved for the transformed output,

$$\hat{y}(s) = \frac{r(s)}{p(s)} + \frac{q(s)}{p(s)}\,\hat{u}(s) \tag{6}$$

Consider the two terms which make up $\hat{y}(s)$. Since the coefficients of r(s) involve only $y(0)$, $\dot{y}(0)$, ..., $y^{(n-1)}(0)$, we see that the first term on the right-hand side of (6) is the *natural response*,

$$\frac{r(s)}{p(s)} \tag{7}$$

which vanishes when the initial conditions are all zero.

The second term is proportional to the input $\hat{u}(s)$ and is known as the *forced response*, and when the initial conditions are all zero, Eq. (6) reduces to

$$\hat{y}(s) = \frac{q(s)}{p(s)}\,\hat{u}(s) \tag{8}$$

Thus the output transform (in the absence of initial conditions) is the product of the input transform and the function q(s)/p(s). Since this function completely characterizes the input-output behavior of (1), the following terminology is appropriate.

1. DEFINITIONS. The rational function

$$\hat{h}(s) = \frac{q(s)}{p(s)} = \frac{q_{n-1}s^{n-1} + \dots + q_1 s + q_0}{s^n + p_{n-1}s^{n-1} + \dots + p_1 s + p_0} \tag{9}$$

is called the *transfer function* of the system (1), and it has the property that the input and output of (1) are related by

$$\hat{y}(s) = \hat{h}(s)\,\hat{u}(s) = \frac{q(s)}{p(s)}\,\hat{u}(s) \tag{10}$$

whenever the initial conditions are all zero. The transfer function may be written in factored form as

$$\hat{h}(s) = \frac{(s-\mu_1)(s-\mu_2)\ \cdots\ (s-\mu_m)}{(s-\lambda_1)(s-\lambda_2)\ \cdots\ (s-\lambda_n)} \tag{11}$$

The roots $\lambda_1, \lambda_2, \ldots \lambda_n$ of the denominator polynomial p(s) are known as the *poles of the system* (1), and the roots $\mu_1, \mu_2, \ldots, \mu_m$ of the numerator polynomial q(s) are known as its *zeros*. These definitions assume that any common factors in the numerator and denominator have not yet been cancelled out. After cancellation, the remaining factors are known as *poles and zeros of the transfer function*.

ΔΔΔ

Recall from Sec. B.2 that $\hat{y}(s)$ in (6) can be expressed as a partial fractions expansion in terms of the roots of its denominator polynomial (assuming that $\hat{u}$ is a rational function of s). These are clearly the system poles [roots of p(s)] plus any additional roots in the denominator of $\hat{u}$. Using the table of Laplace transforms in Sec. B.4 to perform the inverse Laplace transformation, y(t) may be expressed as a sum of terms of the form

$$e^{\lambda t} \qquad e^{\sigma t}\sin\omega t \qquad e^{\sigma t}\cos\omega t \tag{12}$$

possibly multiplied by powers of t, where λ and $\sigma \pm j\omega$ are either poles of the system or roots of the denominator of $\hat{u}$.

Thus we see that the shape of the output time function y(t) is determined principally by two factors: the poles of the system and the shape of the input u(t). This observation leads naturally into the concept of system stability, which is the subject of Chapter 5. The system (1) will be called *stable* if for a zero input, y(t) and all of its derivatives eventually decay to zero, regardless of the initial conditions. It should be clear that this condition will hold when the real parts (λ, σ, etc.) of all the poles are negative, so that the corresponding exponentials approach zero. The role of the zeros is illustrated in Problem 8.

2. *EXAMPLE*. Consider a scalar system governed by the differential equation

$$\dddot{y}(t) + 5\ddot{y}(t) + 17\dot{y}(t) + 13y(t) = 2\dot{u}(t) - 4u(t) \tag{13}$$

with initial conditions

$$\ddot{y}(0) = -7 \qquad \dot{y}(0) = 2 \qquad y(0) = 0 \tag{14}$$

and a unit step function applied to the input,

$$u(t) = 1(t) = \begin{cases} 0 & (t \leq 0) \\ 1 & (t > 0) \end{cases} \tag{15}$$

This differential equation has the form (1), so to determine its Laplace transform we may substitute the above values into (4),

$$(s^3 + 5s^2 + 17s + 13)\,\hat{y}(s) = (2s + 3) + (2s - 4)\,\hat{u}(s) \tag{16}$$

or

$$\hat{y}(s) = \frac{2s + 3}{s^3 + 5s^2 + 17s + 13} + \frac{2s - 4}{s^3 + 5s^2 + 17s + 13}\,\hat{u}(s) \tag{17}$$

The transfer function in this case is

$$\hat{h}(s) = \frac{q(s)}{p(s)} = \frac{2s - 4}{s^3 + 5s^2 + 17s + 13} \tag{18}$$

The roots of the denominator polynomial may be shown to be -1 and $-2 \pm j3$, so (18) can be written in factored form as

$$\hat{h}(s) = 2\,\frac{s - 2}{[(s + 2)^2 + 3^2]\,(s + 1)} \tag{19}$$

We see that $\hat{h}(s)$ has a real pole at $s = -1$, two complex poles at $s = -2 \pm j3$, and a real zero at $s = 2$.

Now we shall perform an inverse Laplace transformation on $\hat{y}(s)$ in order to determine the output time function. Recall from the table of transforms in Sec. B.4 that the transform of the step input (15) is

$$\hat{u}(s) = \frac{1}{s} \tag{20}$$

which may be substituted into (17) to yield

$$\hat{y}(s) = \frac{2s+3}{s^3+5s^2+17s+13} + \frac{2s-4}{s^3+5s^2+17s+13}\,\frac{1}{s}$$

$$= \frac{(2s^2+3s)+(2s-4)}{(s^3+5s^2+17s+13)s} \tag{21}$$

$$= \frac{2s^2+5s-4}{[(s+2)^2+3^2]\,(s+1)\,s}$$

Expanding this in partial fractions (Theorem B.2.2 and Corollary B.2.3),

$$\hat{y}(s) = \frac{C_1(s+2) + S_1 3}{(s+2)^2+3^2} + \frac{R_3}{s+1} + \frac{R_4}{s} \tag{22}$$

where

$$C_1 = \frac{1}{3}\,\mathrm{Im}\left(\frac{2s^2+5s-4}{(s+1)s}\right)_{s=-2+j3} = \frac{-51}{130} = -0.39$$

$$S_1 = \frac{1}{3}\,\mathrm{Re}\left(\frac{2s^2+5s-4}{(s+1)s}\right)_{s=-2+j3} = \frac{83}{130} = 0.64$$

$$R_3 = \left(\frac{2s^2+5s-4}{[(s+2)^2+3^2]\,s}\right)_{s=-1} = \frac{-7}{-10} = 0.70$$

$$R_4 = \left(\frac{2s^2+5s-4}{(s+1)[(s+2)^2+3^2]}\right)_{s=0} = \frac{-4}{13} = -0.31$$

Using the table of Laplace transforms in Sec. B.4, this becomes

$$y(t) = -0.39e^{-2t}\cos 3t + 0.64e^{-2t}\sin 3t + 0.70e^{-t} - 0.31 \tag{23}$$

which is plotted in Fig.1

ΔΔΔ

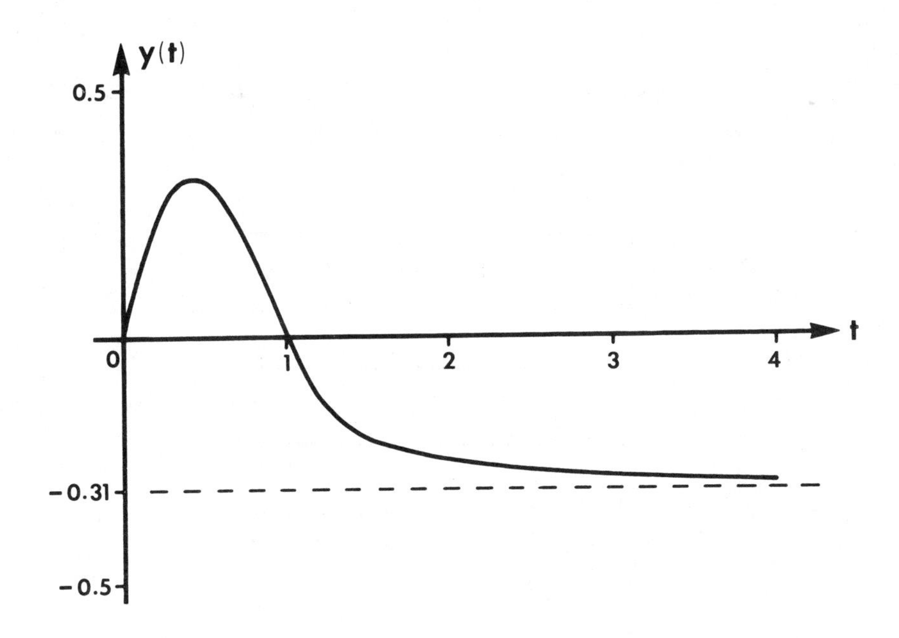

FIG. 1. Output y(t) for Example 2.

The use of block diagrams was introduced in Sec. 2.1 for a system expressed in state-variable format. When working with Laplace transforms, and particularly with input-output relationships, it is often convenient to suppress the natural response and simply write transfer functions in the blocks. Figure 2(a) shows two scalar systems with transfer functions $\hat{h}_1$ and $\hat{h}_2$, respectively, and input-output relationships given by

$$\hat{y}_1(s) = \hat{h}_1(s)\,\hat{u}_1(s) \tag{24}$$

$$\hat{y}_2(s) = \hat{h}_2(s)\,\hat{u}_2(s) \tag{25}$$

for zero initial conditions.

In Fig. 2(b) the two systems are connected in *series* (or *cascade*). Setting $\hat{u}_2 = \hat{y}_1$, substituting (24) into (25), and dropping the subscripts from $\hat{u}_1$ and $\hat{y}_2$, we see that

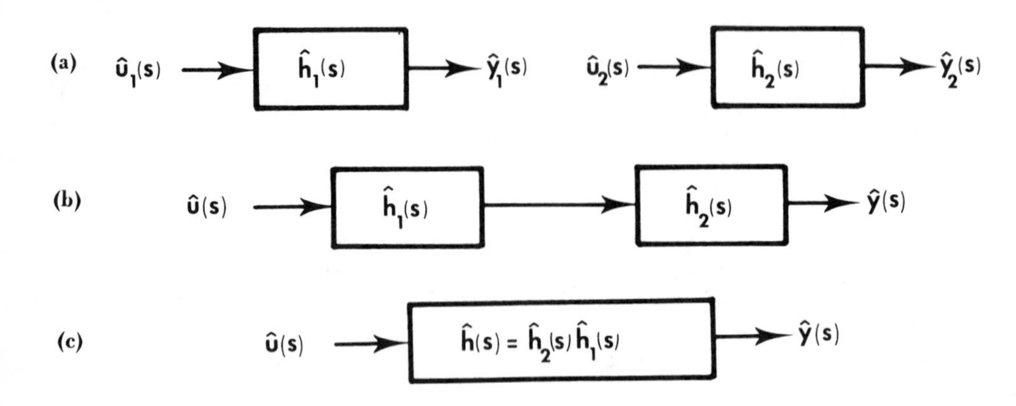

FIG. 2. Block diagrams: (a) independent systems, (b) series connection, and (c) equivalent system.

$$\hat{y}(s) = \hat{h}_2(s)\,\hat{y}_1(s) = \hat{h}_2(s)\,\hat{h}_1(s)\,\hat{u}(s) \tag{26}$$

This establishes the following basic property of transfer functions.

3. *LEMMA*. A series connection of two systems with transfer functions $\hat{h}_1(s)$ and $\hat{h}_2(s)$ is equivalent (in terms of forced response) to a single system whose transfer function is the product

$$\hat{h}(s) = \hat{h}_2(s)\,\hat{h}_1(s) \tag{27}$$

as indicated in Fig. 2(c).

ΔΔΔ

This deceptively simple property is very useful for manipulating transfer functions, particularly in dealing with several interconnected subsystems. The next two examples illustrate its utility. A similar result holds for systems connected in *parallel* (see Problem 5) and for multivariable systems (see Lemma 4.2.6).

A word of caution is in order here. Application of Lemma 3 involves the implicit but critical assumption that connecting the two systems does not alter the equations governing their behavior. To use a circuit analogy, this means that the second system must have a very high input impedance relative to the output impedance of the first system. If this assumption is not valid, then the interaction must be taken into account (see Problem 7). This will be illustrated in Example 5, where the motion of a motor's armature induces a "back emf" in the windings. This appears as a feedback loop in the block diagram (Fig. 6).

4. *EXAMPLE.* Recall the field-controlled d.c. motor which we encountered in Example 1.1.1, shown schematically in Fig. 3. It may be considered as a series connection of three subsystems:

(a) The field circuit, whose input is the applied voltage v, and whose output is the field current i,

(b) The magnetic fields of the motor with input i and output m, the torque developed by the motor,

(c) The armature inertia, whose inputs are m and the external load torque m_L, and whose output is θ, the angular position of the motor shaft.

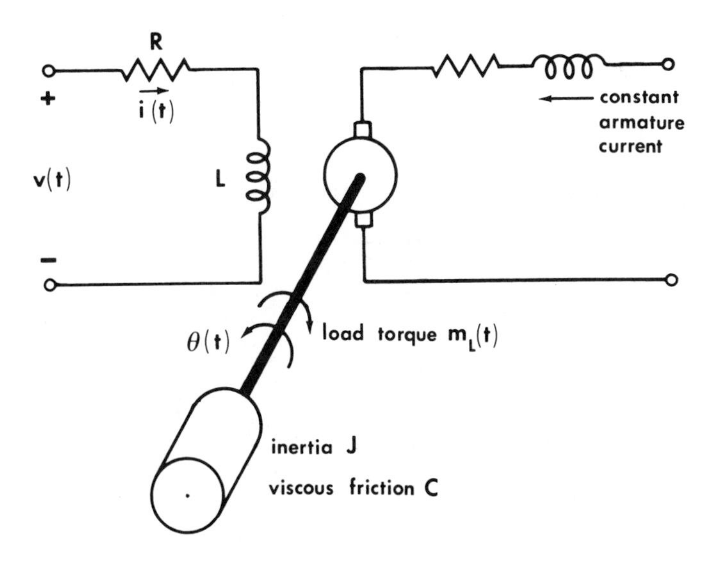

FIG. 3. Schematic diagram of field-controlled d.c. motor.

Using the convention, introduced in Chapter 2, that these variables represent deviations from normal equilibrium values, the equations describing each of the subsystems are

$$\dot{i}(t) = -\frac{R}{L}i(t) + \frac{1}{L}v(t) \tag{28}$$

$$m(t) = k\,i(t) \tag{29}$$

$$\ddot{\theta}(t) = -\frac{C}{J}\dot{\theta}(t) + \frac{1}{J}m(t) - \frac{1}{J}m_L(t) \tag{30}$$

Assuming zero initial conditions and taking Laplace transforms of these equations, we arrive at the block diagram shown in Fig.4(a) as a model for the motor. If variations in load torque can be neglected ($m_L = 0$), application of Lemma 3 reduces the diagram to a single block containing the product of the three transfer functions, as shown in Fig. 4(b).

ΔΔΔ

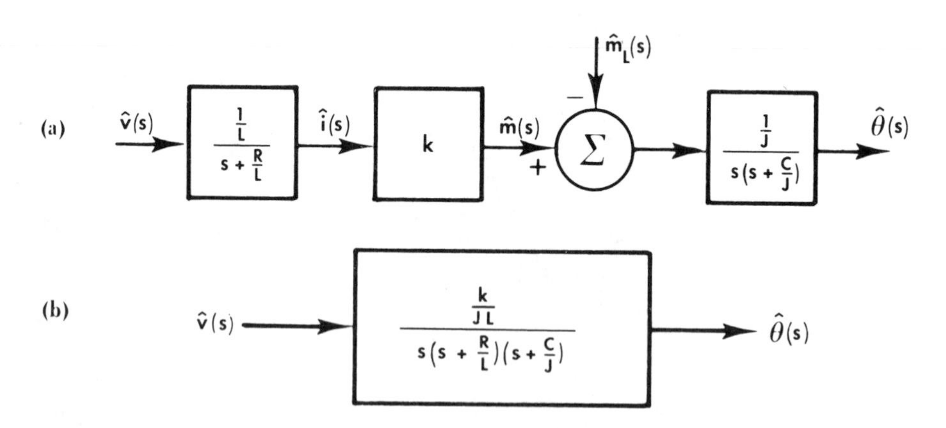

FIG. 4. Equivalent block diagrams of field-controlled d.c. motor.

5. *EXAMPLE*. As another illustration of the use of block diagrams, consider the armature-controlled motor of Example 2.3.1, shown in Fig. 5. The motor inputs are the voltage v applied to the armature (field voltage being constant) and the load torque m_L. Summing voltages around the armature circuit yields an equation for the armature current i,

$$L\dot{i}(t) + Ri(t) = v(t) - v_b(t) \tag{31}$$

where v_b (the back emf) opposes v and is proportional to the angular velocity ω,

$$v_b(t) = k\omega(t) \tag{32}$$

The torque m developed by the motor is proportional to i,

$$m(t) = k\,i(t) \tag{33}$$

and determines the motor speed via

$$J\dot{\omega}(t) + C\omega(t) = m(t) - m_L(t) \tag{34}$$

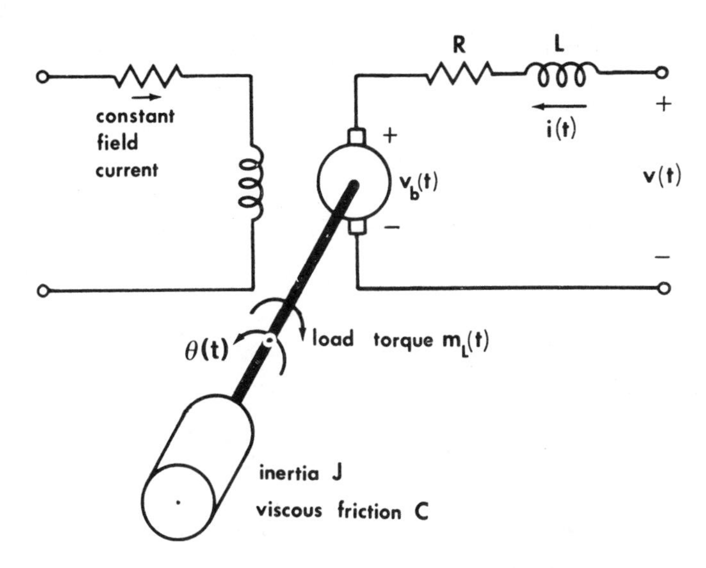

FIG. 5. Schematic representation of armature-controlled d.c. motor.

where J is the moment of inertia and C the coefficient of friction. The output of the system in this case will be angular position θ, which is just the integral of ω,

$$\dot{\theta}(t) = \omega(t) \tag{35}$$

Using the parameter values given in (2.3-7) and taking Laplace transforms, we obtain the equations

$$\hat{i}(s) = \frac{2}{s+2}\,[\hat{v}(s) - \hat{v}_b(s)] \tag{36}$$

$$\hat{v}_b(s) = 3\hat{\omega}(s) \tag{37}$$

$$\hat{m}(s) = 3\hat{i}(s) \tag{38}$$

$$\hat{\omega}(s) = \frac{2}{s+4}\,[\hat{m}(s) - \hat{m}_L(s)] \tag{39}$$

$$\hat{\theta}(s) = \frac{1}{s}\,\hat{\omega}(s) \tag{40}$$

which are shown as a block diagram in Fig. 6(a). If we now assume $m_L = 0$ (i.e., the load torque remains constant at its equilibrium value), Lemma 3 may be used to reduce this diagram to the one shown in Fig. 6(b). Note that this differs from the previous example in that it contains an internal "feedback loop" from $\hat{\omega}$ to the input. Thus, the equation relating $\hat{\omega}$ to $\hat{v}$ is

$$\hat{\omega}(s) = \frac{12}{(s+2)(s+4)}\,[\hat{v}(s) - 3\hat{\omega}(s)] \tag{41}$$

or

$$(s^2 + 6s + 8)\,\hat{\omega}(s) = 12\hat{v}(s) - 36\hat{\omega}(s) \tag{42}$$

which may be solved for $\hat{\omega}(s)$:

$$\hat{\omega}(s) = \frac{12}{s^2 + 6s + 44}\,\hat{v}(s) \tag{43}$$

Substituting this into (40) (or applying Lemma 3 once more) yields the block diagram of Fig. 6(c).

ΔΔΔ

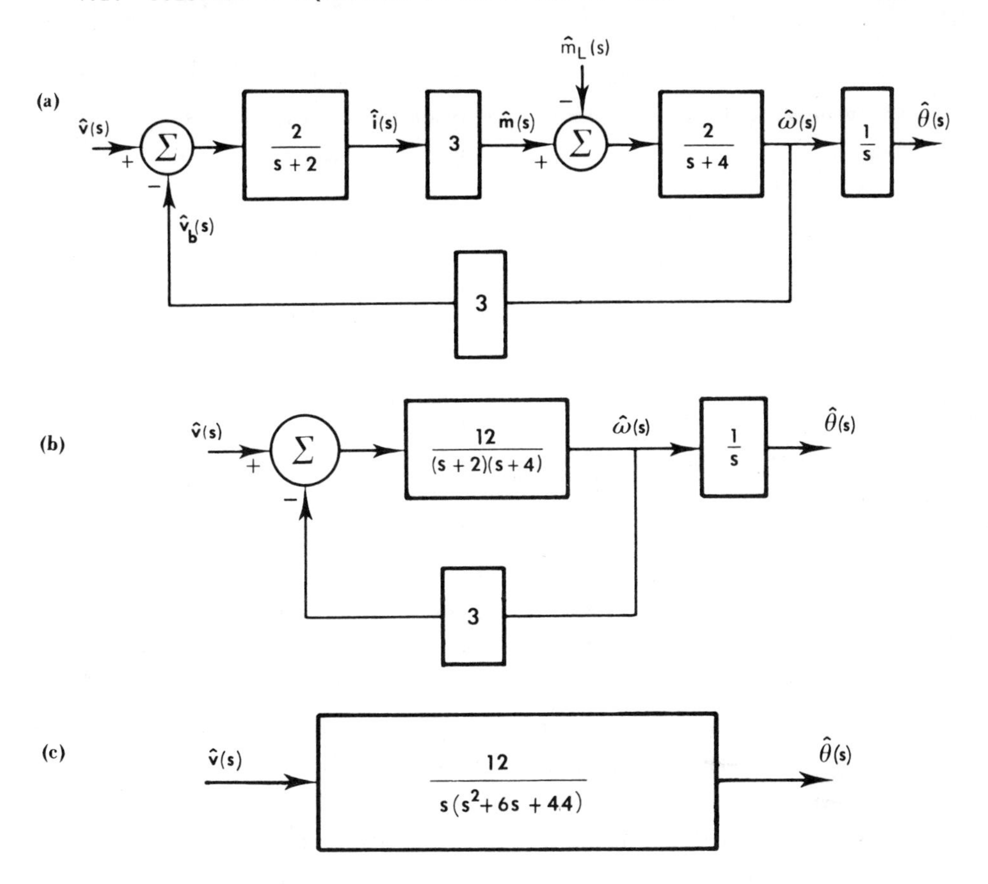

FIG. 6. Equivalent block diagram of armature-controlled d.c. motor.

Impulse Response

We have seen that in the absence of initial conditions, the transfer function $\hat{h}(s)$ completely characterizes the input-output behavior of the system (in the transform domain) via the relation (10),

$$\hat{y}(s) = \hat{h}(s)\,\hat{u}(s) \tag{44}$$

Now suppose the input u(t) is a *unit impulse* at t = 0, as described in Sec. 3.4. In practice, this will really be a finite pulse of unit area and very short time duration, but for mathematical convenience we call it an impulse (or delta function). It is shown in Example B.1.9 that the Laplace transform of such an impulse is constant, that is,

$$\hat{u}(s) = 1 \qquad \text{(for all } s\text{)} \tag{45}$$

Substituting (45) into (44), we find that the output of the system is

$$\hat{y}(s) = \hat{h}(s) \tag{46}$$

for an impulsive input, or

$$y(t) = h(t) = \mathcal{L}^{-1}(\hat{h}(s)) \tag{47}$$

Thus $h(t)$, the inverse transform of the transfer function $\hat{h}(s)$, is appropriately called the *impulse response* of the system (1). In the next section we shall verify that this is the impulse response introduced in Definition 3.4.2.

In order to further illuminate the relationship (44), we may make use of Theorem B.1.14, which states that a product of two transforms corresponds in the time domain to a convolution of the respective time functions. Thus (44) is equivalent to

$$y(t) = \int_0^t h(t - \tau)\, u(\tau)\, d\tau \tag{48}$$

This formula for the output, also derived in Sec. 3.4, is known as a *convolution* (or *superposition*) *integral*.

6. *EXAMPLE*. If the initial conditions in Example 2 are all zero,

$$\ddot{y}(0) = \dot{y}(0) = y(0) = 0 \tag{49}$$

then the output transform $\hat{y}$ may be found directly from the transfer function $\hat{h}$ and the input transform $\hat{u}$ via (44). In the case of a step function input, as in (15), this is

$$\hat{y}(s) = \hat{h}(s)\,\hat{u}(s) = \frac{2s - 4}{s^3 + 5s^2 + 17s + 13}\,\frac{1}{s}$$
$$= \frac{2s - 4}{[(s+2)^2 + 3^2]\,(s+1)\,s} \tag{50}$$

which may be expanded in partial fractions and transformed to yield y(t) (Problem 1).

When the input is a unit impulse, the output is the impulse response,

$$y(t) = h(t) = \mathcal{L}^{-1}(\hat{h}(s))$$
$$= \mathcal{L}^{-1}\left\{\frac{2s - 4}{[(s+2)^2 + 3^2]\,(s+1)}\right\} \tag{51}$$

To determine y(t) explicitly, we expand $\hat{h}(s)$ in partial fractions,

$$\hat{h}(s) = \frac{C_1(s+2) + S_1(3)}{(s+2)^2 + 3^2} + \frac{R_3}{s+1} \tag{52}$$

where

$$C_1 = \frac{1}{3}\,\mathrm{Im}\left(\frac{2s-4}{s+1}\right)\Bigg|_{s=-2+j3} = \frac{1}{3}\,\frac{18}{10} = 0.60$$

$$S_1 = \frac{1}{3}\,\mathrm{Re}\left(\frac{2s-4}{s+1}\right)\Bigg|_{s=-2+j3} = \frac{1}{3}\,\frac{26}{10} \simeq 0.87$$

$$R_3 = \left(\frac{2s-4}{(s+2)^2 + 3^2}\right)\Bigg|_{s=-1} = \frac{-6}{10} = -0.60$$

Using the Laplace transform table in Sec. B.4, we have

$$y(t) = h(t) = 0.60e^{-2t}\cos 3t + 0.87e^{-2t}\sin 3t - 0.60e^{-t} \tag{53}$$

which is shown in Fig. 7.

ΔΔΔ

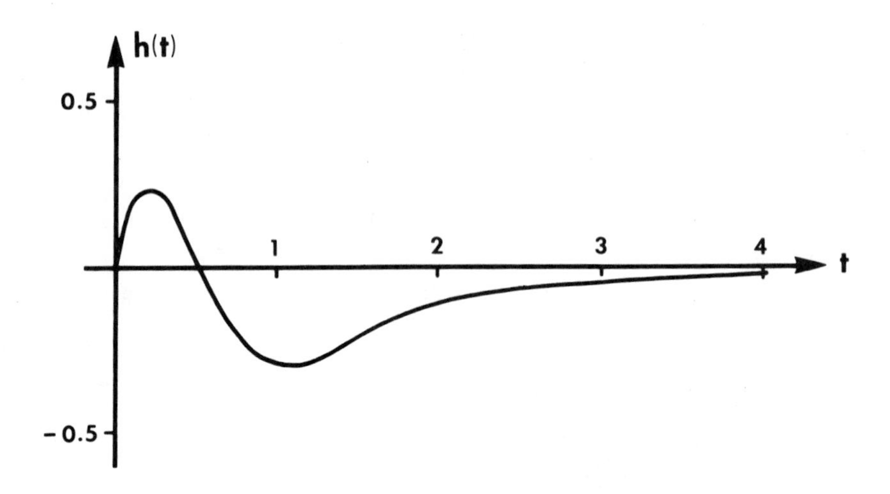

FIG. 7. Impulse response h(t) for Example 6.

4.2 SOLUTION OF EQUATIONS IN STATE-VARIABLE FORM

In this section we shall be concerned with differential equations in the state-variable format introduced in Sec. 2.1,

$$\begin{aligned} \dot{\underline{x}}(t) &= \underline{A}\underline{x}(t) + \underline{B}\underline{u}(t) \\ \underline{y}(t) &= \underline{C}\underline{x}(t) \end{aligned} \tag{1}$$

where $\underline{x}$, $\underline{u}$, and $\underline{y}$ are the state, input, and output vectors, respectively, and the initial state (at $t = 0$) is

$$\underline{x}(0) = \underline{x}_0 \tag{2}$$

The Laplace transform of a vector quantity is defined in Sec. B.3, and using Lemma B.3.2, (1) can be transformed to

$$\begin{aligned} s\hat{\underline{x}}(s) - \underline{x}(0) &= \underline{A}\hat{\underline{x}}(s) + \underline{B}\hat{\underline{u}}(s) \\ \hat{\underline{y}}(s) &= \underline{C}\hat{\underline{x}}(s) \end{aligned} \tag{3}$$

Rearranging the first equation and using (2), we have

$$(s\underline{I} - \underline{A})\hat{\underline{x}}(s) = \underline{x}_0 + \underline{B}\hat{\underline{u}}(s) \tag{4}$$

or

$$\hat{\underline{x}}(s) = (s\underline{I} - \underline{A})^{-1}\underline{x}_0 + (s\underline{I} - \underline{A})^{-1}\underline{B}\hat{\underline{u}}(s) \tag{5}$$ †

and the output transform is

$$\hat{\underline{y}}(s) = \underline{C}(s\underline{I} - \underline{A})^{-1}\underline{x}_0 + \underline{C}(s\underline{I} - \underline{A})^{-1}\underline{B}\hat{\underline{u}}(s) \tag{6}$$

The first term on the right-hand side, due to the initial state $\underline{x}_0$, is called the *natural* (or *zero-input*) *response*, while the second term is the *forced* (or *zero-state*) *response* and depends upon the input $\hat{\underline{u}}(s)$. This terminology is identical to that introduced in Corollary 3.3.3, and the following results verify that the two are consistent.

1. THEOREM. The matrix $(s\underline{I} - \underline{A})^{-1}$ is known as a *resolvent matrix*, and its inverse Laplace transform is the matrix exponential,

$$L^{-1}((s\underline{I} - \underline{A})^{-1}) = e^{\underline{A}t} \tag{7}$$

This formula is often the most efficient means of computing $e^{\underline{A}t}$ by hand (see Example 4).

Proof: First we expand the resolvent in a series†† about $|s| = \infty$,

$$\begin{aligned}(s\underline{I} - \underline{A})^{-1} &= \frac{(\underline{I} - \underline{A}/s)^{-1}}{s} \\ &= \frac{(\underline{I} + \underline{A}/s + \underline{A}^2/s^2 + \ldots)}{s} \\ &= \underline{I}/s + \underline{A}/s^2 + \underline{A}^2/s^3 + \ldots\end{aligned} \tag{8}$$

†Note that the inverse of $(s\underline{I} - \underline{A})$ exists for all values of s except the n eigenvalues of $\underline{A}$. This is analogous to the situation in Sec. 4.1, where the transfer function $\hat{h}(s) = q(s)/p(s)$ was well-defined for all s except the n roots of p(s). No difficulties arise from this since there is always a *region of convergence* to the right of the n singularities in which the transform in question is analytic (see Sec. B.3).

††It can be shown [109,151] that this infinite series converges for all s in the region of convergence of the Laplace transform $(s\underline{I} - \underline{A})^{-1}$. This corresponds, in the scalar case, to the well-known result that $1/(1 - z) = 1 + z + z^2 + \ldots$ for $|z| < 1$.

This equation may be transformed term by term, using the table in Sec. B.4, to yield

$$\mathcal{L}^{-1}((s\underline{I} - \underline{A})^{-1}) = \underline{I}\ \mathcal{L}^{-1}(1/s) + \underline{A}\ \mathcal{L}^{-1}(1/s^2) + \underline{A}^2\mathcal{L}^{-1}(1/s^3) + \cdots$$

$$= \underline{I} + \underline{A}t + \underline{A}^2t^2/2! + \cdots \tag{9}$$

$$= e^{\underline{A}t}$$

(For an alternative proof, see Problem 1.)

ΔΔΔ

2. *COROLLARY.* The output of the system (1) is given by

$$\underline{y}(t) = \underline{C}e^{\underline{A}t}\ \underline{x}_0 + \int_0^t \underline{C}e^{\underline{A}(t - \tau)}\ \underline{B}\underline{u}(\tau)\ d\tau \tag{10}$$

(This is identical to the result of Corollary 3.3.3 with $t_0 = 0$.)

Proof: The transform of the output is given by (6). By Theorem 1, we have

$$\mathcal{L}^{-1}\ (\underline{C}(s\underline{I} - \underline{A})^{-1}\ \underline{x}_0) = \underline{C}e^{\underline{A}t}\underline{x}_0 \tag{11}$$

and

$$\mathcal{L}^{-1}\ (\underline{C}(s\underline{I} - \underline{A})^{-1}\ \underline{B}) = \underline{C}e^{\underline{A}t}\underline{B} \tag{12}$$

If we extend Theorem B.1.14 to matrices, the product of $\underline{C}(s\underline{I} - \underline{A})^{-1}\underline{B}$ and $\hat{\underline{u}}(s)$ in the last term corresponds to a convolution integral of $\underline{C}e^{\underline{A}t}\underline{B}$ and $\underline{u}(t)$. Thus the inverse transform of (6) is (10).

ΔΔΔ

In the case of a zero initial state, the natural response vanishes in (6) and we see that the input-output behavior of the system is determined entirely by the matrix $\underline{C}(s\underline{I} - \underline{A})^{-1}\underline{B}$.

3. *DEFINITION.* The matrix

$$\hat{\underline{H}}(s) \triangleq \underline{C}(s\underline{I} - \underline{A})^{-1}\underline{B} \tag{13}$$

is called the *transfer function matrix* of the system (1), and it has the property that the input and output of (1) are related by

$$\hat{\underline{y}}(s) = \hat{\underline{H}}(s)\,\hat{\underline{u}}(s) \tag{14}$$

whenever $\underline{x}_0 = \underline{0}$. By Theorem 1, the inverse transform of $\hat{\underline{H}}(s)$ is

$$\underline{H}(t) = \underline{C}e^{\underline{A}t}\underline{B} \tag{15}$$

which was previously encountered in Sec. 3.3 and Sec. 3.4 and is known as the *impulse response matrix* of the system. The inverse transform of (14) expresses the output of the system as a *convolution integral* of the impulse response matrix and the input,

$$\underline{y}(t) = \int_0^t \underline{H}(t-\tau)\underline{u}(\tau)\,d\tau \tag{16}$$

ΔΔΔ

4. *EXAMPLE.* Consider a system

$$\dot{\underline{x}} = \underline{A}\underline{x} + \underline{B}\underline{u} = \begin{bmatrix} 3 & 5 \\ 0 & -2 \end{bmatrix}\underline{x} + \begin{bmatrix} 3 & -4 \\ 0 & 4 \end{bmatrix}\underline{u}$$

$$y = \underline{c}'\underline{x} = \begin{bmatrix} 1 & 2 \end{bmatrix}\underline{x} \tag{17}$$

having two inputs and one output. The resolvent matrix in this case is

$$(s\underline{I} - \underline{A})^{-1} = \begin{bmatrix} s-3 & -5 \\ 0 & s+2 \end{bmatrix}^{-1} = \frac{1}{(s-3)(s+2)}\begin{bmatrix} s+2 & 5 \\ 0 & s-3 \end{bmatrix}$$

$$= \begin{bmatrix} \dfrac{1}{(s-3)} & \dfrac{1}{(s-3)(s+2)} \\ 0 & \dfrac{1}{(s+2)} \end{bmatrix} \tag{18}$$

Thus we have

$$\underline{c}'(s\underline{I} - \underline{A})^{-1} = \begin{bmatrix} \frac{1}{s-3} & \frac{5}{(s-3)(s+2)} + \frac{2}{s+2} \end{bmatrix}$$
$$= \begin{bmatrix} \frac{1}{s-3} & \frac{1}{s+2} + \frac{1}{s-3} \end{bmatrix} \tag{19}$$

which transforms (using the table again) to

$$\underline{c}'e^{\underline{A}t} = \begin{bmatrix} e^{3t} & e^{-2t} + e^{3t} \end{bmatrix} \tag{20}$$

Similarly, the transfer function matrix is

$$\underline{\hat{H}}(s) = \underline{c}'(s\underline{I} - \underline{A})^{-1}\underline{B} = \begin{bmatrix} \frac{1}{s-3} & \frac{1}{s+2} + \frac{1}{s-3} \end{bmatrix} \begin{bmatrix} 3 & -4 \\ 0 & 4 \end{bmatrix}$$
$$= \begin{bmatrix} \frac{3}{s-3} & \frac{4}{s+2} \end{bmatrix} \tag{21}$$

which transforms (using the table again) to the impulse response matrix

$$\underline{H}(t) = \underline{c}'e^{\underline{A}t}\underline{B} = \begin{bmatrix} 3e^{3t} & 4e^{-2t} \end{bmatrix} \tag{22}$$

Now suppose the initial state and inputs are

$$\underline{x}_0 = \begin{bmatrix} 1 \\ 1 \end{bmatrix} \qquad u_1(t) = 0 \qquad u_2(t) = 1(t) \tag{23}$$

From (6), the output transform is

$$\underline{\hat{y}}(s) = \underline{c}'(s\underline{I} - \underline{A})^{-1}\underline{x}_0 + \underline{c}'(s\underline{I} - \underline{A})^{-1}\underline{B}\underline{u}(s)$$
$$= \begin{bmatrix} \frac{1}{s-3} & \frac{1}{s+2} + \frac{1}{s-3} \end{bmatrix} \begin{bmatrix} 1 \\ 1 \end{bmatrix} + \begin{bmatrix} \frac{3}{s-3} & \frac{4}{s+2} \end{bmatrix} \begin{bmatrix} 0 \\ 1/s \end{bmatrix}$$
$$= \frac{2}{s-3} + \frac{1}{s+2} + \frac{4}{(s+2)s} \tag{24}$$

When the last term is expanded in partial fractions, this becomes

$$\hat{y}(s) = \frac{2}{s-3} + \frac{1}{s+2} + \frac{-2}{s+2} + \frac{2}{s}$$
$$= \frac{2}{s-3} - \frac{1}{s+2} + \frac{2}{s} \tag{25}$$

and transforms to

$$y(t) = 2e^{3t} - e^{-2t} + 2 \tag{26}$$

Alternatively, the same result could have been derived using (10),(20), and (22),

$$y(t) = \underline{c}'e^{\underline{A}t}\underline{x}_0 + \int_0^t \underline{c}'e^{\underline{A}(t-\tau)}\underline{Bu}(\tau)\, d\tau$$

$$= \begin{bmatrix} e^{3t} & e^{-2t} + e^{3t} \end{bmatrix} \begin{bmatrix} 1 \\ 1 \end{bmatrix} + \int_0^t \begin{bmatrix} 3e^{3(t-\tau)} & 4e^{-2(t-\tau)} \end{bmatrix} \begin{bmatrix} 0 \\ 1 \end{bmatrix} d\tau$$

$$= 2e^{3t} + e^{-2t} + 4e^{-2t} \int_0^t e^{2\tau}\, d\tau \tag{27}$$

$$= 2e^{3t} + e^{-2t} + 4e^{-2t} \left(\frac{1}{2}e^{2t} - \frac{1}{2}e^{0}\right)$$

$$= 2e^{3t} - e^{-2t} + 2$$

ΔΔΔ

We now note, as we did at the end of Sec. 3.4, that input-output behavior is not dependent upon the particular state vector chosen to characterize a system. Theorem 3.4.5 verified that the impulse response matrix $\underline{H}(t)$ is invariant under a state transformation. Since the impulse response and transfer function are Laplace transforms of each other, an equivalent result must hold for the transfer function matrix $\hat{\underline{H}}(s)$.

5. *THEOREM.* The two systems

$$\dot{\underline{x}} = \underline{A}\underline{x} + \underline{B}\underline{u} \qquad \underline{y} = \underline{C}\underline{x} \tag{28}$$

and

$$\dot{\tilde{\underline{x}}} = \underline{Q}\underline{A}\underline{Q}^{-1}\tilde{\underline{x}} + \underline{Q}\underline{B}\underline{u} \qquad \underline{y} = \underline{C}\underline{Q}^{-1}\tilde{\underline{x}} \tag{29}$$

which are related by a state transformation

$$\tilde{\underline{x}} = \underline{Q}\underline{x} \qquad \underline{x} = \underline{Q}^{-1}\tilde{\underline{x}} \tag{30}$$

have the same transfer function matrix

$$\hat{\underline{H}}(s) = \underline{C}(s\underline{I} - \underline{A})^{-1}\underline{B} \tag{31}$$

Proof: As stated above, this is equivalent to Theorem 3.4.5. It is easily proved, however, by noting that

$$\underline{C}\underline{Q}^{-1}(s\underline{I} - \underline{Q}\underline{A}\underline{Q}^{-1})^{-1}\underline{Q}\underline{B} = \underline{C}\underline{Q}^{-1}\underline{Q}(s\underline{I} - \underline{A})^{-1}\underline{Q}^{-1}\underline{Q}\underline{B} = \underline{C}(s\underline{I} - \underline{A})^{-1}\underline{B} \tag{32}$$

ΔΔΔ

Finally, it is useful to observe that Lemma 4.1.3 extends to multivariable systems. Figure 1(a) shows two systems with transfer function matrices $\hat{\underline{H}}_1$ and $\hat{\underline{H}}_2$, their inputs and outputs related by

$$\hat{\underline{y}}_1(s) = \hat{\underline{H}}_1(s)\,\hat{\underline{u}}_1(s) \tag{33}$$

$$\hat{\underline{y}}_2(s) = \hat{\underline{H}}_2(s)\,\hat{\underline{u}}_2(s) \tag{34}$$

for $\underline{x}_0 = \underline{0}$. If the dimensions of $\hat{\underline{y}}_1$ and $\hat{\underline{u}}_2$ are the same, the two systems may be connected in series, as shown in Fig. 1(b). Setting $\hat{\underline{u}}_2 = \hat{\underline{y}}_1$, substituting (33) into (34), and dropping the subscripts from $\hat{\underline{u}}_1$ and $\hat{\underline{y}}_2$, we have

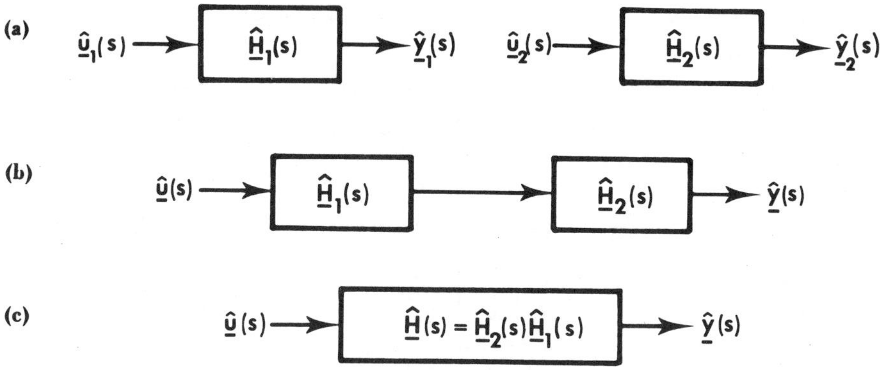

FIG. 1. Series connection of multivariable systems: (a) independent systems, (b) series connection, and (c) equivalent system.

$$\hat{\underline{y}}(s) = \hat{\underline{H}}_2(s)\,\hat{\underline{y}}_1(s) = \hat{\underline{H}}_2(s)\,\hat{\underline{H}}_1(s)\,\hat{\underline{u}}(s) \tag{35}$$

which establishes the following lemma.

6. *LEMMA.* A series connection of two systems with transfer function matrices $\hat{\underline{H}}_1(s)$ and $\hat{\underline{H}}_2(s)$ in the order shown in Fig. 1(b) is equivalent (in terms of forced response) to a single system whose transfer function matrix is the product

$$\hat{\underline{H}}(s) = \hat{\underline{H}}_2(s)\,\hat{\underline{H}}_1(s) \tag{36}$$

as shown in Fig. 1(c). Note that in this case the order of multiplication is critical.

ΔΔΔ

4.3 PROPERTIES OF THE RESOLVENT MATRIX

In Sec. 4.2 we saw that the principal ingredient in the transfer function matrix is the resolvent matrix $(s\underline{I} - \underline{A})^{-1}$, whose inverse Laplace transform is the matrix exponential $e^{\underline{A}t}$. The problem of evaluating the resolvent is a nontrivial one except in relatively simple second- and third-order cases. We shall now investigate an algorithm for determining $(s\underline{I} - \underline{A})^{-1}$, and derive several of its important properties.

A fundamental property is that the resolvent is a proper, rational function of the complex variable s.

1. LEMMA. For any $n \times n$ matrix $\underline{A}$, the elements of the resolvent matrix $(s\underline{I} - \underline{A})^{-1}$ may always be written as a strictly proper ratio of polynomials in s. In particular,

$$(s\underline{I} - \underline{A})^{-1} = \frac{\underline{Q}(s)}{p(s)} = \frac{\underline{Q}_{n-1}s^{n-1} + \dots + \underline{Q}_1 s + \underline{Q}_0}{s^n + p_{n-1}s^{n-1} + \dots + p_1 s + p_0} \tag{1}$$

where

$$p(s) = s^n + p_{n-1}s^{n-1} + \dots + p_1 s + p_0 \tag{2}$$

is the *characteristic polynomial* of $\underline{A}$, and

$$\underline{Q}(s) = \underline{Q}_{n-1}s^{n-1} + \dots + \underline{Q}_1 s + \underline{Q}_0 \tag{3}$$

is an $n \times n$ matrix polynomial, the $\underline{Q}_i$ being constant matrices.

Proof: By Theorem A.2.5, the inverse of a matrix can always be expressed as

$$(s\underline{I} - \underline{A})^{-1} = \frac{\underline{\Delta}'(s\underline{I} - \underline{A})}{\det(s\underline{I} - \underline{A})} \tag{4}$$

where $\underline{\Delta}(s\underline{I} - \underline{A})$ is the matrix of cofactors of $(s\underline{I} - \underline{A})$. It is established in Problem A.3-3 that $\det(s\underline{I} - \underline{A})$ is an n-th order polynomial p(s) called the characteristic polynomial of $\underline{A}$. The cofactors of $(s\underline{I} - \underline{A})$ are determinants of $(n - 1) \times (n - 1)$ submatrices of $(s\underline{I} - \underline{A})$, and so by the same argument they must be polynomials in s of order $n - 1$ or less. This establishes that the numerator $\underline{\Delta}'(s\underline{I} - \underline{A})$ must have the form indicated in (3).

ΔΔΔ

Lemma 1 establishes that the transfer function matrix has the form

$$\hat{H}(s) = C(sI - A)^{-1} B = \frac{CQ(s)\, B}{p(s)} \tag{5}$$

and the roots of the polynomial p(s) (i.e., the eigenvalues of the system matrix A) are known as the *poles* of the system. Occasionally, however, p(s) and $Q(s)$ may have common factors which cancel, so that the transfer function matrix

$$\hat{H}(s) = \frac{C\tilde{Q}(s)\, B}{\tilde{p}(s)} \tag{6}$$

has degree less than the dimension of A. In this case, the roots of $\tilde{p}(s)$ are said to be the poles of the transfer function matrix, and not all eigenvalues of A are poles of $\hat{H}(s)$.

Computation of the resolvent is straightforward and easily carried out on a digital computer using the following algorithm. However, the procedure is rather sensitive to numerical round-off errors, so double-precision arithmetic is usually required.

2. *THEOREM (Resolvent algorithm).*[†] The coefficients of the characteristic polynomial p(s) and the matrix polynomial $Q(s)$ in (1) may be determined sequentially as follows:

$$Q_{n-1} = I \tag{7}$$

$$p_{n-1} = -\mathrm{tr}(Q_{n-1}A) = -\mathrm{tr}(AQ_{n-1}) \tag{8}$$

$$Q_{n-2} = Q_{n-1}A + p_{n-1}I = AQ_{n-1} + p_{n-1}I \tag{9}$$

$$p_{n-2} = -\frac{1}{2}\mathrm{tr}(Q_{n-2}A) = -\frac{1}{2}\mathrm{tr}(AQ_{n-2}) \tag{10}$$

....

$$Q_1 = Q_2A + p_2I = AQ_2 + p_2I \tag{11}$$

$$p_1 = -\frac{1}{n-1}\mathrm{tr}(Q_1A) = -\frac{1}{n-1}\mathrm{tr}(AQ_1) \tag{12}$$

$$Q_0 = Q_1A + p_1I = AQ_1 + p_1I \tag{13}$$

$$p_0 = -\frac{1}{n}\mathrm{tr}(Q_0A) = -\frac{1}{n}\mathrm{tr}(AQ_0) \tag{14}$$

$$0 = Q_0A + p_0I = AQ_0 + p_0I \tag{15}$$

[†] In the literature, this is usually attributed to some combination of Leverrier, Souriau, and Faddeeva.

where the *trace* of a matrix is defined as the sum of its diagonal elements, that is,

$$\operatorname{tr} \underline{A} = \sum_{i=1}^{n} a_{ii} \tag{16}$$

Equation (15) may be used as a check on the accuracy of the preceding computations, and it will be used to prove some useful corollaries.

Proof: First multiply (1) by $(s\underline{I} - \underline{A})$ and p(s) to get

$$\underline{Q}(s)\ (s\underline{I} - \underline{A}) = \underline{I}p(s)$$

or

$$(\underline{Q}_{n-1}s^{n-1} + \ldots + \underline{Q}_1 s + \underline{Q}_0)(s\underline{I} - \underline{A}) = \underline{I}s^n + p_{n-1}\underline{I}s^{n-1} + \ldots + p_1\underline{I}s + p_0\underline{I} \tag{17}$$

Since coefficients of like powers of s must be equal, we have

$$\begin{aligned}
\underline{Q}_{n-1} &= \underline{I} \\
\underline{Q}_{n-2} &= \underline{Q}_{n-1}\underline{A} + p_{n-1}\underline{I} \\
&\cdots\cdots\cdots\cdots \\
\underline{Q}_0 &= \underline{Q}_1\underline{A} + p_1\underline{I} \\
\underline{0} &= \underline{Q}_0\underline{A} + p_0\underline{I}
\end{aligned} \tag{18}$$

which proves (7), (9), (11), (13), and (15).

Establishing (8), (10), (12), and (14) is somewhat more complicated. The interested reader will find a proof in Refs. 61, 62, 67, 149, and 185.

ΔΔΔ

In Theorem 3.2.13, it was stated that the matrix exponential could always be written as a finite sum of powers of $\underline{A}$. The proof of this fact follows easily from the resolvent algorithm.

3. *COROLLARY.* The resolvent and exponential matrices may be expressed as

$$(s\underline{I} - \underline{A})^{-1} = \sum_{i=0}^{n-1} \underline{A}^i \hat{\alpha}_i(s) \tag{19}$$

and

$$e^{\underline{A}t} = \sum_{i=0}^{n-1} \underline{A}^i \alpha_i(t) \tag{20}$$

where

$$\begin{aligned}
\hat{\alpha}_{n-1}(s) &= \frac{1}{p(s)} \\
\hat{\alpha}_{n-2}(s) &= \frac{s + p_{n-1}}{p(s)} \\
\hat{\alpha}_{n-3}(s) &= \frac{s^2 + p_{n-1}s + p_{n-2}}{p(s)} \\
&\cdots\cdots\cdots\cdots\cdots\cdots \\
\hat{\alpha}_0(s) &= \frac{s^{n-1} + p_{n-1}s^{n-2} + \cdots + p_2 s + p_1}{p(s)}
\end{aligned} \tag{21}$$

and

$$\alpha_i(t) = \mathcal{L}^{-1}(\hat{\alpha}_i(s)) \qquad (i = 0,1,\ldots, n-1) \tag{22}$$

Proof: By successively substituting each equation (7), (9), (11), and (13) into the next, we see that

$$\begin{aligned}
\underline{Q}_{n-1} &= \underline{I} \\
\underline{Q}_{n-2} &= \underline{A} + p_{n-1}\underline{I} \\
\underline{Q}_{n-3} &= \underline{A}^2 + p_{n-1}\underline{A} + p_{n-2}\underline{I} \\
&\cdots\cdots\cdots\cdots\cdots\cdots \\
\underline{Q}_1 &= \underline{A}^{n-2} + p_{n-1}\underline{A}^{n-3} + \cdots + p_3\underline{A} + p_2\underline{I} \\
\underline{Q}_0 &= \underline{A}^{n-1} + p_{n-1}\underline{A}^{n-2} + \cdots + p_3\underline{A}^2 + p_2\underline{A} + p_1\underline{I}
\end{aligned} \tag{23}$$

Substitution of these expressions into (1) yields (19) and (21), and an inverse Laplace transformation yields (20) and (22).

ΔΔΔ

If the above proof is carried one step further, it is an easy matter to prove the Cayley-Hamilton theorem (Problem 1). The algorithm also provides us with the determinant and the inverse of $\underline{A}$.

4. *COROLLARY.* The determinant and the inverse of $\underline{A}$ are also computed by the resolvent algorithm,

$$\det \underline{A} = (-1)^n p_0 \tag{24}$$

$$\underline{A}^{-1} = -\frac{\underline{Q}_0}{p_0} \tag{25}$$

Proof: This formula for the determinant is derived in Problem A.3-9. Whenever $\det \underline{A} \neq 0$, $\underline{A}^{-1}$ exists and (25) follows from (15).

ΔΔΔ

Another result which follows from Theorem 2 is that for any two $n \times n$ matrices $\underline{A}$ and $\underline{B}$, $\underline{AB}$ and $\underline{BA}$ have the same eigenvalues (Problem 2).

Recall the *spectral representations* of $e^{\underline{A}t}$ which were derived in Sec. 3.2. In order to illustrate that these also follow directly from a partial fractions expansion of the resolvent matrix, we shall rederive Theorem 3.2.11 (and hence Theorem 3.2.10).

5. *THEOREM.* If $\underline{A}$ has distinct eigenvalues

$$\left.\begin{aligned} \lambda_i &= \sigma_i + j\omega_i \\ \lambda_{i+1} &= \sigma_i - j\omega_i = \overline{\lambda}_i \end{aligned}\right\} (i = 1, 3, 5, \ldots, m-1) \qquad \lambda_i = \overline{\lambda}_i \quad (i = m+1, m+2, \ldots, n) \tag{26}$$

then the resolvent matrix can be expanded as

$$(s\underline{I} - \underline{A})^{-1} = \sum_{i=1,3,\ldots}^{m-1} \frac{\underline{C}_i(s - \sigma_i) + \underline{S}_i\omega_i}{(s - \sigma_i)^2 + \omega_i^2} + \sum_{i=m+1}^{n} \frac{\underline{R}_i}{s - \lambda_i} \tag{27}$$

and the matrix exponential is

$$e^{\underline{A}t} = \sum_{i=1,3,\ldots}^{m-1} e^{\sigma_i t}(\underline{C}_i \cos\omega_i t + \underline{S}_i \sin\omega_i t) + \sum_{i=m+1}^{n} \underline{R}_i e^{\lambda_i t} \tag{28}$$

The case of repeated eigenvalues is left to the reader (Problem 5).

Proof: Recall from Lemma 1 that

$$(s\underline{I} - \underline{A})^{-1} = \frac{\underline{Q}(s)}{p(s)} = \frac{\underline{Q}_{n-1}s^{n-1} + \ldots + \underline{Q}_1 s + \underline{Q}_0}{s^n + p_{n-1}s^{n-1} + \ldots + p_1 s + p_0} \tag{29}$$

where $p(s)$ is the characteristic polynomial of $\underline{A}$. Since the eigenvalues (26) are the roots of $p(s)$, it follows from Corollary B.2.3 that each element of the matrix $\underline{Q}(s)/p(s)$ has a scalar partial fractions expansion. Collecting all of these expansions, we arrive at the matrix partial fractions expansion (27), and an inverse Laplace transformation yields the decomposition (28) of $e^{\underline{A}t}$.

ΔΔΔ

Finally, consider the case of a scalar system

$$\begin{aligned} \dot{\underline{x}}(t) &= \underline{A}\underline{x}(t) + \underline{b}u(t) \\ y(t) &= \underline{c}'\underline{x}(t) \end{aligned} \tag{30}$$

The transfer function matrix of Definition 4.2.3 is thus a scalar

$$\hat{h}(s) = \underline{c}'(s\underline{I} - \underline{A})^{-1}\underline{b} \tag{31}$$

which relates the input and output via

$$\hat{y}(s) = \hat{h}(s)\,\hat{u}(s) \tag{32}$$

when the initial state is zero. A transfer function for scalar systems was also introduced in Definition 4.1.1, and we shall now verify that these two definitions are consistent.

6. *LEMMA*. If the scalar system (30) has a transfer function

$$h(s) = \underline{c}'(s\underline{I} - \underline{A})^{-1}\underline{b} = \frac{q_{n-1}s^{n-1} + \ldots + q_1 s + q_0}{s^n + p_{n-1}s^{n-1} + p_1 s + p_0} \tag{33}$$

then the input and output of (30) satisfy the differential equation

$$\begin{aligned} y^{(n)}(t) + p_{n-1}y^{(n-1)}(t) + \ldots + p_1\dot{y}(t) + p_0 y(t) \\ = q_{n-1}u^{(n-1)}(t) + \ldots + q_1\dot{u}(t) + q_0 u(t) \end{aligned} \tag{34}$$

which is the same as (4.1-1).

Proof: The input and output transforms are related by

$$y(s) = \frac{q_{n-1}s^{n-1} + \ldots + q_1 s + q_0}{s^n + p_{n-1}s^{n-1} + \ldots + p_1 s + p_0}\ \hat{u}(s) \tag{35}$$

or

$$(s^n + p_{n-1}s^{n-1} + \ldots + p_1 s + p_0)\ \hat{y}(s) = (q_{n-1}s^{n-1} + \ldots + q_1 s + q_0)\ \hat{u}(s) \tag{36}$$

when the initial state $\underline{x}(0) = \underline{0}$. The initial conditions for y are also zero, since

$$\begin{aligned} y(0) &= \underline{c}'\underline{x}(0) = 0 \\ \dot{y}(0) &= \underline{c}'\dot{\underline{x}}(0) = \underline{c}'\underline{A}\underline{x}(0) + \underline{c}'\underline{b}u(0) = 0 \\ &\cdots\cdots\cdots\cdots\cdots\cdots \\ y^{(n-1)}(0) &= \underline{c}'\underline{x}^{(n-1)}(0) \\ &= \underline{c}'\underline{A}^{n-1}\underline{x}(0) + \underline{c}'\underline{A}^{n-2}\underline{b}u^{(n-2)}(0) + \ldots + \underline{c}'\underline{b}u(0) = 0 \end{aligned} \tag{37}$$

and we have adopted the convention that u(t) and all of its derivatimes are zero for $t \leq 0$. Thus the inverse transform of (36) is (34).

ΔΔΔ

4.4 SUMMARY

In this chapter we have examined the Laplace transform approach to solving differential equations. The first section dealt with scalar systems, and a *transfer function* relating the input and output transforms (in the absence of initial conditions) was defined. The inverse Laplace transform of the transfer function was termed the *impulse response*.

These terms were generalized in Sec. 4.2 to multivariable systems, and the impulse response matrix was found to be the same as that defined in Sec. 3.4. The principal factor in the transfer function matrix was seen to be the *resolvent matrix* $(s\underline{I} - \underline{A})^{-1}$, whose inverse transform is the exponential matrix $e^{\underline{A}t}$.

An algorithm for determining the resolvent was presented in Sec. 4.3, and several properties of the resolvent were discussed. The consistency of the two definitions of a transfer function in the scalar case was also established.

PROBLEMS

Section 4.1

1. (a) Determine the output response to a step input in Example 4.1.6 by performing an inverse Laplace transformation on (4.1-50).
   (b) Verify your result by recomputing it with the convolution integral (4.1-48).

2. (a) Determine the transfer function and impulse response of the system

$$\dddot{y} + 10\ddot{y} + 37\dot{y} + 52y = \ddot{u} - 3u$$

   Hint: One of the poles is $s = -4$.
   (b) Determine the output time function when the input is a unit step and

$$y(0) = \dot{y}(0) = 1 \qquad \ddot{y}(0) = 0$$

   Identify the natural and forced responses.
   (c) Determine and sketch the response as a function of time when

$$u(t) = \sin t \qquad (t > 0)$$

   and

$$y(0) = \dot{y}(0) = \ddot{y}(0) = 0$$

3. Verify that the definition of system poles following (4.1-11) is consistent with the definition following Theorem 3.2.8.
4. What happens to Lemma 4.1.3 in the case of nonzero initial conditions?

5. Derive the equivalent transfer function for a parallel connection of two systems with transfer functions $h_1(s)$ and $h_2(s)$.

6. Determine the transfer function of a system which consists of a pure time delay of duration Δ sec.

7. (a) Consider the RC and RL circuits shown in Fig. P1, and determine the transfer functions $\hat{h}_1$ and $\hat{h}_2$ relating inputs to outputs.

   (b) Draw a block diagram with blocks containing $\hat{h}_1$ and $\hat{h}_2$ connected in series $(\hat{u}_2 = \hat{y}_1)$, so that the overall transfer function relating $\hat{u}_1$ to $\hat{y}_2$ is $\hat{h}_2\hat{h}_1$. Show that this does *not* correctly represent the series connection of the two circuits in general, by computing the overall transfer function directly.

   (c) Under what conditions will the block diagram in (b) be approximately correct?

   (d) What must be added to the block diagram in (b) so that it will be correct in general?

   (e) Repeat the exercise for two RC circuits connected in series.

8. (a) Suppose z is a real zero of the transfer function $\hat{h}(s)$ in (4.1-11), and consider the input function $u(t) = 1(t)\, e^{zt}$. Show that there is a set of initial conditions of the system (4.1-1) for which the response y(t) to this input function is identically zero.

   (b) Formulate and prove a similar result for pairs of complex zeros. Interpret your result in the case where the zeros lie on the imaginary axis.

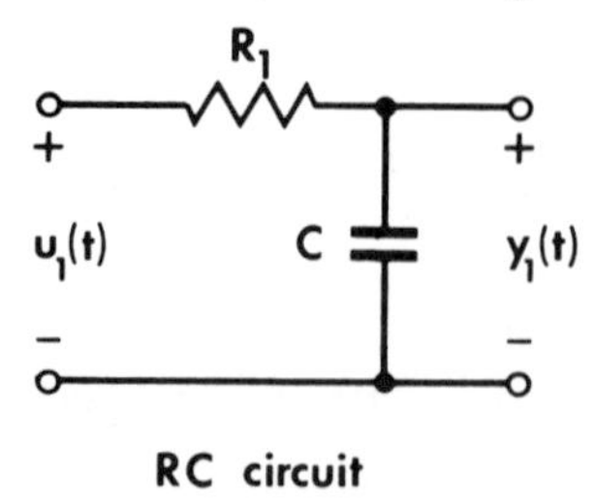

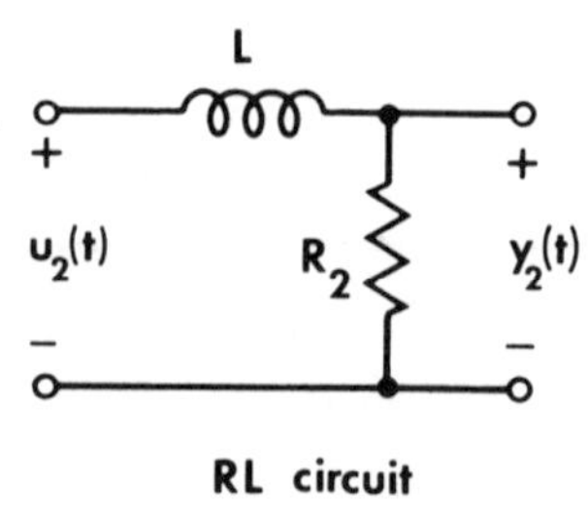

FIG. P1.

Section 4.2

1. Re-prove Theorem 4.2.1 by transforming Eq. (3.1-18) and using (3.1-19).

2. Determine the transfer function matrix and impulse response matrix for the 4-th order system describing the equatorial motion of a satellite in circular orbit, as given in equation (2.4-46).

3. Determine $e^{\underline{A}t}$ for Example 4.2.4 and use this to verify (4.2-20) and (4.2-22).

4. (a) Determine the matrix exponential and the transfer function for the scalar system

$$\dot{\underline{x}} = \begin{bmatrix} 3 & -4 \\ 1 & -2 \end{bmatrix} \underline{x} + \begin{bmatrix} 0 \\ 1 \end{bmatrix} u$$

$$y = [1 \quad 0] \, \underline{x}$$

   (b) Calculate and sketch the response of the system when $\underline{x}_0' = [1 \quad 1]$ and $u(t)$ is a unit step function.

5. Show that the two scalar systems

$$\dot{\underline{x}} = \underline{A}\underline{x} + \underline{b}u \qquad \text{and} \qquad \dot{\tilde{\underline{x}}} = \underline{A}'\tilde{\underline{x}} + \underline{c}u$$

$$y = \underline{c}'\underline{x} \qquad\qquad\qquad y = \underline{b}'\tilde{\underline{x}}$$

   have the same transfer function and impulse response.

6. Determine the transfer function and impulse response matrices for a system of the form

$$\dot{\underline{x}}(t) = \underline{A}\underline{x}(t) + \underline{B}\underline{u}(t)$$

$$\underline{y}(t) = \underline{C}\underline{x}(t) + \underline{D}\underline{u}(t)$$

7. This problem refers to Example 2.3.3 in which a model for the spread of an epidemic disease was developed. Assume that no migration takes place ($\underline{u} \equiv \underline{0}$), and that

$$\underline{x}(0) = \underline{x}_0 = [x_{10} \quad x_{20} \quad x_{30} \quad x_{40}]'$$

   (a) Determine the resolvent matrix $(s\underline{I} - \underline{A})^{-1}$ for this system.

(b) Using your *physical intuition*, state what you expect

$$\lim_{t\to\infty} [x_1(t) + x_2(t)] \quad \text{and} \quad \lim_{t\to\infty} [x_3(t) + x_4(t)]$$

to be.

(c) The state trajectory $\underline{x}(t)$ is the inverse Laplace transform of $\hat{\underline{x}}(s) = (s\underline{I} - \underline{A})^{-1} \underline{x}_0$. Use a *physical* argument to establish that the Final Value theorem (B.1.12) can be applied, and then use the theorem to determine the limits of $x_1(t)$, $x_2(t)$, $x_3(t)$, and $x_4(t)$ as $t \to \infty$.

(d) If your answers to (b) and (c) disagree, go back and determine what went wrong.

(e) Suppose the initial population distribution is

$$x_0 = [1000 \quad 1000 \quad 100 \quad 50]'$$

and the transition parameters are

$$\alpha = 0.15 \qquad \beta = 0.05 \qquad \gamma = \delta = 0.10$$

How many people will end up in categories 3 and 4, respectively, as $t \to \infty$?

(f) Suppose a new medical team arrives with the capacity to *either* increase α to 0.20 *or* increase δ to 0.15 (β and γ cannot be changed). Which course of action does your *intuition* indicate will be more beneficial in the long run?

(g) Check your answer to (f) by recomputing

$$\lim_{t\to\infty} x_3(t)$$

for both cases.

8. Using Laplace transforms, verify the superposition property of linear systems (Corollary 3.3.4).

9. In many textbooks, the transfer function of a scalar system is defined as $\hat{h}(s) \triangleq \hat{y}(s)/\hat{u}(s)$. Is there an analogous definition for multivariable systems?

Section 4.3

1. Prove the Cayley-Hamilton theorem (A.3.9) by extending the proof of Corollary 4.3.3.

2. Let $\underline{A}$ and $\underline{B}$ be $n \times n$ matrices with $\underline{AB} \neq \underline{BA}$, and prove that $\underline{AB}$ and $\underline{BA}$ have the same eigenvalues, as follows. Prove that $\text{tr}(\underline{AB}) = \text{tr}(\underline{BA})$ and use this fact in the resolvent algorithm to show that $\underline{AB}$ and $\underline{BA}$ have the same characteristic polynomial.

3. Use the resolvent algorithm to determine $(s\underline{I} - \underline{A})^{-1}$, $\underline{A}^{-1}$, and det $\underline{A}$ for

$$\underline{A} = \begin{bmatrix} 1 & 2 \\ -3 & -4 \end{bmatrix} \quad \text{and} \quad \underline{A} = \begin{bmatrix} 1 & 3 & -2 \\ -1 & 0 & 4 \\ -2 & 3 & 1 \end{bmatrix}$$

   Verify directly that your answers are correct.

4. Prove Theorem 2.2.1 using Laplace transforms.

5. Generalize Theorem 4.3.5 to include an $\underline{A}$ with repeated eigenvalues.

6. Consider the linearized equatorial motion of a satellite in circular orbit described by (2.4-62).

   (a) Use the resolvent algorithm to determine $(s\underline{I} - \underline{A})^{-1}$. Be sure to check your calculation with (4.3-15).

   (b) Compute $e^{\underline{A}t}$.

   (c) Write $(s\underline{I} - \underline{A})^{-1}$ and $e^{\underline{A}t}$ in the forms specified by Corollary 4.3.3.

   (d) Write $(s\underline{I} - \underline{A})^{-1}$ and $e^{\underline{A}t}$ in the form specified by Theorem 4.3.5 (and Problem 4.3-5), and then write $e^{\underline{A}t}$ in the form specified by Theorem 3.2.11.

   (e) Compute the transfer function matrix.

Chapter 5

STABILITY AND TRANSIENT ANALYSIS

5.0. INTRODUCTION

In this chapter we shall examine questions of *stability*. Generally speaking, a system is said to be *stable* if it returns to equilibrium after a disturbance occurs. More specifically, a linear differential equation is said to be *stable* if its natural response decays to zero with increasing time, for any initial conditions. An *unstable* system, on the other hand, has variables which grow without bound.

In the case of a linear, time-invariant system (or more accurately, a linear, time-invariant *model* of a physical system), we shall see that conditions for stability are almost transparent. Since the response of such a system consists entirely of terms involving exponentials, it will decay if and only if all of the exponents have negative real parts.

Although certain variables in an unstable mathematical model may grow without bound, the physical variables which they represent rarely do so. In such cases the accuracy of the model usually breaks down, and the variables in question are bounded by saturation, overheating, or a myriad of other nonlinear effects. Nevertheless, linear stability criteria are extremely useful tools in the analysis and design of control systems.

5.1. STABILITY OF LINEAR SYSTEMS

We shall consider a linear, time-invariant dynamic system described in state-variable format by

$$\dot{\underline{x}}(t) = \underline{A}\underline{x}(t) + \underline{B}\underline{u}(t)$$
$$\underline{y}(t) = \underline{C}\underline{x}(t) \tag{1}$$

where $\underline{x}(0) = \underline{x}_0$. The state $\underline{x}(t)$ consists of all the relevant dynamic variables in the system; questions of stability naturally hinge on the behavior of $\underline{x}(t)$ for large t. Thus we say that the system is stable if the state always returns to zero when there is no input.

1. DEFINITION. The system (1) is said to be *(asymptotically) stable*† if whenever $\underline{u}(t) \equiv \underline{0}$,

$$\| \underline{x}(t) \| \xrightarrow[t\to\infty]{} 0 \qquad \text{for any initial state } \underline{x}_0. \tag{2}$$

ΔΔΔ

In later chapters we shall usually refer to systems satisfying this condition as simply *stable*. However, various other definitions of stability may be found in the literature, and the reader should bear in mind that this one is intended. The case of a nonzero input is taken up in Definition 5 and Theorem 6.

An alternative description for scalar systems is the input-output format,

$$y^{(n)}(t) + p_{n-1}y^{(n-1)}(t) + \dots + p_1\dot{y}(t) + p_0y(t)$$
$$= q_{n-1}u^{(n-1)}(t) + \dots + q_1\dot{u}(t) + q_0u(t) \tag{3}$$

---

†This is also known as *global asymptotic stability*, *stability in the large*, or *Liapunov stability*. The definition can easily be extended to time-varying and nonlinear systems. In the latter case, stability may be with respect to an equilibrium point other than origin.

which may always be converted to state-variable format (see Theorem 2.2.1 or Problem 2.2-2). In this case we define stability in terms of y and its derivatives.

2. *DEFINITION.* The system (3) is said to be ***(asymptotically) stable*** if whenever $u(t) \equiv 0$,

$$\lim_{t\to\infty} y(t) = \lim_{t\to\infty} \dot{y}(t) = \ldots = \lim_{t\to\infty} y^{(n-1)}(t) = 0 \quad (4)$$

for any set of initial conditions. Problem 2 asks for a verification that this is consistent with Definition 1.

ΔΔΔ

Having characterized stability, we turn to the problem of determining whether a particular system is stable. Recalling that the system response was expressed in Chapters 3 and 4 as a sum of exponentials involving the system poles, we may deduce the following conditions for asymptotic stability.

3. *THEOREM.*

(a) The system (1) is (asymptotically) stable if and only if all of its poles lie strictly in the left half-plane, that is,

$$\text{Re } \lambda_i < 0 \qquad (i = 1,2,\ldots,n) \quad (5)$$

where λ_i are the eigenvalues† of $\underline{A}$.

(b) Similarly, the system (3) is (asymptotically) stable if and only if all of its poles [i.e., the roots of its characteristic polynomial $p(s) = s^n + p_{n-1}s^{n-1} + \ldots + p_1 s + p_0$] lie strictly in the left half-plane.

†Since the eigenvalues of $\underline{QAQ}^{-1}$ are the same as those of $\underline{A}$ (Theorem A.3.2), conclusions regarding stability are unaffected by a state transformation.

Proof: (a) For the case $\underline{u}(t) \equiv \underline{0}$, recall from Theorem 3.1.4 that the solution of (1) is given by

$$\underline{x}(t) = e^{\underline{A}t}\underline{x}_0 \tag{6}$$

If we suppose for a moment that the eigenvalues of $\underline{A}$ are all distinct, then Theorem 3.2.10 establishes that

$$\begin{aligned} \underline{x}(t) &= \left(\sum_{i=1}^{n} \underline{R}_i e^{\lambda_i t} \right) \underline{x}_0 \\ &= \underline{R}_1 \underline{x}_0 e^{\lambda_1 t} + \underline{R}_2 \underline{x}_0 e^{\lambda_2 t} + \dots + \underline{R}_n \underline{x}_0 e^{\lambda_n t} \end{aligned} \tag{7}$$

It is clear that if (5) holds, then all the exponentials in (7) will approach zero, and $\underline{x}(t) \xrightarrow[t\to\infty]{} \underline{0}$ for any $\underline{x}_0$. Conversely, if $\operatorname{Re} \lambda_i > 0$ for some i, the corresponding term of (7) will grow without bound for some $\underline{x}_0$ since $\underline{R}_i \neq \underline{0}$ (Problem 3.2-10). If $\operatorname{Re} \lambda_i = 0$, (7) contains terms which are constant or proportional to $\sin \omega_i t$ and $\cos \omega_i t$ and, hence, never decay to zero.

If $\underline{A}$ has repeated eigenvalues, then

$$\underline{x}(t) = e^{\underline{A}t}\underline{x}_0 = \left(\sum_{i=1}^{p} \sum_{j=1}^{\eta_i} \underline{R}_i^j \, t^{j-1} e^{\lambda_i t} \right) \underline{x}_0 \tag{8}$$

and the same argument applies (note that $\underline{R}_i^1 \neq \underline{0}$ by Problem 3.2-10).

(b) This may be proved in an analogous fashion by using Theorem B.2.2 and B.2.5 to expand the solution of (3) in partial fractions. Details are left to the reader (Problem 3).

ΔΔΔ

Note that a system of the form (1) which is not asymptotically stable could nevertheless have a bounded state, that is,

$$\| \underline{x}(t) \| < K < \infty \qquad 0 < t < \infty \tag{9}$$

for any initial state $\underline{x}_0$ (and for $\underline{u}(t) \equiv \underline{0}$). We shall refer to this state of affairs as *marginal (or neutral) stability*, and we deduce from the proof of Theorem 3 that it may occur if one or more of the poles has a real part equal to zero. This corresponds to terms in the transient response which are constant (pole at the origin) or which oscillate with constant amplitude (poles on the $j\omega$ axis).

If a system is neither asymptotically stable nor marginally stable, then there must be certain initial states $\underline{x}_0$ for which $\|\underline{x}(t)\|$ grows without bound (even though $\underline{u}(t) \equiv \underline{0}$). Such a system is said to be *unstable*. Precise conditions for marginal stability and instability are developed in Problems 7 and 8.

4. EXAMPLE. Consider a scalar system

$$\ddot{y}(t) + 7\dot{y}(t) + 10y(t) = u(t) \tag{10}$$

which may also be described in state-space format as

$$\begin{aligned} \dot{\underline{x}}(t) &= \begin{bmatrix} 0 & 1 \\ -10 & -7 \end{bmatrix} \underline{x}(t) + \begin{bmatrix} 0 \\ 1 \end{bmatrix} u(t) \\ y(t) &= \begin{bmatrix} 1 & 0 \end{bmatrix} \underline{x}(t) \end{aligned} \tag{11}$$

where

$$\underline{x} \triangleq \begin{bmatrix} y & \dot{y} \end{bmatrix}'$$

According to Theorem 3(a), this system is asymptotically stable if the poles of the system lie in the left half-plane. These are the roots of the characteristic polynomial,

$$\det(\lambda \underline{I} - \underline{A}) = \lambda^2 + 7\lambda + 10 \tag{12}$$

Alternatively, Theorem 3(b) reminds us that this polynomial may be read off directly from (10). Either way, we see that its roots are -2 and -5, and since these are both negative, we conclude that the system is asymptotically stable.

To check this result, we may set $u(t) \equiv 0$ and compute

$$\underline{x}(t) = e^{\underline{A}t}\underline{x}_0 = \frac{1}{3}\begin{bmatrix} 5e^{-2t} - 2e^{-5t} & e^{-2t} - e^{-5t} \\ 10e^{-5t} - 10e^{-2t} & 5e^{-5t} - 2e^{-2t} \end{bmatrix} \underline{x}_0 \tag{13}$$

which clearly approaches zero for any initial state $\underline{x}_0$. ΔΔΔ

So far, we have dealt with an "internal" characterization of stability, in terms of the natural response of the system state with no input applied. Alternatively, if we are only interested in "external" behavior, we may set the initial state to zero and describe stability in terms of input-output relationships.

5. *DEFINITION.* A system is said to be *bounded-input, bounded-output (BIBO) stable* if for zero initial conditions, a bounded input always evokes a bounded output. In terms of (1), this means that if $\underline{x}_0 = \underline{0}$ and

$$\| \underline{u}(t) \| \leq K_1 < \infty \qquad (0 \leq t < \infty) \tag{14}$$

for some constant K_1, then there exists another constant K_2 such that

$$\| \underline{y}(t) \| \leq K_2 < \infty \qquad (0 \leq t < \infty) \tag{15}$$

In terms of (3), an equivalent statement is that if

$$y(0) = \dot{y}(0) = \dots = y^{(n-1)}(0) = 0 \tag{16}$$

and

$$|u(t)| \leq K_1 \qquad (0 \leq t < \infty) \tag{17}$$

then

$$|y(t)| \leq K_2 \qquad (0 \leq t < \infty) \tag{18}$$

ΔΔΔ

6. *THEOREM.* A system is BIBO stable if and only if the poles of its *transfer function* (assuming common factors in the numerator and denominator have been cancelled) all lie strictly in the left half-plane.

Proof: The proof, which is straightforward but somewhat tedious, is left to the reader (Problem 4).

ΔΔΔ

If one or more of a system's poles are cancelled by zeros in the transfer function, the transfer function poles will comprise only a subset of the full set of system poles. Thus it is clear from Theorems 3 and 6 that, in general, *asymptotic stability implies BIBO stability,* but that a BIBO-stable system is not necessarily asymptotically stable.

One might be tempted to conclude that the concept of BIBO stability is not particularly useful since a system can have a perfectly well-behaved input-output relation and still be suffering from an "internal hemorrhage" ($\|\underline{x}(t)\| \to \infty$). Nevertheless, under certain conditions BIBO and asymptotic stability are completely equivalent,[†] and BIBO stability can be useful in dealing with systems containing time-varying parameters, nonlinear elements, or pure delays, where it is sometimes easier to establish than asymptotic stability.

Having briefly explored the relationship between internal and external characterizations of system stability, we shall have little further use for the BIBO concept. The reader may assume from this point on that the word "stable," unless otherwise qualified, refers to Definitions 1 and 2.

5.2. TESTS FOR STABILITY

Theorems 5.1.3 and 5.1.6 in the preceding section established conditions for determining the stability or instability of a linear system. To make use of these conditions, one must check whether the roots of a particular polynomial lie strictly in the left half of the complex plane, and if the degree of the polynomial is three or more this becomes a nontrivial task.

Fortunately, straightforward procedures have been devised to test these conditions for a polynomial of arbitrary degree,[††]

$$p(s) = s^n + p_{n-1}s^{n-1} + \ldots + p_1 s + p_0 \tag{1}$$

[†] This will be cleared up in Problem 10.3-8. A sufficient condition for equivalence is clearly that no pole-zero cancellations occur, but we shall see that this condition is necessary only for systems with a single input and/or a single output.

[††] Without loss of generality, we assume that the polynomial is *monic* (the coefficient of s^n is equal to 1).

We denote the roots of p(s) as $\lambda_1, \lambda_2, \ldots, \lambda_n$, and by a slight abuse of terminology, we shall refer to p(s) as a *stable polynomial* (or a *Hurwitz polynomial*) whenever

$$\operatorname{Re}\lambda_i < 0 \qquad (i = 1, 2, \ldots, n) \tag{2}$$

First, we note a useful necessary condition for stability.

1. *LEMMA.* If p(s) is stable, then all of its coefficients must be positive,

$$p_i > 0 \qquad (i = 0, 1, 2, \ldots, n-1) \tag{3}$$

Proof: The polynomial p(s) may be written as a product,

$$p(s) = [(s-\sigma_1)^2 + \omega_1^2] \ldots [(s-\sigma_{m-1})^2 + \omega_{m-1}^2]\ (s-\lambda_{m+1})\ldots(s-\lambda_n) \tag{4}$$

where $\sigma_i \pm j\omega_i$, $i = 1, 3, \ldots, m-1$, and λ_i, $i = m+1, m+2, \ldots, n$, are its roots. If p(s) is stable, then

$$\begin{aligned} \sigma_i &< 0 \qquad (i = 1, 3, 5, \ldots, m-1) \\ \lambda_i &< 0 \qquad (i = m+1, m+2, \ldots, n) \end{aligned} \tag{5}$$

and hence all the constants ($-\sigma_i$, ω_i^2, $-\lambda_k$, etc.) occurring in (4) must be positive. This means that when (4) is multiplied out to get the form (1), all the resulting coefficients must be positive.

ΔΔΔ

This lemma is very useful as an initial stability check: If any of the coefficients p_i are negative or zero, the polynomial cannot be stable. If the coefficients are all positive, however, no conclusion may be drawn (except for $n = 2$) and we must test further (e.g., with the *Hurwitz criterion*).

2. *THEOREM (Hurwitz criterion).* The polynomial p(s) is stable if and only if the *Hurwitz determinants*

$$\Delta_1 = \det \left[p_{n-1} \right] = p_{n-1}$$

$$\Delta_2 = \det \begin{bmatrix} p_{n-1} & p_{n-3} \\ 1 & p_{n-2} \end{bmatrix}$$

$$\Delta_3 = \det \begin{bmatrix} p_{n-1} & p_{n-3} & p_{n-5} \\ 1 & p_{n-2} & p_{n-4} \\ 0 & p_{n-1} & p_{n-3} \end{bmatrix}$$

....

$$\Delta_n = \det \begin{bmatrix} p_{n-1} & p_{n-3} & p_{n-5} & \cdots & 0 & 0 \\ 1 & p_{n-2} & p_{n-4} & \cdots & 0 & 0 \\ 0 & p_{n-1} & p_{n-3} & \cdots & 0 & 0 \\ 0 & 1 & p_{n-2} & \cdots & 0 & 0 \\ 0 & 0 & p_{n-1} & \cdots & 0 & 0 \\ \cdots & \cdots & \cdots & \cdots & \cdots & \cdots \\ 0 & 0 & 0 & \cdots & p_0 & 0 \\ 0 & 0 & 0 & \cdots & p_1 & 0 \\ 0 & 0 & 0 & \cdots & p_2 & p_0 \end{bmatrix} \tag{6}$$

are all positive,

$$\Delta_i > 0 \qquad (i = 1,2,\ldots,n) \tag{7}$$

Proof: The proof is beyond the scope of this book, and the reader is referred to Ref. 68 for a detailed exposition of this and related results.

ΔΔΔ

The effort involved in applying the Hurwitz criterion can be reduced by recognizing that half the conditions (7) are redundant.

3. *THEOREM (Liénard-Chipart).* Suppose the coefficients of p(s) are all positive,[†]

$$p_i > 0 \qquad (i = 0, 1, 2, \ldots, n-1) \tag{8}$$

Then p(s) is stable if and only if

$$\Delta_i > 0 \qquad (i = 2, 4, 6, \ldots, n-1) \quad \text{(for n odd)} \tag{9}$$

or

$$\Delta_i > 0 \qquad (i = 3, 5, 7, \ldots, n-1) \quad \text{(for n even)} \tag{10}$$

Proof: See Ref. 68.

ΔΔΔ

We shall now illustrate these results with two examples.

4. *EXAMPLE.* Consider a system described by

$$\dot{\underline{x}}(t) = \begin{bmatrix} 0 & 1 & 0 & 0 \\ 0 & 0 & 1 & 0 \\ 0 & 0 & 0 & 1 \\ -5 & -4 & -3 & -2 \end{bmatrix} x(t) + \begin{bmatrix} 0 \\ 0 \\ 0 \\ 1 \end{bmatrix} u(t) \tag{11}$$

$$y(t) = \begin{bmatrix} 2 & -6 & 4 & 0 \end{bmatrix} x(t)$$

with characteristic polynomial

$$p(s) = s^4 + 2s^3 + 3s^2 + 4s + 5 \tag{12}$$

To check the stability of this system, we first make use of Lemma 1 and check that all the coefficients of p(s) are positive. They are, so we turn next to the Liénard-Chipart criterion. Since $n = 4$ is even, we use (10) and find that

---

[†]In fact, it is only necessary to check every other coefficient: See Ref. 68.

$$\Delta_3 = \det \begin{bmatrix} 2 & 4 & 0 \\ 1 & 3 & 5 \\ 0 & 2 & 4 \end{bmatrix} = -12 \tag{13}$$

is negative, so we conclude from Theorem 3 that the system (11) is unstable.

If we had wished to check the other Hurwitz determinants, we would have found

$$\Delta_1 = \det \begin{bmatrix} 2 \end{bmatrix} = 2 \tag{14}$$

$$\Delta_2 = \det \begin{bmatrix} 2 & 4 \\ 1 & 3 \end{bmatrix} = 2 \tag{15}$$

$$\Delta_4 = \det \begin{bmatrix} 2 & 4 & 0 & 0 \\ 1 & 3 & 5 & 0 \\ 0 & 2 & 4 & 0 \\ 0 & 1 & 3 & 5 \end{bmatrix} = -60 \tag{16}$$

ΔΔΔ

5. *EXAMPLE.* Consider a system described by

$$\dddot{y}(t) + 7\ddot{y}(t) + 14\dot{y}(t) + 8y(t) = \dot{u}(t) + 3u(t) \tag{17}$$

whose characteristic polynomial is

$$p(s) = s^3 + 7s^2 + 14s + 8 \tag{18}$$

Since the coefficients are all positive, we need only check

$$\Delta_2 = \begin{bmatrix} 7 & 8 \\ 1 & 14 \end{bmatrix} = 90 \tag{19}$$

which is also positive, and conclude from Theorem 3 that the system (17) is stable.

ΔΔΔ

In the case of a second-, third-, or fourth-order system, the stability test can be expressed even more compactly as follows.

6. COROLLARY.

(a) A second-order polynomial

$$p(s) = s^2 + p_1 s + p_0 \tag{20}$$

is stable if and only if the coefficients p_1 and p_0 are both positive.

(b) A third-order polynomial

$$p(s) = s^3 + p_2 s^2 + p_1 s + p_0 \tag{21}$$

is stable if and only if the coefficients are all positive and

$$p_2 p_1 > p_0 \tag{22}$$

(c) A fourth-order polynomial

$$p(s) = s^4 + p_3 s^3 + p_2 s^2 + p_1 s + p_0 \tag{23}$$

is stable if and only if the coefficients are all positive and

$$p_3 p_2 p_1 > p_3^2 p_0 + p_1^2 \tag{24}$$

Proof: This follows directly from Theorem 3. The details are left to the reader.

ΔΔΔ

The results obtained above are very useful since they allow a rapid check as to whether a system is stable. Alternative formulas using *inners* may be found in Refs. 97 and 275. If a digital computer is available, another alternative is to solve directly for the roots of p or the eigenvalues of $\underline{A}$ with an iterative algorithm.† Such programs are readily available in the subroutine libraries of most major computer installations.

†See, for instance, Refs. 2, 80, 81, 83, 93, 146 and 179.

Occasionally one needs to determine not only whether a polynomial is stable, but if not, *how many* of its roots have a positive real part. In particular, this question will arise in Chapter 8, when we introduce the Nyquist stability criterion for systems with feedback. It can be answered by using the following alternative form of Theorem 2, known as the *Routh criterion*.

7. *THEOREM (Routh criterion)*. The number of roots of

$$p(s) = s^n + p_{n-1}s^{n-1} + \dots + p_1 s + p_0 \tag{25}$$

which have strictly positive real parts is equal to the number of sign changes in the left-hand column of the *Routh array*,

$$\begin{array}{lllll}
\text{row 1:} & 1 & p_{n-2} & p_{n-4} & \cdots \\
\text{row 2:} & p_{n-1} & p_{n-3} & p_{n-5} & \cdots \\
\text{row 3:} & b_{n-1} & b_{n-3} & b_{n-5} & \cdots \\
\text{row 4:} & c_{n-1} & c_{n-3} & c_{n-5} & \cdots \\
\text{row 5:} & d_{n-1} & d_{n-3} & d_{n-5} & \cdots \\
\vdots & & & & \\
\text{row } (n+1): & \cdots\cdots\cdots & & &
\end{array} \tag{26}$$

where $p_{n-i} \triangleq 0$ for $i > n$, and for $j = 3, 4, \dots, n+1$, row j is generated from rows $j-1$ and $j-2$ as follows:

$$b_{n-1} = -\frac{1}{p_{n-1}} \det \begin{bmatrix} 1 & p_{n-2} \\ p_{n-1} & p_{n-3} \end{bmatrix}$$

$$b_{n-3} = -\frac{1}{p_{n-1}} \det \begin{bmatrix} 1 & p_{n-4} \\ p_{n-1} & p_{n-5} \end{bmatrix}$$

$$b_{n-5} = -\frac{1}{p_{n-1}} \det \begin{bmatrix} 1 & p_{n-6} \\ p_{n-1} & p_{n-7} \end{bmatrix}$$

....

$$c_{n-1} = -\frac{1}{b_{n-1}} \det \begin{bmatrix} p_{n-1} & p_{n-3} \\ b_{n-1} & b_{n-3} \end{bmatrix}$$

$$c_{n-3} = -\frac{1}{b_{n-1}} \det \begin{bmatrix} p_{n-1} & p_{n-5} \\ b_{n-1} & b_{n-5} \end{bmatrix} \qquad (27)$$

....

$$d_{n-1} = -\frac{1}{c_{n-1}} \det \begin{bmatrix} b_{n-1} & b_{n-3} \\ c_{n-1} & c_{n-3} \end{bmatrix}$$

....

The polynomial is stable if and only if all the elements in the left-hand column are strictly positive. All the elements of a row may be multiplied by a positive constant, in order to simplify the arithmetic, without affecting the result. If zeros occur in the left-hand column, they may be handled as follows:

A. If the row contains other terms which are nonzero, the zero in the left-hand column is replaced by ε, the array is completed as above, and ε is assumed to be very small for the purpose of counting sign changes (see Example 10). This procedure is based on the observation that introducing the parameter ε amounts to a small perturbation in some or all of the coefficients of the polynomial. Since the roots of a polynomial are continuous functions of its coefficients as long as its degree is unchanged, the perturbed polynomial will have the same number of roots in the right half-plane as the original one, provided the perturbation is sufficiently small, and provided also that the original polynomial has no pure imaginary roots which are moved into the right half-plane by the perturbation. This is indicated if the number of sign changes in the left-hand column of the Routh array depends on the sign of ε. In this case, the Routh criterion breaks down, and it is necessary first to remove all factors of the form $s^2 + \omega^2$ from the polynomial. This can be done without determining the imaginary roots explicitly, but the details are somewhat involved; we refer the reader to Refs. 68, 236a and 237a for a discussion.

B. If row j contains all zeros,[†] the entries in row $j - 1$ are used as the coefficients of an *auxiliary polynomial* of order $n - j + 2$, containing only even powers of s. The auxiliary polynomial is differentiated, and the coefficients of the derivative replace the row of zeros (see Example 11). It may be necessary to repeat this procedure several times.

Proof: See Ref. 68.

ΔΔΔ

8. EXAMPLE. The characteristic polynomial in Example 5 was

$$p(s) = s^4 + 2s^3 + 3s^2 + 4s + 5 \tag{28}$$

for which the Routh array is

$$\begin{array}{rrrrl} 1 & 3 & 5 & 0 & \cdots \\ 2 & 4 & 0 & 0 & \cdots \\ 1 & 5 & 0 & 0 & \cdots \\ -6 & 0 & 0 & 0 & \cdots \\ 5 & 0 & 0 & 0 & \cdots \end{array} \tag{29}$$

Since there are two sign changes in the left-hand column, we conclude that two roots of p(s) lie in the right half-plane.

ΔΔΔ

9. EXAMPLE. The characteristic polynomial in Example 5 was

$$p(s) = s^3 + 7s^2 + 14s + 8 \tag{30}$$

[†]In this case, provided that p(s) has no roots at $s = 0$, $n - j$ must be even.

for which the Routh array is

$$\begin{array}{cccc} 1 & 14 & 0 & \cdots \\ 7 & 8 & 0 & \cdots \\ \frac{90}{7} & 0 & 0 & \cdots \\ 8 & 0 & 0 & \cdots \end{array} \tag{31}$$

Since there are no sign changes and no zero elements in the left-hand column, we conclude that p(s) is stable.

ΔΔΔ

10. EXAMPLE. The Routh array for

$$p(s) = s^4 + s^3 + 2s^2 + 2s + 1 \tag{32}$$

is

$$\begin{array}{cccc} 1 & 2 & 1 & 0 \\ 1 & 2 & 0 & 0 \\ \not{0} & \not{1} & \not{0} & \not{0} \\ \varepsilon & 1 & 0 & 0 \\ 2 - \frac{1}{\varepsilon} & 0 & 0 & 0 \\ 1 & 0 & 0 & 0 \end{array} \tag{33}$$

For any ε such that $|\varepsilon| < \frac{1}{2}$ there are two sign changes in the left-hand column and hence two roots in the right half-plane.

ΔΔΔ

11. EXAMPLE. The Routh array for

$$p(s) = s^5 + 2s^4 + 3s^3 + 6s^2 - 4s - 8 \tag{34}$$

is

$$\begin{array}{cccc} 1 & 3 & -4 & 0 \\ 2 & 6 & -8 & 0 \\ \not{0} & \not{0} & \not{0} & \not{0} \\ 8 & 12 & 0 & 0 \\ 3 & -8 & 0 & 0 \\ 33.3 & 0 & 0 & 0 \\ -8 & 0 & 0 & 0 \end{array} \qquad \begin{array}{l} \longrightarrow \quad p_a(s) = 2s^4 + 6s^2 - 8 \\ \qquad\qquad \downarrow \\ \longleftarrow \quad p_a'(s) = 8s^3 + 12s \end{array} \tag{35}$$

There is one sign change in the left-hand column and hence one root in the right half-plane.

ΔΔΔ

It is often necessary to determine not only whether all the rows of a polynomial lie to the left of the $j\omega$ axis, but also whether they lie to the left of a vertical line through $s = -\bar{\sigma}$ for some real number $\bar{\sigma} > 0$. In the next section, a system whose characteristic polynomial satisfies this condition [see Fig. 5.3-5(a)] will be said to have *stability margin* $\bar{\sigma}$.

Testing this condition is quite easy. If the polynomial $p(s)$ has roots λ_i, $i = 1,2,\ldots,n$, then substitution of $s - \bar{\sigma}$ for s yields a new polynomial

$$\bar{p}(s) = p(s - \bar{\sigma}) \tag{36}$$

with roots $\lambda_i + \bar{\sigma}$, $i = 1,2,\ldots,n$. The Hurwitz and Routh criteria, applied to $\bar{p}(s)$, test whether $\lambda_i + \bar{\sigma}$, $i = 1,2,\ldots,n$, all have negative real parts, i.e., whether

$$\operatorname{Re} \lambda_i < -\bar{\sigma} \qquad (i = 1,2,\ldots,n) \tag{37}$$

and the Routh criterion will determine how many roots lie to the right of the vertical line.

12. EXAMPLE. Returning to Example 5, we wish to determine whether all the roots of

$$p(s) = s^3 + 7s^2 + 14s + 8 \tag{38}$$

have real parts less than -1. Substituting $s - 1$ for s, we have

$$\begin{aligned} \bar{p}(s) &= p(s-1) = (s-1)^3 + 7(s-1)^2 + 14(s-1) + 8 \\ &= s^3 + 4s^2 + 3s + 0 \end{aligned} \tag{39}$$

Since the last coefficient is zero, $\bar{p}(s)$ is not stable (using Lemma 1), and we conclude that one or more roots of $p(s)$ have real parts between -1 and 0.

ΔΔΔ

It is also useful to be able to determine whether all the roots of a polynomial lie within a sector of angle 2Θ centered on the negative real axis. In the next section, a system whose characteristic polynomial satisfies this condition [see Fig. 5.3-5(b)] will be said to have *damping margin* $\zeta = \cos\Theta$.

Whereas testing for a given stability margin required only a simple coordinate shift, testing this condition is more difficult. It can be done by rotating coordinates [159], which results in a new polynomial with complex coefficients, and then using a generalized version of the Hurwitz criterion [68, p.248].

A very different stability test will be introduced in Sec. 8.4. This is the *Nyquist criterion*, which checks the stability of a closed-loop (feedback) system using only open-loop information.

5.3. RESPONSE OF FIRST- AND SECOND-ORDER SYSTEMS

We have seen in Chapters 3 and 4 that the response of any linear system can be written in terms of exponential functions such as

$$e^{\lambda t} \qquad e^{\sigma t}\cos\omega t \qquad e^{\sigma t}\sin\omega t \tag{1}$$

possibly multiplied by powers of t, where λ and $\sigma \pm j\omega$ are poles of the system. This fact was used in the two preceding sections to draw conclusions about the stability of a system: The terms in (1) decay to zero or grow without bound, respectively, depending on whether λ and σ are negative or positive.

It is often necessary to know more about the system than simply whether it is stable; the shape and speed of the response to a change of input or other disturbance can also be of great importance. For a system of even moderate size, of course, exact determination of the response will often require either a substantial amount of calculation or a computer simulation. Fortunately, however, a great deal may be learned from analyses of terms such as those in (1); by examining in detail the responses associated with individual poles, we gain valuable insight and intuition into the behavior of the system as a whole.

Since an unstable system is quite useless for most practical purposes, we shall restrict our attention in this section to stable systems. We shall consider only first- and second-order systems, bearing in mind that the same analysis may be applied (with some care) to the response associated with a first- or second-order pole in a higher-order system.

A common procedure for analyzing control systems is to apply a standard input, such as a step, impulse, or sinusoid, and then determine the output response. In a stable system, this response will consist of a *transient* component, including all the terms which decay to zero, and a *steady-state* component, depending upon the form of the input. We shall concern ourselves here with the transient, and postpone a more thorough steady-state analysis until Chapter 6.

1. EXAMPLE (First-order system). Consider a first-order linear system described by

$$\dot{y}(t) = -\lambda y(t) + u(t) \tag{2}$$

We assume stability ($\lambda > 0$), and note that there is a single real pole at $s = -\lambda$, indicated by × in Fig. 1(a). If the system is initially at rest $[y(0) = 0]$, its response is given by

$$y(t) = \int_0^t e^{-\lambda(t-\tau)} u(\tau)\, d\tau \tag{3}$$

or in terms of Laplace transforms as

$$\hat{y}(s) = \frac{1}{s+\lambda} \hat{u}(s) \tag{4}$$

If a unit impulse is applied at the input to (2), the output is simply the impulse response

$$y(t) = e^{-\lambda t} \qquad (t > 0) \tag{5}$$

shown in Fig. 1(b); the steady-state component in this case is zero. The parameter $1/\lambda$ is known as the *time constant* of the system (or of the pole), and the response is characterized by the property that in any time interval of length $1/\lambda$, it drops by a factor of $1/e$ ($\simeq 0.37$).

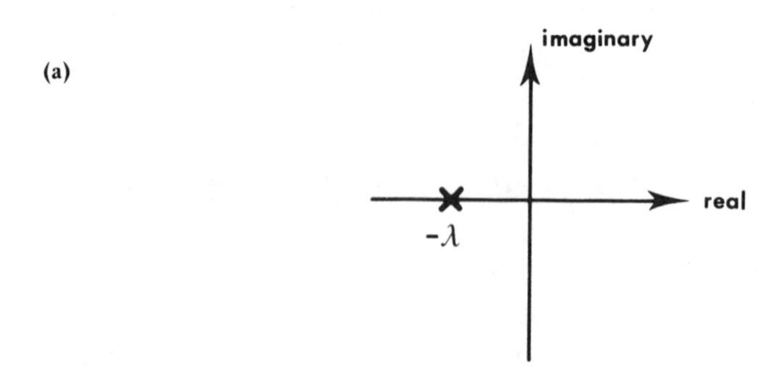

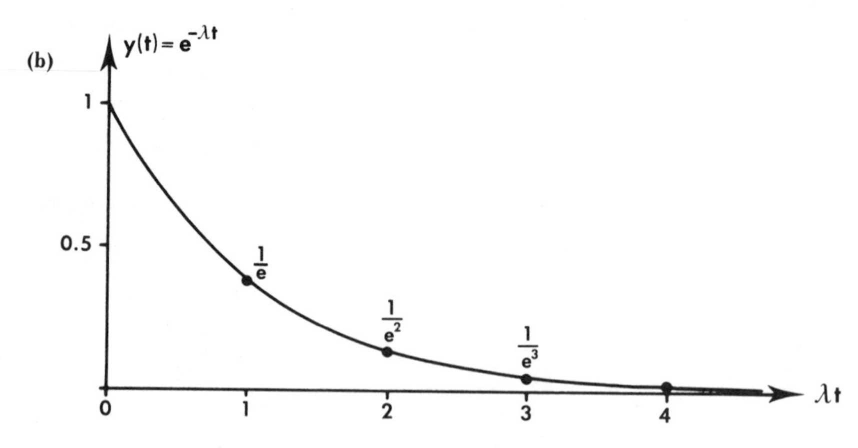

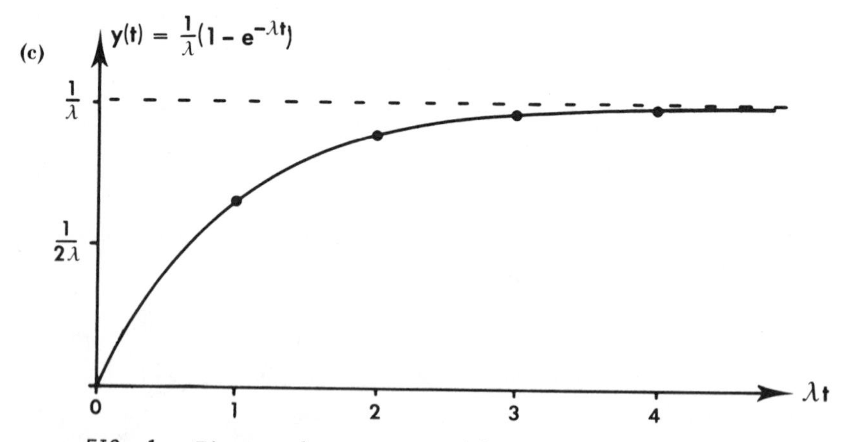

FIG. 1. First-order system: (a) pole location, (b) impulse response, and (c) step response.

Thus when 5 time constants have elapsed, the magnitude of the response has decreased to less than 1% of its original value.

If another standard input, the unit step, is applied to (2), the output may easily be found from (3) or (4) to be

$$y(t) = \frac{1}{\lambda} - \frac{1}{\lambda} e^{-\lambda t} \qquad (t > 0) \tag{6}$$

as shown in Fig. 1(c). In this case the response approaches a steady-state value of $1/\lambda$ as $t \to \infty$, and the transient component decays in proportion to $e^{-\lambda t}$.

It is common to define the *settling time* for such a system to be the time which elapses before the step response settles to within some tolerance (say 5%) of its steady-state value. Using the exponential property mentioned above, we observe that $1/e^3 \simeq 0.05$ and conclude that the settling time (to within 5%) is approximately 3 time constants.

ΔΔΔ

2. *EXAMPLE (Second-order system).* We shall now examine the second-order linear system,

$$\dot{\underline{x}}(t) = \begin{bmatrix} -\sigma & \omega \\ -\omega & -\sigma \end{bmatrix} \underline{x}(t) + \begin{bmatrix} 0 \\ 1 \end{bmatrix} u(t) \tag{7}$$

$$y(t) = \begin{bmatrix} 1 & 0 \end{bmatrix} \underline{x}(t)$$

with a pair of complex poles at $s = -\sigma \pm j\omega$, indicated by × in Fig. 2(a). Again, we assume stability ($\sigma > 0$). If the system is initially at rest [$\underline{x}(0) = \underline{0}$], its response is given by

$$y(t) = \int_0^t e^{-\sigma(t-\tau)} \sin[\omega(t - \tau)]\, u(\tau)\, d\tau \tag{8}$$

or in terms of Laplace transforms as

$$\hat{y}(s) = h(s)\hat{u}(s) = \frac{\omega}{(s + \sigma)^2 + \omega^2} \hat{u}(s) \tag{9}$$

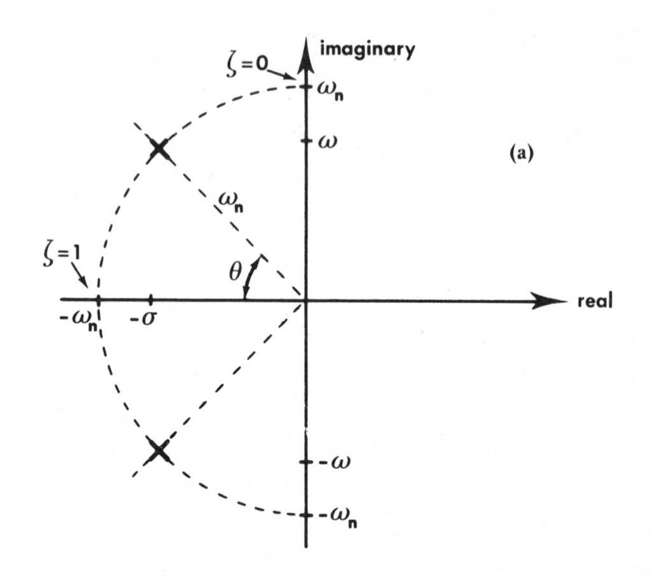

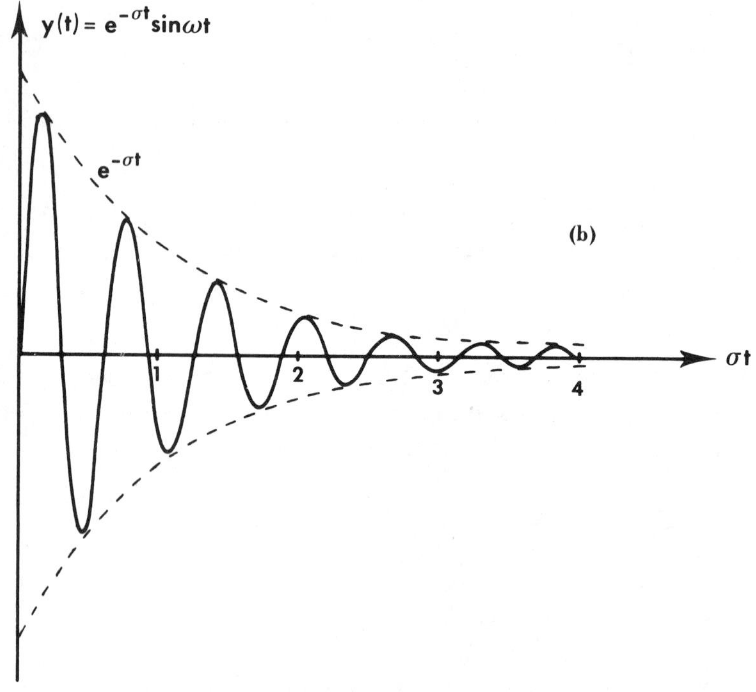

FIG. 2. Second-order system: (a) pole locations and (b) impulse response for $\omega = 10\sigma (\zeta \approx 0.1)$.

The impulse response of (7) is

$$y(t) = e^{-\sigma t} \sin \omega t \qquad (t > 0) \tag{10}$$

Although this response is oscillatory (its exact shape depends upon the ratio σ/ω), the magnitude of the sinusoidal oscillations decays in proportion to $e^{-\sigma t}$. As in the first-order case, we shall refer to $1/\sigma$ as the *time constant* of the system (or of the pair of complex poles). The response (10) is plotted in Fig. 2(b) for the case $\omega = 10\sigma$, with the envelopes $\pm e^{-\sigma t}$ shown as dotted lines.

It is a common practice in the study of control systems to rewrite the transfer function in (9) as

$$h(s) = \frac{\omega}{(s+\sigma)^2 + \omega^2} = \frac{\omega}{s^2 + 2\zeta\omega_n s + \omega_n^2} \tag{11}$$

Here the *natural frequency* ω_n and the *damping ratio* ζ are defined by

$$\omega_n^2 = \sigma^2 + \omega^2 \qquad \zeta = \frac{\sigma}{\omega_n} \tag{12}$$

or, conversely,

$$\omega = \omega_n\sqrt{1-\zeta^2} \qquad \sigma = \omega_n\zeta \tag{13}$$

Referring to Fig. 2(a), we see that ω_n and Θ are polar coordinates of the complex poles, and $\zeta = \cos\Theta$. The motivation for defining these parameters becomes clear if we fix ω_n and let ζ vary between 0 and 1, so that the pair of complex poles moves along the dashed semicircle. When $\zeta = 0$ (no damping), the poles lie on the imaginary axis and the impulse response (10) is an unattenuated oscillation ($\sin\omega_n t$) at the natural frequency ω_n. As ζ increases from 0 to 1, the response becomes less and less oscillatory until at $\zeta = 1$, the system has two identical real poles at $s = -\omega_n = -\sigma$. Since σ and ω are both real, $\zeta \leq 1$ for this example; the reader should verify that a system with $\zeta > 1$ would have two real poles.

In these three cases, the system (or the pair of poles) is said to be *underdamped* ($0 \leq \zeta < 1$), *critically damped* ($\zeta = 1$), or *overdamped* ($\zeta > 1$), respectively. The terms arise from the step response of the system, a standard test for servo or tracking systems (see Example 1.1.1 and Sec. 9.1) in which the output is supposed to follow (or track) the input. For $0 \leq \zeta < 1$, the response of this system to a unit step input can easily be shown to be[†]

$$y(t) = \frac{\omega}{\omega_n^2} \left[1 - \frac{\omega_n}{\omega} e^{-\sigma t} \sin(\omega t + \Theta) \right] \tag{14}$$

where $\Theta = \cos^{-1}\zeta$ was defined above. The reader is asked in Problem 1 to verify this and to derive step response formulas for a transfer function with $\zeta \geq 1$.

The step response is plotted in Fig. 3 for fixed ω_n and various values of ζ. For small ζ, the response rises sharply to a peak and

[†] In this example, the input and output differ by a factor of ω/ω_n^2 in steady state which will be defined in Corollary 6.1.2 as the *static gain* of the system. This has no effect on the shape of the transient response.

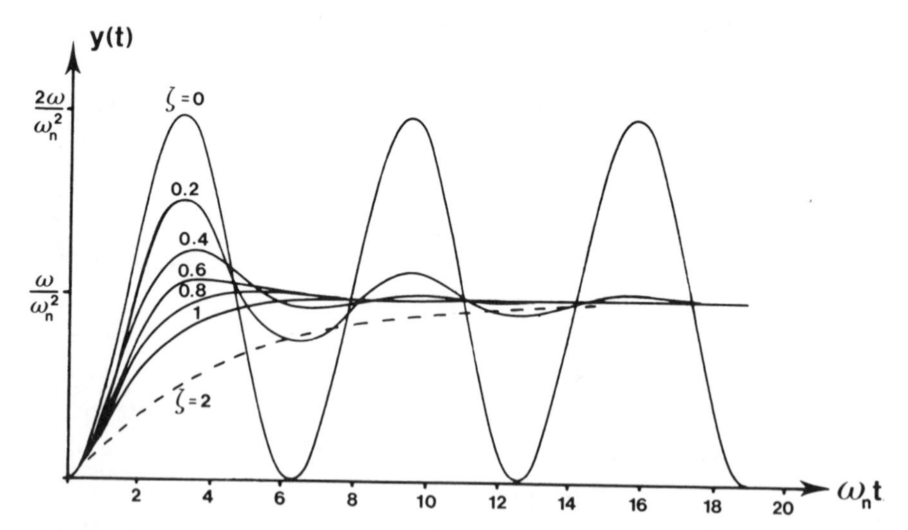

FIG. 3. Step response of second-order system for various ζ

then oscillates before settling down to its ultimate steady-state value. *Overshoot* is defined as the percentage by which the peak exceeds the steady-state value. As the damping ratio increases, there is less overshoot and the oscillations decrease until for $\zeta = 1$, the response becomes monotonic. When $\zeta > 1$, the response is even slower, as indicated by the dotted line in Fig. 3 [see Problem 1(b)]. However, this conclusion does not necessarily carry over to transfer functions with zeros as well as poles (see Problem 3).

In addition to overshoot, *settling time* is an important consideration; it was defined in Example 1 as the time which elapses before the response settles to within some tolerance (say 5%) of its final value. A third useful parameter of the step response is its *rise time*, often defined as the time which elapses while it rises from 10% to 90% of its final value. An alternative to rise time is *delay time*, which will be defined in Sec. 9.1 as the time required for the response to reach half its final value.

In Problem 2, the reader is asked to show that *for this example*, the amount of overshoot depends only upon the damping ratio ζ (the angle of the complex poles) while the settling time is determined solely by $\sigma = \zeta\omega_n$ (the real part of the complex poles). The rise time depends upon both ζ and ω_n, or equivalently, upon σ and ω.

ΔΔΔ

Rise time, overshoot, and settling time, the three parameters most commonly used to characterize step response, are illustrated in Fig. 4. Generally speaking, a large value of $\sigma = \zeta\omega_n$ in a second-order system with complex poles indicates a fast response, i.e., a small rise time and settling time. Overshoot and oscillations, on the other hand, are large for small values of ζ (see Problem 2).

The value $\zeta = 0.707$ (poles at 45°) is widely accepted as a sort of "optimal" damping ratio. This is because *for the particular system of Example 2*, it corresponds to about as fast a response as can be achieved (for fixed ω_n) without exceeding 5% overshoot. However, the reader should beware of the temptation to draw general

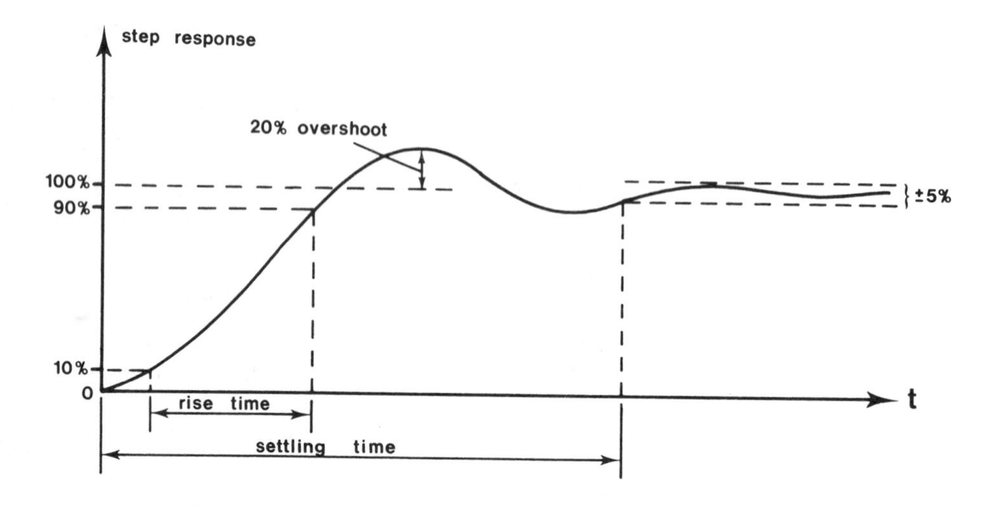

FIG. 4. Step response parameters.

conclusions about the relation between damping ratio and overshoot from this example since the introduction of a zero in the numerator of the transfer function can radically alter the situation (see Problem 3).

Since the response of an n-th-order linear system is made up of terms like (5) and (10), it is standard practice to apply the analysis of Examples 1 and 2 to the general case and draw various conclusions about system behavior from the pattern of poles in the complex plane. This is, at best, an approximation, complete with the pitfalls illustrated in Problem 3. Moreover, the relative weighting of different modes in the response is determined in part by the locations of zeros, which reduce the residues of any nearby poles. Nevertheless, the practice is a very useful one, provided we bear in mind its limitations.

Accordingly, we will refer to the time constant associated with each pole of the system and to the damping ratio of each pair of complex poles. If all the poles lie in the shaded half-plane shown in Fig. 5(a), then we conclude that all modes in the response must decay at least as fast as $e^{-\sigma t}$ and say that the system has *stability margin* σ. Similarly, if all the poles lie in the shaded

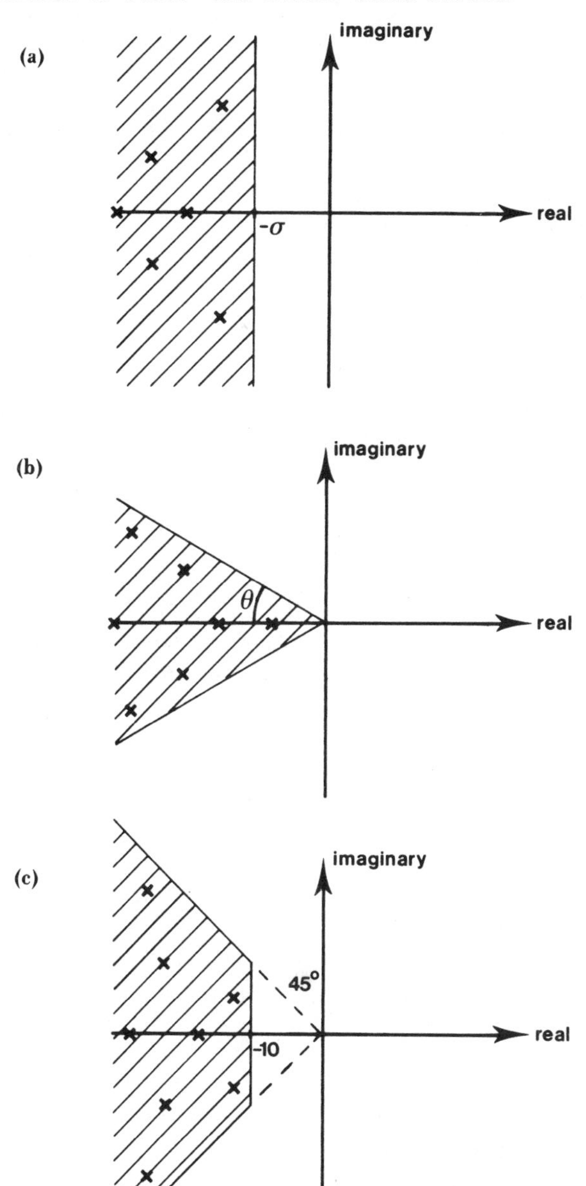

FIG. 5. Pole patterns of n-th-order systems: (a) system with stability margin σ, (b) system with damping margin $\zeta = \cos\Theta$, and (c) system with stability margin 10 and damping margin $0.7 = \cos 45°$.

sector of Fig. 5(b), the system will be said to have *damping margin* $\zeta = \cos \Theta$, and we may draw tentative conclusions about the tendency of the step response to overshoot. In practice, both of these criteria must be considered: Fig. 5(c) shows the pole pattern of a system with stability margin 10 and damping margin 0.7.

It is often true that one or more poles *dominate* the others in determining the response of a system. This situation occurs when some of the poles lie a considerable distance to the left of the dominant ones, as illustrated in Fig. 6. For a system with the pole pattern shown, the *dominant* pair of complex poles at $s = -1 \pm j1$ gives rise to a term of the form $e^{-t} \sin(t + \phi)$ in the response. The other poles all give rise to terms which decay in proportion to e^{-5t} or faster, and hence nearly vanish before 1 time constant of the dominant term elapses.

The notion of dominance must also be used with care since the occurrence of a zero near an apparently dominant pole can alter the situation considerably (see Problem 4). Nevertheless, it is a useful concept which often allows us to approximate a high-order system with one of first- or second-order.

5.4. LIAPUNOV STABILITY THEORY†

In Sec. 5.1 we examined stability from a *spectral* or *modal* point of view, summarized by Theorem 5.1.3: If all the poles of a system have negative real parts, then every mode contains a decaying exponential factor and all unforced state trajectories must eventually approach the origin.

In this section we shall adopt a somewhat different point of view which originated in the late 19th century when A. M. Liapunov borrowed some ideas from classical dynamics and generalized the concept of scalar *energy* of a system. This approach has proved to be

---

†This section assumes familiarity with Sec. A.6.

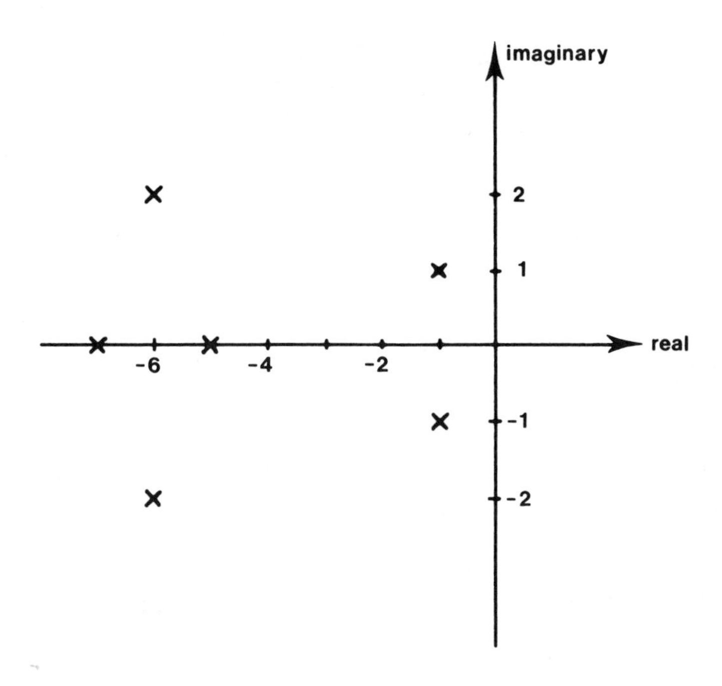

FIG. 6. System with two dominant poles.

extremely useful, particularly for the analysis of time-varying and nonlinear systems, and here it provides an interesting supplement to the results of Sec. 5.1.

Being limited to linear, time-invariant systems and quadratic Liapunov functions, this brief discussion will be somewhat transparent; the interested reader may wish to pursue the subject further in Refs. 32, 52, 77, 107, 165, and 181.

We consider the system

$$\dot{\underline{x}}(t) = \underline{\underline{A}}\underline{x}(t) \tag{1}$$

and observe that the positive scalar function of the state vector $\underline{x}$ defined by its norm,

$$f(\underline{x}(t)) = \|\underline{x}(t)\|^2 = \underline{x}'(t)\underline{x}(t) \tag{2}$$

can be viewed as a sort of "energy measure" for the system. If it can be established that this scalar function eventually approaches zero, then of course the system is stable (Definition 5.1.1).

Indeed, the same observations apply to any quadratic function†

$$v(\underline{x}(t)) = \underline{x}'(t)\underline{Q}\underline{x}(t) \tag{3}$$

for which $\underline{Q}$ is symmetric and positive definite since by (A.6-10) we have

$$\lambda_{min} \|\underline{x}\|^2 \le \underline{x}'\underline{Q}\underline{x} \le \lambda_{max}\|\underline{x}\|^2 \tag{4}$$

where λ_{min} and λ_{max} are the smallest and largest eigenvalues of $\underline{Q}$, respectively. Rather than simply requiring that v decay to zero, it will be useful to adopt a slightly stronger definition.

1. DEFINITION. The scalar function v defined by (3) is said to be a *(quadratic) Liapunov function* for the system (1) if $\underline{Q} > \underline{0}$ and there exists a constant $k > 0$ such that

$$\dot{v}(\underline{x}(t)) \le -k\|\underline{x}(t)\|^2 \qquad \text{(for all t)} \tag{5}$$

along all trajectories $\underline{x}$ of (1).

ΔΔΔ

In other words, a Liapunov function is a generalized "energy measure" for a system with the additional property (5) that it decreases monotonically with time. One suspects that this must imply that the function eventually approaches zero, and we shall see shortly that this is indeed the case. The requirement of monotonicity is somewhat restrictive, of course, and for some stable systems there will always be functions of the form (3) which oscillate as they decay (see Problem 1). Nevertheless, the definition is extremely useful, and the basic Liapunov stability result may be stated.

2. THEOREM. If a quadratic Liapunov function exists for the system (1), then (1) is (asymptotically) stable.

---

†A state transformation $\hat{\underline{x}} = \underline{Q}^{1/2}\underline{x}$ will turn (2) into (3).

Proof: Using (4), we may bound the right-hand side of (5) by

$$-k \, \|\underline{x}\|^2 \leq -\frac{k}{\lambda_{max}} v(\underline{x}) \tag{6}$$

and conclude that

$$\dot{v}(\underline{x}) \leq -\frac{k}{\lambda_{max}} v(\underline{x}) \tag{7}$$

If we were to replace the inequality sign with an equality, we would have the first-order linear differential equation

$$\dot{v}(\underline{x}) = -\frac{k}{\lambda_{max}} v(\underline{x}) \tag{8}$$

whose solution is a decaying exponential,

$$v(\underline{x}(t)) = e^{-(k/\lambda_{max})t} \, v(\underline{x}_0) \tag{9}$$

The effect of the inequality (since $v(\underline{x}) \geq 0$) is that v may in fact decay *faster* than the solution of (8), that is,

$$v(\underline{x}(t)) \leq e^{-(k/\lambda_{max})t} \, v(\underline{x}_0) \tag{10}$$

Using (4) once again, we have

$$\|\, \underline{x}(t) \,\|^2 \leq \frac{v(\underline{x}(t))}{\lambda_{min}}$$

$$\leq \frac{e^{-(k/\lambda_{max})t} \, v(\underline{x}_0)}{\lambda_{min}} \xrightarrow[t \to \infty]{} 0 \tag{11}$$

which implies that (1) is stable.

ΔΔΔ

It will be useful at this point to make use of (1) and expand the derivative $\dot{v}(\underline{x})$ into

$$\begin{aligned} \dot{v}(\underline{x}) &= \frac{d}{dt} (\underline{x}'\underline{Q}\underline{x}) = \dot{\underline{x}}'\underline{Q}\underline{x} + \underline{x}'\underline{Q}\dot{\underline{x}} \\ &= \underline{x}'\underline{A}'\underline{Q}\underline{x} + \underline{x}'\underline{Q}\underline{A}\underline{x} = \underline{x}'(\underline{A}'\underline{Q} + \underline{Q}\underline{A})\underline{x} \end{aligned} \tag{12}$$

If we define

$$\underline{A}'\underline{Q} + \underline{Q}\underline{A} = -\underline{M} \tag{13}$$

then we have

$$\dot{v}(\underline{x}) = -\underline{x}'\underline{M}\underline{x} \tag{14}$$

and the condition in (5) becomes

$$-\underline{x}'(t)\underline{M}\underline{x}(t) \leq -k\underline{x}'(t)\underline{x}(t) \qquad \text{(for all t)} \tag{15}$$

In particular, (15) must hold for $t = 0$ and for any initial state $\underline{x}_0$, which is equivalent to requiring that $\underline{M}$ be positive definite ($\underline{M} \geq k\underline{I} > \underline{0}$).

Thus we have a straightforward test to determine whether any particular function is a Liapunov function: Choose any $\underline{Q} > \underline{0}$, compute $\underline{M}$ via (13), and check whether $\underline{M} > \underline{0}$. If so, then $v(\underline{x}) = \underline{x}'\underline{Q}\underline{x}$ is a Liapunov function for (1), implying stability. If not, we must choose another $\underline{Q}$ and try again; we cannot conclude instability just because one particular v is not a Liapunov function. This illustrates one potential, practical flaw in using the method above to check stability: it may involve a good deal of trial and error.

A better approach is to choose some $\underline{M} > \underline{0}$ first, thus specifying the derivative (14), and then to solve the linear equation (13) for $\underline{Q}$ and test whether $\underline{Q} > \underline{0}$, implying stability. The following theorem assures us that this procedure need only be carried out once: if $\underline{Q} \not> \underline{0}$, then the system cannot be stable.

3. THEOREM. Suppose $\underline{M} > \underline{0}$. Then $\underline{A}$ is stable if and only if (13) has a unique solution $\underline{Q} > \underline{0}$.

Proof: The if part is implied by Theorem 2. Next we shall establish that for *any* $\underline{M}$, the matrix

$$\underline{Q} = \int_0^\infty e^{\underline{A}'t}\underline{M}e^{\underline{A}t}\, dt \tag{16}$$

is a solution of (13). The integral must be convergent for stable $\underline{A}$ since the integrand consists of terms proportional to $t^k e^{\lambda_i t}$ with

$\operatorname{Re}\lambda_i < 0$. Substituting (16) into the left-hand side of (13), we have

$$\underline{A}' \int_0^\infty e^{\underline{A}'t}\underline{M}e^{\underline{A}t}\,dt + \int_0^\infty e^{\underline{A}'t}\underline{M}e^{\underline{A}t}\,dt\,\underline{A} = \int_0^\infty (\underline{A}'e^{\underline{A}'t}\underline{M}e^{\underline{A}t} + e^{\underline{A}'t}\underline{M}e^{\underline{A}t}\underline{A})\,dt$$
$$= \left[e^{\underline{A}'t}\underline{M}e^{\underline{A}t} \right]_0^\infty = -\underline{M} \tag{17}$$

as required.

We may look upon (13) as defining a linear mapping $\underline{L}(\underline{Q}) = \underline{A}'\underline{Q} + \underline{Q}\underline{A}$ from R^{n^2} into R^{n^2}. Since we have just shown that $\underline{L}(\underline{Q}) = \underline{M}$ has a solution for all $\underline{M}$, the rank of the mapping is n^2 (i.e., it is invertible), and hence the solution must be unique.

Finally, to verify that $\underline{Q} > \underline{0}$, we compute

$$\begin{aligned} \underline{x}_0'\underline{Q}\underline{x}_0 &= \int_0^\infty \underline{x}_0' e^{\underline{A}'t}\underline{M}e^{\underline{A}t}\underline{x}_0\,dt \\ &= \int_0^\infty \underline{x}'(t)\underline{M}\underline{x}(t)\,dt \end{aligned} \tag{18}$$

where $\underline{x}(t)$ can be zero only if $\underline{x}_0 = \underline{0}$ because $e^{\underline{A}t}$ is nonsingular. Since $\underline{M} > \underline{0}$, this means that the integrand is positive, and $\underline{x}_0'\underline{Q}\underline{x}_0 > 0$ unless $\underline{x}_0 = \underline{0}$.

ΔΔΔ

4. *EXAMPLE*. Suppose we are seeking a Liapunov function for the system

$$\dot{\underline{x}}(t) = \begin{bmatrix} 1 & -3 \\ 1 & -2 \end{bmatrix} \underline{x}(t) \tag{19}$$

in order to determine whether it is stable. Choosing $\underline{M} = \underline{I}$, we must then solve

$$\begin{bmatrix} 1 & 1 \\ -3 & -2 \end{bmatrix}\begin{bmatrix} q_{11} & q_{12} \\ q_{12} & q_{22} \end{bmatrix} + \begin{bmatrix} q_{11} & q_{12} \\ q_{12} & q_{22} \end{bmatrix}\begin{bmatrix} 1 & -3 \\ 1 & -2 \end{bmatrix} = \begin{bmatrix} -1 & 0 \\ 0 & -1 \end{bmatrix} \tag{20}$$

for $\underline{Q}$ (note that $\underline{Q}$ must be symmetric because $\underline{M}$ is). Expanding, this becomes

$$\begin{aligned} -2q_{11} \quad -2q_{12} \quad\quad\quad &= -1 \\ -3q_{11} \quad - q_{12} + q_{22} &= 0 \\ -6q_{12} \quad -4q_{22} &= -1 \end{aligned} \tag{21}$$

which may readily be solved for

$$\underline{Q} = \begin{bmatrix} q_{11} & q_{12} \\ q_{12} & q_{22} \end{bmatrix} = \begin{bmatrix} 3 & -3.5 \\ -3.5 & 5.5 \end{bmatrix} \tag{22}$$

We observe that $\underline{Q} > \underline{0}$, and conclude that the system must be stable. As a check, note that the poles of this system are $-1/2 \pm j\sqrt{3/2}$.

ΔΔΔ

The Liapunov function approach is sometimes useful for establishing stability (or instability) in general classes of time-invariant systems, but for particular examples it is usually easier to check the pole locations, as in Secs. 5.1 and 5.2. If the system is time varying or nonlinear, however, the latter approach breaks down, and Liapunov methods have found many important applications in such cases.

5.5. SUMMARY

In this chapter we have looked into the stability of linear, time-invariant systems. We began with (equivalent) definitions of asymptotic stability for systems described in state-variable and input-output format, and then we observed that the whole question of stability amounts to asking whether all of a system's poles have negative real parts. In Sec. 5.2 the Hurwitz and Routh tests were described, allowing us to check the stability of a system without actually determining its poles.

In Sec. 5.3 we saw how the behavior of a system could be qualitatively analyzed without actually computing a solution, i.e., by examining the time constants and damping ratios of its set of poles. Step response characteristics such as rise time, overshoot, and settling time were defined, and their relationship to the pole locations was discussed.

In the last section we considered an alternative approach to stability analysis, using the Liapunov function.

PROBLEMS

Section 5.1

1. Suppose the system

$$\dot{\underline{x}}(t) = \underline{A}\underline{x}(t) + \underline{B}\underline{u}(t) \qquad \underline{x}(0) = \underline{x}_0$$

$$\underline{y}(t) = \underline{C}\underline{x}(t)$$

is asymptotically stable. Prove that if

$$\underline{u}(t) = \underline{u}_c = \text{constant}$$

then

$$\underline{x}(t) \xrightarrow[t\to\infty]{} \underline{x}_c = \text{constant}$$

2. Use Theorem 2.2.1 and Lemma 4.3.6 to verify that Definitions 5.1.2 and 5.1.1 are consistent.

3. Complete the proof of Theorem 5.1.3(b) for the case of distinct poles. To prove necessity ("only if"), you will need to establish that no residue can be zero for all initial conditions (Problem B.2-2 may be helpful in this regard).

4. (a) Prove Theorem 5.1.6 for a scalar transfer function with distinct roots. *Hint:* Expand the transfer function in partial fractions using Theorem B.2.2 and Corollary B.2.3, obtain the impulse response, and then use the

convolution integral (4.1-48). As in Problem 5.1-3, you will need to verify that the residues are nonzero.

(b) Indicate how your proof can be extended to multivariable systems and to nondistinct poles.

5. Consider the system

$$\dot{\underline{x}} = \begin{bmatrix} 0 & 1 & 0 \\ 0 & 0 & 1 \\ 250 & 0 & -5 \end{bmatrix} \underline{x} + \begin{bmatrix} 0 \\ 0 \\ 50 \end{bmatrix} u$$

$$\underline{y} = \begin{bmatrix} -5 & 1 & 0 \end{bmatrix} \underline{x}$$

(a) Determine the transfer function $\hat{h}(s)$ relating the input and output transforms $\hat{u}(s)$ and $\hat{y}(s)$.

(b) Show that the input u(t) and output y(t) are related by the following differential equation:

$$\ddot{y}(t) + 10\dot{y}(t) + 50y(t) = 50u(t)$$

Is this consistent with your answer to (a)? Why or why not?

(c) Determine the output y(t) when

$$y(0) = C_1 \qquad \dot{y}(0) = C_2 \qquad u(t) = \text{unit step}$$

(d) Is the system BIBO stable? If so, explain how you know. If not, find a bounded input u(t) which leads to an unbounded output y(t).

(e) Determine the state trajectory $\underline{x}(t)$ when

$$\underline{x}(0) = \begin{bmatrix} 1 \\ -5 \\ 0 \end{bmatrix} \qquad u(t) = 0$$

(f) Is the system asymptotically stable? If so, explain how you know. If not, find an initial state $\underline{x}_0$ which results in an unbounded state trajectory $\underline{x}(t)$.

(g) Can you find an initial state $\underline{x}_0$ which results in an unbounded output y(t)? Explain.

6. Consider a scalar system described by

$$\ddot{y}(t) + 9y(t) = \dot{u}(t) + 3u(t)$$

Is this system BIBO stable? Determine the output y(t) when

$$y(0) = \dot{y}(0) = 0 \qquad u(t) = \sin 3t$$

7. (a) Prove the following theorem:
The system (5.1-1) is marginally stable if and only if the eigenvalues $\lambda_1, \lambda_2, \ldots, \lambda_n$ of $\underline{A}$ satisfy the following conditions:
(1) $\text{Re}\,\lambda_i \leq 0$, $i = 1,2,\ldots,n$.
(2) If $\text{Re}\,\lambda_i = 0$, and λ_i is a repeated eigenvalue, then the block containing λ_i in the Jordan canonical form of $\underline{A}$ has no 1's above the main diagonal.

(b) State conditions on the eigenvalues of $\underline{A}$ which are necessary and sufficient for the system (5.1-1) to be unstable.

8. Consider the scalar system

$$y^{(n)}(t) = u(t)$$

(a) What are the roots of the characteristic polynomial? Is the system marginally stable?
(b) Can you draw any general conclusions about the marginal stability of systems described in input-output format which have repeated roots on the $j\omega$ axis?[†]

9. Prove that a system which is only marginally stable cannot be BIBO stable.

Section 5.2

1. Use the Hurwitz, Liénard-Chipart, and Routh tests to determine whether the following polynomials are stable, and if not, how many roots lie in the right half-plane.
(a) $s^3 + 2s^2 + s + 3$
(b) $s^4 + 2s^3 + 2s^2 + 4s + 2$

[†]This question will be taken up again in Problem 10.4-20.

(c) $s^7 + 3s^6 + 11s^5 + 19s^4 + 36s^3 + 38s^2 + 36s + 24$

(d) $s^6 + 2s^5 + 4s^4 + 8s^3 + 6s^2 + 8s + 4$

2. A system has the characteristic polynomial

$$s^5 + 8s^4 + 24s^3 + 32s^2 + as + ab = 0$$

where a and b are unspecified parameters. Use the Routh criterion to determine what constraints must be placed on a and b in order to ensure that the system is stable.

Section 5.3

1. (a) Verify that the unit step response in Example 5.3.2 is given by (5.3-14) for $0 \leq \zeta < 1$.
   (b) Determine the step response formulas for $\zeta = 1$ and $\zeta > 1$. What happens as $\zeta \to \infty$?

2. (a) For the step response of Example 5.3.2 (where $0 \leq \zeta \leq 1$), determine the rise time and overshoot as functions of ζ and ω_n, and find an upper bound on the settling time (to within 5% of steady state) as a function of σ.
   (b) Determine these three parameters (approximately) for the same transfer function with $\zeta > 1$.
   (c) Do you expect your formulas in (a) and (b) to hold if the numerator of the transfer function is changed? Explain.

3. Consider two second-order systems with transfer functions

$$h_1(s) = \frac{1}{s^2 + 1.42s + 1}$$

$$h_2(s) = \frac{3.34s + 1}{s^2 + 1.42s + 1}$$

   (a) What are the damping ratios of these transfer functions?
   (b) Sketch the response of each system to a unit step input, and determine (approximately) the respective rise times, overshoots, and settling times. Is your answer consistent with the solution to Problem 5.3-2?

(c) Repeat (a) and (b) for the transfer functions

$$h_3(s) = \frac{0.1}{s^2 + 1.1s + 0.1}$$

$$h_4(s) = \frac{1.9s + 0.1}{s^2 + 1.1s + 0.1}$$

4. Consider the transfer functions

$$h_5(s) = \frac{30}{(s + 1)(s + 5)(s + 6)}$$

$$h_6(s) = \frac{27s + 30}{(s + 1)(s + 5)(s + 6)}$$

Which poles appear to be dominant? Check your answer by carrying out a partial fractions expansion and sketching the step response in each case.

Section 5.4

1. (a) Does the choice $\underline{Q} = \underline{I}$ provide a Liapunov function for the system

$$\dot{\underline{x}}(t) = \begin{bmatrix} -1 & 1 \\ -4 & -1 \end{bmatrix} \underline{x}(t)$$

(b) Determine the poles of the system and explain why this choice of $\underline{Q}$ fails.

(c) Find a Liapunov function for the system.

Chapter 6

STEADY-STATE AND FREQUENCY RESPONSE

6.0 INTRODUCTION

In the previous chapter we were concerned with the short-term behavior of linear systems; we examined some tests for stability and saw how the transient response could be analyzed qualitatively from a knowledge of the system's poles. In this chapter, we shall consider the long-term behavior of stable systems and see what response to certain inputs remains after the transient has died away.

The reader will not be surprised to learn that the response to a constant (i.e., step) input approaches a constant, and that a sinusoidal input eventually evokes an output with the same shape. However, we shall see that this relatively simple property enables us to extract a wealth of information about a system simply by applying sinusoidal inputs and observing the outputs. In principle, the entire transfer function can be determined in this manner.

In later chapters, this approach will prove to be very useful for the analysis and design of single-loop feedback control systems.

6.1 STEADY-STATE RESPONSE

Throughout this chapter (except for Theorem 5) we shall restrict our attention to a stable scalar system described either by

$$\dot{\underline{x}} = \underline{A}\underline{x} + \underline{b}u \qquad y = \underline{c}'\underline{x} \tag{1a}$$

or, equivalently, by

$$y^{(n)} + p_{n-1}y^{(n-1)} + \dots + p_0 y = q_{n-1}u^{(n-1)} + \dots + q_0 u \tag{1b}$$

Recall from Chapter 4 that the transfer function of this system is

$$h(s) = \underline{c}'(s\underline{I} - \underline{A})^{-1}\underline{b} = \frac{q(s)}{p(s)} = \frac{q_{n-1}s^{n-1} + \dots + q_0}{s^n + p_{n-1}s^{n-1} + \dots + p_0} \tag{2}$$

and that the output consists of a natural response (determined by initial conditions) plus a forced response (determined by the input),

$$\hat{y}(s) = \underline{c}'(s\underline{I} - \underline{A})^{-1}\underline{x}_0 + \underline{c}'(s\underline{I} - \underline{A})^{-1}\underline{b}\hat{u}(s) = \frac{r(s)}{p(s)} + h(s)\hat{u}(s) \tag{3}$$

Recall also from Sec. 5.1 that our assumption of stability means that all the system poles [roots of $p(s)$] have negative real parts.

In Examples 5.3.1 and 5.3.2 we observed that a constant (step) input resulted in a transient output component plus a constant, and we referred to the latter as the steady-state response. More generally, suppose the input is any function with a rational Laplace transform,

$$\hat{u}(s) = \frac{n(s)}{d(s)} \tag{4}$$

then the output transform (3) becomes

$$\hat{y}(s) = \frac{r(s)}{p(s)} + \frac{q(s)}{p(s)}\frac{n(s)}{d(s)} \tag{5}$$

$$= \hat{y}_1(s) + \hat{y}_2(s) + \hat{y}_3(s) \tag{6}$$

Here $\hat{y}_1(s)$ denotes the partial fractions expansion[†] of the natural response $r(s)/p(s)$, and consists of various terms having factors of $p(s)$ in their denominators. The forced response $q(s)n(s)/p(s)d(s)$ may also be expanded in partial fractions, the terms involving factors of $p(s)$ denoted collectively by $\hat{y}_2(s)$ and those involving factors of $d(s)$ denoted collectively by $\hat{y}_3(s)$.

Taking the inverse transform,

$$y(t) = y_1(t) + y_2(t) + y_3(t) \qquad (t > 0) \tag{7}$$

where $y_1(t)$ and $y_2(t)$ consist entirely of terms corresponding to the system poles. Since we have assumed that these poles are all stable, y_1 and y_2 must decay to zero, leaving

$$y(t) \xrightarrow[t\to\infty]{} y_3(t) \tag{8}$$

To summarize the above discussion, we have the following theorem.

1. THEOREM. If an input having a transform of the form (4) is applied to the stable system (1), the output is given by (7), where $y_1(t) + y_2(t)$ is known as the *transient response* and decays to zero, leaving the *steady-state response* $y_3(t)$.

ΔΔΔ

To be more specific, consider the case of a constant input.

2. COROLLARY. If a step function input

$$u(t) = \alpha 1(t) \tag{9}$$

is applied to the stable system (1), the steady-state response is

$$\lim_{t\to\infty} y(t) = y_3(t) = h(0)\alpha = \text{constant} \tag{10}$$

Thus in the limit the output differs from the input only by a constant factor $h(0) = -\underline{c}'\underline{A}^{-1}\underline{b}$, which will be known as the *static*

[†]See Sec. B.2.

gain† (or *d.c. gain*, or often just the *gain*) of the system.

Proof: The Laplace transform of the input (9) is

$$\hat{u}(s) = \frac{\alpha}{s} \tag{11}$$

and using (3), the output transform is

$$\hat{y}(s) = \frac{r(s)}{p(s)} + h(s)\frac{\alpha}{s} \tag{12}$$

When this expression is expanded in partial fractions, the residue associated with the term 1/s is found by multiplying (12) by s and then evaluating it for $s = 0$. This yields the steady-state response

$$\hat{y}_3(s) = \frac{\alpha h(0)}{s} \tag{13}$$

whose inverse transform is (10).

ΔΔΔ

3. *EXAMPLE.* Consider the system

$$\ddot{y} + 3\dot{y} + 2y = 6u \tag{14}$$

where $y(0) = 1$, $\dot{y}(0) = 0$, and the input is a unit step $[\hat{u}(s) = 1/s]$. The transfer function in this case is

$$h(s) = \frac{6}{s^2 + 3s + 2} \tag{15}$$

with a static gain of $h(0) = 3$, so according to Corollary 2, the steady-state response will have the constant value 3.

To verify this, we evaluate the output transform using (4.1-4),

$$\begin{aligned}\hat{y}(s) &= \frac{r(s)}{p(s)} + \frac{q(s)}{p(s)} \cdot \frac{n(s)}{d(s)} \\ &= \frac{s+3}{s^2+3s+2} + \frac{6}{s^2+3s+2} \cdot \frac{1}{s}\end{aligned} \tag{16}$$

---

†Note that $h(0) = \infty$ if h(s) has any poles at the origin, and $\underline{c}A^{-1}\underline{b} = \infty$ if the system has any poles at the origin. We have avoided this difficulty by assuming a stable system. In Sec. 8.1 a more general definition of gain will be adopted which avoids this singularity.

Expanding in partial fractions, we have

$$\begin{aligned}\hat{y}(s) &= \left(\frac{2}{s+1} - \frac{1}{s+2}\right) + \left(\frac{3}{s+2} - \frac{6}{s+1}\right) + \frac{3}{s} \\ &= \qquad \hat{y}_1(s) \qquad + \qquad \hat{y}_2(s) \qquad + \hat{y}_3(s)\end{aligned} \tag{17}$$

or

$$\begin{aligned}y(t) &= (2e^{-t} - e^{-2t}) + (3e^{-2t} - 6e^{-t}) + 3 \\ &= \qquad y_1(t) \qquad + \qquad y_2(t) \qquad + y_3(t) \xrightarrow[t\to\infty]{} 3\end{aligned} \tag{18}$$

As predicted, the transient $y_1 + y_2$ decays to zero, leaving the steady-state term y_3.

ΔΔΔ

The existence of a steady-state response is not restricted to constant inputs, of course; Theorem 1 applies to any input with a rational transform. The case of a *ramp* input (proportional to t) is considered in Problem 2. Oscillatory inputs are the subject of the next corollary, which establishes a very important property of stable, linear, time-invariant systems: A sinusoidal input induces a sinusoidal steady-state output, unchanged except for its amplitude and phase.

4. COROLLARY. If a sinusoidal input with (angular) frequency ω,

$$u(t) = \alpha 1(t) \sin \omega t \tag{19}$$

is applied to the stable system (1), the steady-state response is a sinusoid of the same frequency,

$$\lim_{t\to\infty} y(t) = y_3(t) = M\alpha \sin(\omega t + \phi) \tag{20}$$

where

$$M(\omega) \triangleq |h(j\omega)| \qquad \phi(\omega) \triangleq \angle h(j\omega) \tag{21}$$

The *gain* (or *amplitude ratio*, or *magnification*) $M(\omega)$ and *phase shift* $\phi(\omega)$ at the frequency ω are given by the magnitude and angle of the complex number[†] $h(j\omega) = M(\omega)e^{j\phi(\omega)}$, which is obtained by evaluating the transfer function $h(s)$ at $s = j\omega$.

Proof: The Laplace transform of the input (19) is

$$\hat{u}(s) = \frac{\alpha\omega}{s^2 + \omega^2} \tag{22}$$

and using (3), the output transform is

$$\hat{y}(s) = \frac{r(s)}{p(s)} + h(s)\frac{\alpha\omega}{s^2 + \omega^2} \tag{23}$$

When this expression is expanded in partial fractions, the residues C and S in the steady-state response term

$$\hat{y}_3(s) = \frac{Cs + S\omega}{s^2 + \omega^2} \tag{24}$$

are found via

$$S + jC = \frac{1}{\omega}(s^2 + \omega^2)\left[\frac{r(s)}{p(s)} + h(s)\frac{\alpha\omega}{s^2 + \omega^2}\right]_{s = j\omega} \tag{25}$$

$$= h(j\omega)\alpha = M\alpha e^{j\phi}$$

and the inverse transform of (24) is seen to be (20).

ΔΔΔ

Corollary 4 is one of the foundations of linear systems theory. It says that the steady-state response to a sinusoidal input of frequency ω is easily determined by substituting $s = j\omega$ in the system transfer function. The output approaches a sinusoid of the same frequency with its amplitude scaled by $M(\omega) = |h(j\omega)|$ and its phase shifted by $\phi(\omega) = \angle h(j\omega)$, as shown in Fig.1. Thus $h(j\omega)$, regarded as a complex-valued function of ω, completely specifies the steady-state response of the system to sinusoids of all frequencies. The function

---

[†] If $\phi > 0$, the output is said to have a *phase lead* relative to the input; if $\phi < 0$, it is said to have a *phase lag*. Note that our assumption of a stable system ensures $h(j\omega) \neq \infty$.

(a)

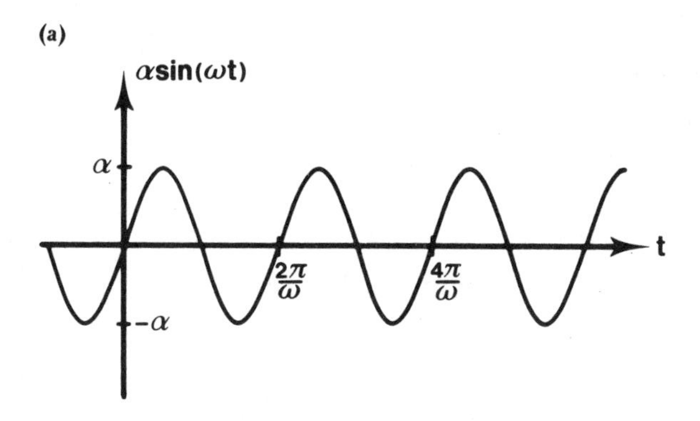

(b)

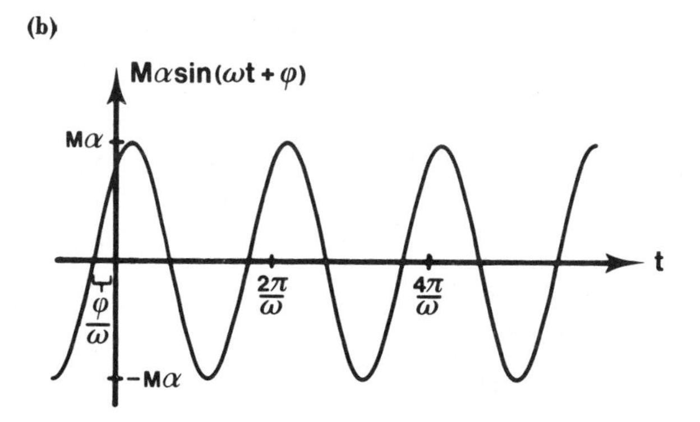

FIG. 1. Sinusoidal steady-state: (a) input and (b) output.

$h(j\omega)$ is known as the *frequency response* of the system, and in the next section we shall see how it is commonly displayed in graphical form.

It is worthwhile noting that the frequency response is in fact another type of transform. Recall (Definition B.1.1) that the Laplace transform of a time function $f(t)$ is defined by

$$\hat{f}(s) = \int_0^\infty f(t)e^{-st}\,dt \tag{26}$$

where $s = \sigma + j\omega$ and the integral converges for all $\sigma > \sigma_c$. In particular, if $\sigma_c < 0$, then we may choose $\sigma = 0$ and (26) becomes the standard definition of a *Fourier transform* or *Fourier integral*,[†]

$$\hat{f}(j\omega) = \int_0^\infty f(t)e^{-j\omega t}\,dt \tag{27}$$

which may be viewed as an extension to aperiodic functions of the concept of Fourier series. Except for the restriction that σ_c be negative [which simply means that f(t) must die away exponentially as $t \to \infty$], the Fourier transform is just as general as the Laplace transform and may be used and manipulated in the same way.

Recalling from (4.1-47) that the transfer function h(s) is the Laplace transform of the system's impulse response, we see that the frequency response $h(j\omega)$ is its Fourier transform. The condition $\sigma_c < 0$ requires that the system be stable, but this can be circumvented by using the Analytic Continuation theorem (see Example B.1.2), and the frequency response of an unstable system can be a useful, if slightly fictitious, concept.

The results of this section have so far been confined to scalar systems, but extension to the stable system

$$\begin{aligned} \dot{\underline{x}} &= \underline{A}\underline{x} + \underline{B}\underline{u} \\ \underline{y} &= \underline{C}\underline{x} \end{aligned} \tag{28}$$

with $p \times m$ transfer function matrix

$$\underline{H}(s) = \underline{C}(s\underline{I} - \underline{A})^{-1}\underline{B} \tag{29}$$

is mainly a matter of bookkeeping. The multivariable results are summarized in the following theorem.

5. *THEOREM.*

(a) If an input $\underline{u}(t)$ with a rational transform is applied to the stable system (28), the output $\underline{y}(t)$ consists of a

[†]This would normally be denoted $\hat{f}(\omega)$ with ω interpreted as angular frequency. Additional material on Fourier transforms may be found in Refs. 31, 96, and 141.

transient response which decays to zero and a *steady-state response* whose form depends upon the input.

(b) If the input is a step function,

$$\underline{u}(t) = \underline{\alpha}1(t) \tag{30}$$

then the steady-state response is

$$\lim_{t\to\infty} \underline{y}(t) = \underline{H}(0)\underline{\alpha} = \text{constant} \tag{31}$$

where $\underline{H}(0) = -\underline{C}\underline{A}^{-1}\underline{B}$ is called the *static gain matrix*.†

(c) If the input is sinusoidal with (angular) frequency ω,

$$\underline{u}(t) = \underline{\alpha}1(t) \sin \omega t \tag{32}$$

then the steady state response is

$$\lim_{t\to\infty} y_i(t) = \sum_{k=1}^{m} M_{ik}\, \alpha_k \sin(\omega t + \phi_{ik}) \tag{33}$$

where

$$M_{ik}(\omega) \triangleq |H_{ik}(j\omega)| \qquad \phi_{ik}(\omega) \triangleq \angle H_{ik}(j\omega) \tag{34}$$

for $i = 1,2,\ldots,p$ and $k = 1,2,\ldots,m$. The *gains* $M_{ik}(\omega)$ and *phase shifts* $\phi_{ik}(\omega)$ at the frequency ω are given by the magnitudes and phases of the complex *frequency response matrix* $\underline{H}(j\omega)$, which is obtained by evaluating the transfer function matrix $\underline{H}(s)$ at $s = j\omega$.

Proof: Left to the reader (Problem 5).

ΔΔΔ

6.2 FREQUENCY RESPONSE

In the previous section we saw that a sinusoidal signal of frequency ω passes through a linear system unaltered in the steady state except for its amplitude and phase. The changes in these two parameters are

---

†This is also called a *d.c. gain matrix*. As in Corollary 2, we have avoided singularities by assuming a stable system.

specified by the magnitude and angle of the system's *frequency response* $h(j\omega)$, $0 \leq \omega < \infty$, which may be obtained by evaluating the transfer function $h(s)$ at $s = j\omega$.

The two most common graphical forms for displaying $h(j\omega) = M(\omega)e^{j\phi(\omega)}$ are the *Bode* (or *logarithmic*) *diagram*, in which $\log_{10} M(\omega)$ and $\phi(\omega)$ are plotted separately against $\log_{10} \omega$, and the *Nyquist* (or *polar*) *diagram*, which plots the locus of $h(j\omega)$ in the complex plane as ω varies from $-\infty$ to ∞. We shall illustrate these with some examples.

1. EXAMPLE. Consider a first-order system described by

$$\dot{y}(t) = -\lambda y(t) + \lambda u(t) \tag{1}$$

which has a single real pole at $s = -\lambda$, assumed stable ($\lambda > 0$), and a transfer function

$$h(s) = \frac{\lambda}{s + \lambda} \tag{2}$$

The transient response of this system (without the extra factor of λ multiplying u) was examined in Example 5.3.1. Evaluating $h(s)$ at $s = j\omega$, the frequency response is

$$h(j\omega) = \frac{\lambda}{j\omega + \lambda} = \frac{\lambda^2 - j\lambda\omega}{\lambda^2 + \omega^2} = M(\omega)e^{j\phi(\omega)} \tag{3}$$

where

$$M(\omega) = |h(j\omega)| = \frac{\lambda}{(\lambda^2 + \omega^2)^{1/2}} \tag{4}$$

$$\phi(\omega) = \angle h(j\omega) = \tan^{-1}\frac{-\lambda\omega}{\lambda^2} = -\tan^{-1}\frac{\omega}{\lambda} \tag{5}$$

The Nyquist diagram in this case is easy to construct and is shown in Fig. 1(a). For $\omega = 0$, we have $h(j\omega) = 1$, and as ω increases, $h(j\omega)$ traces out a semicircle in the fourth quadrant, approaching the origin from below as $\omega \to \infty$. It is conventional in such diagrams to also plot $h(j\omega)$ for $\omega < 0$ even though this is just the complex

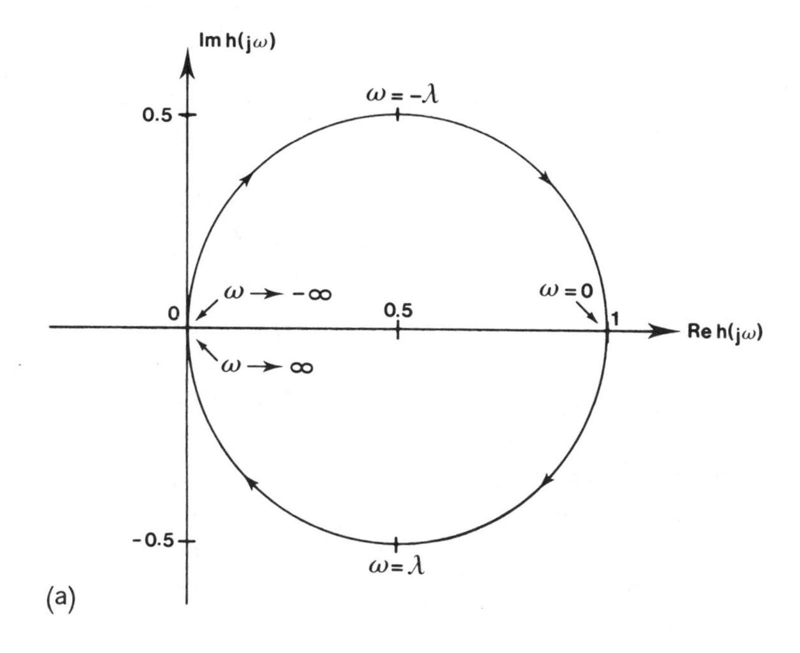

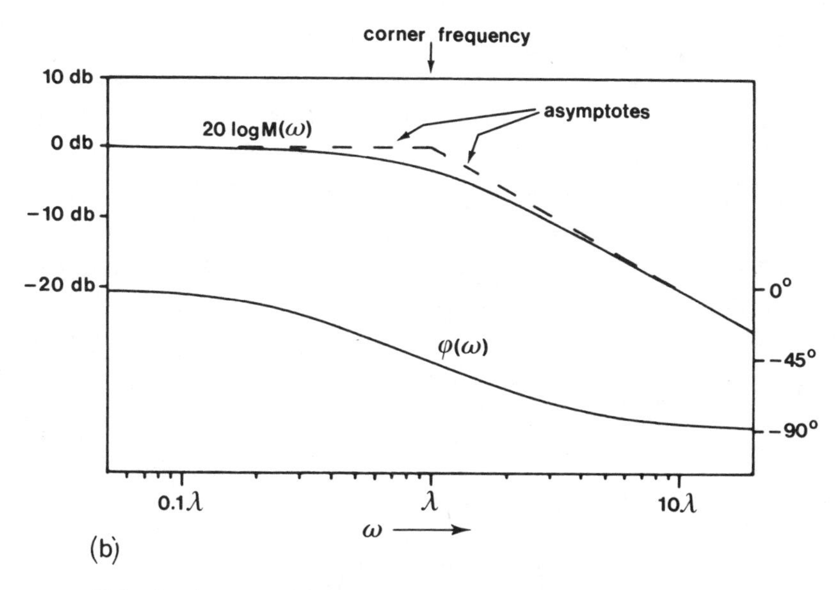

FIG. 1. Frequency response of first-order system: (a) Nyquist diagram and (b) Bode diagram.

conjugate of the plot for $\omega > 0$, and hence redundant (see Problem 2). (Arrowheads on the curve indicate the direction of increasing ω.)

The Bode diagram shown in Fig. 1(b) is an alternative representation of the same information, with the gain $M(\omega)$ and phase $\phi(\omega)$ plotted separately against ω. The phase is plotted linearly, but the gain is expressed in logarithmic units known as *decibels* (db), and defined by

$$\text{gain (db)} = 10 \log_{10} M^2(\omega) = 20 \log_{10} M(\omega) \tag{6}$$

Thus 0 db represents a gain of 1, and each 20-db interval corresponds to a factor of 10 in gain. A logarithmic scale is also used for ω, which is then conveniently measured in *octaves* (factors of 2 in frequeny) and *decades* (factors of 10).

We see from (4) that the gain $M(\omega)$ is very close to 1 (0 db) for $\omega \ll \lambda$, and for $\omega \gg \lambda$, its logarithm is approximated by

$$20 \log M(\omega) \approx 20 \log \frac{\lambda}{\omega} = -20 \log \frac{\omega}{\lambda} \tag{7}$$

which is a straight line on the plot with a slope of -20 db/decade (-6 db/octave). Both of these asymptotes are shown as dotted lines, and their point of intersection is known as the *corner* or *break frequency* $\omega = \lambda$, where $M(\omega)$ takes on the value $1/\sqrt{2}$ (-3 db). The curve converges very rapidly to its asymptotes within one decade on either side of the corner frequency, where the largest deviation (3 db) occurs. For this reason, it is often convenient to simply approximate the curve by its asymptotes.

Using (5), it is easy to see that the phase shift $\phi(\omega)$ starts out at 0° for small ω, passes through -45° at the corner frequency $\omega = \lambda$, and then approaches -90° for large ω.

ΔΔΔ

2. *EXAMPLE*. The second-order system

$$\begin{aligned} \dot{\underline{x}}(t) &= \begin{bmatrix} -\bar{\sigma} & \bar{\omega} \\ -\bar{\omega} & -\bar{\sigma} \end{bmatrix} \underline{x}(t) + \begin{bmatrix} 0 \\ \bar{\sigma}^2 + \bar{\omega}^2 \end{bmatrix} u(t) \\ y(t) &= \begin{bmatrix} 1/\bar{\omega} & 0 \end{bmatrix} \underline{x}(t) \end{aligned} \tag{8}$$

was considered in Example 5.3.2 (except that the nonzero entries in the input and output matrices were simply 1). We again assume stability ($\bar{\sigma} > 0$) and write the transfer function as

$$h(s) = \frac{\bar{\sigma}^2 + \bar{\omega}^2}{(s+\bar{\sigma})^2 + \bar{\omega}^2} = \frac{\omega_n^2}{s^2 + 2\zeta\omega_n s + \omega_n^2} \tag{9}$$

in terms of the natural frequency $\omega_n = \sqrt{\bar{\sigma}^2 + \bar{\omega}^2}$ and damping ratio $\zeta = \bar{\sigma}/\omega_n$. The step response of (8) was plotted for constant ω_n and various values of ζ in Fig. 5.3-3. The frequency response in this case is

$$h(j\omega) = \frac{\omega_n^2}{(j\omega)^2 + 2\zeta\omega_n(j\omega) + \omega_n^2} = \frac{\omega_n^2[(\omega_n^2 - \omega^2) - j(2\zeta\omega_n\omega)]}{(\omega_n^2 - \omega^2)^2 + (2\zeta\omega_n\omega)^2} \tag{10}$$

so that

$$M(\omega) = |h(j\omega)| = \frac{\omega_n^2}{[(\omega_n^2 - \omega^2)^2 + (2\zeta\omega_n\omega)^2]^{\frac{1}{2}}} \tag{11}$$

$$\phi(\omega) = \angle h(j\omega) = -\tan^{-1}\frac{2\zeta\omega_n\omega}{\omega_n^2 - \omega^2} \tag{12}$$

The Nyquist diagram of this transfer function is shown in Fig. 2(a) for various values of ζ. Once again we have $h(0) = 1$, but this time the curve traverses the fourth quadrant as ω increases, entering the third quadrant and approaching the origin from the left as $\omega \to \infty$. Reflecting about the real axis, we get the other half of the curve $(-\infty < \omega \leq 0)$.

The corresponding Bode plots are shown in Fig. 2(b). Using (11), it is clear that the gain approaches 1 (0 db) for $\omega \ll \omega_n$, and for $\omega \gg \omega_n$, we have

$$20 \log M(\omega) \simeq 20 \log \frac{\omega_n^2}{\omega^2} = -40 \log \frac{\omega}{\omega_n} \tag{13}$$

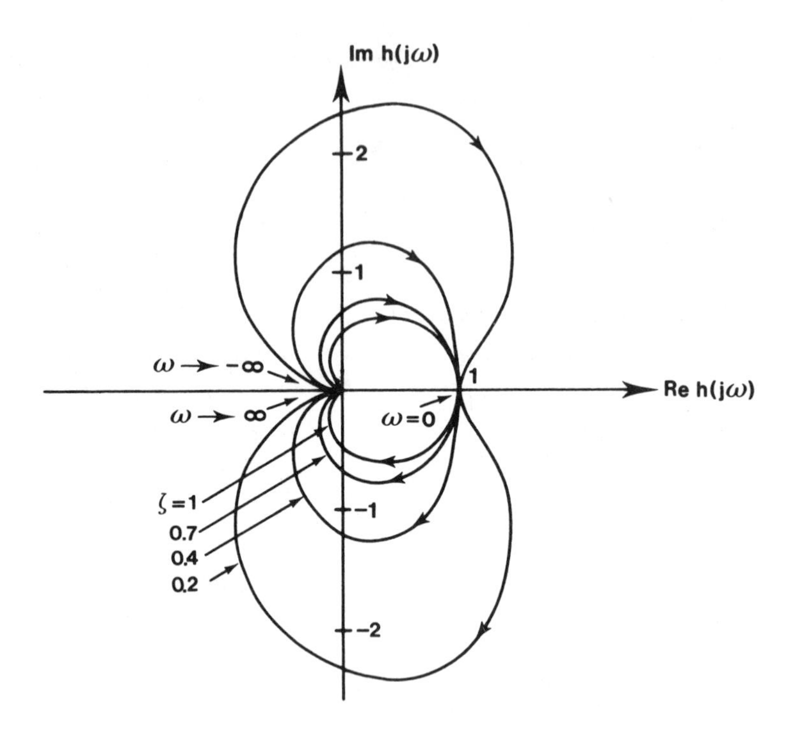

FIG. 2(a). Nyquist diagram of second-order system.

which is a straight line with a slope of -40 db/decade (-12 db/octave). These asymptotes are shown as dotted lines which intersect at the corner frequency $\omega = \omega_n$. In this case the behavior of the curve in the vicinity of the corner frequency depends upon ζ, and several different values are included in the figure. For a heavily damped system ($1/\sqrt{2} \leq \zeta \leq 1$), the curve simply drops away, as in the previous example, but with light damping ($0 \leq \zeta < 1/\sqrt{2}$), there is a definite *resonant peak* (see Problems 4 and 12). This implies that the amplitude of the output is greatest for sinusoidal inputs near the resonant frequency.

The phase shift (12) is 0° for small ω, passing through -90° at $\omega = \omega_n$ and ultimately approaching -180°. The transition is steeper

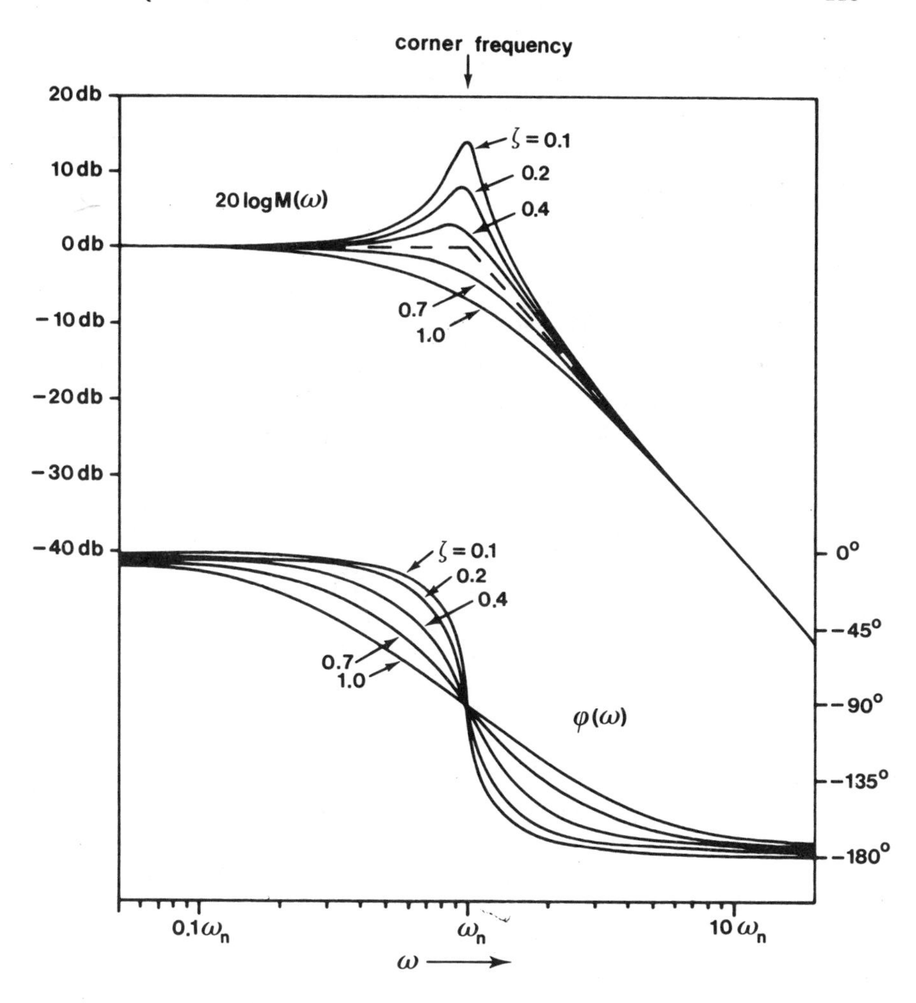

FIG. 2(b). Bode diagram of second-order system.

for smaller values of the damping ratio ζ.

ΔΔΔ

Note that in both of these examples the static gain was $h(0) = 1$, so that the low-frequency asymptote was a horizontal line at $20 \log 1 = 0$ db. If $h(0) \neq 1$, the whole curve simply shifts up or down by an amount $20 \log h(0)$.

The reader will no doubt already have observed one reason for plotting the frequency response in logarithmic units on a Bode diagram:

once the corner frequency is known, the gain curve can be sketched rapidly using little more than the high- and low-frequency straight-line asymptotes. A more important reason becomes evident when a complicated transfer function is expressed as a product of lower-order transfer functions

$$h(s) = h_1(s)\, h_2(s) \ldots h_p(s) \tag{14}$$

This will be the case if the system in question consists of a series connection of various subsystems, as in Example 4.1.4, or if the denominator and numerator polynomials of h(s) are factored into their first- and second-order pole and zero components. The frequency response is then

$$\begin{aligned} h(j\omega) &= h_1(j\omega)h_2(j\omega) \ldots h_p(j\omega) \\ &= M_1 e^{j\phi_1} M_2 e^{j\phi_2} \ldots M_p e^{j\phi_p} = Me^{j\phi} \end{aligned} \tag{15}$$

where

$$M = M_1 M_2 \ldots M_p \tag{16}$$

$$\log M = \log M_1 + \log M_2 + \ldots + \log M_p \tag{17}$$

$$\phi = \phi_1 + \phi_2 + \ldots + \phi_p \tag{18}$$

If the Bode diagrams for $h_1(j\omega)$, $h_2(j\omega)$, ..., $h_p(j\omega)$ are all available, then we see from (17) and (18) that constructing the composite plot for $h(j\omega)$ is simply a matter of summing the components. This will be illustrated with an example.

3. *EXAMPLE*. Suppose we require a frequency response plot for a system with the transfer function

$$h(s) = \frac{15s + 30}{s^3 + 10s^2 + 100s} = \frac{15(s + 2)}{s(s^2 + 10s + 100)} \tag{19}$$

This can conveniently be factored into three distinct transfer functions as follows:

$$\begin{aligned} h(s) &= \frac{0.3}{s}\;\frac{s + 2}{2}\;\frac{100}{s^2 + 10s + 100} \\ &= h_1(s)h_2(s)h_3(s) \end{aligned} \tag{20}$$

Since $h_3(s)$ is just the second-order transfer function of Example 2, with $\omega_n = 10$ and $\zeta = 0.5$, we can plot $M_3(\omega)$ and $\phi_3(\omega)$ immediately. It is also easy to plot $h_2(s)$ since it is just the reciprocal of Example 1 with $\lambda = 2$. This means that $20 \log M_2(\omega)$ and $\phi_2(\omega)$ are obtained by negating the gain and phase found in Example 1. For $h_1(s)$ we observe that

$$h_1(j\omega) = \frac{0.3}{j\omega} = -j\frac{0.3}{\omega} \tag{21}$$

$$M_1(\omega) = |h_1(j\omega)| = \frac{0.3}{\omega} \tag{22}$$

$$\phi_1(\omega) = -90° \tag{23}$$

Thus the phase is constant at -90° and the gain curve is

$$20 \log M_1(\omega) = 20 \log \frac{0.3}{\omega} = -20 \log \frac{\omega}{0.3} \tag{24}$$

which is a straight line with a slope of -20 db/decade passing through 0 db at the frequency $\omega = 0.3$.

All three of these Bode plots are shown together in Fig. 3(a). The composite diagram for $h(j\omega)$, according to (17) and (18), is easily obtained by adding them together, as shown in Fig. 3(b). Finally, the information contained in this latter plot may be transferred to a Nyquist diagram, as shown in Fig. 4.

ΔΔΔ

The above examples illustrate the ease with which frequency response plots may be constructed from a known transfer function, and further examples may be found in the problems. Much of the value of these plots, however, lies in the fact that they may also be determined experimentally when the transfer function of a system is *not* known: sinusoidal signals at various frequencies are applied to the input of the system and relative amplitudes and phases are recorded for the resulting steady-state outputs. Even in situations where such direct measurements are not feasible, such as large industrial processes which cannot be shut down, it is often possible to apply small perturbations to the input and recover the required information at the output using correlation techniques.

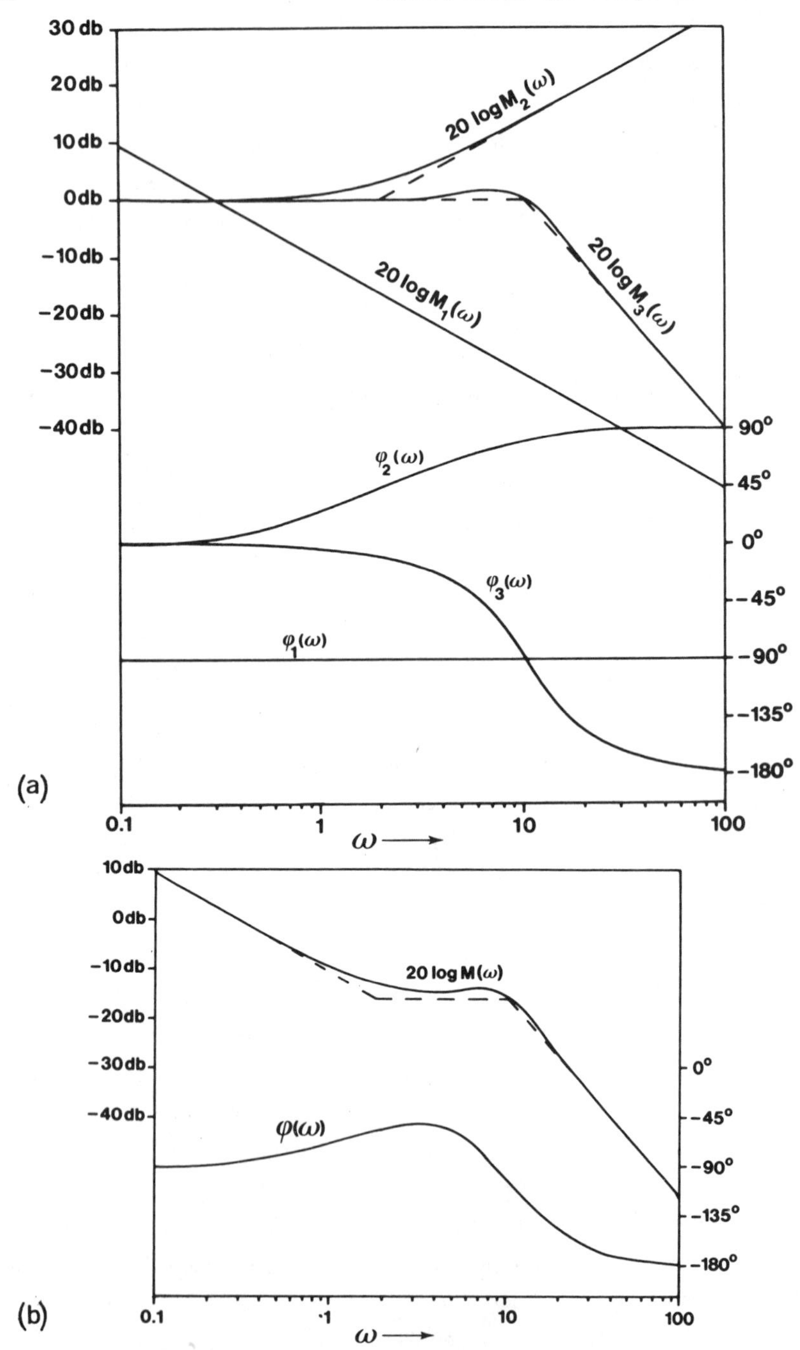

FIG. 3. Bode diagrams for Example 3: (a) subsystems and (b) composite system.

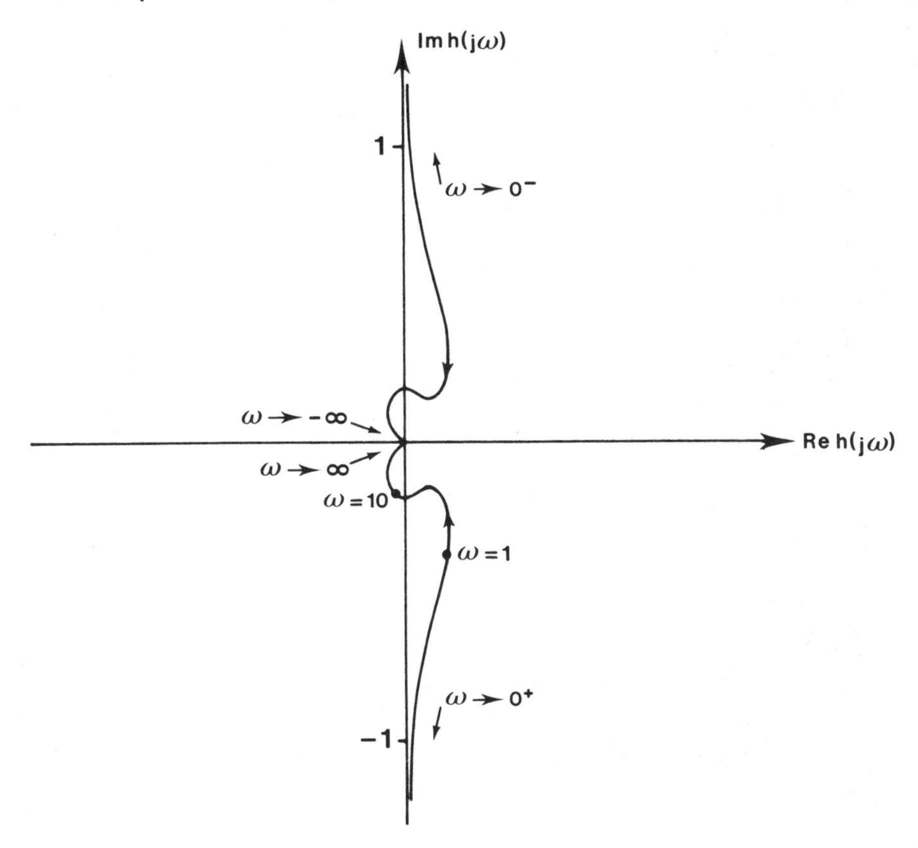

FIG. 4. Nyquist diagram for Example 3.

The Bode diagram provides a wealth of information about a system's characteristics, the most obvious being those associated with the concept of *bandwidth*. The bandwidth of a system is (roughly speaking) the frequency region beyond which signals are attenuated significantly. More precise definitions are available, such as the frequency at which the gain drops to some fraction (usually -3 db) of its peak or low-frequency value, or the second moment of the gain curve (but we will not make use of them here).

A system whose frequency response gain curve remains at about the same level from $\omega = 0$ out to some value $\omega = \bar{\omega}$ and then drops off sharply is usually said to be a *low-pass* system because sinusoidal signals at lower frequencies pass through relatively unattenuated while those at higher frequencies are blocked. One may also encounter *band-pass* systems, which attenuate both low and high frequencies, passing only signals whose frequencies lie in a particular interval or band [see Problem 5(b)].

Similarly, a system whose gain is small at low frequencies and then approaches some constant nonzero value at high frequencies is called *high-pass* [see Problem 5(e)]. Note that such behavior cannot really occur (it would require infinite bandwidth) because every system has some inertia. Nevertheless, it is a useful fiction, provided we bear in mind that it is valid only for frequencies out to the limit of our model (or of our measuring equipment). The same comments apply to so-called *all-pass* systems.

The relationship between frequency response and transient behavior can be illustrated by means of Fourier series. In particular, a step function input to a system can be viewed as one section of a square wave with a very long period, which in turn can be expressed in a Fourier series as a sum of sinusoids at various frequencies. A low-pass system filters out the high-frequency components, which has the effect of both slowing and smoothing the response to the step input (in the extreme case, the system acts as an integrator). The higher the bandwidth of the system, the more high-frequency components remain and the more faithfully the step input is reproduced at the output. In general, high bandwidth corresponds to fast transient response.† Conversely, a high-pass system removes the d.c. and low-frequency components, making the output appear more "noisy" than the input (in the extreme case, the system acts as a differentiator).

The first-order frequency response shown in Fig. 1(b) exhibits low-pass behavior with a bandwidth of about λ. Comparison with Fig. 5.3-1 illustrates the relation between transient and frequency responses: the speed of the step response is directly proportional to λ.

The bandwidths of the second-order systems whose frequency responses are plotted in Fig. 2(b) are all roughly ω_n. Comparison with Fig. 5.3-3 highlights another useful relationship: A sharp

†However, this statement must be qualified in the case of *nonminimum phase* systems (see Definition 5 and the discussion following it).

resonant peak in the frequency response corresponds to an oscillatory transient response.

The usefulness of the Nyquist plot will become clear in Chapter 8, when the *Nyquist stability criterion* is introduced. This famous result makes possible the design of single-loop feedback control systems using only frequency response information.

Mathematically speaking, the frequency response completely characterizes a system's input-output behavior; in principle, both the transfer function and impulse response may be derived from it. For example, the impulse response could be determined approximately from experimental frequency response data by numerically performing an inverse Fourier transform. Another approach is to reverse the procedure used in Example 3, where a composite Bode diagram was constructed from the individual pole and zero factors of the transfer function. By carefully studying the gain curve, one may attempt to identify various slopes, corner frequencies, and resonant peaks and thereby determine the poles and zeros of the transfer function (see Problem 6).

Determining a model from experimental data is known as the *system identification* problem. In practice, the procedure mentioned above can be useful, but if the system is of high order or the data is noisy, it often breaks down.[†]

Another problem which complicates system identification using frequency response information is that the gain plot alone does not uniquely determine the system. This is best illustrated with an example.

4. EXAMPLE. Consider a system whose transfer function is

$$h(s) = \frac{s + 3}{(s + 1)(s + 10)} \tag{25}$$

with frequency response

$$h(j\omega) = \frac{j\omega + 3}{(j\omega + 1)(j\omega + 10)} \tag{26}$$

[†]Various time-domain algorithms for system identification have been proposed in recent years, some of which have proved to be quite effective. These will be discussed briefly in Chapter 16.

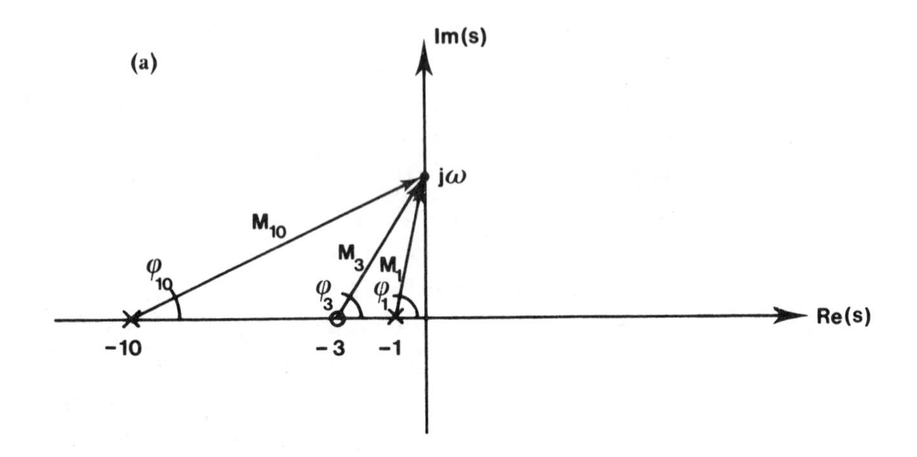

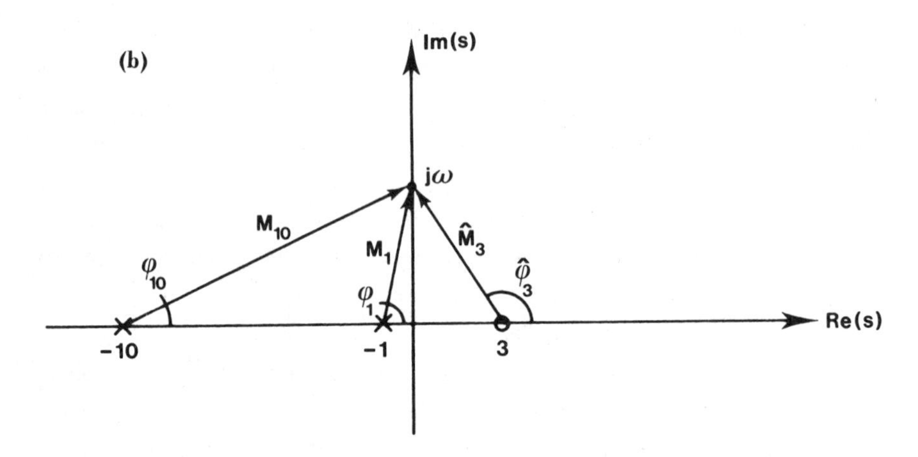

FIG. 5. Component vectors of frequency response: (a) minimum phase system $h(s)$ and (b) non-minimum phase system $\hat{h}(s)$.

At any given frequency ω, $h(j\omega)$ may be evaluated using the vectors $j\omega + 1$, $j\omega + 3$, and $j\omega + 10$ drawn from the poles and the zero to the point $j\omega$, as shown in Fig. 5(a). Denoting the magnitudes of these three vectors by M_1, M_3, M_{10}, and the angles as ϕ_1, ϕ_3, ϕ_{10}, we see that

$$M(\omega) = |h(j\omega)| = \frac{M_3(\omega)}{M_1(\omega)M_{10}(\omega)} \tag{27}$$

$$\phi(\omega) = \angle h(j\omega) = \phi_3(\omega) - \phi_1(\omega) - \phi_{10}(\omega) \tag{28}$$

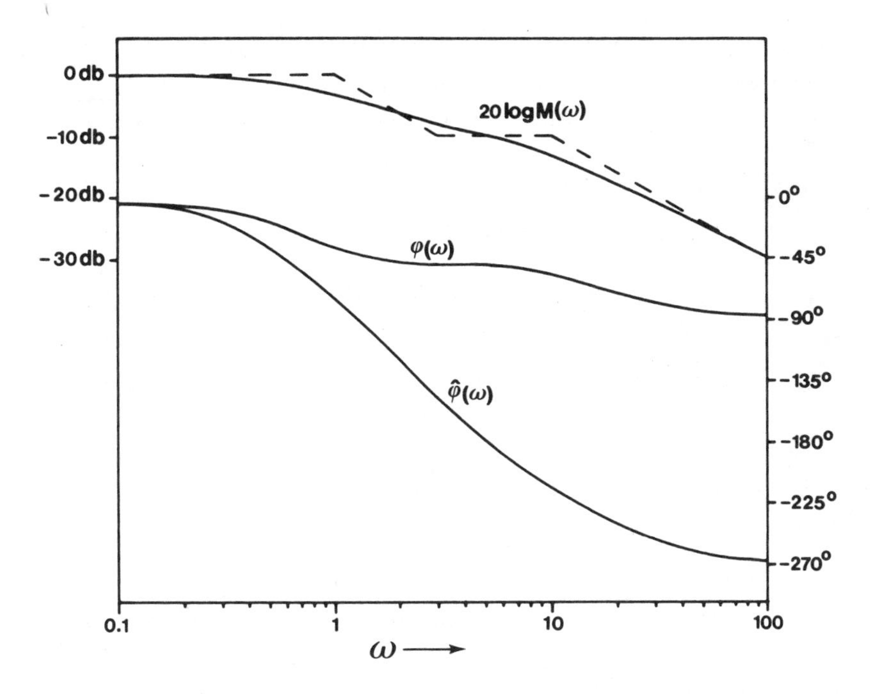

FIG. 6. Frequency responses for Example 4.

The Bode plot for this system is shown in Fig. 6. Note that the phase varies from 0° ($\phi_1 = \phi_3 = \phi_{10} = 0°$) at $\omega = 0$ to -90° ($\phi_1 = \phi_3 = \phi_{10} = 90°$) as $\omega \to \infty$.

Next, consider a system with transfer function

$$\hat{h}(s) = -\frac{s - 3}{(s + 1)(s + 10)} \tag{29}$$

and frequency response

$$\hat{h}(j\omega) = -\frac{j\omega - 3}{(j\omega + 1)(j\omega + 10)} \tag{30}$$

which differs from (25) only in that the zero lies at +3 instead of -3 [the factor of -1 is included to make $\hat{\phi}(0) = \phi(0)$]. The vector components of $\hat{h}(j\omega)$ are shown in Fig. 5(b). The two associated with the poles are the same as for $h(j\omega)$, but the vector associated with the zero is $j\omega - 3$ with magnitude and angle

$$\hat{M}_3(\omega) = M_3(\omega) \tag{31}$$

$$\hat{\phi}_3(\omega) = 180° - \phi_3(\omega) \tag{32}$$

The Bode plot for this system is also shown in Fig. 6. Since the magnitudes of the vectors are the same for both systems, their gain curves are identical $[\hat{M}(\omega) = M(\omega)]$. The phase, however, is given by

$$\begin{aligned} \hat{\phi}(\omega) &= 180° + \hat{\phi}_3(\omega) - \phi_1(\omega) - \phi_{10}(\omega) \\ &= -\phi_3(\omega) - \phi_1(\omega) - \phi_{10}(\omega) \end{aligned} \tag{33}$$

where the extra 180° comes from the factor of -1 in (30). This varies from 0° (or 360°) at $\omega = 0$ to -270° as $\omega \to \infty$.

Thus we see that although the gain curves coincide for the two systems considered, the phase curves are quite distinct. Indeed, the phase lag in the second case is greater at all frequencies, which motivates the next definition.

ΔΔΔ

5. *DEFINITION*. A stable linear system having all of its zeros in the left half-plane is said to be *minimum phase*; if one or more zeros have positive real parts the system is designated *nonminimum phase*. This terminology refers to the fact that a minimum phase system always produces less phase lag than its nonminimum phase counterpart (see Problem 7).

ΔΔΔ

We have already observed in Example 4 (also see Problem 9) that the correspondence between gain and phase plots is not one-to-one in general. For minimum phase systems, however, the gain uniquely determines the phase via the following formula:†

$$\phi(\omega_1) = \frac{2\omega_1}{\pi} \int_0^\infty \frac{\ln M(\omega) - \ln M(\omega_1)}{\omega^2 - \omega_1^2}\, d\omega \qquad (0 \le \omega_1 < \infty) \tag{34}$$

This means that in principle, when dealing with minimum phase systems, one need only be concerned with gain information. In practice, of course, the phase information is also useful, particularly if the frequency response is constructed from noisy experimental data.

---

†This is sometimes known as *Bode's equation*. For a derivation, see pp. 428-436 of Ref. 185.

Nonminimum phase systems, having greater phase lag than their minimum phase counterparts, are invariably slower in response to an input. In many cases (see Problem 10), this manifests itself in a curious way: the step response of a nonminimum phase system may move initially in the direction *opposite* to its ultimate steady-state value. Some interesting aeronautical examples of this effect may be found in Sec. 8.5.4 of Ref. 71.

Finally, it should be noted that many systems contain *pure time delays* or *transportation lags*. These lie outside the framework of our analysis (such systems are linear, but infinite-dimensional) and in practice are often ignored or approximated by finite-dimensional elements. In the Laplace transform domain, a pure delay of Δ sec. becomes a multiplicative factor of $e^{-s\Delta}$ (Theorem B.1.4). Since $|e^{-j\omega\Delta}| = 1$, this does not affect the gain plot, but on the phase curve it shows up as an additional phase lag of $\omega\Delta$ rad at each frequency ω. For this reason, systems with delay are often included in the nonminimum phase category.

6.3. SUMMARY

In this chapter we have considered the behavior of stable linear systems under steady-state conditions, that is, after any initial transients have died out. The response to a constant (step) input was seen to approach the same constant, multiplied by the static gain h(0). Similarly, sinusoidal inputs were seen to produce sinusoidal outputs in steady state with amplitude and phase changes at each frequency ω determined by the magnitude and angle of $h(j\omega)$.

In Sec. 6.2 we computed the frequency response $h(j\omega)$, $0 \leq \omega < \infty$, of a first- and a second-order system, displaying it on both Bode and Nyquist diagrams, and saw how to build up Bode diagrams for more complex systems from their lower-order components. We also observed that frequency response data, which is relatively easy to obtain experimentally, completely characterizes the input-output behavior of a system. Finally, nonminimum phase systems were briefly discussed.

PROBLEMS

Section 6.1

1. Determine the static gains of the systems in Examples 5.3.1 and 5.3.2 and verify that the steady-state values given there are correct.

2. Suppose the input to the system of Example 6.1.3 is a *ramp function*

$$u(t) = \begin{cases} 0 & (t \leq 0) \\ 2t & (t > 0) \end{cases}$$

Determine the entire system response, identifying the transient and steady-state components. How does the shape of the steady-state output differ from that of the input?

3. (a) Extend Corollary 6.1.4 to allow an input of the form $u(t) = \alpha 1(t) \sin(\omega t + \Theta)$.

   (b) Do the same for the input $u(t) = \alpha 1(t) \cos \omega t$.

4. (a) Re-prove Corollary 6.1.2 by using the Final Value theorem (B.1.12).

   (b) Can you do the same for Corollary 6.1.4? Explain.

5. (a) Prove Theorem 6.1.5 by extending Theorem 6.1.1 and its corollaries.

   (b) Determine the steady-state response of the system (6.1-28) to the input

$$\underline{u}(t) = \begin{bmatrix} \alpha_1 \sin(\omega_1 t) \\ \alpha_2 \sin(\omega_2 t) \\ \vdots \\ \alpha_m \sin(\omega_m t) \end{bmatrix} 1(t)$$

Section 6.2

1. Construct Nyquist and Bode diagrams for the unstable system

$$\dot{x} = \lambda x + u \qquad (\lambda > 0)$$

$$y = \lambda x$$

2. Show that the negative frequency portion $(-\infty < \omega \leq 0)$ of a Nyquist plot is just a reflection about the real axis of the positive frequency portion $(0 \leq \omega < \infty)$. In other words, show that for real, rational h(s),

$$h(-j\omega) = \overline{h(j\omega)}$$

3. For the case of $\zeta > 1$, show how to derive the second-order frequency response (Example 6.2.2) from the first-order case (Example 6.2.1).

4. Referring to Example 6.2.2, show that if $0 \leq \zeta < 1/\sqrt{2}$, the maximum gain occurs at the resonant frequency

$$\omega_r = \omega_n \sqrt{1 - 2\zeta^2}$$

and determine $M(\omega_r)$. Show also that if $\zeta \geq 1/\sqrt{2}$, no resonance occurs.

5. Construct Bode and Nyquist diagrams for the following transfer functions:

(a) $h_1(s) = \dfrac{20(s + 20)}{(s + 10)(s + 40)}$

(b) $h_2(s) = \dfrac{25(s + 1)}{s^2 + 20s + 75}$

(c) $h_3(s) = \dfrac{1000}{(s + 10)(s^2 + 8s + 100)}$

(d) $h_4(s) = \dfrac{s - 5}{s^2 - s}$

(e) $h_5(s) = \dfrac{(s + 2)^2}{(s + 8)^2}$

6. Identify the transfer function of the system whose Bode diagram is shown in Fig. P1.

7. (a) Suppose $h_1(s)$ is a minimum phase transfer function and $h_2(s)$ is a nonminimum phase version with one or more zeros moved to their "mirror-image" locations in the right half-plane.

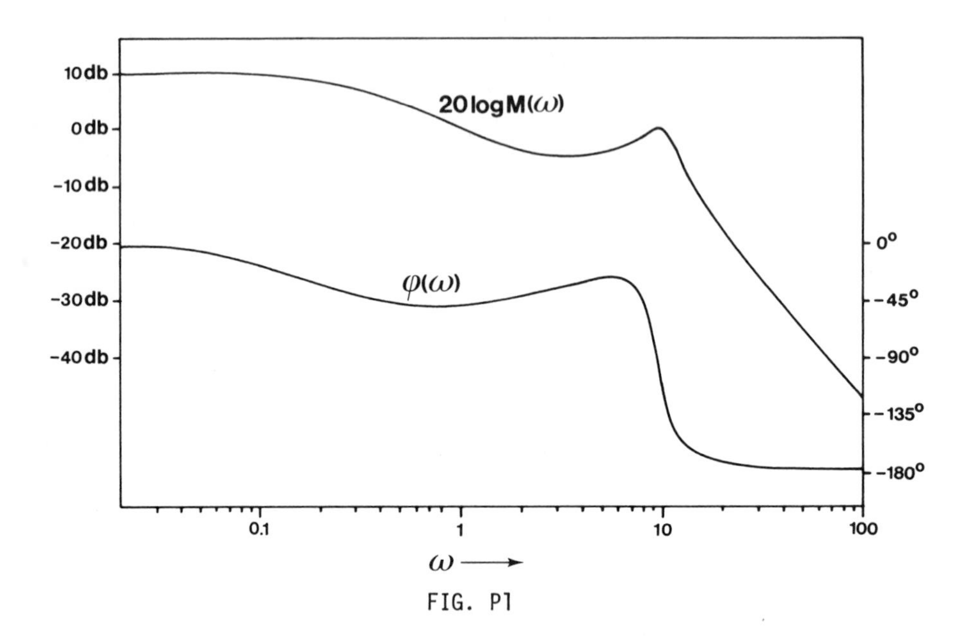

FIG. P1

Assuming that the sign of $h_2(s)$ is adjusted to make $\phi_2(0) = \phi_1(0)$, show that

$$M_1(\omega) = M_2(\omega) \qquad 0 \leq \omega < \infty$$

and

$$\phi_1(\omega) > \phi_2(\omega) \qquad 0 < \omega < \infty$$

which justifies the terminology "minimum phase."

(b) Some textbooks define a nonminimum phase system to be one with zeros *or* poles in the right half-plane. Is the terminology appropriate in this case?

8. (a) Suppose a Bode diagram is constructed for the transfer function

$$h(s) = \frac{s^m + q_{m-1}s^{m-1} + \ldots + q_0}{s^n + p_{n-1}s^{n-1} + \ldots + p_0}$$

which is assumed to be both stable and minimum phase (i.e., all of its poles and zeros lie in the left half-plane). What will be the *slope of the high-frequency asymptote* ($\omega \to \infty$) of the gain curve $M(\omega)$? What will be the *range of phase angle* $\phi(\infty)-\phi(0)$?

(b) How do your answers in (a) change for a nonminimum phase system?

(c) Can you suggest a test for determining whether a stable system is minimum or nonminimum phase by looking *only* at the Bode diagram [i.e., assuming that no other information about h(s) is available]?

9. Example 6.2.4 concerns two systems whose frequency responses have identical gain curves but different phase curves. Devise an example of two systems with identical phase curves but different gain curves.

10. (a) Plot the step response of each system in Example 6.2.4. What distinctive characteristic is exhibited in the non-minimum phase case?

(b) Prove that this is a characteristic of *every* nonminimum phase system with a single zero in the right half-plane.

(c) What happens in the case of multiple zeros in the right half-plane?

11. Under what conditions does a low-pass filter approximate an integrator? Under what conditions does a high-pass filter approximate a differentiator?

12. (a) Attach a small mass ($\simeq$ 100 g) to one end of a rubber band about 15 cm in length. Hold the other end in your hand with the mass suspended. Move your hand up and down through a distance of 2-3 cm at a very low frequency and observe the motion of the mass. Next, slowly increase your frequency of oscillation without changing amplitude

and observe the changes in the motion of the mass. Describe its asymptotic behavior at high and low frequencies. Do you detect a resonance?

(b) Sketch a rough Bode diagram (both magnitude and phase, if possible) of the frequency response of this system. Try to estimate the transfer function relating input (position of your hand) to output (position of the mass) from the Bode plot.

(c) Write down the equations of motion of this system, using rough estimates of the parameters. Compute the transfer function and compare it with your answer in (b).

(d) Is your hand motion perfectly sinusoidal? Does it matter? Why?

Chapter 7

COMPUTER SOLUTION OF DIFFERENTIAL EQUATIONS

7.0. INTRODUCTION

In the preceding chapters, we investigated analytical solutions of differential equations and found that a good deal of qualitative information about the behavior of a system can be deduced rather directly from its coefficients. This has laid the foundation for the design techniques we shall develop in later chapters when we address the problem of controlling a system. These methods will enable us to determine what basic structure a control system must have if it is to meet given performance specifications, but not all questions about system behavior can be answered without an explicit solution of the differential equations. For example, we will use short-cut methods to find the system poles for specified parameter values, and hence determine what exponential functions will appear in the transient response, but this is not sufficient to determine the exact response to, say, a step input. Thus, the final process of tuning parameter values in order to meet quantitative transient response specifications will often require the complete solution of the system equations.

This, of course, is exceedingly laborious if done manually; in all but the simplest cases, the aid of a computer will be necessary.

Differential equations can be solved on a computer in three fundamentally different ways:

(i) The coefficients in a closed-form, analytic solution can be determined by some algorithm, and the resulting functions evaluated at the required values of time.

(ii) The equations can be solved by wholly numerical methods which do not use a closed-form solution.

(iii) The equations can be solved by simulation on an analog computer.

Of these alternatives, the first is by far the least efficient in all but very simple cases. It requires computation of all the eigenvalues, evaluation of the resulting combination of exponential functions for each state variable at many values of time, and some very complex programming. Moreover, it cannot be generalized to handle nonlinear and time-varying cases. We therefore dismiss methods requiring an analytic solution as impractical and concentrate in this chapter on the remaining methods: analog computation and direct numerical solution.

7.1. ANALOG COMPUTATION

An analog computer is a physical system constructed so that the quantities in a mathematical problem are represented by physical variables, and the relations between them are known to obey the same law as in the mathematical problem. The problem is solved by appropriately manipulating the physical system and making measurements on it. A slide rule, for example, is an analog computer: the mathematical problem of finding the product of two numbers is solved by determining the sum of two lengths, each representing the logarithm of one of the numbers. However, this is a rather broad definition of analog computing; by common usage, an analog computer is now understood to be an electronic machine built specifically to solve ordinary differential equations. It contains a range of electronic devices from which a circuit can be constructed which is known to be described by the differential equation to be solved.

The solution is then obtained by appropriately exciting the circuit and observing or recording its behavior over time. The dependent variables of the problem are represented by voltages appearing at various points in the circuit; the independent variable is always time. Thus, an analog computer circuit is a physical model or analog of the system from which the equations were originally derived.

The basic building blocks for constructing computer models are operational amplifiers, potentiometers or voltage dividers, multipliers, and diode function-generating circuits. The input and output terminals of each device are brought out to a patch panel, allowing them to be connected by patch cords. The following mathematical operations can be implemented with these devices:

(i) Multiplication of a voltage by a constant
(ii) Algebraic addition of voltages
(iii) Integration with respect to time of a voltage
(iv) Continuous multiplication of two voltages
(v) Generation of arbitrary functions of a voltage

Other equipment in an analog computer comprises power supplies, reference voltage supplies, and control circuitry allowing the simultaneous switching of computing elements necessary to start and stop a solution.

The basis of nearly all analog computing circuits is the operational amplifier. This is a direct-coupled electronic feedback amplifier of very high negative gain ($\simeq 2 \times 10^8$), often equipped with a separate a.c. amplifier in the feedback path for drift compensation. By connecting various impedances around the amplifier, a number of mathematical operations can be performed. Consider the circuit shown in Fig. 1(a) whose equations are

$$\begin{aligned}
e_0(t) &= -A\, e_G(t) \\
e_0(t) - e_G(t) &= R_f\, i_2(t) \\
e_i(t) - e_G(t) &= R_i\, i_1(t) \\
i_1(t) + i_2(t) &= i_G(t)
\end{aligned} \tag{1}$$

The amplifier is designed to draw an extremely small input current, so

(a)

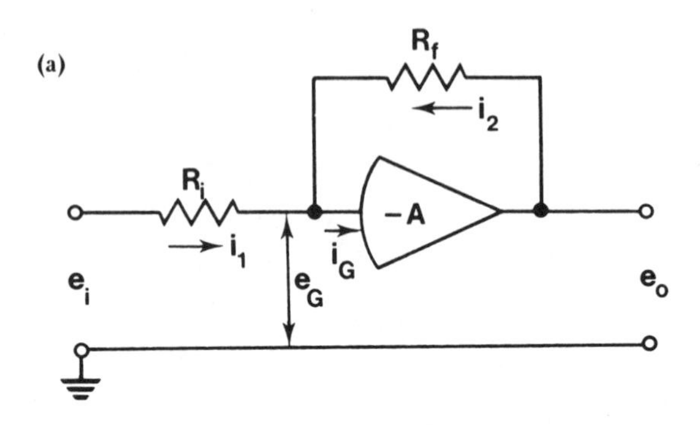

(b)

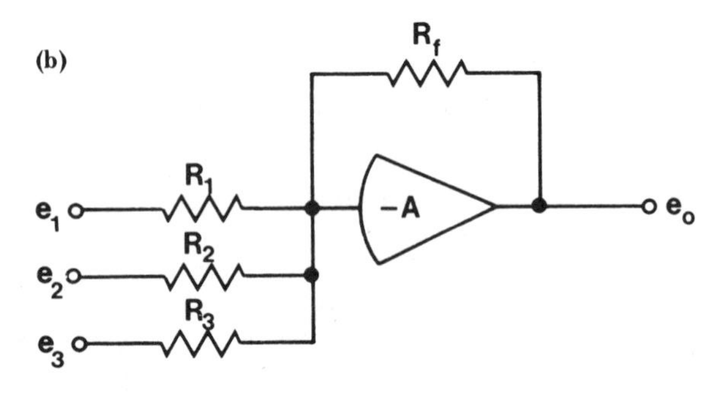

(c)

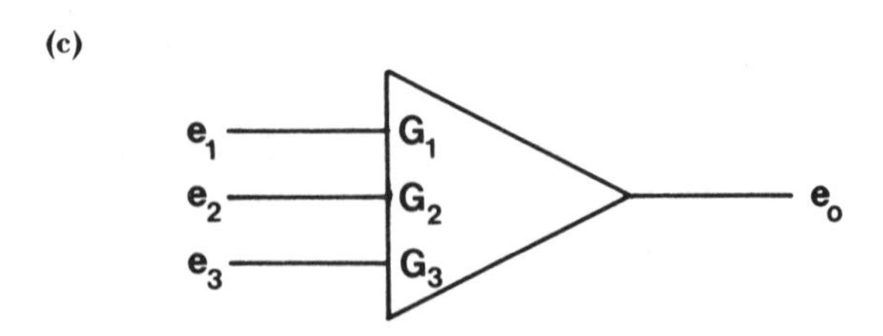

FIG. 1. Summing amplifier circuits.

that $i_G \approx 0$ and these equations yield

$$e_0(t) \approx \frac{-G\, e_i(t)}{1 + (1 + G)/A} \tag{2}$$

where $G \triangleq R_f/R_i$. For practical circuits, $G \le 100$, so that $(1 + G)/A$ is negligible and

$$e_0(t) \approx -G\, e_i(t) \tag{3}$$

Several input voltages can be applied simultaneously to the input of one amplifier. The reader may show, for example, that the circuit of Fig. 1(b) is described by the equation

$$e_0(t) = -[G_1 e_1(t) + G_2 e_2(t) + G_3 e_3(t)] \tag{4}$$

where $G_j = R_f/R_j$, $j = 1,2,3$. Amplifiers used in this way with input and feedback resistors are called *summing amplifiers*; a standard way of representing them graphically is shown in Fig. 1(c).

The resistors required for such summing circuits are built into the computer and connected internally around the amplifiers. Different manufacturers have different arrangements; one fairly typical circuit is shown in Fig. 2. The circles indicate patch panel terminals. A feedback resistor is inserted by patching a wire from an appropriate input terminal to one of the output terminals. With the range of resistors provided in this case, amplifier gains G $(=R_f/R_i)$ of 0.1, 1.0, and 10 can be obtained.

If a voltage is to be multiplied by a constant not obtainable with the built-in amplifier resistors, a *coefficient potentiometer* or adjustable voltage divider is required in conjunction with a summing amplifier. For example, the circuit of Fig. 3(a) is described by

$$e_0(t) = -2.47\, e_i(t) \tag{5}$$

if the potentiometer is set at 0.247 and an amplifier gain of 10 is used. The standard symbol for a potentiometer on a computer circuit diagram is a circle, as shown in Fig. 3(b).

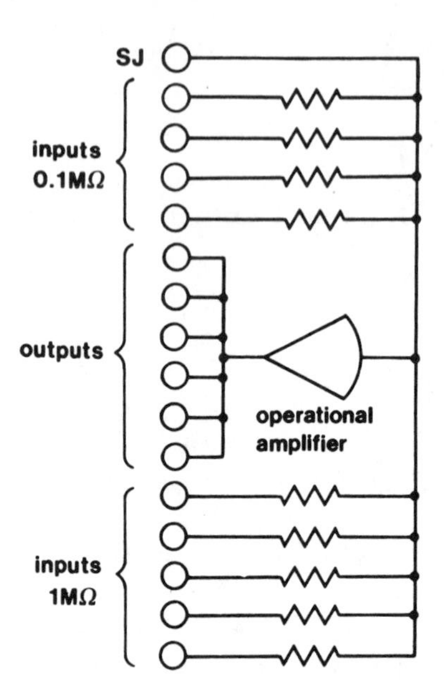

FIG. 2. Summing amplifier circuit.

(a)

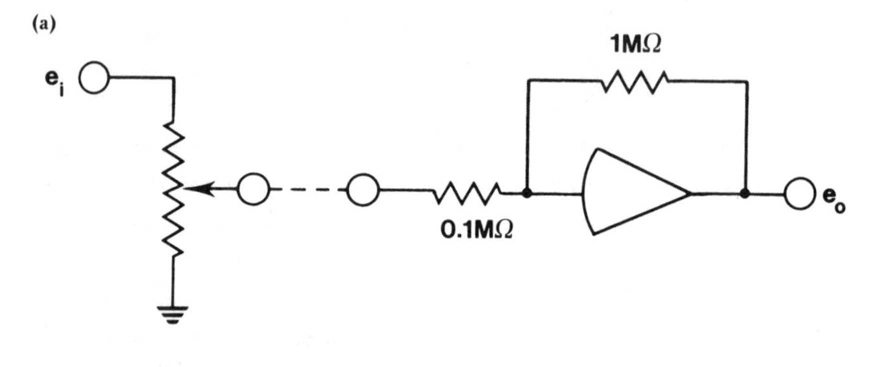

(b)

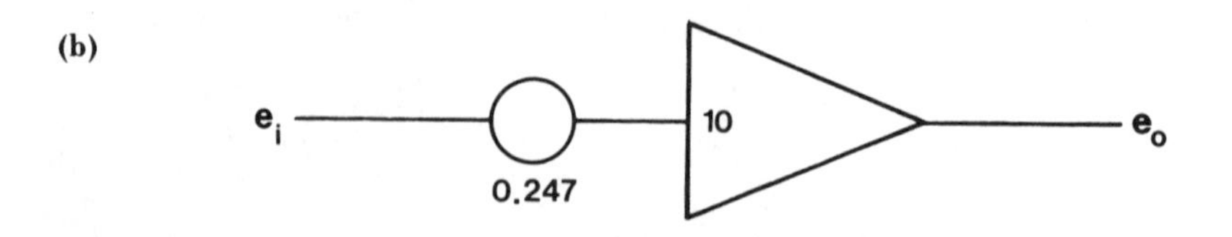

FIG. 3. Use of coefficient potentiometers: (a) Circuit diagram (circles are patch panel terminals, broken line is a patch panel connection) and (b) symbolic diagram.

By replacing the feedback resistor of the basic amplifier circuit with a capacitor, as shown in Fig. 4(a), integration of the input voltage with respect to time is achieved. As before, the amplifier input current is extremely small, so that the circuit

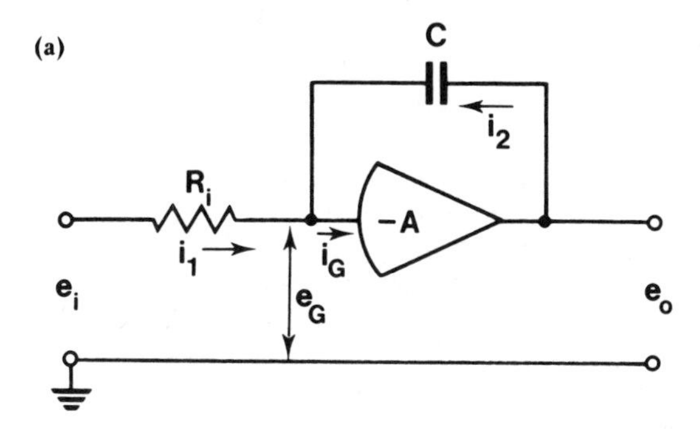

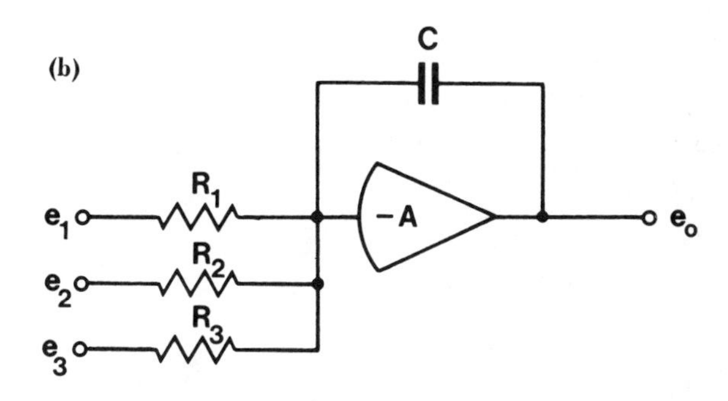

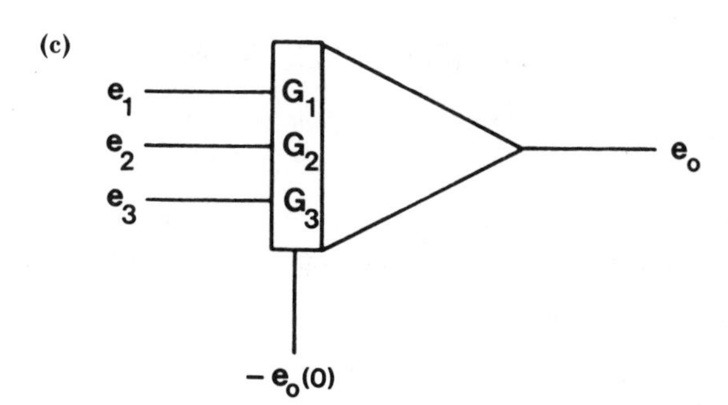

FIG. 4. Integrating amplifiers.

equations are:

$$
\begin{aligned}
e_0(t) &= -A\,e_G(t) \\
e_0(t) - e_G(t) &= e_0(t_0) - e_G(t_0) + \frac{1}{C}\int_{t_0}^{t} i_2(\tau)\,d\tau \\
e_i(t) - e_G(t) &= R_i i_i(t) \\
i_2(t) + i_1(t) &= i_G(t) \approx 0
\end{aligned}
\tag{6}
$$

Elimination of e_G, i_1 and i_2 yields

$$(1 + \frac{1}{A})e_0(t) = (1 + \frac{1}{A})e_0(t_0) - \frac{1}{R_i C}\int_{t_0}^{t} [e_i(\tau) - \frac{e_0(\tau)}{A}]\,d\tau \tag{7}$$

Since A is very large, and e_0 is limited by the capacity of the amplifier to not much more than the computer reference voltage, typically 10 V, this simplifies to

$$e_0(t) \simeq e_0(t_0) - G\int_{t_0}^{t} e_i(\tau)\,d\tau \tag{8}$$

where $G \triangleq 1/R_i C$. Thus the circuit performs pure time integration of a voltage. Again, several input voltages can be connected to the amplifier: the circuit of Fig. 4(b) solves the equation

$$e_0(t) = e_0(t_0) - \int_{t_0}^{t} [G_1 e_1(\tau) + G_2 e_2(\tau) + G_3 e_3(\tau)]\,d\tau \tag{9}$$

where $G_j = 1/R_j C$, $j = 1,2,3$. A standard schematic representation for integrating amplifiers is shown in Fig. 4(c). As in the case of pure summing amplifiers, the resistor and capacitor networks required to construct such integrating circuits are already built into the computer. To be useful, the circuits need to contain additional equipment which allows suitable "initial-condition" voltages to be placed on the feedback capacitors before integration is started, and which allows starting and stopping of the integration process. A typical simplified circuit diagram of an integrating amplifier is shown in Fig. 5. As before, circles indicate patch panel terminals. To use the amplifier as a "summer", a patch cord is used to connect the SJ and G terminals; the amplifier input is then always connected to the input resistor "summing junction." To use it as an integrator,

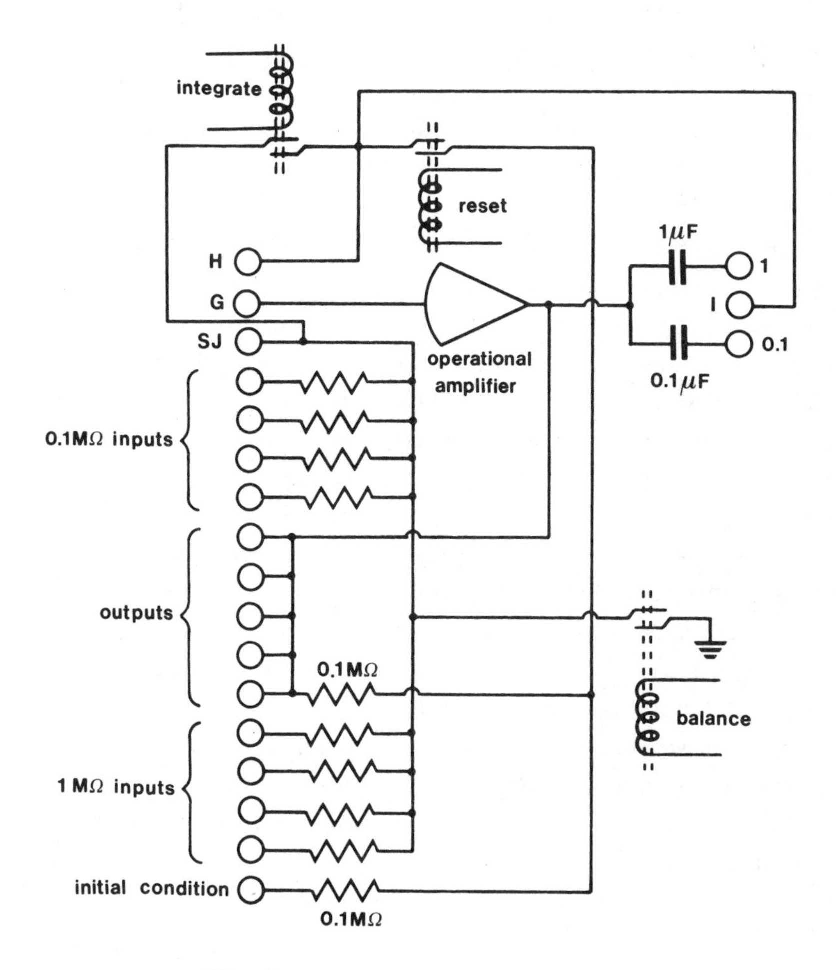

FIG. 5. Integrating amplifier circuit.

patch cords connecting G to H and I to 1 or 0.1 are required; the choice of the latter depends on whether a 1-μF or 0.1-μF feedback capacitor is required. Again, the details of integrating circuits and of patch panel terminals differ among manufacturers, but the basic principles are much the same.

Computer control is obtained by a master switch which simultaneously energizes the appropriate relays or solid-state switches in each of the amplifier circuits. The computer can be switched to four modes of operation: *balance* (or *potset*), *reset* (or *initial condition*), *integrate*, and *hold*. In the *balance* mode, all amplifiers are disconnected from their patch panel terminals, so that their "drift" or state of balance can be corrected. At the same time, the amplifier summing junctions are grounded by energizing the balance relay in Fig. 5, allowing coefficient potentiometers to be set with their correct loading. In the *reset* mode, the amplifiers are reconnected to their patch panel terminals, and the summing amplifiers operate normally. Note that when the reset relay is energized, integrating amplifiers are connected as summers, producing at their output the negative of the voltage connected to the initial-condition terminal. This voltage is also applied to the integrating capacitor. Switching the computer to *integrate* energizes the corresponding relay and starts the problem solution. At any time, the computer can be switched to *hold*, interrupting the integration process and "freezing" the capacitor voltages. They can then be measured at leisure. Returning to *integrate* continues the problem solution.

We now consider how to construct an analog computer circuit which will solve a given set of differential equations. Recall that the dependent variables of the problem are represented by voltages in the computer circuit while the independent variable is represented by (computer) time. In particular, if the differential equations are in state-variable format

$$\dot{\underline{x}}(t) = \underline{f}(\underline{x},\underline{u},t) \qquad \underline{x}(t_0) = \underline{x}_0 \tag{10}$$

or alternatively,

$$x_i(t) = x_i(0) + \int_{t_0}^{t} f_i(\underline{x}(\tau),\underline{u}(\tau),\tau)\, d\tau \qquad (i = 1,2,\dots,n) \quad (11)$$

then it follows from (8) that each state variable $x_i(t)$ may be represented by the output voltage of an integrating amplifier. The rest of the computer circuit produces voltages representing the forcing functions $\underline{u}(t)$, and the integrator gets $-f_i$ from the forcing functions and the state variables. The computer variables can be, but need not be, numerically equal to the problem variables; we shall consider scaling below.

Nonlinear and time-varying functions f_i require nonlinear computing devices such as multipliers and function-generating networks; constant-coefficient linear equations can be solved using only summing and integrating amplifiers and potentiometers. Only the latter will be considered here. Circuits for linear equations are most conveniently constructed from the state-variable form of the equations, particularly if the input-output form contains derivatives of the inputs.

1. EXAMPLE. Consider a system described by

$$\dddot{y}(t) + 2.2\ddot{y}(t) + 5.4\dot{y}(t) + y(t) = 2\dot{u}(t) + u(t) \quad (12)$$

or, alternatively, by the transfer function

$$h(s) = \frac{2s + 1}{s^3 + 2.2s^2 + 5.4s + 1} \quad (13)$$

and suppose that we require its response for two cases:

(a) *Zero input and initial conditions* $\ddot{y}(0) = \dot{y}(0) = 0$, $y(0) = 4$.

We first use Theorem 2.2.1 to obtain an equivalent set of state equations,

$$\begin{aligned}
\dot{x}_1(t) &= x_2(t) \\
\dot{x}_2(t) &= x_3(t) \\
\dot{x}_3(t) &= -x_1(t) - 5.4x_2(t) - 2.2x_3(t) + u(t) \\
y(t) &= x_1(t) + 2x_2(t)
\end{aligned} \quad (14)$$

The initial values of the state variables can be found by solving

$$y(0) = x_1(0) + 2x_2(0) = 4$$

$$\dot{y}(0) = \dot{x}_1(0) + 2\dot{x}_2(0) = x_2(0) + 2x_3(0) = 0$$

$$\begin{aligned} \ddot{y}(0) &= \ddot{x}_1(0) + 2\ddot{x}_2(0) = \dot{x}_2(0) + 2\dot{x}_3(0) \\ &= x_3(0) + 2[-x_1(0) - 5.4x_2(0) - 2.2x_3(0) + u(0)] \\ &= -2x_1(0) - 10.8x_2(0) - 3.4x_3(0) + u(0) = 0 \end{aligned} \tag{15}$$

With our convention that $u(0) = 0$, this has the solution $x_1(0) = 7.14$, $x_2(0) = -1.57$, $x_3(0) = 0.784$. The first state equation is therefore

$$\dot{x}_1(t) = x_2(t) \qquad x_1(0) = 7.14 \tag{16}$$

Assuming that 1 unit of "problem time" t is represented by 1 sec. on the computer and that a voltage is available such that 1 V represents 1 unit of the variable x_2, then (16) is solved by an integrating amplifier provided an initial-condition voltage of $x_1(0) = 7.14$ V is connected. This can be done by passing the computer reference voltage, assumed to be 10 V, through a coefficient potentiometer set at 0.714. The variable $-x_1(t)$ then appears as the output voltage of the amplifier shown in Fig. 6(a); recall that amplifiers change the sign of input as well as initial-condition voltages.

The second state equation

$$\dot{x}_2(t) = x_3(t) \qquad x_2(0) = -1.57 \tag{17}$$

is solved by a similar circuit, shown in Fig. 6(b). The third equation is

$$\dot{x}_3(t) = -x_1(t) - 5.4x_2(t) - 2.2x_3(t) \qquad x_3(0) = 0.784 \tag{18}$$

since $u(t) \equiv 0$ for part (a). Assuming that voltages representing $-x_1, -x_2$ and $-x_3$ are available, this equation can [by (9)] also be solved by an integrating amplifier. The constants 5.4 and 2.2

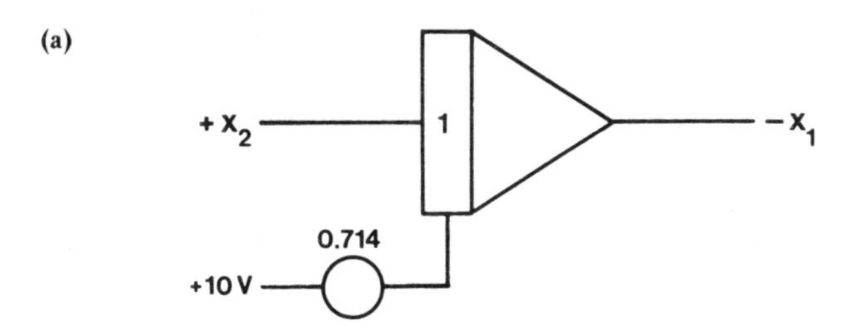

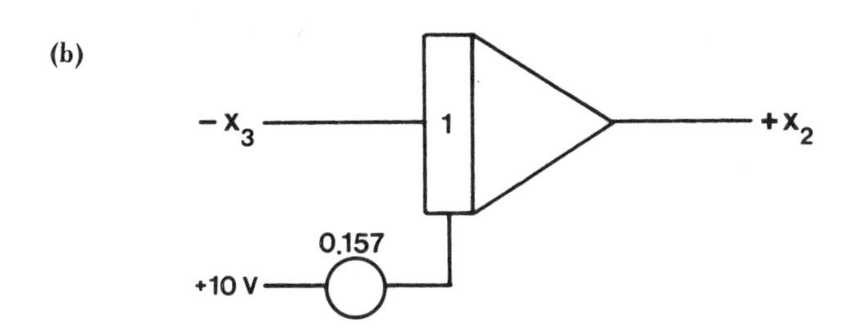

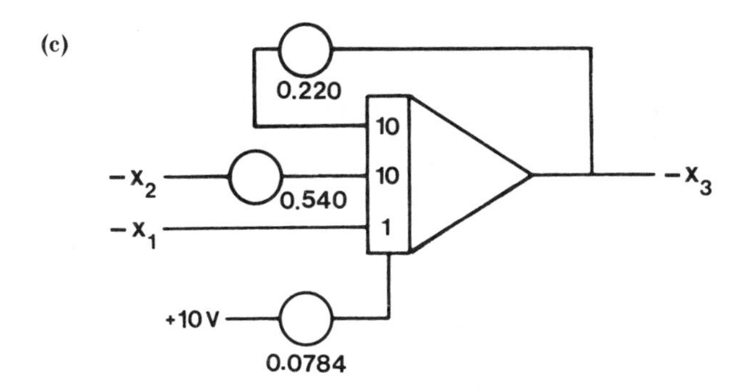

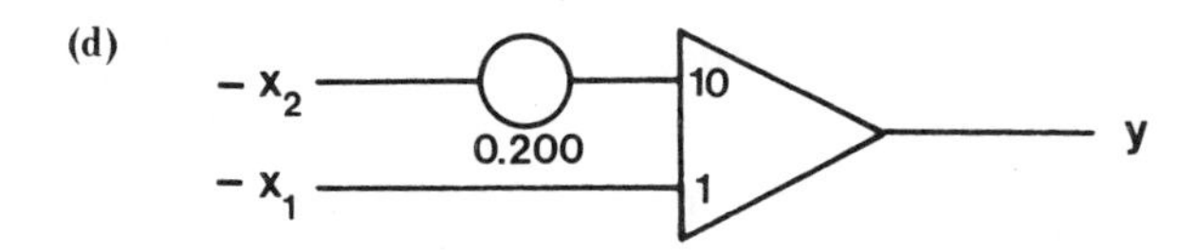

FIG. 6. Amplifier circuits for Example 1.

are obtained by potentiometers set at 0.54 and 0.22, followed by amplifier gains of 10. The circuit solving (18) is shown in Fig. 6(c). The output equation is solved by a summing amplifier and a potentiometer [see Fig. 6(d)].

The complete computer circuit is now obtained by joining the individual amplifier circuits in the appropriate way. Since the circuit for (18) requires $-x_2$ as an input, while the circuit for (17) produces $+x_2$, a summing amplifier with unit gain needs to be inserted to change the sign of the voltage. The complete circuit is shown in Fig. 7(a).

(b) *Zero initial conditions and a step input* $u(t) = 4 \times 1(t)$

In this case, (18) is replaced by

$$\dot{x}_3(t) = -x_1(t) - 5.4x_2(t) - 2.2x_3(t) + 4 \times 1(t) \tag{19}$$

The only modifications to the circuit required for (a) are to remove all initial condition voltages, and to provide a voltage representing the step input to a fourth input terminal of the left-hand amplifier. The simplest way to do this is to connect a constant voltage of 4 V to an input terminal having a gain of 1. Since the integrating amplifier is not connected to its inputs until the computer is switched from *reset* to *integrate*, the effect on the circuit is the same as if a step function were actually applied at that instant. The required circuit is shown in Fig. 7(b).

ΔΔΔ

We have been assuming that a unit of each of the dependent variables of the problem can be represented by 1 V of the corresponding voltage in the computer model, but this is often impossible. The computer circuit cannot operate with voltages outside the range given by the computer reference voltage (typically -10 V to +10 V), nor can the computer model give good accuracies with very small voltages. The latter is due to the limited precision with which attenuators, resistors, and capacitors can be made, and with which voltages can be measured. *Amplitude scaling* must then be used.

(a)

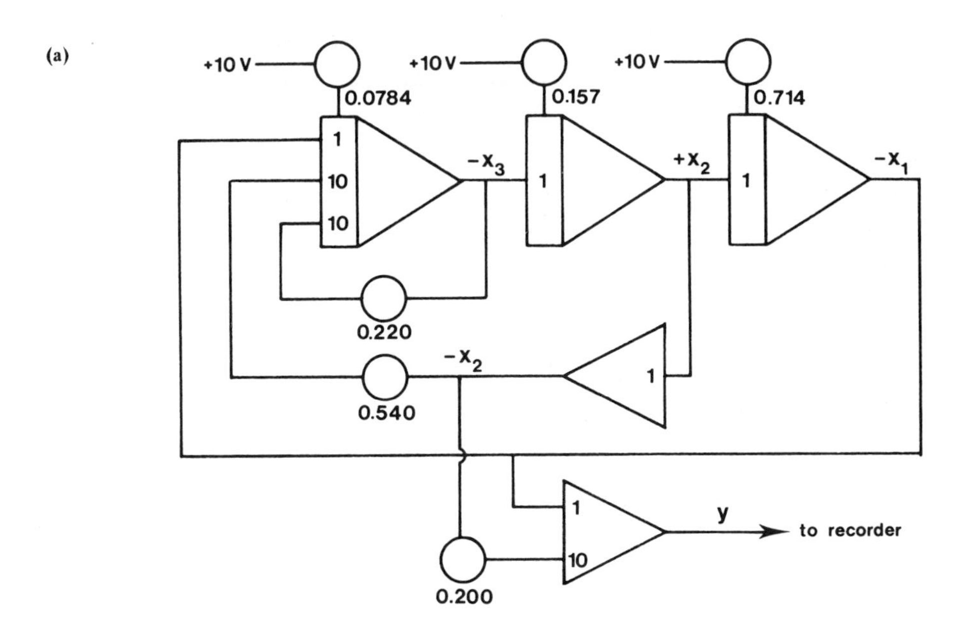

(b)

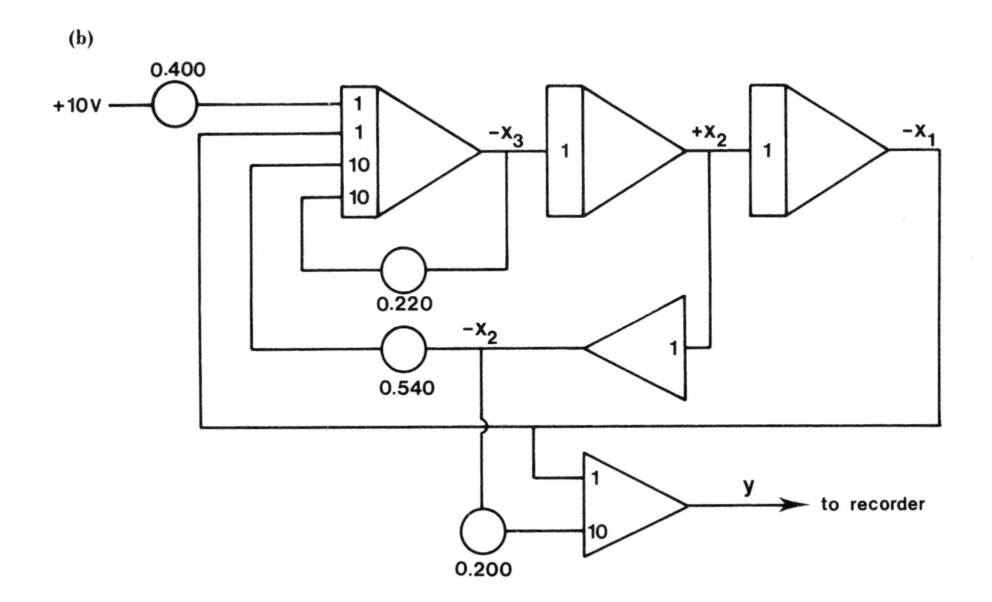

FIG. 7. Complete analog computer circuits for Example 1.

Suppose that from physical considerations, or from a run with an unscaled computer model, it is known that the maximum absolute values of each of the problem state variables x_i, inputs u_j, and outputs y_k are approximately $\bar{x}_i$, $\bar{u}_j$, and $\bar{y}_k$, respectively, and that these maximum values are each to be represented by, say, 8 V on the computer. Define the computer variables

$$\begin{aligned} \hat{x}_i &= \alpha_i x_i = \frac{8}{\bar{x}_i} x_i \qquad (i = 1,2,\ldots,n) \\ \hat{u}_j &= \beta_j u_j = \frac{8}{\bar{u}_j} u_j \qquad (j = 1,2,\ldots,m) \\ \hat{y}_k &= \gamma_k y_k = \frac{8}{\bar{y}_k} y_k \qquad (k = 1,2,\ldots,p) \end{aligned} \tag{20}$$

The original equations

$$\begin{aligned} \dot{x}_\ell(t) &= \sum_{i=1}^{n} a_{\ell i} x_i(t) + \sum_{j=1}^{m} b_{\ell j} u_j(t) \\ y_k(t) &= \sum_{i=1}^{n} c_{ki} x_i(t) \end{aligned} \tag{21}$$

may be rewritten as

$$\begin{aligned} \dot{\hat{x}}_\ell(t) &= \sum_{i=1}^{n} \hat{a}_{\ell i} \hat{x}_i(t) + \sum_{j=1}^{m} \hat{b}_{\ell j} \hat{u}_j(t) \\ \hat{y}_k(t) &= \sum_{i=1}^{n} \hat{c}_{ki} \hat{x}_i(t) \end{aligned} \tag{22}$$

where

$$\begin{aligned} \hat{a}_{\ell i} &= a_{\ell i} \frac{\alpha_\ell}{\alpha_i} \\ \hat{b}_{\ell j} &= b_{\ell j} \frac{\alpha_\ell}{\beta_j} \\ \hat{c}_{ki} &= c_{ki} \frac{\gamma_k}{\alpha_i} \end{aligned} \tag{23}$$

A computer circuit is then constructed to solve (22), instead of (21); numerical values of the original variables can be recovered from computer measurements via (20).

It may also happen that one unit of time cannot be conveniently represented by 1 sec. on the computer. The solution may take too long, or be too rapid for accurate recording, so that *time scaling* is required. This simply consists of replacing problem time t by "computer time" $\tau = \delta t$, where the scale factor $\delta > 1$ if the solution is to be slowed down on the computer, and $\delta < 1$ if it is to be speeded up. Functions of t are replaced by functions of τ, and derivatives with respect to t by derivatives with respect to τ, using the chain rule of differentiation:

$$\dot{x}_\ell = \frac{dx_\ell}{dt} = \frac{dx_\ell}{d\tau} \times \frac{d\tau}{dt} = \delta \frac{dx_\ell}{d\tau} \tag{24}$$

Hence (21) becomes

$$\frac{dx_\ell(\tau)}{d\tau} = \frac{1}{\delta}\left[\sum_{i=1}^{n} a_{\ell i} x_i(\tau) + \sum_{j=1}^{m} b_{\ell j} u_j(\tau)\right]$$

$$y_k(\tau) = \sum_{i=1}^{n} c_i x_i(\tau)$$

This shows that time scaling merely involves multiplying the inputs to *all* integrating amplifiers by the same factor $1/\delta$; the magnitudes of the dependent variables, i.e., their initial and maximum values, remain unchanged if the equations are in state-variable form.

Most analog computers are equipped with a facility for repetitive operation, which switches the computer automatically between initial condition and integrate modes at a frequency determined by the user. At the same time, a substantial time scale compression, typically by a factor of 100 or 500, is achieved by internally replacing all integrating capacitors selected on the patch panel by capacitors 100 or 500 times smaller, so that the integrating amplifier gains $G = 1/R_jC$ are all increased by this factor. This permits solutions to be displayed on an oscilloscope, and by varying the appropriate coefficient potentiometers, to investigate very rapidly how the solution is affected by changes in the coefficients of the system equation.

This capacity to provide the solution to a complex problem very rapidly is one of the attractive features of analog computers. On a digital computer, any mathematical problem must be solved by a sequence of elementary arithmetic operations; the solution time depends very much on the size and complexity of the problem. An analog computer, on the other hand, is a pure "parallel" machine: All operations occurring simultaneously in the real system are performed simultaneously on the computer, and the solution time is independent of the size of the problem. This, together with the possibility of time scale compression, gives the analog computer a speed advantage which is significant in the analysis and design of very complex systems, where large sets of differential equations must be solved many times before a final design evolves.

For large simulations, *hybrid computers* are frequently used. These machines are combinations of digital and analog computers. The analog portion is equipped with motor driven potentiometers or digital coefficient attenuators which can be remotely set by applying appropriate voltages. Interface equipment allows the mode of operation of the analog machine to be completely controlled by the digital computer, and voltages generated in the analog computer model can be returned to the digital machine by analog-digital converters. Thus an algorithm to determine an optimal set of system parameters can be implemented by the digital computer; at each iteration, the new set of parameters is determined from the previous solutions of the differential equations on the analog machine. Essentially, the analog computer is used as a special-purpose peripheral device in much the same way as, for example, a floating-point arithmetic processor.

For the solution of small-to-medium-size sets of differential equations, the speed advantage offered by analog computing is no longer significant enough to outweigh its main drawbacks: the comparatively high cost of precision analog devices, and the need for careful amplitude scaling if good accuracy is to be achieved. The latter can be a tedious and time-consuming process. Thus, for all but very large simulation tasks, analog computing is gradually being replaced by numerical methods which can be implemented efficiently and economically on any of the modern general purpose digital computers. Such methods are the subject of the next section.

7.2. NUMERICAL METHODS

As the name implies, a numerical solution of an ordinary differential equation consists of a table of values of the dependent variables at discrete values of the independent variable (usually time). The table is computed by arithmetic operations directly from the equation, without making use of a closed formula for its solution. The basic strategy is to approximate the solution of the *differential* equation

$$\dot{\underline{x}} = \underline{f}(\underline{x}(t),\underline{u}(t),t) \qquad \underline{x}(t_0) = \underline{x}_0$$

by the solution of a *difference* equation, usually a linear one, in which the solution at time $t_i = t_0 + ih$, $i = 1,2,\ldots$, is computed from one or more of the previously computed values $\underline{x}(t_i),\underline{x}(t_{i-1}), \ldots$, and from values of $\underline{f}$ at $t_i,t_{i-1},\ldots$, or even at some intermediate points. The difference equation is so chosen that the error between the exact and approximate solutions has a known upper bound, low enough to guarantee the required accuracy.

A numerical approach is essential if an analytic solution of the differential equation is not known, as is usually the case for nonlinear and time-varying equations. For a linear, time-invariant equation, on the other hand, it is possible to find an analytic solution using the methods of Chapters 3 and 4, and then to evaluate the resulting expression at the required values of time. However, if the solution is required at many points (e.g., for graph plotting) and if the state vector is at all large, it is simpler and usually much more efficient to compute the solution by a direct numerical method.

The development of general purpose methods for solving ordinary differential equations, and a study of their accuracy and efficiency, form a large branch of numerical analysis. The interested reader will find good elementary introductions in Refs. 51 and 125, and more thorough treatments in Refs. 80, 81, 83, 93, and 146. These methods are generally applicable to time-varying and nonlinear systems such as (1), and examples may be found in Problems 5 and 6. Here we shall restrict our attention to the linear, time-invariant case; this will enable us to obtain efficient methods more directly and still illustrate the basic principles involved.

Consider first the homogeneous differential equation

$$\dot{\underline{x}}(t) = \underline{A}\underline{x}(t) \tag{2}$$

where $\underline{x}(0) = \underline{x}_0^*$. We will denote the exact solution of (2) by $\underline{x}^*(t)$, while $\underline{x}(t)$ will refer to an approximate solution. In Theorem 3.1.4, it was shown that

$$\underline{x}^*(t) = e^{\underline{A}t}\,\underline{x}_0^* \tag{3}$$

From the composition property of matrix exponentials (Theorem 3.2.1), it is clear that at discrete instants of time $t_i = ih$, $i = 1, 2, \ldots$, the exact solution $\underline{x}_i^* = \underline{x}^*(t_i)$ can also be computed by means of the difference equation[†]

$$\underline{x}_{i+1}^* = e^{\underline{A}h}\,\underline{x}_i^* \qquad (i = 0,1,2,\ldots) \tag{4}$$

where

$$e^{\underline{A}h} = \underline{I} + \underline{A}h + \frac{\underline{A}^2h^2}{2!} + \frac{\underline{A}^3h^3}{3!} + \cdots \tag{5}$$

[†]Difference equations will be the subject of Chapter 15.

Now suppose we approximate $e^{\underline{A}h}$ by the first $r+1$ terms in its power series, and replace (4) by

$$\underline{x}_{i+1} = \underline{F}_r(h)\, \underline{x}_i \qquad (i = 0, 1, 2, \ldots) \tag{6}$$

where $\underline{x}_0 = \underline{x}_0^*$ and

$$\underline{F}_r(h) \triangleq \underline{I} + \underline{A}h + \frac{\underline{A}^2 h^2}{2!} + \cdots + \frac{\underline{A}^r h^r}{r!} \approx e^{\underline{A}h} \tag{7}$$

The solution $\underline{x}_i$ of (6) is then an approximate solution of (4) whose accuracy depends on how well $\underline{F}_r(h)$ approximates $e^{\underline{A}h}$. For identical starting points $\underline{x}_i^*$, the exact and approximate values after one time interval differ by a *truncation error*†

$$\underline{x}_{i+1}^* - \underline{x}_{i+1} = \delta\underline{F}_r(h)\, \underline{x}_i^* \tag{8}$$

where

$$\delta\underline{F}_r(h) \triangleq e^{\underline{A}h} - \underline{F}_r(h) = \frac{\underline{A}^{r+1} h^{r+1}}{(r+1)!} + \cdots \tag{9}$$

Assuming that the first term in the power series (9) dominates the others, we see that the truncation error is roughly proportional to h^{r+1}, and hence depends upon both the *step size* h and the *order* r of the approximation. Taking norms (see Definitions A.1.1 and A.6.6), we have

$$\begin{aligned} \| \underline{x}_{i+1}^* - \underline{x}_{i+1} \| &= \| \delta\underline{F}_r(h)\, \underline{x}_i^* \| = \| \delta\underline{F}_r(h)\, e^{-\underline{A}h}\, \underline{x}_{i+1}^* \| \\ &\le \| \delta\underline{F}_r(h)\, e^{-\underline{A}h} \| \; \| \underline{x}_{i+1}^* \| \end{aligned} \tag{10}$$

and the *relative truncation error* introduced in one step is bounded by

$$\frac{\| \underline{x}_{i+1}^* - \underline{x}_{i+1} \|}{\| \underline{x}_{i+1}^* \|} \le \| \delta\underline{F}_r(h)\, e^{-\underline{A}h} \| \tag{11}$$

---

†In practice, r is often chosen sufficiently large so that $\delta\underline{F}$ actually represents computer *roundoff error* rather than truncation error.

From this point on, we will denote $\underline{F}_r(h)$ by $\underline{F}$ and $\delta\underline{F}_r(h)$ by $\delta\underline{F}$ for simplicity.

It is not sufficient, of course, to merely bound the single-step error; we must also determine how this error propagates. First note from (4) and (6) that

$$\underline{x}_n^* = e^{\underline{A}hn}\,\underline{x}_0^* \qquad \underline{x}_n = \underline{F}^n\,\underline{x}_0^* \tag{12}$$

Next, the reader should verify that $\underline{F}$, $\delta\underline{F}$, and $e^{\underline{A}h}$ all commute with one another, and hence that

$$\begin{aligned}\underline{x}_n &= \underline{F}^n\underline{x}_0 = (e^{\underline{A}h} - \delta\underline{F})^n\,\underline{x}_0^* \\ &= [e^{\underline{A}hn} - ne^{\underline{A}h(n-1)}\delta\underline{F} + n(n-1)\,e^{\underline{A}h(n-2)}\delta\underline{F}^2 - \ldots \pm \delta\underline{F}^n]\,\underline{x}_0^*\end{aligned} \tag{13}$$

Since $\delta\underline{F}$ is small, we neglect the terms involving powers of $\delta\underline{F}$ and write

$$\underline{x}_n \simeq e^{\underline{A}hn}\,\underline{x}_0^* - n\delta\underline{F}e^{-\underline{A}h}e^{\underline{A}hn}\underline{x}_0^* = \underline{x}_n^* - n\delta\underline{F}e^{-\underline{A}h}\,\underline{x}_n^* \tag{14}$$

Subtracting and taking norms, we have

$$\|\,\underline{x}_n^* - \underline{x}_n\| \simeq \|\,n\delta\underline{F}e^{-\underline{A}h}\,\underline{x}_n^*\| \le n\|\,\delta\underline{F}e^{-\underline{A}h}\|\,\|\,\underline{x}_n^*\,\| \tag{15}$$

and the *relative error* accumulated in n steps is bounded approximately by

$$\frac{\|\,\underline{x}_n^* - \underline{x}_n\,\|}{\|\,\underline{x}_n^*\,\|} \le n\,\|\delta\underline{F}e^{-\underline{A}h}\| \simeq \frac{nh^{r+1}}{(r+1)!}\,\|\underline{A}^{r+1}\underline{F}_r(-h)\,\| \tag{16}$$

In other words, the bound (11) on the relative truncation error accumulates linearly with n. The absolute error in (15) can often be bounded independently of n (see Problem 2).

To achieve a specified accuracy, we may guess a step size h and order r, compute the bound in (16), and compare the result with our required level of accuracy. Then h and/or r may be adjusted accordingly and the process repeated until the desired accuracy is reached. In practice, the bound (16) is generally quite conservative, and the actual error can be one or two orders of magnitude smaller, as we shall see in Example 1. Indeed, it is often safe to omit the factor of n when (16) is used to select a step size.

The bound in (16) becomes somewhat more tractable if (A.6-19) is used to replace $\|\cdot\|$ by the Euclidian matrix norm $\|\cdot\|_E$, which does not involve eigenvalues. If $\|\underline{A}h\| \ll 1$, then (16) may be simplified further with the approximation $\underline{F}_r(-h) \approx \underline{I} - \underline{A}h$.

Forced Equations

Now we shall extend the discussion above to the case where an input (forcing function) is present,

$$\dot{\underline{x}}(t) = \underline{A}\underline{x}(t) + \underline{B}\underline{u}(t) \tag{17}$$

with $\underline{x}(0) = \underline{x}_0^*$. From Theorem 3.3.2, the exact solution of (17) is

$$\underline{x}^*(t) = e^{\underline{A}t}\underline{x}_0^* + \int_0^t e^{\underline{A}(t-\tau)}\underline{B}\underline{u}(\tau)\, d\tau \tag{18}$$

and at discrete instants of time $t_i = ih$, $i = 1, 2, \ldots$, this satisfies the difference equation

$$\underline{x}_{i+1}^* = e^{\underline{A}h}\underline{x}_i^* + \int_0^h e^{\underline{A}(h-\tau)}\underline{B}\underline{u}(t_i+\tau)\, d\tau \tag{19}$$

The convolution integral in (19) can be evaluated approximately by means of a numerical integration (or quadrature) formula, and we shall choose one whose accuracy is consistent with that of the approximation (7) for $e^{\underline{A}h}$. More specifically, since the truncation error in (7) is roughly proportional to h^{r+1}, we need a quadrature formula whose truncation error also varies as h^{r+1}.

There are many numerical integration schemes that can be used; among the simplest are the trapezoidal rule, Simpson's rule, and the parabolic rule with four subintervals (see Refs. 51, 80, 81, 83, 93, 125, and 146). For a scalar function f, these approximate the definite integral

$$I = \int_a^{a+h} f(\tau)\, d\tau \tag{20}$$

by

$$I_T = \frac{h}{2}[f(a) + f(a+h)] \tag{21}$$

$$I_S = \frac{h}{6}[f(a) + 4f(a+\frac{h}{2}) + f(a+h)] \tag{22}$$

$$I_p = \frac{h}{12}[f(a) + 4f(a+\frac{h}{4}) + 2f(a+\frac{h}{2}) + 4f(a+\frac{3h}{4}) + f(a+h)] \tag{23}$$

respectively. The truncation error is proportional to h^3 for I_T and h^5 for I_S; that of I_P is comparable to a five-point Newton-Cotes formula having an error proportional to h^7. We conclude that the trapezoidal rule is a suitable choice for $r = 2$, that Simpson's rule matches $r = 4$, and that the five-point parabolic rule can be used for $r = 6$.

Generalizing these quadrature formulas to the integration of vector-valued functions and using $\underline{F}_r(h)$ as defined in (7), we obtain the following three schemes for the approximate solution of (19):

$r = 2$:
$$\underline{x}_{i+1} = \underline{F}_2(h)\,\underline{x}_i + \underline{G}_1\underline{u}(t_i) + \underline{G}_2\underline{u}(t_i + h) \tag{24}$$

where

$$\underline{G}_1 = \frac{h}{2}\underline{F}_2(h)\,\underline{B}$$

$$\underline{G}_2 = \frac{h}{2}\underline{B}$$

$r = 4$:
$$\underline{x}_{i+1} = \underline{F}_4(h)\,\underline{x}_i + \underline{G}_3\underline{u}(t_i) + \underline{G}_4\underline{u}(t_i + \frac{h}{2}) + \underline{G}_5\underline{u}(t_i + h) \tag{25}$$

where

$$\underline{G}_3 = \frac{h}{6}\underline{F}_4(h)\,\underline{B}$$

$$\underline{G}_4 = \frac{4h}{6}\underline{F}_4(\frac{h}{2})\,\underline{B}$$

$$\underline{G}_5 = \frac{h}{6}\underline{B}$$

r = 6:
$$\underline{x}_{i+1} = \underline{F}_6(h)\underline{x}_i + \underline{G}_6\underline{u}(t_i) + \underline{G}_7\underline{u}(t_i + \frac{h}{4}) + \underline{G}_8\underline{u}(t_i + \frac{h}{2}) + \underline{G}_9\underline{u}(t_i + \frac{3h}{4}) + \underline{G}_{10}\underline{u}(t_i + h) \tag{26}$$

where

$$\underline{G}_6 = \frac{h}{12}\underline{F}_6(h)\,\underline{B}$$

$$\underline{G}_7 = \frac{4h}{12}\underline{F}_6(\frac{3h}{4})\underline{B}$$

$$\underline{G}_8 = \frac{2h}{12}\underline{F}_6(\frac{h}{2})\underline{B}$$

$$\underline{G}_9 = \frac{4h}{12}\underline{F}_6(\frac{h}{4})\underline{B}$$

$$\underline{G}_{10} = \frac{h}{12}\underline{B}$$

The most efficient way to compute the matrices $\underline{F}_r(h), \underline{F}_r(h/2)$, etc., is to compute and store the required powers of $\underline{A}$ first; this minimizes the number of full matrix multiplications, each of which requires $2n^3 - n^2$ arithmetic operations. Appropriate values of r and h can be determined using (16) once the powers of $\underline{A}$ have been computed.

We shall not discuss the propagation of truncation errors introduced in approximating (19) by one of the difference equations (24)-(26). However, it can be shown that whenever the homogeneous equation $\dot{\underline{x}} = \underline{A}\underline{x}$ is strictly stable and the input $\underline{u}(\cdot)$ in (17) is bounded, then any of the approximate methods is *stable* for sufficiently small h, in the sense that the norm of the accumulated error $\underline{\varepsilon}_i \triangleq \underline{x}_i^* - \underline{x}_i$ has a bound depending only on $\underline{x}(0)$ and $\underline{u}(\cdot)$, but not on the length of the solution interval. We ask the reader to prove this in Problem 2.

For a given step size h, higher-order methods are more accurate. Conversely, for a given accuracy, a higher-order method allows a greater step size, so that the solution is evaluated at fewer points. Up to a point, this saving can outweigh the slight increase in computational effort per step and the considerable increase in the effort required to compute the coefficients of the difference equation. If K is the number of points t_i at which the solution is computed, and n is the dimension of the system, then a count of the total number N of floating-point arithmetic operations required for each of the methods above yields the following expressions:

$$r = 2: \quad N = 2n^3 + 2n^2 + K(2n^2 + 3n) \tag{27}$$

$$r = 4: \quad N = 6n^3 + 12n^2 + K(2n^2 + 5n) \tag{28}$$

$$r = 6: \quad N = 10n^3 + 32n^2 + K(2n^2 + 7n) \tag{29}$$

As an example, suppose that a system of dimension $n = 20$ is to be solved for $0 \le t \le 10$. Suppose further that a step size $h = 0.1$ is necessary to give the required accuracy with the sixth-order method (26), but that the fourth-order method (25) requires half that spacing for the same accuracy. Then the total number of floating-point arithmetic operations is 233,000 for $r = 4$ and 187,000 for $r = 6$, a saving of 20%. As a general rule, the use of methods of order higher than four or six, with correspondingly larger time steps, will give little or no advantage, unless an exceptionally large set of equations is to be solved over a very large time interval. If (26) does not give sufficient accuracy, it will usually be more economical to reduce the step size than to increase the order of the method.

1. EXAMPLE. Consider the differential equations

$$\dot{\underline{x}}(t) = \begin{bmatrix} 0 & 1 & 0 & 0 & 0 \\ 0 & 0 & 1 & 0 & 0 \\ 0 & 0 & 0 & 1 & 0 \\ 0 & 0 & 0 & 0 & 1 \\ -8.08 & -21 & -20.16 & -10.64 & -4.4 \end{bmatrix} \underline{x}(t) + \begin{bmatrix} 0 \\ 0 \\ 0 \\ 0 \\ 8.08 \end{bmatrix} u(t) \tag{30}$$

where $\underline{x}(0) = \underline{0}$ and $u(t) = t \times 1(t)$.

Suppose that we require a solution accurate throughout the interval $0 \le t \le 14$ to within 0.1%. Evaluating (16) several times yields the following step sizes:

$$\begin{aligned} &\text{Second-order method } (r = 2): \quad h \approx 0.001 \\ &\text{Fourth-order method } (r = 4): \quad h \approx 0.05 \\ &\text{Sixth-order method } (r = 6): \quad h \approx 0.2 \end{aligned} \tag{31}$$

Results obtained by these methods for various step sizes are shown in Table 1. Only $x_1(t)$ is tabulated. The results indicate that the largest time steps h that can be taken for 0.1% accuracy throughout the interval are approximately 0.0625 for $r = 2$, 0.25 for $r = 4$, and 0.5 for $r = 6$. These values are substantially larger than the ones given in (31).

There are a number of important special cases of the differential equation (17) in which substantial computational savings are possible by exploiting special features of the input or the coefficients of the equation. We shall consider one of these, and leave the others as exercises for the reader (see Problems 3 and 4).

As we shall see in Chapters 8 and 9, automatic control systems are frequently designed so that their responses to step, or constant, inputs satisfy certain criteria. The simulation of such systems on a digital computer thus calls for the solution of (17) with a constant input $\underline{u}(t) = \bar{\underline{u}}$. The convolution integral in the exact difference equation (19) can then be evaluated analytically and does not require a numerical integration formula:

$$\int_0^h e^{\underline{A}(h-\tau)} \underline{B}\bar{\underline{u}} \, d\tau = \underline{Q}(h) \underline{B}\bar{\underline{u}} \tag{32}$$

where

$$\underline{Q}(h) = h\left[\underline{I} + \frac{\underline{A}h}{2!} + \frac{\underline{A}^2h^2}{3!} + \frac{\underline{A}^3h^3}{4!} + \cdots\right] \tag{33}$$

Then the vector

TABLE 1

Numerical Solution of Example 1

Order r	Time t	Exact solution $x_1^*(t)$	Solutions $x_1(t)$				Relative error $\|x_1^*-x_1\|/\|x_1^*\|$			
			h = .5	h = .25	h = .125	h = .0625	h = .5	h = .25	h = .125	h = .0625
2	2	0.16486	0.17328	0.16667	0.16555	0.16508	.051	.011	.0042	.0013
	6	3.41469	3.60166	3.42972	3.41785	3.41548	.055	.0044	.00093	.00023
	10	7.41019	7.56714	7.41674	7.41124	7.41043	.021	.00088	.00014	.000032
	14	11.4036	11.7313	11.4034	11.4023	11.4032	.029	.000018	.00011	.000035
4	2	0.16486	0.16836	0.16497	0.16486		.021	.00067	0	
	6	3.41469	3.42203	3.41479	3.41469		.0021	.000029	0	
	10	7.41019	7.42334	7.41054	7.41021		.0018	.000047	.0000027	
	14	11.4036	11.4253	11.4042	11.4036		.0019	.000053	0	
6	2	0.16486	0.16502	0.16486			.00097	0		
	6	3.41469	3.41561	3.41470			.00027	.0000029		
	10	7.41019	7.41179	7.41022			.00022	.0000041		
	14	11.4036	11.4058	11.4036			.00019	0		

$$\underline{q}_r(h) = h\left(\underline{I} + \frac{\underline{A}h}{2!} + \ldots + \frac{\underline{A}^{r-1}h^{r-1}}{r!}\right)\overline{\underline{B}\underline{u}} \tag{34}$$

gives an approximation to (32) whose accuracy is consistent with that of approximating $e^{\underline{A}h}$ by $\underline{F}_r(h)$; both approximations have truncation errors roughly proportional to h^{r+1}. Thus (17) can be solved by the difference equation

$$\underline{x}_{i+1} = \underline{F}_r(h)\,\underline{x}_i + \underline{q}_r(h) \tag{35}$$

and this is clearly more efficient than (24), (25), or (26). Again, the most economical way to compute $\underline{F}_r(h)$ and $\underline{q}_r(h)$ is to precompute and store the required powers of $\underline{A}$.

7.3 SUMMARY

This chapter has provided a brief introduction to the use of computers in solving ordinary differential equations. Our aim was to enable a reader with access to an analog or digital computer to supplement the analytical study of a system with a computer solution of its equations. As elsewhere in this book, only linear differential equations with constant coefficients were considered in detail.

In Sec. 7.1 we described how the operations in solving such equations could be implemented by analog electronic circuits based on the operational amplifier. A systematic procedure for constructing an analog computer circuit for a given equation was described by an example. Computer scaling of both independent and dependent variables was discussed.

The solution of linear differential equations by numerical methods was considered in Sec. 7.2. We saw that the solution of the differential equation at discrete, equally spaced values of time could be approximated by that of a linear difference equation, solved recursively. Difference equations of various degrees of accuracy were derived from the exact solution of the differential equation over one time step by truncating the power series for the matrix exponential and by using a numerical integration formula for the convolution integral. A bound on the relative error was derived and used to determine the order and step size required to achieve any given level of accuracy.

PROBLEMS

Section 7.1

1. Construct analog computer circuits having the following transfer functions:

   (i) $h(s) = \dfrac{0.5}{s + 0.5}$

   (ii) $h(s) = \dfrac{3s + 0.5}{s + 0.5}$

   (iii) $h(s) = \dfrac{1}{s^2 + 1}$

   (iv) $h(s) = \dfrac{s - 1}{s^2 + s + 5}$

2. A circuit which solves the equation

$$\dddot{y} + 0.15\ddot{y} + 0.2\dot{y} + 0.05y = 0.5\dot{u} + 0.1u$$

   where

$$y(0) = 5 \qquad \dot{y}(0) = 0 \qquad \ddot{y}(0) = 0.1 \qquad u(t) = 0.1 \times 1(t)$$

   can be constructed

   (a) By using Theorem 2.2.1 to first obtain an equivalent set of state equations

   or

   (b) By defining $v = \ddot{y} - 0.5u$, solving the equation

   $\dot{v} = -0.15\ddot{y} - 0.2\dot{y} - 0.05y + 0.1u$

   by an integrator, recovering $\ddot{y}$ from v with a summer, and integrating $\ddot{y}$ and $\dot{y}$.

Derive analog computer circuits using both methods. Which is simpler? Show that in (b) the integrator outputs form an alternative state vector.

3. What differential equation can be solved by the circuit shown in Fig. P1?

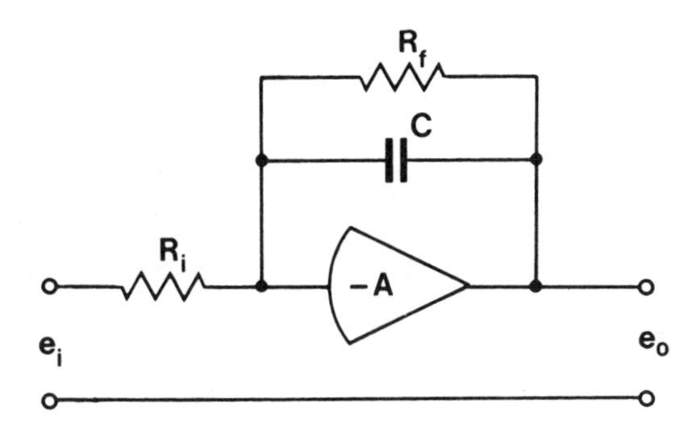

FIG. P1

4. A feedback system with the block diagram shown in Fig. P2 is to be simulated on an analog computer. Initial conditions are all zero. The parameter ξ lies between 0.1 and 1.

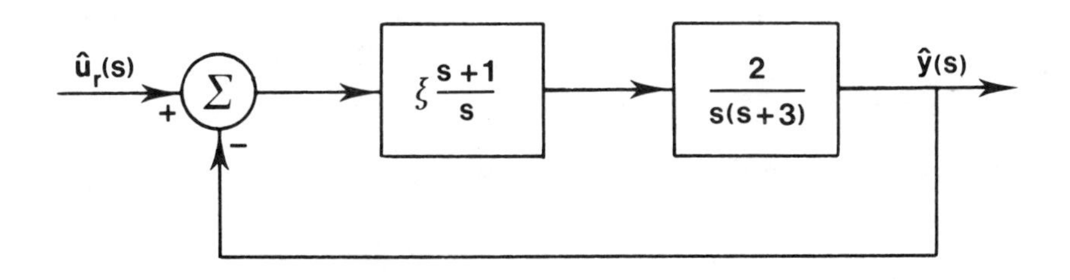

FIG. P2

(a) Derive an analog computer circuit which reflects the feedback structure of the system. Use separate circuits for each of the subsystems and then connect them as in the block diagram.

(b) Derive the transfer function relating y and u_r and then proceed as in Example 7.1.1.

Which circuit is more convenient to (i) derive and (ii) use?

5. The inputs or forcing functions for analog computer circuits must be supplied either by external signal generators or by separate analog computer circuits. Construct circuits whose outputs are the functions

(i) $u(t) = \alpha t\, 1(t) \qquad (\alpha > 0)$

(ii) $u(t) = (\alpha + \beta t + \gamma t^2)\, 1(t) \qquad (\alpha, \beta, \gamma > 0)$

(iii) $u(t) = \alpha(\cos \beta t)\, 1(t) \qquad (\alpha, \beta > 0)$

Hint: What differential equations do these functions solve?

Section 7.2

1. Draw a detailed flowchart and write a Fortran program implementing the algorithm suggested in the text for solving (7.2-17) numerically.

2. It was shown in the text that the *relative* error in solving $\dot{\underline{x}} = \underline{\underline{A}}\underline{x}$; $\underline{x}(0) = \underline{x}_0$ by the difference equation (6) has a bound which grows linearly as the solution proceeds. Prove that the *actual* error $\underline{\varepsilon}_i = \underline{x}_i^* - \underline{x}_i$ has a bound which is independent of the length of the solution interval, provided

(i) The differential equation is stable
(ii) $\underline{F}_r(h)$ is a sufficiently "close" approximation to $e^{\underline{A}h}$.

Show that the result can be extended to the forced equation $\dot{\underline{x}} = \underline{\underline{A}}\underline{x} + \underline{\underline{B}}\underline{u}$, solved by the methods in (24)-(26), provided that $\underline{u}(\cdot)$ is bounded.

3. Suppose that the coefficient matrix $\underline{A}$ in (17) is in companion form, such as in Example 7.2.1. For a general $n \times n$ matrix $\underline{A}$, the product $\underline{\underline{A}}\underline{x}$ requires $2n^2$ arithmetic operations, while for a companion matrix, it can be evaluated with only 2n operations.

The product $\underline{F}_r(h)\,\underline{x}_i$ required in the numerical solution of (17) by one of the methods (24)-(26) can be evaluated recursively without pre-computing $\underline{F}_r(h)$. For $r = 4$,

$$\underline{F}_4(h)\,\underline{x}_i = \underline{x}_i + h\underline{A}\,(\underline{x}_i + \frac{1}{2}\,h\underline{A}\,[\underline{x}_i + \frac{1}{3}h\underline{A}\,(\underline{x}_i + \frac{1}{4}h\underline{A}\,\underline{x}_i)])$$

and if $\underline{A}$ is in companion form, the right-hand side takes fewer operations (if n is sufficiently large) than the left-hand side. Investigate under what circumstances, if any, it pays to exploit the companion form of $\underline{A}$ in the numerical solution of (17) by the methods of the text.

4. Consider the numerical solution of a large set of n equations in which only a few of the state variables or output variables are needed. Thus, in the set of equations

$$\begin{aligned} \dot{\underline{x}}(t) &= \underline{A}\underline{x}(t) + \underline{b}u(t) \\ \underline{y}(t) &= \underline{C}\underline{x}(t) \end{aligned} \qquad \text{(P1)}$$

only the p-vector $\underline{y}$ is actually required, and $p \ll n$. Assume that the matrix $\underline{A}$ does not have a simple form, so that no shortcuts are possible in computing the coefficients of the difference equations (24), (25), or (26), or in solving them. There exists a state transformation $\underline{z}(t) = \underline{T}\underline{x}(t)$ such that (P1) takes the form

$$\begin{aligned} \dot{\underline{z}}(t) &= \hat{\underline{A}}\underline{z}(t) + \hat{\underline{b}}u(t) = \underline{T}\underline{A}\underline{T}^{-1}\underline{z}(t) + \underline{T}\underline{b}u(t) \\ \underline{y}(t) &= \hat{\underline{C}}\underline{z}(t) = \underline{C}\underline{T}^{-1}\underline{z}(t) \end{aligned} \qquad \text{(P2)}$$

where $\hat{\underline{A}}$ is in *tri-diagonal* form, that is, the entries of $\hat{\underline{A}}$ are all zero except on the main diagonal, the first superdiagonal and the first subdiagonal. Thus the i-th equation in (P2) has the form

$$\dot{z}_i = \hat{a}_{i,i-1}z_{i-1}(t) + \hat{a}_{i,i}z_i(t) + \hat{a}_{i,i+1}z_{i+1}(t) + \hat{b}_i u(t)$$

Details of methods of reducing $\underline{A}$ to tri-diagonal form are given in Ref. 179.

(a) Let K be the number of points t_i at which the solution is required. Show that provided n and K are sufficiently large, it is more efficient not to precompute the matrix $\underline{F}_r(h)$ in the solution of (P2) by the methods in the text, but to exploit the tri-diagonal form of $\underline{\hat{A}}$ in computing $\underline{F}_r(h)\,\underline{x}_i$.

(b) Assume that the effort required to compute $\underline{T}$ and $\underline{\hat{A}}$ is roughly comparable to that required to determine the coefficients of (25) or (26) if the original form (P1) were used. Investigate under what circumstances, if any, it is advantageous to solve (P1) by first transforming the equations to (P2) and then to use the methods in the text.

5. One standard general purpose method of numerically solving the set of differential equations

$$\dot{\underline{x}}(t) = \underline{f}(\underline{x}(t),\underline{u}(t),t) \qquad \underline{x}(t_0) = \underline{x}_0 \tag{P3}$$

is the *fourth-order Runge-Kutta method* [80, 81, 93, 146]:

$$\underline{s}_1 = \underline{f}(\underline{x}_i,\underline{u}(t_i),t_i)$$

$$\underline{s}_2 = \underline{f}(\underline{x}_i + \frac{h}{2}\underline{s}_1 , \underline{u}(t_i + \frac{h}{2}), t_i + \frac{h}{2})$$

$$s_3 = \underline{f}(\underline{x}_i + \frac{h}{2}\underline{s}_2 , \underline{u}(t_i + \frac{h}{2}), t_i + \frac{h}{2})$$

$$s_4 = \underline{f}(\underline{x}_i + h\underline{s}_3 , \underline{u}(t_i + h), t_i + h)$$

$$\underline{x}_{i+1} = \underline{x}_i + \frac{h}{6}(\underline{s}_1 + 2\underline{s}_2 + 2\underline{s}_3 + \underline{s}_4)$$

(a) Show that for the linear constant-coefficient equations (17), this method is almost identical with the fourth-order method (25). [The difference is in the accuracy with which the integrand $\exp[\underline{A}(h - \tau)]\underline{B}\underline{u}(\tau)$ is evaluated at $\tau = 0$, $h/2$, h.]

(b) Show that unless the solution is required at only a small number of points, then (25) is substantially more efficient than the Runge-Kutta method.

6. The Runge-Kutta method described in the previous problem is normally used only for starting the solution. Once the first few values are computed, it is customary to switch to a *predictor-corrector method.* One of the most popular of these is the fourth-order method of Hamming [146]:

An initial estimate of $\underline{x}_{i+1}$, $\underline{x}_{i+1}^0$, is obtained by an extrapolation formula, the "predictor":

$$\underline{x}_{i+1}^0 = \underline{x}_{i-3} + \frac{112(\underline{x}_i - \underline{x}_i^0)}{121} + \frac{4h}{3}(2\underline{f}_i - \underline{f}_{i-1} + 2\underline{f}_{i-2})$$

where $\underline{f}_i \triangleq \underline{f}(\underline{x}_i, \underline{u}(t_i), t_i)$.

This estimate of $\underline{x}_{i+1}$ is then refined by several iterations of the "corrector" formula

$$\underline{x}_{i+1}^{k+1} = \frac{9\underline{x}_i - \underline{x}_{i-2}}{8} + \frac{3h}{8}(\underline{f}_{i+1}^k + 2\underline{f}_i - \underline{f}_{i-1})$$

where

$$\underline{f}_{i+1}^k \triangleq \underline{f}(\underline{x}_{i+1}^k, \underline{u}(t_i + h), t_i + h).$$

The step size h is adjusted as the solution proceeds so that $\underline{x}_{i+1}^2$ and $\underline{x}_{i+1}^1$ agree to the level of accuracy required in the solution; then $\underline{x}_{i+1} = \underline{x}_{i+1}^2$. Thus the corrector formula is never applied more than twice at each time step. This adjustment of h provides an internal accuracy check not available with the Runge-Kutta method.

Apply this method to the linear constant-coefficient equation (17) and compare the computational effort of solving the resulting difference equation with that of solving (25). Under what circumstances is the predictor-corrector method more efficient? In particular, consider the solution of the set of differential equations describing a series connection of p scalar subsystems of low order:

$$\dot{\underline{x}}_1(t) = \underline{A}_1\underline{x}_1(t) + \underline{b}_1 u(t) \qquad y_1(t) = \underline{c}_1'\underline{x}_1(t)$$

$$\dot{\underline{x}}_2(t) = \underline{A}_2\underline{x}_2(t) + \underline{b}_2 y_1(t) \qquad y_2(t) = \underline{c}_2'\underline{x}_2(t)$$

....

$$\dot{\underline{x}}_p(t) = \underline{A}_p\underline{x}_p(t) + \underline{b}_p y_{p-1}(t) \qquad y_p(t) = \underline{c}_p'\underline{x}_p(t)$$

Part III

CONTROL OF LINEAR SYSTEMS

Chapter 8

SINGLE-LOOP FEEDBACK SYSTEMS

8.0 INTRODUCTION

In the previous chapters we have developed a number of the basic mathematical tools necessary for the analysis of linear, time-invariant dynamic systems. We have described some of the most important general features of their behavior and have discussed various methods of computing their responses when excited by known inputs.

We now turn our attention to problems of control. In the next five chapters, our aim will be to determine what inputs should be applied to a given system to achieve certain specified objectives. These objectives can take many forms. For example, one might wish to know what inputs should be applied to a system to drive it from some initial state at time t_0 to another state a short time later; this brings up the question of *controllability*, which will be examined in Chapter 10. Interesting *optimal control* problems arise if the state transfer must be achieved with a minimum expenditure of control energy,† or if it must occur in a minimum time subject to constraints on the magnitudes of the inputs. A brief introduction to optimal control concepts may be found in Chapter 13.

†See Corollary 10.1.5.

In this chapter and the next we shall deal with more traditional design objectives, which are often conflicting and somewhat less precise mathematically than those mentioned above, but which nevertheless are extremely useful. These include stability, fast transient response, steady-state tracking capabilities, immunity to disturbances, and insensitivity to parameter variations. We shall also restrict our attention for the present to scalar systems, i.e., those having only one output and one control input; extensions to multivariable systems will be taken up in Chapters 10-12.

8.1 FEEDBACK CONTROL

The class of control problems to be examined here is one of considerable engineering interest. We shall consider systems with several inputs, some known as *controls* because they may be manipulated and others called *external disturbances*, which are quite unpredictable. For example, in an industrial furnace we may consider the fuel flow, the ambient temperature, and the loading of material into the furnace to be inputs. Of these, the fuel flow is accessible and can readily be controlled, while the latter two are usually unpredictable disturbances.

In such situations, one aspect of the control problem is to determine how the controls should be manipulated so as to counteract the effects of the external disturbances on the state of the system. One possible approach to the solution of this problem is to use a continuous measurement of the disturbances, and from this and the known system equations to determine what the control inputs should be as functions of time to give appropriate control of the system state.

A different approach is to construct a *feedback system*, that is, rather than measure the disturbances directly and then compute their effects on the system from the model or system equations, we compare direct and continuous measurements of the accessible system states with signals representing their "desired values" to form an error signal, and use this signal to produce inputs to the system which will drive the error as close to zero as possible. Diagrams representing these two basic strategies of control are shown in Fig. 1.

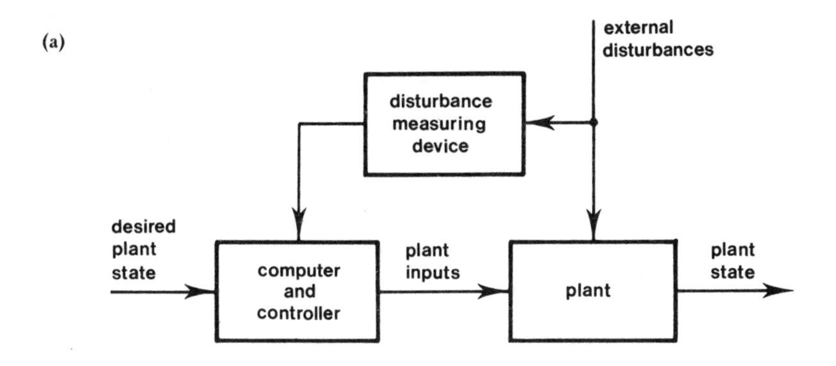

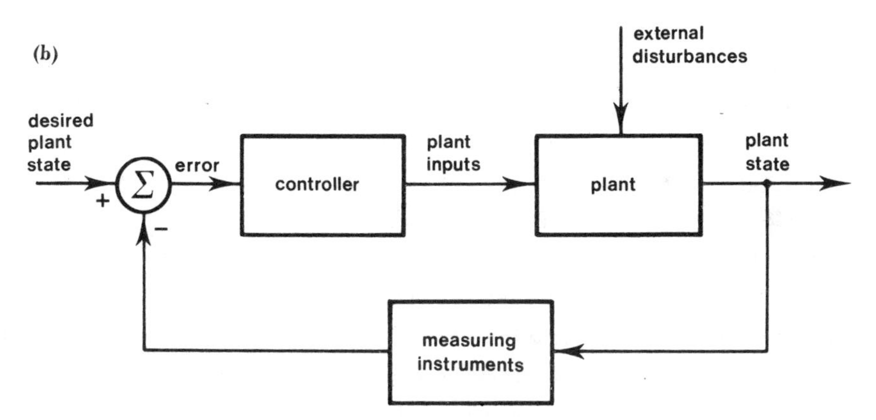

FIG. 1. Schematic representation of (a) open-loop and (b) closed-loop control strategies.

By some abuse of terminology, the former approach has come to be known as *open-loop* control, and the latter as *closed-loop* control. At first sight, the two approaches might appear to be essentially equivalent. Indeed, one might surmise that an open-loop control scheme is preferable since it is not necessary to wait until the disturbances have produced an undesirable change in the system state before corrective inputs can be computed and applied.

However, this advantage is more than outweighed by the disadvantages of open-loop control and the inherent advantages of feedback systems. First, in many cases the implementation of the open-loop control suggested above would require a very sophisticated (and hence expensive) computing device to determine the inputs required to counteract the predicted disturbance effects. Second, a feedback system turns out to be inherently far less sensitive to the accuracy with which a mathematical model of the system has been determined. Put another way, a properly designed feedback system will still operate satisfactorily even when the internal properties of the system change by significant amounts.

Another major advantage of the feedback approach is that by placing a "feedback loop" around a system which initially has quite unsatisfactory performance characteristics, one can in many cases construct a system with satisfactory behavior. Consider, for example, a rocket in vertical flight. This is essentially an inverted pendulum (see Example 2.3.2), balancing on the gas jet produced by the engine, and inherently unstable (any deviation of the rocket axis from the vertical will cause the rocket to topple over). It can, however, be kept stable in vertical flight by appropriate changes in the direction of the exhaust jet, which may be achieved by rotating the engine on its gymbal mountings. The only satisfactory way of achieving these variations in jet direction is to use a feedback strategy in which continuous measurements of the angular motions of the rocket in two mutually perpendicular vertical planes cause a controller to make appropriate adjustments to the direction of the rocket engine. Stabilization of an inherently unstable system could not be achieved in practice by an open-loop control strategy.

The mathematical tools required for the analysis and design of feedback systems differ according to the structural complexity of the systems to be controlled and according to the objectives the feedback control is meant to achieve.

In the simplest situation, one controls a single plant state variable, called the *output*, by means of adjustments to a single plant input. The problem is to design a feedback loop around the system which will ensure that the output changes in response to certain specified time functions or trajectories with an acceptable degree of accuracy. In either case, the transients which are inevitably excited should not be too "violent" or persist for too long.

In later chapters we will consider feedback controls for more complex multivariable systems, but for now let us focus upon the "simple" problem outlined above. Such problems, in fact, comprise the great majority of current applications of automatic control. In a typical situation, shown in Fig. 2, we are given a system, or *plant*, with *control input* u, *external disturbance* d, and *output* y, all scalars. The problem is to design a feedback system around the plant consisting of (a) a device which produces a continuous measurement y_m of the output, (b) a comparator in which this signal is subtracted from a *reference input* (or *set point*, or *desired output*) y_r, representing the desired value of the output, to produce an error signal e, and (c) a controller which uses the error signal e to produce an appropriate input u to the plant. We shall call this configuration a *single-loop feedback system*, a term which is meant to convey the essential feature that just one of the plant states (the output y) is to be controlled using only one input. The objective of the feedback system is to make the output y(t) follow its desired value $y_r(t)$ as closely as possible even in the presence of nonzero disturbances d(t). The ability of a system to do so under steady-state conditions is known as *static accuracy*.

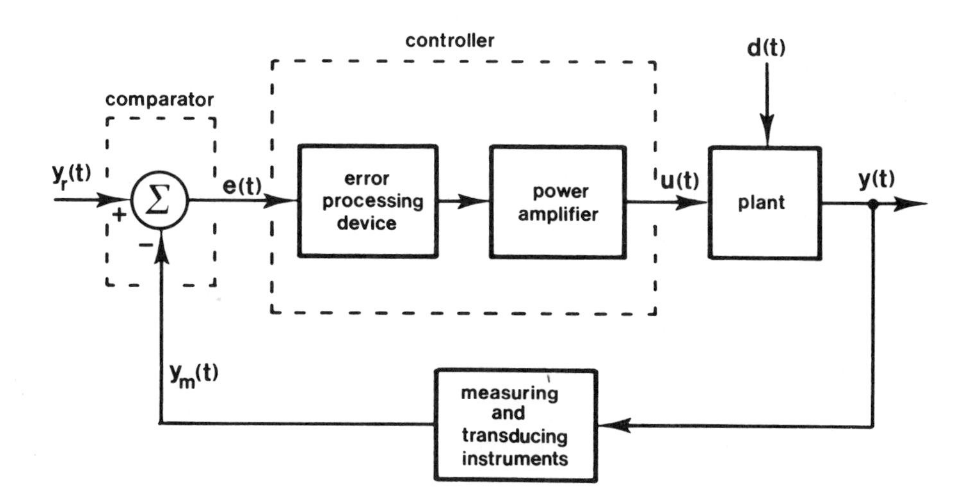

FIG. 2. Structure of single-loop feedback system.

Frequently y_r is a constant, in which case we call the feedback system a *regulator system*. An example is the speed control system of a turbine-generator set in a power station, whose main purpose is to maintain the generator speed as nearly constant as possible. Sometimes y_r is a prescribed nonconstant function of time, such as a ramp function:

$$y_r(t) = \alpha t \times 1(t) = \begin{cases} 0 & (t \leq 0) \\ \alpha t & (t > 0) \end{cases} \tag{1}$$

An example of this would be the control system for a radar antenna whose axis is to be kept aligned with the line of sight to an aircraft flying past with constant angular velocity. In this case, we refer to the system as a *tracking system*.

Single-loop feedback systems with the structure of Fig. 2 are often called *servomechanisms* because the controller usually includes a device giving considerable power amplification. For instance, in the control system of a hydroelectric turbine-generator set, the signals representing measured speed and desired speed might be voltages at a power level of milliwatts, while several hundred horsepower might be required to operate the main turbine valve regulating the water flow. This example also illustrates an important engineering constraint in the design of feedback control systems. In many applications, the plant and the activating device immediately preceding it operate at comparatively high power levels, and their dynamic properties, if unsatisfactory for some reason, can be changed only at the expense of changing large, powerful, and hence costly components. Therefore the design of a feedback system is preferably done in the low-power components of the feedback system, i.e., in the measuring elements and the controller.

We have stated above that a single-loop feedback system arises if the objective is to control just one of the state variables of the plant, which we have called the output. It is worth pointing out that such an objective is appropriate only if it is not really necessary to explicitly control all the other plant state variables. In such circumstances we usually know from our physical understanding of the structure of the system that the plant output cannot have satisfactory transient and steady-state behavior unless the other state variables behave similarly. For example, if we observe that a change in the voltage applied to an armature-controlled d.c. motor (Example 2.3.1) results in a smooth change of the armature velocity from one value to another, we can be certain that the other state variable of the motor, the armature current, does not undergo violent fluctuations in this period; an explicit knowledge of this current would be superfluous.

This suggests that for the single-loop feedback systems in this chapter, input-output relations will be a simpler and more appropriate description of the major components of the system than equations in the state-variable format. Indeed, most of the standard techniques for analyzing and designing single-loop systems are based on a transfer function description of the system components. In later chapters, when more stringent performance criteria are imposed or when several of the plant state variables are to be controlled simultaneously, sometimes by controlling several input variables, a system description using transfer functions is no longer adequate. It will then be necessary to consider the internal dynamics of the system and to use equations in state-variable form.

The use of block diagrams to display system structure was illustrated in Examples 4.1.4 and 4.1.5; there we also saw how the block diagrams could be simplified by combining transfer functions. In this way, any single-loop feedback system can usually be represented by a block diagram of the form of Fig. 3(a), where the $h_i(s)$ are the transfer functions of the various components or subsystems. A system model of this form will be the starting point for the analysis techniques we shall develop in this chapter. For most of our purposes, however, a somewhat simpler version will suffice. First, it is usually convenient to eliminate the block containing $h_2(s)$ by simply

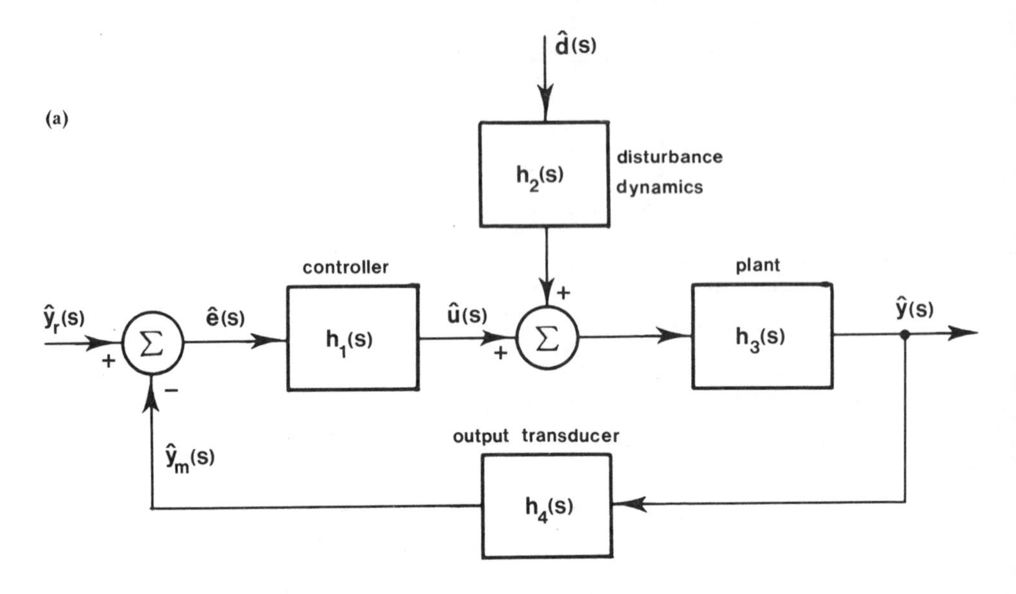

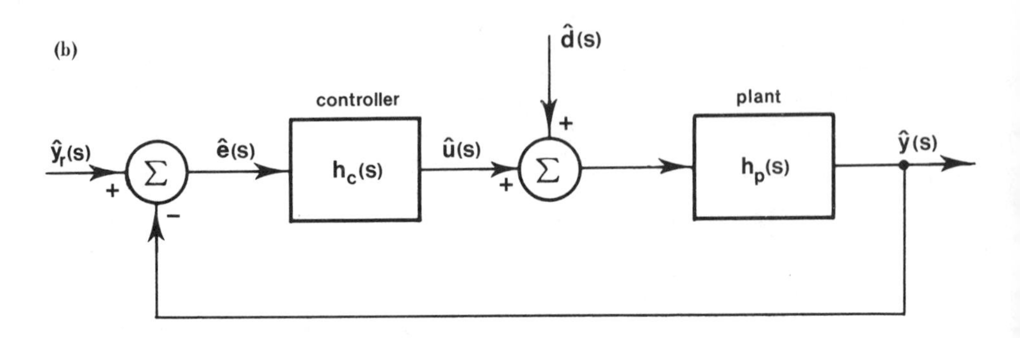

FIG. 3. Block diagrams of single-loop feedback systems.

agreeing to call its output the disturbance input. Second, it is also convenient in most cases to ignore the distinction between the output y(t) and its measured value $y_m(t)$; this means that the transfer function $h_4(s)$ in the feedback path can be combined into a single block with that of the plant, $h_3(s)$. With these alterations, our basic system model becomes the *unity feedback* arrangement shown in

Fig. 3(b), where $h_p(s)$ and $h_c(s)$ will simply be called the *plant* and *controller* transfer functions, respectively.

Before proceeding to analyze the behavior of such a system, it will be useful to establish some more nomenclature and a few preliminary results. In Corollary 6.1.2, application of a constant (step) input of amplitude α to a system with transfer function h(s) was seen to evoke a constant steady-state response of amplitude $\alpha h(0)$, and h(0) was consequently called the *static gain* of the system. However, if h(s) contains one or more factors of 1/s (i.e., if the system contains one or more *pure integrators* in series), then the static gain becomes $h(0) = \infty$. It will be convenient at this point to generalize the definition of gain so that it remains finite in the presence of integrators, and to adopt a transfer function notation which exhibits the gain explicitly.

1. DEFINITION. Suppose a transfer function h(s) has a pole at $s = 0$ of multiplicity m. We say that the transfer function is *normalized* if it is written in the form kg(s), where k is a constant and $[s^m g(s)]_{s=0} = 1$. The constant k is called the *gain* of the transfer function.[†]

ΔΔΔ

We will often need to determine the transfer function of a feedback system from the transfer functions of its components. This can be accomplished with the following formula, which is fundamental to the study of single-loop feedback systems.

2. LEMMA. Consider the single-loop feedback system shown in Fig. 4(a) with *feedforward transfer function* $h_1(s)$ and *feedback transfer function* $h_2(s)$. The *closed-loop transfer function* relating $\hat{y}$ to $\hat{u}$ is

---

[†]Such a transfer function is often said to be of *type m* since it represents a system containing m pure integrators in series. Also, k is often called a steady-state *error coefficient*, for reasons which will become clear in Sec. 8.2.

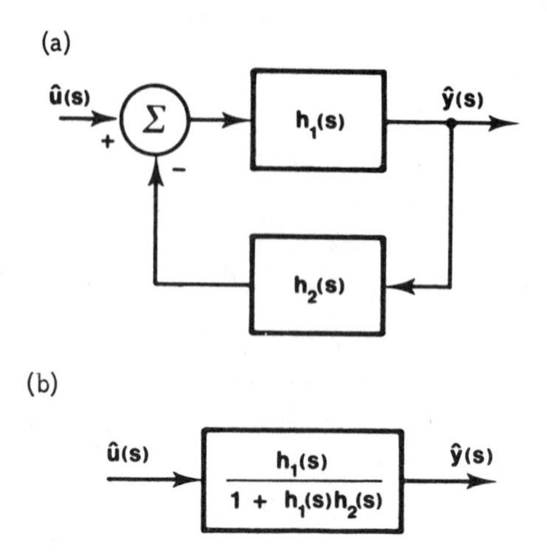

FIG. 4. Reduction of feedback system: (a) single-loop feedback system and (b) equivalent system.

$$h(s) = \frac{h_1(s)}{1 + h_1(s)\, h_2(s)} \tag{2}$$

as indicated in Fig. 4(b). (Also see Problem 7.)

Proof: From Fig. 4(a), we see that

$$\hat{y}(s) = h_1(s)[\hat{u}(s) - h_2(s)\hat{y}(s)] \tag{3}$$

and solving for $\hat{y}$, we have

$$\hat{y}(s) = \frac{h_1(s)}{1 + h_1(s)\, h_2(s)}\, \hat{u}(s) = h(s)\, \hat{u}(s) \tag{4}$$

ΔΔΔ

We return now to the feedback system of Fig. 3(b) with inputs y_r and d. Setting $d = 0$, the system fits into the framework of Lemma 2 with $h_1(s) = h_c(s)\, h_p(s)$ and $h_2(s) = 1$; so we conclude that the closed-loop *reference transfer function* relating y_r to the output y is

$$h_r(s) = \frac{h_c(s)\,h_p(s)}{1 + h_c(s)\,h_p(s)} \tag{5}$$

Similarly, if $y_r = 0$, then Lemma 2 again applies with $h_1(s) = h_p(s)$ and $h_2(s) = h_c(s)$, so that the closed-loop *disturbance transfer function* relating d to y is

$$h_d(s) = \frac{h_p(s)}{1 + h_c(s)\,h_p(s)} \tag{6}$$

Using the superposition property (Corollary 3.3.4), we see that the complete closed-loop response is

$$\begin{aligned} \hat{y}(s) &= h_r(s)\,\hat{y}_r(s) + h_d(s)\,\hat{d}(s) \\ &= \frac{h_c(s)\,h_p(s)}{1 + h_c(s)\,h_p(s)}\,\hat{y}_r(s) + \frac{h_p(s)}{1 + h_c(s)\,h_p(s)}\,\hat{d}(s) \end{aligned} \tag{7}$$

This may also be written as

$$\hat{y}(s) = [h_r(s) \quad h_d(s)]\begin{bmatrix} \hat{y}_r(s) \\ \hat{d}(s) \end{bmatrix} = \underline{H}(s)\begin{bmatrix} \hat{y}_r(s) \\ \hat{d}(s) \end{bmatrix} \tag{8}$$

where $\underline{H}(s)$ is the closed-loop transfer function matrix relating the inputs y_r and d to the output y.

We have assumed that the plant and controller transfer functions may always be expressed as ratios of polynomials,

$$h_p(s) = \frac{q_p(s)}{p_p(s)} \qquad h_c(s) = \frac{q_c(s)}{p_c(s)} \tag{9}$$

Substituting these expressions into (5) and (6) and clearing fractions, we find that

$$h_d(s) = \frac{q_p(s)\,p_c(s)}{p_p(s)\,p_c(s) + q_p(s)\,q_c(s)} \tag{10}$$

$$h_r(s) = \frac{q_p(s)\,q_c(s)}{p_p(s)\,p_c(s) + q_p(s)\,q_c(s)} \tag{11}$$

In Sec. 5.3, we saw that the basic characteristics of the transient response of a system are determined by its poles. It is clear, therefore, that the poles of the closed-loop transfer functions are of major importance in determining the dynamic properties of a feedback system. In fact, most of the analysis and design techniques we shall discuss in the sequel revolve around obtaining rapid estimates of the values of the poles, and then making changes in various parts of the feedback system in order to improve the pole locations.

Examining (10) and (11), we see that both closed-loop transfer functions have the same denominator polynomial and hence the same set of poles (except for possible cancellations). Therefore the same set of exponential functions, or modes, will appear in the transient response to a change in either the disturbance d or the reference input y_r, even though the relative weighting of the modes [which depends upon the closed-loop zeros in (10) and (11)] will in general be different. We conclude, therefore, that the basic characteristics of the transient response are determined by the feedback system itself and do not depend on just where the input enters the feedback loop. We shall see that this conclusion does not carry over to the steady-state behavior of the system; here the point at which the inputs enter the loop has an important bearing on the response.

With these observations, some additional terminology is in order.

3. *DEFINITION*. The denominator of the closed-loop transfer functions in (10) and (11) is called the *closed-loop characteristic polynomial*, and its roots are the *closed-loop poles* of the system. If the feedback loop is broken, the transfer function relating y_r to y is simply the product $h_c(s)\,h_p(s)$, which is called the *open-loop transfer function*; its poles and zeros are the *open-loop poles and zeros* of the system. If the open-loop transfer function is written in normalized form,

$$h_c(s)\,h_p(s) = kg(s) \tag{12}$$

then k is called the *loop gain* of the feedback system. The quantity

$$1 + h_c(s)\,h_p(s) = 1 + kg(s) \tag{13}$$

which appears in the denominator of (5) and (6), is known as the *return difference*.

ΔΔΔ

We also observe that the closed-loop zeros of the reference transfer function (11) are the same as the open-loop zeros of the system (i.e., the open-loop zeros of the plant and controller transfer functions). The closed loop zeros of the disturbance transfer function (10), on the other hand, are the open-loop zeros of the plant transfer function and the open-loop poles of the controller transfer function.

We conclude this section by studying in some detail two simple examples of feedback systems, obtained by placing feedback loops around first- and second-order plants. In particular, we shall illustrate the effect of varying one of the system parameters, the loop gain, on the transient and steady-state response of the feedback system, and on its sensitivity to changes in one of the plant parameters.

4. EXAMPLE.[†] Figure 5(a) shows a first-order feedback system utilizing *proportional control*, in which the plant input u is directly proportional to the error e, the controller being simply an amplifier of gain k_c. We assume for convenience that the plant is stable ($\lambda > 0$).

Consider first the response of the system to a reference input $y_r(t)$ when the disturbance input $d(t) \equiv 0$. From (7), the transform of the response is

$$\begin{aligned} \hat{y}(s) &= \frac{(k\lambda)/(s+\lambda)}{1 + (k\lambda)/(s+\lambda)}\,\hat{y}_r(s) \\ &= \frac{k}{1+k}\;\frac{\lambda(1+k)}{s+\lambda(1+k)}\,\hat{y}_r(s) \end{aligned} \tag{14}$$

[†]Also see Example 8.2.1 and Problem 8.2-1.

(a)

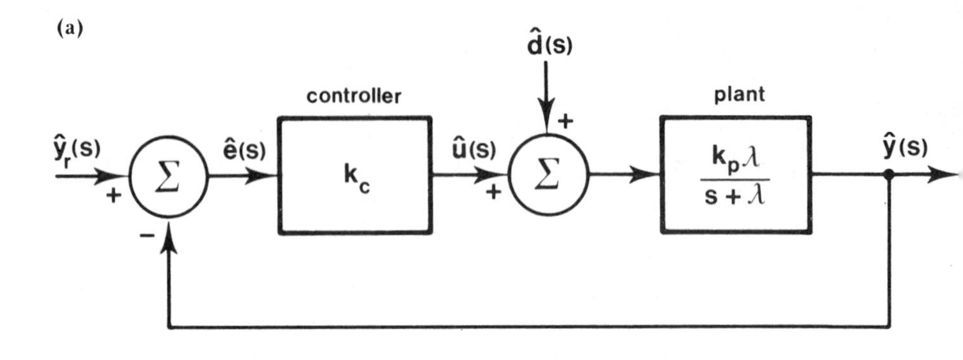

(b)

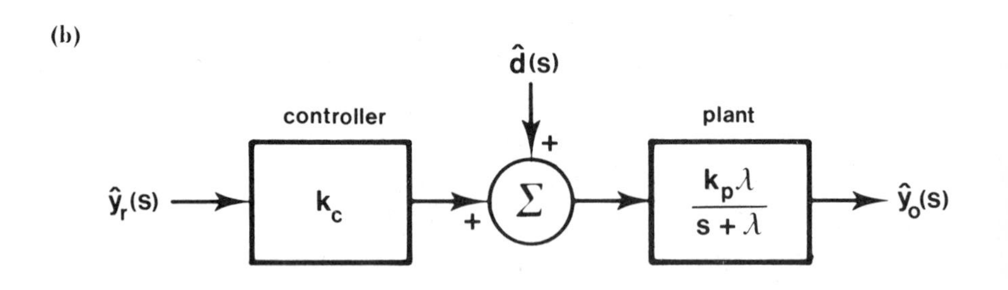

FIG. 5. System of Example 4: (a) closed-loop system with proportional control, and (b) open-loop system.

where the loop gain is

$$k = k_c k_p \tag{15}$$

If $y_r(t)$ is a step function of magnitude α, then $\hat{y}_r(s) = \alpha/s$ and

$$\hat{y}(s) = \frac{k}{1+k} \frac{\lambda(1+k)}{s+\lambda(1+k)} \frac{\alpha}{s} \tag{16}$$

The inverse transform yields the closed-loop step response

$$y(t) = \frac{k\alpha}{1+k}(1 - e^{-\lambda(1+k)t}) \tag{17}$$

which is shown in Figure 6(a) for various values of loop gain k.

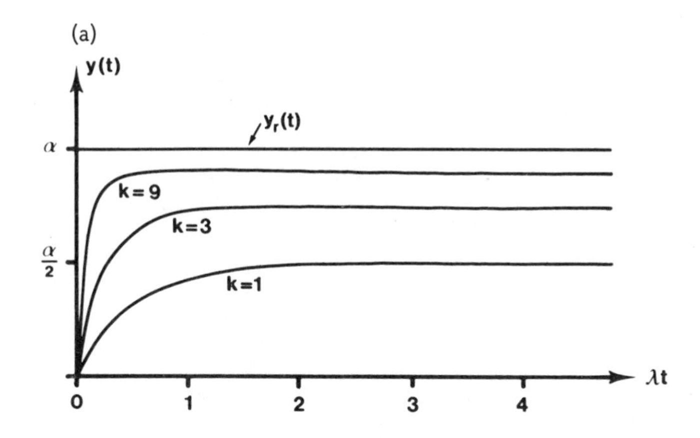

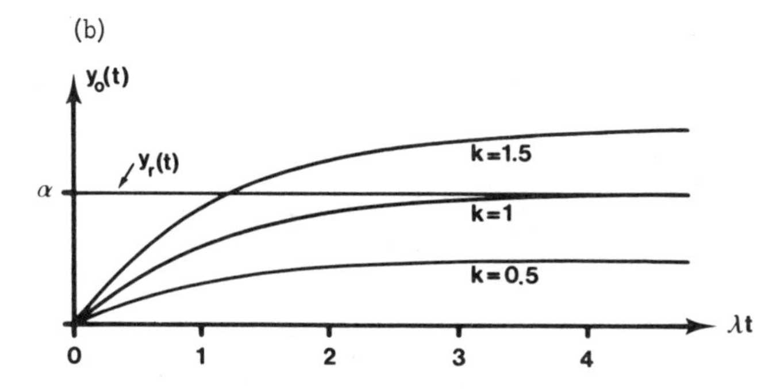

FIG. 6. Step responses for Example 4: (a) closed-loop responses and (b) open-loop response.

The corresponding open-loop system is shown in Fig. 5(b). With no disturbance input and a step of magnitude α applied to the reference input, the output transform is

$$\hat{y}_0(s) = \frac{k\lambda}{s + \lambda}\,\hat{y}_r(s) = \frac{k\lambda}{s + \lambda}\,\frac{\alpha}{s} \tag{18}$$

Thus the open-loop step response is

$$y_0(t) = k\alpha(1 - e^{-\lambda t}) \tag{19}$$

which is shown in Fig.6(b) for various values of loop gain k.

Comparing these step responses, we see that the speed of the closed-loop response can always be improved by increasing k, while the speed of the open-loop response is fixed by the plant parameter λ. Note also that the closed-loop system always has a steady-state error ($y \neq y_r$) which can be reduced by increasing k, while the open-loop system has no steady-state error if $k = 1$.

We will now show that the effect of the external disturbance is also reduced by feedback. Assume that $y_r(t) \equiv 0$ and that $d(t) = \delta 1(t)$. The response of the open-loop system in this case has the transform

$$\hat{y}_0(s) = \frac{k_p \lambda}{s + \lambda} \frac{\delta}{s} \tag{20}$$

and hence, from the Final Value theorem (B.1.12), the steady-state error caused by the disturbance is

$$\lim_{t \to \infty} [y_0(t)] = \delta k_p \tag{21}$$

By contrast, the response of the closed-loop system to the same disturbance is found from (7) to be

$$\hat{y}(s) = \frac{k_p}{1 + k} \left[\frac{\lambda(1 + k)}{s + \lambda(1 + k)} \right] \frac{\delta}{s} \tag{22}$$

so that the steady-state error is

$$\lim_{t \to \infty} y(t) = \frac{\delta k_p}{1 + k} \tag{23}$$

Comparing (21) and (23), we conclude that the steady-state error of the closed-loop system is less than that of the open-loop system, and can be made as small as desired by increasing the controller gain k_c, to which the loop gain k is directly proportional.

Finally, we shall demonstrate that the feedback system has a lower sensitivity to parameter changes than the open-loop system. Assume that the disturbance input d(t) is zero, and that the reference input is a nonzero constant, $y_r(t) = \alpha 1(t)$. Assume further that over an interval of time the plant gain changes from k_p to $k_p + \Delta k_p$. This causes a change in loop gain of

$$\Delta k = k_c \, \Delta k_p \tag{24}$$

or a percentage change of

$$\frac{\Delta k}{k} = \frac{\Delta k_p}{k_p} \tag{25}$$

Now consider the resulting change in the steady-state response of the open-loop system. From (19), this response is

$$\bar{y}_0 \triangleq \lim_{t\to\infty} y_0(t) = k\alpha \tag{26}$$

and clearly

$$\frac{\Delta \bar{y}_0}{\bar{y}_0} = \frac{\Delta k}{k} \tag{27}$$

If we now define a *steady-state sensitivity* S as the ratio of the percentage change in the steady-state response to the percentage change in a system parameter, it follows from (25) and (27) that the open-loop system has a steady-state sensitivity to plant gain changes of

$$S_0 = 1 \tag{28}$$

The steady-state response of the closed-loop system to the same input may be found from (17),

$$\bar{y} \triangleq \lim_{t\to\infty} y(t) = \frac{k\alpha}{1 + k} \tag{29}$$

and its steady-state sensitivity is

$$S = \frac{(k + \Delta k)/(1 + k + \Delta k) - k/(1 + k)}{k/(1 + k)} \frac{k}{\Delta k} = \frac{1}{1 + k + \Delta k} \tag{30}$$

This is clearly less than the open-loop sensitivity for any positive plant or controller gain and, moreover, can be made arbitrarily small by choosing a sufficiently large loop gain.

ΔΔΔ

To summarize, we have shown that the introduction of a feedback loop around a simple first-order plant results in

(i) Faster response to a step change in the reference input

(ii) Reduced effect of the external disturbance

(iii) Reduced sensitivity to a change in the plant gain

Moreover, we have seen that in all three respects the performance of the feedback system can be improved by increasing the controller gain k_c. However, it is important to stress that this conclusion is based on a model of the real system and is valid only for limited ranges of the system variables. Clearly, there is a limit to the torque a motor can develop or to the increase in steam flow one can obtain by opening a valve. Moreover, in deriving the system equations, the assumption has undoubtedly been made that some system components act so fast as to be practically instantaneous by comparison with others. If the loop gain is increased sufficiently to give a very fast overall system response, this assumption is no longer valid, and the model ceases to be an accurate representation of the real system.

Therefore, while it is true that the performance of the feedback system we have discussed can be improved by increasing the loop gain, knowledge of the actual system hardware and a measure of experience and judgment are necessary to decide the extent of the improvement possible in the real system.

5. *EXAMPLE*. The feedback system shown in Fig. 7 again uses proportional control, and differs from the previous example only in that the plant is now described by a second-order transfer function, with natural frequency $\bar{\omega}_n$ and damping ratio $\bar{\zeta}$. We shall confine our discussion to the effect of varying the loop gain on the transient response of the system. The reader is asked in Problem 4 to show that the steady-state sensitivity of this system to the external disturbance and to changes in plant gain varies with loop gain exactly as in the previous example.

Letting $d(t) \equiv 0$ and using (5), we find that the closed-loop transfer function relating the output y to the reference input y_r is

$$h_r(s) = \frac{k_c k_p \bar{\omega}_n^2}{s^2 + 2\bar{\zeta}\bar{\omega}_n s + \bar{\omega}_n^2 + k_c k_p \bar{\omega}_n^2} = \frac{k}{1+k} \frac{\omega_n^2}{s^2 + 2\zeta\omega_n s + \omega_n^2} \tag{31}$$

where $k = k_c k_p$ is the loop gain, and the closed-loop natural frequency

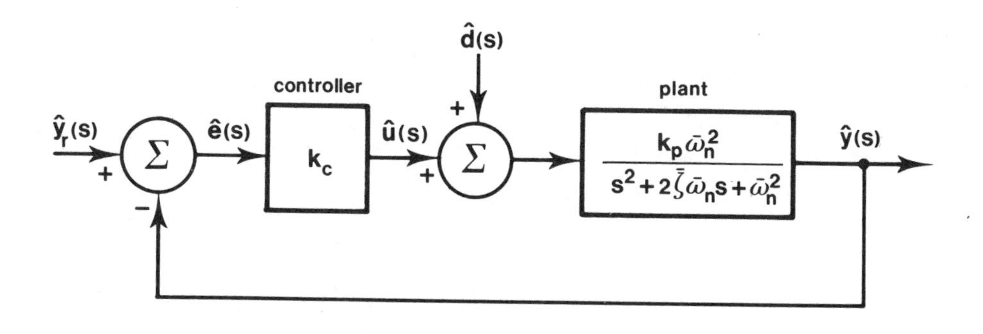

FIG. 7. Block diagram of system for Example 5.

and damping ratio are

$$\omega_n \triangleq \bar{\omega}_n \sqrt{1 + k} \tag{32}$$

$$\zeta \triangleq \frac{\bar{\zeta}}{\sqrt{1 + k}} \tag{33}$$

We see that application of proportional feedback to a second-order plant results in a second-order closed-loop system with poles

$$\begin{aligned} \lambda_1, \lambda_2 &= -\zeta\omega_n \pm \omega_n \sqrt{\zeta^2 - 1} \\ &= -\bar{\zeta}\bar{\omega}_n \pm \bar{\omega}_n \sqrt{\bar{\zeta}^2 - 1 - k} \end{aligned} \tag{34}$$

These closed-loop poles, which determine the basic characteristics of the closed-loop transient response, can be varied by adjusting the loop gain k. Note that this can be accomplished without altering the plant by simply changing the controller gain k_c.

Assume now that the plant itself is overdamped ($\bar{\zeta} > 1$), so that its open-loop poles are real and distinct:

$$\lambda_1, \lambda_2 = -\bar{\zeta}\bar{\omega}_n \pm \bar{\omega}_n \sqrt{\bar{\zeta}^2 - 1} \tag{35}$$

Each of these is indicated by an × in Fig. 8. Using (34), we can draw the path traced out on the complex plane by the closed-loop poles as the loop gain varies from zero to very large positive values. Such a graph, shown in Fig. 8, is called a *root-locus diagram*. The arrows indicate the direction in which the closed-loop poles move as k is increased.

For $k < \bar{\zeta}^2 - 1$ (that is, $\zeta > 1$), the closed-loop poles are real and distinct and lie between the plant poles. The response of the system to a step input of unit magnitude can then easily be computed from (31),

$$y(t) = \frac{k}{1 + k} \left(1 + \frac{\lambda_2}{\lambda_1 - \lambda_2} e^{\lambda_1 t} - \frac{\lambda_1}{\lambda_1 - \lambda_2} e^{\lambda_2 t} \right) \tag{36}$$

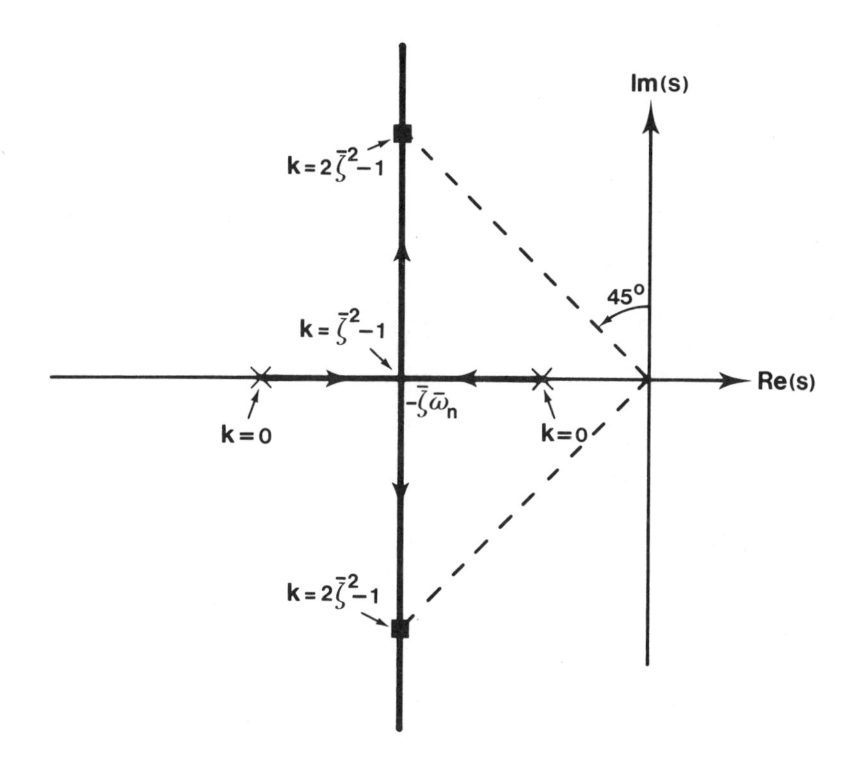

FIG. 8. Root-locus diagram for Example 5.

For $k > \bar{\zeta}^2 - 1$ (that is, $\zeta < 1$), the closed-loop poles are complex conjugates,

$$\begin{aligned} \lambda_1, \lambda_2 &= -\sigma \pm j\omega \\ &= -\zeta\omega_n \pm j\omega_n\sqrt{1-\zeta^2} = -\bar{\zeta}\bar{\omega}_n \pm j\bar{\omega}_n\sqrt{k+1-\bar{\zeta}^2} \end{aligned} \tag{37}$$

and the response to a unit step input is

$$y(t) = \frac{k}{1+k}\left[1 - \frac{\omega_n}{\omega} e^{-\sigma t} \sin(\omega t + \phi)\right] \tag{38}$$

where

$$\phi = \tan^{-1}\frac{\omega}{\sigma} = \cos^{-1}\zeta \tag{39}$$

This step response is plotted in Fig. 9 for $\bar{\zeta} = 1.34$ and various values of the loop gain k. The reader may verify, by computing the open-loop step response of the plant, that the introduction of proportional feedback increases the speed of response in all cases. Examining (36) and (38), we also see that the gain of the closed-loop system is again $k/(1+k)$, so that the steady-state error can only be made small (gain $\simeq$ 1) by choosing a large value of k.

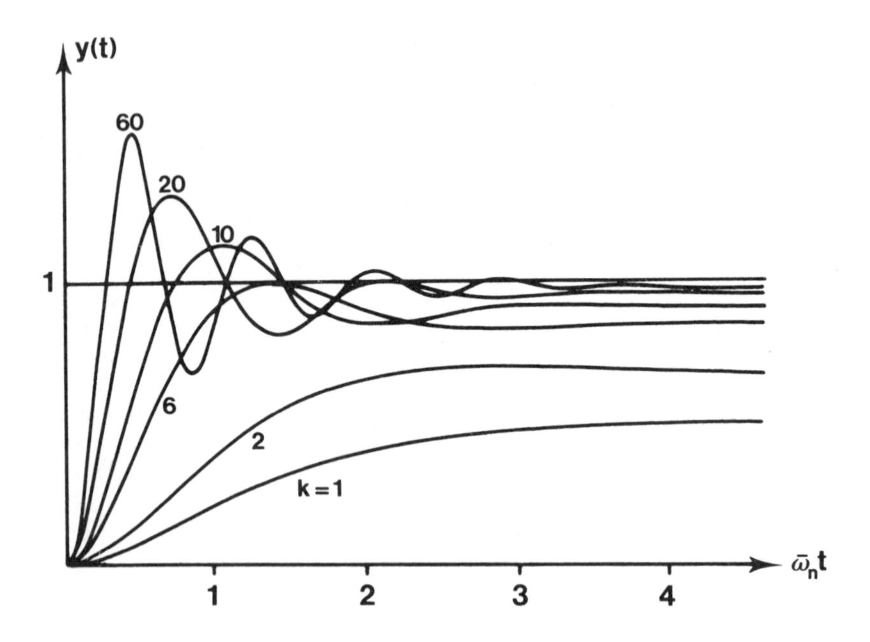

FIG. 9. Step responses for Example 5.

A large loop gain, however, results via (33) in a small damping ratio ζ and hence a very oscillatory response with a fast initial rise accompanied by large overshoots and poorly damped high-frequency oscillations about the final steady-state value. Thus, while the system always remains stable, large loop gains result in a poor transient response. We observed in Sec. 5.3 that a reasonably smooth step response with about 5% overshoot may be achieved with a damping ratio of 0.7, or $k = 2\bar{\zeta}^{-2} - 1$ in this example. This value of gain may or may not provide a satisfactory speed of response and steady-state error, depending on the original plant damping ratio $\bar{\zeta}$.

In Problem 4, the reader is asked to verify for this example that very low steady-state sensitivities to external disturbances and changes in plant gain also require the selection of a high loop gain. It is clear from the discussion above that in some cases very low sensitivity can be achieved only at the expense of poorly damped, oscillatory transients.

These are common problems in the design of feedback systems. With few exceptions, such as the first-order system considered in Example 4, the simultaneous achievement of high steady-state accuracy and satisfactory transient response merely by an adjustment of the loop gain are conflicting requirements, and the optimum choice of loop gain usually results from a compromise between them.

ΔΔΔ

8.2. STEADY-STATE PERFORMANCE

Most feedback control systems are required to perform satisfactorily as regulators. This means that a change in disturbance or plant load should not permanently drive the system output from its desired value, and that the error should tend to zero for any one of a whole range of reference inputs (set points). In some cases, it is also required that the system output accurately track simple nonconstant reference inputs, with a zero or small constant steady-state error.

In Examples 8.1.4 and 8.1.5, good static accuracy with constant reference inputs and disturbances was achievable by choosing a large value of the loop gain k although perfect static accuracy (zero steady-state error) was impossible except in the limit as $k \to \infty$. Moreover, we saw that there may be difficulties associated with a large loop gain, such as (in Example 8.1.5) excessive oscillations in the transient response. In this section, we will examine the use of integrators to eliminate steady-state errors without requiring excessive values of loop gain.

We saw in Chapter 6 that the steady-state response to a sinusoidal input is itself sinusoidal, and because of the phase shift between input and output, the steady-state error has the same shape. Therefore, we will not regard the steady-state response to sinusoids as giving an indication of the static accuracy of the system.

Before analyzing the steady-state performance of feedback systems in general, it will be useful to discuss a specific example.

1. EXAMPLE. Consider a simplified industrial continuous annealing furnace through which items to be heat-treated are carried continuously by means of a conveyor system. Let the furnace output y be its temperature; the external disturbance or load d is the rate of flow of material to be heat treated; the furnace input is the rate of oil flow u delivered to the burners and is regulated by means of metering pumps. Assume that the furnace temperature is measured by means of a thermocouple, whose voltage is compared with a voltage representing the required temperature, y_r. The difference or error voltage is amplified and used to change the oil flow rate delivered by the metering pumps by an amount directly proportional to the error voltage. If we assume that the rate of change of furnace temperature is proportional to the net rate of heat flow into the furnace, the system may be described by the block diagram shown in Fig.1. In accordance with our usual convention, all variables represent deviations from nominal values; they are zero if the furnace operates in its normal equilibrium condition. The reader will find it

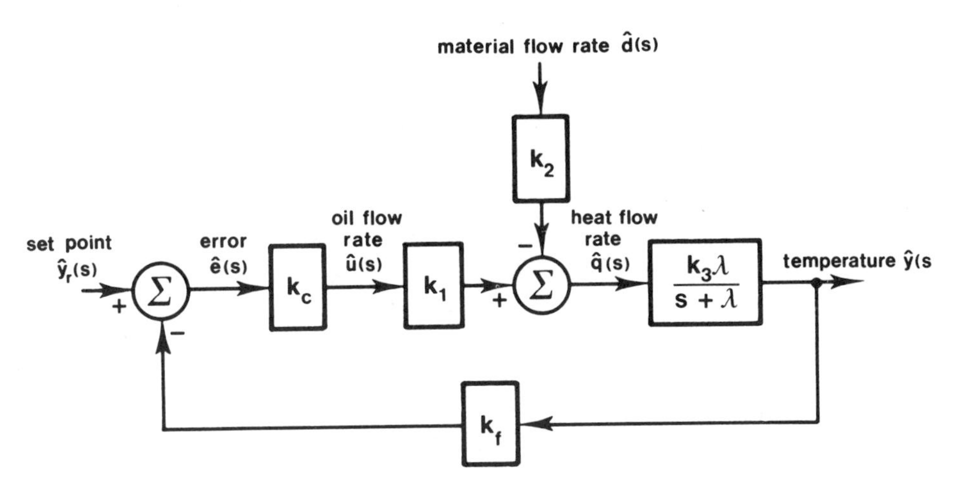

FIG. 1. Block-diagram model for annealing furnace of Example 1.

instructive to show how this model is derived, reduce it to the form of Fig. 8.1.3(b), and compare it with Example 8.1.4 (see Problem 1).

Now consider what happens if it is decided to change the set point from its equilibrium value by, say -100°, without any change in the rate of material flow through the furnace. This clearly requires a reduction in the rate of oil flow from the metering pumps, so that in the new steady-state condition the furnace must operate with a nonzero constant value of u. However, the change in oil flow rate from its original equilibrium value is directly proportional to the error voltage, the constant of proportionality being the controller gain. It follows that a reduction in oil flow rate and hence furnace temperature can be achieved only if there is a nonzero error. In other words, in its new equilibrium state the furnace cannot ever reach the new desired temperature. The error can be made small by choosing a large controller gain, but, as we have argued in Examples 8.1.4 and 8.1.5, there are physical limitations to this. Ideally, the error in the furnace temperature should tend to zero after the disappearance of the initial transients; its graph should

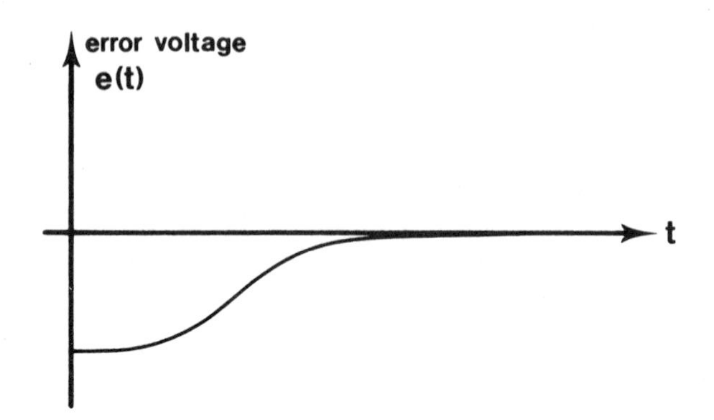

FIG. 2. Graph of error voltage for perfect static accuracy.

look somewhat like Fig. 2. Now this can happen only if the controller produces a constant reduction in fuel flow when excited by an error voltage of the form of Fig. 2. Looked at in this way, the requirement for perfect static accuracy is obvious: the controller must be such that its output contains a component proportional to the *time integral* of the error; perfect static accuracy cannot, in this case, be achieved with simple proportional control.

We reach a similar conclusion if we consider the response of the furnace to a change in the disturbance, i.e., the rate of flow of items to be heat treated, the set point or desired temperature remaining unchanged. An increase in the throughput will result in a drop in temperature, and if the temperature is to be restored to its original value, the fuel flow rate must be increased and maintained at a higher value, i.e., u(t) must become a positive constant. With a proportional control, this requires a nonzero steady-state error, so that the furnace temperature cannot ever return to its original equilibrium value [y(t) = 0]. As before, the only way to achieve perfect steady-state accuracy is to have the controller produce an increase in oil flow rate from an error signal which is positive for a time and then returns to zero; the controller must have an integrating action.

ΔΔΔ

We shall now extend these observations to the more general case of a single-loop feedback system of the form shown in Fig. 8.1.3(b), which is repeated here as Fig. 3 with the transfer functions expressed in normalized form (see Definition 8.1.1). The feedback system will be assumed stable although stability and static accuracy are really separate issues.

Generalizing the discussion in Example 1 to Fig. 3, we expect that perfect static accuracy for a nonzero constant reference input y_r can be achieved if y contains a term proportional to the time integral of e; in other words, *either* the controller or the plant transfer function should contain a factor of 1/s. With a nonzero constant disturbance input, perfect static accuracy results if u contains a term proportional to the integral of e; this requires a factor 1/s in the controller transfer function (integrating action in the plant does not help in this case). We shall now state and prove these qualitative conclusions more precisely, also allowing for nonconstant inputs.

2. *THEOREM*. In the absence of an external disturbance [$d(t) \equiv 0$], the response of the feedback system of Fig. 3 to a constant reference input

$$y_r(t) = \alpha 1(t) \tag{1}$$

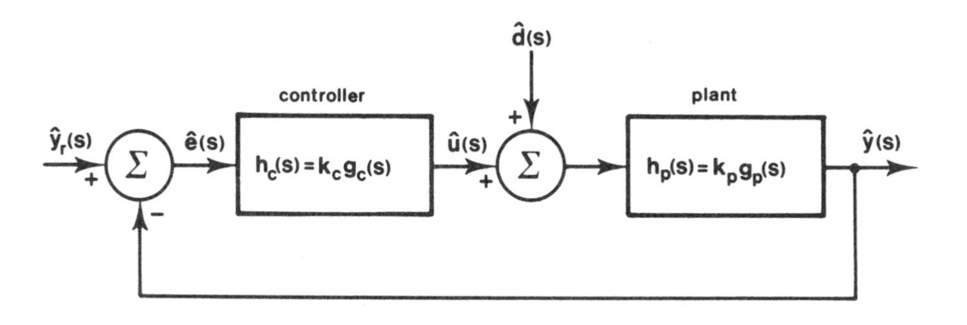

FIG. 3. Single-loop feedback system.

has a zero steady-state error if and only if the open-loop transfer function

$$kg(s) = k_c k_p g_c(s)\ g_p(s) \tag{2}$$

has at least one pole at $s = 0$. If not, there is a constant steady-state error

$$\lim_{t\to\infty} e(t) = \frac{\alpha}{1 + k} \tag{3}$$

The steady-state error in response to a reference input of the form

$$y_r(t) = \alpha t^n 1(t) \qquad (n = 1,2,3,\ldots) \tag{4}$$

is zero if and only if $kg(s)$ has a pole at $s = 0$ of multiplicity $m > n$. If $m = n$, the steady-state error is

$$\lim_{t\to\infty} e(t) = \frac{\alpha n!}{k} \qquad (n = m = 1,2,3,\ldots) \tag{5}$$

and if $m < n$, the error grows without bound.

Proof: It is easy to show (using Lemma 8.1.2, for instance, with input y_r and output e) that the transform of the error is given by

$$\hat{e}(s) = \frac{1}{1 + kg(s)}\,\hat{y}_r(s) \tag{6}$$

Since (1) is just (4) with $n = 0$, the input transform may be written as

$$\hat{y}_r(s) = \frac{\alpha n!}{s^{n+1}} \qquad (n = 0,1,2,\ldots) \tag{7}$$

Further, $kg(s)$ may be expressed in terms of its numerator and denominator polynomials as

$$kg(s) = \frac{kq(s)}{s^m p(s)} \tag{8}$$

with a pole of multiplicity m at $s = 0$ shown explicitly.

Substituting (7) and (8) into (6), we have

$$\hat{e}(s) = \frac{\alpha n!}{s^{n+1}[1 + kq(s)/s^m p(s)]} = \frac{s^m p(s)\alpha n!}{s^{n+1}[s^m p(s) + kq(s)]} \tag{9}$$

Recall from Definition 8.1.3 and the discussion preceding it that the quantity $1 + kg(s)$ occurs in the denominator of the system's closed-loop transfer function. Clearing fractions, the closed-loop characteristic polynomial is $s^m p(s) + kq(s)$, all of whose roots are stable by our earlier assumption.

Examining (9), we see that for $m \geq n + 1$, all factors of s in the denominator cancel out, leaving only the stable closed-loop poles, and we conclude that $e(t) \xrightarrow[t\to\infty]{} 0$.

If $m = n$, the transform of the error is

$$\hat{e}(s) = \frac{p(s)\alpha n!}{s[s^m p(s) + kq(s)]} \tag{10}$$

so that e(t) contains a constant term. Since $q(0)/p(0) = 1$ [recall that kg(s) was expressed in normalized form], we may use either a partial fractions expansion or the Final Value theorem (B.1.12) to conclude that

$$\lim_{t\to\infty} e(t) = \begin{cases} \dfrac{\alpha}{1 + k} & (n = m = 0) \\[2ex] \dfrac{\alpha n!}{k} & (n = m \geq 1) \end{cases} \tag{11}$$

If $m < n$, it is clear that e(t) contains a term proportional to t^{n-m}, and hence grows without bound.

ΔΔΔ

A feedback system is said to be of *type m* according to the number m of open-loop poles at the origin. Thus, according to Theorem 2, a type-1 system will follow a constant (step) reference input with perfect static accuracy, whereas a type-0 system exhibits steady-state error. More generally, a type-m system can follow a polynomial reference input of order up to $m - 1$ with no steady-state error (see Problem 6).

We turn now to the case of a nonzero disturbance input.

3. *THEOREM.* In the absence of a reference input $[y_r(t) \equiv 0]$, the response of the feedback system of Fig. 3 to a constant disturbance input

$$d(t) = \beta 1(t) \tag{12}$$

has a zero steady-state error if and only if the controller transfer function $k_c g_c(s)$ has at least one pole at $s = 0$. If not, there is a constant steady-state error

$$\lim_{t\to\infty} e(t) = \begin{cases} \dfrac{-\beta k_p}{1+k} = \dfrac{-\beta k_p}{1+k_c k_p} & \left(\text{if } \lim_{s\to\infty} g_p(s) = 1\right) \\[2ex] -\dfrac{\beta}{k_c} & \left(\text{if the } \textit{plant} \text{ has one or more poles at the origin}\right) \end{cases} \tag{13}$$

The steady-state error in response to a disturbance input of the form

$$d(t) = \beta t^n 1(t) \qquad (n = 1,2,3,\ldots) \tag{14}$$

is zero if and only if $k_c g_c(s)$ has a pole at $s = 0$ of multiplicity $m > n$. If $m = n$, the steady-state error is

$$\lim_{t\to\infty} e(t) = -\frac{\beta n!}{k_c} \qquad (n = m = 1,2,3,\ldots) \tag{15}$$

and if $m < n$, the error grows without bound.

Proof: The proof is similar to that of Theorem 2 and is left as an exercise for the reader (Problem 2).

ΔΔΔ

Note that a classification of feedback systems by type numbers is also valid with respect to disturbance inputs, provided that only open-loop controller poles at the origin are counted. In practice, however, the type number is nearly always associated with a reference input, and we will maintain this convention.

Theorems 2 and 3 verify the qualitative conclusions reached in Example 1: The static accuracy of a feedback system depends on the number of integrating devices through which the error signal passes in traversing the loop (each contributing a pole at $s = 0$ to the transfer function), and on the gain if the number of integrators is not sufficient to give a zero steady-state error. Moreover, the location of integrating devices in the loop is immaterial if static accuracy in following the *reference* input $y_r(t)$ is considered. In contrast, the static insensitivity to *disturbance* inputs depends not only on the presence of integrating devices but also on their location in the loop. Using the principle of superposition, it is easy to see that these conclusions remain valid when y_r and d are nonzero simultaneously (Problem 3).

Thus we see that good static accuracy in a feedback system requires integrators in the loop. Problem 4 asks for a comparison (with respect to steady-state performance) of an integral controller and one with both proportional and integral terms. These types of control will arise again in Example 8.3.5 and will be considered in more detail in the next chapter.

8.3. TRANSIENT ANALYSIS BY ROOT-LOCUS METHODS

Apart from achieving good steady-state accuracy, the main objective in the design of most feedback systems is to ensure a satisfactory transient response. The feedback system should be stable; it should have an adequate "margin of stability," so that small changes in system parameters do not make the system unstable; and transitions from one steady-state condition to another should neither be very slow nor very oscillatory.

The transient response of a linear system is a weighted sum of exponential functions. The set of functions appearing in the response depends on the system itself: each real pole $s = -\lambda$ of the transfer function contributes a real exponential $e^{-\lambda t}$; each complex conjugate pair of poles $s = -\sigma \pm j\omega$ contributes a damped sinusoid of the form $e^{-\sigma t} \sin(\omega t + \phi)$; in the case of repeated poles, these functions are multiplied by polynomials in t. The weighting factors of the various terms are not independent of the input to the system, but we shall see that a good deal of information about the relative magnitude of the weighting factors can be directly determined from the pattern of poles and zeros.

It follows that the problem of analyzing the transient behavior of a feedback system is principally a matter of determining its set of closed-loop poles. However, this is not a trivial task in any but the simplest systems and, moreover, does not give much guidance to the changes that should be made in the system if the pattern of closed-loop poles turns out to be unsatisfactory. A common procedure for designing single-loop feedback systems is one of trial and error: the plant and elements in the feedback path are given, and the designer assumes the simplest type of controller with which the steady-state accuracy requirements can be met, leaving the values of one or two parameters, such as controller gain, to be determined by transient response specifications. In this situation, it is obviously desirable to have a direct method of adjusting the free parameters and obtaining the "best" closed-loop pole configuration without having to repeatedly solve for all the roots of the characteristic polynomial.

The root-locus method has been developed for this purpose. It consists of a set of rules which allow for a rapid graphical determination of the paths (or "root-loci") traced out in the complex plane by the closed-loop poles as one of the system parameters is varied through a range of values. Most often, the variable parameter is the loop gain of the system, and for the sake of clarity, the details of obtaining root-locus plots will be developed for this specific case.

Recall that a root-locus diagram was constructed in Example 8.1.5 by explicitly determining the closed-loop poles of a second-order system as a function of the loop gain. More generally, we define the root-locus plot of a feedback system to be the directed graph traced out by its closed-loop poles as the loop gain varies from 0 to $+\infty$. The modifications required if a parameter other than loop gain is to be varied, or if negative loop gains are to be considered, will be pointed out in Example 8 and Problem 8, respectively.

We will again be considering the feedback system shown in Fig. 1. Recall from (8.1-7) that the closed-loop reference and disturbance transfer functions for this system both have the return difference

$$1 + h_c(s)h_p(s) = 1 + k_c k_p g_c(s)\, g_p(s) = 1 + kg(s) \tag{1}$$

as their denominator, where $k = k_c k_p$ is the loop gain and $kg(s)$ is the normalized open-loop transfer function. If the open-loop poles p_i, $i = 1, 2, \ldots, n$, and zeros z_i, $i = 1, 2, \ldots, r$, are known, then this can be written in factored form as

$$1 + kg(s) = 1 + \xi \frac{(s - z_1)(s - z_2) \cdots (s - z_r)}{(s - p_1)(s - p_2) \cdots (s - p_n)} \tag{2}$$

Note that ξ is *not* the loop gain, but is proportional to it,

$$k = \xi \frac{(-z_1)(-z_2) \cdots (-z_r)}{(-p_{\ell+1})(-p_{\ell+2}) \cdots (-p_n)} \tag{3}$$

where the first ℓ poles are assumed to be at the origin. Clearing fractions in (2), the closed-loop characteristic equation is

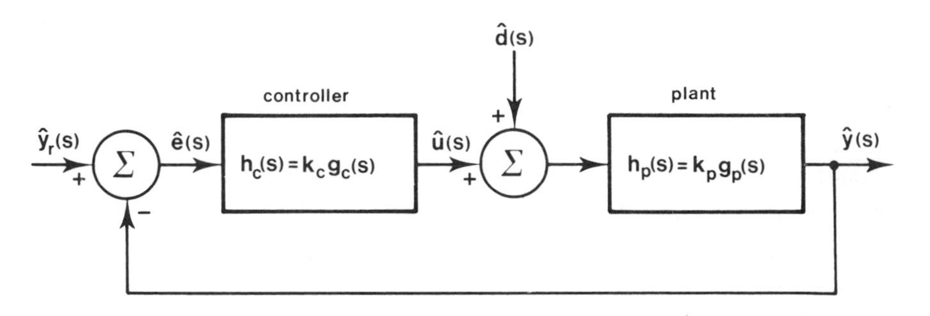

FIG. 1. Feedback control system.

$$(s - p_1)(s - p_2) \ldots (s - p_n) + \xi (s - z_1)(s - z_2) \ldots (s - z_r) = 0 \quad (4)$$

As usual, we assume that kg(s) is strictly proper $(r < n)$, so that (4) is an n-th order polynomial with real coefficients, some of which depend on the parameter ξ. It is well known that such a polynomial has exactly n roots, real or complex (the latter occurring in conjugate pairs), and that the roots are continuous functions of the polynomial coefficients. These two facts immediately give the first property of root-locus diagrams.

1. PROPERTY. As the parameter ξ in (4) varies between 0 and $+\infty$, the closed-loop poles trace out n continuous curves, or *branches* of the root-locus, and the branches form a pattern symmetrical about the real axis.

ΔΔΔ

The points on the root-loci are the roots of (4) for $0 \leq \xi < \infty$. Equivalently, they are the values of s for which (2) vanishes, that is,

$$\frac{(s - p_1)(s - p_2) \ldots (s - p_n)}{(s - z_1)(s - z_2) \ldots (s - z_r)} = -\xi \quad (5)$$

for some value of ξ. Writing each factor as a complex number in polar form, (5) becomes

$$\frac{|s - p_1|e^{j\theta_1}\ |s - p_2|e^{j\theta_2} \ldots |s - p_n|e^{j\theta_n}}{|s - z_1|e^{j\phi_1}\ |s - z_2|e^{j\phi_2} \ldots |s - z_r|e^{j\phi_r}} = \xi e^{\pm jm\pi} \quad (6)$$

$$(m = 1,3,5,\ldots)$$

Note that the factors $(s - p_i)$ and $(s - z_i)$ can be interpreted as vectors in the complex plane originating at p_i and z_i, respectively, and terminating at s, as illustrated in Fig. 2. Equating angles and magnitudes in (6) yields

$$\sum_{i=1}^{n} \theta_i - \sum_{i=1}^{r} \phi_i = \pm m\pi \qquad (m = 1,3,5,\ldots) \quad (7)$$

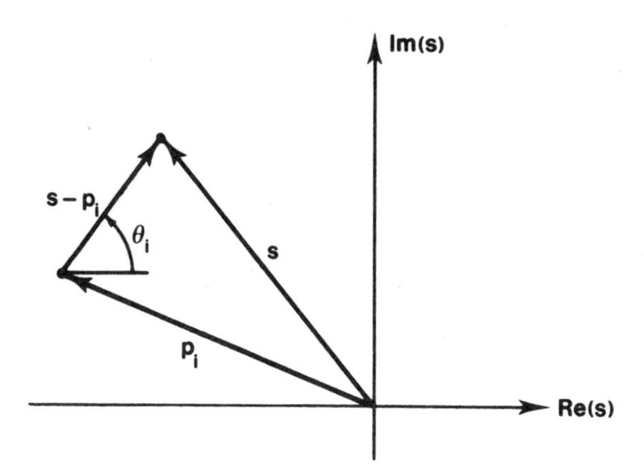

FIG. 2. Vectors in the complex plane.

and

$$\frac{|s - p_1|\ |s - p_2|\ \cdots\ |s - p_n|}{|s - z_1|\ |s - z_2|\ \cdots\ |s - z_r|} = \xi \tag{8}$$

Equation (7), known as the *angle condition*, is the key to the construction of root-locus plots. It states that a point s is on the root-locus, i.e., it is a closed-loop pole for some positive value of the parameter ξ, if and only if the sum of the angles of the vectors drawn from the poles to s less the sum of the angles of the vectors drawn from the zeros to s is an odd multiple of π.

Once it has been established that a particular point s is on a branch of the root-locus using (7), the corresponding value of the parameter ξ can be computed from the *magnitude condition* (8): it is equal to the product of the lengths of the vectors drawn from the poles to s, divided by the product of the lengths of the vectors drawn from the zeros to s. If there are no zeros, the denominator of the left-hand side of (8) is unity. The required value of loop gain can then be computed from (3).

The precise shape and location of the n branches of the root-locus must in general be determined by trial and error using (7).

This would be a formidable task were it not for the fact that the location of root-locus branches corresponding to very small and very large values of ξ can be very simply established from (7) and (8). In particular, we have the following property.

2. *PROPERTY.* Each branch of the root-locus originates at an open-loop pole, and each open-loop zero is a terminating point for one branch. The remaining $n - r$ branches go to infinity along straight-line asymptotes whose angles are given by

$$\phi_a = \frac{\pm m\pi}{n-r} \qquad (m = 1, 3, 5, \ldots) \tag{9}$$

and which intersect on the real axis at the point given by

$$\sigma_a = \frac{1}{n-r}\left(\sum_{i=1}^{n} p_i - \sum_{i=1}^{r} z_i\right) \tag{10}^\dagger$$

Proof: Clearly, the only points which can satisfy (5) when $\xi = 0$ are $s = p_i$, $i = 1, 2, \ldots, n$. Hence the n branches of the root-locus originate at the poles. As $\xi \to \infty$, (5) can be satisfied if $s \to z_i$, $i = 1, 2, \ldots, r$. It is also necessary to prove that every zero of multiplicity q is approached by precisely q branches of the root-locus. [See Problem 1(a).]

From (8), it is clear that there are no other finite points on the complex plane which can satisfy (5) as $\xi \to \infty$; since $n > r$, $n - r$ branches of the root-locus must go to infinity. To verify this, write (5) as

$$\frac{s^n + \alpha_{n-1} s^{n-1} + \ldots + \alpha_1 s + \alpha_0}{s^r + \beta_{r-1} s^{r-1} + \ldots + \beta_1 s + \beta_0} = -\xi \tag{11}$$

and carry out long division on the left-hand side to obtain

$$s^{n-r} + \gamma_1 s^{n-r-1} + \gamma_2 s^{n-r-2} + \ldots = -\xi \tag{12}$$

---

† This may be interpreted as the *center of gravity* of the open-loop pole-zero diagram, poles having mass +1 and zeros -1.

Writing $s = \zeta e^{j\psi}$, and considering large values of $s(\zeta \to \infty)$, we have

$$\zeta^{n-r} e^{j(n-r)\psi} \approx -\xi = \xi e^{\pm jm\pi} \qquad (m = 1,3,5,\ldots) \tag{13}$$

because the first term on the left-hand side of (13) dominates the others. Thus it is clear that on the branches which go to infinity, $s = \zeta e^{j\psi}$ must satisfy $\zeta^{n-r} \to \xi \to \infty$ and

$$\psi \to \phi_a = \frac{\pm m\pi}{n - r} \qquad (m = 1,3,5,\ldots) \tag{14}$$

This equation specifies the angles of the straight-line asymptotes, and the reader is asked in Problem 1(b) to show that these all intersect the real axis at the point specified by (10).

ΔΔΔ

We next establish several additional properties which are useful for the construction of root-locus diagrams, and then illustrate them with some examples.

3. PROPERTY. A point on the real axis lies on a branch of the root-locus if and only if it lies to the left of an *odd* number of *real* poles and zeros.

Proof: Let s be real. At s, the angle of the vector drawn from any real pole or zero to the left of s is zero; the sum of the angles of the two vectors from each complex conjugate pair of poles or zeros is precisely 2π; the angle of the vector from any real pole or zero to the right of s equals π; and by (7), the sum of the angles of all such vectors must be an odd multiple of π.

ΔΔΔ

4. PROPERTY. The location of *breakpoints* on the real axis, where two or more branches of the root-locus meet, turn, and then diverge, may be determined by treating ξ as a rational function (5) of s and solving the equation

$$\frac{d\xi}{ds} = 0 \tag{15}$$

for real s. All real breakpoints will be solutions of this equation, but some of them will correspond to negative values of ξ (or k).

In practice, solving (15) can be difficult if the order of the open-loop transfer function is greater than 3 or 4. In such cases it is often easier to consider a few trial points a short distance above the real axis, and then use the angle condition (7).

Proof: On the real axis, ξ may be considered to be a function (5) of s *with s constrained to be real* [Im(s) = 0]. With this restriction, each breakpoint must be a local maximum or minimum of $\xi(s)$, and hence a solution of (15).

ΔΔΔ

5. *PROPERTY.* Points where the root-loci cross the imaginary axis may be determined by substituting $s = j\omega$ into (5), separating real and imaginary parts, and solving for ω and ξ. Similarly, points of intersection with any vertical or horizontal line may be determined by substituting $s = \sigma + j\omega$ with either σ or ω fixed.

As with Property 4, this procedure becomes intractable for systems of even moderate complexity. Thus it is often easier to determine such points by trial and error, utilizing the angle condition (7).

Proof: All points on the root-locus are solutions of (5).

ΔΔΔ

6. *PROPERTY.* The angle at which the j-th branch of the root-locus departs from the open-loop pole p_j is given by

$$\psi_j = \pi - \sum_{\substack{i=1 \\ i \neq j}}^{n} \angle (p_j - p_i) + \sum_{i=1}^{r} \angle (p_j - z_i) \tag{16}$$

Similarly, the angle at which the k-th branch of the root-locus approaches the open-loop zero z_k is given by

$$\eta_k = \sum_{i=1}^{n} \angle(z_k - p_i) - \sum_{\substack{i=1 \\ i \neq k}}^{r} \angle(z_k - z_i) \tag{17}$$

Note that these formulae are needed only for complex open-loop poles and zeros; those on the real axis and at infinity are covered by Properties 2 and 3.

Proof: This is easily proved (Problem 2) by applying the angle condition to points on a circle of very small radius about the pole or zero in question. Indeed, it is often easier to simply follow this procedure than it is to remember Eqs. (16) and (17).

ΔΔΔ

We shall now illustrate the use of these properties by constructing root-locus diagrams for several simple systems, and show how the diagrams can be used to investigate the transient response obtainable by loop gain adjustment.

7. *EXAMPLE.* Consider the usual feedback system, shown in Fig. 3. Using root-locus diagrams, we shall evaluate how the transient response varies with controller gain k_c using three different types of controller:

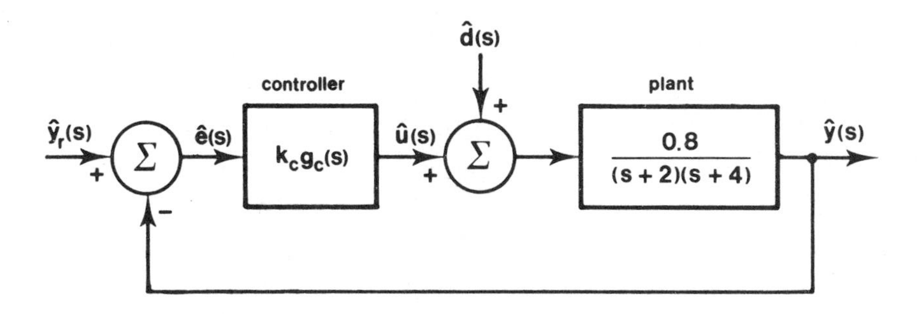

FIG. 3. Feedback system of Example 7.

(a) Proportional (P): $g_c(s) = 1$

(b) Integral (I): $g_c(s) = \frac{1}{s}$

(c) Proportional-plus-integral (PI):

$$g_c(s) = \frac{5}{8} + \frac{1}{s} = \frac{s + 1.6}{1.6s}$$

From (8.1-5) and (8.1-6), the closed-loop transfer functions are

$$h_r(s) = \frac{kg(s)}{1 + kg(s)} \qquad h_d(s) = \frac{k_p g_p(s)}{1 + kg(s)} \tag{18}$$

where

$$kg(s) = \frac{0.8}{(s + 2)(s + 4)} \; k_c g_c(s) \tag{19}$$

is the open-loop transfer function. Referring to (2) - (5), the root-loci consist of points where

$$1 + kg(s) = 1 + (0.1\ k_c)\frac{8g_c(s)}{(s + 2)(s + 4)} = 0 \tag{20}$$

and the loop gain is $k = 0.1k_c$.

(a) *Proportional control:* $k_c g_c(s) = k_c$.

For this control law the system is a particular case of Example 8.1.5 with $\bar{\omega}_n = \sqrt{8}$ and $\bar{\zeta} = 3/\sqrt{8}$. The closed-loop poles were found explicitly in (8.1-34), and the root-locus diagram is repeated in Fig. 4. The reader should verify that this diagram has the properties derived above, although they were not actually required for its construction.

For loop gains large enough to place the poles in the complex portion of the locus, the transient response is a damped sinusoid of the form $e^{-3t} \sin(\omega t + \phi)$. The reader will recall from Section 5.3 that a reasonably good response results if $\omega = 3$, so that the closed-loop poles lie at angles of 45° from the negative real axis, as indicated by ■ in Fig. 4. Using (8.1-37), we see that this requires

$$\bar{\omega}_n \sqrt{k + 1 - \bar{\zeta}^2} = 2\sqrt{2}\,\sqrt{k + 1 - \frac{9}{8}} = 3 \tag{21}$$

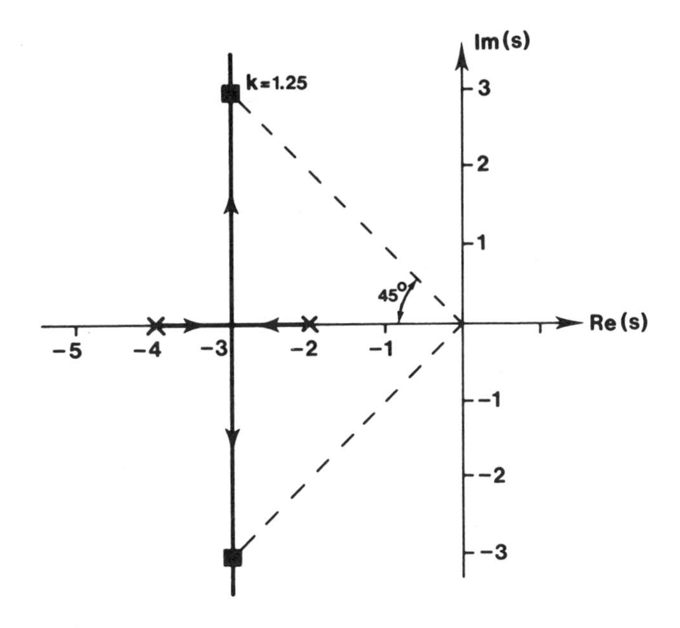

FIG. 4. Root-locus diagram for Example 7(a).

or a loop gain of $k = 1.25$. Thus $\xi = 10$, and the controller gain is $k_c = 12.5$.

(b) *Integral control:* $k_c g_c(s) = k_c/s$.

In this case u is the time integral of $y_r - y$, and (2) becomes

$$1 + kg(s) = 1 + \frac{0.8k_c}{s(s+2)(s+4)} = 0 \tag{22}$$

or

$$s(s+2)(s+4) = -0.8\,k_c = -\xi \tag{23}$$

The system now has three open-loop poles, at $s = 0$, -2, and -4, and no open-loop zeros. According to Theorems 8.2.2 and 8.2.3, the closed-loop system will therefore be capable of following constant reference and disturbance inputs with no steady-state error.

Properties 1 and 2 establish that the root-locus has three branches originating at the poles and going to infinity along straight-line asymptotes with angles $\pm\pi/3$ and π. These are the only distinct angles which can be obtained from (9) by letting m take on all possible integer values. By (10), the asymptotes intersect the real axis at

$$\sigma_a = \frac{0 - 2 - 4}{3 - 0} = -2 \tag{24}$$

By Property 3, root-locus branches exist on the real axis between the poles at 0 and -2, and everywhere to the left of the pole at -4, but not on the positive real axis or between the poles at -2 and -4 (see Fig. 5).

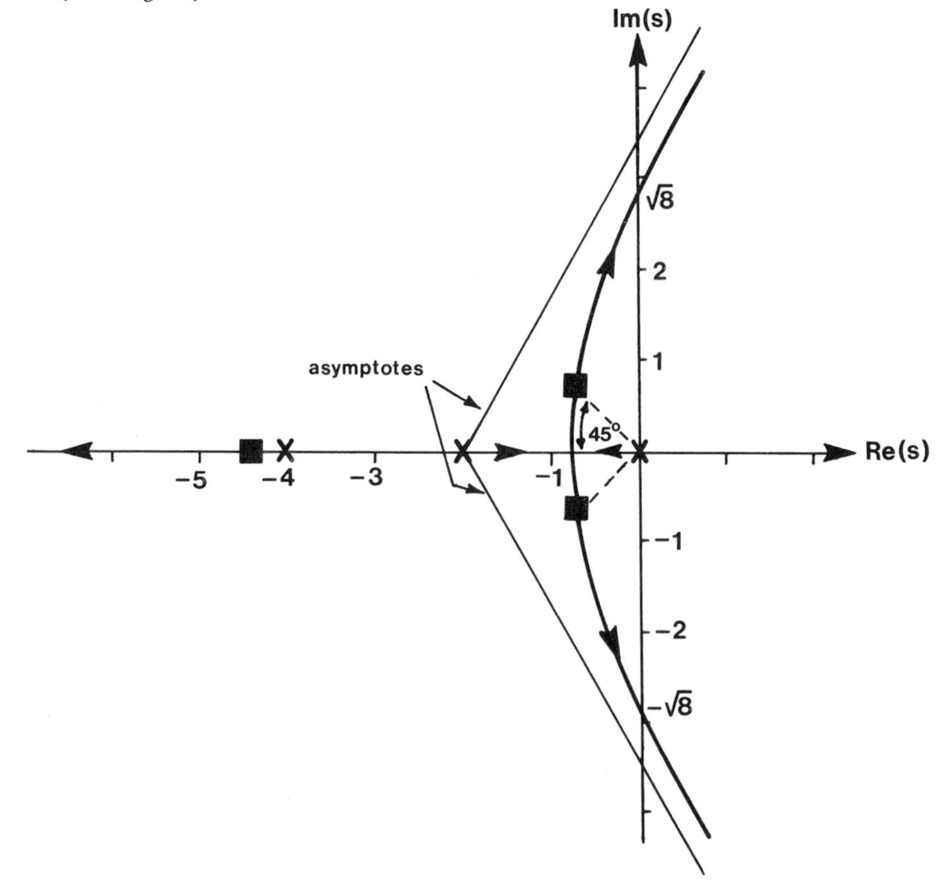

FIG. 5. Root-locus diagram for Example 7(b).

It is clear that the two branches originating at 0 and -2 must meet and then leave the real axis at some point in between. According to Property 4, this breakpoint may be found by differentiating (23) and setting

$$\frac{d\xi}{ds} = -3s^2 - 12s - 8 = 0 \tag{25}$$

One solution of this equation is $s \approx -0.85$, which is the required breakpoint (the other is $s \approx -3.15$, which is a breakpoint for negative ξ). This could also be determined by trial and error, drawing vectors from the open-loop poles to a series of trial points above the real axis and checking the angle condition (7).

At this stage, only one or two additional points on the root-locus are required before quite an accurate graph can be drawn. One such point can be found by checking which of a number of trial points on the horizontal line $s = \sigma + j1$ satisfies (7). From the pattern of the asymptotes, it is also clear that the two branches leaving the real axis at -0.85 must pass into the right half-plane. The points at which they cross the imaginary axis may be found as indicated in Property 5 by substituting $s = j\omega$ into (23),

$$j\omega(j\omega + 2)(j\omega + 4) = -j\omega^3 - 6\omega^2 + 8j\omega = -\xi \tag{26}$$

Separating real and imaginary parts yields

$$-\omega^3 + 8\omega = -\omega(\omega^2 - 8) = 0 \tag{27}$$

$$-6\omega^2 = -\xi \tag{28}$$

The required points are the solutions[†] $\omega = \pm\sqrt{8}$ of (27), for which substitution in (28) yields $\xi = 48$. Since $\xi = 0.8k_c$, it follows that the feedback system is unstable for values of controller gain k_c greater than 60.

The complete root-locus is shown in Fig. 5. The value of ξ (or of controller gain $k_c = 1.25\xi$), at any point on the locus may be determined graphically using the magnitude condition (8).

[†]The other solution, $\omega = \xi = 0$, is simply the open-loop pole at the origin.

A comparison of Figs. 4 and 5 now allows a number of qualitative conclusions to be drawn about the effect on the transient response of using an integrating device rather than a simple amplifier as the controller. To begin with, the system has been changed from one which is stable for all values of loop gain to one which is stable for only a limited range of gains. Moreover, if the gain is adjusted as in (a) to give a pair of complex conjugate closed-loop poles with equal real and imaginary parts, so that the damped sinusoid in the transient response decays rapidly without excessive oscillation, then the complex conjugate poles are much closer to the origin than in (a). [The reader may verify that a value of $\xi = 5.2$, or $k_c = 6.5$, places the roots of (23) at $s = -0.76 \pm j0.76$ and -4.47, as indicated by ■ in Fig. 5.] The damped sinusoid contributed by them dominates the transient since the real exponential term due to the real closed-loop pole decays much more rapidly. Therefore, while the integrating controller greatly improves the static accuracy of the system (see Theorems 8.2.2 and 8.2.3), it also results in a considerable reduction in the maximum speed of response obtainable by loop-gain adjustment.

(c) *Proportional-plus-integral control:* $k_c g_c(s) = k_c(\frac{5}{8} + \frac{1}{s})$

In this case u contains terms proportional to both e and its time integral, and (20) becomes

$$1 + kg(s) = 1 + \frac{k_c(s + 1.6)}{2s(s + 2)(s + 4)} = 0 \tag{29}$$

or

$$\frac{s(s + 2)(s + 4)}{(s + 1.6)} = -\frac{k_c}{2} = -\xi \tag{30}$$

The system now has three open-loop poles, at $s = 0, -2,$ and -4, and one zero at $s = -1.6$; so it will still follow a constant input with no steady-state error.

By Property 3, branches of the root-locus exist on the real axis between the pole at 0 and the zero at -1.6, and between the two poles at -2 and at -4. From this, and from Properties 1 and 2, it follows that one branch originates at $s = 0$ and terminates at $s = -1.6$, while the other two branches originating at $s = -2$ and -4 meet on the real axis between the poles, break away from it, and go to infinity, approaching asymptotes with angles $\pm\pi/2$. According to (10), the asymptotes intersect the real axis at

$$\sigma_a = \frac{(0 - 2 - 4) - (-1.6)}{3 - 1} = -2.2 \tag{31}$$

Using Property 4, the breakpoint may be found by differentiating (30) and setting

$$\frac{d\xi}{ds} = -2\,\frac{s^3 + 5.4s^2 + 9.6s + 6.4}{(s + 1.6)^2} = 0 \tag{32}$$

The numerator polynomial has a real root at $s \approx -2.8$, which is the required breakpoint.

Locating one or two additional points, for instance by testing (7) along the horizontal line $s = \sigma + 2j$, completes the root-locus diagram shown in Fig. 6. Values of $\xi(= k_c/2)$ at various points on the root-loci may be determined using the magnitude condition (8).

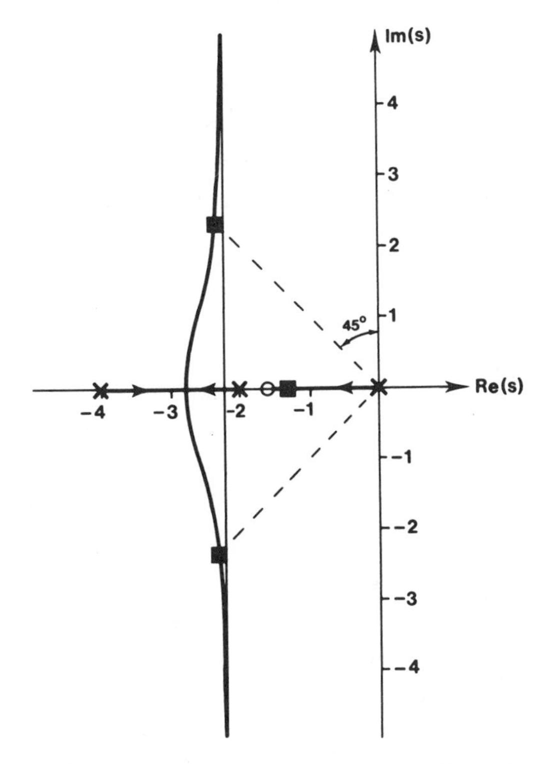

FIG. 6. Root-locus diagram for Example 7(c).

The reader should verify that choosing $\xi = 9$ ($k_c = 18$) places the closed-loop poles at $s \simeq -2.3 \pm j2.3$ and -1.34, indicated by ■ on the diagram.

Comparing Figs. 4 and 6, we see that by an appropriate choice of ξ, and hence of controller gain, the system with PI controller can be given a pair of complex closed-loop poles which are fairly close to those selected in (a), where simple proportional control was used. Since these are slightly to the right of the closed-loop poles in (a), and since an additional real closed-loop pole at $s = -1.34$ has been introduced, the transient response must be slightly slower. The steady-state response, of course, is greatly improved because of the integral term in the control law.

Step responses for the closed-loop systems in (a), (b), and (c) are shown in Fig. 7.

ΔΔΔ

A very interesting and important conclusion can be drawn from a comparison of the step responses and root-locus plots of systems (b) and (c) above. It will be noted that the introduction of the zero at $s = -1.6$ brings about a dramatic improvement in the behavior of the system. The proportional-plus-integral control gives just as good a steady-state accuracy as pure integral control (see Theorems 8.2.2 and 8.2.3), while the speed with which the transient decays is considerably greater. Moreover, system (c) is stable for all values of gain.

To reiterate, the introduction of a pole at $s = 0$ improves steady-state performance but has an adverse effect on stability and speed of response [(compare systems (a) and (b)]; the introduction of a zero makes the system more stable and improves the speed of response [compare systems (b) and (c)]. We shall exploit these ideas systematically in the next chapter, where we consider the problem of improving an unsatisfactory system by compensation (the introduction of additional dynamic elements).

At this point it is useful to note that the relative weighting of different modes in the transient response of a closed-loop system

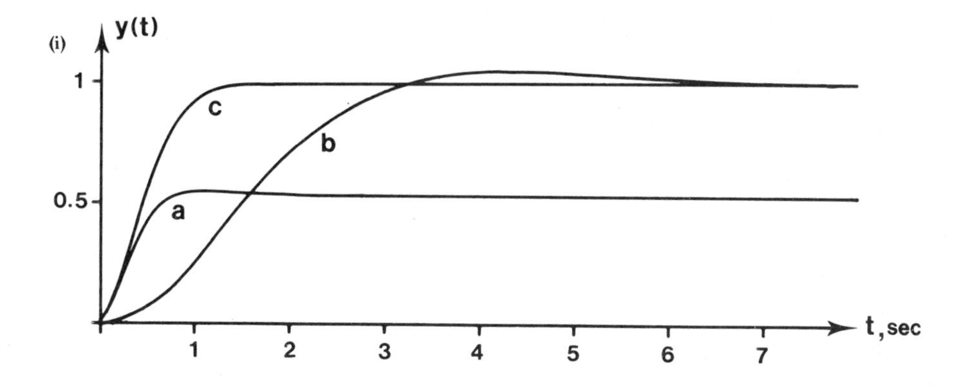

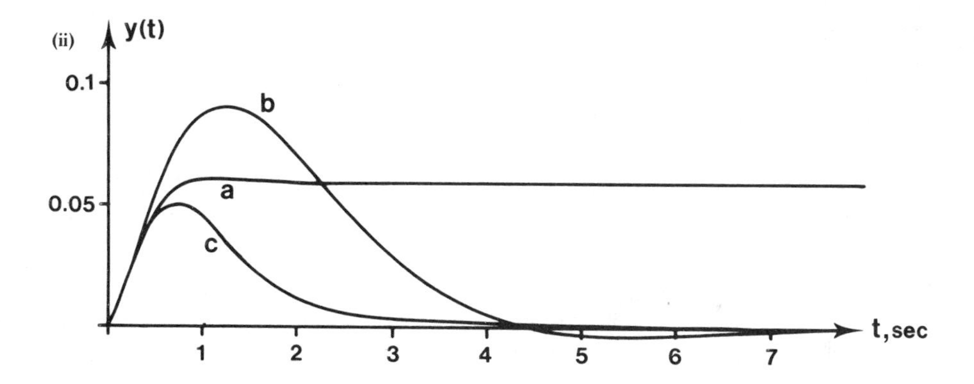

FIG. 7. Step responses for feedback system of Example 7: (a) proportional control, (b) integral control, and (c) proportional-plus-integral control; (i) unit step reference input $y_r(t) = 1(t)$, and (ii) unit step disturbance input $d(t) = 1(t)$.

can also be estimated graphically from the root-locus diagram. Once a value of ξ (and hence of loop gain) has been chosen, the closed-loop poles λ_j, $j = 1, 2, \ldots, n$, are fixed at certain points on the root-loci. The closed loop zeros η_i, $i = 1, 2, \ldots, m$, are also known, as indicated at the end of Definition 8.1.3. Thus the closed-loop (reference or disturbance) transfer function may be expressed as

$$h(s) = k \frac{(s - \eta_1)(s - \eta_2) \ldots (s - \eta_m)}{(s - \lambda_1)(s - \lambda_2) \ldots (s - \lambda_n)} \tag{33}$$

and the transform of the response to a unit step input is

$$\hat{y}(s) = \frac{h(s)}{s} = k\,\frac{(s-\eta_1)(s-\eta_2)\dots(s-\eta_m)}{s(s-\lambda_1)(s-\lambda_2)\dots(s-\lambda_n)} \tag{34}$$

(the numerators are unity if there are no closed-loop zeros).

Referring to Theorem B.2.2, we recall that $\hat{y}(s)$ in (34) may be expressed as a partial fractions expansion,

$$\hat{y}(s) = k\left(\frac{R_0}{s} + \sum_{\ell=1}^{n} \frac{R_\ell}{s-\lambda_\ell}\right) \tag{35}$$

where the residues R_ℓ are given by

$$R_0 = [s\,\hat{y}(s)]_{s=0} = \frac{\prod_{i=1}^{m} -\eta_i}{\prod_{j=1}^{n} -\lambda_j} \tag{36}$$

$$R_\ell = [(s-\lambda_\ell)\hat{y}(s)]_{s=\lambda_\ell} = \frac{\prod_{i=1}^{m}(\lambda_\ell - \eta_i)}{\lambda_\ell \prod_{\substack{j=1 \\ j\neq \ell}}^{n}(\lambda_\ell - \lambda_j)} \qquad (\ell = 1,2,\dots,n) \tag{37}$$

An inverse transform yields

$$\mathcal{L}^{-1}\left(\frac{R_\ell}{s-\lambda_\ell}\right) = R_\ell e^{\lambda_\ell t} \tag{38}$$

or in the case of complex poles (Corollary B.3.2),

$$\mathcal{L}^{-1}\left(\frac{R_\ell}{s-\lambda_\ell} + \frac{\overline{R}_\ell}{s-\overline{\lambda}_\ell}\right) = 2\,|R_\ell|\,e^{\lambda_\ell t}\sin(\omega_\ell t + \phi_\ell) \tag{39}$$

where $\phi_\ell = \tan^{-1}(-\mathrm{Re}\,R_\ell / \mathrm{Im}\,R_\ell)$.

Thus the relative weighting of the ℓ-th mode is determined by

$$|R_\ell| = \frac{\prod_{i=1}^{m} |\lambda_\ell - n_i|}{|\lambda_\ell| \prod_{\substack{j=1 \\ j\neq\ell}}^{n} |\lambda_\ell - \lambda_j|} \tag{40}$$

This can easily be calculated on the root-locus diagram by measuring the lengths of the vectors drawn to λ_ℓ from the other *closed-loop* poles and zeros and from the origin.

This procedure is illustrated in Problem 3 and explored further in Problem 4.

Extensions

The discussion of the root-locus method has been concerned so far with the specific problem of how the closed-loop poles depend on the loop gain of the system. It is often of interest to investigate what closed-loop pole configurations are possible by varying some other system parameter, keeping the loop gain fixed. Let this parameter be denoted by α and suppose that it is possible to rearrange the closed-loop characteristic equation to

$$a(s) + \alpha b(s) = 0 \tag{41}$$

where a and b are monic polynomials in s. This has the same form as (4), so that for $\alpha \geq 0$ all the properties derived previously are directly applicable to the new problem. The only changes necessary are to redefine the root-locus as the path traced out by the closed-loop poles as α varies from 0 to large positive values, and to replace the poles and zeros of the system by the roots of a and b, respectively. To illustrate, we return to the previous example.

8. EXAMPLE. In the PI feedback controller of Example 7(c), proportional and integral control terms were applied in the ratio of 5:8. Now suppose this ratio is allowed to vary,

$$h_c(s) = k_c g_c(s) = 18\left(\frac{5}{8} + \frac{\beta}{s}\right) = 11.25\,\frac{s + 1.6\beta}{s} \tag{42}$$

where $0 \leq \beta < \infty$ and $k_c = 18\beta$. Choosing $\beta = 1$ yields the closed-loop pole configuration indicated by ■ in Fig. 6.

From (42) and (20), the closed-loop poles must be solutions of

$$1 + kg(s) = 1 + 9\,\frac{(s + 1.6\beta)}{s(s + 2)(s + 4)} = 0 \tag{43}$$

Clearing fractions, this becomes

$$(s^3 + 6s^2 + 17s) + 14.4\beta = 0 \tag{44}$$

which has the form (41) with $\alpha = 14.4\beta$, $b(s) = 1$, and

$$a(s) = s^3 + 6s^2 + 17s = s[(s + 3)^2 + 8] \tag{45}$$

Equation (44) can be viewed as the characteristic equation of a system having three "poles," at $s = 0$ and $s = -3 \pm j\sqrt{8}$, and no "zeros." Of course, since in this case the parameter being varied is not the loop gain, the "poles" are not the open-loop poles as previously defined but the values to which the closed-loop poles tend as α becomes small. With this understanding, the previously derived results on the construction of the root-locus can be directly applied. In particular, by Properties 1 and 2 the root-locus consists of three branches, originating at $s = 0$ and $s = -3 \pm j\sqrt{8}$, and ultimately going to infinity along asymptotes at angles of $\pm\pi/3$ and π which intersect the real axis at

$$\sigma_a = \frac{0 - 3 - j\sqrt{8} - 3 + j\sqrt{8}}{3 - 0} = -2 \tag{46}$$

as shown in Fig. 8.

In this example it is useful to know the angles of departure of the two branches from the complex conjugate "poles". Using Property 6, the angle of departure from $-3 + j\sqrt{8}$ is given by

$$\pi - \angle(-3 + j\sqrt{8} + 3 + j\sqrt{8}) - \angle(-3 + j\sqrt{8} - 0) \approx \pi - \frac{\pi}{2} - \frac{3\pi}{4}$$
$$\approx -\frac{\pi}{4} \tag{47}$$

and because of symmetry, the angle of departure from $-3 - j\sqrt{8}$ must be $\approx \pi/4$.

Finally, we may substitute $s = j\omega$ into (44) and find that two branches cross the imaginary axis at $\omega = \pm\sqrt{17}$ where $\alpha = 102$ and hence $\beta = 7.1$ (Property 5).

The complete root-locus plot is shown in Fig. 8. It is clear that increasing α (and hence k_c) decreases the damping ratio of the complex closed-loop poles, so that the transient response becomes increasingly oscillatory. This illustrates once again a common

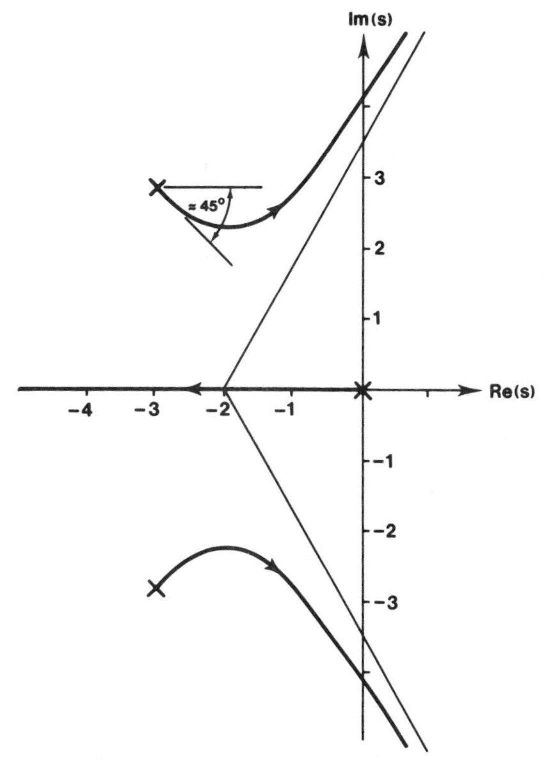

FIG. 8. Root-locus diagram for Example 8.

difficulty in the design of single-loop feedback systems: large loop gains, although desirable for good static accuracy, usually result in poor transient performance.

ΔΔΔ

Once a satisfactory configuration of closed-loop poles has been determined in the design of a feedback system, it is frequently of some interest to determine its sensitivity to variations in plant parameters other than loop gain. We have just seen that the root-locus method can readily be extended to such investigations. However, in sensitivity studies it is sometimes necessary to determine the behavior of the closed-loop poles when the parameter is allowed to take on negative values. Indeed, this situation can even arise when the parameter is the loop gain: it is not difficult to contrive systems which can be stabilized only by negative values of loop gain, i.e., by *positive* feedback (see Problems 8.1-2, 8.1-3, and 8.3-10).

The rules developed so far for constructing root-loci do not cover this situation. We ask the reader in Problem 8 to determine how these rules must be changed when the loop gain k varies between 0 and $-\infty$.

Finally, we note that construction of root-locus diagrams for systems of even moderate complexity can be a very time-consuming chore. Some efficiency can be gained by using a commercially-available device known as a *Spirule*, which assists in rapid application of the angle and magnitude conditions. An alternative, if one has a computer and display scope or plotter available, is to make use of one of the many programs available for plotting root-loci. This eliminates the most laborious part of the job and allows the designer to test a greater range of controller structures and parameter variations than he otherwise could.

8.4. THE NYQUIST STABILITY CRITERION

In Chapter 6, we saw that a knowledge of the steady-state response of linear systems to sinusoidal inputs at various frequencies (the *frequency response*) provides a good deal of qualitative information about transient behavior, both for nonzero initial states and for other inputs, such as impulses and step functions. In particular, we observed that the bandwidth of the frequency response is an indication of the speed of transient response, and that resonant peaks are an indication of oscillatory transients. Moreover, frequency response information is readily available; it can be derived from a model of the system or, if the system is stable, obtained experimentally.

The use of frequency response methods will now be extended to analyze the transient response of feedback systems. Given the open-loop transfer function, the root-locus method described in Sec. 8.3 allows a rapid qualitative analysis of the main features of the transient response of the feedback system. Moreover, it permits some designing to be done as well: a value of one parameter, usually the loop gain, can be selected which will yield a good or even a "best possible" configuration of closed-loop poles, given the open-loop poles and zeros. Much of the same information about transient performance can be obtained from a knowledge of the open-loop frequency response. The criterion we shall develop provides tools which are both complementary and alternative to those based on root-locus methods.

The basis of most frequency response methods for analyzing feedback systems is the Nyquist stability criterion, which provides a condition on the open-loop frequency response under which the closed-loop system will be stable.

Consider the usual feedback system, shown in Fig. 1, and recall from (8.1-7) that both the reference and disturbance closed-loop transfer functions have the return difference

$$w(s) = 1 + k_c k_p g_c(s)\, g_p(s) = 1 + kg(s) \tag{1}$$

as their denominator, where $kg(s)$ is the normalized open-loop transfer function and $k = k_c k_p$ is the loop gain. Note that the "poles" of $w(s)$ (values of s where $w = \infty$) are just the poles of $g(s)$[†], that is, the

[†] This assumes that no poles and zeros are "cancelled" in the product $g_c g_p$ (see Example 9.3.4 and the ensuing discussion).

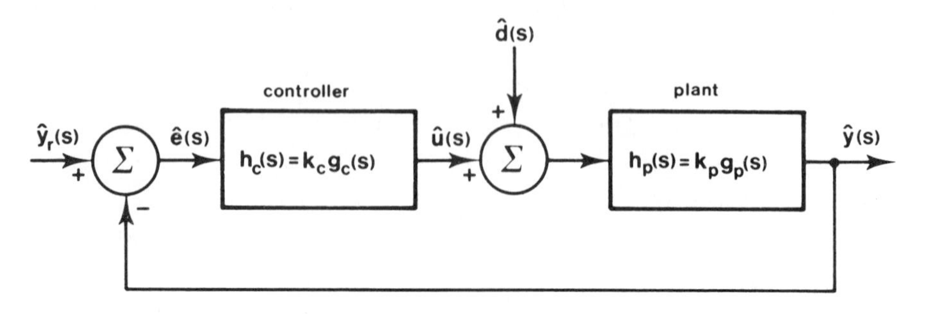

FIG. 1. Feedback control system.

open-loop poles of the system, denoted $p_1, p_2, \ldots, p_n$. The "zeros" of w(s), on the other hand, are those values of s for which the denominator of the closed-loop transfer functions vanishes, and hence are precisely the *closed-loop poles* of the feedback system, denoted $\lambda_1, \lambda_2, \ldots, \lambda_n$. These closed-loop poles depend, of course, upon the value of the loop gain k.

This means that we could, if we knew $\lambda_1, \lambda_2, \ldots, \lambda_n$, write

$$w(s) = \frac{(s - \lambda_1)(s - \lambda_2) \ldots (s - \lambda_n)}{(s - p_1)(s - p_2) \ldots (s - p_n)} \tag{2}$$

Nyquist's stability test involves letting s vary along a closed path in the complex plane, observing the path (also closed) which is consequently traced out in the complex plane by w(s), and noting that the angle traversed by w(s) depends upon the number of p_i's and λ_i's encircled by the path of s. Since our aim is to test for closed-loop stability, we choose a path for s which encloses the entire right half-plane.

1. DEFINITION. The *Nyquist contour* is a closed path in the complex s-plane, traversed in the clockwise direction and consisting of the imaginary axis together with a semicircle of very large radius in the right half-plane. If the system in question has any open-loop poles on the imaginary axis, the path is modified by making a semicircular "detour" of very small radius to the right of each such pole. ΔΔΔ

As an example, a Nyquist contour is shown in Fig. 2 for a system whose open-loop poles are marked by ×.

Now consider what happens to w(s) as s traverses the Nyquist contour. On the imaginary axis (ignoring any "detours" for the moment), we have

$$w(s) = w(j\omega) = 1 + kg(j\omega) \tag{3}$$

which is just the constant 1 plus the open-loop frequency response $kg(j\omega)$. Assuming (as usual) that the open-loop transfer function is strictly proper (see Problem 1), we have

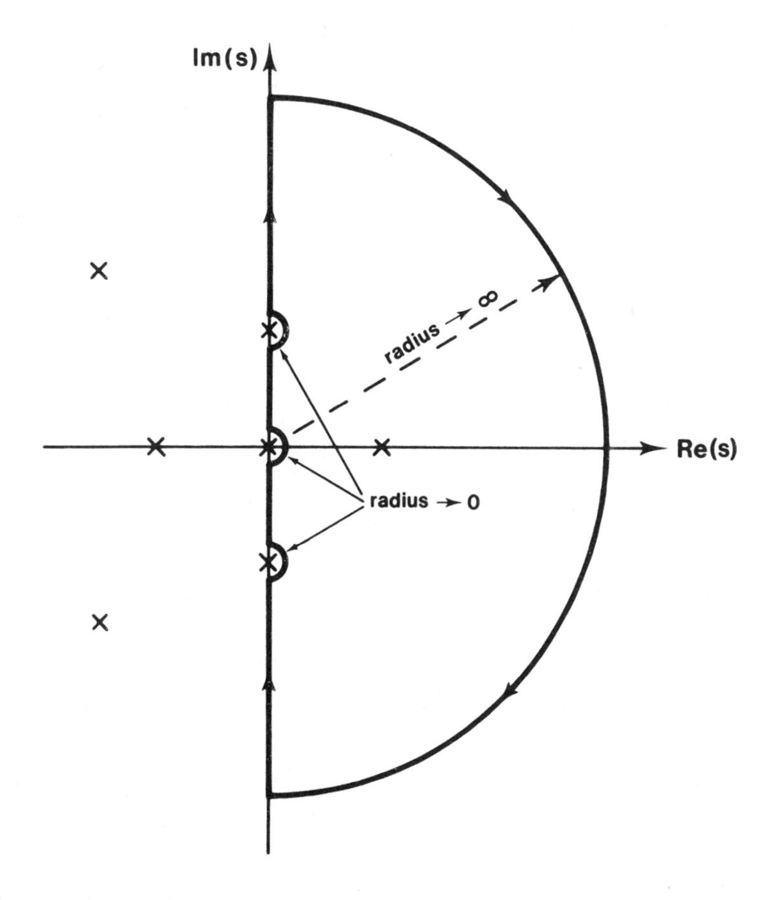

FIG. 2. Typical Nyquist contour.

$$\lim_{\omega\to\infty} w(j\omega) = 1 + \lim_{\omega\to\infty} kg(j\omega) = 1 \tag{4}$$

so that $w(j\omega) \approx 1$ on the large semicircle. This means that the plot of $w(s)$ as s traverses the Nyquist contour is simply a Nyquist (polar) diagram of the open-loop frequency response (see Sec. 6.2), with the constant 1 added to it. Some caution is necessary if there are open-loop poles on the imaginary axis: see Examples 4(b) and (c).

Nyquist's criterion tells us how to use this open-loop frequency response information to test for stability of the closed-loop system:

2. *THEOREM (Nyquist).* Suppose the feedback system of Fig. 1 has ℓ open-loop poles in the right half-plane (not counting poles on the imaginary axis). Then the closed-loop system is stable if and only if the path of $w(s)$ encircles the origin exactly ℓ times in the counterclockwise direction as s traverses the Nyquist contour.

If the path of $w(s)$ encircles the origin $\ell - m$ times in the counterclockwise direction (or $m - \ell$ times in the clockwise direction), then the closed-loop system has m unstable closed-loop poles.

Proof: Rather than attempt a rigorous proof, which would require the theory of functions of a complex variable and conformal mappings, we will simply outline the main argument.

From (2), $w(s)$ may be written in polar form as

$$w(s) = |w(s)| \angle w(s) \tag{5}$$

where

$$|w(s)| = \prod_{i=1}^{n} \frac{|s - \lambda_i|}{|s - p_i|} \tag{6}$$

$$\angle w(s) = \sum_{i=1}^{n} \angle (s - \lambda_i) - \angle (s - p_i) \tag{7}$$

The complex numbers $(s - \lambda_i)$ and $(s - p_j)$ may be represented as vectors in the complex plane drawn from λ_i and p_j, respectively, to s, as illustrated in Fig. 3 for a second-order system.

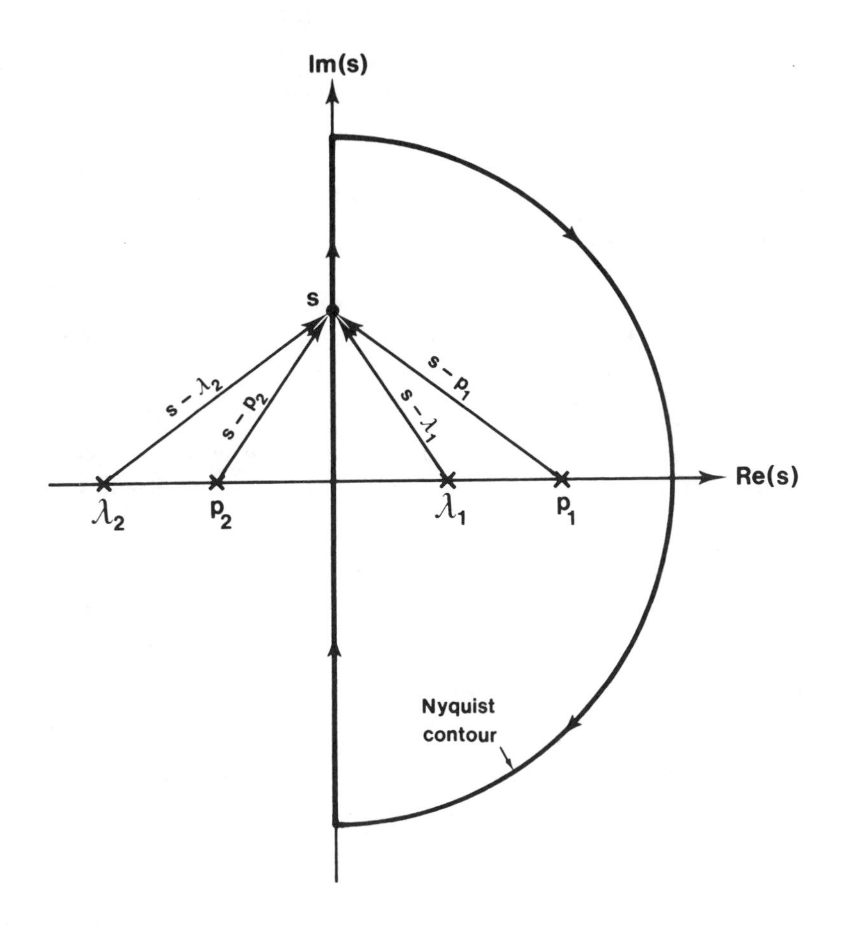

FIG. 3. Nyquist contour for second-order system.

As s makes one complete trip around the Nyquist contour in Fig. 3, the vectors $s - \lambda_1$ and $s - p_1$ each undergo a net angle change of -2π, while the net angle change of $s - p_2$ and $s - \lambda_2$ is zero. Referring to (7), the net angle change of w(s) is $2\pi - 2\pi = 0$.

Generalizing this observation, we conclude that every open-loop pole within the Nyquist contour contributes a net angle change of 2π to w(s), while each closed-loop pole within the contour contributes -2π. Poles outside the contour always contribute 0. Thus for a system with ℓ unstable open-loop poles and m unstable closed-loop poles, w(s) will undergo an angle change of $2\pi(\ell - m)$ as s traverses the Nyquist contour. Since a 2π angle change is the same as one counterclockwise encirclement of the origin, this establishes the theorem.

ΔΔΔ

This result becomes even more useful when restated in terms of the frequency response.

3. COROLLARY (Nyquist stability criterion). Suppose the feedback system of Fig. 1 has ℓ open-loop poles in the right half-plane (not counting poles on the imaginary axis):

(a) Then the closed-loop system is stable if and only if the Nyquist plot of $kg(j\omega)$ encircles the point -1 exactly ℓ times in the counterclockwise direction.

(b) Equivalently, the closed-loop system is stable if and only if the Nyquist plot of $g(j\omega)$ encircles the point $-1/k$ exactly ℓ times in the counterclockwise direction.

Proof: These are simply coordinate changes, and the details are left to the reader (Problem 2).

ΔΔΔ

The Nyquist criterion is an extremely powerful tool for control system analysis and design; closed-loop stability of a feedback system is determined using only a plot of the open-loop frequency response and a knowledge of the number of unstable open-loop poles. Moreover, the range of loop gains for which stability holds is determined explicitly by Corollary 3(b).

Strictly speaking, the term frequency response is meaningful only in the case of a stable open-loop system, since it is defined in terms of the steady-state response to a sinusoidal input. Nevertheless, a Nyquist plot of $g(j\omega)$ may be constructed just as readily for an unstable system (although it cannot be determined experimentally), and we shall also refer to it as a frequency response diagram. The Nyquist criterion, of course, applies regardless of whether the open-loop system is stable or unstable.

We shall now illustrate the use of the Nyquist criterion with some examples.

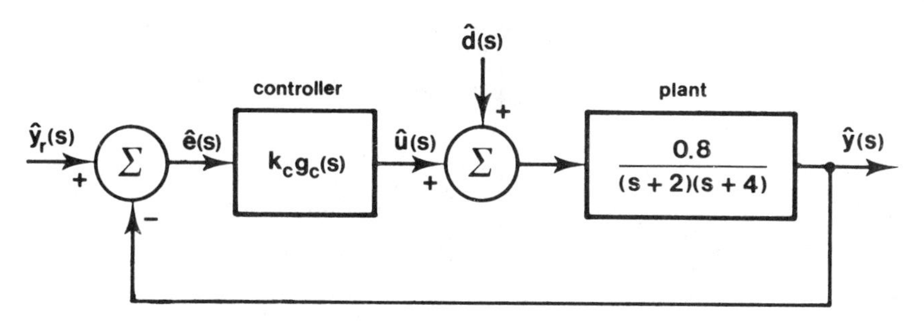

FIG. 4. Feedback system of Example 4.

4. *EXAMPLE*. Consider once again the feedback system of Example 8.3.7, shown here in Fig. 4. The open-loop transfer function is

$$kg(s) = k_p k_c g_p(s) g_c(s) = 0.1k_c \frac{8}{(s+2)(s+4)} g_c(s) \tag{8}$$

where the loop gain is $k = 0.1k_c$. We again consider three different types of controller: proportional, integral, and proportional plus integral.

(a) *Proportional control:* $k_c g_c(s) = k_c$.

For this control law we have

$$kg(s) = k\frac{8}{(s+2)(s+4)} \tag{9}$$

and the Nyquist plot of $g(j\omega)$ is shown in Fig. 5. As in Example 8.3.7(a), the loop gain is chosen to be $k = 1.25$, and the point $-1/k = -0.8$ is indicated on the plot.

We note that the open-loop system has no unstable poles, and observe that the Nyquist plot does not encircle the point $-1/k$. Thus Corollary 3(b) implies that the closed-loop system is stable. Indeed, it is clear that the Nyquist plot cannot encircle the point $-1/k$ for any finite value of $k > 0$. This verifies our conclusion reached earlier using the root-locus diagram: the feedback system is stable for all positive values of loop gain.

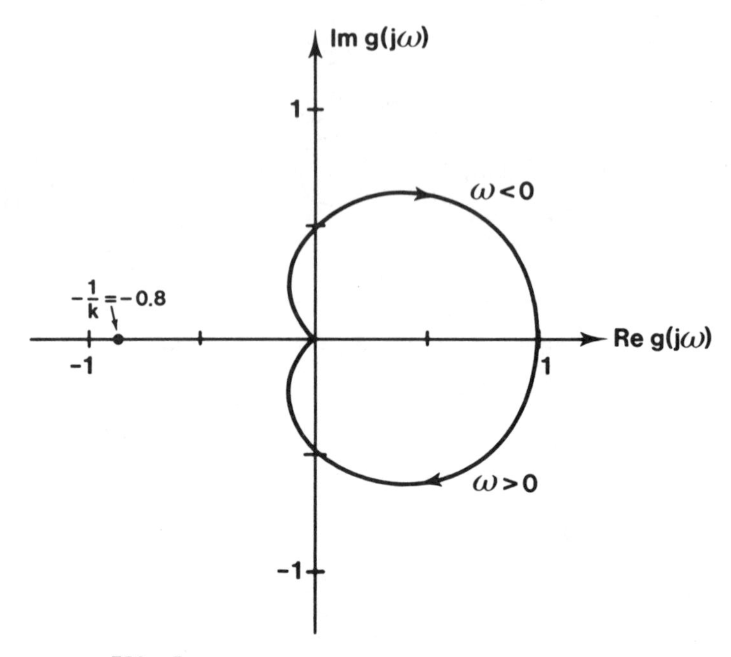

FIG. 5. Nyquist plot for Example 4(a).

(b) *Integral control:* $k_c g_c(s) = k_c/s$.

The open-loop transfer function now becomes

$$kg(s) = k \frac{8}{s(s + 2)(s + 4)} \tag{10}$$

and the Nyquist plot of g(s) is shown in Fig. 6. In this case there is an open-loop pole at the origin, so that the plot of $g(j\omega)$ approaches $\pm j\infty$ as ω approaches 0 from above and below. Referring to Fig. 2, we recall that the Nyquist contour must make a small semicircular detour to the right of the pole at the origin. Along this path, the factor s in the denominator of g(s) will have a very small constant magnitude and its angle will *increase* from $-\pi/2$ to $\pi/2$ while the other factors all remain nearly constant. This means that the angle of g(s) must *decrease* by π, as indicated by the large semicircle on the plot.

The loop gain, as in Example 8.3.7(b), is chosen to be k = 0.65, and the point -1/k = -1.54 is indicated on the plot. Since this point

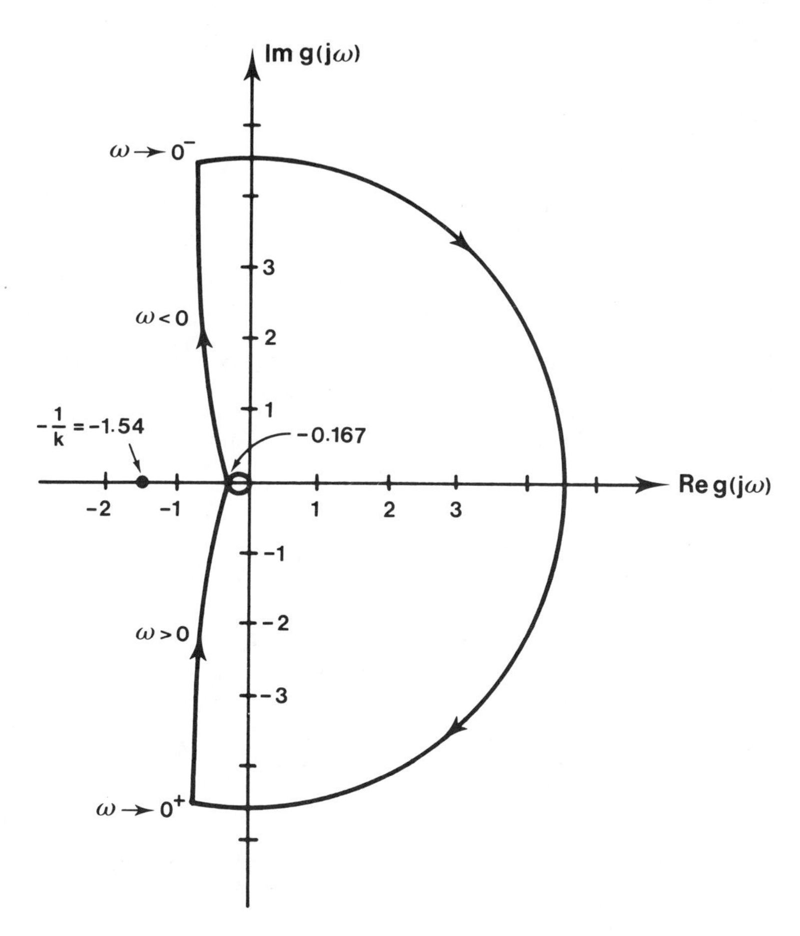

FIG. 6. Nyquist plot for Example 4(b).

is not encircled, and the open-loop system has no unstable poles, we again conclude from the Nyquist criterion that the closed-loop system is stable.

In this case stability is *not* maintained for all positive values of loop gain. Examination of the Nyquist plot near the origin reveals that $-1/k$ will be encircled if $-1/k > -0.167$, or $k > 6$. This was also evident, of course, from the root-locus diagram in Fig. 8.3-5.

(c) *Proportional-plus-integral control:* $k_c g_c(s) = k_c(\frac{5}{8} + \frac{1}{s})$.

The open-loop transfer function in this case is

$$kg(s) = k \frac{5(s + 1.6)}{s(s + 2)(s + 4)} \tag{11}$$

and the Nyquist plot of g(s) is shown in Fig. 7. Again, the Nyquist contour detours to the right of the open-loop pole at the origin, and g(s) follows a large semicircle in the right half-plane.

Using the loop gain $k = 1.8$ chosen in Example 8.3.7(c), we see that the point $-1/k = -0.556$ is not encircled by the Nyquist plot; indeed, the closed-loop system is stable (as expected) for all values of $k > 0$.

ΔΔΔ

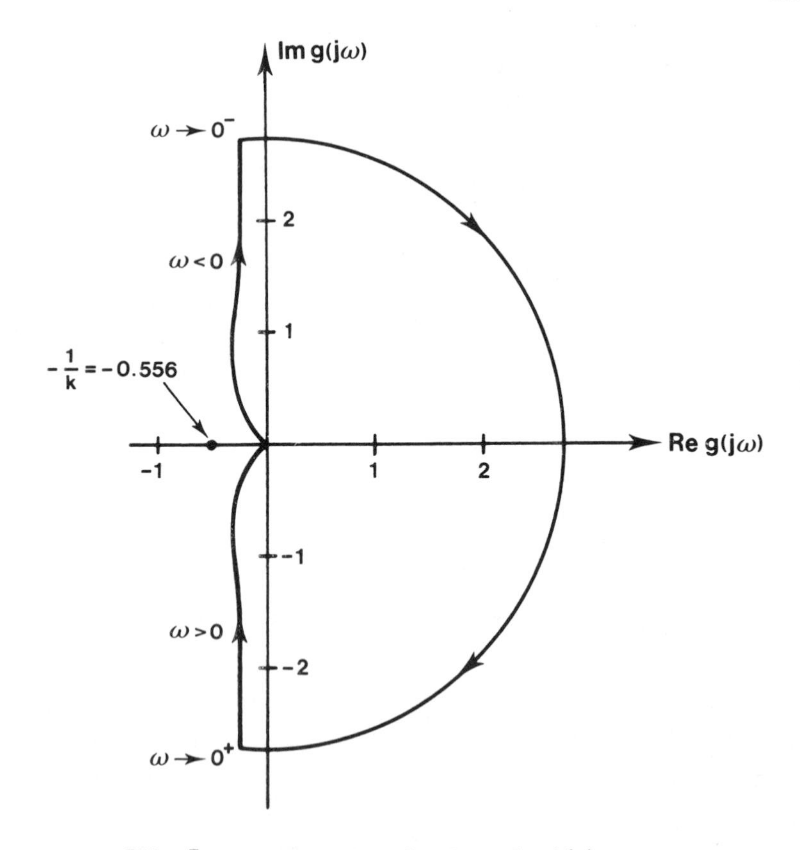

FIG. 7. Nyquist plot for Example 4(c).

In Example 4(b), the gain was chosen to be $k = 0.65$ and we saw that this could be increased by a factor of about 9.2, to $k = 6$, before instability would result. This feedback system is thus said to have a margin of safety, or *gain margin*, of 9.2. The gain margins in Examples 4(a) and (c) are infinite. More generally, we adopt the following terminology.

5. *DEFINITION*. The *gain margin* of a stable closed-loop system is the factor by which the loop gain can be increased before instability results.

The *phase margin* of a stable closed-loop system is the angle of phase lag (negative phase shift) that can be introduced in the system before instability results.

ΔΔΔ

These definitions are standard, and are illustrated below with a further example. However, some caution must be exercised, as there are systems for which a *decrease* in gain [see Problem 8(a)] or a *positive* phase shift will cause instability, and the terminology in such cases can be misleading.

6. *EXAMPLE*. Suppose the plant transfer function in Fig. 1 is given by

$$k_p g_p(s) = \frac{10}{(s+2)[(s+1)^2 + 2^2]} \tag{12}$$

and a proportional controller is used:

$$k_c g_c(s) = k_c = 2 \tag{13}$$

This means that the open-loop transfer function is

$$kg(s) = 2\,\frac{10}{(s+2)[(s+1)^2 + 2^2]} \tag{14}$$

and the loop gain is $k = 2$.

The Nyquist plot of $g(j\omega)$ is shown in Fig. 8. Since there are no unstable open-loop poles and the point $-1/k = -0.5$ is not encircled, the closed-loop system is stable. Stability holds for all k such

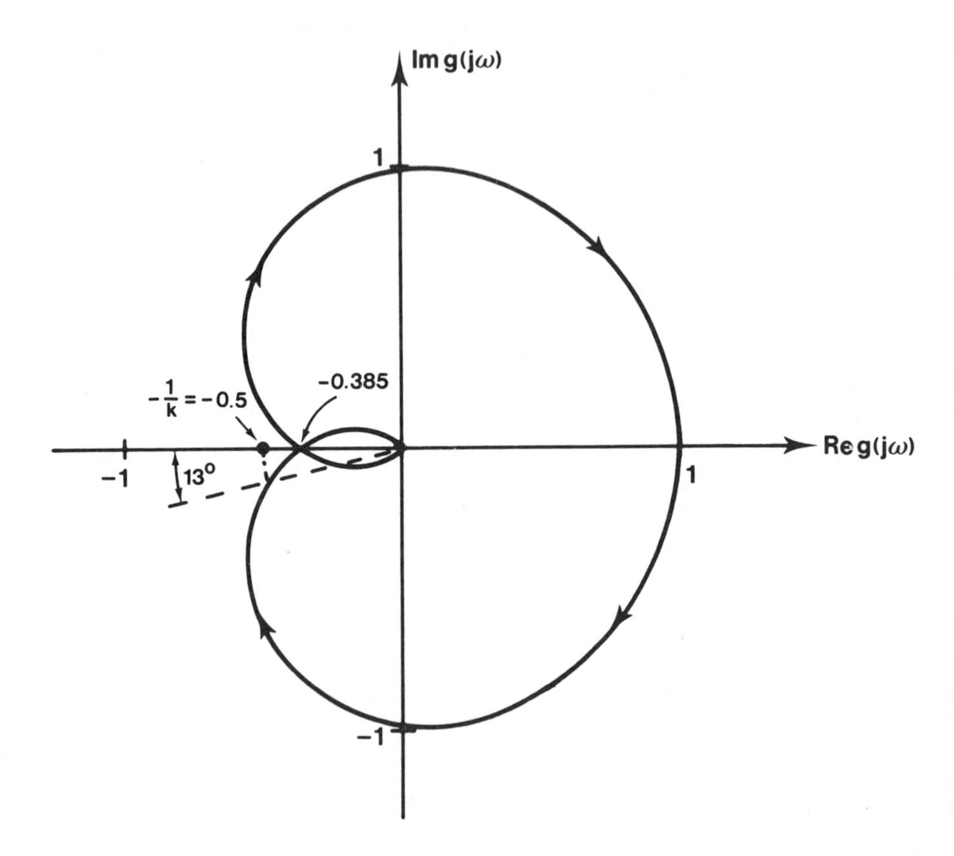

FIG. 8. Nyquist plot for Example 6.

that $1/k > 0.385$, or $k < 2.6$, so the gain margin in this case is $2.6/2 = 1.3$. Similarly, we observe that a phase lag of 13° or more will shift the plot[†] so that $-1/k = -0.5$ is encircled, and conclude that the phase margin is 13°.

ΔΔΔ

The concepts of gain and phase margin have a great deal of practical significance, since they provide a quantitative measure of how much deviation can be tolerated from the mathematical model

[†]Note that phase lag implies a clockwise shift in the $\omega > 0$ portion of the plot, and a corresponding counterclockwise shift in the $\omega < 0$ portion.

assumed for the system. Such deviations may occur as a result of thermal effects, nonlinearities, amplifier drift, component variations, or just from inaccuracies in modelling.

One very common source of phase shift is the presence of a pure time delay in the feedback loop, which introduces a phase lag proportional to frequency. A delay of Δ sec. multiplies the open-loop transfer function $g(j\omega)$ by $e^{-j\omega\Delta}$ (Theorem B.1.4), and the Nyquist plot may be altered accordingly. Although it is a common and useful practice to incorporate time delays in this manner, the resulting closed-loop transfer function is not rational, and some of the foregoing mathematical statements do not hold. The Nyquist approach can be made rigorous in this case [53a], but we will not pursue the subject any further here.

This section contains only the briefest introduction to the use of frequency response methods in the analysis and design of single-loop control systems. A great deal of additional material is available in the literature, and the interested reader is referred to Refs. 48, 71, 89, and 139.

8.5. SUMMARY

This chapter has dealt extensively with single-loop feedback control. The concept of feedback control was developed in Sec. 8.1, and several of its uses and advantages were pointed out. Attention was focused upon systems with a single output, a single control or reference input, and a single unpredictable disturbance input. A standard format and terminology for single-loop feedback systems was adopted and illustrated with examples. The question of steady-state performance was taken up in Sec. 8.2, where we saw that elimination of steady-state errors in the presence of nonzero reference and disturbance inputs requires the presence of integrators at appropriate points in the feedback loop.

Root-locus methods were the subject of Sec. 8.3. There we saw how to plot the poles of a closed-loop system as a function of the loop gain (or some other system parameter). This technique was used to investigate in some detail the behavior of a feedback system using proportional, integral, and proportional-plus-integral control on a second-order plant.

A somewhat different approach was developed in Sec. 8.4. The Nyquist criterion determines closed-loop stability of a feedback system using only the open-loop frequency response diagram and a knowledge of any open-loop instabilities. Indeed, this method also provides a measure of just *how stable* a closed-loop system is in terms of the gain and phase margins.

Use of the techniques described here has traditionally been restricted to single-loop systems. The root-locus concept can be extended to include a second parameter by means of *parameter-plane* methods [159], but additional parameters render this technique intractable. Extensions of the Nyquist approach to multivariable systems have also been explored in the recent literature: the interested reader is referred to Refs. 39, 53a, 141, and 297.

Another approach to single-loop design, which we have not mentioned, is to adjust parameters so as to minimize some scalar integral of the error (see Problem 8.2-5). This touches upon the theory of optimal control (Chapter 13), and the subject is discussed at greater length in Refs. 57, 137, and 139.

PROBLEMS

Section 8.1

1. (a) Consider the system shown in Fig. P1, and determine the closed-loop transfer functions relating each of the inputs (y_r, d_1, and d_2) to y.

   (b) Define the input and output vectors

$$\underline{u} = \begin{bmatrix} y_r \\ d_1 \\ d_2 \end{bmatrix} \qquad \underline{y} = \begin{bmatrix} y \\ e \end{bmatrix}$$

   and determine the closed-loop transfer function matrix relating them.

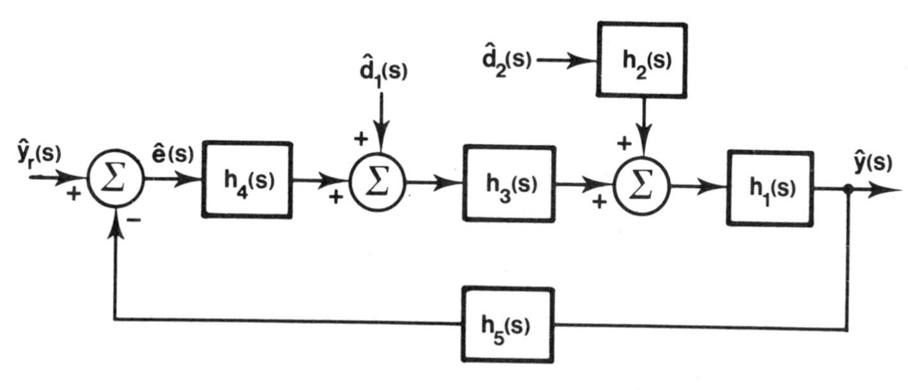

FIG. P1

2. (a) Draw a complete root-locus diagram for Example 8.1.4, allowing for both positive and negative values of λ and k. If the plant is unstable ($\lambda < 0$), are there any values of loop gain for which the closed-loop system is stable?

   (b) Discuss what happens to the transient response in Example 8.1.5 if the loop gain is allowed to be negative (extend the root-locus plot to include $-\infty < k < 0$).

3. (a) Suppose the plant in Example 8.1.5 is unstable, with transfer function

$$k_p g_p(s) = \frac{-3k_p}{s^2 + 2s - 3}$$

   Draw a root-locus plot for the closed-loop system ($-\infty < k < \infty$) and determine whether there are any values of k for which the closed-loop system is stable.

   (b) Repeat for the unstable plant transfer function

$$k_p g_p(s) = \frac{-3k_p}{s^2 - 2s - 3}$$

4. Show that the system of Example 8.1.5 has the same steady-state sensitivity to external disturbances and to changes in plant gain as the system of Example 8.1.4.

5. For the open- and closed-loop systems of Example 8.1.4 [Fig. 1(a) and 1(b)], derive expressions for the steady-state response to a ramp input, $y_r(t) = \alpha t1(t)$. Is it true that the closed-loop system can be made to have a better steady-state accuracy than the open-loop system?

6. Consider a second-order feedback system similar to that of Example 8.1.5 but with a plant transfer function given by

$$k_p g_p(s) = \frac{k_p \lambda}{s(s+\lambda)}$$

   (a) Compute the steady-state error of this feedback system for a zero disturbance and a unit step input $y_r(t) = 1(t)$ and compare with the steady-state error of the system of Example 8.1.5.

   (b) Draw a root-locus plot and determine the gain required for a closed-loop damping ratio of 0.7.

   (c) Compare the steady-state response of this system and that of Example 8.1.5 for a ramp input, $y_r(t) = \alpha t1(t)$.

7. How can you account for initial conditions in Lemma 8.1.2? Are the two systems in Fig. 8.1-4 equivalent? In what sense? Explain.

Section 8.2

1. (a) Show how the system described in Example 8.2.1 may be represented by the block diagram of Fig. 8.2-1.

   (b) Show how this block diagram may be reduced to one of the form shown in Fig. 8.1-3(b).

   (c) How does this system compare with that of Example 8.1.4?

2. Prove Theorem 8.2.3.

3. State and prove a theorem which specifies the static accuracy of a feedback system in the case where *both* y_r and d are nonzero.

Verify that your result reduces to Theorem 8.2.2 when $d = 0$ and to Theorem 8.2.3 when $y_r = 0$.

4. For the system shown in Fig. 8.2-3, show that equally good steady-state performance results if the controller is a pure integrator,

$$k_c g_c(s) = \frac{k_c}{s}$$

of if proportional-plus-integral control is chosen

$$k_c g_c(s) = k_1 + \frac{k_2}{s}$$

In the latter case, determine an expression for the loop gain k of the system.

5. For the system of Problem 8.1-6, assume that $d = 0$ and that y_r is a unit step function, and apply Theorem 2 to verify that the steady-state error is zero. Derive an expression for the error as a function of time and use it to determine the loop gains k_a^* and k_b^* for which the integrals

$$a = \int_0^\infty e^2(t)\,dt \qquad\qquad b = \int_0^\infty |e(t)|\,dt$$

take on their minimum values (you may do this analytically, by digital computation, or with an analog computer simulation). Compare the gains required with that giving a closed-loop damping ratio of 0.7, plot the unit step responses for the three values of gain and discuss the suitability of the three criteria for evaluating the "best" loop gain. For simplicity, choose $k_p = \lambda = 1$.

6. Show that a system of type m will follow a reference input of the form

$$y_r(t) = [\alpha_0 + \alpha_1 t + \dots + \alpha_{m-1} t^{m-1}]\ 1(t)$$

with zero steady-state error.

Section 8.3

1. Complete the proof of Property 8.3.2.

   (a) Show that a zero of multiplicity q is approached by precisely q branches of the root-locus as $\xi \to \infty$. *Hint:* Express a point s in the immediate neighborhood of a zero z_i as

   $$s = z_i + \varepsilon e^{j\psi}$$

   where ε is small, and solve for ψ.

   (b) Verify that the asymptotes of the root-locus branches which tend to infinity intersect the real axis at the point σ_a given in (10). *Hint:* Letting σ_a be the point of intersection, show that the left-hand side of (5) [and hence of (12)] approximates $(s - \sigma_a)^{n-r}$ as s grows large. Determine γ_1 and then solve for σ_a by equating like powers of s.

2. Prove Property 8.3.6.

3. (a) Estimate the relative weighting of the various modes in the response of the system of Example 8.3.7(c) to a unit step reference input. How well do your estimates correlate with the actual step response shown in Fig. 8.3-7?

   (b) Do the same for a unit step disturbance input.

   (c) Repeat (a) for unit impulse inputs.

   (d) Repeat (b) for unit impulse inputs.

4. Can you suggest a few rules of thumb for estimating the relative weighting of transient response modes? In particular, what may be concluded if a closed-loop pole lies very near a closed-loop zero, or very far from all the other closed-loop poles? What if it also lies near the origin? Do your conclusions depend upon the form of the input (step, impulse, etc.)? Explain.

5. (a) Construct a root-locus diagram for the system of Example 8.3-7(c) using the controller

$$g_c(s) = 10 + \frac{1}{s}$$

Estimate the value of controller gain required to place the complex closed-loop poles at 45°. Estimate the speed of response to unit step reference and disturbance inputs, respectively. How does this controller compare with the one in the example?

(b) Repeat for the controller

$$g_c(s) = \frac{1}{2} + \frac{1}{s}$$

6. Consider the feedback system shown in Fig. 8.3-1, construct root-locus diagrams ($0 < k \leq \infty$) for the following cases, and determine the range of k for which the closed-loop system is stable.

(a) $k_p g_p(s) = \dfrac{s + 6}{s^2 + 4s}$ $\qquad k_c g_c(s) = k_c$

(b) $k_p g_p(s) = \dfrac{1}{s^2 + 10s + 24}$ $\qquad k_c g_c(s) = \dfrac{k_c}{s}$

(c) $k_p g_p(s) = \dfrac{s + 5}{(s + 1)(s + 4)}$ $\qquad k_c g_c(s) = k_c(1 + \dfrac{1.1}{s})$

(d) $k_p g_p(s) = \dfrac{1}{(s - 1)(s^2 + 4s + 20)}$ $\qquad k_c g_c(s) = k_c(1 + \dfrac{1}{s})$

(e) $k_p g_p(s) = \dfrac{1}{(s - 1)(s^2 + 4s + 16)}$ $\qquad k_c g_c(s) = k_c(1 + \dfrac{1}{s})$

7. Show that the root-locus plot for a system with two open-loop poles p_1, p_2 and one open-loop zero $z_1 < \min(p_1, p_2)$ always consists of a real-axis portion plus a circle of radius $\sqrt{(p_1 - z_1)(p_2 - z_1)}$ centered at z_1.

8. Suppose the loop gain in Fig. 8.3-1 is negative ($-\infty < k < 0$), and determine how the rules for constructing root-loci must be altered:

(a) Restate Property 1 for this case.

(b) Make any necessary alterations to the angle and magnitude conditions (7) and (8).

(c) Revise Properties 2-6 accordingly.

9. Extend the root-locus diagrams of Example 8.3.7 to include negative loop gains $(-\infty \leq k < 0)$.

10. Consider the feedback system shown in Fig. 8.3-1, with

$$k_p g_p(s) = \frac{s^2 - s - 2}{s^3 + 3s^2 + s - 5} = \frac{s^2 - s - 2}{(s - 1)(s^2 + 4s + 5)}$$

and

$$k_c g_c(s) = k_c \qquad \text{(proportional control)}$$

(a) Construct a root-locus diagram for positive values of loop gain $(0 \leq k < \infty)$. Can the system be stabilized?

(b) Repeat for negative loop gains $(-\infty < k \leq 0)$.

11. Sketch out a *moderately* detailed flow chart for a computer program to plot or display root-locus diagrams.

Section 8.4

1. Explain how the discussion following Definition 8.4.1 must be modified if the open-loop transfer function has a numerator and denominator of the same degree.

2. (a) Prove Corollary 8.4.3.

   (b) Are any modifications to the Nyquist criterion required to allow for negative loop gains $(k < 0)$?

3. What can you conclude about the closed-loop poles of a system if the frequency response plot of $g(j\omega)$ passes through the point $-1/k$?

4. Show how the Nyquist contour can be modified so that the Nyquist criterion will test whether all the closed loop poles

   (a) Have stability margin σ.

   (b) Have damping margin ζ.

Is the frequency response information of any use in constructing the plot of $g(s)$ corresponding to your modified Nyquist contours?

5. Construct a root-locus diagram for the feedback system of Example 8.4.6 and determine the points on the loci corresponding to $k = 2$. Verify the gain margin which was determined using the Nyquist criterion.

6. Construct a Nyquist plot for each system in Problem 8.3-6, and verify the stability ranges of k determined there.

7. Construct a Nyquist plot for the system of Problem 8.3-10, and verify the stability range of k determined there.

8. Consider the feedback system shown in Fig. P2, and construct Nyquist plots for the following cases. Determine whether or not each system is stable, and if so, find the gain and phase margins.

(a) $k_p g_p(s) = \dfrac{(s+3)}{(s-1)}$ $\qquad k_c g_c(s) = \dfrac{2}{s}$

(b) $k_p g_p(s) = \dfrac{2}{s(s+2)}$ $\qquad k_c g_c(s) = \dfrac{s+1}{s}$

(c) $k_p g_p(s) = \dfrac{2}{s(s+2)}$ $\qquad k_c g_c(s) = \dfrac{s+3}{s}$

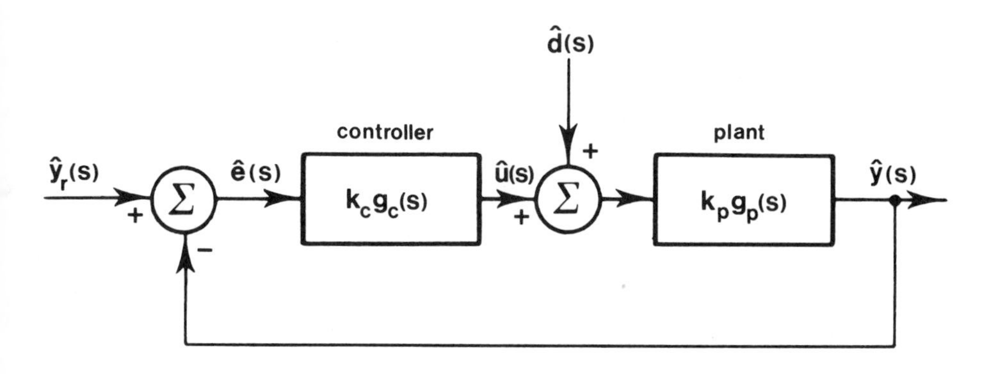

FIG. P2

9. How must Nyquist's Theorem (8.4.2) be modified if the Nyquist contour detours to the *left* of all open-loop poles on the imaginary axis?

10. Translate the statement of the Nyquist stability criterion into a form suitable for use with a Bode plot. Must you apply any restrictions in order for this to make sense?

11. Suppose the open-loop frequency response of a feedback system has been plotted as a Bode diagram. From the Nyquist stability criterion, deduce what properties the magnitude and phase curves must have to guarantee stability of the feedback system. Consider three cases: (i) the phase lag never reaches +180°, (ii) the phase lag equals +180° at only one frequency, and is greater than 180° at high frequencies, and (iii) the phase lag is less than 180° between two frequencies, and greater than 180° outside this range. In the latter case, how should the term gain margin used in the text be defined?

Chapter 9

COMPENSATION OF SINGLE-LOOP SYSTEMS

9.0. INTRODUCTION

The previous chapter introduced the most useful tools available for the analysis of scalar feedback systems. Once the open-loop system is described, either by its set of poles and zeros or by its frequency response, the root-locus method or Nyquist's stability criterion will indicate whether the feedback system can be given an acceptable transient response by adjustment of the loop gain. The static accuracy of the system can then be determined from Theorems 8.2.2 and 8.2.3.

It frequently happens that good transient response and high static accuracy cannot both be achieved simply by adjusting a single parameter such as a loop gain. The speed of response depends, roughly speaking, on the distance of the dominant groups of closed-loop poles from the origin, which in turn depends on the loop gain. For any system in which two or more root-locus branches tend to infinity, this distance can be increased beyond some limit only at the cost of an unacceptable reduction in the degree of damping of at least one sinusoid in the transient response. Thus gain adjustment may not give sufficiently fast as well as sufficiently smooth transients. In other cases, the required static accuracy may not be achievable for any value of loop gain, or the high gain necessary to achieve it may result in inadequate damping of the transients.

It is then necessary to modify the system dynamics: either the dynamic properties of some components in the loop need to be altered, or additional components need to be inserted into the loop. The process of modifying the properties of one or more system components so as to allow the performance specifications to be met by subsequent loop gain adjustment is known as *compensation,* and is the subject of this chapter.

We begin with a description of the performance specifications most commonly used in the design of scalar feedback systems. In subsequent sections the main types of system compensation are discussed.

In this chapter, as in the previous one, only scalar feedback systems will be considered. Terms such as "response" and "performance" refer to the *output* of the system, unless qualified by a reference to the internal system *state*.

9.1. PERFORMANCE SPECIFICATIONS

The design of any engineering system or component requires a specification of the job to be done, or of certain essential properties the system must have. This specification should be fairly precise; it is useful to know when a given design is just good enough for the job at hand since better quality almost invariably results in more complex, difficult, and expensive designs. Automatic control systems are no exception.

We observed in Chapter 8 that the behavior of scalar feedback systems involves both steady-state and transient responses. The same two categories are traditionally used for the performance specifications of feedback systems.

The majority of such systems must perform in steady state as regulators; that is, they must keep the system output at or near any one of a range of setpoints for a range of constant values of load or disturbance. In this case the design criteria will include a specification of maximum permissible steady-state error. Sometimes a system is required to follow a time-varying input with an acceptable steady-state accuracy. It has become customary to specify static accuracy by the maximum permissible steady-state errors in the response to polynomial inputs of the form

$$y_r(t) = \alpha t^i \, 1(t) \tag{1}$$

$$d(t) = \beta t^j \, 1(t) \tag{2}$$

where i and j are given nonnegative integers.

In the case of transient response, it is somewhat more difficult to state meaningful quantitative specifications since the relative weighting of the modes in the transient depends on the input, which can rarely be predicted in advance. One commonly used set of specifications puts constraints on the step response of the system. By specifying the three parameters *delay time*, *overshoot*, and *settling time*, the response is confined to within the shaded borders of Fig. 1. It is then *assumed* that a system whose step response satisfies these constraints will have an acceptable transient response to any kind of input. Another popular design parameter is the *rise time*, defined in Sec. 5.3.

Alternatively, constraints can be put on the frequency response of the system. It was shown in Chapter 6 that there is a qualitative

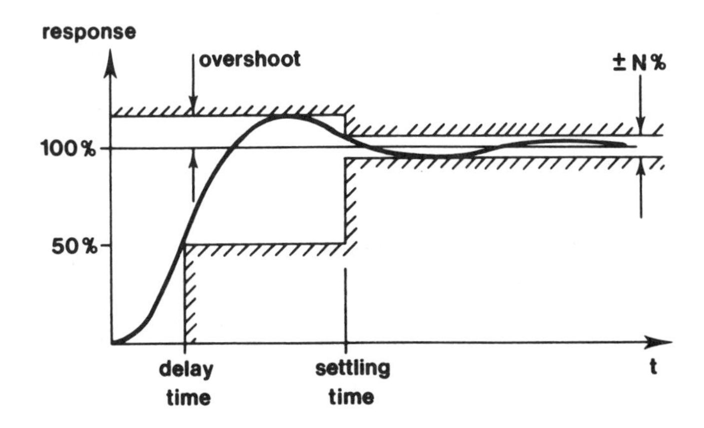

FIG. 1. Unit step response specifications.

connection between the transient response and the steady-state frequency response. Roughly speaking, a large bandwidth means that a system can follow rapidly changing inputs (i.e., signals containing high frequency components in their Fourier series representations), while the presence of a large resonance peak in the frequency response is usually an indication of at least one poorly damped sinusoid in the transient response. Thus, *bandwidth* B and *resonance peak height* M_p of the closed-loop frequency response are figures of merit roughly corresponding to delay time and overshoot, respectively. Specification of these parameters constrains the magnitude of the closed-loop frequency response to the region shown in Fig. 2. The closed-loop bandwidth is not a very convenient figure of merit (see Problem 2), and the resonant frequency ω_r is often used in its place. An alternative method of constraining the transient response by frequency domain criteria is to stipulate the smallest acceptable gain and phase margins; these can be determined using only the *open-loop* frequency response (see Sec. 8.5).

There are thus three alternative sets of transient response specifications in common use:

(a)	Closed-loop step response:	delay time (or rise time), overshoot, settling time
(b)	Closed-loop frequency response:	resonance peak height, bandwidth or resonant frequency
(c)	Open-loop frequency response:	gain margin, phase margin

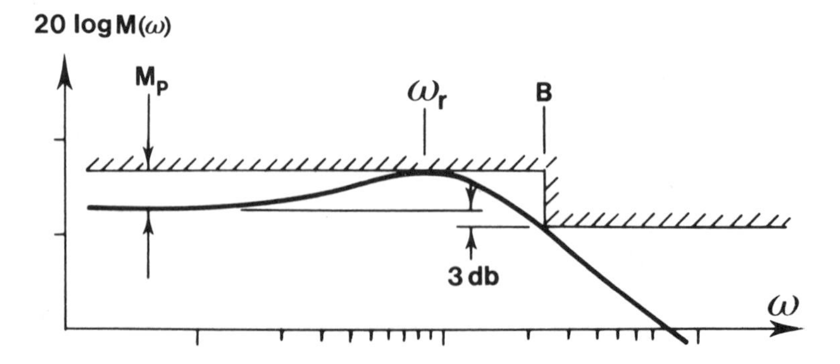

FIG. 2 Closed-loop frequency response specifications.

At first glance, at least the first two of these sets of specifications seem quite reasonable ways of ensuring an acceptable transient. Difficulties can arise, however, when the transient contains poorly damped, high frequency sinusoids of relatively low weighting.

1. EXAMPLE. Consider a system with the open-loop transfer function

$$kg(s) = k\,\frac{5(s+2)}{s(s^2+2s+10)} \tag{3}$$

Suppose that the closed-loop system is to have

(i) Steady-state error in response to a ramp input of unit slope ($y_r(t) = t1(t)$) not exceeding 0.75 units.
(ii) A delay time of not more than 1 time unit.
(iii) An overshoot of not more than 5%.
(iv) A settling time (to within 3%) of not more than 4 time units.

Using (8.2-5), (i) requires a loop gain k of not less than (1/0.75) = 1.33. The step response of the system with $k = 1.4$ is shown in Fig. 3. It clearly satisfies (ii)-(iv), yet would be considered too oscillatory for many practical applications. If the output of the system were the velocity of a large motor or turbine, for instance,

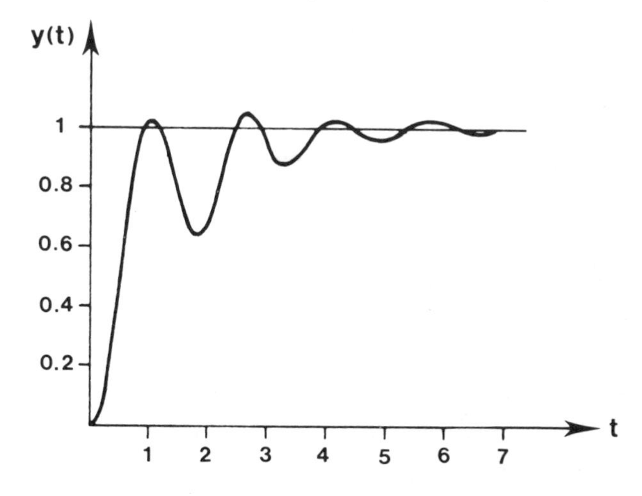

FIG. 3. Unit step response for Example 1.

the oscillations would indicate considerable stress fluctuations in the shaft which might well exceed fatigue safety limits. A Bode diagram of the closed-loop frequency response is shown in Fig. 4. The magnitude curve exhibits a pronounced resonance peak M_p of about 4.5 db, or an amplitude ratio of 1.66. This is quite large: a peak amplitude ratio of about 1.2 is a more usual limit in servomechanism design. The open-loop frequency response is shown in Fig. 5. The system clearly has an infinite gain margin, and a phase margin of about 40°; these would be considered satisfactory in most cases.

ΔΔΔ

The example above is meant to illustrate some of the difficulties associated with transient response design specifications. Constraints on the step response are appealing at first sight, and yet are clearly not always sufficient; no wholly satisfactory quantitative criterion has yet been found which guarantees smoothness without sacrificing speed. Moreover, the required step response properties cannot be determined directly from the system poles and zeros or from its frequency response; usually the system equations must be solved numerically or by an analog simulation. Frequency domain specifications are more convenient if computer assistance is unavailable. Gain and phase margins are clearly the simplest criteria with which to work, but they give little information about the shape of the transients; for example, in Example 1, the presence of an inadequately damped term cannot be deduced directly from Fig. 5. Thus, even a good margin of stability does not always rule out the presence of excessive oscillations. Figure 4 suggests that a constraint on the closed-loop resonance peak is the best way to ensure smoother transients. However, the use of closed-loop frequency response criteria alone is also awkward due to the lack of a quantitative connection between bandwidth or resonant frequency and speed of response.

The reader might suggest that with these reservations about step and frequency response specifications the most direct and satisfactory way of ensuring good transients would be to constrain the closed-loop poles to lie in an appropriate region of the complex

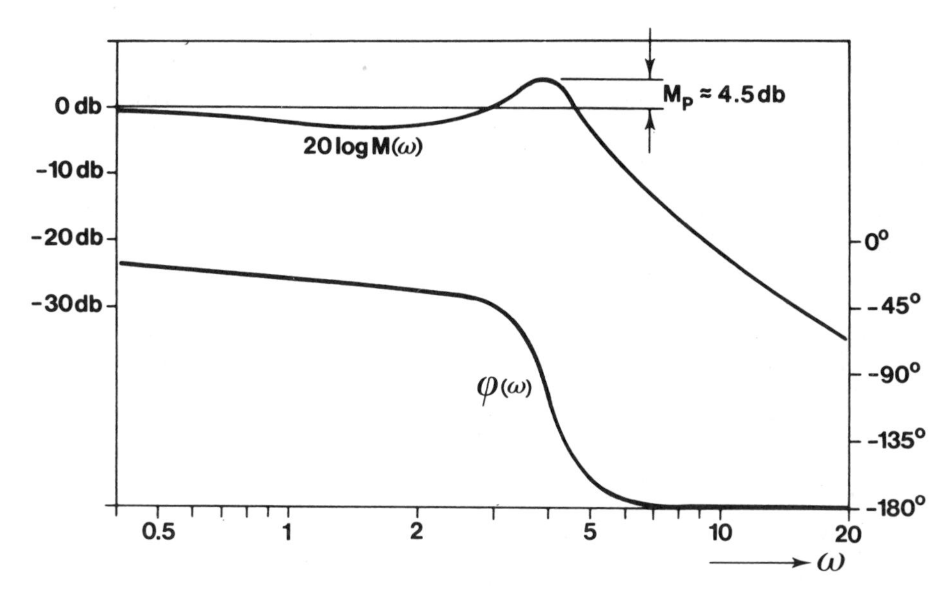

FIG. 4. Closed-loop frequency response of Example 1.

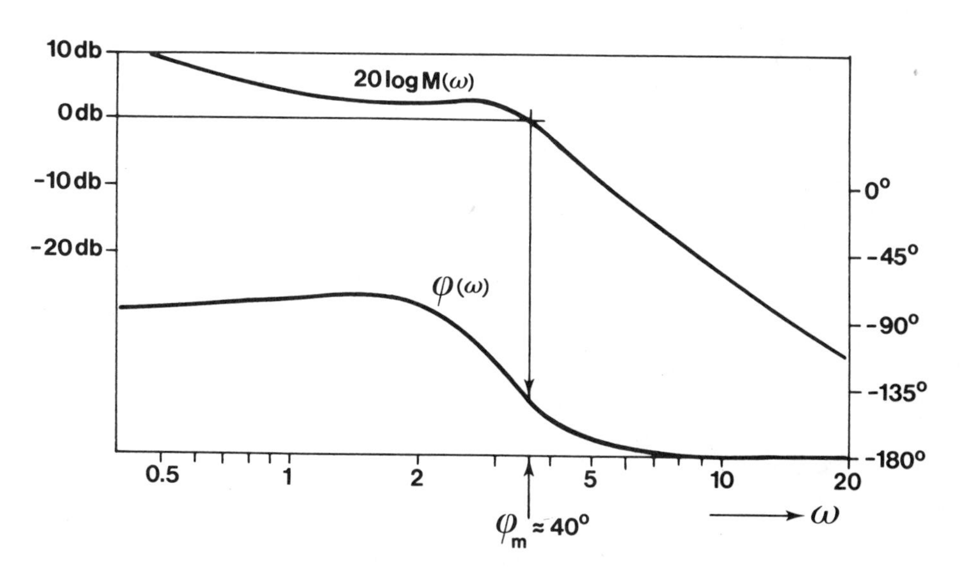

FIG. 5. Open-loop frequency response of Example 1.

plane by specifying a minimum acceptable damping margin ζ and stability margin σ (see Fig. 5.3-5). If all natural modes of the system decay sufficiently rapidly and are well enough damped, then the output, being a linear combination of them, should also be acceptable. Unfortunately, this is not always true, as the following example shows.

2. *EXAMPLE*. Consider a feedback system with the open-loop transfer function

$$kg(s) = \frac{1.9s + 0.1}{s(s - 0.8)} \tag{4}$$

Its closed-loop transfer function (which was used in Problem 5.3-3) is

$$\bar{h}_c(s) = \frac{1.9s + 0.1}{(s + 1)(s + 0.1)} \tag{5}$$

The transient response is a linear combination of e^{-t} and $e^{-0.1t}$, yet the step response of the system (shown in Fig. 6) has an overshoot of nearly 65%, and the magnitude curve of the closed-loop frequency response (shown in Fig. 7) has a resonance peak of about 4.7 db.

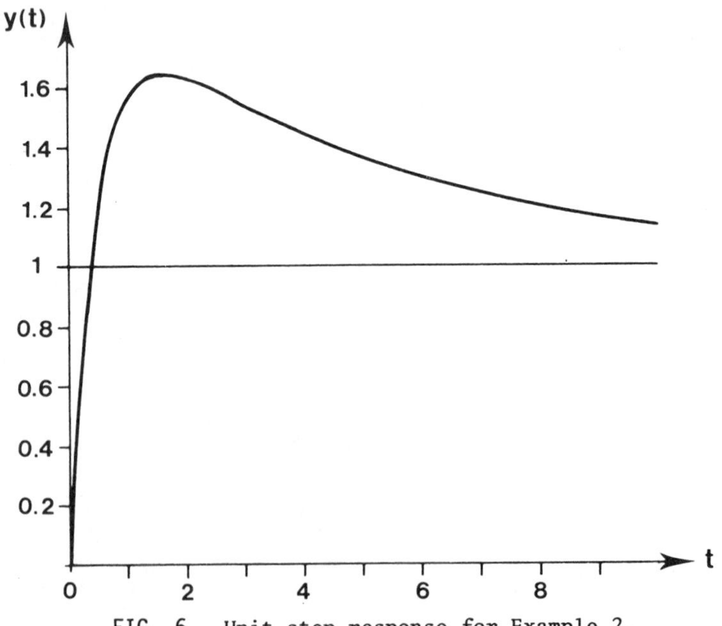

FIG. 6. Unit step response for Example 2.

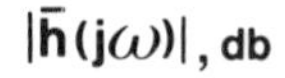

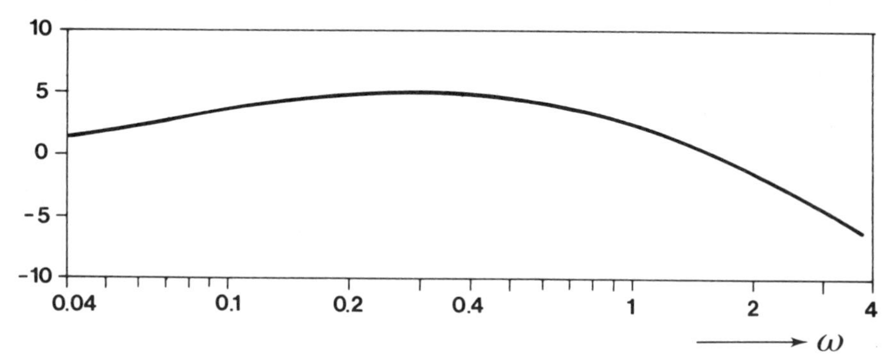

FIG. 7. Magnitude of closed-loop frequency response for Example 2.

If the reader considers this example to be somewhat unusual, due to the open-loop pole at +0.8, he should plot the step response of the feedback system with open-loop transfer function $(6s + 5)/s^2$; this has an overshoot of nearly 9%, with closed-loop poles at $s = -1$ and $s = -5$.

ΔΔΔ

We must conclude that none of the three commonly used sets of transient response specifications are wholly satisfactory by themselves. A good deal of judgment and experience is required to decide not only what values should be given to the different performance measures, but also just what combination of them should be used so that a system will perform satisfactorily in its operating environment.

A further difficulty is also apparent. The design constraints which are most convenient for the analysis techniques we have developed so far, e.g., constraints on the closed-loop pole locations, or on gain and phase margins, have no direct quantitative relation with "physically meaningful" constraints such as step response specifications. For this reason, the compensation techniques to be described in the sequel are trial-and-error methods requiring a good deal of intuition. We shall return to this point later.

9.2. IMPROVEMENT OF STATIC ACCURACY

Consider a system which fails to meet a specification calling for zero steady-state error in response to a reference or disturbance input of the form (9.1-1) or (9.1-2). By Theorems 8.2.2 and 8.2.3, there is then no choice but to insert the required number of integrators into the loop. A consideration of the properties of root-locus diagrams (see Sec. 8.3) will show that this invariably results in a degradation of the transient response, and may even make the system unstable. Addition of an open-loop pole at $s = 0$ introduces another root-locus branch starting there, increases the number of branches going to infinity by one, and shifts the intersection of the asymptotes with the real axis closer to the origin. Thus, all branches tending to infinity now do so along asymptotes which have been moved further towards the right half-plane, so that at least some of the closed-loop poles are forced closer to the origin if the previous degree of damping is to be maintained. The system is slower, or more oscillatory, and certainly less stable. We shall not attempt to make this rather loose statement more precise, but leave it to the reader to satisfy himself of its validity by sketching the effect of adding a pole at $s = 0$ on some typical root-locus diagrams (see Problem 1).

If, after the insertion of the necessary number of integrators, the system can be given a satisfactory transient response by loop gain adjustment, the design is complete. If not, we have a problem in improving the transient performance, which will be discussed in Sec. 9.3.

Sometimes the performance specifications permit a nonzero steady-state error, so that additional integrators can be avoided, but the loop gain cannot be given a sufficiently high value without an unacceptable deterioration in transient performance. In such cases, the insertion of a passive network known as a *lag compensator* may provide a simple solution. Such a device has a transfer function of the form

$$h_c(s) = \alpha \frac{s + a}{s + \alpha a} \tag{1}$$

where $0 < \alpha < 1$ and $a > 0$.

Inserting the network into the loop amounts to adding a real pole at $s = -\alpha a$ and a real zero at $s = -a$ to the open-loop transfer function, with the pole closer to the origin than the zero, as shown in Fig. 1. The open-loop gain is unchanged since the compensator has unity gain. If the uncompensated system has no root-locus branch between $-\alpha a$ and $-a$, then the compensator introduces a new branch starting at $-\alpha a$ and terminating at $-a$.

To see how a lag compensator works, consider a scalar feedback system with open-loop transfer function

$$kg(s) = \xi \frac{(s - z_1)(s - z_2) \ldots (s - z_r)}{(s - p_1)(s - p_2) \ldots (s - p_n)} \tag{2}$$

where ξ is related to the loop gain k by (8.3-3),

$$k = \xi \frac{(-z_1)(-z_2) \ldots (-z_r)}{(-p_{\ell+1})(-p_{\ell+2}) \ldots (-p_n)} \tag{3}$$

and the first ℓ open-loop poles are assumed to be at the origin. Now recall that a point s in the complex plane lies on the root-locus of the closed-loop system if and only if the *angle condition* (8.3-7) is satisfied,

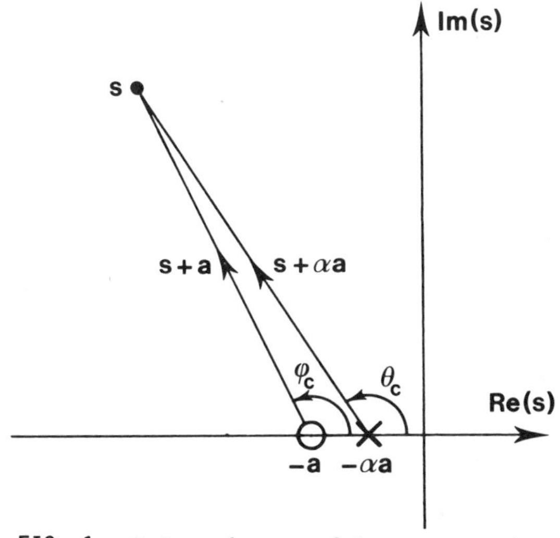

FIG. 1. Pole and zero of lag compensator.

$$\sum_{i=1}^{n} \Theta_i - \sum_{i=1}^{r} \phi_i = \pm m\pi \qquad (m = 1,3,5,\ldots) \tag{4}$$

where Θ_i are the angles of the vectors drawn from the open-loop poles p_i to s and ϕ_i are the angles of the vectors drawn from the open-loop zeros z_i to s. Moreover, for any such point on the root-locus, ξ is specified by the *magnitude condition* (8.3-8),

$$\xi = \frac{|s - p_1|\ |s - p_2| \cdots |s - p_n|}{|s - z_1|\ |s - z_2| \cdots |s - z_r|} \tag{5}$$

The basic assumption regarding a lag compensator is that its zero -a and pole $-\alpha a$ lie close together and relatively far away from all the original closed-loop poles. Thus for a point s on the root-locus in the vicinity of an existing closed-loop pole, the vectors drawn from -a and $-\alpha a$ to s will be nearly the same (see Fig. 1). This means that introduction of the additional compensator pole and zero alters the angle condition (4) by adding

$$\Theta_c - \phi_c \approx 0 \tag{6}$$

and the shape of the root-locus is essentially unchanged in the vicinity of the original closed-loop poles. Moreover, in this vicinity the magnitude condition (5) is also changed very little since it is multiplied by the factor

$$\frac{|s + \alpha a|}{|s + a|} \approx 1 \tag{7}$$

Note, however, that (3) becomes

$$k = \xi \frac{(-z_1)(-z_2) \cdots (-z_r)(a)}{(-p_{\ell+1})(-p_{\ell+2}) \cdots (-p_n)(\alpha a)} \tag{8}$$

so that the loop gain k is *increased* by a factor of $1/\alpha$.

Thus a lag compensator allows an increase in loop gain and a consequent improvement in static accuracy, without an appreciable change in the original configuration of closed-loop poles. Some degradation of the transient response is inevitable since an additional closed-loop pole in the vicinity of $-a$ and $-\alpha a$ is introduced. However, the loss of performance can be kept sufficiently small in many cases. An example will help to illustrate this.

1. EXAMPLE. Consider the proportional control system of Example 8.3.7(a), whose block diagram is reproduced in Fig. 2 (the disturbance input is omitted). Suppose the system is to have a step response with

(i) An overshoot not exceeding 6%

(ii) A settling time (to within 5%) of not more than 4 units

(iii) A steady-state error not exceeding 2.5%

Unit step responses of the uncompensated system, obtained by a computer simulation, are shown in Fig. 3. In A, a loop gain of 1.5 gives an overshoot of 5.8%, a settling time of 1.2, and a steady-state error of 40%. To meet specification (iii), a gain of 40 is required, giving an overshoot of 56% and a settling time of about 1.2, as shown in B.

Clearly, the specifications cannot be met by gain adjustment alone, so we insert a lag compensator in series with the controller amplifier, making the controller transfer function

$$k_c \hat{g}_c(s) = k_c \alpha \cdot \frac{s + a}{s + \alpha a} \tag{9}$$

Choosing $\alpha = 0.1$, $a = 0.5$, and a loop gain of 40 gives the response C shown in Fig. 3. This has an overshoot of 2.5%, a settling time of about 4 and a steady-state error of 2.5%, and thus meets all the specifications. Root-locus diagrams of the original and compensated

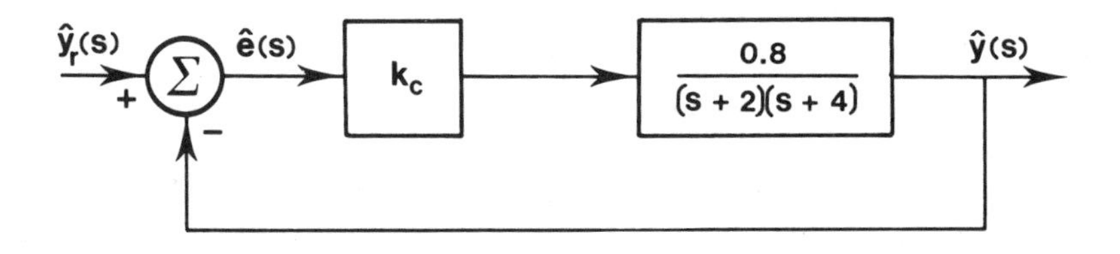

FIG. 2. Block diagram for Example 1.

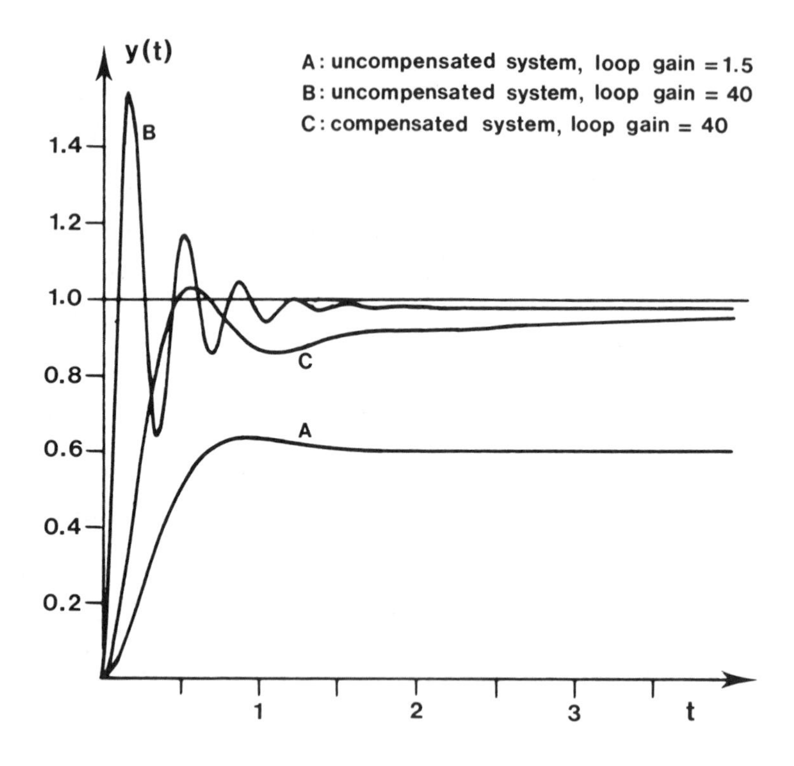

FIG. 3. Unit step responses for Example 1.

systems are shown in Fig. 4; the closed-loop poles corresponding to step responses A and C of Fig. 3 are indicated by dark squares.

ΔΔΔ

Several features in this example deserve emphasis. First, the reader should note the effect of the slowly decaying real exponential contributed by the closed-loop pole between the compensator pole and zero in Fig. 4. This considerably increases the system settling time. However, since the sinusoid due to the pair of complex conjugate closed-loop poles now contributes only a part of the transient, its damping can be reduced without violating the overshoot constraint, i.e., the original closed-loop poles can be moved further out. Thus the loop gain can be increased by a factor of $40/1.5 = 26.7$, considerably greater than $1/\alpha = 10$. The resulting reduction in delay time partly compensates for the much longer settling time of the response.

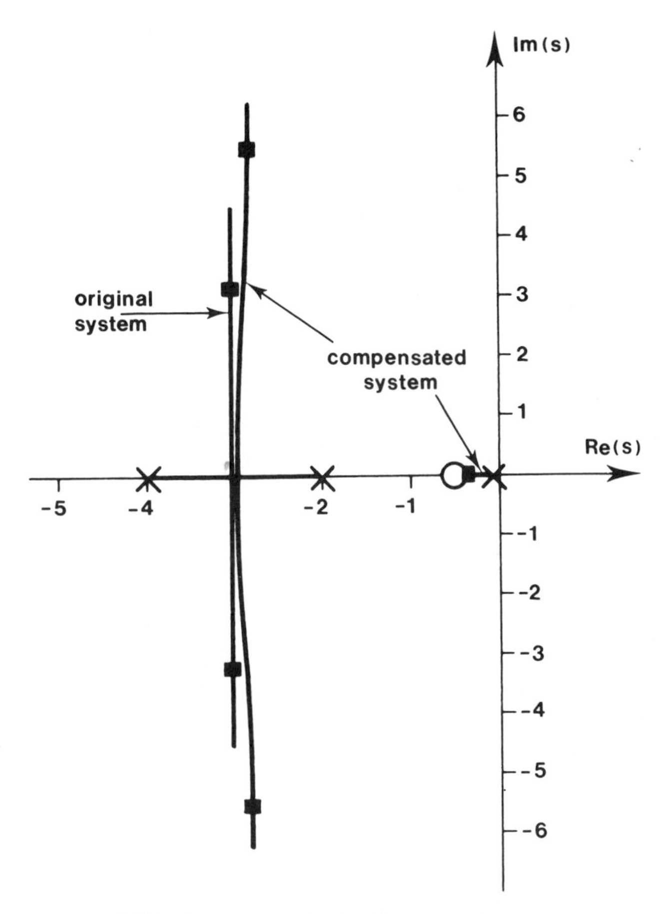

FIG. 4. Root-loci for Example 1.

A second point worth noting concerns the design of the compensator, i.e., the choice of suitable values for a and α. There is no simple way of computing the best values of these parameters directly from specifications on the step response of the system. In the example above, the compensator was "designed" by trial and error, using a computer simulation of the system. A fairly typical figure for α is about 0.1; smaller values are often difficult to implement in practical circuits. In addition, the parameter a should itself be fairly small so as to keep the compensator pole and zero close to the origin and minimize the change in the root-locus near the original closed-loop poles. These should be viewed as guidelines rather than definite rules for compensator design.

We have described the effect of integral and lag compensation in terms of the poles and zeros of the system. The theory can also be quite readily developed using the open-loop frequency response of the system. Consider the lag compensator (1). Its frequency response is $h_c(j\omega) = \alpha(j\omega + a)/(j\omega + \alpha a)$, where $\alpha < 1$. Clearly, the magnitude is close to unity for $\omega << \alpha a$, while for frequencies greater than a, the magnitude decreases to a limiting value of α. The compensator phase shift can readily be shown to be $\tan^{-1}\{\omega a(\alpha - 1) / (\alpha a^2 + \omega^2)\}$. Since $\alpha < 1$, this is negative for all frequencies, varying from 0 at low frequencies to a minimum of $\sin^{-1}\{(\alpha - 1)/(\alpha + 1)\}$ at a frequency of $a\sqrt{\alpha}$, and then increasing to 0 again. On a Bode diagram, the phase curve is symmetrical about the frequency $a\sqrt{\alpha}$. Incidentally, the negative phase shift or lag gives the compensator its name.

For frequencies greater than αa, the compensator reduces the magnitude of the open-loop frequency response by a factor which decreases from unity to a limiting value of α, while leaving the loop gain unchanged. Now suppose that the break points αa and a of the compensator can be placed in a range where the original phase lag of the system is comparatively small, so that the open-loop phase curve is not materially altered near the frequency where it crosses -180°. If the original gain and phase margins are satisfactory, then to restore them it is necessary only to *increase* the loop gain sufficiently to make up for the attenuation introduced by the compensator near the critical frequencies where the gain and phase margins are measured, i.e., by a factor of approximately $1/\alpha$. This increase in loop gain, possible without reducing the phase margin or open-loop bandwidth, is the point of the compensation. However, no compensation which increases the phase lag of the open-loop frequency response can possibly increase the phase margin or open-loop bandwidth. It is then an immediate consequence of the Nyquist stability theorem that neither a lag compensator nor a pure integrator can increase the degree of stability or speed of response of a system. We invite the reader to work out the details of this argument, and of compensator design in the frequency domain, by solving Problems 5-8.

There are thus two methods for improving the static accuracy of a system when loop gain adjustment by itself will not work: inserting pure integrators into the loop or inserting a passive lag network. Of these, the former is more effective, at the cost of a more severe degradation of the transient response, so that a further step of compensation is usually necessary to improve transients. There are several ways to implement time integration of a variable by analog (as distinct from digital) devices. Some of these have already been described in earlier chapters. One is a combination of a hydraulic four-way valve and power cylinder, in which the piston displacement is the time integral of the valve spool displacement (see Problem 2.4-1). Another device is a frictionless rotor whose mass moment of inertia can be thought of as converting an angular acceleration (proportional to the applied torque) into its integral, angular velocity. Probably the most common integrating device is the basic integrating circuit of the electronic analog computer, consisting of an operational amplifier with an input resistor and a feedback capacitor (see Sec. 7.1). With the development of integrated circuits, such devices have become very compact, and quite cheap compared with systems employing mechanical displacements. In many industrial control systems, signals are realized electrically because of the ease of transmitting, comparing, amplifying, and summing voltages, so that electronic integrators can be readily implemented. Nevertheless, all integrators are "active" devices requiring a power supply. By contrast, lag compensating networks can be implemented by very simple passive circuits. Figure 5 shows an electric network whose input and output voltages are related by the transfer function (1), with $a = 1/R_2C$ and $\alpha = R_2/(R_1 + R_2)$.

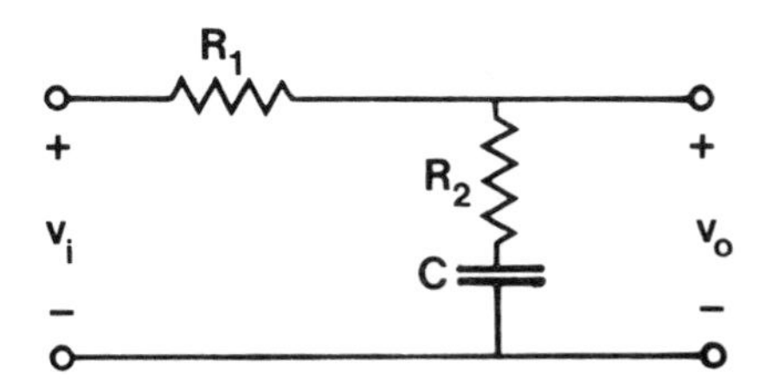

FIG. 5. Lag compensating network.

9.3. STABILIZATION AND IMPROVEMENT OF TRANSIENT RESPONSE

We have seen that the transient response of a feedback system can be unsatisfactory for one or more of the following reasons:

(a) It is unstable, with closed-loop poles in the right half-plane.

(b) The response is too oscillatory, usually because of complex closed-loop poles with insufficient damping.

(c) The response is too slow since one or more closed-loop poles lie too close to the origin.

We assume that the system has a root-locus diagram of such a nature that satisfactory placement of the closed-loop poles cannot be achieved by loop gain adjustment. The problem therefore is to modify the system so that those branches of its root-locus which contain the troublesome closed-loop poles are either removed or moved sufficiently far into the left half-plane. A consideration of the rules for constructing the root-locus shows that this can be achieved by increasing the number of open-loop zeros in the left half-plane. Adding a negative real zero, for example, reduces by one the number of branches of the root-locus which go to infinity, increases the angle of the asymptotes, and may shift the intersection of the asymptotes to the left. The reader should compare the root-locus diagrams of the systems (b) and (c) of Example 8.3.7 (Figs. 8.3-5 and 8.3-6); they illustrate well the stabilizing effect of introducing a zero into the loop.

Multiplying the open-loop transfer function by a factor $s + a$, $a > 0$, means that somewhere in the loop a signal must be replaced by a weighted sum of itself and its *time derivative*. Now the differentiation of a physical variable with respect to time is often difficult: the noise which is invariably present is drastically accentuated in differentiation. Devices can be built which produce a good approximation to the time rate of change of some signals such as voltages, and which also act as noise filters. However, they are usually complex and expensive. Generally, differentiation of a physical variable is considered only as a last resort. It follows, then, that introduction of a real zero into the open-loop transfer function can be readily implemented only when the derivative of a signal is already available in the system. This requires that the open-loop system contains at least one pure integrator (pole at the origin). Moreover, the required derivative must be a signal which can be measured continuously and added to its integral. This is illustrated in the block diagrams of Fig. 1. In (a) the system transfer function is

$$h_1(s) = \frac{k_1}{s} g_1(s)g_2(s) \tag{1}$$

while the compensated system in (b) has the transfer function

$$h_2(s) = k_2 \frac{s + k_1/k_2}{s} g_1(s)g_2(s) \tag{2}$$

We have already encountered two examples of this technique; it will be useful to consider them in more detail.

1. EXAMPLE. In the introductory example in Chapter 1, we attempted to design an automatic controller for the angular position of a radar

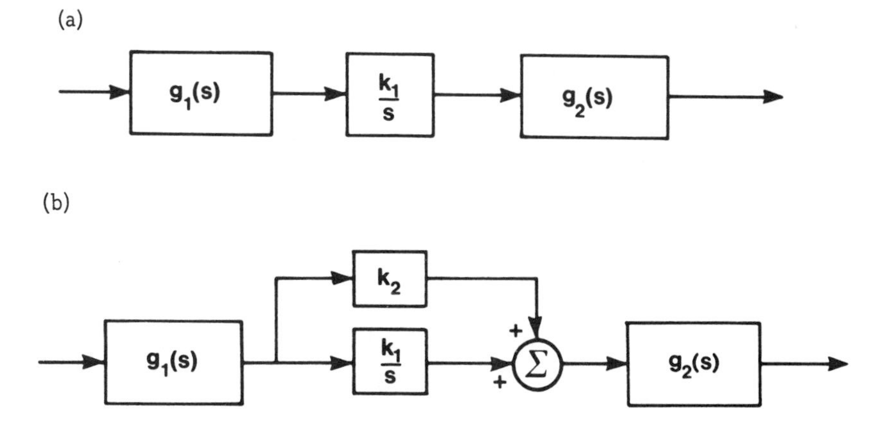

FIG. 1. Adding a zero to a system: (a) original system containing an integrator, and (b) compensated system with a new zero at $-k_1/k_2$.

antenna by applying to the antenna shaft a torque proportional to the error in antenna position. This strategy produced an unstable (or marginally stable) system: with the assumption that the electric motor driving the antenna could be modelled as a simple amplifier (1.1.3), it turned out that any change in desired antenna position or any change in wind torque resulted in oscillations of constant amplitude. Figure 2 shows a block diagram and root-locus for this proportional control system. In Chapter 1 it was argued on physical grounds that replacing the error by a signal depending not only on antenna position but also on antenna velocity

$$e(t) = y_r(t) - y(t) - k_v\dot{y}(t) \tag{3}$$

would have a stabilizing effect: the system can apply corrective torques without having to wait for position errors to develop. If

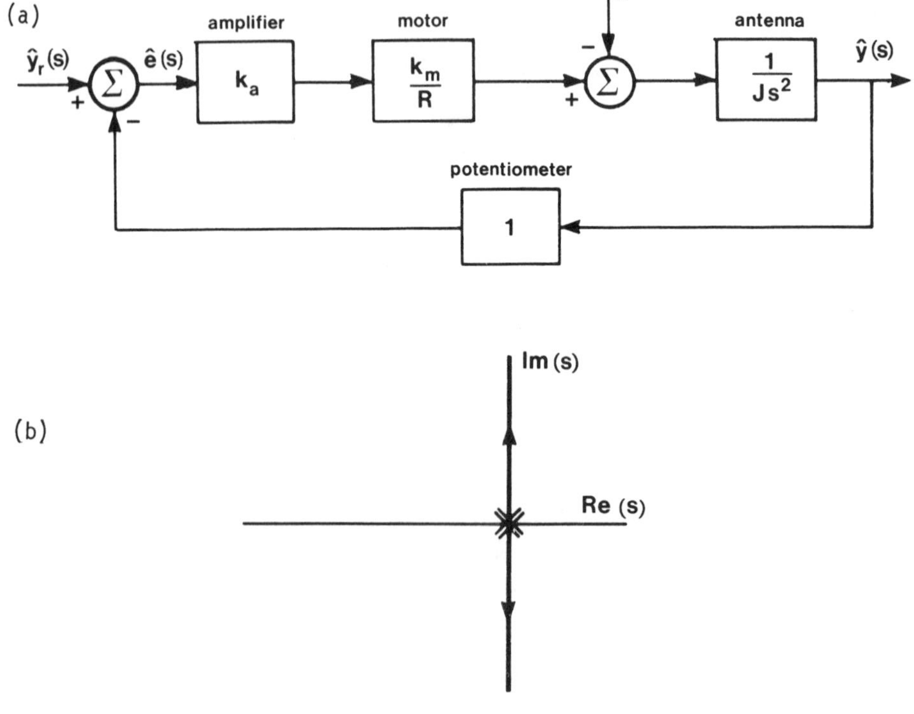

FIG. 2. Uncompensated system of Example 1: (a) block diagram and (b) root-locus diagram.

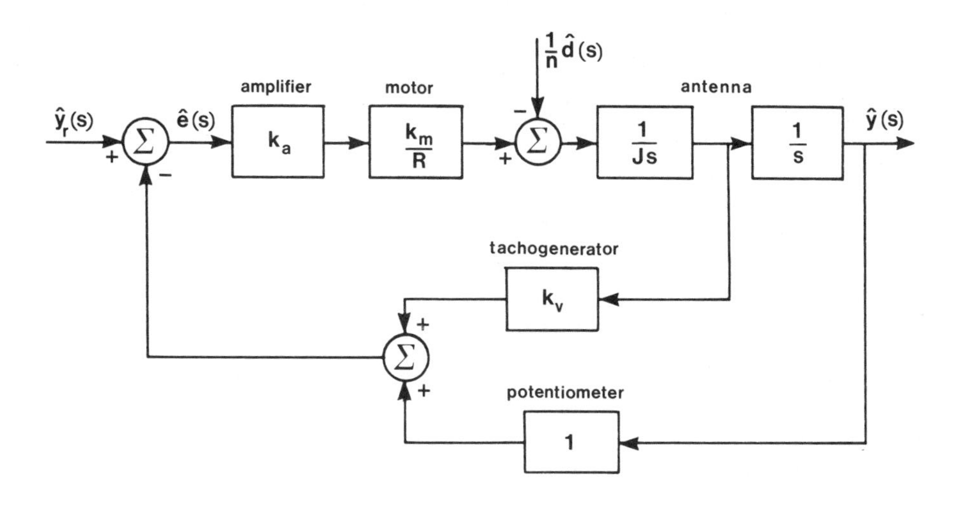

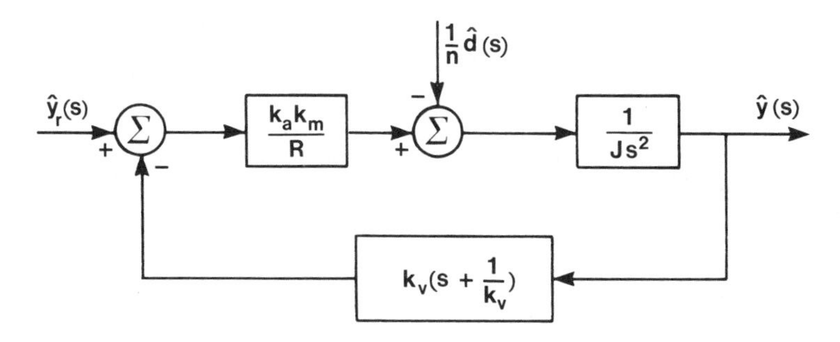

FIG. 3. Block diagrams of a compensated antenna system (Example 1): (top) introduction of derivative feedback and (bottom) equivalent block diagram.

we use (3) and take Laplace transforms, the block diagrams of Fig. 3 result, and we see that a zero at $s = -1/k_v$ has been introduced into the system. This modification is possible since a device for continuously measuring the derivative of the output is readily available in the form of a tachogenerator, and the voltage signals produced by the potentiometer for output feedback and the tachogenerator for derivative or rate feedback can be easily added together.

The characteristic equation of the compensated system is

$$s^2 + \omega^2 k_v s + \omega^2 = 0 \tag{4}$$

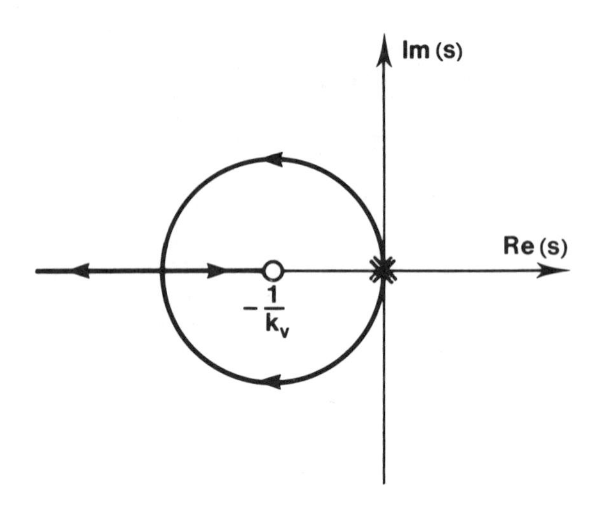

FIG. 4. Root locus of the compensated system of Example 1.

where $\omega^2 \triangleq k_a k_m/RJ$. A typical root-locus diagram is shown in Fig. 4: the system is clearly stable for all positive loop gains. In fact, it is easy to see from (4) that by a suitable choice of k_a and k_v the closed-loop poles can be placed *anywhere* in the complex plane (except that complex roots must occur in conjugate pairs). For example, if a repeated closed-loop pole at $-\lambda$ is required, then the characteristic equation should be

$$(s + \lambda)^2 = s^2 + 2\lambda s + \lambda^2 = 0 \tag{5}$$

Equating coefficients in (4) and (5), we find that $\omega^2 k_v = 2\lambda$ and $\omega^2 = \lambda^2$, from which it follows that $k_v = 2/\lambda$ and $k_a = \lambda^2 RJ/k_m$.

Of course, the conclusion about arbitrary placement of closed-loop poles is valid only for the system *model*; poles of large magnitude correspond to very fast transients, and at sufficiently high speeds the model will no longer be an adequate representation of the real antenna system. For instance, the assumption that the motor is a pure amplifier, i.e., that it can produce an instantaneous change in motor torque from a step change in applied voltage, will break down. Thus appropriate values of k_a and k_v must be chosen on the basis of physical knowledge of the real system, and an understanding of the extent to which the model can be expected to represent it. In this case, theory alone gives little help regarding the best choice of the design parameters.

ΔΔΔ

It is interesting to look at the compensation procedure of Example 1 from a slightly different point of view. The reader should show, using Theorem 2.2.1, that the output and its derivative in this case form a complete state vector for the open-loop antenna system. Thus we have established that with a weighted sum of both states in the feedback loop, the closed-loop poles can be placed arbitrarily. On the other hand, if only the output is fed back, then the closed-loop poles are restricted to a one-dimensional region: the root-locus. In a first-order system (see Example 8.1.4), the output is also the sole state variable, and it may be fed back to place the closed-loop pole anywhere on the real axis.

This suggests that as more information about the system states is included in the feedback signal, the designer's freedom in specifying closed-loop behavior increases. In fact, this is a manifestation of one of the most important results in linear control theory, which holds under certain controllability conditions: If all the states of an open-loop system can be directly measured, then a feedback signal exists in the form of a linear combination of the states to give the closed-loop system any preassigned configuration of closed-loop poles, subject to complex conjugate pairing (Theorem 11.2.1). For the special case of a scalar open-loop system with no zeros, the result can be established quite simply using Laplace transforms (Problem 1).

2. *EXAMPLE.* Another instance of compensation by adding a zero has already been given in Example 8.3.7. There, system (b) has poles at 0, -2, and -4; its root-locus is shown in Fig. 8.3-5. A block diagram of the system is repeated in Fig. 5(a) for convenience. Replacing the integrator by a proportional-plus-integral controller as shown in Fig. 5(b) results in a system having a zero at $-a = -k_2/k_1$. The

(a)

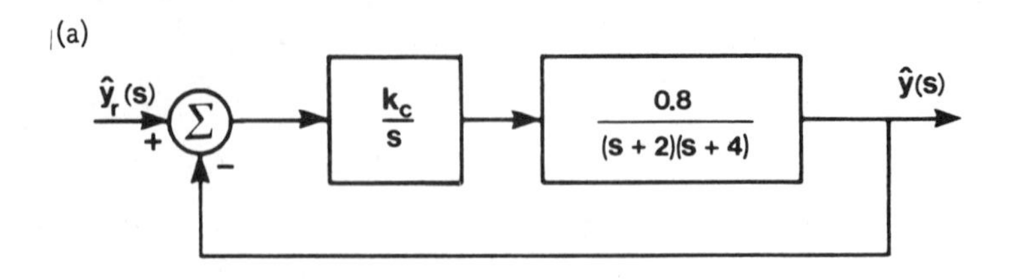

(b)

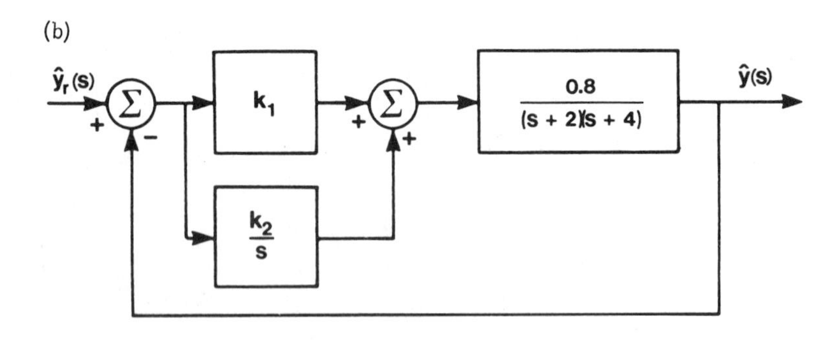

FIG. 5. Block diagrams for Example 2: (a) uncompensated systems and (b) compensated system.

root-locus of the compensated system (with $a = 1.6$) is shown in Fig. 8.3-6. The reader should again note the stabilizing effect of the zero: only two root-locus branches now tend to infinity along vertical asymptotes in Re $s < 0$, so that the system is stable for all values of gain. A new root-locus branch now exists between 0 and $-a$, but provided the closed-loop pole on this branch is placed close to the zero, the real exponential term due to it will, by (8.3-40), account for only a small fraction of the transient response. Thus the compensated system not only has much better stability properties, it can also be expected to be considerably faster for the same degree of damping. This is illustrated by Fig. 6, showing the responses to a unit step input of both the original and compensated systems for

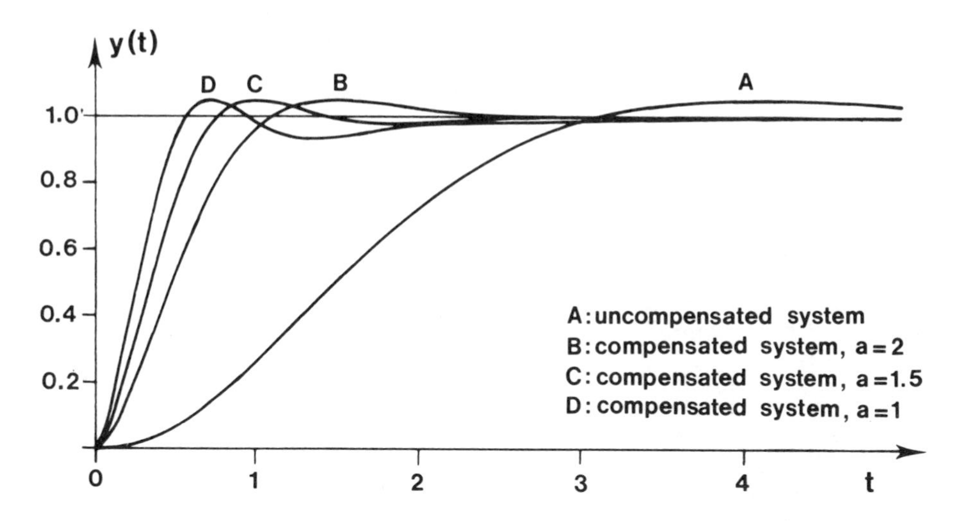

FIG. 6. Unit step responses for Example 2 (all have 5% overshoot).

several values of the zero at -a. The curves were obtained by a computer solution of the system equations, and in all cases the loop gain was adjusted to give an overshoot of 5%. If the settling time to within 5% is taken as the criterion, then the speed of response of the system has been tripled by introducing the zero. As in the previous example, the best values of the zero and of loop gain cannot be directly obtained from the root-locus diagrams; these give no more than a general guide. Moreover, the "best" parameter values depend in quite a complicated fashion on the design criteria adopted. For example, a limit on the peak overshoot in the step response cannot be precisely translated into constraints on the location of the closed-loop poles; in such a situation, there is usually no alternative to a computer simulation of the system and a determination of the parameter values by trial and error. Thus, from Fig. 6 it appears that a value of $a = 1.5$ is near optimal if it is desired to minimize the step response settling time to within 5% subject to a peak overshoot of not more than 5%.

ΔΔΔ

The methods described so far for improving the transient response of a system cannot always be used. The system may contain no integrator, for example, or it may not be possible to measure, amplify, and add the derivative of a signal to the signal itself (the reader may try to devise a means for adding a signal proportional to the applied net torque to the angular velocity of a rotor). In such cases, insertion of a so-called passive *lead compensator* into the loop may provide the solution.

Such a device has the transfer function

$$h_c(s) = \frac{s + a}{s + \alpha a} \tag{6}$$

where $\alpha > 1$ and $a > 0$, and it adds a zero at $-a$ and a pole at $-\alpha a$ to the open-loop transfer function. Its static gain is clearly $1/\alpha$, so that an increase of gain by a factor α is necessary elsewhere in the system if the previous value of loop gain is to be maintained. If in some parts of the system the signal is a voltage at low power level, then a very simple and cheap realization of the compensator is given by the network of Fig. 7; we leave it as an exercise for the reader to show that the transfer function relating input and output voltages has the form (6). This will only be true if the device driven by the output voltage of the circuit draws a negligible current, i.e., has a high input impedance. To ensure this, it is customary to insert an isolating amplifier after the network, if it is not already present. Giving this amplifier a gain of α then provides a convenient means of compensating for the attenuation of the network. Another method of implementing a lead compensator, very common in control systems using pneumatic components, is given in Problem 14.

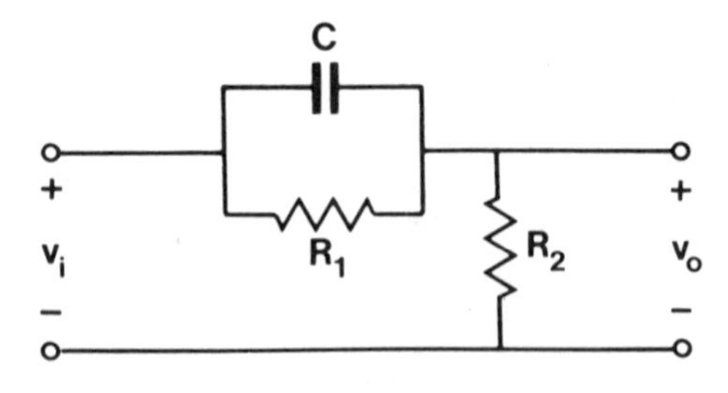

FIG. 7. Passive lead compensator.

3. EXAMPLE. To illustrate how well this compensation works, consider again the uncompensated system of Example 2. The block diagrams of the uncompensated and compensated systems are shown in Fig. 8. Root-locus diagrams of the systems are plotted in Figure 9 with $a = 1.5$. These show that the passive lead compensator is not quite as effective a stabilizing device as the introduction of a pure zero: the compensated system will still become unstable for very large loop gains. However, in the region of the complex plane in which the closed-loop poles have to be located to give adequate damping, the root-locus of the passively compensated system is quite similar to that of the system compensated by adding a zero. [Compare Fig. 9(b) with Fig. 8.3-6.] We can expect therefore that the actual transients will also be similar. This is indeed the case: the step responses of the passively compensated system, plotted in Fig. 10 for values of $a = 1$, 1.5, and 2 and with loop gains adjusted to give 5% overshoot, are slightly slower, but otherwise differ little from those of the system compensated by adding just a zero (compare with Fig. 6).

ΔΔΔ

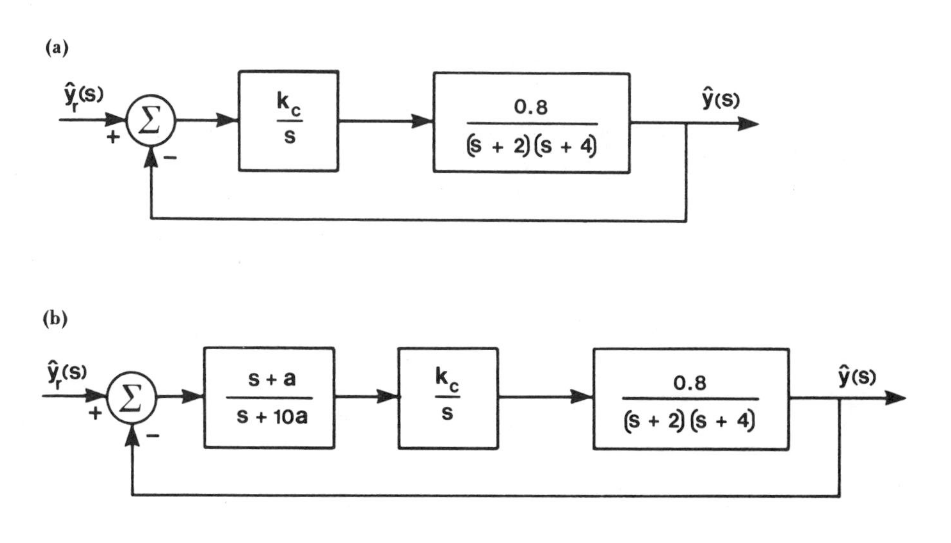

FIG. 8. Block diagrams for Example 3: (a) uncompensated system and (b) system with passive lead compensation.

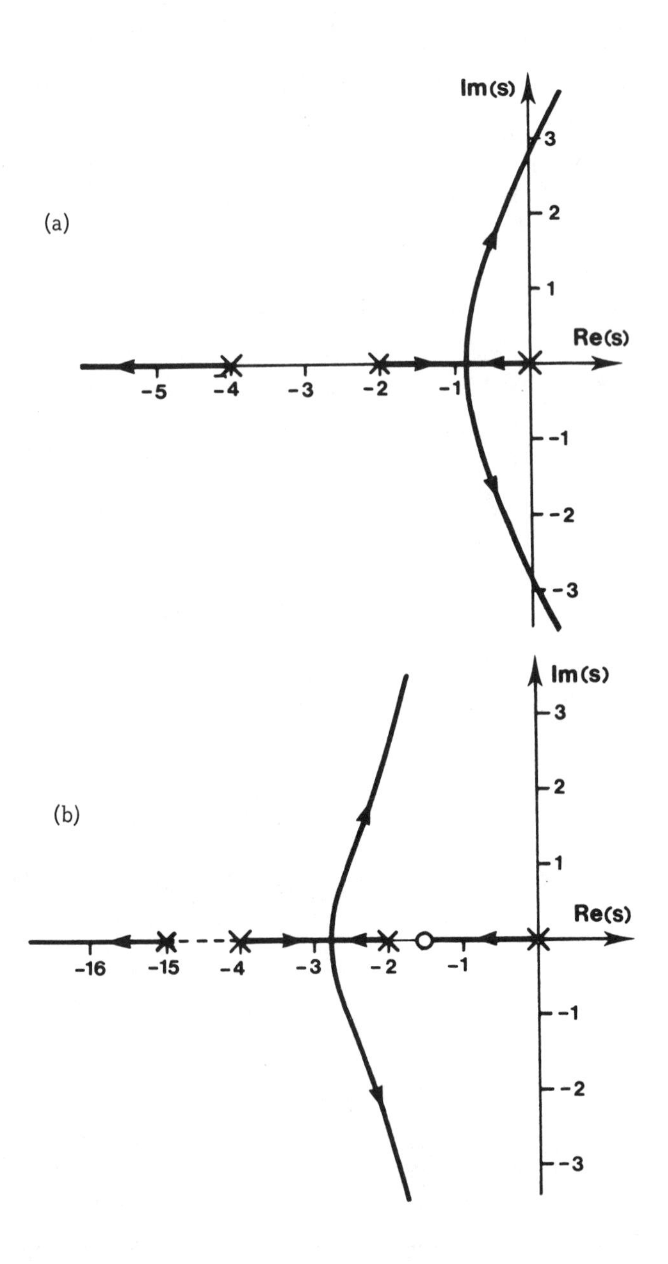

FIG. 9. Root-locus diagrams for Example 3 : (a) uncompensated system and (b) system with passive lead compensation.

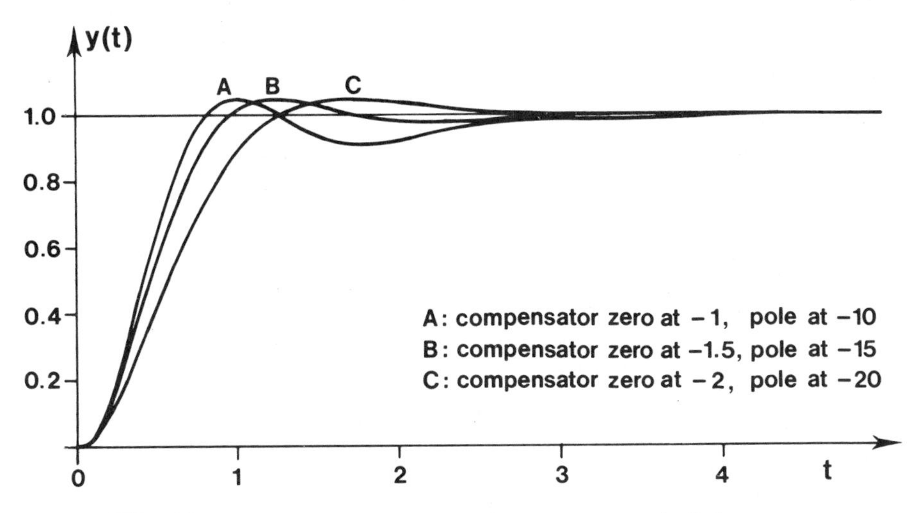

FIG. 10. Unit step responses for Example 3 (all have 5% overshoot).

As in the case of the lag compensator, the effect of lead compensation can also be discussed using frequency domain arguments. If the compensator has unit gain, i.e., if it consists of a network with the transfer function (6) in series with an amplifier of gain α, then its frequency response is $h_c(j\omega) = \alpha(j\omega + a)/(j\omega + \alpha a)$, where $\alpha > 1$. For frequencies less than a, the compensator amplitude ratio is unity, while for frequencies greater than αa, it increases to a limiting value of α. It is not hard to show that the compensator phase shift is positive (hence the term "lead" compensator), increasing from 0 at low frequencies to a maximum of $\sin^{-1}\{(\alpha - 1)/(\alpha + 1)\}$ at a frequency $a\sqrt{\alpha}$ and decreasing to zero again at high frequencies. If the corner frequencies a and αa of the compensator are placed in the range where the phase shift of the original open-loop frequency response is close to -180°, then the effect of the compensator is to give either a greater open-loop bandwidth for roughly the same gain and phase margins, or greater gain and phase margins for the same bandwidth. In other words, the system has faster transients without becoming more oscillatory, or has a greater margin of stability without a reduction in speed. Indeed, in some cases a simple lead compensator can stabilize an inherently unstable system by reducing the open-loop phase lag to less than 180° over a finite frequency range. Again, we invite the reader to work out details by solving Problems 4-7 and 10.

There are some systems whose transient performance cannot be significantly improved by the techniques described so far. Such systems have degree higher than two with a pair of poorly damped complex conjugate open-loop poles, and the root-locus branches originating at these poles never move into a region of sufficient damping. If the configuration of the remaining poles and zeros is such that the damped sinusoid due to this pair of closed-loop poles has a fairly large weighting, then the transient cannot be damped well by gain adjustment alone. Moreover, a little experimentation using root-locus sketches (see Problem 9) will convince the reader that with the techniques of introducing real zeros, or real zero-pole pairs by passive compensators, it is quite difficult to move the troublesome root-locus branches sufficiently far into the left half-plane. Of course, it is necessary to keep in mind our earlier argument that we cannot add more real zeros to the system than there are poles at the origin.

In such a situation it will be necessary to use a compensator which places a pair of complex conjugate zeros on top of, or at least close to, the poorly damped pair of open-loop poles. In the former case, the poles are said to be *cancelled*, thus removing the root-locus branches altogether; in the second case, the zeros are placed close to the closed-loop poles on the troublesome root-locus branches, giving the poorly damped sinusoid due to them a low weighting [see (8.3-40)]. This is best illustrated by an example.

4. EXAMPLE. Consider a feedback system having open-loop poles at 0, -3, and $-0.5 \pm j2$, and no zeros; its block diagram and root-locus are shown in Fig. 11. Suppose that static accuracy specifications require a minimum loop gain of $k = \xi/12.75 = 0.75$, which places the

(a)

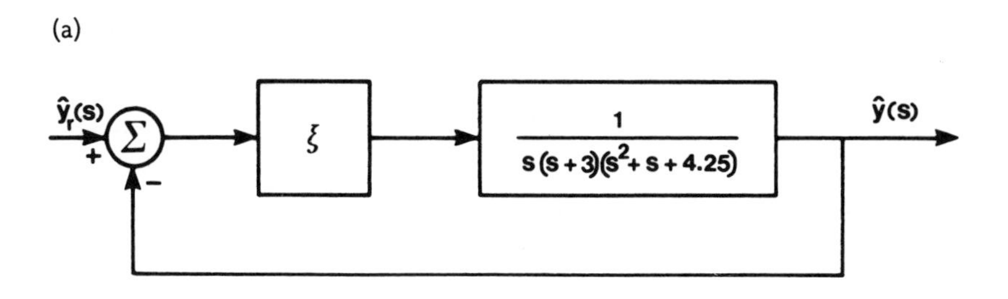

(b)

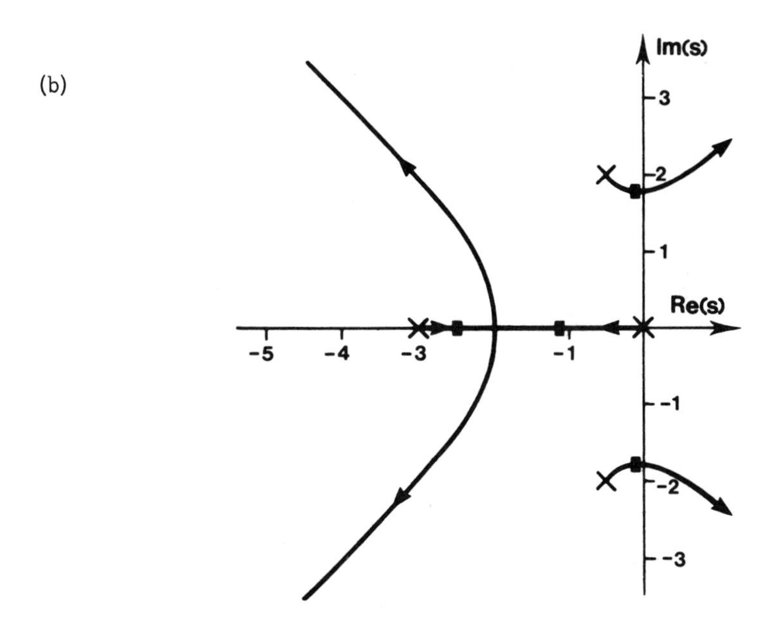

FIG. 11. Uncompensated system of Example 4 : (a) block diagram and (b) root-locus.

closed-loop poles in the positions indicated by ■ on the root-locus and results in a response to a unit step input shown by curve A in Fig. 12. Clearly, the transient response is very poor, with a peak overshoot of about 27% and a settling time of nearly 15 units. The root-locus shows that any increase in gain will only make matters worse: the oscillations in the response will increase in frequency and by (8.3-40), their amplitude will certainly not get smaller.

Now suppose it is possible to introduce a compensating device into the loop having a transfer function

$$h_c(s) = \frac{(s + 0.5 + j2)(s + 0.5 - j2)}{(s + 6)(s + 10)} = \frac{s^2 + s + 4.25}{(s + 6)(s + 10)} \tag{7}$$

A block diagram of the compensated system is shown in Fig. 13(a), and its root-locus in Fig. 13(b). The compensator cancels the system poles at $-0.5 \pm j2$, and introduces new ones at -6 and -10. With a

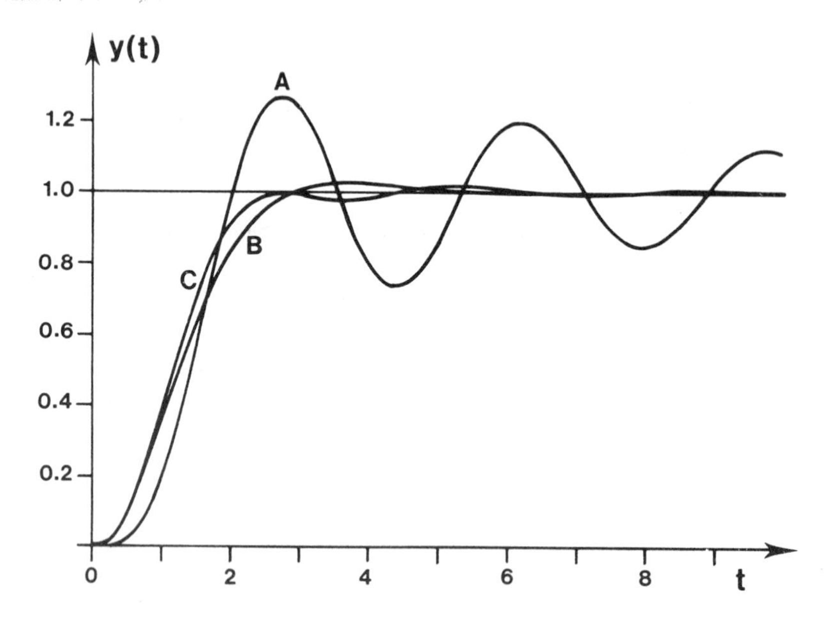

FIG. 12. Unit step responses for Example 4 : (A) uncompensated system, loop gain 0.75; (B) compensated system, compensator zeros at $-0.5 \pm j2$, loop gain 0.75; and (C) compensated system, compensator zeros at $-0.7 \pm j2$; loop gain 0.79.

(a)

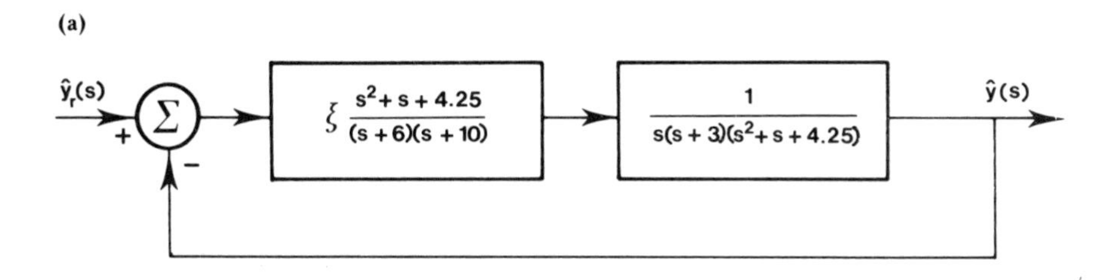

(b)

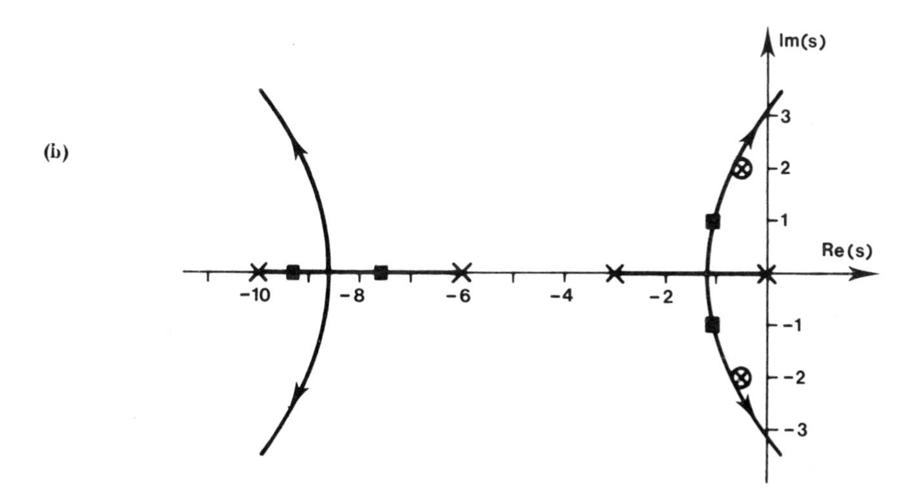

FIG. 13. Compensation by cancelling plant poles in Example 4: (a) block diagram and (b) root-locus.

loop gain of 0.75 as before, the compensated system has closed-loop poles shown by ■ in Figure 13(b), and the unit step response shown by curve B in Fig. 12. The improvement in the transient response is quite dramatic: the settling time has been reduced to about 2.7. The compensated system is more than five times as fast and has only a barely perceptible overshoot of about 3% with no deterioration in the static accuracy.

It is important to check how critically this improvement depends on the placement of the compensator zeros; above we assumed them to exactly cancel the original poles at $-0.5 \pm j2$. Now suppose that the compensator transfer function is given by

$$h_c(s) = \frac{(s+0.7+j2)(s+0.7-j2)}{(s+6)(s+10)} = \frac{s^2+1.4s+4.49}{(s+6)(s+10)} \tag{8}$$

instead of by (7), and that a loop gain of 0.79 is chosen instead of 0.75. The compensated system will then have the root-locus shown in Fig. 14 with the closed-loop poles again indicated by ■. Now the compensator zeros are also zeros of the closed-loop transfer function relating y(t) and $y_r(t)$ (see the remarks following Definition 8.1.3) and since the distance between either of them and the corresponding closed-loop pole closest to the imaginary axis is quite small, we may expect from (8.3-40) that the weighting factor of the poorly damped sinusoid in the transient response is also small. This is borne out by the actual step response, shown as curve C in Fig. 12. We conclude that small changes in the positioning of the poles and zeros, and in loop gain, lead to small changes in the transient response.

In Problem 15 the reader is asked to show that these conclusions do not hold in the case of a disturbance input.

ΔΔΔ

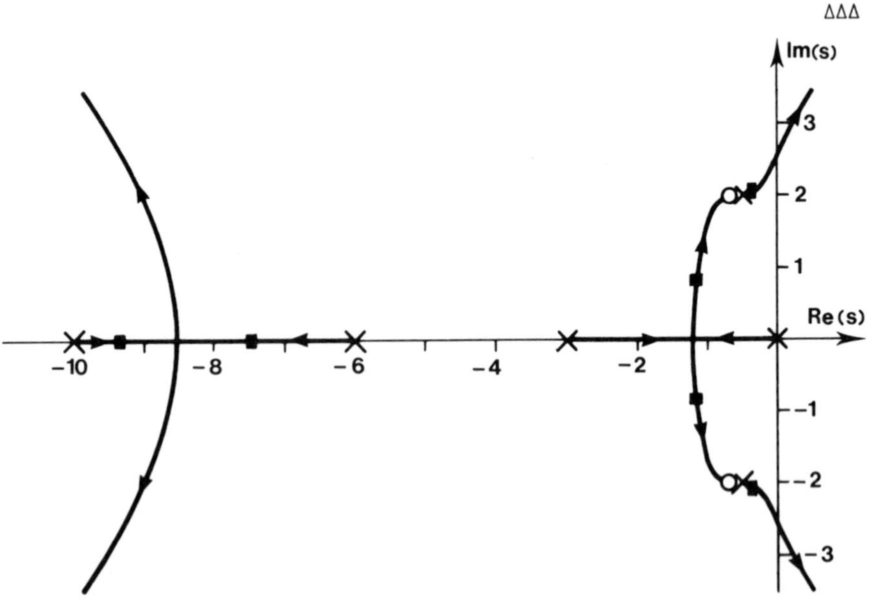

FIG. 14. Root-locus of compensated system in Example 4 with imperfect pole cancellation.

So far in our analysis of single-loop feedback systems and their compensation, we have considered only what happens to the output of the system. It has been tacitly assumed that satisfactory output behavior implies a satisfactory internal transient response, but this is not always valid. In particular, the technique we have just described of cancelling undesirable system poles leads to a system in which certain internal modes are masked from the output. The reader may verify with the help of Theorem 2.2.1 that a state-variable model for the compensated system of Example 4, shown in Fig. 13(a), is given by

$$\dot{\underline{x}} = \begin{bmatrix} 0 & 1 & 0 & 0 & 0 & 0 \\ 0 & 0 & 1 & 0 & 0 & 0 \\ 0 & 0 & 0 & 1 & 0 & 0 \\ 0 & 0 & 0 & 0 & 1 & 0 \\ 0 & 0 & 0 & 0 & 0 & 1 \\ 0 & -765 & -639 & -369 & -131 & -20 \end{bmatrix} \underline{x} + \begin{bmatrix} 0 \\ 0 \\ 0 \\ 0 \\ 0 \\ 1 \end{bmatrix} (y_r - y) \tag{9}$$

$$y = \begin{bmatrix} 4.25\xi & \xi & \xi & 0 & 0 & 0 \end{bmatrix} \underline{x} \tag{10}$$

Substituting (10) into (9) shows that

$$\dot{\underline{x}} = \underline{A}\underline{x}(t) + \underline{b}y_r(t) \tag{11}$$

where $\underline{A}$ is a 6×6 matrix whose characteristic polynomial is given by

$$\det(\lambda\underline{I} - \underline{A}) = \lambda^6 + 20\lambda^5 + 131\lambda^4 + 369\lambda^3 + (639 + \xi)\lambda^2 + (765 + \xi)\lambda + 4.25\xi \tag{12}$$

A simple division shows that

$$(\lambda + 0.5 + j2)(\lambda + 0.5 - j2) = \lambda^2 + \lambda + 4.25 \tag{13}$$

is a factor of (12) for *any* value of the gain constant ξ. This means that at least one of the six state variables of the system must contain the function $e^{-0.5t} \cos(2t + \phi)$ in its transient response. The output does not; removal of this mode was the whole point of the compensation. The response of the six state variables to a unit step input, i.e., a solution of (11) with $y_r(t) = 1(t)$ and zero initial conditions, is shown in Fig.15. These curves, particularly x_5 and

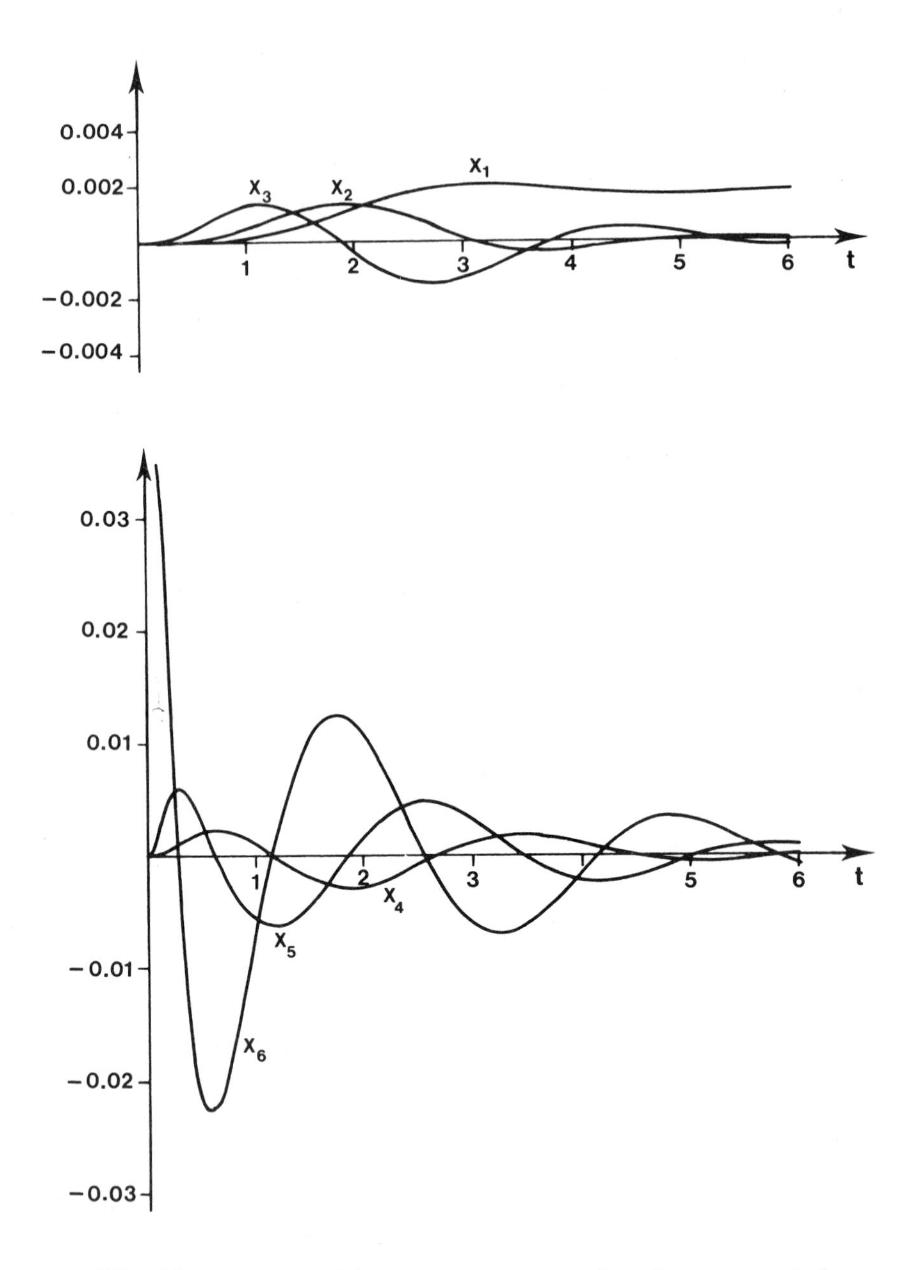

FIG. 15. State-variable step responses for the compensated system of Example 4.

x_6, show clearly that the poorly damped sinusoid is still very much present among the system states, although it does not appear in the three components which make up the output (x_1, x_2, and x_3).

We must conclude, therefore, that the cancellation of the open-loop poles at $-0.5 \pm j2$ has merely suppressed a poorly damped mode from the output without also removing it from the system state. Stating this another way, the presence of a particular mode in the system can no longer be deduced from an observation of the output: we have made part of the system *unobservable*.†

This conclusion has two immediate consequences. The first is that for a compensation by cancelling system poles, an input-output description of the system is inadequate. It is essential to consider in detail what happens to the system states before one can conclude that the compensation is successful. In particular, it is necessary to check where oscillatory transients (such as that exhibited by x_6 in the example above) appear in the system, and whether they can be tolerated.

The second consequence is that the pole-cancellation technique cannot be used to stabilize an inherently unstable system. Cancelling an unstable open-loop pole by placing a zero on top of it does not make the system stable: there will still be at least one state variable which will grow without bound (See Problems 12 and 13, and Example 10.4.8).

We conclude that the improvement of the transient behavior of a system by pole cancellation, i.e., by removing certain system modes from its output, requires care. In particular, the internal behavior of the compensated system should always be analyzed with a state-variable model.

To complete our discussion of Example 4, we consider how a compensating device with a transfer function having complex conjugate zeros can be constructed. To begin with, we point out that the realization of a system having a prescribed mathematical model of other than the simplest form is best done by electric devices operating on

---

†This concept will be discussed in detail in Chapter 10 (see Problem 10.2-12).

currents and voltages; devices using pneumatic or hydraulic pressures or mechanical displacements as variables have an unfortunate tendency to become cumbersome, elaborate, and very costly. Thus even if the signals in the system are not in electrical form, it will often be advantageous to transduce them into this form and use an electric circuit for compensation.

The design of circuits having a prescribed transfer function is a classic problem in network synthesis. A number of circuit configurations exist which have a pair of real poles and a pair of complex conjugate zeros in their transfer function and which use only passive elements, i.e., resistors, capacitors, and inductors. We refer the reader to any standard text on network theory for details, e.g., Refs. 14 and 176.

The simplest such circuits, however, do not give complete freedom in the location of poles and zeros. We shall therefore illustrate a more general method using active elements, i.e., amplifiers. In fact, we use the standard technique of constructing an electronic analog computer model to realize (7).

One way to do this is to begin with a partial fractions expansion of (7)

$$h_c(s) = \frac{s^2 + s + 4.25}{(s+6)(s+10)} = 1 + \frac{8.56}{s+6} - \frac{23.6}{s+10} \tag{9}$$

or

$$h_c(s) = \frac{1.43}{1 + 0.167s} - \frac{2.36}{1 + 0.1s} + 1 \tag{10}$$

In Problem 7.1-3 the reader was asked to show that a circuit consisting of an operational amplifier with an input resistor R_i and a feedback network having a capacitor C in parallel with a resistor R_f, has the transfer function $-(R_f/R_i)/(1 + R_f Cs)$. It follows that (10) can be realized by the circuit shown in Fig. 16, where $R_2/R_1 = 1.43$, $R_2C_1 = 0.167$, $R_4/R_3 = 2.36$, and $R_4C_2 = 0.1$.

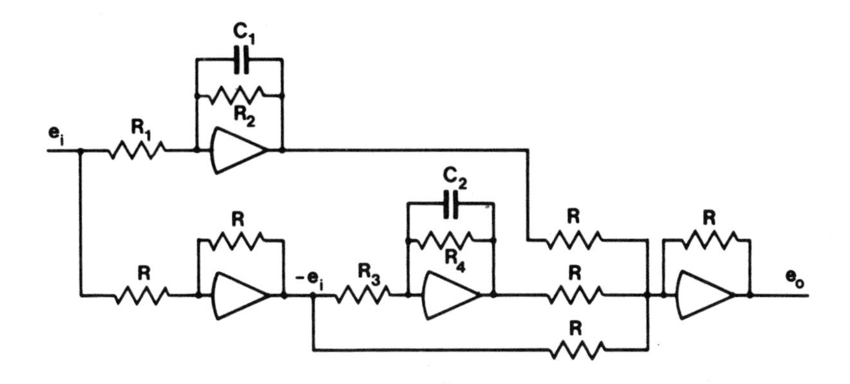

FIG. 16. Active compensator for Example 4.

We remarked earlier that integrated circuits have made it possible to build a circuit such as this one from standard and fairly cheap components. Many control systems use solid-state electronic devices elsewhere, so that no additional power supplies are required, and the compensator can be incorporated at little additional expense. This has given the system designer much more freedom than he had in the past when there was an economic premium on the use of passive devices.

9.4. LAG-LEAD COMPENSATORS AND THREE-TERM CONTROLLERS

In the preceding sections, improvement of static accuracy and transient response were discussed as separate problems, each requiring its own type of compensation. It often happens, of course, that a system design will require both types of compensation. Provided that the static performance requirements do not call for the insertion of pure integrators, it will often be possible to satisfy the performance specifications by inserting both a lag and a lead network in series, separated by an isolating amplifier. The combined transfer function of the compensating networks will then be

$$h_c(s) = \frac{(s + a)(s + b)}{(s + \alpha a)(s + \beta b)} \tag{1}$$

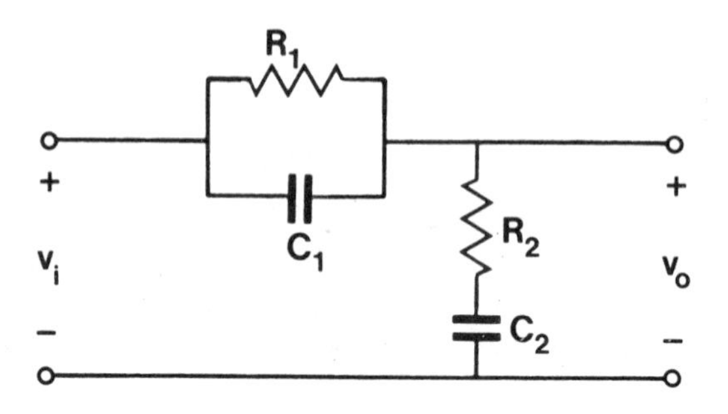

FIG. 1. Lag-lead network.

where $b > \alpha a$, $\beta b > a$, and $0 < \alpha < 1 < \beta$. This transfer function can also be realized directly by a single passive circuit called a *lag-lead network*. The circuit shown in Fig. 1 has the transfer function given by Eq. (1) if $a = 1/R_1C_1$, $b = 1/R_2C_2$, $\alpha\beta = 1$, and $\alpha a + \beta b = a + b + 1/R_2C_1$. In this case, the two zeros of (1) always lie between the two poles, and the gain of the network is unity.

1. EXAMPLE. Consider the feedback system shown in Fig. 2(a), and suppose that the overshoot in the response to a reference step input must not exceed 15%, while the steady-state error for a unit step

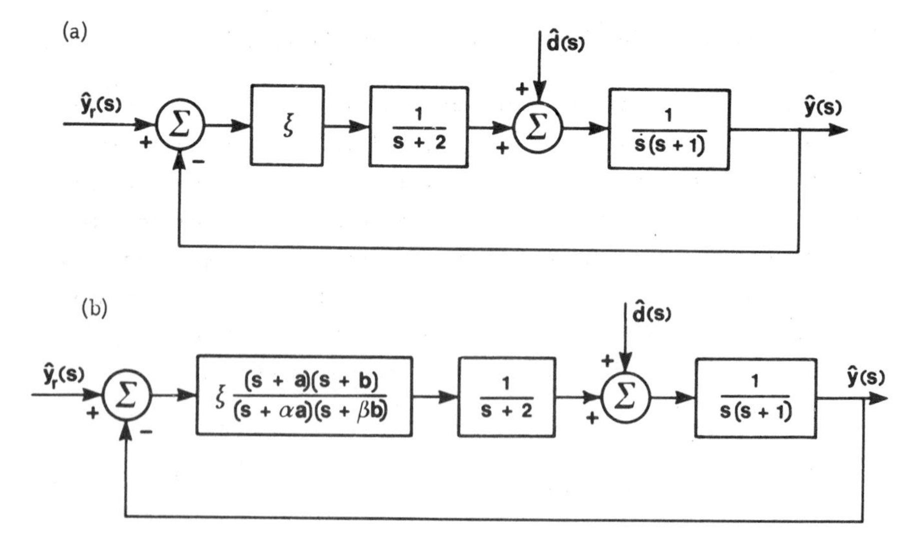

FIG. 2. Block diagrams for Example 1: (a) original system and (b) compensated system.

(a)

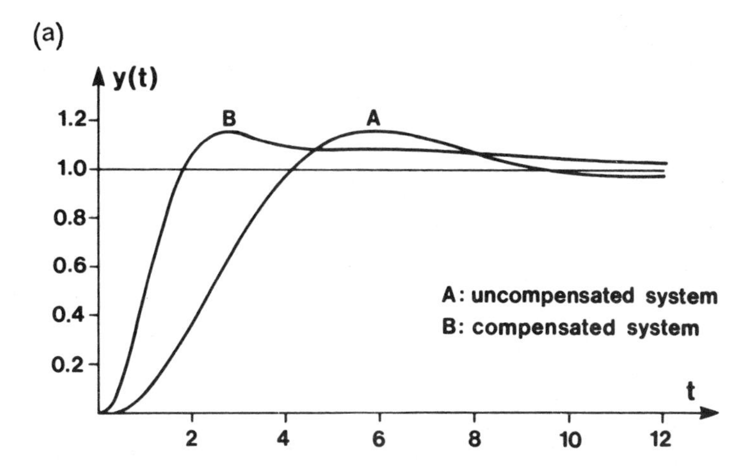

(b)

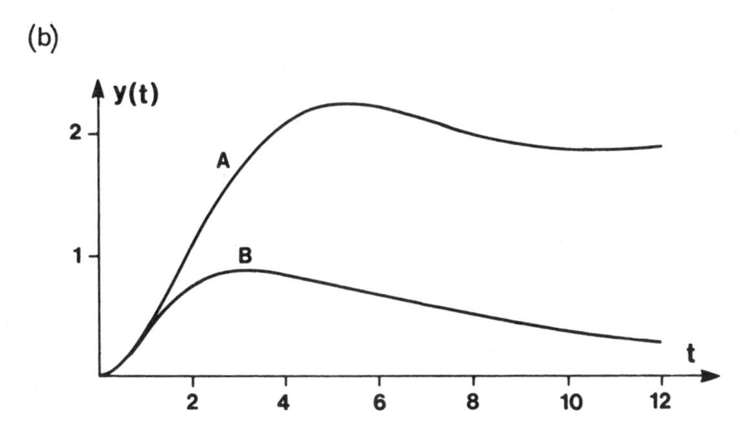

FIG. 3. Step responses for Example 1: (a) response to step reference input and (b) response to step disturbance input.

disturbance must be less than 0.15 units. A computer simulation shows that the largest controller gain ξ possible without violating the overshoot constraint is $\xi = 1$. The step responses for this gain are shown as A in Fig. 3; the steady-state error for a unit step disturbance is 2. Conversely, to give the required steady-state accuracy, a controller gain of $\xi = 13.3$ is needed. Unfortunately, the Routh criterion shows that the system is unstable for $\xi \geq 6$.

A little experimenting, either with root-locus plots or with a computer model, will show that a lag compensator by itself is insufficient. This can also be deduced from the open-loop frequency response (see Problem 9.2-7): if the compensated system is to have a phase margin similar to that of the uncompensated system with $\xi = 1$, then a lag compensator alone will not permit a sufficiently large increase in loop gain. However, a lag-lead network, inserted as shown in Fig. 2(b), works very well. A choice of $\alpha = 1/\beta = 0.1$, $a = 0.2$, $b = 0.6$ and $\xi = 16$ gives the responses shown as B in Fig. 3. The steady-state error for a unit step disturbance has been reduced to $2/\xi = 0.125$ and the rise time has been more than halved, while the overshoot has been maintained at 15%. However, the settling time (to within 5%) has increased somewhat due to the influence of the slowly decaying real exponentials contributed by the two closed-loop poles between -a and -b. Thus, if settling time is not an important consideration, the compensator has simultaneously yielded a significant improvement in both static accuracy and response speed without sacrificing smoothness of the transients.

As in the previous examples, the compensator parameters were determined by trial and error using a computer simulation of the system. Such a simulation shows that the appropriate selection of the parameters is essentially a matter of compromise between the different performance specifications. For a given value of α, greater static accuracy (i.e., a higher value of ξ) without an increase in overshoot can be obtained by smaller values of a, α, and b, but only at the cost of considerably greater settling time. Conversely, for a given static accuracy, a shorter settling time can be achieved by larger values of a and b, but at the expense of increased overshoot. Moreover, the choice of parameters is constrained by practical network considerations. For example, if the compensator is implemented with the circuit of Fig. 1, the parameters selected above can be obtained with $R_1 = 4.3$ MΩ, $C_1 = 1.2$ μF, $R_2 = 0.17$ MΩ and $C_2 = 10$ μF. Much smaller values of a or α, to obtain greater static accuracy, could be achieved only with much larger resistance and capacitance values, making the circuit very sensitive to noise and requiring expensive components.

Lag-lead circuits provide a convenient means of compensating servomechanisms, particularly if low weight and compactness are important, as in instrument servomechanisms or aircraft control systems. As a general rule, such servomechanisms are "tailor-made" for each application.

This is quite different from industrial process control, the other large area of application of automatic control systems, which includes such tasks as the control of flow rates in pipelines, temperatures in furnaces and chemical reactors, concentrations of reagents, and levels in tanks. These are almost invariably carried out with standard off-the-shelf controllers, either electronic or pneumatic, which have provision for manual or remote entry of the setpoint y_r, and often include a recorder which produces a continuous plot of y. In its simplest form, such a controller provides only proportional control action: a power amplifier in the controller produces a signal proportional to the error which can then be used to drive an actuator such as a flow regulating valve. More elaborate controllers include compensating networks which give proportional-plus-derivative (PD), proportional-plus-integral (PI), or proportional-plus-integral-plus-derivative (PID) control action. The latter is also known as a *three-term controller,* and is described by

$$u(t) = k_p e(t) + k_d \frac{de}{dt} + k_i \int_0^t e(\tau)\, d\tau \tag{2}$$

where u is the controller output and $e = y_r - y$ is the error signal. Thus the ideal transfer function of the controller is

$$h_c(s) = k_p + k_d s + \frac{k_i}{s} \tag{3}$$

This, of course, cannot be achieved perfectly in practice; we have already commented on the difficulty of differentiating signals. In electronic controllers, the component proportional to the derivative

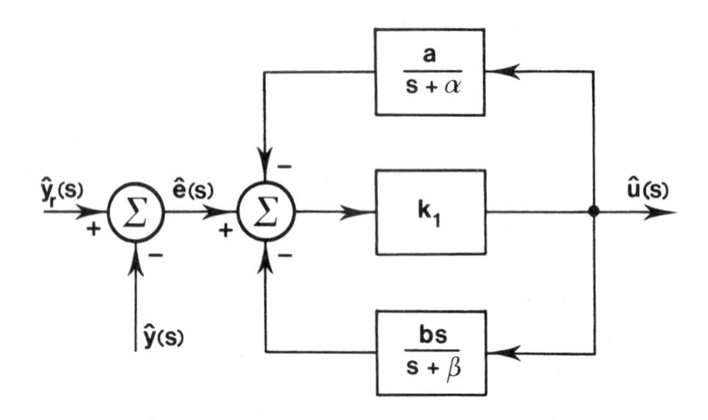

FIG. 4. Block diagram of a pneumatic PID controller.

of the error is usually produced approximately by a lead network; in pneumatic devices, both integral and derivative components are produced by passive (lag and lead) networks. (See Problems 9.2-3 and 9.3-14.) A block diagram of a typical pneumatic PID controller is shown in Fig. 4; the reader should verify that its transfer function is just that of a lag-lead compensator in series with an amplifier.

In all such controllers, the constants k_p, k_d, and k_i are manually adjustable, and the process of designing the control system, or compensating it, is usually carried out by trial-and-error "tuning" when the equipment is installed and the plant is first started up.

9.5. SUMMARY

This chapter has given a brief introduction to methods of improving the performance of single-loop feedback systems, useful if only the system output is of interest. The most common forms of performance specifications for both steady-state and transient behavior were discussed in Sec. 9.1. The improvement of static accuracy was taken up next in Sec. 9.2. It was shown that if the system does not already contain a sufficient number of integrators in the appropriate part of the feedback loop, then a specification of zero steady-state error requires additional integrators, which usually leads to a fairly drastic degradation of transient performance. If finite steady-state errors are permitted, and the systems have acceptable transients when the loop gain is too low for the required static accuracy, then insertion of a passive lag compensator will often allow a sufficient increase in gain without a serious degradation of the transient.

The problem of improving systems which cannot be given a satisfactory transient response by loop gain adjustment was discussed in Sec. 9.3. It was shown, by root-locus arguments, that introducing an additional zero (derivative compensation) or a pole-zero pair (passive lead compensation) usually allows troublesome closed-loop system poles to be moved further into the left half-plane. This results in a more stable system or one with better speed or damping in the step response. Another compensation technique known as pole-zero cancellation was then discussed with an example. We saw that its use requires care, since the unwanted transient component is merely suppressed from the system output but not removed from the internal system dynamics.

Finally, simultaneous improvement of static accuracy and transient response by lag-lead compensation was considered, and industrial process controllers were discussed.

PROBLEMS

Section 9.1

1. Suppose a system has a frequency response whose magnitude curve is similar to that shown in Fig. P1. There is no resonance peak

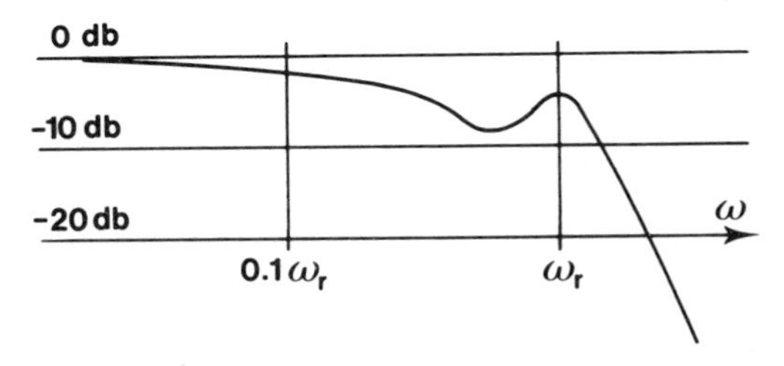

FIG. P1

as defined in Fig. 9.1-2, but the system does exhibit resonance. Is it possible to conclude from the presence of the peak at frequency ω_r that the transient response of the system contains poorly damped sinusoids?

2. Suggest a method of deducing the closed-loop bandwidth of a feedback system from its open-loop frequency response. Will an increase in open-loop bandwidth always lead to an increase in closed-loop bandwidth?

3. Analytically, or by an analog computer simulation, or by digital solution of the system equations, determine the loop gains k necessary to minimize

(i) $$I_1 = \int_0^\infty [e(t)]^2 \, dt \qquad \text{(ISE criterion)}$$

(ii) $$I_2 = \int_0^\infty t \, |e(t)| \, dt \qquad \text{(ITAE criterion)}$$

where e(t) is the error in the response to a unit step input of the system having the open-loop transfer function

$$h(s) = k \frac{2}{s(s+2)}$$

Plot the corresponding step responses and comment on the utility of these integral criteria in this case.

Section 9.2

1. Study the effect on the root-locus and open-loop frequency response of introducing an additional integrator into the feedback systems characterized by the following sets of open-loop poles and zeros:

(a) Poles at 0, -1, -2 ; no zeros
(b) Poles at 0, 0, -2 ; zero at -1
(c) Poles at $-1 \pm j2$, -2 ; no zeros
(d) Poles at 0, -1 ; zero at -1

Discuss how the stability margin and the quality of transient performance achievable by gain adjustment are affected by the new integrator (define what you understand by these terms, and indicate whether they have any meaning in each case).

2. Show that the transfer function relating output and input voltages of the circuit of Fig. 9.2-5 has the form (9.2-1).

3. Figure P2(a) shows a schematic arrangement of a common pneumatic controller claimed to give "proportional-plus-integral" control action. The input to the device is the displacement e(t) of the right hand end of the floating link L which acts as the flapper of a pneumatic flapper-nozzle amplifier (see Example 2.4.1). The pressure produced by the flapper-nozzle amplifier is itself amplified in a pneumatic power amplifier, producing the controller output, a pressure u(t). An increase in u(t) drives air through the adjustable resistance R_i into the upper bellows until its pressure, $p_b(t)$, equals u(t). The output pressure u(t) always acts in the lower bellows; pressure drops in connecting pipes can be neglected. The two bellows have equal effective area. The difference in the two bellows forces is balanced by the spring of stiffness k, and the resulting displacement determines the position f(t) of the left hand end of the floating flapper.

Assume

(i) Both pneumatic amplifiers act instantaneously, that is, u(t) is always directly proportional to the position of the midpoint of the flapper link.

(ii) u(t) is independent of the air flow rate drawn from the controller, i.e., the power amplifier has zero output impedance.

(iii) The mass flow rate of air through the restriction R_i is directly proportional to the pressure drop across it, that is, to $u(t) - p_b(t)$.

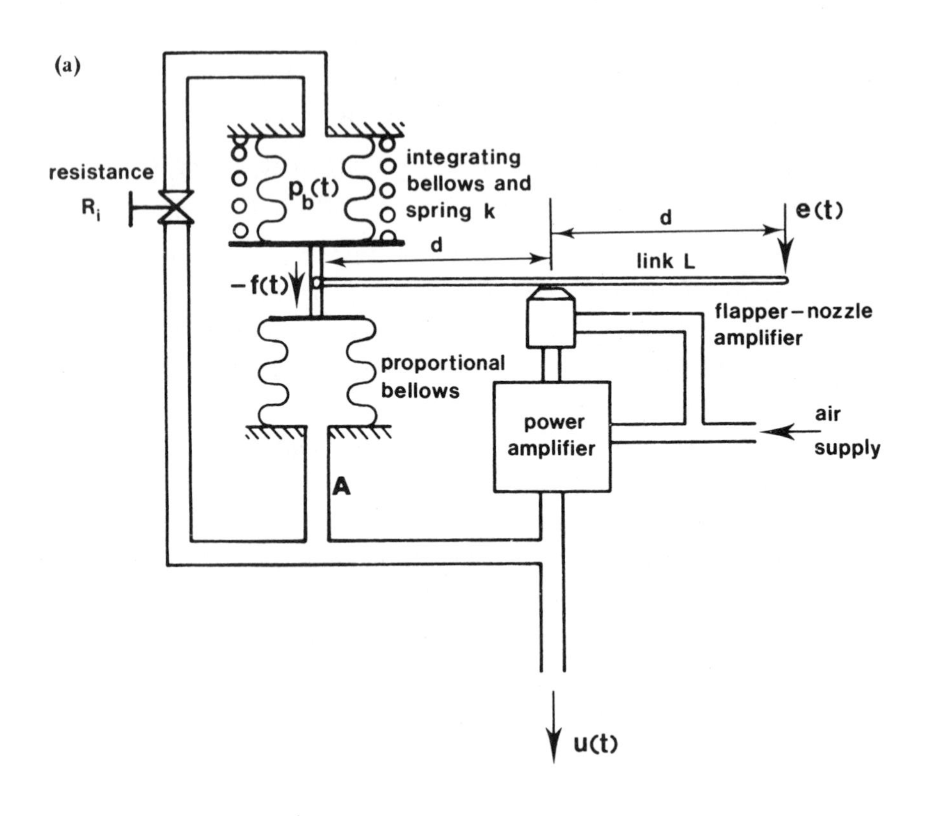

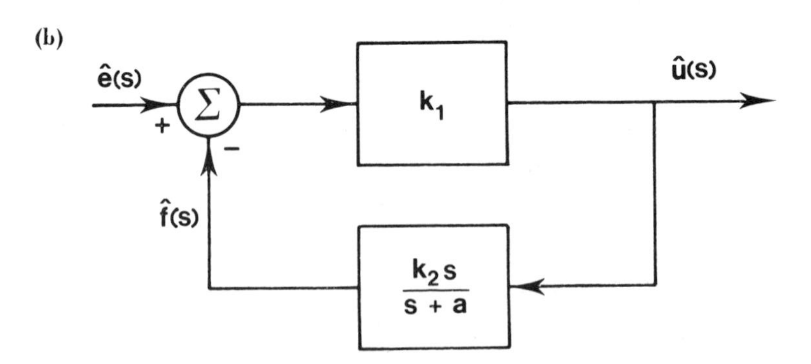

FIG. P2. Pneumatic proportional-plus-integral controller: (a) schematic arrangement, and (b) block diagram.

(iv) The pressure in the bellows is directly proportional to the mass of air in it and depends on nothing else; in particular, the change in bellows volume is too small to noticeably affect the pressure.

(v) The flapper has negligible mass.

Answer the following:

(a) Show that with these assumptions the controller can be represented by the block diagram of Fig. P2(b) and give expressions for k_1, k_2, and a.

(b) Hence show that the controller transfer function is that of a lag compensator in series with an amplifier.

(c) How must the parameters be chosen to give a good approximation to proportional-plus-integral control?

(d) What is the point of having R_i adjustable? In particular, how will the controller behave if R_i is completely closed?

4. A feedback system has open-loop poles at -2 and $-1 \pm j$ and no zeros. The system must not have a steady-state error in the unit step response of more than 0.1, and the complex conjugate closed-loop poles must have a degree of damping of at least 0.5. Can these specifications be met with the use of a lag compensator of the form (9.2-1)? If so, specify suitable values of loop gain and of the parameters a and α. Plot the unit step responses of the original and compensated systems and comment.

5. (a) Draw Nyquist and Bode diagrams of the frequency response of a lag compensator with a transfer function given by (9.2-1). Show that the maximum phase shift of the compensator is given by

$$\phi_m = \sin^{-1}\frac{\alpha - 1}{\alpha + 1}$$

and that this occurs at a frequency given by $a\sqrt{\alpha}$.

(b) Show by means of Bode diagrams and the Nyquist stability criterion that the introduction of a lag compensator into the loop of a feedback system cannot

(i) Stabilize a system which is unstable for all gains.

(ii) Give an increased margin of stability for a given speed of response or, more precisely, give increased gain or phase margins without lowering the open-loop bandwidth.

(iii) Give an increased speed of response for a given margin of stability, i.e., increase the open-loop bandwidth without reducing the phase margin.

6. Draw Bode diagrams of the original and compensated systems in Example 9.2.1, and determine the gain and phase margins corresponding to the values of gain chosen in the example.

7. Consider a system with the open-loop transfer function

$$h(s) = \frac{\xi}{s(s+1)(s+2)}$$

(a) Draw a Bode diagram of the open-loop frequency response using a loop gain k (= $\xi/2$) of unity. Hence determine the maximum permissible value of gain consistent with a phase margin of at least 30°. What is the corresponding gain margin of the system?

(b) Suppose a lag compensator with the transfer function (1) is introduced. If the compensator parameter α cannot be less than 0.1, and if the compensated system must have a phase margin of at least 30°, determine values of the parameters a and α which allow the greatest possible increase in loop gain.

8. Suppose that in Problem 4 above the specification of the location of the closed-loop poles is replaced by the requirement that the phase margin of the system must not be less than 60°. Design a suitable lag compensator, plot the step responses of the original and compensated systems, and comment on both the effectiveness of the compensator and the utility of the phase margin constraint as a transient performance specification.

9. If the static accuracy in the response to both reference and disturbance inputs is improved by adding pure integrators, then the integrators cannot be placed just anywhere in the loop (see Theorems 8.2.2 and 8.2.3). Investigate whether a similar restriction applies to the placement of a lag compensator.

Section 9.3

1. A feedback system has the open-loop transfer function relating the output y(t) and the error u(t)

$$h(s) = \frac{q_0}{s^n + p_{n-1}s^{n-1} + \ldots + p_1 s + p_0}$$

Prove the pole-placement theorem (11.2.1) for this case, i.e., show that there exist real constants α_i such that the feedback system obtained by setting

$$u(t) = y_r(t) - \alpha_0 y(t) - \alpha_1 \dot{y}(t) - \ldots - \alpha_{n-1} y^{(n-1)}(t)$$

has any arbitrarily specified set of closed-loop poles, subject only to the constraint that complex poles must occur in conjugate pairs. *Hint:* Use Theorem 2.2.1.

2. Consider the antenna positioning system of Chapter 1 and Example 9.3.1. Assume as before that the antenna, its gearbox and the motor armature can be represented by a frictionless inertia, but that the torque developed by the motor is related to the applied field voltage by a first-order differential equation. A feedback system is required which will give a zero steady-state error for a step change in wind torque on the antenna.

(a) Draw a block diagram of the simplest system which satisfies the steady-state specification. Draw its root-locus and show that it is inherently unstable.

(b) By means of root-locus diagrams, investigate whether the system can be stabilized by introducing

(i) A zero as in Example 9.2.1

(ii) Two zeros (How could this be implemented in practice?)

(iii) A single lead compensator of the form of (9.3-6)

(iv) A zero and a lead compensator

(v) Two lead compensators in series

Which of the feasible schemes would you recommend, and why?

3. The purpose of both pure derivative and passive lead compensation is to improve transients, which may be excited by both reference and disturbance inputs. Investigate whether the position of the compensator in the loop is of any consequence in this regard.

4. Draw a Bode diagram of the frequency response of a passive lead compensator with the transfer function (9.3-6) and show that the maximum phase shift, and the frequency at which it occurs, are again given by the formulas stated in Problem 9.2-5.

5. Sketch the open-loop frequency response of both the original and compensated systems of Example 9.3.3, and determine the open loop bandwidth, gain and phase margins for the gains which give the complex closed-loop poles a damping ratio of 0.7.

6. Consider the system and performance specification of Problem 9.2-7. Is it possible to meet the specifications with a *lead* compensator? If so, give suitable values of a and α in (9.3-6) and of loop gain.

7. Show that the circuit of Fig. 9.3-7 is a lead network.

8. Devise mechanical networks which may consist of (but need not contain all of) masses, springs, massless levers, and dashpots in which the output and input are displacements related by the transfer functions

(i) $\dfrac{a}{1 + bs}$

(ii) $a\,\dfrac{s + b}{s + c} \qquad b > c$

(iii) $a\,\dfrac{s + b}{s + c} \qquad b < c$

9. (a) Consider the systems characterized by the following sets of open-loop poles and zeros:

   (i) Poles at $-1/2 \pm j3$, -3, 0 ; no zeros

   (ii) Poles as in (i) ; zero at -2

   (iii) Poles at 0, 0, $-1/2 \pm j3$, -5 ; zero at -2

   By means of root-locus sketches, discuss the likelihood of successfully compensating these systems by introducing additional zeros (where possible) or passive lead compensators.

   (b) Repeat using frequency response methods, i.e., the Nyquist stability criterion and Bode or Nyquist diagrams of the open-loop frequency response.

10. Compute the weighting factors of the transient terms in the unit step responses of Example 9.3.4, for

   (i) The uncompensated system

   (ii) The system compensated by approximate pole cancellation.

11. It is known that a certain furnace can be quite accurately modelled by a first-order differential equation

$$\dot{y}(t) = -ay(t) + bm(t) - cd(t)$$

where $y(t)$ is the furnace temperature, $m(t)$ the rate of fuel flow into the furnace and $d(t)$ the ambient temperature.

   (a) Discuss the claim that a proportional-plus-derivative controller (whose transfer function is usually that of a proportional controller in series with a lead compensator) will give better control of the furnace than a simple proportional controller.

   (b) Suppose that a proportional-plus integral controller is necessary for static accuracy. Will the addition of derivative action now be advantageous?

(c) Repeat (a) and (b) if the furnace has the transfer function

$$\frac{k}{(s+a)(s+b)}$$

where a and b are positive.

12. A unity feedback system has open-loop poles at +2, 0, and -2. It is proposed to stabilize it by introducing a zero at +2, thus cancelling the unstable open-loop pole there. Use Theorem 2.2.1 to construct a state-equation model of the compensated system (which still has three state variables!) and hence show that the system is still unstable.

13. Assume a unity feedback system has a pair of open-loop poles $\sigma \pm j\omega$, which are to be cancelled by inserting into the loop a compensating device whose transfer function has the factor $(s - \sigma - j\omega)(s - \sigma + j\omega)$ in its numerator. Show that the closed-loop system will always have at least one state variable whose transient response contains the term

$$e^{\sigma t} \cos (\omega t + \phi)$$

14. Consider the pneumatic controller described in Problem 9.2-3, and shown in Fig. P2(a). Show that the controller behaves as a lead network in series with an amplifier, provided that the resistance R_i is closed completely, and that another adjustable resistance R_d is installed at the position labelled A. How must the device be proportioned to give a good approximation to proportional-plus-derivative control?

15. Suppose the system of Example 9.3.4 has both reference and disturbance inputs. Does the pole-zero cancellation also improve the transient response to a step disturbance? Explain.

Section 9.4

1. Sketch the Bode diagram of a lag-lead compensator.

2. (a) Draw the Bode diagrams of the original and compensated systems of Example 9.4.1 [use your results for Problem 9.2-7(a)]. For the controller gains ξ used in the example, determine the gain and phase margins before and after compensation.

   (b) From the curves obtained in (a), you will see that the lag-lead compensator has two effects on the open-loop frequency response: it lowers the magnitude curve over the critical range of frequencies without changing the loop gain, and it increases the frequency at which the phase shift equals -180°. Both changes together permit a somewhat greater loop gain, for an unchanged phase margin, than is possible with a lag compensator alone. For the criterion that the compensated system should have the same phase margin as that of the uncompensated system with $\xi = 1$, the lag-lead compensator used in Example 9.4.1 does not give the greatest possible increase of loop gain. Design a better compensator, still using $\alpha = 1/\beta = 0.1$, and determine the corresponding loop gain.

   (c) By means of an analog or digital computer solution of the system equations, plot the responses to unit step reference and disturbance inputs. In this case, is a specification of phase margin sufficient to ensure a satisfactory transient response?

3. (a) Using your results for Problem 2, suggest general rules for designing lag-lead compensators, using Bode diagrams and specifications of gain and phase margins. In particular, comment on the appropriate placement of the corner frequencies of the lag and lead sections of the compensator, relative to the frequencies at which the original system has a magnitude of 0 db and a phase shift of -180°. Bear in mind that the pole and zero in each section of the

network cannot in practice be separated by more than a factor of about 10 or 15.

(b) In the text it is stated that a lag-lead compensator has its two zeros located between its poles, and this is illustrated in Example 9.4.1. Using frequency response arguments, discuss whether a compensator with an alternating pattern of poles and zeros would be more or less effective. Can you suggest a system for which it would make sense to use a "lead-lag" compensator, having its poles located between its zeros, so that $b < \alpha a$ and $\beta b < a$ in 9.4-1?

4. Consider the pneumatic controller discussed in Problems 9.2-3 and 9.3-14. Show that with adjustable restrictions R_i and R_d in the lines to both bellows, the device has the block diagram of Fig. 9.4-4 and hence behaves as a lag-lead compensator in series with an amplifier.

Chapter 10

STRUCTURE OF MULTIVARIABLE SYSTEMS†

10.0 INTRODUCTION

In Chapters 8 and 9 our attention was restricted to dynamic systems with only one control input and one output, and we investigated the use of feedback control in some detail. In all cases the objective was to improve in some way the behavior of the output, and this was done by feeding back the output (either directly or through another dynamic system) and subtracting it from the input.

Now we shall consider systems which have multiple inputs and outputs and take into account the behavior of the entire state vector. More specifically, the *controllability* of the system is of interest: Given the state of the system at some particular time, can a control function be found which will drive the state to zero a short time later? The *observability* of the system is of equal interest: Given the value of the system's output over a finite time interval, is it possible to deduce the value of the state vector?

These concepts lead to some very important conclusions regarding the basic structure of linear systems. Indeed, we shall find that any system may be *canonically decomposed* into various parts, and that input-output characterizations such as the transfer function and

---

†An understanding of Appendix A through Sec. A.5 is assumed from this point on.

impulse response describe *only* that portion which is both controllable and observable.

The definitions of controllability and observability will be seen to be open loop in nature. A time function is sought, in one case to control the state from the input and in the other to determine it from the output, and subsequent information is not used to alter the control or the observation. Nevertheless, these concepts have important implications for closed-loop systems. In Chapter 11 we shall see that the eigenvalues of a controllable system can be placed arbitrarily using *state-variable feedback*. In Chapter 12 we shall see that an *observer* can be constructed to estimate the state of any observable system, and the eigenvalues of the observer can also be placed arbitrarily.

10.1 CONTROLLABILITY

Questions of controllability concern the extent to which the input to a linear system influences the behavior of the state. To illustrate this concept, consider the following simple circuit.

1. EXAMPLE. The state equations governing the circuit shown in Fig. 1(a) are

$$\dot{\underline{x}} = \begin{bmatrix} -2 & 1 \\ 1 & -2 \end{bmatrix} \underline{x} + \begin{bmatrix} 1 \\ 1 \end{bmatrix} u = \underline{A}\underline{x} + \underline{b}u \qquad (1)$$

$$y = \begin{bmatrix} 0 & 1 \end{bmatrix} \underline{x} = \underline{c}'\underline{x}$$

and the matrix exponential in this case is easily seen to be

$$e^{\underline{A}t} = \frac{1}{2}\begin{bmatrix} e^{-t} + e^{-3t} & e^{-t} - e^{-3t} \\ e^{-t} - e^{-3t} & e^{-t} + e^{-3t} \end{bmatrix} \qquad (2)$$

If the initial state is $\underline{x}_0 = \underline{0}$, then the state trajectory $\underline{x}(t)$ depends only upon the input voltage u(t) and is given (Corollary 3.3.3) by

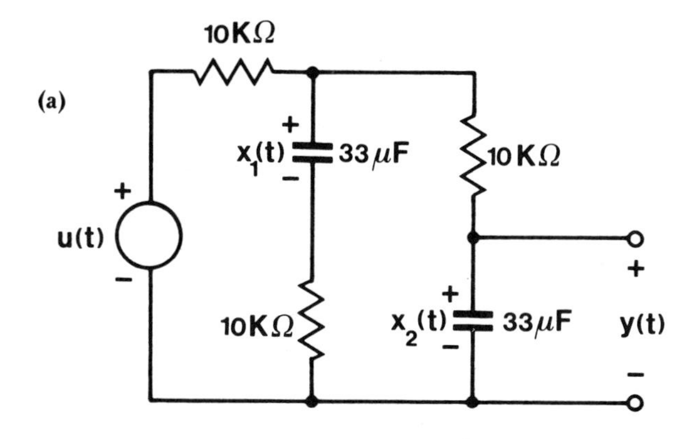

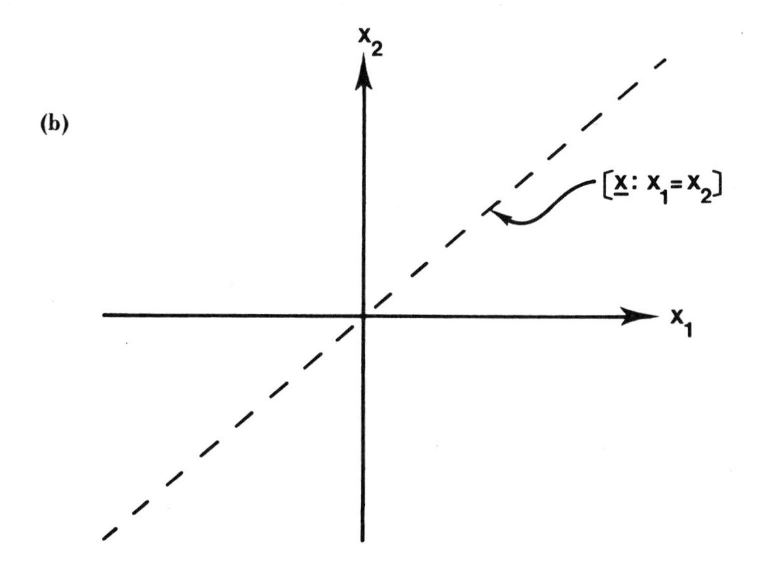

FIG. 1. System of Example 1: (a) circuit and (b) subspace of controllable states.

$$\underline{x}(t) = \int_0^t e^{\underline{A}(t-\tau)} \underline{b}u(\tau)\, d\tau$$
$$= \int_0^t \frac{1}{2} \begin{bmatrix} e^{-(t-\tau)} + e^{-3(t-\tau)} & e^{-(t-\tau)} - e^{-3(t-\tau)} \\ e^{-(t-\tau)} - e^{-3(t-\tau)} & e^{-(t-\tau)} + e^{-3(t-\tau)} \end{bmatrix} \begin{bmatrix} 1 \\ 1 \end{bmatrix} u(\tau)\, d\tau$$
$$= \int_0^t \begin{bmatrix} e^{-(t-\tau)} \\ e^{-(t-\tau)} \end{bmatrix} u(\tau)\, d\tau = \begin{bmatrix} 1 \\ 1 \end{bmatrix} \int_0^t e^{-(t-\tau)} u(\tau)\, d\tau \qquad (3)$$

From this expression it is clear that no matter what control function $u(t)$ is applied, $\underline{x}(t)$ will always be proportional to the vector $[1 \;\; 1]'$, so that the state trajectory remains in the subspace shown in Fig. 1(b).

Since there is no control which will drive the state $\underline{x}(t)$ to a point where $x_1(t) \neq x_2(t)$, we say that this system is (partially) *uncontrollable* and refer to $\{\underline{x} : x_1 = x_2\}$ as the subspace of *controllable states.*

Note that this state of affairs is obvious from the topology of the circuit: any voltage applied must affect the two branches identically, so that if $x_1(0) = x_2(0)$, then $x_1(t) = x_2(t)$ for all t.

ΔΔΔ

We now consider the usual system,

$$\begin{aligned} \dot{\underline{x}}(t) &= \underline{A}\underline{x}(t) + \underline{B}\underline{u}(t) \\ \underline{y}(t) &= \underline{C}\underline{x}(t) \end{aligned} \qquad (4)$$

where $\underline{x}$, $\underline{u}$, and $\underline{y}$ have dimensions n, m, and p, respectively. In view of Example 1, the following definitions are appropriate.

2. *DEFINITION.* A particular state $\overline{\underline{x}}$ of the system (4) is said to be *controllable* if for every $T > 0$, there exists a control function

$$\underline{u}(t) \qquad (0 < t \leq T) \qquad (5)$$

which drives the system (4) from the initial state $\underline{x}(0) = \overline{\underline{x}}$ to $\underline{x}(T) = \underline{0}$. If so, $\underline{u}$ is not unique; other controls will also drive the state to zero via different trajectories. One such state trajectory (for a two-dimensional state vector) is illustrated in Fig. 2. A state $\bar{\underline{x}}$ is also said to be *reachable* if a control exists which drives the system from $\underline{x}(0) = \underline{0}$ to $\underline{x}(T) = \bar{\underline{x}}$. In this case controllability and reachability are completely equivalent† (see Problem 3).

ΔΔΔ

3. *DEFINITION*. The system (4) is said to be *(completely) controllable* if every state is controllable. Equivalently, every state is reachable and the system is said to be *(completely) reachable*. It follows that given any two states $\underline{x}_0$ and $\underline{x}_1$, a control exists which will drive the system from $\underline{x}(0) = \underline{x}_0$ to $\underline{x}(T) = \underline{x}_1$ (see Problem 4). If any uncontrollable states exist, then the system is *uncontrollable*.

ΔΔΔ

†The distinction becomes important only when we generalize to time-varying and/or discrete-time systems.

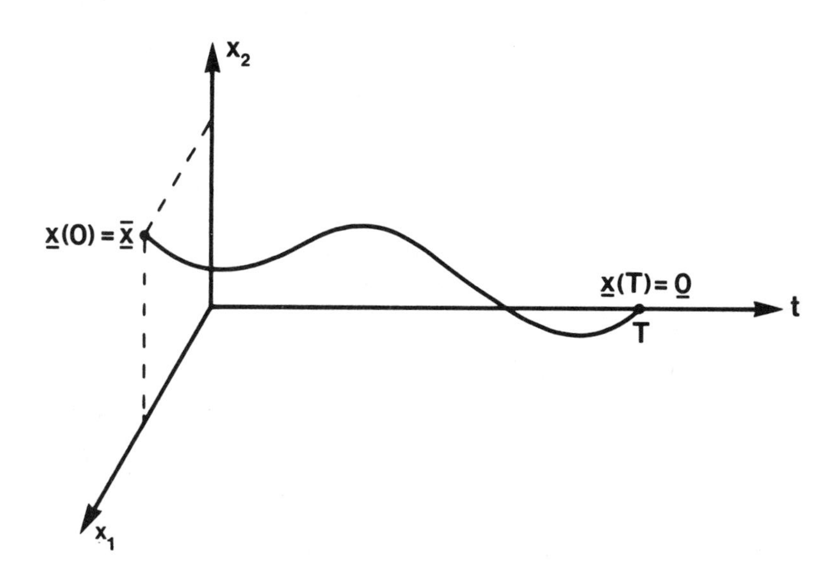

FIG. 2. State trajectory driven to zero.

If these definitions are to be useful, we shall need some conditions with which to test for controllability. These are provided by the following theorems.

4. THEOREM. A state $\bar{\underline{x}}$ of (4) is controllable if and only if it belongs to the range space† $R(\underline{W}_c(0,T))$, where

$$\underline{W}_c(0,T) \triangleq \int_0^T e^{-\underline{A}\tau}\underline{B}\underline{B}'\, e^{-\underline{A}'\tau}\, d\tau \tag{6}$$

is a symmetric $n \times n$ matrix known as the *controllability Gramian*. Consequently, the system is (completely) controllable if and only if $\underline{W}_c(0,T)$ is nonsingular.

Proof: Assume $\bar{\underline{x}} \in R(\underline{W}_c)$. Then the equation

$$\underline{W}_c\underline{z} = \bar{\underline{x}} \tag{7}$$

has a solution $\underline{z} \in R^n$. Using the control

$$\bar{\underline{u}}(t) = -\underline{B}'e^{-\underline{A}'t}\underline{z} \qquad (0 < t \leq T) \tag{8}$$

in the forced response formula (Corollary 3.3.3), we find that

$$\begin{aligned} \underline{x}(t) &= e^{\underline{A}T}\bar{\underline{x}} + \int_0^T e^{\underline{A}(T-\tau)}\underline{B}\underline{u}(\tau)\, d\tau \\ &= e^{\underline{A}T}\bar{\underline{x}} - e^{\underline{A}T}\int_0^T e^{-\underline{A}\tau}\underline{B}\underline{B}'\, e^{-\underline{A}'\tau}\underline{z}\, d\tau \qquad (9) \\ &= e^{\underline{A}T}\bar{\underline{x}} - e^{\underline{A}T}\underline{W}_c\underline{z} = \underline{0} \end{aligned}$$

Conversely, suppose a control $\underline{u}$ exists which drives $\bar{\underline{x}} \neq \underline{0}$ to $\underline{0}$. We shall prove that $\bar{\underline{x}} \in R(\underline{W}_c)$ by showing that the opposite conclusion

---

†We shall see in Theorem 6 that $R[\underline{W}_c(0,T)]$ is the same for all $T > 0$, due to the time-invariance of the system.

leads to a contradiction. First, if $\bar{\underline{x}} \notin R(\underline{W}_c)$, then $\bar{\underline{x}} \not\perp N(\underline{W}_c')$ by (A.5-4). Since $\underline{W}_c' = \underline{W}_c$, we deduce that there exists a vector $\underline{w} \in N(\underline{W}_c)$ such that

$$\underline{w}'\bar{\underline{x}} \neq 0 \tag{10}$$

Next, the assumption that $\underline{u}$ drives $\bar{\underline{x}}$ to $\underline{0}$ means that

$$\underline{x}(T) = \underline{0} = e^{\underline{A}T}\bar{\underline{x}} + \int_0^T e^{\underline{A}(T-\tau)}\underline{B}\underline{u}(\tau)\, d\tau \tag{11}$$

and this may be multiplied on the left by $\underline{w}'e^{-\underline{A}T}$ to yield

$$0 = \underline{w}'\bar{\underline{x}} + \int_0^T \underline{w}'e^{-\underline{A}\tau}\underline{B}\underline{u}(\tau)\, d\tau \tag{12}$$

From (10) and (12) we see that

$$\int_0^T \underline{w}'e^{-\underline{A}\tau}\underline{B}\underline{u}(\tau)\, d\tau \neq \underline{0} \tag{13}$$

However, $\underline{w} \in N(\underline{W}_c)$ means that $\underline{W}_c\underline{w} = \underline{0}$, so that

$$\begin{aligned} 0 = \underline{w}'\underline{W}_c\underline{w} &= \int_0^T \underline{w}'e^{-\underline{A}\tau}\underline{B}\underline{B}'e^{-\underline{A}'\tau}\underline{w}\, d\tau \\ &= \int_0^T \|(\underline{w}'e^{-\underline{A}\tau}\underline{B})'\|^2\, d\tau \end{aligned} \tag{14}$$

and[†]

$$\underline{w}'e^{-\underline{A}\tau}\underline{B} = \underline{0} \qquad (0 < \tau \leq T) \tag{15}$$

which contradicts (13). Thus $\bar{\underline{x}} \in R(\underline{W}_c)$.

Finally, we note that the system is completely controllable if and only if $R(\underline{W}_c) = R^n$, that is, $\underline{W}_c$ is nonsingular.

ΔΔΔ

[†]The reader should recall that if the integral of a nonnegative, continuous function is zero, then the function itself must be identically zero.

This theorem stipulates that a given state $\bar{\underline{x}}$ can be driven to zero if and only if it lies in $R(\underline{W}_c(0,T))$, which is known as the *subspace of controllable states* of the system (4). From Definition A.5.2 and the fact that $\underline{W}_c' = \underline{W}_c$, the orthogonal complement of this subspace is

$$R(\underline{W}_c(0,T))^{\perp} = N(\underline{W}_c(0,T)) \tag{16}$$

which will be known as the *subspace of uncontrollable states.*

This decomposition of the state space (see Definition A.4.4) into

$$R^n = R(\underline{W}_c(0,T)) \oplus N(\underline{W}_c(0,T)) \tag{17}$$

means that every state $\underline{x} \in R^n$ can be uniquely expressed as

$$\underline{x} = \underline{x}_c + \underline{x}_{uc} \qquad (\underline{x}_c \perp \underline{x}_{uc}) \tag{18}$$

where $\underline{x}_c \in R(\underline{W}_c(0,T))$ is called the *controllable component* of $\underline{x}$, and $\underline{x}_{uc} \in N(\underline{W}_c(0,T))$ is its *uncontrollable component.*

A particular control function (8) was used in the proof of Theorem 4, and now we shall see that this control not only drives the state to zero, but does so with a minimum expenditure of "energy."

5. *COROLLARY.* Suppose that $\underline{W}_c(0,T)$ is nonsingular.† Then the control function

$$\bar{\underline{u}}(t) = -\underline{B}'e^{-\underline{A}'t}\underline{W}_c^{-1}(0,T)\,\bar{\underline{x}} \tag{19}$$

transfers the state of the system (4) from $\underline{x}(0) = \bar{\underline{x}}$ to $\underline{x}(T) = \underline{0}$.

---

†If not, the corollary still holds whenever $\bar{\underline{x}} \in R(\underline{W}_c)$. The control to be used in this case is given by (8) (see Problem 5). A similar control function may be used to reach a nonzero final state $\underline{x}(T)$ (See Problems 3 and 4).

Moreover, this control minimizes the "energy" expended in the transfer in the sense that

$$\int_0^T \|\underline{u}(t)\|^2 \, dt \geq \int_0^T \|\bar{\underline{u}}(t)\|^2 \, dt = \bar{\underline{x}}' \underline{W}_c^{-1}(0,T)\bar{\underline{x}} \tag{20}$$

for any other control $\underline{u}$ which drives $\underline{x}(0) = \bar{\underline{x}}$ to $\underline{x}(T) = \underline{0}$.

Proof: That the control (19) accomplishes the transfer was established by direct substitution in (9). The expression on the right in (20) follows by substitution of (19) in the integral.

Suppose $\underline{u}$ is another control which transfers $\bar{\underline{x}}$ to $\underline{0}$, that is,

$$\begin{aligned} \underline{x}(T) = \underline{0} &= e^{\underline{A}T}\bar{\underline{x}} + \int_0^T e^{\underline{A}(T-\tau)}\underline{B}\bar{\underline{u}}(\tau)\, d\tau \\ &= e^{\underline{A}T}\bar{\underline{x}} + \int_0^T e^{\underline{A}(T-\tau)}\underline{B}\underline{u}(\tau)\, d\tau \end{aligned} \tag{21}$$

Subtracting, we have

$$\int_0^T e^{\underline{A}(T-\tau)}\underline{B}[\underline{u}(\tau) - \bar{\underline{u}}(\tau)]\, d\tau = \underline{0} \tag{22}$$

and multiplying on the left by $\bar{\underline{x}}'\underline{W}_c^{-1} e^{-\underline{A}T}$,

$$\int_0^T \bar{\underline{x}}'\underline{W}_c^{-1} e^{-\underline{A}\tau}\underline{B}[\underline{u}(\tau) - \bar{\underline{u}}(\tau)]d\tau = -\int_0^T \bar{\underline{u}}'(\tau)[\underline{u}(\tau) - \bar{\underline{u}}(\tau)]\, d\tau = \underline{0} \tag{23}$$

or

$$\int_0^T \bar{\underline{u}}'(\tau)\underline{u}(\tau)\, d\tau = \int_0^T \|\bar{\underline{u}}(\tau)\|^2 \, d\tau \tag{24}$$

Next, we expand the obvious inequality

$$\begin{aligned} 0 &\leq \int_0^T \|\underline{u}(\tau) - \bar{\underline{u}}(\tau)\|^2 \, d\tau \\ &= \int_0^T [\|\underline{u}(\tau)\|^2 - 2\bar{\underline{u}}'(\tau)\underline{u}(\tau) + \|\bar{\underline{u}}(\tau)\|^2]\, d\tau \end{aligned} \tag{25}$$

and substitute (24) to find

$$0 \leq \int_0^T [\| \underline{u}(\tau) \|^2 - \| \underline{\bar{u}}(\tau) \|^2]\, d\tau \tag{26}$$

or†

$$\int_0^T \| \underline{\bar{u}}(\tau) \|^2 \, d\tau \leq \int_0^T \| \underline{u}(\tau) \|^2 \, d\tau \tag{27}$$

ΔΔΔ

The condition established in Theorem 4 for controllability is rather awkward to use in practice since it requires evaluation of the integral (6) to determine $\underline{W}_c(0,T)$. The following theorem provides more convenient characterizations of the subspaces of controllable and uncontrollable states.

6. *THEOREM.* If we define the $n \times nm$ *controllability matrix* for the system (4) to be

$$\underline{P}_c \triangleq \left[\underline{B} \quad \underline{AB} \quad \underline{A}^2\underline{B} \quad \cdots \quad \underline{A}^{n-1}\underline{B} \right] \tag{28}$$

then the subspace of controllable states is

$$R(\underline{W}_c(0,T)) = R(\underline{P}_c) \tag{29}$$

and the subspace of uncontrollable states is

$$N(\underline{W}_c(0,T)) = N(\underline{P}_c') = N \begin{bmatrix} \underline{B}' \\ \underline{B}'\underline{A}' \\ \vdots \\ \underline{B}'\underline{A}'^{n-1} \end{bmatrix} \tag{30}$$

†The interested reader may wish to note that this proof amounts to an application of the *projection theorem* in Hilbert space. In (23) we have established that $\underline{\bar{u}} \perp (\underline{\bar{u}} - \underline{u})$, where orthogonality is defined by the inner product

$$\langle \underline{x}, \underline{y} \rangle = \int_0^T \underline{x}'(t)\underline{y}(t)\, dt$$

and this implies that $\| \underline{u} \|^2 = \langle \underline{u}, \underline{u} \rangle$ is minimal. See Ref. 117 for further details.

for any $T > 0$. Thus the system is completely controllable if and only if the controllability matrix has full rank, that is,

$$\operatorname{rank}(\underline{P}_c) = n \tag{31}$$

Proof: From (A.5-4) and the fact that $\underline{W}_c = \underline{W}_c'$, we have

$$\mathcal{R}(\underline{P}_c)^{\perp} = \mathcal{N}(\underline{P}_c') \qquad \mathcal{R}(\underline{W}_c)^{\perp} = \mathcal{N}(\underline{W}_c) \tag{32}$$

Thus (29) and (30) are equivalent and it will suffice to prove the latter,

$$\mathcal{N}(\underline{W}_c) = \mathcal{N}(\underline{P}_c') = \mathcal{N}\begin{bmatrix} \underline{B}' \\ \underline{B}'\underline{A}' \\ \vdots \\ \underline{B}'\underline{A}'^{n-1} \end{bmatrix} \tag{33}$$

First, we shall establish that $\mathcal{N}(\underline{W}_c) \subseteq \mathcal{N}(\underline{P}_c')$. If $\underline{z} \in \mathcal{N}(\underline{W}_c)$, then $\underline{W}_c\underline{z} = \underline{0}$ and

$$\underline{z}'\underline{W}_c\underline{z} = \int_0^T (\underline{z}'e^{-\underline{A}\tau}\underline{B})(\underline{B}'e^{-\underline{A}'\tau}\underline{z})\, d\tau = \int_0^T \|\underline{B}'e^{-\underline{A}'\tau}\underline{z}\|^2\, d\tau = 0 \tag{34}$$

Since the integrand in (34) is continuous, it must be identically zero, that is

$$\underline{B}'e^{-\underline{A}'\tau}\underline{z} = \underline{B}'\left(\underline{I} - \underline{A}'\tau + \frac{\underline{A}'^2\tau^2}{2!} - \cdots\right)\underline{z} = \underline{0} \qquad (0 < \tau \le T) \tag{35}$$

and since this power series is identically zero on a finite interval, each term must be zero,

$$\underline{B}'\underline{z} = \underline{B}'\underline{A}'\underline{z} = \underline{B}'\underline{A}'^2\underline{z} = \cdots = \underline{0} \tag{36}$$

This means that

$$\underline{P}_c'\underline{z} = \begin{bmatrix} \underline{B}'\underline{z} \\ \underline{B}'\underline{A}'\underline{z} \\ \vdots \\ \underline{B}'\underline{A}'^{n-1}\underline{z} \end{bmatrix} = \begin{bmatrix} \underline{0} \\ \underline{0} \\ \vdots \\ \underline{0} \end{bmatrix} \tag{37}$$

so that $\underline{z} \in N(\underline{P}_c')$.

Conversely, we shall establish that $N(\underline{W}_c) \supseteq N(\underline{P}_c')$. If $\underline{z} \in N(\underline{P}_c')$, then (37) holds and

$$\underline{B}'\underline{z} = \underline{B}'\underline{A}'\underline{z} = \ldots = \underline{B}'\underline{A}'^{n-1}\underline{z} = \underline{0} \tag{38}$$

Using the expression (4.3-20),

$$e^{\underline{A}t} = \sum_{i=0}^{n-1} \underline{A}^i \alpha_i(t) \tag{39}$$

for the matrix exponential, we have

$$\begin{aligned} \underline{B}'e^{-\underline{A}'\tau}\underline{z} &= \underline{B}'\left[\sum_{i=0}^{n-1} \underline{A}'^i\alpha_i(-\tau)\right]\underline{z} \\ &= \sum_{i=0}^{n-1} (\underline{B}'\underline{A}'^i\underline{z})\alpha_i(-\tau) = \underline{0} \end{aligned} \tag{40}$$

from which it follows that

$$\underline{W}_c\underline{z} = \int_0^T e^{-\underline{A}\tau}\underline{B}(\underline{B}'e^{-\underline{A}'\tau}\underline{z})\, d\tau = \underline{0} \tag{41}$$

and we see that $\underline{z} \in N(\underline{W}_c)$.

Having shown that each of the spaces $N(\underline{W}_c)$ and $N(\underline{P}_c')$ is contained in the other, we conclude that they must be identical.

ΔΔΔ

Note that the controllability matrix $\underline{P}_c$ in (28) does not depend upon the terminal time T (assuming $T > 0$), which verifies that a controllable state may be driven to zero over *any* time interval [0,T]. Some caution is necessary, however, since to accomplish the transfer in an arbitrarily small time interval may require an arbitrarily large control function (see Problem 12(d)).

Theorem 6 provides a more useful test for controllability than Theorem 4 does since $\underline{P}_c$ is relatively easy to calculate. In the case of a multiple-input system ($m > 1$), a further improvement results from noting that the $n \times nm$ controllability matrix $\underline{P}_c$ can be replaced (provided $\underline{B}$ has rank m) by the $n \times (n-m+1)m$ matrix

$$\begin{bmatrix} \underline{B} & \underline{AB} & \underline{A}^2\underline{B} & \cdots & \underline{A}^{n-m}\underline{B} \end{bmatrix} \tag{42}$$

(see Problem 7). A useful necessary condition for controllability is that the $n \times (n+m)$ matrix $[\underline{B} \quad \underline{A}]$ have full rank (see Problem 8).

It may also be convenient in some multiple-input cases to make use of Lemma A.5.3, which says that

$$R(\underline{P}_c) = R(\underline{P}_c\underline{P}_c') \qquad N(\underline{P}_c') = N(\underline{P}_c\underline{P}_c') \tag{43}$$

and implies that the system is controllable if and only if

$$\det\,(\underline{P}_c\underline{P}_c') \neq 0 \tag{44}$$

The advantage is that $\underline{P}_c\underline{P}_c'$ is $n \times n$, whereas $\underline{P}_c$ is $n \times nm$. This must be weighed, of course, against the extra calculations involved in carrying out the matrix multiplication.

Because the system (4) is time invariant, the definitions and results in this section are all valid for an initial time $t_0 \neq 0$ (see Problems 2 and 4). This leads us to ask the following question: if the state of a system is controllable at one time, will it still be controllable at some future time? The answer, which is always in the affirmative in the case of a time-invariant system, is provided by the following lemma (also see Problem 9).

7. *LEMMA*.[†] If the initial state $\underline{x}(0) = \overline{\underline{x}}$ of the system (4) is controllable, then the state

$$\underline{x}(t) = e^{\underline{A}t}\overline{\underline{x}} + \int_0^t e^{\underline{A}(t-\tau)}\underline{Bu}(\tau)\, d\tau \tag{45}$$

remains controllable for all time, regardless of what input function $\underline{u}$ is applied.

Proof: The assumption that $\overline{\underline{x}}$ is controllable means, by Corollary 7, that

---

[†]This result is a manifestation of the fact that the subspace $R(\underline{P}_c)$ is *invariant* under $\underline{A}$, and hence under $e^{\underline{A}t}$.

$$\bar{\underline{x}} \in R(\underline{P}_c) = R[\underline{B} \quad \underline{AB} \quad \underline{A}^2\underline{B} \quad \ldots \quad \underline{A}^{n-1}\underline{B}] \tag{46}$$

In other words, $\bar{\underline{x}}$ may be expressed as a linear combination of the columns of $\underline{P}_c$,

$$\bar{\underline{x}} = \sum_{j=0}^{n-1} \underline{A}^j \underline{Bd}_j \tag{47}$$

for some set of m-vectors $\{\underline{d}_0, \underline{d}_1, \ldots, \underline{d}_{n-1}\}$. Using (39), we may then write the natural response as

$$\begin{aligned} e^{\underline{A}t}\bar{\underline{x}} &= \sum_{i=0}^{n-1} \underline{A}^i \alpha_i(t) \sum_{j=0}^{n-1} \underline{A}^j \underline{Bd}_j \\ &= \sum_{i,j=0}^{n-1} \underline{A}^{i+j} \underline{Bd}_j \alpha_i(t) = \sum_{k=0}^{2n-2} \underline{A}^k \underline{Be}_k(t) \end{aligned} \tag{48}$$

where

$$\underline{e}_k(t) \triangleq \sum_{i+j=k} \underline{d}_j \alpha_i(t) \tag{49}$$

Equation (48) establishes that $e^{\underline{A}t}\bar{\underline{x}}$ can be expressed as a linear combination of the columns of $\underline{A}^k\underline{B}$ for $k = 0,1,2,\ldots, 2n-2$, or

$$e^{\underline{A}t}\bar{\underline{x}} \in R\left[\underline{B} \quad \underline{AB} \quad \underline{A}^2\underline{B} \quad \ldots \quad \underline{A}^{2n-2}\underline{B}\right] \quad \text{(for all t)} \tag{50}$$

It then follows from the Cayley-Hamilton Theorem (see Problem 10.1-6) that

$$e^{\underline{A}t}\bar{\underline{x}} \in R(\underline{P}_c) \qquad \text{(for all t)} \tag{51}$$

i.e., that the natural response remains controllable for all time.

We shall now prove the same for the forced response, by using (39) and writing it as

$$\begin{aligned} \int_0^t e^{\underline{A}(t-\tau)} \underline{Bu}(\tau)\, d\tau &= \int_0^t \sum_{i=0}^{n-1} \underline{A}^i \alpha_i(t-\tau) \underline{Bu}(\tau)\, d\tau \\ &= \sum_{i=0}^{n-1} \underline{A}^i \underline{Bf}_i(t) \end{aligned} \tag{52}$$

where

$$\underline{f}_i(t) \triangleq \int_0^t \alpha_i(t-\tau)\underline{u}(\tau)\, d\tau \tag{53}$$

Thus we see that the forced response (52) is also a linear combination of the columns of $\underline{B}$, $\underline{AB}$, ..., $\underline{A}^{n-1}\underline{B}$, that is,

$$\int_0^t e^{\underline{A}(t-\tau)}\underline{B}\underline{u}(\tau)\, d\tau \in R(\underline{P}_c) \qquad \text{(for all } t \text{ and all } \underline{u}) \tag{54}$$

Combining (51) and (54), we see that $\underline{x}(t)$ remains controllable for all t and all $\underline{u}$.

ΔΔΔ

We shall now illustrate the above results by means of a few examples.

8. EXAMPLE. In Example 1, using Theorem 6, we can compute

$$\underline{P}_c = [\ \underline{b}\quad \underline{Ab}\ ] = \begin{bmatrix} 1 & -1 \\ 1 & -1 \end{bmatrix} \tag{55}$$

and

$$R(\underline{P}_c) = \left\{ \underline{x} : \underline{x} \propto \begin{bmatrix} 1 \\ 1 \end{bmatrix} \right\} = R(\underline{W}_c(0,T)) \tag{56}$$

$$N(\underline{P}_c') = R(\underline{P}_c)^{\perp} = \left\{ \underline{x} : \underline{x} \propto \begin{bmatrix} 1 \\ -1 \end{bmatrix} \right\} = N(\underline{W}_c(0,T)) \tag{57}$$

Thus the subspace of controllable states consists of all those for which $x_1 = x_2$, as indicated in Example 1, and any state with $x_1 \neq x_2$ cannot be reached from or driven to the origin.

The same information may be obtained from Theorem 4, with a slightly greater computational effort. In Example 1 it was found that

$$e^{-\underline{A}\tau}\underline{b} = \begin{bmatrix} e^{\tau} \\ e^{\tau} \end{bmatrix} \tag{58}$$

so that

$$\begin{aligned} \underline{W}_c(0,T) &= \int_0^T e^{-\underline{A}\tau}\underline{b}\underline{b}'e^{-\underline{A}'\tau}\, d\tau = \int_0^T \begin{bmatrix} e^{\tau} \\ e^{\tau} \end{bmatrix} \begin{bmatrix} e^{\tau} & e^{\tau} \end{bmatrix} d\tau \\ &= \int_0^T e^{2\tau} \begin{bmatrix} 1 & 1 \\ 1 & 1 \end{bmatrix} d\tau = \frac{e^{2T} - 1}{2} \begin{bmatrix} 1 & 1 \\ 1 & 1 \end{bmatrix} \end{aligned} \tag{59}$$

which verifies (56) and (57).

Now suppose the initial state is $\underline{x}(0) = \overline{\underline{x}} = [1 \quad 1]'$. Then (7) becomes

$$\frac{e^{2T} - 1}{2} \begin{bmatrix} 1 & 1 \\ 1 & 1 \end{bmatrix} \underline{z} = \begin{bmatrix} 1 \\ 1 \end{bmatrix} \tag{60}$$

and has the (nonunique) solution

$$z = \begin{bmatrix} 2/(e^{2T} - 1) \\ 0 \end{bmatrix} \tag{61}$$

Substituting into (8), we conclude that the control

$$\overline{u}(t) = -\underline{b}'e^{-\underline{A}'t}\underline{z} = -\begin{bmatrix} e^{t} & e^{t} \end{bmatrix} \begin{bmatrix} 2/(e^{2T} - 1) \\ 0 \end{bmatrix} = \frac{-2e^{t}}{e^{2T} - 1} \tag{62}$$

will drive the state to $\underline{x}(T) = \underline{0}$. Moreover, this control accomplishes the transfer from $\overline{\underline{x}}$ to $\underline{0}$ in T seconds with a minimal expenditure of energy $\int_0^T u^2(t)\, dt$ (see the footnote in Corollary 5). The reader should verify that the same control is obtained for every possible solution $\underline{z}$ of (60).

9. *EXAMPLE*. Recall the *broom-balancer* of Example 2.3.2, shown in Fig. 3, in which the object is to apply a force u(t) to the cart so that the small mass m remains balanced in the vertical position.

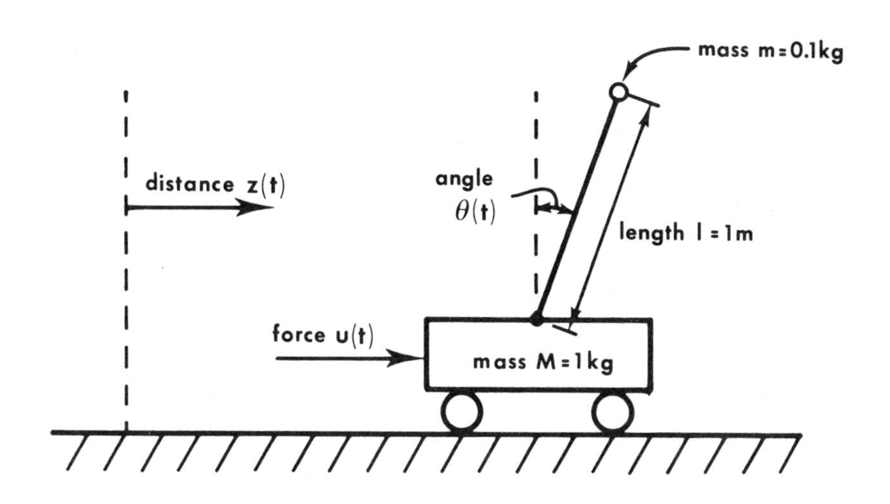

FIG. 3. Broom-balancing system.

This has obvious applications to the stabilization of a rocket booster which "balances" on its own thrust vector.

Defining the state vector for this system to be

$$\underline{x}(t) = \begin{bmatrix} z(t) & \dot{z}(t) & \Theta(t) & \dot{\Theta}(t) \end{bmatrix}' \tag{63}$$

and assuming that $\Theta(t)$ remains close to zero, we found the linearized equations governing the system to be

$$\dot{\underline{x}} = \underline{A}\underline{x} + \underline{b}u = \begin{bmatrix} 0 & 1 & 0 & 0 \\ 0 & 0 & -1 & 0 \\ 0 & 0 & 0 & 1 \\ 0 & 0 & 11 & 0 \end{bmatrix} \underline{x} + \begin{bmatrix} 0 \\ 1 \\ 0 \\ -1 \end{bmatrix} u \tag{64}$$

$$z = \underline{c}'\underline{x} = \begin{bmatrix} 1 & 0 & 0 & 0 \end{bmatrix} \underline{x}$$

To check the controllability of this system, we compute

$$\underline{P}_c = \begin{bmatrix} \underline{b} & \underline{A}\underline{b} & \underline{A}^2\underline{b} & \underline{A}^3\underline{b} \end{bmatrix} = \begin{bmatrix} 0 & 1 & 0 & 1 \\ 1 & 0 & 1 & 0 \\ 0 & -1 & 0 & -11 \\ -1 & 0 & -11 & 0 \end{bmatrix} \tag{65}$$

Since det $\underline{P}_c = 100$, $\underline{P}_c$ has full rank, and by Theorem 6 the system is completely controllable. Thus if the angle Θ departs from equilibrium by a small amount, a control always exists which will drive it back to zero,[†] just as one's intuition suggests. Moreover, a control also exists which will drive *both* Θ and z, as well as their derivatives, to zero.

The reader may wish to attempt balancing a broom or similar object in order to verify these conclusions empirically.

ΔΔΔ

10. EXAMPLE. In Example 2.4.3 we considered a satellite in a circular, equatorial orbit about the earth, and found that small perturbations about this orbit (in the equatorial plane) would be governed by the linearized equations

$$\dot{\underline{x}} = \underline{A}\underline{x} + \underline{B}\underline{u} = \begin{bmatrix} 0 & 1 & 0 & 0 \\ 3\omega^2 & 0 & 0 & 2\omega \\ 0 & 0 & 0 & 1 \\ 0 & -2\omega & 0 & 0 \end{bmatrix} \underline{x} + \begin{bmatrix} 0 & 0 \\ 1 & 0 \\ 0 & 0 \\ 0 & 1 \end{bmatrix} \underline{u} \tag{66}$$

where

$$\underline{x} = \begin{bmatrix} r \\ \dot{r} \\ \Theta \\ \dot{\Theta} \end{bmatrix} \qquad \underline{u} = \begin{bmatrix} u_r \\ u_\Theta \end{bmatrix} \tag{67}$$

To determine controllability, we make use of (42) and compute

$$\underline{P}_c = [\underline{B} \quad \underline{A}\underline{B} \quad \underline{A}^2\underline{B}] = \begin{bmatrix} 0 & 0 & 1 & 0 & 0 & 2\omega \\ 1 & 0 & 0 & 2\omega & -\omega^2 & 0 \\ 0 & 0 & 0 & 1 & -2\omega & 0 \\ 0 & 1 & -2\omega & 0 & 0 & -4\omega^2 \end{bmatrix} \tag{68}$$

[†]This justifies the assumption that $\Theta(t) \approx 0$, provided we choose an appropriate control strategy.

The first four columns are independent since

$$\det \begin{bmatrix} 0 & 0 & 1 & 0 \\ 1 & 0 & 0 & 2\omega \\ 0 & 0 & 0 & 1 \\ 0 & 1 & -2\omega & 0 \end{bmatrix} = -1 \neq 0 \tag{69}$$

so rank $(\underline{P}_c) = 4$ and the system is completely controllable.

Now suppose that the tangential thruster becomes inoperable, so that $u_\Theta = 0$. In this situation we need to know whether the system can be controlled from u_r alone, so we rewrite (66) as

$$\dot{\underline{x}} = \underline{\underline{A}}\underline{x} + \underline{b}_r u_r = \begin{bmatrix} 0 & 1 & 0 & 0 \\ 3\omega^2 & 0 & 0 & 2\omega \\ 0 & 0 & 0 & 1 \\ 0 & -2\omega & 0 & 0 \end{bmatrix} \underline{x} + \begin{bmatrix} 0 \\ 1 \\ 0 \\ 0 \end{bmatrix} u_r \tag{70}$$

and compute

$$\underline{P}_{cr} = \left[\underline{b}_r \quad \underline{\underline{A}}\underline{b}_r \quad \underline{A}^2\underline{b}_r \quad \underline{A}^3\underline{b}_r \right] = \begin{bmatrix} 0 & 1 & 0 & -\omega^2 \\ 1 & 0 & -\omega^2 & 0 \\ 0 & 0 & -2\omega & 0 \\ 0 & -2\omega & 0 & 2\omega^3 \end{bmatrix} \tag{71}$$

Since $\det \underline{P}_{cr} = 0$, the satellite is *not* completely controllable using only u_r. Indeed, the fourth column of $\underline{P}_{cr}$ is proportional to the second (and the first and third columns are orthogonal to these two), so the subspace of controllable states is

$$R(\underline{P}_{cr}) = \{ \underline{x} : x_4 = -2\omega x_1 \} \tag{72}$$

Thus any initial state for which $\dot{\Theta} \neq -2\omega r$ cannot be returned to zero with only radial thrusting.

Finally, suppose instead that the radial thruster becomes inoperable, leaving us with

$$\dot{\underline{x}} = \underline{\underline{A}}\underline{x} + \underline{b}_\Theta \underline{u}_\Theta = \begin{bmatrix} 0 & 1 & 0 & 0 \\ 3\omega^2 & 0 & 0 & 2\omega \\ 0 & 0 & 0 & 1 \\ 0 & -2\omega & 0 & 0 \end{bmatrix} \underline{x} + \begin{bmatrix} 0 \\ 0 \\ 0 \\ 1 \end{bmatrix} u_\Theta \tag{73}$$

The controllability matrix is now

$$\underline{P}_{c\theta} = \left[\underline{b}_\Theta \quad \underline{A}\underline{b}_\Theta \quad \underline{A}^2\underline{B}_\Theta \quad \underline{A}^3\underline{b}_\Theta\right] = \begin{bmatrix} 0 & 0 & 2\omega & 0 \\ 0 & 2\omega & 0 & -2\omega^3 \\ 0 & 1 & 0 & -4\omega^2 \\ 1 & 0 & -4\omega^2 & 0 \end{bmatrix} \tag{74}$$

and det $\underline{P}_{c\Theta} = -12\omega^4 \neq 0$, so the satellite is completely controllable using the tangential thruster by itself.

ΔΔΔ

This section has introduced the concept of controllability, which is fundamental in the theory of linear systems. Before we turn to the dual (and equally fundamental) concept of observability, a word of caution is in order.

Controllability is concerned with the problem of transferring the state of a system to a particular value *at a particular instant of time*. However, controllability of a system does *not* imply that the state can be made to follow some arbitrary trajectory over an interval of time (the servomechanism problem). Problem 13 investigates this question further, and Problem 12(d) is also relevant.

Moreover, the controls discussed in this section have all been *open-loop:* an input function is precomputed and then applied to the system over some time interval, with no feedback to adjust for variations and errors. The subject of closed-loop controls will be taken up in the next chapter.

Note also that the above results have all dealt with controllability of the state vector $\underline{x}$. If one is concerned only with controlling the output $\underline{y}$, then the definitions can be altered accordingly. In particular, a system is said to be *output controllable* if for any two outputs $\underline{y}_0$ and $\underline{y}_1$, and $T > 0$, a control exists which will drive the system from $\underline{y}(0) = \underline{y}_0$ to $\underline{y}(T) = \underline{y}_1$. It can be shown that the system is output controllable if and only if the $p \times p$ matrix

$$\underline{C}\ \underline{W}_c(0,T)\ \underline{C}' = \int_0^T \underline{H}(-\tau)\underline{H}'(-\tau)\ d\tau \tag{75}$$

is nonsingular, where $\underline{H}(t) = \underline{C}e^{\underline{A}t}\underline{B}$ is the impulse response of the system. An equivalent condition is that the $p \times nm$ matrix

$$\underline{CP}_c = [\underline{CB} \quad \underline{CAB} \quad \cdots \quad \underline{CA}^{n-1}\underline{B}] \tag{76}$$

have full rank p. This issue is explored further in Problem 16, and in Refs. 32, 39, and 47.

Finally, we remark that uncontrollability is a "singular" condition, in the sense that if the system

$$\dot{\underline{x}} = \underline{Ax} + \underline{Bu} \tag{77}$$

is uncontrollable then almost any small perturbations of the elements of $\underline{A}$ and $\underline{B}$ will cause it to become controllable [108]. To put it another way, if the elements of $\underline{A}$ and $\underline{B}$ are chosen at random, then the probability of obtaining a controllable system is equal to one. We must bear this fact in mind whenever we model a physical system, since some of the parameters are bound to be only approximate. Other parameters (usually zeros and ones) are fixed by the structure of the system and may be specified exactly. An interesting discussion of these issues may be found in Refs. 243a, 287 and 334a, where the concept of *structural controllability* is introduced to deal with them.

10.2 OBSERVABILITY

Observability concerns the extent to which the state of a linear system influences the output. To illustrate this concept, consider the following simple circuit.

1. EXAMPLE. The state equations governing the circuit shown in Fig. 1(a) are

$$\begin{aligned} \dot{\underline{x}} &= \begin{bmatrix} -2 & 1 \\ 1 & -2 \end{bmatrix} \underline{x} + \begin{bmatrix} 1 \\ 0 \end{bmatrix} u = \underline{Ax} + \underline{b}u \\ y &= \begin{bmatrix} 1 & -1 \end{bmatrix} \underline{x} = \underline{c}'\underline{x} \end{aligned} \tag{1}$$

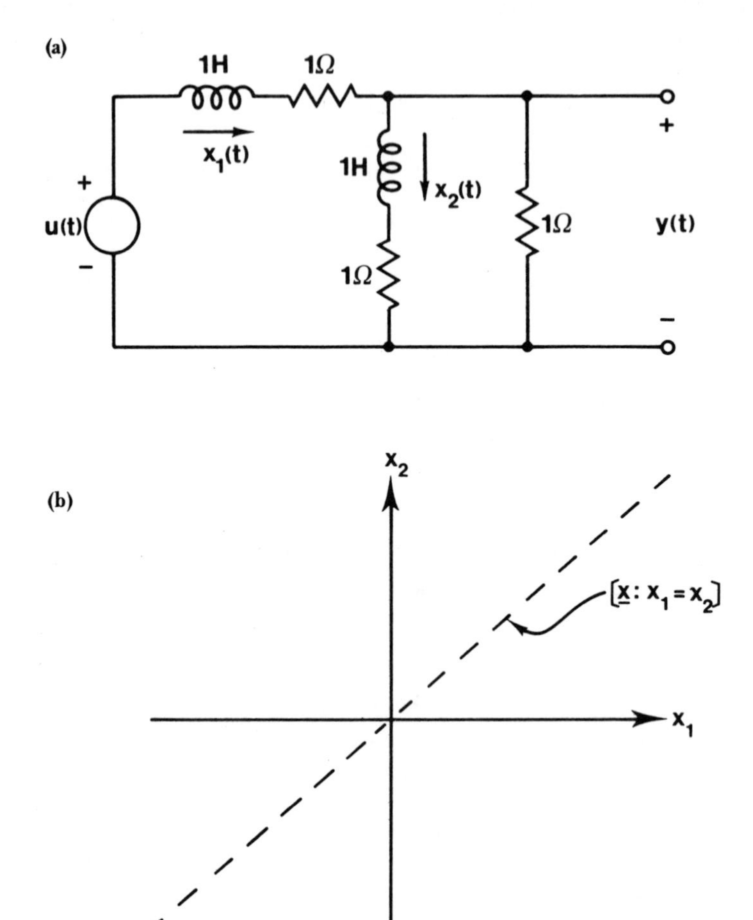

FIG. 1. System of Example 1: (a) circuit and (b) subspace of unobservable states.

and the matrix exponential in this case is easily seen to be

$$e^{\underline{A}t} = \frac{1}{2}\begin{bmatrix} e^{-t} + e^{-3t} & e^{-t} - e^{-3t} \\ e^{-t} - e^{-3t} & e^{-t} + e^{-3t} \end{bmatrix} \tag{2}$$

If the input is $u(t) \equiv 0$, then the output voltage depends only upon the initial state $\underline{x}(0) = \bar{\underline{x}}$ and is given (Corollary 3.3.3) by

$$y(t) = \underline{c}'e^{\underline{A}t}\bar{\underline{x}} = \begin{bmatrix} 1 & -1 \end{bmatrix} \frac{1}{2} \begin{bmatrix} e^{-t}+e^{-3t} & e^{-t}-e^{-3t} \\ e^{-t}-e^{-3t} & e^{-t}+e^{-3t} \end{bmatrix} \begin{bmatrix} \bar{x}_1 \\ \bar{x}_2 \end{bmatrix}$$

$$= \begin{bmatrix} e^{-3t} & -e^{-3t} \end{bmatrix} \begin{bmatrix} \bar{x}_1 \\ \bar{x}_2 \end{bmatrix} = (\bar{x}_1 - \bar{x}_2)\, e^{-3t} \tag{3}$$

From this expression it is clear that no matter what the initial state $\bar{\underline{x}}$ may be, the output $y(t)$ depends only upon the difference $\bar{x}_1 - \bar{x}_2$. Indeed, if $\bar{x}_1 = \bar{x}_2$, so that $\bar{\underline{x}}$ belongs to the subspace shown in Fig.1(b), then the output is identically zero.

Since an initial state with $\bar{x}_1 = \bar{x}_2$ produces no response at the output, we say that the system is (partially) *unobservable* and refer to $\{\underline{x} : x_1 = x_2\}$ as the subspace of *unobservable states*.

Note that this state of affairs is obvious from the topology of the circuit: the two inductive branches have identical time constants, so if $u(t) = 0$ and $x_1(0) = x_2(0)$ then $x_1(t) = x_2(t)$ for all t. Since the output $y(t)$ is the difference of these two inductor currents, it remains identically zero.

ΔΔΔ

We now consider the usual system,

$$\begin{aligned} \dot{\underline{x}}(t) &= \underline{A}\underline{x}(t) + \underline{B}\underline{u}(t) \\ \underline{y}(t) &= \underline{C}\underline{x}(t) \end{aligned} \tag{4}$$

where $\underline{x}$, $\underline{u}$ and $\underline{y}$ have dimensions n, m, and p, respectively. In view of Example 1 above, the following definitions are appropriate.

2. *DEFINITION.* A particular state $\bar{\underline{x}}$ of the system (4) is said to be *unobservable* if for $\underline{u}(t) \equiv \underline{0}$ and any $T > 0$, the initial state $\underline{x}(0) = \bar{\underline{x}}$ produces an output response

$$\underline{y}(t) = \underline{0} \qquad (0 \leq t \leq T) \tag{5}$$

A state $\bar{\underline{x}}$ is also said to be *unreconstructible* if the output vanishes for $-T \leq t \leq 0$, but in this case observability and reconstructibility are completely equivalent[†] (see Problem 3).

ΔΔΔ

3. *DEFINITION.* The system (4) is said to be *(completely) observable* if no state (except $\underline{0}$) is unobservable. Equivalently, no nonzero state is unreconstructible and the system is said to be *(completely) reconstructible*. If any nonzero unobservable states exist, then the system is *unobservable*.

ΔΔΔ

If these definitions are to be useful, we shall need some conditions with which to test for observability. These are provided by the following theorems.

4. *THEOREM.* A state $\bar{\underline{x}}$ of (4) is unobservable if and only if it belongs to the null space[††] $N(\underline{W}_o(0,T))$, where

$$\underline{W}_o(0,T) \triangleq \int_0^T e^{\underline{A}'\tau}\underline{C}'\underline{C}e^{\underline{A}\tau}\, d\tau \tag{6}$$

is a symmetric $n \times n$ matrix known as the *observability Gramian*. Consequently, the system is (completely) observable if and only if $\underline{W}_o(0,T)$ is nonsingular.

Proof: Assume that $\bar{\underline{x}} \in N(\underline{W}_o)$. Then $\underline{W}_o\bar{\underline{x}} = \underline{0}$ and

---

[†]The distinction becomes important only when we generalize to time-varying and/or discrete-time systems.

[††]We shall see in Theorem 6 that $N(\underline{W}_o(0,T))$ is the same for all $T > 0$, due to the time invariance of the system. $\underline{W}_o(0,T)$ has an interesting interpretation as an *information matrix* [95,280].

$$
\begin{aligned}
0 = \underline{\bar{x}}'\underline{W}_o\underline{\bar{x}} &= \int_0^T \underline{\bar{x}}' e^{\underline{A}'\tau}\underline{C}'\underline{C}e^{\underline{A}\tau}\underline{\bar{x}}\, d\tau \\
&= \int_0^T \| \underline{C}e^{\underline{A}\tau}\underline{\bar{x}} \|^2 \, d\tau
\end{aligned} \tag{7}
$$

The output of the system (4), when $\underline{u} = \underline{0}$ and $\underline{x}(0) = \underline{\bar{x}}$, is given (Corollary 3.3.3) by

$$
\underline{y}(\tau) = \underline{C}e^{\underline{A}\tau}\underline{\bar{x}} \qquad (\tau \geq 0) \tag{8}
$$

and we conclude[†] from (7) that

$$
\underline{y}(\tau) = \underline{C}e^{\underline{A}\tau}\underline{\bar{x}} = \underline{0} \qquad (0 \leq \tau \leq T) \tag{9}
$$

so that $\underline{\bar{x}}$ is unobservable.

Conversely, if $\underline{\bar{x}}$ is unobservable then (9) holds, and

$$
\underline{W}_o\underline{\bar{x}} = \int_0^T e^{\underline{A}'\tau}\underline{C}'\underline{C}e^{\underline{A}\tau}\underline{\bar{x}}\, d\tau = \underline{0} \tag{10}
$$

or $\underline{\bar{x}} \in N(\underline{W}_o)$.

Finally, we note that the system is completely observable if and only if $N(\underline{W}_o) = \{\underline{0}\}$, that is, $\underline{W}_o$ is nonsingular.

ΔΔΔ

This theorem states that a given initial state $\underline{\bar{x}}$ produces zero output if and only if it lies in $N(\underline{W}_o(0,T)) \subseteq R^n$, which is known as the *subspace of unobservable states* of the system (4). Using Definition A.5.2 and the fact that $\underline{W}_o' = \underline{W}_o$, the orthogonal complement of this subspace is

[†]The reader should recall that if the integral of a nonnegative, continuous function is zero, then the function itself must be identically zero.

$$N(\underline{W}_o(0,T))^{\perp} = R(\underline{W}_o(0,T)) \tag{11}$$

which will be known as the *subspace of observable states.*

This decomposition of the state-space (see Definition A.4.4) into

$$R^n = R(\underline{W}_o(0,T)) \oplus N(\underline{W}_o(0,T)) \tag{12}$$

means that every state $\underline{x} \in R^n$ can be uniquely expressed as

$$\underline{x} = \underline{x}_o + \underline{x}_{uo} \qquad (\underline{x}_0 \perp \underline{x}_{uo}) \tag{13}$$

where $\underline{x}_o \in R(\underline{W}_o(0,T))$ is called the *observable component* of $\underline{x}$, and $\underline{x}_{uo} \in N(\underline{W}_o(0,T))$ is its *unobservable component.*

Next we shall see that a state which is observable can be determined by appropriately processing the information available at the output.

5. *COROLLARY.*[†] Suppose that $\underline{W}_o(0,T)$ is nonsingular, the input $\underline{u}(t)$ to the system (4) is zero, and the initial state $\underline{x}(0) = \overline{\underline{x}}$ is unknown. Then $\overline{\underline{x}}$ can be determined by processing the output function $\underline{y}(t)$, $0 \le t \le T$, as follows:

$$\overline{\underline{x}} = \int_0^T \underline{W}_o^{-1}(0,T)\, e^{\underline{A}'\tau}\underline{C}'\underline{y}(\tau)\, d\tau \tag{14}$$

Proof: Using (8) and (6), the integral in (14) is easily seen to yield

$$\begin{aligned}\int_0^T \underline{W}_o^{-1}(0,T)\, e^{\underline{A}'\tau}\underline{C}'\underline{y}(\tau)\, d\tau &= \underline{W}_o^{-1}(0,T) \int_0^T e^{\underline{A}'\tau}\underline{C}'\underline{C}e^{\underline{A}\tau}\overline{\underline{x}}\, d\tau \\ &= \underline{W}_o^{-1}(0,T)\, \underline{W}_o(0,T)\, \overline{\underline{x}} = \overline{\underline{x}}\end{aligned} \tag{15}$$

ΔΔΔ

[†] If $\underline{W}_o$ is singular, the corollary still holds for $\overline{\underline{x}} \in R(\underline{W}_o)$ (see Problem 4). If the input is a known nonzero function, its effect may simply be subtracted before (14) is used (see Problem 5).

The condition established in Theorem 4 for observability is somewhat awkward to use in practice, since it requires evaluation of the integral (6) to determine $\underline{W}_o(0,T)$. The following theorem provides more convenient characterizations of the subspaces of observable and unobservable states.

6. *THEOREM.* If we define the $np \times n$ *observability matrix* for the system (4) to be

$$\underline{P}_o = \begin{bmatrix} \underline{C} \\ \underline{C}\underline{A} \\ \underline{C}\underline{A}^2 \\ \vdots \\ \underline{C}\underline{A}^{n-1} \end{bmatrix} \tag{16}$$

then the subspace of unobservable states is

$$N(\underline{W}_o(0,T)) = N(\underline{P}_o) \tag{17}$$

and the subspace of observable states is

$$R(\underline{W}_o(0,T)) = R(\underline{P}_o') = R\left[\underline{C}' \quad \underline{A}'\underline{C}' \quad \dots \quad \underline{A}'^{n-1}\underline{C}'\right] \tag{18}$$

for any $T > 0$. Thus the system is completely observable if and only if the observability matrix has full rank, that is,

$$\text{rank } (\underline{P}_o) = n \tag{19}$$

Proof: The proof is virtually identical to that of Theorem 10.1.6 and is left to the reader.

ΔΔΔ

Note that the observability matrix $\underline{P}_o$ in (16) does not depend upon the terminal time T (assuming $T > 0$). According to Corollary 5, this means that the state of an observable system may be determined by processing the output on *any* time interval $[0,T]$. Some caution is necessary, however, since for arbitrarily small T the integrand in (14) may be physically unrealizable (see Problem 13).

Theorem 6 provides a more useful test for observability than Theorem 4 does since $\underline{P}_o$ is relatively easy to calculate. In the case of a multiple-output system ($p > 1$), a further improvement results from noting that the $np \times n$ observability matrix $\underline{P}_o$ can be replaced (provided $\underline{C}$ has rank p) by the $(n-p+1)p \times n$ matrix (see Problem 7)

$$\begin{bmatrix} \underline{C} \\ \underline{C}\underline{A} \\ \underline{C}\underline{A}^2 \\ \vdots \\ \underline{C}\underline{A}^{n-p} \end{bmatrix} \tag{20}$$

It may also be convenient in some multiple-output cases to note, as in (10.1.43) and (10.1.44), that

$$R(\underline{P}_o') = R(\underline{P}_o'\underline{P}_o) \qquad N(\underline{P}_o) = N(\underline{P}_o'\underline{P}_o) \tag{21}$$

and that the system is observable if and only if

$$\det \underline{P}_o'\underline{P}_o \neq 0 \tag{22}$$

Because the system (4) is time-invariant, the definitions and results in this section are all valid for an initial time $t_0 \neq 0$ (see Problem 2). The question of whether an unobservable state remains unobservable for all future time is answered by the following lemma (see also Problems 8 and 9).

7. *LEMMA.*[†] If the initial state $\underline{x}(0) = \overline{\underline{x}}$ of the system (4) is unobservable and the input $\underline{u}(t)$ is identically zero, then the state

$$\underline{x}(t) = e^{\underline{A}t}\overline{\underline{x}} \tag{23}$$

remains unobservable for all time.

Proof: A proof analogous to that of Lemma 10.1.7 could easily be given, but this is unnecessary, since $\overline{\underline{x}}$ unobservable implies

$$\underline{y}(t) = \underline{C}e^{\underline{A}t}\overline{\underline{x}} = \underline{C}\underline{x}(t) = \underline{0} \qquad \text{(for all } t \geq 0) \tag{24}$$

ΔΔΔ

[†]This result is a manifestation of the fact that the subspace $N(\underline{P}_o)$ is *invariant* under $\underline{A}$, and hence under $e^{\underline{A}t}$.

We shall now illustrate the above results by means of a few examples.

8. EXAMPLE. In Example 1, using Theorem 6, we can compute

$$\underline{P}_o = \begin{bmatrix} \underline{c}' \\ \underline{c}'\underline{A} \end{bmatrix} = \begin{bmatrix} 1 & -1 \\ -3 & 3 \end{bmatrix} \tag{25}$$

and

$$\mathcal{N}(\underline{P}_o) = \{\underline{x} : \underline{x} \propto \begin{bmatrix} 1 \\ 1 \end{bmatrix}\} = \mathcal{N}(\underline{W}_o(0,T)) \tag{26}$$

$$\mathcal{R}(\underline{P}_o') = \mathcal{N}(\underline{P}_o)^{\perp} = \{\underline{x} : \underline{x} \propto \begin{bmatrix} 1 \\ -1 \end{bmatrix}\} = \mathcal{R}(\underline{W}_o(0,T)) \tag{27}$$

Thus the subspace of unobservable states consists of all those for which $x_1 = x_2$, as indicated in Example 1, and any initial state with $x_1 \neq x_2$ will result in a nonzero output.

The same information may be obtained from Theorem 4, with a slightly greater computational effort. In Example 1 it was found that

$$\underline{c}'e^{\underline{A}\tau} = \begin{bmatrix} e^{-3\tau} & -e^{-3\tau} \end{bmatrix} \tag{28}$$

so that

$$\underline{W}_o(0,T) = \int_0^T e^{\underline{A}'\tau}\,\underline{c}\,\underline{c}'e^{\underline{A}\tau}\,d\tau = \int_0^T \begin{bmatrix} e^{-3\tau} \\ -e^{-3\tau} \end{bmatrix}\begin{bmatrix} e^{-3\tau} & -e^{-3\tau} \end{bmatrix} d\tau$$

$$= \int_0^T e^{-6\tau}\begin{bmatrix} 1 & -1 \\ -1 & 1 \end{bmatrix} d\tau = \frac{1 - e^{-6T}}{6}\begin{bmatrix} 1 & -1 \\ -1 & 1 \end{bmatrix} \tag{29}$$

which verifies (26) and (27).

ΔΔΔ

9. EXAMPLE. We now return to the broom-balancer of Example 10.1.9, illustrated in Fig. 10.1-3. Assuming that the only output available for measurement is z(t), the position of the cart, then the linearized equations governing the system [for $\Theta(t) \simeq 0$] are

$$\dot{\underline{x}} = \underline{A}\underline{x} + \underline{b}\underline{u} = \begin{bmatrix} 0 & 1 & 0 & 0 \\ 0 & 0 & -1 & 0 \\ 0 & 0 & 0 & 1 \\ 0 & 0 & 11 & 0 \end{bmatrix} \underline{x} + \begin{bmatrix} 0 \\ 1 \\ 0 \\ -1 \end{bmatrix} u \tag{30}$$

$$\underline{z} = \underline{c}'\underline{x} = \begin{bmatrix} 1 & 0 & 0 & 0 \end{bmatrix} \underline{x} \tag{31}$$

To check the observability of this system, we compute

$$\underline{P}_o = \begin{bmatrix} \underline{c}' \\ \underline{c}'\underline{A} \\ \underline{c}'\underline{A}^2 \\ \underline{c}'\underline{A}^3 \end{bmatrix} = \begin{bmatrix} 1 & 0 & 0 & 0 \\ 0 & 1 & 0 & 0 \\ 0 & 0 & -1 & 0 \\ 0 & 0 & 0 & -1 \end{bmatrix} \tag{32}$$

Since det $\underline{P}_o = 1$, we see that $\underline{P}_o$ has full rank and by Theorem 6 the system is completely observable.

Thus, by Corollary 5, the values of $\dot{z}$, Θ, and $\dot{\Theta}$ can all be determined by observing z(t) over an arbitrary time interval. Once again, the reader may wish to verify this conclusion empirically (by attempting to balance a broom with his eyes closed).

ΔΔΔ

10. EXAMPLE. In the orbiting satellite of Example 10.1.10, if the measurable outputs are r(t) and Θ(t), then the system is described by the linearized equations

$$\dot{\underline{x}} = \underline{A}\underline{x} + \underline{B}\underline{u} \quad \begin{bmatrix} 0 & 1 & 0 & 0 \\ 3\omega^2 & 0 & 0 & 2\omega \\ 0 & 0 & 0 & 1 \\ 0 & -2\omega & 0 & 0 \end{bmatrix} \underline{x} + \begin{bmatrix} 0 & 0 \\ 1 & 0 \\ 0 & 0 \\ 0 & 1 \end{bmatrix} \underline{u} \tag{33}$$

$$\underline{y} = \begin{bmatrix} r \\ \Theta \end{bmatrix} = \underline{C}\underline{x} = \begin{bmatrix} 1 & 0 & 0 & 0 \\ 0 & 0 & 1 & 0 \end{bmatrix} \underline{x} \tag{34}$$

To determine observability, we make use of (20) and compute

$$\underline{P}_o = \begin{bmatrix} \underline{C} \\ \underline{CA} \\ \underline{CA}^2 \end{bmatrix} = \begin{bmatrix} 1 & 0 & 0 & 0 \\ 0 & 0 & 1 & 0 \\ 0 & 1 & 0 & 0 \\ 0 & 0 & 0 & 1 \\ 3\omega^2 & 0 & 0 & 2\omega \\ 0 & -2\omega & 0 & 0 \end{bmatrix} \tag{35}$$

The first four rows are obviously independent, so that rank$(\underline{P}_o) = 4$ and the system is completely observable.

Now suppose that the tangential measuring device becomes inoperable, so that only radial measurements are available. In this situation we need to know whether the state of the system can be determined from r(t) alone, so we rewrite (34) as

$$r = \underline{c}_r'\underline{x} = \begin{bmatrix} 1 & 0 & 0 & 0 \end{bmatrix}\underline{x} \tag{36}$$

and compute

$$\underline{P}_{or} = \begin{bmatrix} \underline{c}_r' \\ \underline{c}_r'A \\ \underline{c}_r'A^2 \\ \underline{c}_r'A^3 \end{bmatrix} = \begin{bmatrix} 1 & 0 & 0 & 0 \\ 0 & 1 & 0 & 0 \\ 3\omega^2 & 0 & 0 & 2\omega \\ 0 & -\omega^2 & 0 & 0 \end{bmatrix} \tag{37}$$

Since det $\underline{P}_{or} = 0$, the satellite is *not* completely observable using only r. Indeed, the third column of $\underline{P}_{or}$ is zero, so that any state with $\Theta \neq 0$ will belong to $N(\underline{P}_o)$ and hence be indistinguishable from the corresponding state with $\Theta = 0$.

Finally, suppose instead that radial measurements are lost, leaving us with

$$\Theta = \underline{c}_\Theta'\underline{x} = \begin{bmatrix} 0 & 0 & 1 & 0 \end{bmatrix}\underline{x} \tag{38}$$

The observability matrix is now

$$\underline{P}_{o\Theta} = \begin{bmatrix} \underline{c}_\Theta' \\ \underline{c}_\Theta'A \\ \underline{c}_\Theta'A^2 \\ \underline{c}_\Theta'A^3 \end{bmatrix} = \begin{bmatrix} 0 & 0 & 1 & 0 \\ 0 & 0 & 0 & 1 \\ 0 & -2\omega & 0 & 0 \\ -6\omega^3 & 0 & 0 & -4\omega^2 \end{bmatrix} \tag{39}$$

and det $\underline{P}_{o\Theta} = -12\omega^4 \neq 0$, so that the satellite is completely observable using tangential measurements alone.

ΔΔΔ

10.3. DUALITY AND CANONICAL DECOMPOSITION OF THE STATE SPACE

By now the reader will undoubtedly have noticed striking similarities between the conditions for controllability and those for observability. Indeed, these are *dual* concepts, and the similarities can be summarized as follows.

1. THEOREM (Duality). The subspace of *controllable* states of the system

$$\begin{aligned} \dot{\underline{x}} &= \underline{A}\underline{x} + \underline{B}\underline{u} \\ \underline{y} &= \underline{C}\underline{x} \end{aligned} \tag{1}$$

is identical to the subspace of *observable* states of the system[†]

$$\begin{aligned} \dot{\underline{z}} &= -\underline{A}'\underline{z} + \underline{C}'\underline{v} \\ \underline{w} &= \underline{B}'\underline{z} \end{aligned} \tag{2}$$

Similarly, the subspace of *observable* states of (1) is identical to the subspace of *controllable* states of (2). Hence one system is controllable if and only if the other is observable. The theorem also holds if $-\underline{A}'$ in (2) is replaced by $\underline{A}'$.

Proof: This follows immediately from the respective $\underline{W}$ matrices (Theorems 10.1.4 and 10.2.4):

$$\underline{W}_{c1}(0,T) = \int_0^T e^{-\underline{A}\tau}\underline{B}\underline{B}'e^{-\underline{A}'\tau}\, d\tau = \underline{W}_{o2}(0,T) \tag{3}$$

$$\underline{W}_{o1}(0,T) = \int_0^T e^{\underline{A}'\tau}\underline{C}'\underline{C}e^{\underline{A}\tau}\, d\tau = \underline{W}_{c2}(0,T) \tag{4}$$

Replacement of $-\underline{A}'$ by $\underline{A}'$ is requested in Problem 5.

ΔΔΔ

[†]Note that if $\underline{x}$, $\underline{u}$, and $\underline{y}$ are vectors of dimension n, m, and p, respectively, then $\underline{z}$, $\underline{v}$, and $\underline{w}$ have dimension n, p, and m, respectively.

The duality relation between controllability and observability is quite important. It means, in particular, that for every system property involving controllability, there is a corresponding dual property involving observability. Moreover, once the first property is derived or proved, its dual usually follows with no additional effort. We will observe this phenomenon shortly in Theorems 3 and 3', and again in Theorems 10.4.6 and 10.4.6'. Duality will also prove to be very useful when we take up the subject of multivariable feedback control in Chapter 11.

In Sec. 2.5 we introduced the idea of a state transformation (i.e., a change of basis), which allows us to determine alternative but equivalent representations for a system and to study it from different "points of view." Because systems related by state transformations behave identically (recall Theorems 3.4.5 and 4.2.5), we conclude that their controllability and observability properties must coincide:

2. *LEMMA.* The system

$$\begin{aligned} \dot{\underline{x}} &= \underline{A}\underline{x} + \underline{B}\underline{u} \\ \underline{y} &= \underline{C}\underline{x} \end{aligned} \tag{5}$$

is controllable (observable) if and only if the system

$$\begin{aligned} \dot{\hat{\underline{x}}} &= \underline{Q}\underline{A}\underline{Q}^{-1}\hat{\underline{x}} + \underline{Q}\underline{B}\underline{u} \\ \underline{y} &= \underline{C}\underline{Q}^{-1}\underline{x} \end{aligned} \tag{6}$$

is controllable (observable), where $\underline{Q}$ is any nonsingular matrix. Moreover, the subspaces of controllable (observable) states have the same dimension in both cases.

Proof: Left to the reader (Problem 1).

ΔΔΔ

Next we shall use two specific state transformations to reveal the underlying structure imposed upon a system by the concepts of controllability and observability.

Recall the discussion following Theorem 10.1.4, where we remarked that every state $\underline{x}$ of an n-dimensional system could be uniquely expressed as

$$\underline{x} = \underline{x}_c + \underline{x}_{uc} \qquad (\underline{x}_c \perp \underline{x}_{uc}) \tag{7}$$

where $\underline{x}_c$ and $\underline{x}_{uc}$ belong to the orthogonal subspaces of controllable and uncontrollable states, respectively.

To be more specific, if the controllability matrix $\underline{P}_c$ has rank k then we may find a set of vectors $\underline{p}_1, \underline{p}_2, \ldots, \underline{p}_k$ which form a basis for $R(\underline{P}_c)$, the subspace of controllable states. We may also find additional vectors† $\underline{p}_{k+1}, \underline{p}_{k+2}, \ldots, \underline{p}_n$ which form a basis for $R(\underline{P}_c)^{\perp} = N(\underline{P}_c')$, the subspace of uncontrollable states. Thus the $n \times n$ matrix

$$\underline{Q}^{-1} \triangleq [\, \underline{p}_1 \;\; \underline{p}_2 \cdots \;\; \underline{p}_n \,] \tag{8}$$

must be nonsingular (why?).

Next we use the state transformation

$$\underline{Q}\underline{x} = \hat{\underline{x}} = \begin{bmatrix} \hat{x}_1 \\ \vdots \\ \hat{x}_k \\ \hline \hat{x}_{k+1} \\ \vdots \\ \hat{x}_n \end{bmatrix} = \begin{bmatrix} \hat{\underline{x}}_1 \\ \hline \hat{\underline{x}}_2 \end{bmatrix} \tag{9}$$

so that

$$\underline{x} = \underline{Q}^{-1} \begin{bmatrix} \hat{\underline{x}}_1 \\ \hat{\underline{x}}_2 \end{bmatrix} = \sum_{i=1}^{k} \hat{x}_i \underline{p}_i + \sum_{i=k+1}^{n} \hat{x}_i \underline{p}_i = \underline{x}_c + \underline{x}_{uc} \tag{10}$$

---

†The vectors $\underline{p}_{k+1}, \underline{p}_{k+2}, \ldots, \underline{p}_n$ do not necessarily have to be orthogonal to $\underline{p}_1, \underline{p}_2, \ldots, \underline{p}_k$. Theorem 3 will still hold as long as they are chosen so that the matrix in (8) is nonsingular.

This means that in terms of the new state vector $\hat{\underline{x}}$, the subvectors $\hat{\underline{x}}_1$ and $\hat{\underline{x}}_2$ represent the controllable and uncontrollable components of the state, respectively, and the decomposition (7) becomes

$$\hat{\underline{x}} = \hat{\underline{x}}_c + \hat{\underline{x}}_{uc} = \begin{bmatrix} \hat{\underline{x}}_1 \\ \underline{0} \end{bmatrix} + \begin{bmatrix} \underline{0} \\ \hat{\underline{x}}_2 \end{bmatrix} \tag{11}$$

The effect upon the system equations is given in the following theorem.

3. *THEOREM*. If the state transformation (9) is applied to the system

$$\begin{aligned} \dot{\underline{x}} &= \underline{A}\underline{x} + \underline{B}\underline{u} \\ \underline{y} &= \underline{C}\underline{x} \end{aligned} \tag{12}$$

then the resulting state equations have the form

$$\begin{aligned} \begin{bmatrix} \dot{\hat{\underline{x}}}_1 \\ \dot{\hat{\underline{x}}}_2 \end{bmatrix} &= \begin{bmatrix} \hat{\underline{A}}_{11} & \hat{\underline{A}}_{12} \\ \underline{0} & \hat{\underline{A}}_{22} \end{bmatrix} \begin{bmatrix} \hat{\underline{x}}_1 \\ \hat{\underline{x}}_2 \end{bmatrix} + \begin{bmatrix} \hat{\underline{B}}_1 \\ \underline{0} \end{bmatrix} \underline{u} = \underline{Q}\underline{A}\underline{Q}^{-1} \hat{\underline{x}} + \underline{Q}\underline{B}\underline{u} \\ \underline{y} &= \begin{bmatrix} \hat{\underline{C}}_1 & \hat{\underline{C}}_2 \end{bmatrix} \begin{bmatrix} \hat{\underline{x}}_1 \\ \hat{\underline{x}}_2 \end{bmatrix} = \underline{C}\underline{Q}^{-1}\hat{\underline{x}} \end{aligned} \tag{13}$$

where the k-dimensional subsystem

$$\dot{\hat{\underline{x}}}_1 = \hat{\underline{A}}_{11}\hat{\underline{x}}_1 + \hat{\underline{B}}_1\underline{u} + \hat{\underline{A}}_{12}\hat{\underline{x}}_2 \tag{14}$$

is controllable[†] (from $\underline{u}$) and the (n - k)-dimensional subsystem

$$\dot{\hat{\underline{x}}}_2 = \hat{\underline{A}}_{22}\hat{\underline{x}}_2 \tag{15}$$

is entirely uncontrollable. The set of poles of the overall system consists of *controllable* and *uncontrollable* subsets (the eigenvalues of $\hat{\underline{A}}_{11}$ and $\hat{\underline{A}}_{22}$, respectively).

[†]The reader is asked in Problem 7 to verify that the additional driving term $\hat{\underline{A}}_{12}\hat{\underline{x}}_2$ has no effect on controllability.

Proof: It is easy to verify (using Problem 10.1-6) that the subspace $R(\underline{P}_c)$ is *invariant* under $\underline{A}$, that is,

$$\underline{z} \in R(\underline{P}_c) \implies \underline{A}\underline{z} \in R(\underline{P}_c) \tag{16}$$

This means that the first k columns of

$$\underline{A}\underline{Q}^{-1} = [\ \underline{A}\underline{p}_1 \quad \underline{A}\underline{p}_2 \quad \cdots \quad \underline{A}\underline{p}_n\ ] \tag{17}$$

all belong to $R(\underline{P}_c)$.

If we write $\underline{Q}$ in terms of its rows as

$$\underline{Q} = \begin{bmatrix} \underline{q}_1' \\ \underline{q}_2' \\ \vdots \\ \underline{q}_n' \end{bmatrix} \tag{18}$$

then the equation $\underline{Q}\underline{Q}^{-1} = \underline{I}$ means that

$$\underline{q}_i'\underline{p}_j = 0 \qquad \text{(for } i \neq j\text{)} \tag{19}$$

Moreover, since $\underline{p}_1, \underline{p}_2, \ldots, \underline{p}_k$ are a basis for $R(\underline{P}_c)$ we have

$$\underline{q}_i'\underline{z} = 0 \qquad \text{(for } i \geq k+1,\ \underline{z} \in R(\underline{P}_c)\text{)} \tag{20}$$

and more specifically

$$\underline{q}_i'\underline{A}\underline{p}_j = 0 \qquad \text{(for } i \geq k+1,\ \ j \leq k\text{)} \tag{21}$$

Since (21) is an expression for the ij-th element of $\underline{Q}\underline{A}\underline{Q}^{-1}$, we see that the lower left-hand $(n - k) \times k$ block of this matrix vanishes, as indicated in (13).

Since the columns of $\underline{B}$ all belong to $R(\underline{P}_c)$, it also follows from (20) that the last $(n - k)$ rows of

$$\underline{Q}\underline{B} = \begin{bmatrix} \underline{q}_1'\underline{B} \\ \underline{q}_2'\underline{B} \\ \vdots \\ \underline{q}_n'\underline{B} \end{bmatrix} \tag{22}$$

all vanish, as indicated in (13).

Finally, we may compute the controllability matrix of (13),

$$\hat{\underline{P}}_c = \begin{bmatrix} \hat{\underline{B}}_1 & \hat{\underline{A}}_{11}\hat{\underline{B}}_1 & \cdots & \hat{\underline{A}}_{11}^{n-1}\hat{\underline{B}}_1 \\ \underline{0} & \underline{0} & \cdots & \underline{0} \end{bmatrix} = \underline{Q}\underline{P}_c\underline{Q}^{-1} \tag{23}$$

which has rank k by assumption and by Lemma 2. But the zero rows contribute nothing, so the matrix

$$\begin{bmatrix} \hat{\underline{B}}_1 & \hat{\underline{A}}_{11}\hat{\underline{B}}_1 & \cdots & \hat{\underline{A}}_{11}^{n-1}\hat{\underline{B}}_1 \end{bmatrix} \tag{24}$$

must also have rank k, and it follows from the Cayley-Hamilton theorem (see Problem 10.1-6) that

$$\hat{\underline{P}}_{c1} = \begin{bmatrix} \hat{\underline{B}}_1 & \hat{\underline{A}}_{11}\hat{\underline{B}}_1 & \cdots & \hat{\underline{A}}_{11}^{k-1}\hat{\underline{B}}_1 \end{bmatrix} \tag{25}$$

has rank k, i.e., that (14) is controllable.

The uncontrollability of (15) is obvious, since $\hat{\underline{P}}_{c2} = \underline{0}$. The breakdown of poles into two subsets follows from Problem A.3-11.

ΔΔΔ

With the system equations expressed in the form (13), the decomposition into controllable and uncontrollable states has become transparent. The controllable component ($\hat{\underline{x}}_1$) of the state belongs to a controllable subsystem (14) and the uncontrollable component belongs to an uncontrollable subsystem (15) upon which the control $\underline{u}$ has no effect. This state of affairs is shown schematically in Fig. 1.

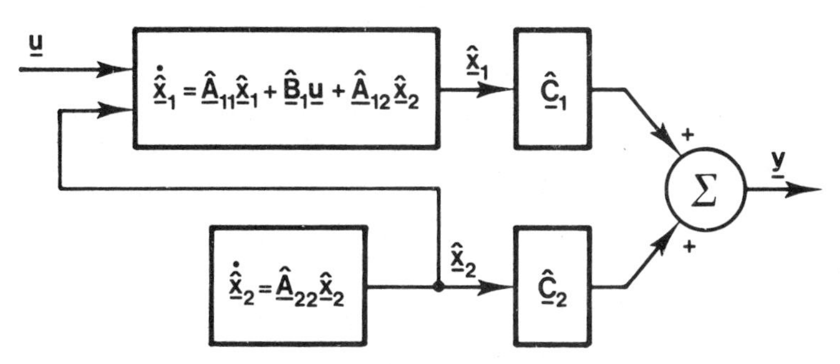

FIG. 1. Decomposition into controllable and uncontrollable components.

In the case of observability, the state $\underline{x}$ of an n-dimensional system can always be decomposed into

$$\underline{x} = \underline{x}_o + \underline{x}_{uo} \qquad \underline{x}_o \perp \underline{x}_{uo} \tag{26}$$

where $\underline{x}_o$ and $\underline{x}_{uo}$ belong to the subspace of observable and unobservable states, respectively.

If the observability matrix $\underline{P}_o$ has rank $n - \ell$, we may find a basis $\{\underline{p}_1, \underline{p}_2, \ldots, \underline{p}_\ell\}$ for $N(\underline{P}_o)$ and a basis $\{\underline{p}_{\ell+1}, \underline{p}_{\ell+2}, \ldots, \underline{p}_n\}$ for $N(\underline{P}_o)^\perp = R(\underline{P}_o')$, to form the nonsingular matrix

$$\underline{Q}^{-1} = [\underline{p}_1 \quad \underline{p}_2 \quad \cdots \quad \underline{p}_n] \tag{27}$$

This defines a state transformation

$$\underline{Qx} = \hat{\underline{x}} = \begin{bmatrix} \hat{x}_1 \\ \vdots \\ \hat{x}_\ell \\ \hline \hat{x}_{\ell+1} \\ \vdots \\ \hat{x}_n \end{bmatrix} = \begin{bmatrix} \hat{\underline{x}}_1 \\ \hline \hat{\underline{x}}_2 \end{bmatrix} \tag{28}$$

and we have the dual of Theorem 3.

3'. THEOREM. If the state transformation (28) is applied to the system

$$\begin{aligned} \dot{\underline{x}} &= \underline{Ax} + \underline{Bu} \\ \underline{y} &= \underline{Cx} \end{aligned} \tag{29}$$

then the resulting state equations have the form

$$\begin{aligned} \begin{bmatrix} \dot{\hat{\underline{x}}}_1 \\ \dot{\hat{\underline{x}}}_2 \end{bmatrix} &= \begin{bmatrix} \hat{\underline{A}}_{11} & \hat{\underline{A}}_{12} \\ \underline{0} & \hat{\underline{A}}_{22} \end{bmatrix} \begin{bmatrix} \hat{\underline{x}}_1 \\ \hat{\underline{x}}_2 \end{bmatrix} + \begin{bmatrix} \hat{\underline{B}}_1 \\ \hat{\underline{B}}_2 \end{bmatrix} \underline{u} = \underline{QAQ}^{-1}\hat{\underline{x}} + \underline{QBu} \\ \underline{y} &= \begin{bmatrix} \underline{0} & \hat{\underline{C}}_2 \end{bmatrix} \begin{bmatrix} \hat{\underline{x}}_1 \\ \hat{\underline{x}}_2 \end{bmatrix} = \underline{CQ}^{-1}\hat{\underline{x}} \end{aligned} \tag{30}$$

where the ℓ-dimensional subsystem

$$\dot{\hat{\underline{x}}}_1 = \hat{\underline{A}}_{11}\hat{\underline{x}}_1 + \hat{\underline{A}}_{12}\hat{\underline{x}}_2 + \hat{\underline{B}}_1\underline{u} \tag{31}$$

is entirely unobservable and the $(n - \ell)$-dimensional subsystem

$$\begin{aligned} \dot{\hat{\underline{x}}}_2 &= \hat{\underline{A}}_{22}\hat{\underline{x}}_2 + \hat{\underline{B}}_2\underline{u} \\ \underline{y} &= \hat{\underline{C}}_2\hat{\underline{x}}_2 \end{aligned} \tag{32}$$

is observable. The set of poles of the overall system consists of *unobservable* and *observable* subsets (the eigenvalues of $\hat{\underline{A}}_{11}$ and $\hat{\underline{A}}_{22}$, respectively).

Proof: Left to the reader (Problem 3). ΔΔΔ

Once again, the decomposition of $\underline{x}$ has become transparent since the observable component ($\underline{x}_2$) belongs to an observable subsystem and the unobservable component ($\underline{x}_1$) belongs to an unobservable subsystem which has no effect upon the output $\underline{y}$. This is shown schematically in Fig. 2.

Theorems 3 and 3' may be combined as follows (also see Problem 6(b)).

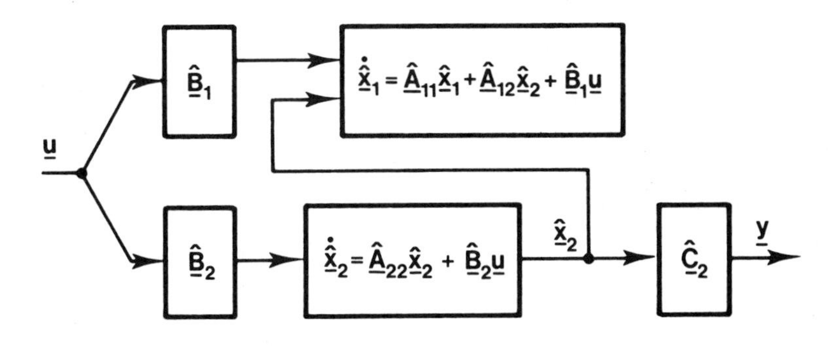

FIG. 2. Decomposition into observable and unobservable components.

4. THEOREM (Canonical decomposition). A linear system can always be represented, by means of a state transformation, in the form

$$\begin{bmatrix} \dot{\underline{x}}_1 \\ \dot{\underline{x}}_2 \\ \dot{\underline{x}}_3 \\ \dot{\underline{x}}_4 \end{bmatrix} = \begin{bmatrix} \underline{A}_{11} & \underline{A}_{12} & \underline{A}_{13} & \underline{A}_{14} \\ \underline{0} & \underline{A}_{22} & \underline{0} & \underline{A}_{24} \\ \underline{0} & \underline{0} & \underline{A}_{33} & \underline{A}_{34} \\ \underline{0} & \underline{0} & \underline{0} & \underline{A}_{44} \end{bmatrix} \begin{bmatrix} \underline{x}_1 \\ \underline{x}_2 \\ \underline{x}_3 \\ \underline{x}_4 \end{bmatrix} + \begin{bmatrix} \underline{B}_1 \\ \underline{B}_2 \\ \underline{0} \\ \underline{0} \end{bmatrix} \underline{u}$$

$$\underline{y} = \begin{bmatrix} \underline{0} & \underline{C}_2 & \underline{0} & \underline{C}_4 \end{bmatrix} \begin{bmatrix} \underline{x}_1 \\ \underline{x}_2 \\ \underline{x}_3 \\ \underline{x}_4 \end{bmatrix} \tag{33}$$

where the controllability and/or observability of the four subsystems is indicated in Fig. 3. The set of system poles breaks up into four corresponding sets.

Proof: Left to the reader (Problem 4).

ΔΔΔ

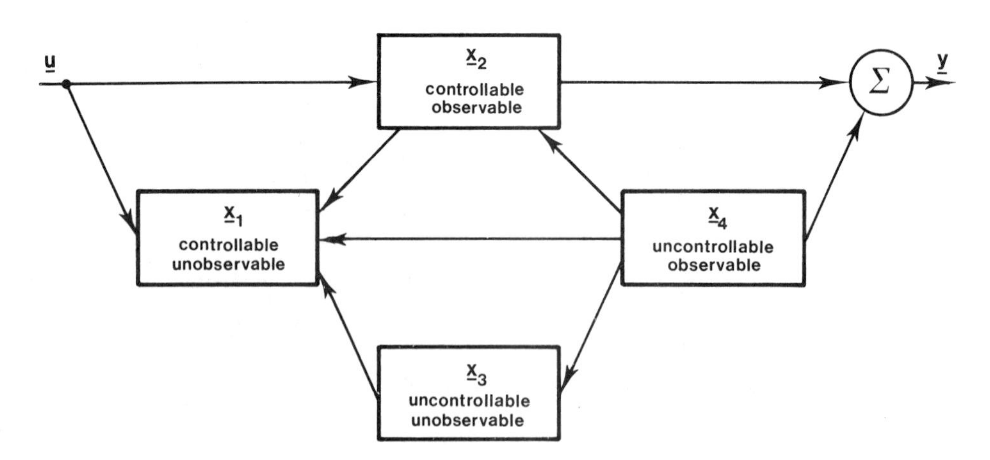

FIG. 3. Canonical decomposition of a linear system.

Theorem 4 and Fig. 3 make explicit the controllability/observability structure of a linear system. The input $\underline{u}$ influences only the state subvectors $\underline{x}_1$ and $\underline{x}_2$, and only $\underline{x}_2$ and $\underline{x}_4$ appear at the output $\underline{y}$. Indeed, we see that the only path from input to output is through the subsystem containing $\underline{x}_2$. This leads us to deduce that input-output descriptions of the system, such as the transfer function and impulse response matrices, represent only the controllable and observable subsystem:

5. *COROLLARY.* The transfer function and impulse response matrices of the system (33) are given by

$$\hat{\underline{H}}(s) = \underline{C}_2(s\underline{I} - \underline{A}_{22})^{-1}\underline{B}_2 \tag{34}$$

and

$$\underline{H}(t) = \underline{C}_2 e^{\underline{A}_{22}t}\underline{B}_2 \tag{35}$$

Proof: This follows by multiplying out

$$\hat{\underline{H}}(s) = \begin{bmatrix} \underline{0} & \underline{C}_2 & \underline{0} & \underline{C}_4 \end{bmatrix} \begin{bmatrix} (s\underline{I} - \underline{A}_{11}) & -\underline{A}_{12} & -\underline{A}_{13} & -\underline{A}_{14} \\ \underline{0} & (s\underline{I} - \underline{A}_{22}) & \underline{0} & -\underline{A}_{24} \\ \underline{0} & \underline{0} & (s\underline{I} - \underline{A}_{33}) & -\underline{A}_{34} \\ \underline{0} & \underline{0} & \underline{0} & (s\underline{I} - \underline{A}_{44}) \end{bmatrix}^{-1} \begin{bmatrix} \underline{B}_1 \\ \underline{B}_2 \\ \underline{0} \\ \underline{0} \end{bmatrix}$$

$$= \begin{bmatrix} \underline{0} & \underline{C}_2 & \underline{0} & \underline{C}_4 \end{bmatrix} \begin{bmatrix} (s\underline{I} - \underline{A}_{11})^{-1} & \times & \times & \times \\ \underline{0} & (s\underline{I} - \underline{A}_{22})^{-1} & \times & \times \\ \underline{0} & \underline{0} & (s\underline{I} - \underline{A}_{33})^{-1} & \times \\ \underline{0} & \underline{0} & \underline{0} & (s\underline{I} - \underline{A}_{44})^{-1} \end{bmatrix} \begin{bmatrix} \underline{B}_1 \\ \underline{B}_2 \\ \underline{0} \\ \underline{0} \end{bmatrix}$$

$$= \begin{bmatrix} \underline{0} & \underline{C}_2 & \underline{0} & \underline{C}_4 \end{bmatrix} \begin{bmatrix} \times \\ (s\underline{I} - \underline{A}_{22})^{-1}\underline{B}_2 \\ \underline{0} \\ \underline{0} \end{bmatrix} = \underline{C}_2(s\underline{I} - \underline{A}_{22})^{-1}\underline{B}_2 \tag{36}$$

Equation (35) is the inverse Laplace transform of (34).

ΔΔΔ

We now illustrate these results with two examples.

6. *EXAMPLE*. Example 10.1.1 had the form

$$\dot{\underline{x}} = \begin{bmatrix} -2 & 1 \\ 1 & -2 \end{bmatrix} \underline{x} + \begin{bmatrix} 1 \\ 1 \end{bmatrix} u = \underline{A}\underline{x} + \underline{b}u$$

$$y = \begin{bmatrix} 0 & 1 \end{bmatrix} \underline{x} = \underline{c}'\underline{x} \tag{37}$$

and we later found that

$$\underline{P}_c = \begin{bmatrix} \underline{b} & \underline{A}\underline{b} \end{bmatrix} = \begin{bmatrix} 1 & -1 \\ 1 & -1 \end{bmatrix} \tag{38}$$

Thus the vectors

$$\underline{p}_1 = \begin{bmatrix} 1 \\ 1 \end{bmatrix} \qquad \underline{p}_2 = \begin{bmatrix} 1 \\ -1 \end{bmatrix} \tag{39}$$

are bases for $R(\underline{P}_c)$ and $N(\underline{P}_c')$, respectively. We form the matrix

$$\underline{Q}^{-1} = \begin{bmatrix} 1 & 1 \\ 1 & -1 \end{bmatrix} \tag{40}$$

and use the state transformation $\hat{\underline{x}} = \underline{Q}\underline{x}$ to write (37) as

$$\dot{\hat{\underline{x}}} = \underline{Q}\underline{A}\underline{Q}^{-1}\hat{\underline{x}} + \underline{Q}\underline{b}u$$

$$y = \underline{c}'\underline{Q}^{-1}\hat{\underline{x}} \tag{41}$$

where

$$\underline{Q}\underline{A}\underline{Q}^{-1} = \begin{bmatrix} \frac{1}{2} & \frac{1}{2} \\ \frac{1}{2} & -\frac{1}{2} \end{bmatrix} \begin{bmatrix} -2 & 1 \\ 1 & -2 \end{bmatrix} \begin{bmatrix} 1 & 1 \\ 1 & -1 \end{bmatrix} = \begin{bmatrix} -1 & 0 \\ 0 & -3 \end{bmatrix}$$

$$\underline{Q}\underline{b} = \begin{bmatrix} \frac{1}{2} & \frac{1}{2} \\ \frac{1}{2} & -\frac{1}{2} \end{bmatrix} \begin{bmatrix} 1 \\ 1 \end{bmatrix} = \begin{bmatrix} 1 \\ 0 \end{bmatrix}$$

$$\underline{c}'\underline{Q}^{-1} = \begin{bmatrix} 0 & 1 \end{bmatrix} \begin{bmatrix} 1 & 1 \\ 1 & -1 \end{bmatrix} = \begin{bmatrix} 1 & -1 \end{bmatrix}$$

This yields the form predicted† by Theorem 3,

$$\dot{\hat{\underline{x}}} = \begin{bmatrix} -1 & 0 \\ 0 & -3 \end{bmatrix} \hat{\underline{x}} + \begin{bmatrix} 1 \\ 0 \end{bmatrix} u \tag{42}$$

$$y = \begin{bmatrix} 1 & -1 \end{bmatrix} \hat{\underline{x}}$$

which is shown in Fig. 4. We may also verify that the transfer function is

$$\begin{aligned} \hat{h}(s) &= \begin{bmatrix} 0 & 1 \end{bmatrix} \begin{bmatrix} s+2 & -1 \\ -1 & s+2 \end{bmatrix}^{-1} \begin{bmatrix} 1 \\ 1 \end{bmatrix} \\ &= \begin{bmatrix} 0 & 1 \end{bmatrix} \begin{bmatrix} \frac{s+2}{(s+1)(s+3)} & \frac{1}{(s+1)(s+3)} \\ \frac{1}{(s+1)(s+3)} & \frac{s+2}{(s+1)(s+3)} \end{bmatrix} \begin{bmatrix} 1 \\ 1 \end{bmatrix} = \frac{1}{s+1} \end{aligned} \tag{43}$$

as can easily be seen from Fig. 4.

ΔΔΔ

---

†This is also the form predicted by Theorem 4, since the system is easily seen to be observable.

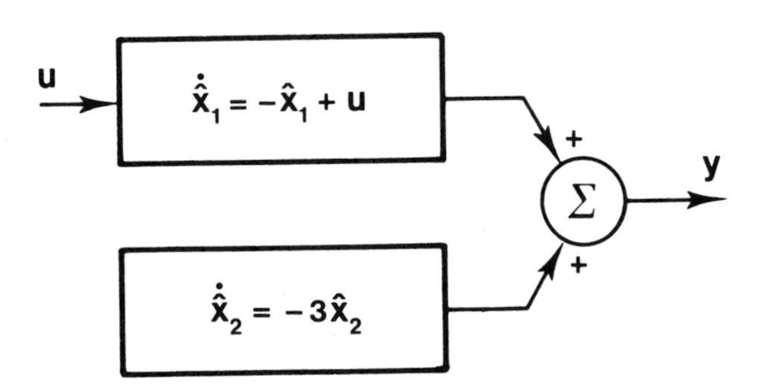

FIG. 4. Decomposition of Example 6.

7. *EXAMPLE*. Example 10.2.1 had the form

$$\dot{\underline{x}} = \begin{bmatrix} -2 & 1 \\ 1 & -2 \end{bmatrix} \underline{x} + \begin{bmatrix} 1 \\ 0 \end{bmatrix} u = \underline{A}\underline{x} + \underline{b}u$$
$$y = \begin{bmatrix} 1 & -1 \end{bmatrix} \underline{x} = \underline{c}'\underline{x} \tag{44}$$

and we later found that

$$\underline{P}_o = \begin{bmatrix} \underline{c}' \\ \underline{c}'\underline{A} \end{bmatrix} = \begin{bmatrix} 1 & -1 \\ -3 & 3 \end{bmatrix} \tag{45}$$

Thus the vectors $\underline{p}_1$ and $\underline{p}_2$ in (39) are bases for $N(\underline{P}_o)$ and $R(\underline{P}_o')$, respectively. Using the same state transformation $\hat{\underline{x}} = \underline{Q}\underline{x}$ as in the previous example, we arrive at the form predicted† by Theorem 3',

$$\dot{\hat{\underline{x}}} = \underline{Q}\underline{A}\underline{Q}^{-1}\hat{\underline{x}} + \underline{Q}\underline{b}u = \begin{bmatrix} -1 & 0 \\ 0 & -3 \end{bmatrix} \hat{\underline{x}} + \begin{bmatrix} \frac{1}{2} \\ \frac{1}{2} \end{bmatrix} u$$
$$y = \underline{c}'\underline{Q}^{-1}\hat{\underline{x}} = \begin{bmatrix} 0 & 2 \end{bmatrix} \hat{\underline{x}} \tag{46}$$

which is shown in Fig. 5. We may also verify that the transfer

†This is also the form predicted by Theorem 4, since the system is easily seen to be controllable.

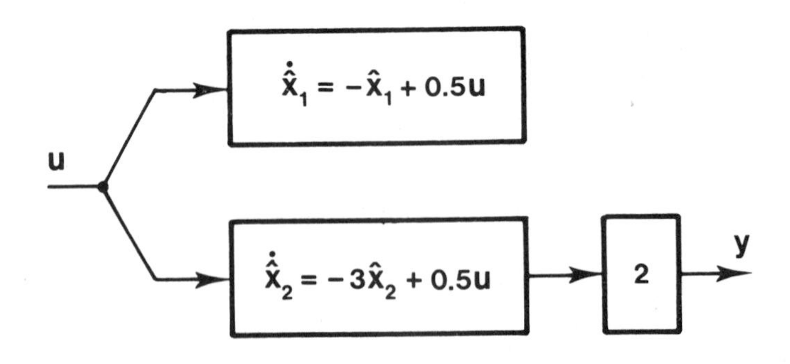

FIG. 5. Decomposition of Example 7.

function is

$$\hat{h}(s) = \begin{bmatrix} 1 & -1 \end{bmatrix} \begin{bmatrix} s+2 & -1 \\ -1 & s+2 \end{bmatrix}^{-1} \begin{bmatrix} 1 \\ 0 \end{bmatrix}$$

$$= \begin{bmatrix} 1 & -1 \end{bmatrix} \begin{bmatrix} \frac{s+2}{(s+1)(s+3)} & \frac{1}{(s+1)(s+3)} \\ \frac{1}{(s+1)(s+3)} & \frac{s+2}{(s+1)(s+3)} \end{bmatrix} \begin{bmatrix} 1 \\ 0 \end{bmatrix} = \frac{1}{s+3} \tag{47}$$

as can easily be seen from Fig. 5.

ΔΔΔ

Another circuit example is given in Problem 6. Note that Theorem 4 and Corollary 5 clarify the relationship between asymptotic stability and BIBO stability, which were defined in Sec. 5.1 (see Problem 8).

10.4. REALIZATIONS AND CANONICAL FORMS

In most of our investigations thus far, the starting point has been a set of differential equations governing a system, such as the state-variable form

$$\begin{aligned} \dot{\underline{x}}(t) &= \underline{A}\underline{x}(t) + \underline{B}\underline{u}(t) \\ \underline{y}(t) &= \underline{C}\underline{x}(t) \end{aligned} \tag{1}$$

where $\underline{x}$, $\underline{u}$, and $\underline{y}$ have dimensions n, m, and p, respectively. We have then proceeded to analyze the system in various ways, often making use of the $p \times m$ transfer function matrix

$$\hat{\underline{H}}(s) = \underline{C}(s\underline{I} - \underline{A})^{-1}\underline{B} \tag{2}$$

or, equivalently, the impulse response matrix

$$\underline{H}(t) = \mathcal{L}^{-1}\{\hat{\underline{H}}(s)\} = \underline{C}e^{\underline{A}t}\underline{B} \tag{3}$$

both of which characterize completely the input-output behavior of the system.

Now suppose instead that we begin with a transfer function $\hat{H}(s)$ (perhaps obtained from measurements of the frequency response of an unknown system) and seek a system of the form (1) which correctly models this input-output behavior. To be more precise, we adopt the following terminology:

1. *DEFINITION*. Given a transfer function matrix $\hat{H}(s)$, a system

$$\dot{x} = Ax + Bu \qquad (4)$$
$$y = Cx$$

is said to be a *realization* of $\hat{H}(s)$ if

$$C(sI - A)^{-1}B = \hat{H}(s) \qquad (5)$$

Equivalently, such a system is said to be a *realization* of an impulse response matrix $H(t)$ if

$$Ce^{At}B = H(t) \qquad (6)$$

ΔΔΔ

Now suppose we seek to realize a $p \times m$ transfer function matrix of the form

$$\hat{H}(s) = \frac{Q_0 + Q_1 s + \dots + Q_{n-1}s^{n-1}}{p_0 + p_1 s + \dots + p_{n-1}s^{n-1} + s^n} \qquad (7)$$

that is, $\hat{H}(s)$ is both *rational* (a ratio of polynomials in s) and *strictly proper* (the degree of the denominator exceeds that of the numerator).

In the scalar case $(m = p = 1)$, this reduces to

$$\hat{h}(s) = \frac{q_0 + q_1 s + \dots + q_{n-1}s^{n-1}}{p_0 + p_1 s + \dots + p_{n-1}s^{n-1} + s^n} \qquad (8)$$

and our problem is equivalent to finding a state-variable description of the n-th order differential equation

$$y^{(n)}(t) + p_{n-1}y^{(n-1)}(t) + \dots + p_1\dot{y}(t) + p_0y(t)$$
$$= q_{n-1}u^{(n-1)}(t) + \dots + q_1\dot{u}(t) + q_0u(t) \tag{9}$$

This was the subject of Sec. 2.2, where two such realizations were presented. One of these (Theorem 2.2.1) is known as the *controllable canonical form* of the state equations,

$$\dot{\underline{x}} = \begin{bmatrix} 0 & 1 & 0 & \dots & 0 & 0 \\ 0 & 0 & 1 & \dots & 0 & 0 \\ 0 & 0 & 0 & \dots & 0 & 0 \\ \dots & \dots & \dots & \dots & \dots & \dots \\ 0 & 0 & 0 & \dots & 1 & 0 \\ 0 & 0 & 0 & \dots & 0 & 1 \\ -p_0 & -p_1 & -p_2 & \dots & -p_{n-2} & -p_{n-1} \end{bmatrix} \underline{x} + \begin{bmatrix} 0 \\ 0 \\ 0 \\ \vdots \\ 0 \\ 0 \\ 1 \end{bmatrix} u \triangleq \underline{A}_c\underline{x} + \underline{b}_c u$$
$$y = \begin{bmatrix} q_0 & q_1 & q_2 & \dots & q_{n-2} & q_{n-1} \end{bmatrix} \underline{x} \triangleq \underline{c}_c'\underline{x} \tag{10}$$

The other (Problem 2.2-2) is the dual (or "transpose") of (10),

$$\dot{\underline{x}} = \begin{bmatrix} 0 & 0 & 0 & \dots & 0 & 0 & -p_0 \\ 1 & 0 & 0 & \dots & 0 & 0 & -p_1 \\ 0 & 1 & 0 & \dots & 0 & 0 & -p_2 \\ \dots & \dots & \dots & \dots & \dots & \dots & \dots \\ 0 & 0 & 0 & \dots & 1 & 0 & -p_{n-2} \\ 0 & 0 & 0 & \dots & 0 & 1 & -p_{n-1} \end{bmatrix} \underline{x} + \begin{bmatrix} q_0 \\ q_1 \\ q_2 \\ \vdots \\ q_{n-2} \\ q_{n-1} \end{bmatrix} u \triangleq \underline{A}_o\underline{x} + \underline{b}_o u \tag{11}$$
$$y = \begin{bmatrix} 0 & 0 & 0 & \dots & 0 & 0 & 1 \end{bmatrix} \underline{x} \triangleq \underline{c}_o'\underline{x}$$

and is known as the *observable canonical form*. The reader should verify (Problem 1) that each of these forms realizes the transfer function (8). The matrices $\underline{A}_c$ and $\underline{A}_o$ are known as *companion matrices*.

Both (10) and (11) lend themselves readily to interpretation in terms of analog computer models[†] of $\hat{h}(s)$. The reader should verify

[†]Note, however, that an actual analog computer circuit in this form may be overly sensitive to small parameter variations and hence unsuitable for simulation.

(a)

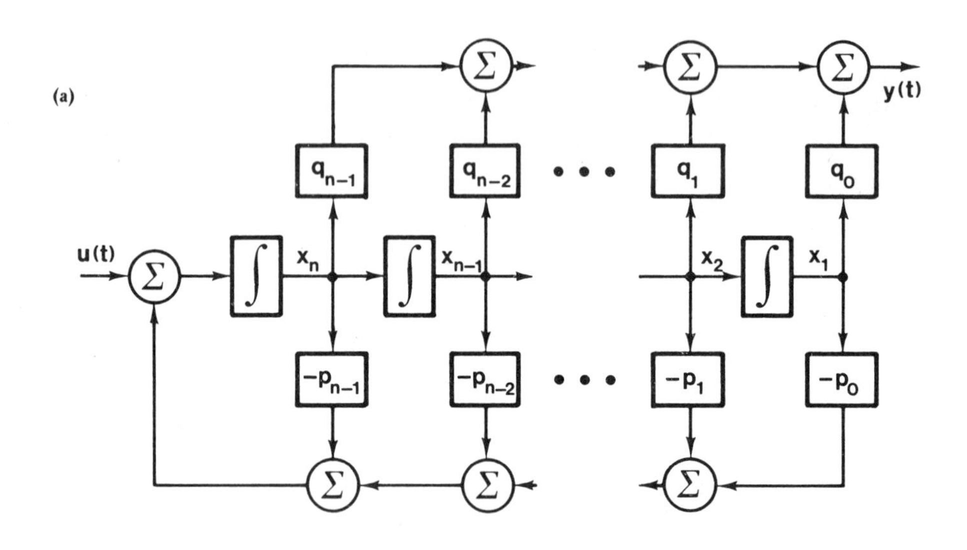

(b)

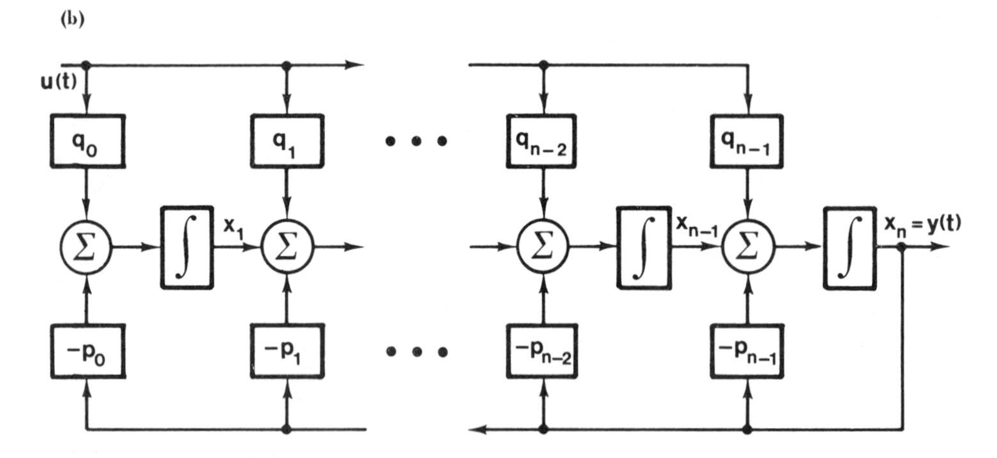

FIG. 1. Analog computer models of standard realizations: (a) controllable canonical form (10) and (b) observable canonical form (11).

that the circuits shown in Fig. 1 correctly model the two canonical forms.

In the multivariable case, we may expand upon (10) and realize $\hat{H}(s)$ with the *controllable canonical form*

$$\dot{x} = \begin{bmatrix} 0_m & I_m & 0_m & \cdots & 0_m \\ 0_m & 0_m & I_m & \cdots & 0_m \\ \cdots & \cdots & \cdots & \cdots & \cdots \\ 0_m & 0_m & 0_m & \cdots & I_m \\ -p_0 I_m & -p_1 I_m & -p_2 I_m & \cdots & -p_{n-1} I_m \end{bmatrix} x + \begin{bmatrix} 0_m \\ 0_m \\ \vdots \\ 0_m \\ I_m \end{bmatrix} u \tag{12}$$

$$y = \begin{bmatrix} Q_0 & Q_1 & Q_2 & \cdots & Q_{n-1} \end{bmatrix} x$$

where 0_m and I_m denote the $m \times m$ zero and identity matrices, respectively. Note that the dimension of this system is nm, where n is the order of the denominator polynomial in (7) and m is the number of inputs. This reduces, of course, to the n-dimensional form (10) when $m = p = 1$.

Similarly, we may expand (11) into the *observable canonical form*

$$\dot{x} = \begin{bmatrix} 0_p & 0_p & 0_p & \cdots & 0_p & -p_0 I_p \\ I_p & 0_p & 0_p & \cdots & 0_p & -p_1 I_p \\ 0_p & I_p & 0_p & \cdots & 0_p & -p_2 I_p \\ \cdots & \cdots & \cdots & \cdots & \cdots & \cdots \\ 0_p & 0_p & 0_p & \cdots & I_p & -p_{n-1} I_p \end{bmatrix} x + \begin{bmatrix} Q_0 \\ Q_1 \\ Q_2 \\ \vdots \\ Q_{n-1} \end{bmatrix} u \tag{13}$$

$$y = \begin{bmatrix} 0_p & 0_p & 0_p & \cdots & 0_p & I_p \end{bmatrix} x$$

Note that (13) is *not* simply the "transpose" of (12); indeed, the dimension in this case is np, where p is the number of outputs, which differs from the dimension of (12) if $p \neq m$. This system reduces to (11) when $p = m = 1$.

The reader should verify (Problem 3) that (13) and (12) both realize $\hat{\underline{H}}(s)$ in (7). This establishes that *a realization exists for any proper, rational transfer function matrix.*

In order to justify the use of the terms controllable and observable to describe the canonical forms (10)-(13) we must verify that these properties in fact hold:

2. *LEMMA.* A system expressed in the controllable canonical form (10) or (12) is always controllable. Similarly, a system expressed in the observable canonical form (11) or (13) is always observable.

Proof: Left to the reader (Problem 4). ΔΔΔ

We have seen that any strictly proper, rational transfer function can be realized with one of the canonical forms (10)-(13). Of course, there are also infinitely many other realizations. Some of these are related to the canonical forms by state transformations, but others are not: indeed, the two realizations (12) and (13) do not even have the same dimension when $m \neq p$.

If two realizations of the same transfer function have different dimensions, then the realization with the smaller dimension is "better" in the sense that it can be modeled (for instance, on an analog computer) with a smaller number of dynamic elements. Thus it will be useful to seek a realization with the smallest possible dimension.

3. *DEFINITION.* A realization

$$\begin{aligned} \dot{\underline{x}} &= \underline{A}\underline{x} + \underline{B}\underline{u} \\ \underline{y} &= \underline{C}\underline{x} \end{aligned} \tag{14}$$

of a transfer function $\hat{\underline{H}}(s)$ is said to be a *minimal realization* if no other realization

$$\begin{aligned} \dot{\tilde{\underline{x}}} &= \tilde{\underline{A}}\tilde{\underline{x}} + \tilde{\underline{B}}\underline{u} \\ \underline{y} &= \tilde{\underline{C}}\tilde{\underline{x}} \end{aligned} \tag{15}$$

of $\hat{\underline{H}}(s)$ exists with $\dim(\tilde{\underline{x}}) < \dim(\underline{x})$. ΔΔΔ

In the previous section (Corollary 10.3.5) we saw that the transfer function represents only the controllable and observable component of any system. This means that any uncontrollable and/or unobservable states can be eliminated without altering the transfer function, and hence the existence of such states implies nonminimality. This link between controllability/observability and minimality works both ways and is fundamental to the theory of linear systems;

4. THEOREM. A realization

$$\begin{aligned} \dot{\underline{x}} &= \underline{A}\underline{x} + \underline{B}\underline{u} \\ \underline{y} &= \underline{C}\underline{x} \end{aligned} \tag{16}$$

of a transfer function $\hat{\underline{H}}(s)$ is minimal if and only if it is both controllable and observable.

Proof: Suppose the system is not both controllable and observable. Then by Corollary 10.3.5, any uncontrollable and/or observable states can be removed without changing the transfer function. Since this results in a realization of lower order, the first realization could not have been minimal.

Conversely, suppose (16) is controllable and observable, with $\underline{A}$ $n \times n$, and let

$$\begin{aligned} \dot{\tilde{\underline{x}}} &= \tilde{\underline{A}}\tilde{\underline{x}} + \tilde{\underline{B}}\underline{u} \\ \underline{y} &= \tilde{\underline{C}}\tilde{\underline{x}} \end{aligned} \tag{17}$$

be another realization of $\hat{\underline{H}}(s)$ with $\tilde{\underline{A}}$ $r \times r$. We shall show that $r \geq n$.

Since (16) and (17) both realize $\hat{\underline{H}}(s)$, they must also realize the same impulse response $\underline{H}(t) = \mathcal{L}^{-1}\{\hat{\underline{H}}(s)\}$, and

$$\begin{aligned} \underline{H}(t-\tau) &= \tilde{\underline{C}}e^{\tilde{\underline{A}}(t-\tau)}\tilde{\underline{B}} = \underline{C}e^{\underline{A}(t-\tau)}\underline{B} \\ &= \tilde{\underline{C}}e^{\tilde{\underline{A}}t}e^{-\tilde{\underline{A}}\tau}\tilde{\underline{B}} = \underline{C}e^{\underline{A}t}e^{-\underline{A}\tau}\underline{B} \end{aligned} \tag{18}$$

for all t and τ. We multiply this equation on the left by $e^{\underline{A}'t}\underline{C}'$ and on the right by $\underline{B}'e^{-\underline{A}'\tau}$, and then integrate with respect to both t and τ to get

$$\begin{aligned}\int_0^T e^{\underline{A}'t}\underline{C}'\tilde{\underline{C}}e^{\tilde{\underline{A}}t}\,dt \int_0^T e^{-\tilde{\underline{A}}\tau}\tilde{\underline{B}}\underline{B}'e^{-\underline{A}'\tau}\,d\tau \\ = \int_0^T e^{\underline{A}'t}\underline{C}'\underline{C}e^{\underline{A}t}\,dt \int_0^T e^{-\underline{A}\tau}\underline{B}\underline{B}'e^{-\underline{A}'\tau}\,d\tau\end{aligned} \tag{19}$$

Recognizing $\underline{W}_o$ and $\underline{W}_c$ (from Theorems 10.1.4 and 10.2.4) on the right, and defining $\underline{V}_o$ and $\underline{V}_c$ on the left, we have

$$\underline{V}_o(0,T)\ \underline{V}_c(0,T) = \underline{W}_o(0,T)\ \underline{W}_c(0,T) \tag{20}$$

Both $\underline{W}_o$ and $\underline{W}_c$ are nonsingular because (16) is controllable and observable, so that the right-hand side is an $n \times n$ matrix of rank n. The matrices $\underline{V}_o$ and $\underline{V}_c$, however, have dimension $n \times r$ and $r \times n$, respectively, and according to Problems A.5-3 and A.5-11 their product must have rank $\leq \min\{n,r\}$. We conclude that $r \geq n$.

ΔΔΔ

With this result, one can easily propose a procedure for determining a minimal realization of any strictly proper, rational $p \times m$ transfer function $\hat{\underline{H}}(s)$. First, construct the nm-dimensional controllable canonical form (12). Next, apply the state transformation of Theorem 10.3.3' and reduce the dimension of the system by eliminating any unobservable states. This results in a controllable and observable realization which, by the theorem above, must be minimal. Alternatively (and preferably, if $p < m$), construct the np-dimensional observable canonical form (13) and use Theorem 10.3.3 to eliminate any uncontrollable states.

Although this procedure is straightforward, the computational requirements can be considerable, and some effort has gone into finding more efficient algorithms. For example, see Refs. 39, 98, 205, 265, 299, 322, 336, and 363.

Having shown that controllability and observability are synonymous with minimality, we now establish that all minimal realizations of the same transfer function are equivalent, in the sense that they are related by state transformations.

5. *THEOREM.*[†] If the systems

$$\dot{\underline{x}} = \underline{A}\underline{x} + \underline{B}u \qquad \dot{\tilde{\underline{x}}} = \tilde{\underline{A}}\tilde{\underline{x}} + \tilde{\underline{B}}u$$
$$\underline{y} = \underline{C}\underline{x} \qquad \underline{y} = \tilde{\underline{C}}\tilde{\underline{x}} \tag{21}$$

are two *minimal* realizations of the same transfer function $\hat{\underline{H}}(s)$, then there exists a state transformation $\tilde{\underline{x}} = \underline{Q}\underline{x}$ relating them, that is,

$$\tilde{\underline{A}} = \underline{Q}\underline{A}\underline{Q}^{-1} \qquad \tilde{\underline{B}} = \underline{Q}\underline{B} \qquad \tilde{\underline{C}} = \underline{C}\underline{Q}^{-1} \tag{22}$$

Proof: Both systems have the same transfer function $\hat{\underline{H}}(s)$, and hence the same impulse response $\underline{H}(t)$, so that

$$\begin{aligned}\underline{H}(\sigma + t - \tau) &= \tilde{\underline{C}}e^{\tilde{\underline{A}}(\sigma + t - \tau)}\tilde{\underline{B}} = \underline{C}e^{\underline{A}(\sigma + t - \tau)}\underline{B} \\ &= \tilde{\underline{C}}e^{\tilde{\underline{A}}\sigma}e^{\tilde{\underline{A}}t}e^{-\tilde{\underline{A}}\tau}\tilde{\underline{B}} = \underline{C}e^{\underline{A}\sigma}e^{\underline{A}t}e^{-\underline{A}\tau}\underline{B}\end{aligned} \tag{23}$$

for all σ, t, and τ. Proceeding as in the previous proof, we multiply on the left by $e^{\underline{A}'\sigma}\underline{C}'$ and on the right by $\underline{B}'e^{-\underline{A}'\tau}$, then integrate with respect to both σ and τ to get

$$\begin{aligned}&\int_0^T e^{\underline{A}'\sigma}\underline{C}'\tilde{\underline{C}}e^{\tilde{\underline{A}}\sigma}\,d\sigma\; e^{\tilde{\underline{A}}t}\int_0^T e^{-\tilde{\underline{A}}\tau}\tilde{\underline{B}}\underline{B}'e^{-\underline{A}'\tau}\,d\tau \\ &\qquad = \int_0^T e^{\underline{A}'\sigma}\underline{C}'\underline{C}e^{\underline{A}\sigma}\,d\sigma\; e^{\underline{A}t}\int_0^T e^{-\underline{A}\tau}\underline{B}\underline{B}'e^{-\underline{A}'\tau}\,d\tau\end{aligned} \tag{24}$$

Recognizing $\underline{W}_o(0,T)$ and $\underline{W}_c(0,T)$, and defining the $n \times n$ matrices $\underline{V}_o(0,T)$ and $\underline{V}_c(0,T)$ as before, we find

$$\underline{V}_o(0,T)\; e^{\tilde{\underline{A}}t}\underline{V}_c(0,T) = \underline{W}_o(0,T)\; e^{\underline{A}t}\underline{W}_c(0,T) \tag{25}$$

Both systems are minimal, and hence (by Theorem 4) controllable and observable, so that $\underline{W}_o$ and $\underline{W}_c$ are invertible and

$$e^{\underline{A}t} = (\underline{W}_o^{-1}\underline{V}_o)\; e^{\tilde{\underline{A}}t}(\underline{V}_c\underline{W}_c^{-1}) = \underline{P}e^{\tilde{\underline{A}}t}\underline{Q} \tag{26}$$

for all t, where $\underline{P} \triangleq \underline{W}_o^{-1}\underline{V}_o$ and $\underline{Q} \triangleq \underline{V}_c\underline{W}_c^{-1}$. At $t = 0$ this yields

[†]Also see Problem 19.

$\underline{P} = \underline{Q}^{-1}$, and from (3.2-33),

$$e^{\underline{A}t} = \underline{Q}^{-1} e^{\tilde{\underline{A}}t} \underline{Q} = e^{\underline{Q}^{-1}\tilde{\underline{A}}\underline{Q}t} \qquad (27)$$

for all t. From this we conclude that

$$\underline{A} = \underline{Q}^{-1}\tilde{\underline{A}}\underline{Q} \qquad (28)$$

Returning to (23), we set $t = \tau = 0$, multiply on the left by $e^{\underline{A}'\sigma}\underline{C}'$, and integrate to get

$$\underline{V}_o(0,T)\tilde{\underline{B}} = \underline{W}_o(0,T)\underline{B} \qquad (29)$$

or

$$\underline{B} = (\underline{W}_o^{-1}\underline{V}_o)\tilde{\underline{B}} = \underline{Q}^{-1}\tilde{\underline{B}} \qquad (30)$$

Similarly, we set $\sigma = t = 0$ in (23), multiply on the right by $\underline{B}'e^{-\underline{A}'\tau}$, and integrate to get

$$\tilde{\underline{C}}\,\underline{V}_c(0,T) = \underline{C}\,\underline{W}_c(0,T) \qquad (31)$$

or

$$\underline{C} = \tilde{\underline{C}}(\underline{V}_c\underline{W}_c^{-1}) = \tilde{\underline{C}}\underline{Q} \qquad (32)$$

ΔΔΔ

Since Theorem 5 has established that all minimal realizations of a transfer function are related by state transformations, we might ask under what conditions a system will be so related to one of the canonical forms. For the single-input and/or single-output cases, the answer is provided by the following two theorems.

6. *THEOREM*. For any controllable single-input system

$$\begin{aligned} \dot{\underline{x}} &= \underline{A}\underline{x} + \underline{b}u \\ \underline{y} &= \underline{C}\underline{x} \end{aligned} \qquad (33)$$

there exists a state transformation $\tilde{\underline{x}} = \underline{Q}\underline{x}$ which puts the system into controllable canonical form,

$$\dot{\tilde{\underline{x}}} = \underline{QAQ}^{-1}\tilde{\underline{x}} + \underline{Qb}u = \underline{A}_c\tilde{\underline{x}} + \underline{b}_c u$$
$$\underline{y} = \underline{CQ}^{-1}\tilde{\underline{x}} = \tilde{\underline{C}}\tilde{\underline{x}} \tag{34}$$

where $\underline{A}_c$ and $\underline{b}_c$ were defined in (10).

Proof: Replace the system (33) temporarily by

$$\dot{\underline{x}} = \underline{Ax} + \underline{b}u$$
$$\underline{y}_1 = \underline{Ix} \tag{35}$$

which must be both controllable and observable, and hence a minimal realization of

$$\hat{\underline{h}}_1(s) = (s\underline{I} - \underline{A})^{-1}\underline{b} \tag{36}$$

The controllable canonical form (12) for $\hat{\underline{h}}_1(s)$ is

$$\dot{\tilde{\underline{x}}} = \underline{A}_c\tilde{\underline{x}} + \underline{b}_c u$$
$$\underline{y}_1 = \underline{C}_1\tilde{\underline{x}} \tag{37}$$

where $\underline{A}_c$ and $\underline{b}_c$ are as in (10) and we are not concerned with the form of $\underline{C}_1$. Since (35) and (37) have the same dimension, they are both minimal realizations of $\hat{\underline{h}}_1(s)$, and by Theorem 5 there is a state transformation $\tilde{\underline{x}} = \underline{Qx}$ relating them, that is,

$$\underline{A}_c = \underline{QAQ}^{-1} \tag{38}$$

$$\underline{b}_c = \underline{Qb} \tag{39}$$

$$\underline{C}_1 = \underline{IQ}^{-1} = \underline{Q}^{-1} \tag{40}$$

Finally, we may reinstate the original system (33) in place of (35) without affecting the validity of (38) and (39), so we simply replace (40) by

$$\tilde{\underline{C}} = \underline{CQ}^{-1} \tag{41}$$

The detailed form of $\tilde{\underline{C}}$ is unimportant.

ΔΔΔ

6'. *THEOREM.* For any observable single-output system

$$\dot{\underline{x}} = \underline{A}\underline{x} + \underline{B}\underline{u}$$
$$y = \underline{c}'\underline{x} \tag{42}$$

there exists a state transformation $\tilde{\underline{x}} = \underline{Q}\underline{x}$ which puts the system into observable canonical form,

$$\dot{\tilde{\underline{x}}} = \underline{Q}\underline{A}\underline{Q}^{-1}\tilde{\underline{x}} + \underline{Q}\underline{B}\underline{u} = \underline{A}_o\tilde{\underline{x}} + \tilde{\underline{B}}\underline{u}$$
$$y = \underline{c}'\underline{Q}^{-1}\tilde{\underline{x}} = \underline{c}_o'\tilde{\underline{x}} \tag{43}$$

where $\underline{A}_o$ and $\underline{c}_o'$ were defined in (11).

Proof: This is the dual of Theorem 6 (see Problem 11).

ΔΔΔ

These theorems establish the *existence* of state transformations to controllable and observable canonical form. Determination of the matrix $\underline{Q}$ for transforming a particular system is another matter, and has generated a large amount of correspondence in the recent literature. Expressions for $\underline{Q}$ are given in Problem 9.

With multiple inputs and outputs, transformation to a canonical form becomes somewhat more complicated (see Problem 10). In particular, the forms (12) and (13) are of little use for this purpose because their dimensions are usually too large. One controllable canonical form which has been used extensively for multi-input systems is shown in (44). Here the system has m inputs, and the m diagonal blocks have dimensions $\nu_1, \nu_2, \ldots, \nu_m$, respectively, which sum to n, the dimension of the system. Nonzero elements are indicated by ×, and we see that each diagonal block is a companion matrix of the same form as $\underline{A}_c$ in (10). Because nonzero elements occur only in the bottom rows of the off-diagonal blocks, this system consists of m single-input subsystems of the form (10) which are coupled only through their inputs, as shown in Fig. 2 (See Problem 16).

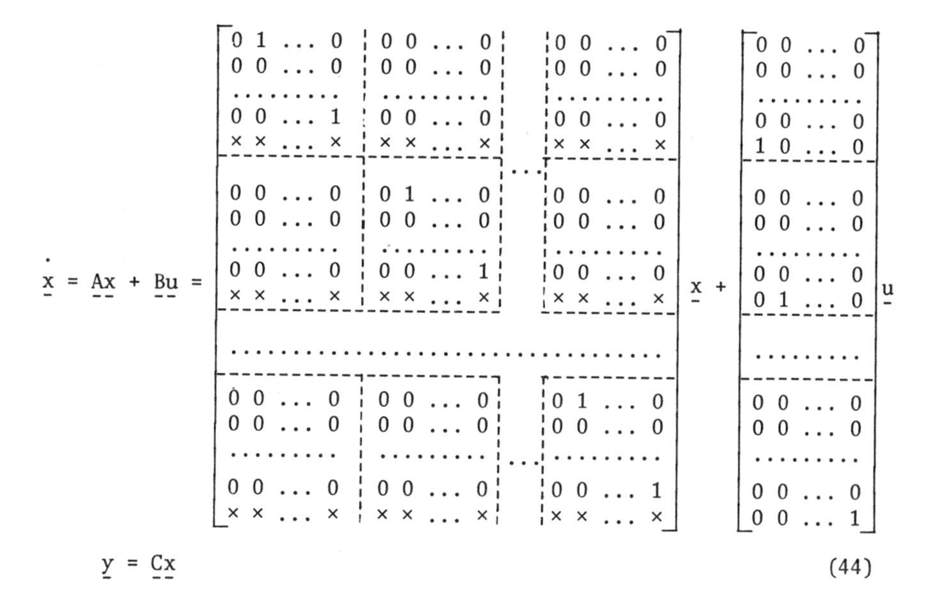

$$\dot{\underline{x}} = \underline{A}\underline{x} + \underline{B}\underline{u} = \left[\begin{array}{ccc|ccc|c|ccc} 0\ 1 & \cdots & 0 & 0\ 0 & \cdots & 0 & & 0\ 0 & \cdots & 0 \\ 0\ 0 & \cdots & 0 & 0\ 0 & \cdots & 0 & & 0\ 0 & \cdots & 0 \\ \cdots & \cdots & \cdots & \cdots & \cdots & \cdots & & \cdots & \cdots & \cdots \\ 0\ 0 & \cdots & 1 & 0\ 0 & \cdots & 0 & & 0\ 0 & \cdots & 0 \\ \times\ \times & \cdots & \times & \times\ \times & \cdots & \times & \cdots & \times\ \times & \cdots & \times \\ \hline 0\ 0 & \cdots & 0 & 0\ 1 & \cdots & 0 & & 0\ 0 & \cdots & 0 \\ 0\ 0 & \cdots & 0 & 0\ 0 & \cdots & 0 & & 0\ 0 & \cdots & 0 \\ \cdots & \cdots & \cdots & \cdots & \cdots & \cdots & & \cdots & \cdots & \cdots \\ 0\ 0 & \cdots & 0 & 0\ 0 & \cdots & 1 & & 0\ 0 & \cdots & 0 \\ \times\ \times & \cdots & \times & \times\ \times & \cdots & \times & & \times\ \times & \cdots & \times \\ \hline \cdots & \cdots & \cdots & \cdots & \cdots & \cdots & \cdots & \cdots & \cdots & \cdots \\ \hline 0\ 0 & \cdots & 0 & 0\ 0 & \cdots & 0 & & 0\ 1 & \cdots & 0 \\ 0\ 0 & \cdots & 0 & 0\ 0 & \cdots & 0 & & 0\ 0 & \cdots & 0 \\ \cdots & \cdots & \cdots & \cdots & \cdots & \cdots & \cdots & \cdots & \cdots & \cdots \\ 0\ 0 & \cdots & 0 & 0\ 0 & \cdots & 0 & & 0\ 0 & \cdots & 1 \\ \times\ \times & \cdots & \times & \times\ \times & \cdots & \times & & \times\ \times & \cdots & \times \end{array}\right] \underline{x} + \left[\begin{array}{ccc} 0\ 0 & \cdots & 0 \\ 0\ 0 & \cdots & 0 \\ \cdots & \cdots & \cdots \\ 0\ 0 & \cdots & 0 \\ 1\ 0 & \cdots & 0 \\ \hline 0\ 0 & \cdots & 0 \\ 0\ 0 & \cdots & 0 \\ \cdots & \cdots & \cdots \\ 0\ 0 & \cdots & 0 \\ 0\ 1 & \cdots & 0 \\ \hline \cdots & \cdots & \cdots \\ \hline 0\ 0 & \cdots & 0 \\ 0\ 0 & \cdots & 0 \\ \cdots & \cdots & \cdots \\ 0\ 0 & \cdots & 0 \\ 0\ 0 & \cdots & 1 \end{array}\right] \underline{u}$$

$$\underline{y} = \underline{C}\underline{x} \qquad (44)$$

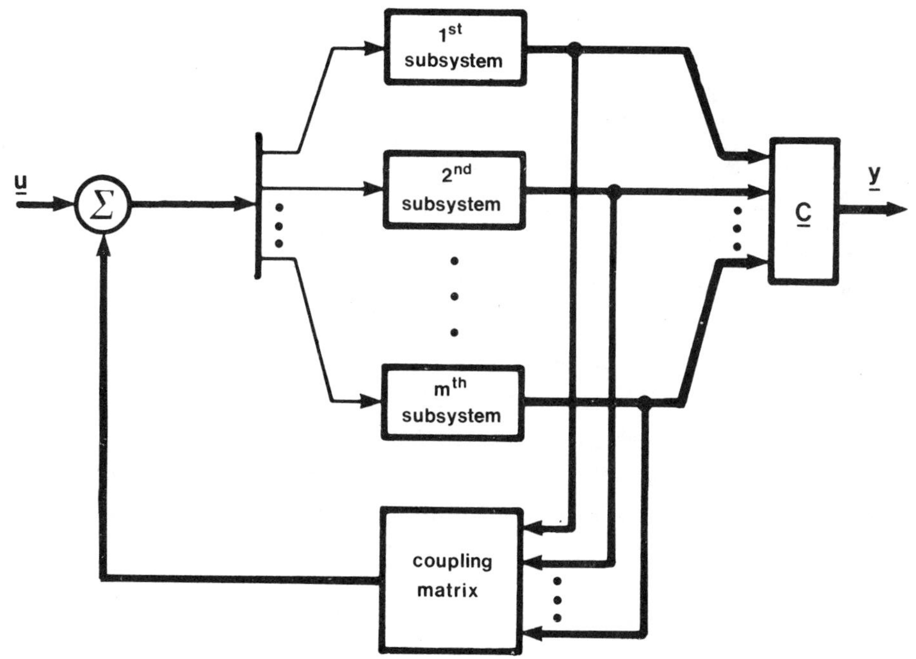

FIG. 2. Multi-input controllable canonical form. (Heavy lines indicate vector quantities; light lines are scalars.)

The corresponding multi-output observable canonical form for a system with p outputs is shown below:

$$\dot{\underline{x}} = \underline{A}\underline{x} + \underline{B}\underline{u} = \left[\begin{array}{c|c|c|c}
0\ 0\ \ldots\ 0\ \times & 0\ 0\ \ldots\ 0\ \times & & 0\ 0\ \ldots\ 0\ \times \\
1\ 0\ \ldots\ 0\ \times & 0\ 0\ \ldots\ 0\ \times & & 0\ 0\ \ldots\ 0\ \times \\
\ldots\ldots\ldots & \ldots\ldots\ldots & & \ldots\ldots\ldots \\
0\ 0\ \ldots\ 1\ \times & 0\ 0\ \ldots\ 0\ \times & & 0\ 0\ \ldots\ 0\ \times \\
\hline
0\ 0\ \ldots\ 0\ \times & 0\ 0\ \ldots\ 0\ \times & \cdots & 0\ 0\ \ldots\ 0\ \times \\
0\ 0\ \ldots\ 0\ \times & 1\ 0\ \ldots\ 0\ \times & & 0\ 0\ \ldots\ 0\ \times \\
\ldots\ldots\ldots & \ldots\ldots\ldots & & \ldots\ldots\ldots \\
0\ 0\ \ldots\ 0\ \times & 0\ 0\ \ldots\ 1\ \times & & 0\ 0\ \ldots\ 0\ \times \\
\hline
\multicolumn{4}{c}{\ldots\ldots\ldots\ldots\ldots\ldots\ldots\ldots\ldots\ldots\ldots\ldots} \\
\hline
0\ 0\ \ldots\ 0\ \times & 0\ 0\ \ldots\ 0\ \times & & 0\ 0\ \ldots\ 0\ \times \\
0\ 0\ \ldots\ 0\ \times & 0\ 0\ \ldots\ 0\ \times & \cdots & 1\ 0\ \ldots\ 0\ \times \\
\ldots\ldots\ldots & \ldots\ldots\ldots & & \ldots\ldots\ldots \\
0\ 0\ \ldots\ 0\ \times & 0\ 0\ \ldots\ 0\ \times & & 0\ 0\ \ldots\ 1\ \times
\end{array}\right] \underline{x} + \underline{B}\underline{u} \tag{45}$$

$$\underline{y} = \underline{C}\underline{x} = \left[\begin{array}{c|c|c|c}
0\ 0\ \ldots\ 0\ 1 & 0\ 0\ \ldots\ 0\ 0 & & 0\ 0\ \ldots\ 0\ 0 \\
0\ 0\ \ldots\ 0\ 0 & 0\ 0\ \ldots\ 0\ 1 & \ldots & 0\ 0\ \ldots\ 0\ 0 \\
\ldots\ldots\ldots & \ldots\ldots\ldots & & \ldots\ldots\ldots \\
0\ 0\ \ldots\ 0\ 0 & 0\ 0\ \ldots\ 0\ 0 & & 0\ 0\ \ldots\ 0\ 1
\end{array}\right] \underline{x}$$

Here the p diagonal blocks are companion matrices of the form $\underline{A}_o$, and the system consists of p single-output subsystems of the form (11) which are coupled only through their outputs, as shown in Fig. 3.

For any controllable system there exists a state transformation which puts it into the form (44). Dually (See Problem 18), any observable system can be put into the form (45). Various transformations to these and other related canonical forms may be found in the literature [39, 210, 235, 292, 363].

We are now in a position to examine the implications of a *pole-zero cancellation* in a transfer function, a subject which was discussed at the end of Sec. 9.3. We begin with a simple example in order to illustrate this phenomenon and the problems which may arise.

7. *EXAMPLE*. Given an unstable system with transfer function

$$\hat{h}_1(s) = \frac{1}{s - 1} \tag{46}$$

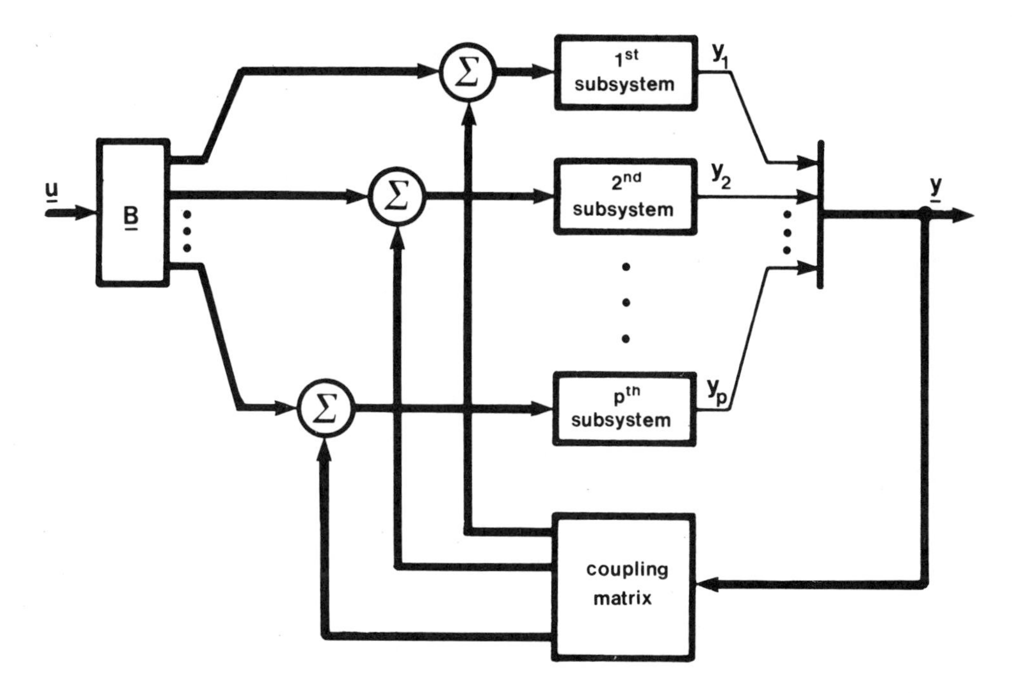

FIG. 3. Multi-output observable canonical form (heavy lines indicate vector quantities; light lines are scalars.)

suppose we attempt to stabilize it by cascading a second system with transfer function

$$\hat{h}_2(s) = \frac{s - 1}{(s + 1)(s + 2)} \tag{47}$$

as shown in Fig. 4(a). The resulting system, shown in Fig. 4(b), has the transfer function

$$\hat{h}_1(s)\hat{h}_2(s) = \frac{1}{(s + 1)(s + 2)} \tag{48}$$

Since the unstable pole at $s = +1$ has been cancelled out, one is tempted to assume that the new system is stable.† Indeed, it is BIBO stable (recall Definition 5.1.5 and Theorem 5.1.6), but we shall see that in a more fundamental sense, such a system can never be stable.

†We ignore the impossibility of constructing a realization of $h_2(s)$ with sufficient precision to insure perfect cancellation.

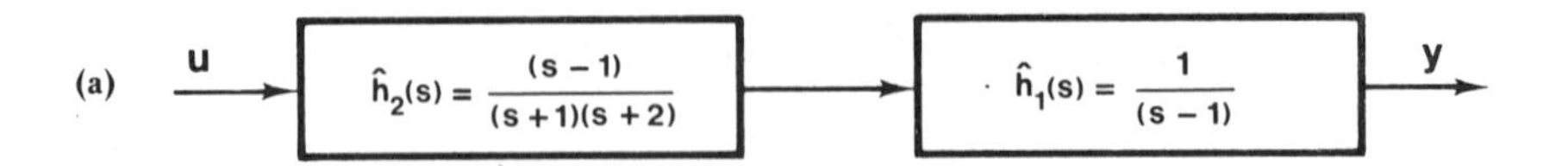

(b)

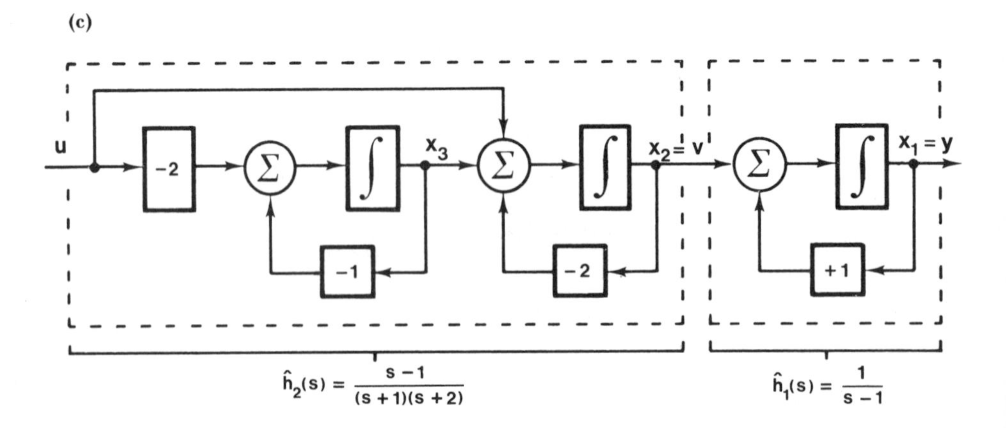

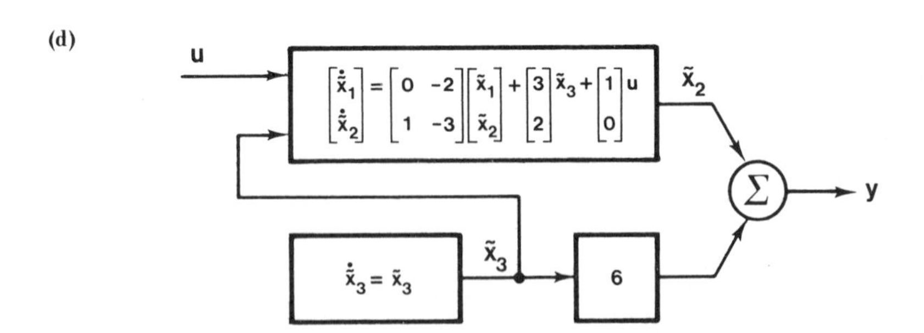

FIG. 4. System of Example 7: (a) cascaded systems, (b) input-output equivalent system, (c) analog computer implementation, and (d) decomposition into controllable and uncontrollable components.

Consider the analog computer implementations of the two cascaded systems shown in Fig. 4(c). The reader should verify that these realize $\hat{h}_2(s)$ and $\hat{h}_1(s)$, respectively (Problem 12). Designating the outputs of the integrators as states, the differential equations governing the system are

$$\begin{aligned} \dot{x}_1(t) &= x_1(t) + v(t) \\ y(t) &= x_1(t) \end{aligned} \tag{49}$$

and

$$\begin{aligned} \dot{x}_2(t) &= -2x_2(t) + x_3(t) + u(t) \\ \dot{x}_3(t) &= -x_3(t) - 2u(t) \\ v(t) &= x_2(t) \end{aligned} \tag{50}$$

Combining these, we have

$$\begin{aligned} \dot{\underline{x}} &= \begin{bmatrix} 1 & 1 & 0 \\ 0 & -2 & 1 \\ 0 & 0 & -1 \end{bmatrix} \underline{x} + \begin{bmatrix} 0 \\ 1 \\ -2 \end{bmatrix} u \\ y &= \begin{bmatrix} 1 & 0 & 0 \end{bmatrix} \underline{x} \end{aligned} \tag{51}$$

and the transfer function is as predicted,

$$\begin{aligned} h(s) &= \begin{bmatrix} 1 & 0 & 0 \end{bmatrix} \begin{bmatrix} s-1 & -1 & 0 \\ 0 & s+2 & -1 \\ 0 & 0 & s+1 \end{bmatrix}^{-1} \begin{bmatrix} 0 \\ 1 \\ -2 \end{bmatrix} \\ &= \begin{bmatrix} 1 & 0 & 0 \end{bmatrix} \begin{bmatrix} \frac{1}{s-1} & \frac{1}{(s-1)(s+2)} & \frac{1}{(s+1)(s-1)(s+2)} \\ 0 & \frac{1}{s+2} & \frac{1}{(s+1)(s+2)} \\ 0 & 0 & \frac{1}{s+1} \end{bmatrix} \begin{bmatrix} 0 \\ 1 \\ -2 \end{bmatrix} \\ &= \begin{bmatrix} \frac{1}{s-1} & \frac{1}{(s-1)(s+2)} & \frac{1}{(s+1)(s-1)(s+2)} \end{bmatrix} \begin{bmatrix} 0 \\ 1 \\ -2 \end{bmatrix} = \frac{1}{(s+1)(s+2)} \end{aligned} \tag{52}$$

Thus an input-output analysis (via the transfer function) seems to imply stability, but the circumstances are somewhat suspicious, as this is a third-order system with only a second-order transfer function. Since we saw in the previous section that uncontrollable and/or unobservable components of a system are always "hidden" from the transfer function, we check both of these conditions:

$$\underline{P}_c = \begin{bmatrix} 0 & 1 & -3 \\ 1 & -4 & 10 \\ -2 & 2 & -2 \end{bmatrix} \qquad \underline{P}_o = \begin{bmatrix} 1 & 0 & 0 \\ 1 & 1 & 0 \\ 1 & -1 & 1 \end{bmatrix} \tag{53}$$

The system is observable since det $\underline{P}_o = 1 \neq 0$, but det $\underline{P}_c = 0$ indicates the existence of uncontrollable states. Indeed, it is not hard to establish that

$$R(\underline{P}_c) = \text{span}\{\begin{bmatrix} 0 \\ 1 \\ -2 \end{bmatrix}, \begin{bmatrix} 1 \\ -4 \\ 2 \end{bmatrix}\} \quad \text{and} \quad R(\underline{P}_c)^{\perp} = \text{span}\{\begin{bmatrix} 6 \\ 2 \\ 1 \end{bmatrix}\} \tag{54}$$

are the subspaces of controllable and uncontrollable states, respectively. Using the transformation (10.3-9)

$$\underline{Q}^{-1} = \begin{bmatrix} 0 & 1 & 6 \\ 1 & -4 & 2 \\ -2 & 2 & 1 \end{bmatrix} \qquad \underline{Q} = \begin{bmatrix} \frac{8}{41} & \frac{-11}{41} & \frac{-26}{41} \\ \frac{5}{41} & \frac{-12}{41} & \frac{-6}{41} \\ \frac{6}{41} & \frac{2}{41} & \frac{1}{41} \end{bmatrix} \tag{55}$$

we may decompose (51) into

$$\begin{aligned} \dot{\tilde{\underline{x}}} &= \underline{QAQ}^{-1}\tilde{\underline{x}} + \underline{QB}u = \begin{bmatrix} 0 & -2 & 3 \\ 1 & -3 & 2 \\ 0 & 0 & 1 \end{bmatrix} \tilde{\underline{x}} + \begin{bmatrix} 1 \\ 0 \\ 0 \end{bmatrix} u \\ y &= \underline{c}'\underline{Q}^{-1} = \begin{bmatrix} 0 & 1 & 6 \end{bmatrix} \tilde{\underline{x}} \end{aligned} \tag{56}$$

which is shown as a block-diagram in Fig. 4(d).

Now the reason for the system's apparent "stability" is clear: the part of the system containing the unstable eigenvalue at +1 has been made uncontrollable, so that it does not show up in the transfer function. Nevertheless, the system is quite unstable, for even an infinitesimal initial condition $\tilde{x}_3(0) \neq 0$ will result in a term $6x_3(0)e^t$ appearing at the output.

ΔΔΔ

This example illustrates that the existence of pole-zero cancellations in a transfer function may indicate a system which is uncontrollable and/or unobservable, that is, a nonminimal realization. Indeed, we shall now show that for scalar systems, the absence of pole-zero cancellations is equivalent to minimality, while for multivariable systems it is sufficient (but not necessary).

8. THEOREM.

(a) The system

$$\dot{\underline{x}} = \underline{A}\underline{x} + \underline{B}\underline{u}$$
$$\underline{y} = \underline{C}\underline{x} \tag{57}$$

is controllable and observable (i.e., a minimal realization) if there are no pole-zero cancellations† between the numerator and denominator of the transfer function matrix

$$\hat{\underline{H}}(s) = \underline{C}(s\underline{I} - \underline{A})^{-1}\underline{B} = \frac{\underline{Q}(s)}{\det(s\underline{I} - \underline{A})} \tag{58}$$

(where $\underline{Q}(s)$ is a matrix polynomial in s). The converse statement is false for systems with both multiple inputs and multiple outputs; the system may still be controllable and observable when there are cancellations.

(b) Suppose (57) has either a scalar input ($\dot{\underline{x}} = \underline{A}\underline{x} + \underline{b}u$) or a scalar output ($y = \underline{c}'\underline{x}$) or both. Then the system is controllable and observable if and only if there are no pole-zero cancellations between the numerator and denominator of the transfer function.

†When the transfer function is a matrix, this means that the same cancellation occurs in every nonzero element of the matrix.

Proof:

(a) If (57) is not minimal, then some system

$$\dot{\tilde{\underline{x}}} = \tilde{\underline{A}}\tilde{\underline{x}} + \tilde{\underline{B}}\underline{u}$$
$$\underline{y} = \tilde{\underline{C}}\tilde{\underline{x}} \tag{59}$$

has smaller dimension and

$$\frac{\underline{Q}(s)}{\det(s\underline{I} - \underline{A})} = \frac{\tilde{\underline{Q}}(s)}{\det(s\underline{I} - \tilde{\underline{A}})} = \hat{\underline{H}}(s) \tag{60}$$

But then the polynomial $\det(s\underline{I} - \tilde{\underline{A}})$ is of lower order than $\det(s\underline{I} - \underline{A})$, and there must be a cancellation on the left-hand side of (60).

A counterexample to the converse may be found in Problem 14.

(b) Conversely, suppose there is a cancellation. Then we can use either the controllable canonical form (single-input system) or the observable canonical form (single-output system) with the reduced-order transfer function to construct a realization of smaller dimension.

ΔΔΔ

Note that the absence of pole-zero cancellations is *not* a necessary condition for minimality in the multiple-input, multiple-output case. This apparent anomaly can be removed by adopting a slightly more general definition of "denominator" for matrix transfer functions. The interested reader should consult Sec. 6-2 of Ref. 39. Some results relating the controllability and observability of interconnected systems to pole-zero cancellations are reported in Refs. 39 and 245. A further discussion of zeros in multivariable systems may be found at the end of Sec. 11.2.

10.5. SUMMARY

In this chapter we have examined two fundamental aspects of linear systems. The first of these was controllability, which concerns the amount of control one can exercise over the state of a system from the input. A precise definition was stated and algebraic conditions for controllability were derived in terms of the matrix $\underline{P}_c$.

Observability concerns the extent to which the state of a system influences the output, and whether the state can be determined from knowledge of the output over a finite interval. Again, a precise definition was stated and algebraic conditions for observability were derived in terms of the matrix $\underline{P}_o$.

In Sec. 10.3 the duality between controllability and observability was made explicit, and the underlying structure of a linear system with regard to these two properties was explored, resulting in the canonical decomposition set out in Theorem 10.3.4 and Fig. 10.3-3. There it became clear that input-output descriptions (such as the transfer function) can only represent the controllable and observable component of any system.

In Sec. 10.4 we turned to the problem of finding a state-space realization for a given transfer function, and two standard canonical forms (one controllable, one observable) were introduced for this purpose. Next we saw that a realization has minimal dimension if and only if it is both controllable and observable, and that all minimal realizations of the same transfer function are equivalent (i.e., related by state transformations). This led us to observe that controllable or observable systems can always be transformed into appropriate canonical forms.

Finally, we investigated the phenomenon of pole-zero cancellations in transfer functions, and saw that these are (usually) brought about by the existence of uncontrollable and/or unobservable states.

PROBLEMS

Section 10.1

1. Verify the state equations and the matrix exponential in Example 10.1.1.

2. (a) Show that Definition 10.1.2 is equivalent to the following: "...for every t_0 and for every $T > 0$, there

exists a control function

$$\underline{u}(t) \qquad (t_0 \leq t < t_0 + T)$$

which drives the system (4) from $\underline{x}(t_0) = \overline{\underline{x}}$ to $\underline{x}(t_0 + T) = \underline{0}$."

(b) Does this equivalence hold for a nonlinear system? What about a time-varying system?

(c) Extend Corollary 10.1.5 to the case where $t_0 \neq 0$.

3. (a) Show that $\overline{\underline{x}} \in R(\underline{W}_c(0,T))$ if and only if $e^{\underline{A}\tau}\overline{\underline{x}} \in R(\underline{W}_c(0,T))$ for all τ. *Hint:* See Lemma 10.1.7.

(b) Assuming that $\overline{\underline{x}}$ is a controllable state of (10.1-4), determine a control which will drive the system from $\underline{x}(0) = \underline{0}$ to $\underline{x}(T) = \overline{\underline{x}}$.

(c) Alter the proof of Theorem 10.1.4 to show that a state is reachable if and only if it belongs to $R(\underline{W}_c(0,T))$, and hence that reachability and controllability are equivalent in this case.

4. (a) Assume that $\underline{W}_c(0,T)$ is nonsingular and determine a control which will drive the system (10.1-4) from $\underline{x}(0) = \underline{x}_0$ to $\underline{x}(T) = \underline{x}_1$.

(b) Does your control in (a) achieve the transfer with a minimum expenditure of energy? Prove or disprove your answer.

(c) Repeat for $\underline{x}(t_0) = \underline{x}_0$ and $\underline{x}(t_0 + T) = \underline{x}_1$, where $t_0 \neq 0$.

5. Generalize the proof of Corollary 10.1.5 to the case where $\underline{W}_c(0,T)$ is singular.

6. Use the Cayley-Hamilton theorem (A.3.9) to prove that

$$R[\underline{B} \;\; \underline{AB} \;\; \underline{A}^2\underline{B} \;\; \cdots \;\; \underline{A}^k\underline{B}] = R[\underline{B} \;\; \underline{AB} \;\; \underline{A}^2\underline{B} \;\; \cdots \;\; \underline{A}^{n-1}\underline{B}]$$

for all $k \geq n = \dim(\underline{A})$. *Hint:* First prove that

$$R[\underline{B} \;\; \underline{AB} \;\; \underline{A}^2\underline{B} \;\; \cdots \;\; \underline{A}^n\underline{B}] = R[\underline{B} \;\; \underline{AB} \;\; \underline{A}^2\underline{B} \;\; \cdots \;\; \underline{A}^{n-1}\underline{B}]$$

and then proceed by induction.

7. If $\underline{A}$ is n × n and $\underline{B}$ is n × m with rank m, prove that

$$R[\underline{B} \quad \underline{AB} \quad \underline{A}^2\underline{B} \quad \ldots \quad \underline{A}^{n-1}\underline{B}] = R[\underline{B} \quad \underline{AB} \quad \underline{A}^2\underline{B} \quad \ldots \quad \underline{A}^{n-m}\underline{B}]$$

You may find it useful to first prove the following:

LEMMA. For any j, if every column of $\underline{A}^j\underline{B}$ is linearly dependent upon the columns of $\underline{B}$, $\underline{AB}$, ..., $\underline{A}^{j-1}\underline{B}$, then so is every column of $\underline{A}^{j+1}\underline{B}$, $\underline{A}^{j+2}\underline{B}$, ...

8. Prove that a necessary condition for controllability is that the matrix $[\underline{B} \quad \underline{A}]$ have full rank, by demonstrating that $R(\underline{P}_c)$ is contained in $R[\underline{B} \quad \underline{A}]$. Verify that this condition is not sufficient.

9. (a) If the initial state $\underline{x}(0) = \overline{\underline{x}}$ of a system is totally uncontrollable, that is, $\overline{\underline{x}} \in N(\underline{P}_c')$, does the natural response $e^{\underline{A}t}\overline{\underline{x}}$ remain uncontrollable for all time?

   (b) Test your answer to (a) on the following example:

$$\dot{\underline{x}}(t) = \begin{bmatrix} -1 & 1 \\ 0 & -2 \end{bmatrix} \underline{x}(t) + \begin{bmatrix} 1 \\ 0 \end{bmatrix} u(t)$$

10. Consider the system

$$\dot{\underline{x}} = \begin{bmatrix} 0 & 2 & -1 \\ 3 & 0 & 1 \\ 0 & 0 & 2 \end{bmatrix} \underline{x} + \begin{bmatrix} 1 & 0 \\ 2 & 1 \\ 0 & 2 \end{bmatrix} \underline{u}$$

   (a) Is this system completely controllable?

   (b) Identify the subspace of states controllable from u_1 only, and the corresponding subspace of uncontrollable states. What are their dimensions?

   (c) Identify the subspace of states controllable from u_2 only, and the corresponding subspace of uncontrollable states. What are their dimensions?

11. Repeat Problem 10 for the system

$$\dot{\underline{x}} = \begin{bmatrix} 3 & 0 & -2 \\ 1 & 2 & -1 \\ 0 & 0 & 1 \end{bmatrix} \underline{x} + \begin{bmatrix} 1 & 1 \\ 1 & 0 \\ 1 & 1 \end{bmatrix} \underline{u}$$

12. Consider the circuit shown in Fig. P1, where the initial voltage on the capacitor is V at time $t = 0$.

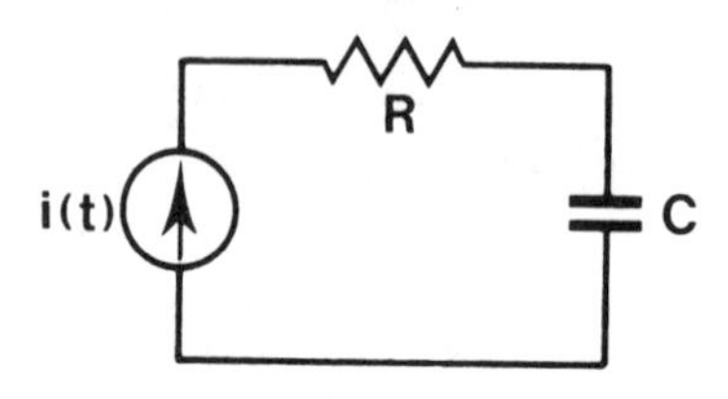

FIG. P1

(a) Determine the input current $i(t)$ which makes the capacitor voltage equal to zero at time T with minimum loss in the resistor.

(b) Determine a different current which also drives the capacitor voltage to zero at time T.

(c) Compare the loss in the resistor for your solutions in (a) and (b).

(d) Explain what happens to the current in (a) as $T \to 0$.

13. Consider the network shown in Fig. P2 and assume that

$$x_1(0) = x_2(0) = 0$$

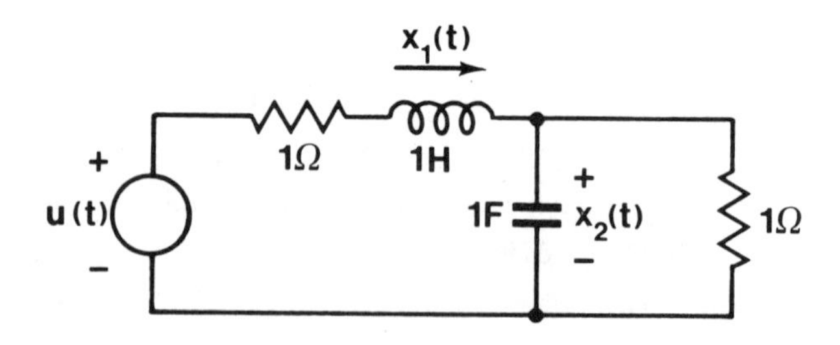

FIG. P2

(a) Does there exist an input voltage $u(t)$, $0 \leq t < \pi$, such that

$$x_1(\pi) = x_2(\pi) = 1 \ ?$$

If so, find one (you may find Problem 3 or 4 useful here).

(b) Suppose we want to maintain

$$x_1(t) = x_2(t) = 1 \qquad (t \geq \pi)$$

Can a suitable $u(t)$, $t \geq \pi$, be found? If so, what is it?

(c) Suppose the system

$$\dot{\underline{x}} = \underline{A}\underline{x} + \underline{B}\underline{u}$$

is controllable and $\tilde{\underline{x}}(t)$, $0 \leq t \leq T$, is a given (differentiable) trajectory. Assuming that $\underline{x}(0) = \tilde{\underline{x}}(0)$, under what conditions on $\underline{A}$, $\underline{B}$, and $\tilde{\underline{x}}(t)$ will there exist a control $\underline{u}(t)$, $0 \leq t < T$, such that

$$\underline{x}(t) = \tilde{\underline{x}}(t) \qquad (0 \leq t \leq T)$$

Check these conditions for part (b) above.

(d) Under what conditions on $\underline{A}$ and $\underline{B}$ can the system be made to follow <u>any</u> arbitrary differentiable trajectory?

14. (a) If the integrand of

$$\underline{W}_c(0,T) = \int_0^T e^{-\underline{A}\tau}\underline{B}\underline{B}'\, e^{-\underline{A}'\tau}\, d\tau$$

is singular, i.e.,

$$\det(e^{-\underline{A}\tau}\underline{B}\underline{B}'\, e^{-\underline{A}'\tau}) = 0 \qquad (0 \leq \tau \leq T)$$

does it follow that $\underline{W}_c(0,T)$ must be singular?

(b) Test out your answer to (a) on the system

$$\dot{\underline{x}} = \begin{bmatrix} 0 & 1 \\ -1 & 0 \end{bmatrix} \underline{x} + \begin{bmatrix} 1 \\ 0 \end{bmatrix} u$$

15. (a) Use *physical intuition* to predict under what conditions on m_1, m_2, ℓ_1, and ℓ_2 the dual-broom balancing system shown in Fig. P3(a) will be controllable (assume that the brooms move in two parallel planes and cannot collide).

(b) Check your answer to (a) by forming a set of linearized equations and testing the controllability matrix.

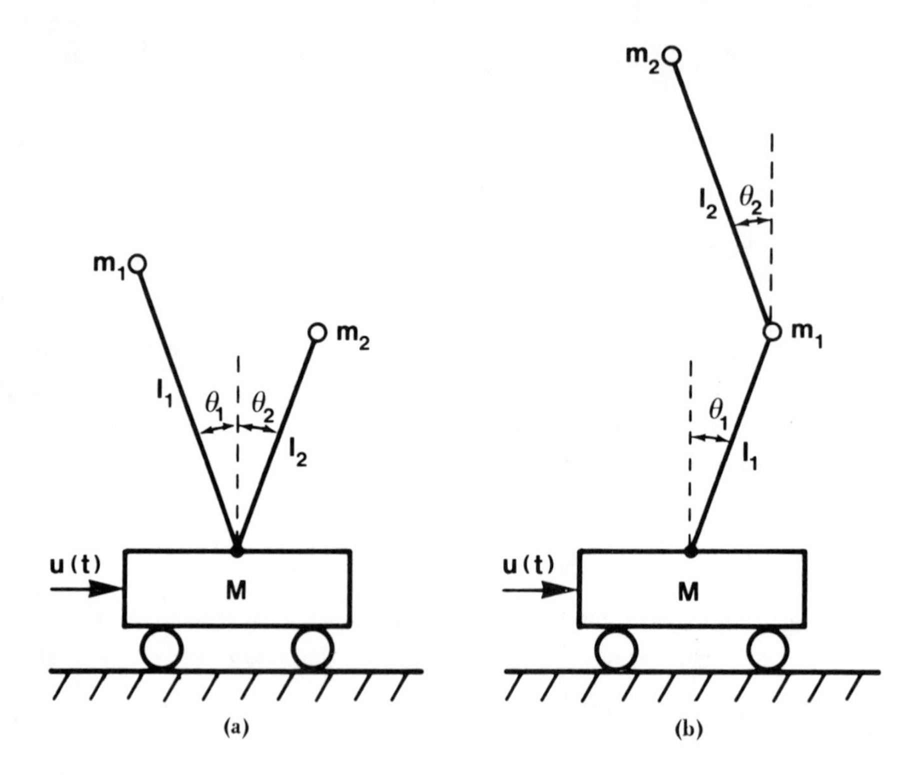

FIG. P3. Dual-broom balancing systems.

(c) Repeat for the dual-broom system shown in Fig. P3(b).

16. *Output controllability:* consider the system

$$\dot{\underline{x}} = \underline{A}\underline{x} + \underline{B}\underline{u}$$

$$\underline{y} = \underline{C}\underline{x}$$

with initial state $\underline{x}(0) = \underline{x}_0$.

(a) Prove that there exists a control which drives the system from its initial output $\underline{y}(0) = \underline{y}_0 = \underline{C}\underline{x}_0$ to $\underline{y}(T) = \underline{y}_1$ if and only if

$$(\underline{y}_1 - \underline{C}e^{\underline{A}T}\underline{x}_0) \in R(\underline{C}e^{\underline{A}T}\underline{W}_c(0,T))$$

Conclude that the system is output controllable if and only if the matrix on the right has full rank.

(b) Suppose the system is output controllable, and find a control function which accomplishes the transfer. Can you find such a control if you know $\underline{y}_0$ and $\underline{y}_1$, but not $\underline{x}_0$?

(c) Use Lemma 10.1.7 to show that $R(\underline{C}e^{\underline{A}T}\,\underline{W}_c(0,T)) = R(\underline{C}\underline{W}_c(0,T))$.

(d) Show that $R(\underline{C}\underline{W}_c(0,T)) = R(\underline{C}\underline{W}_c(0,T)\underline{C}')$. *Hint:* Try using a null space argument.

(e) Show that $R(\underline{C}\underline{W}_c(0,T)) = R(\underline{C}\underline{P}_c)$.

(f) Under what conditions (involving the rank of the $p \times n$ matrix $\underline{C}$) does output controllability imply controllability? Under what conditions does controllability imply output controllability? Can the system be output controllable if $p > n$?

17. (a) Derive controllability conditions for a single-input system described in Jordan form. Illustrate with a block diagram.

(b) Generalize your result to a multiple-input system.

Section 10.2

1. Verify the state equations and the matrix exponential in Example 10.2.1.

2. (a) Show that Definition 10.2.2 is equivalent to the following: ... for $\underline{u}(t) \equiv \underline{0}$, any $T > 0$, and any t_0, the initial state $\underline{x}(t_0) = \overline{\underline{x}}$ produces an output response

$$\underline{y}(t) = \underline{0} \qquad (t_0 \leq t < t_0 + T)$$

(b) Does this equivalence hold for a nonlinear system? What about a time-varying system?

(c) Extend Corollary 10.2.5 to the case where $t_0 \neq 0$.

3. (a) Alter part of the proof of Theorem 10.2.4 to show that the system (10.2-4) is completely reconstructible if and only if it is completely observable.

(b) Assuming the system is completely reconstructible (observable), alter Corollary 10.2.5 to show how the output $\underline{y}(t)$, $0 \leq t < T$, can be processed to determine $\underline{x}(T)$.

(c) How can $\underline{y}(t)$, $-T \leq t < 0$, be processed to determine $\underline{x}(0)$?

4. (a) Suppose, in Corollary 10.2.5, that $\underline{W}_o(0,T)$ is singular but $\bar{\underline{x}}$ is observable, that is, $\bar{\underline{x}} \in R(\underline{W}_o)$. Show that $\underline{x}(0) = \bar{\underline{x}}$ can be determined from $\underline{y}(t)$ via

$$\bar{\underline{x}} = \int_0^T \underline{\Phi}\underline{V}'e^{\underline{A}'\tau}\underline{C}'\underline{y}(\tau)\, d\tau$$

where the columns of $\underline{\Phi}$ are an orthonormal basis for $R(\underline{W}_o)$ and $\underline{V}$ is any solution of the equation

$$\underline{W}_o\underline{V} = \underline{\Phi}$$

(b) What does the above integral yield when $\bar{\underline{x}}$ does not belong to $R(\underline{W}_o)$?

(c) Show that the weighting function in the integral has a minimum energy property analogous to that of Corollary 10.1.5. Suppose that the measurements $\underline{y}(t)$ are corrupted by white noise (see Sec. 12.4) and interpret this property in terms of the variance of the estimate of $\bar{\underline{x}}$.

5. Suppose, in Corollary 10.2.5, that $\underline{u}(t)$ is a known (nonzero) function. Show how $\underline{y}(t)$, $0 \leq t < T$, can be processed to determine $\underline{x}(0)$.

6. Show how the proof of Theorem 10.1.6 can be applied to Theorem 10.2.6.

7. Use the result of Problem 10.1-7 to verify that $\underline{P}_o$ can be replaced by the matrix in (10.2-20), provided $\underline{C}$ has rank p.

8. (a) Does Lemma 10.2.7 hold in general for a nonzero input $\underline{u}(t)$? Give arguments to support your answer.

(b) Under what additional conditions on $\underline{A}$, $\underline{B}$, and $\underline{C}$ does Lemma 10.2.7 hold for any $\underline{u}(t) \neq 0$?

9. (a) If the initial state $\underline{x}(0) = \overline{\underline{x}}$ is observable, that is, $\overline{\underline{x}} \in R(\underline{P}_o')$, does the natural response $e^{\underline{A}t}\overline{\underline{x}}$ remain observable for all time?

   (b) Test your answer to (a) on the following example:

$$\dot{\underline{x}}(t) = \begin{bmatrix} -1 & 1 \\ 0 & -2 \end{bmatrix} \underline{x}(t)$$

$$y(t) = \begin{bmatrix} 0 & 1 \end{bmatrix} \underline{x}(t)$$

10. Consider the system

$$\dot{\underline{x}} = \begin{bmatrix} 0 & 2 & -1 \\ 3 & 0 & 1 \\ 0 & 0 & 2 \end{bmatrix} \underline{x}$$

$$\underline{y} = \begin{bmatrix} 0 & -2 & 1 \\ 0 & 0 & 1 \end{bmatrix} \underline{x}$$

   (a) Is the system completely observable?

   (b) Identify the subspace of states which are unobservable from y_1 only and the corresponding subspace of observable states. What are their dimensions?

   (c) Identify the subspace of states which are unobservable from y_2 only and the corresponding subspace of observable states. What are their dimensions?

11. Repeat Problem 10 for the system

$$\dot{\underline{x}} = \begin{bmatrix} 3 & 0 & -2 \\ 1 & 2 & -1 \\ 0 & 0 & 1 \end{bmatrix} \underline{x}$$

$$\underline{y} = \begin{bmatrix} 1 & 0 & 0 \\ 1 & 1 & 1 \end{bmatrix} \underline{x}$$

12. Investigate the observability of the compensated system in Example 9.3.4.

13. (a) Consider the scalar system $\dot{x}(t) = 0$, with output $y(t) = x(t)$. Show how to determine $x(0)$ from $y(t)$, $0 \leq t \leq T$ using Lemma 10.2.5. Plot $\underline{W}_o^{-1}(0,T)e^{\underline{A}'t}\underline{C}'$ as a function of t and interpret its behavior as $T \to 0$.

    (b) Do the same for the system

$$\dot{\underline{x}}(t) = \begin{bmatrix} 0 & 1 \\ 0 & 0 \end{bmatrix} \underline{x}(t)$$

$$y(t) = \begin{bmatrix} 1 & 0 \end{bmatrix} \underline{x}(t)$$

    Hint: See Problem B.1-3.

14. Formulate and prove a dual version of Problem 10.1-16.

15. Repeat Problems 10.1-15 and 10.1-17 for observability.

Section 10.3

1. Prove Lemma 10.3.2

    (a) directly from the definitions of controllability and observability

    (b) by determining the controllability and observability matrices of the two systems.

2. Explain why the matrix defined by (10.3-8) must be nonsingular.

3. (a) Prove Theorem 10.3.3' directly.

    (b) Show how Theorem 10.3.3' follows from Theorem 10.3.3 using duality (you will need to renumber $\hat{\underline{x}}_1$ and $\hat{\underline{x}}_2$ to get the form shown).

4. Prove Theorem 10.3.4 by choosing basis vectors for the subspaces

$$\mathcal{R}(\underline{P}_c) \cap \mathcal{N}(\underline{P}_o)$$
$$\mathcal{R}(\underline{P}_c) \cap \mathcal{R}(\underline{P}_o')$$
$$\mathcal{N}(\underline{P}_c') \cap \mathcal{N}(\underline{P}_o)$$
$$\mathcal{N}(\underline{P}_c') \cap \mathcal{R}(\underline{P}_o')$$

to be the columns of $\underline{Q}^{-1}$.

5. Show that Theorem 10.3.1 is still valid with $-\underline{A}'$ replaced by $\underline{A}'$ in (10.3-2).

6. (a) Verify that the results of Secs. 10.1 and 10.2 remain valid for the system

$$\dot{\underline{x}} = \underline{A}\underline{x} + \underline{B}\underline{u}$$

$$\underline{y} = \underline{C}\underline{x} + \underline{D}\underline{u}$$

(b) Indicate how Theorem 10.3.4, Fig. 10.3-3, and Corollary 10.3.5 must be altered in this case.

(c) Suppose a section of transmission line is modeled as the circuit shown in Fig. P4(a), and we attempt to compensate for the inductance by introducing a shunt capacitance and resistance, as indicated in Fig. P4(b). Determine the values of R and C required to just cancel the effect of the inductor, i.e., to make the admittance (transfer function from v to i) equal to 1S.

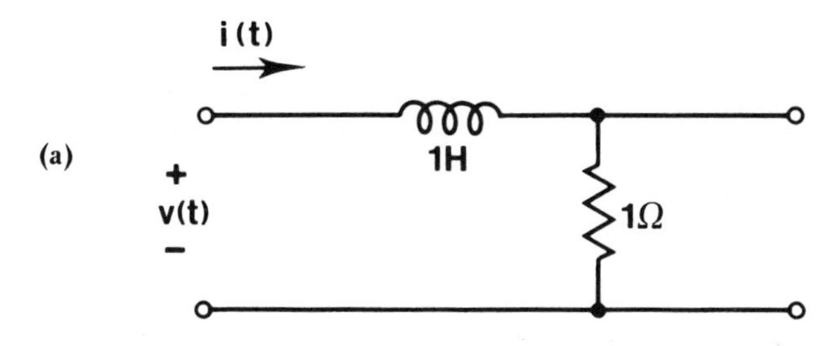

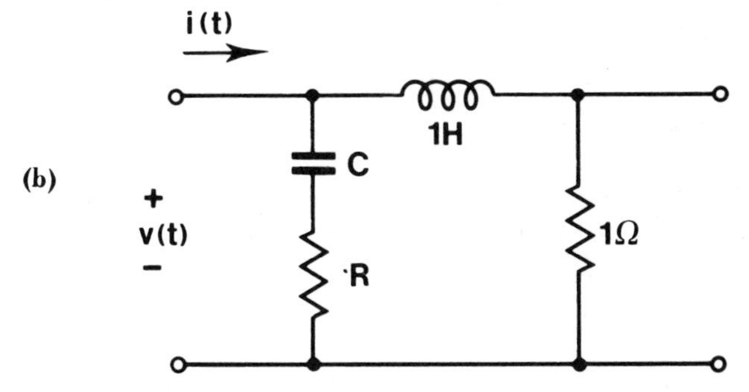

FIG. P4. (a) Transmission line model; (b) compensation with shunt capacitance and resistance.

(d) Using the above values of R and C, derive a set of state equations for this circuit, with v as the input and i as the output. Perform the decomposition of Theorem 10.3.4 on this system and explain its behavior accordingly.

7. In Theorem 10.3.3 a system is decomposed into a subsystem

$$\dot{\underline{x}}_1 = \underline{A}_{11}\underline{x}_1 + \underline{B}_1\underline{u} + \underline{A}_{12}\underline{x}_2$$

which is controllable (from $\underline{u}$), and an uncontrollable subsystem

$$\dot{\underline{x}}_2 = \underline{A}_{22}\underline{x}_2$$

where the $\hat{\underline{x}}$ notation has been dropped for simplicity. Suppose that the initial states $\underline{x}_1(0) = \overline{\underline{x}}_1$ and $\underline{x}_2(0) = \overline{\underline{x}}_2$ are known, and find a control function $\underline{u}(t)$, $0 \le t \le T$, which drives the state of the controllable subsystem to $\hat{\underline{x}}_1(T) = \underline{0}$.

8. (a) Prove that for a controllable and observable system, asymptotic stability (Definition 5.1.1) and BIBO stability (Definition 5.1.5) are equivalent.

(b) Explain why the above result does not hold for systems with uncontrollable and/or unobservable states.

(c) Devise an example of a system which is BIBO stable but not asymptotically stable.

Section 10.4

1. (a) Verify that (10.4-11) realizes the transfer function (10.4-8).

(b) Use the fact that (10.4-10) is the "transpose" of (10.4-11) to show that (10.4-10) also realizes (10.4-8).

2. Verify that the analog computer circuits shown in Fig. 10.4-1(a) and (b) correctly model the state equations in (10.4-10) and (10.4-11), respectively.

3. (a) Extend your solution of Problem 1(a) to show that (10.4-12) realizes the transfer function (10.4-7).

   (b) Can your solution of Problem 1(b) be similarly extended?

4. (a) Show that (10.4-10) is controllable and (10.4-11) is observable.

   (b) Show that (10.4-12) is controllable and (10.4-13) is observable.

5. Suppose we write the transfer function (10.4-8) in the form of a *series* expansion about $|s| = \infty$.

$$\hat{h}(s) = \frac{\ell_0}{s} + \frac{\ell_1}{s^2} + \frac{\ell_2}{s^3} + \ldots$$

(This can easily be done for any proper, rational $\hat{h}(s)$ by long division.)

   (a) Show that the system

$$\dot{\underline{x}} = \begin{bmatrix} 0 & 1 & 0 & \ldots & 0 \\ 0 & 0 & 1 & \ldots & 0 \\ \cdots & \cdots & \cdots & \cdots & \cdots \\ 0 & 0 & 0 & \ldots & 1 \\ -p_0 & -p_1 & -p_2 & \ldots & -p_{n-1} \end{bmatrix} \underline{x} + \begin{bmatrix} \ell_0 \\ \ell_1 \\ \vdots \\ \ell_{n-2} \\ \ell_{n-1} \end{bmatrix} u \triangleq \underline{A}_\ell \underline{x} + \underline{b}_\ell u$$

$$y = \begin{bmatrix} 1 & 0 & 0 & \ldots & 0 \end{bmatrix} \underline{x} \triangleq \underline{c}_\ell' \underline{x}$$

   is a realization of $\hat{h}(s)$, as follows:

   (i) Expand the resolvent matrix as in (4.2-8) to determine the series for $\underline{c}_\ell'(s\underline{I} - \underline{A}_\ell)^{-1}\underline{b}_\ell$

   (ii) Show that the first n terms of this series agree with those in the series for $\hat{h}(s)$.

   (iii) Use the Cayley-Hamilton theorem and the fact that the coefficient of s^{-1} is zero in

$$q(s) = p(s)\hat{h}(s) = (p_0 + p_1 s + \ldots + s^n)\left(\frac{\ell_0}{s} + \frac{\ell_1}{s^2} + \ldots\right)$$

   to show that the (n+1)-th terms also agree.

(iv) Verify that a similar argument establishes that all successive terms must agree.

(v) Conclude that $\underline{c}_\ell'(s\underline{I} - \underline{A}_\ell)^{-1}\underline{b}_\ell = \hat{h}(s)$

(b) Show that the realization is observable.

(c) Show that the system

$$\dot{\underline{x}} = \underline{A}_\ell'\underline{x} + \underline{c}_\ell u$$

$$y = \underline{b}_\ell'\underline{x}$$

is also a realization of $\hat{h}(s)$ and is controllable.

6. Draw analog computer circuits (as in Fig. 10.4-1) for modeling the systems of Problem 5(a) and (c).

7. Use the results of Problem 5 to show that the dimension of a minimal realization of $\hat{h}(s)$ is equal to the rank of the *Hankel matrix*

$$\underline{H}_{n-1} = \begin{bmatrix} \ell_0 & \ell_1 & \ell_2 & \cdots & \ell_{n-1} \\ \ell_1 & \ell_2 & \ell_3 & \cdots & \ell_n \\ \cdots & \cdots & \cdots & \cdots & \cdots \\ \ell_{n-1} & \ell_n & \ell_{n+1} & \cdots & \ell_{2n-2} \end{bmatrix}$$

[This is known as the *McMillan degree* of $\hat{h}(s)$].

8. Extend Problems 5 and 7 to the multiple input-output case, i.e., to the transfer function matrix (10.4-7).

9. (a) Show that the state transformation defined by

$$\underline{Q}^{-1} = [\underline{b}\ \ \underline{A}\underline{b}\ \ \underline{A}^2\underline{b}\ \ldots\ \underline{A}^{n-1}\underline{b}] \begin{bmatrix} p_1 & p_2 & p_3 & \cdots & p_{n-1} & 1 \\ p_2 & p_3 & p_4 & \cdots & 1 & 0 \\ \cdots & \cdots & \cdots & \cdots & \cdots & \cdots \\ p_{n-1} & 1 & 0 & \ldots & 0 & 0 \\ 1 & 0 & 0 & \ldots & 0 & 0 \end{bmatrix}$$

is the one specified in Theorem 10.4.6. This can be done without explicitly computing $\underline{Q}$.

(b) Show that this transformation may also be expressed as

$$\underline{Q}^{-1} = [\underline{Q}_0\underline{b} \quad \underline{Q}_1\underline{b} \quad \cdots \quad \underline{Q}_{n-2}\underline{b} \quad \underline{b}]$$

where $\underline{Q}_0, \underline{Q}_1, \ldots, \underline{Q}_{n-2}$, and $\underline{Q}_{n-1} = \underline{I}$ are the coefficients in the numerator of the resolvent matrix $(s\underline{I} - \underline{A})^{-1}$.

Hint: Use Theorem 4.3.2.

(c) Find corresponding expressions for the transformation of Theorem 10.4.6'.

10. (a) Explain why the proof of Theorem 10.4.6 will not work for the case of multiple inputs.

(b) Test your explanation on the system

$$\dot{\underline{x}} = \begin{bmatrix} 1 & 0 \\ 0 & -1 \end{bmatrix} \underline{x} + \begin{bmatrix} 1 & 0 \\ 0 & 1 \end{bmatrix} \underline{u}$$

$$\underline{y} = \begin{bmatrix} 1 & 0 \\ 0 & 1 \end{bmatrix} \underline{x}$$

11. Use duality to derive Theorem 10.4.6' from Theorem 10.4.6.

12. Verify that Fig. 10.4-4(c) realizes Fig. 10.4-4(a).

13. Repeat the analysis of Example 10.4.7 for the case where the two systems are connected in the opposite order (retain the same numbering of states). Explain why this system can be said to suffer from an "internal hemorrhage."

14. Disprove the converse of Theorem 10.4.8(a) by considering a 2×2 system with $\underline{A} = \underline{B} = \underline{C} = \underline{I}$.

15. (a) Find a realization of the transfer function matrix

$$\underline{h}'(s) = \begin{bmatrix} \dfrac{1}{s+1} & \dfrac{1}{s^2 + 3s + 2} \end{bmatrix}$$

(b) Is your realization minimal? If not, find a minimal one.

16. (a) Show how to construct Fig. 10.4-2 from the controllable canonical form (10.4-44), by explicitly determining the coupling matrix.

    (b) Do the same for Fig. 10.4-3 and the observable canonical form (10.4-45).

17. (a) Verify that the controllable canonical form (10.4-44) is controllable.

    (b) Verify that the observable canonical form (10.4-45) is observable.

18. Assuming that any controllable system can be put into the form (10.4-44) with a state transformation, show how an observable system can be put into the form (10.4-45).

19. Theorem 10.4.5 establishes that two minimal realizations of the same transfer function are related by a state transformation $\tilde{\underline{x}} = \underline{Q}\underline{x}$, but the computation of $\underline{Q}$ is rather awkward. Show that a simpler expression for $\underline{Q}$ is

$$\underline{Q} = \tilde{\underline{P}}_c [\underline{P}_c' (\underline{P}_c \underline{P}_c')^{-1}]$$

where $\underline{P}_c$ and $\tilde{\underline{P}}_c$ are the respective controllability matrices for the two realizations and the matrix in brackets is a *pseudo-inverse* of $\underline{P}_c$ (see Problem A.5-5). Derive a corresponding expression using observability matrices.

20. Prove that a controllable and observable scalar system whose transfer function has repeated poles on the $j\omega$ axis is unstable (not marginally stable). *Hint:* Use Problems 10.1-17 and 5.1-7.

Chapter 11

FEEDBACK CONTROL OF MULTIVARIABLE SYSTEMS

11.0 INTRODUCTION

In Chapter 8 we examined the concept of *feedback control*, wherein the output is measured and used to determine the control input to be applied. Various potential advantages of the feedback mechanism were explored: an ability to track reference inputs and correct for disturbance inputs, faster transient response, and reduced sensitivity to parameter variations and modeling errors. Attention was focused principally upon scalar systems, where the characteristics of the closed-loop system depended upon a single parameter (the loop gain k), and the root-locus and Nyquist techniques were used to exhibit graphically the dependence of transient behavior on the value of loop gain. We also explored the relationship between static (steady-state) accuracy and the placement of integrators in the feedback loop. The use of differentiators was discussed in Chapter 9.

In Chapter 10 we were concerned with multivariable systems and open-loop controls, applying a time function at the input to a system in order to drive the state to a particular value. Systems whose states could all be reached in this manner were termed *controllable* (or *reachable*). We were also concerned with processing the output of a system over an interval of time in order to determine the state; systems for which this could be done were termed *observable* (or *reconstructible*).

In this chapter, the properties of controllability and observability will also be seen to have important implications for closed-loop control. First, we shall specify more precisely the sorts of feedback which may be used in a multivariable system, and determine the effects of these upon the state equations and transfer function. Next, we shall encounter one of the most important results of linear systems theory: If all the states of a controllable system are available for feedback, then the feedback gains may be chosen so as to place the eigenvalues of the closed-loop system *anywhere* in the complex plane (subject to conjugate pairing). This allows the designer a great deal of freedom to specify the transient behavior of the closed-loop system.

Steady-state performance will be taken up in Sec. 11.3 where we shall see how static accuracy may be achieved by placing integrators in a multivariable control loop. Finally, another use of state-variable feedback will be explored in Sec. 11.4 where the object is to *decouple* the system so that each input affects *only* the corresponding output.

11.1 MULTIVARIABLE FEEDBACK

Consider a linear system in the usual form,

$$\begin{aligned} \dot{\underline{x}}(t) &= \underline{A}\underline{x}(t) + \underline{v}(t) \\ \underline{v}(t) &= \underline{B}\underline{u}(t) \\ \underline{y}(t) &= \underline{C}\underline{x}(t) \end{aligned} \tag{1}$$

where the state $\underline{x}$, input $\underline{u}$, and output $\underline{y}$ are vectors of dimension n, m, and p, respectively. For a variety of reasons, most of which were indicated in Chapter 8, we often find it useful to apply linear feedback from the output $\underline{y}$ to the input $\underline{u}$,

$$\underline{u}(t) = \underline{L}\underline{y}(t) + \underline{w}(t) \tag{2}$$

where $\underline{L}$ is a constant $m \times p$ *feedback gain matrix* and $\underline{w}(t)$ is an external input. The resulting *closed-loop system* is shown in Fig. 1(a). Note that each element $u_i(t)$ of the input vector $\underline{u}(t)$ is driven by a linear combination $\underline{l}_i'\underline{y}(t)$ of the outputs, where $\underline{l}_i'$ is the i-th row of $\underline{L}$.

The term $\underline{w}$ can have various interpretations. For example, it can be used to represent a *reference input* (or *desired output*) $\underline{y}_r$ by setting $\underline{w} = -\underline{L}\,\underline{y}_r$ and noting that in (2) $\underline{u}$ is thus proportional to the error vector $\underline{y} - \underline{y}_r$:

$$\underline{u}(t) = \underline{L}[\underline{y}(t) - \underline{y}_r(t)] \tag{3}$$

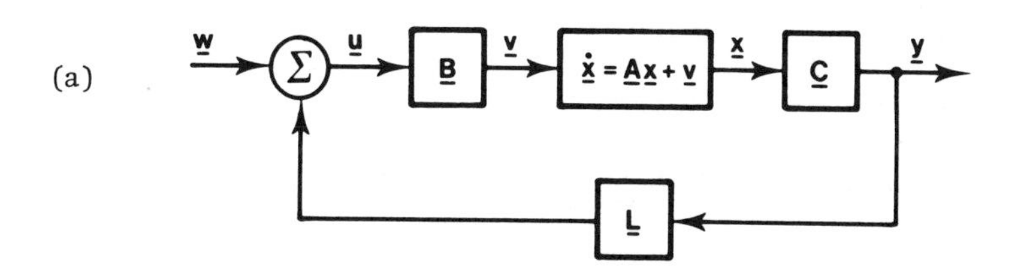

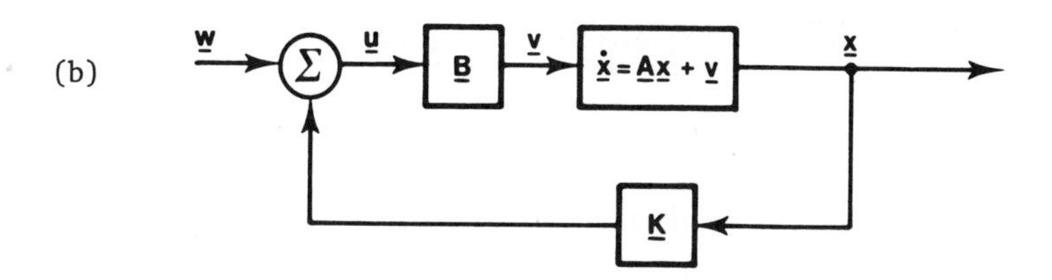

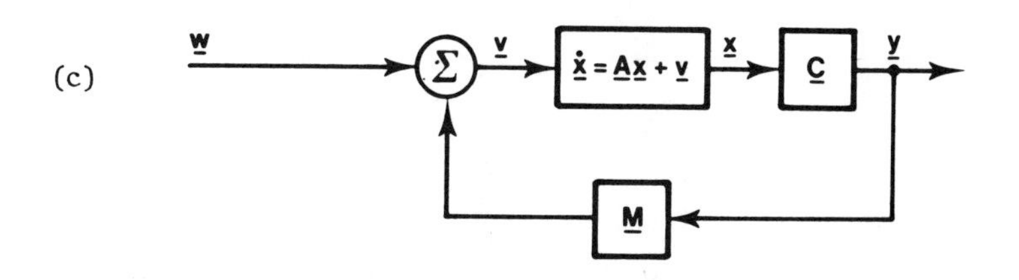

FIG. 1. Multivariable feedback (a) from output to input, (b) from state to input, and (c) from output to state derivative.

Similarly, $\underline{w}$ (or part of $\underline{w}$) can represent an unknown *disturbance input*. These topics will be taken up in Sec. 11.3, and for the moment the interpretation of $\underline{w}$ is left unspecified.

In the literature, (2) is usually called *output feedback*, and its effect on the system is as follows:

1. LEMMA. When the feedback law (2) is applied to the system (1), the resulting closed-loop system is described by

$$\begin{aligned} \dot{\underline{x}} &= \underline{\tilde{A}}\underline{x} + \underline{B}\underline{w} = (\underline{A} + \underline{B}\underline{L}\underline{C})\underline{x} + \underline{B}\underline{w} \\ \underline{y} &= \underline{C}\underline{x} \end{aligned} \tag{4}$$

Moreover, the closed-loop transfer function matrix (relating $\underline{y}$ to $\underline{w}$) may be expressed as

$$\begin{aligned} \underline{\tilde{H}}(s) &= \underline{C}(s\underline{I} - \underline{\tilde{A}})^{-1}\underline{B} \\ &= [\underline{I} - \underline{H}(s)\underline{L}]^{-1}\underline{H}(s) \end{aligned} \tag{5}$$

where $\underline{H}(s) = \underline{C}(s\underline{I} - \underline{A})^{-1}\underline{B}$ is the transfer function matrix of (1), provided $\underline{I} - \underline{H}(s)\underline{L}$ is nonsingular (except possibly at a finite number of points s).

Proof: Equation (4) follows by direct substitution into (1) of

$$\underline{u} = \underline{L}\underline{y} + \underline{w} = \underline{L}\underline{C}\underline{x} + \underline{w} \tag{6}$$

Taking Laplace transforms, we have

$$\hat{\underline{y}}(s) = \underline{H}(s)\hat{\underline{u}}(s) = \underline{H}(s)[\underline{L}\hat{\underline{y}}(s) + \hat{\underline{w}}(s)] \tag{7}$$

or

$$[\underline{I} - \underline{H}(s)\underline{L}]\hat{\underline{y}}(s) = \underline{H}(s)\hat{\underline{w}}(s) \tag{8}$$

If the matrix in brackets is nonsingular,

$$\hat{\underline{y}}(s) = [\underline{I} - \underline{H}(s)\underline{L}]^{-1}\underline{H}(s)\hat{\underline{w}}(s) = \underline{\tilde{H}}(s)\hat{\underline{w}}(s) \tag{9}$$

ΔΔΔ

We have established that the effect of a feedback law of the form (2) is to change the matrix governing the system dynamics from $\underline{A}$ to $\underline{\tilde{A}} = \underline{A} + \underline{B}\underline{L}\underline{C}$, and the behavior of the closed-loop system depends principally upon the eigenvalues of $\tilde{\underline{A}}$ (the closed-loop system poles). We shall thus attempt to choose the feedback gain matrix $\underline{L}$ so that $\tilde{\underline{A}}$ is stable and its eigenvalues lie in "desirable" locations. This problem will be examined shortly, but first we shall look into some other properties of the closed-loop system (4) and two special forms of the feedback law (2).

Suppose for the moment that (1) is a scalar system,

$$\dot{\underline{x}} = \underline{A}\underline{x} + \underline{b}u$$
$$y = \underline{c}'\underline{x} \tag{10}$$

with normalized[†] transfer function

$$h(s) = \underline{c}'(s\underline{I} - \underline{A})^{-1}\underline{b} = kg(s) \tag{11}$$

and gain k. In this case the feedback gain matrix is 1×1 (i.e., a scalar) and (2) becomes

$$u(t) = \ell y(t) + w(t) \tag{12}$$

Thus the closed-loop system (4) becomes

$$\dot{\underline{x}} = (\underline{A} + \underline{b}\ell\underline{c}')\underline{x} + \underline{b}w = \tilde{\underline{A}}\underline{x} + \underline{b}w$$
$$y = \underline{c}'\underline{x} \tag{13}$$

and the closed-loop transfer function (5) is

$$\tilde{h}(s) = \frac{h(s)}{1 - \ell h(s)} = \frac{kg(s)}{1 - \ell kg(s)} \tag{14}$$

The reader should verify that this expression agrees with Lemma 8.1.2, and that the *loop gain* in this case is $-\ell k$.

It is interesting to note that letting $\ell = 1$ in (14) leads to a handy formula for computing the open-loop transfer function in terms of determinants (see Problem 1):

$$h(s) = \underline{c}'(s\underline{I} - \underline{A})^{-1}\underline{b} = \frac{\det(s\underline{I} - \underline{A}) - \det(s\underline{I} - \underline{A} - \underline{b}\underline{c}')}{\det(s\underline{I} - \underline{A})} \tag{15}$$

[†]See Definition 8.1.1.

In (1) and (10) only certain linear combinations of the states (specified by the rows of $\underline{C}$) were assumed to be available for direct observation at the output $\underline{y}$. This reflects the usual situation in practice, but occasionally it will be feasible to install enough sensors for the entire state vector $\underline{x}$ to be available for feedback. In this case, we have feedback *from the state* to the input (commonly known as *state feedback*), and (2) takes on the form

$$\underline{u}(t) = \underline{K}\underline{x}(t) + \underline{w}(t) \tag{16}$$

where the feedback gain matrix $\underline{K}$ is now $m \times n$. This situation (which is equivalent to setting $\underline{C} = \underline{I}$) is illustrated in Fig. 1(b), and the equations (4) of the closed-loop system become

$$\dot{\underline{x}} = (\underline{A} + \underline{B}\underline{K})\underline{x} + \underline{B}\underline{w} \tag{17}$$

Equation (16) is sometimes called *proportional state feedback* in the literature, but it actually includes what is known as *proportional-plus-derivative* feedback in the context of a scalar system. This is because the state vector effectively contains the output and its derivatives (see Theorem 2.2.1). In Sec. 11.3 we will see how *integral* feedback may also be included.

Similarly, in (1) and (10) only certain combinations of the state derivatives (specified by the columns of $\underline{B}$) were assumed to be available for direct control from the input $\underline{u}$. This reflects the usual situation in practice, but occasionally it will be feasible to install enough actuators for the entire state derivative vector $\dot{\underline{x}}$ to be manipulated by feedback. In this case, we have feedback from the output *to the state derivative*, and the feedback law (2) is replaced by

$$\underline{v}(t) = \underline{M}\underline{y}(t) + \underline{w}(t) \tag{18}$$

where the feedback gain matrix $\underline{M}$ is now $n \times p$ and $\underline{w}$ is an n-vector. This situation (which is equivalent to setting $\underline{B} = \underline{I}$) is illustrated in Fig. 1(c), and the equations (4) of the closed-loop system become

$$\begin{aligned} \dot{\underline{x}} &= (\underline{A} + \underline{M}\underline{C})\underline{x} + \underline{w} \\ \underline{y} &= \underline{C}\underline{x} \end{aligned} \tag{19}$$

This type of feedback is not very common. Nevertheless, there are some physical systems, such as distillation columns, for which it appears to be a more natural control structure than state feedback. Moreover, it is extremely useful in the dual context of a state estimator or observer, as we shall see in the next chapter.

Having defined these three forms of feedback laws (output to input, state to input, and output to state derivative), one might legitimately ask what effect they have on such system properties as controllability and observability:

2. *LEMMA*. The subspace of controllable states of a system is unchanged by feedback to the input. The subspace of observable states of a system is unchanged by feedback from the output. Consequently, both controllability and observability are unaffected when feedback from output to input is applied.

Proof:[†] If $\underline{x}_1$ is a controllable state of

$$\dot{\underline{x}} = \underline{A}\underline{x} + \underline{B}\underline{u} \tag{20}$$

then a control $\underline{\bar{u}}(t)$, $0 < t \leq T$, exists such that the state trajectory is $\underline{\bar{x}}(t)$, $0 \leq t \leq T$, with $\underline{\bar{x}}(0) = \underline{x}_1$ and $\underline{\bar{x}}(T) = \underline{0}$. If feedback is applied to the input, the system becomes

$$\dot{\underline{x}} = (\underline{A} + \underline{B}\underline{K})\underline{x} + \underline{B}\underline{w} \tag{21}$$

(where we may have $\underline{K} = \underline{L}\underline{C}$). In this case the control

$$\underline{\bar{w}}(t) = \underline{\bar{u}}(t) - \underline{K}\underline{\bar{x}}(t) \qquad (0 < t \leq T) \tag{22}$$

causes the closed-loop system to follow *precisely the same trajectory* $\underline{\bar{x}}(t)$, $0 \leq t \leq T$, because the term $-\underline{K}\underline{\bar{x}}(t)$ just cancels out the feedback. Thus the control $\underline{\bar{w}}$ drives $\underline{x}_1$ to $\underline{0}$, and $\underline{x}_1$ must be a controllable state of (21). The same argument establishes that any controllable state of (21) is also a controllable state of (20), so we

---

[†]This form of proof is particularly useful because it applies equally well to time-varying and even nonlinear systems. An algebraic proof using the controllability and observability matrices $\underline{P}_c$ and $\underline{P}_o$ may also be devised (Problem 2).

conclude that feedback to the input has no effect on controllability.

Similarly, suppose $\underline{x}_2$ is an unobservable state of

$$\dot{\underline{x}} = \underline{A}\underline{x} + \underline{v} \qquad y = \underline{C}\underline{x} \tag{23}$$

so that for $\underline{v}(t) \equiv \underline{0}$ and $\underline{x}(0) = \underline{x}_2$ we have

$$\underline{y}(t) = \underline{0} \qquad (0 \leq t < T) \tag{24}$$

If feedback is applied from the output, the system becomes

$$\dot{\underline{x}} = (\underline{A} + \underline{M}\underline{C})\underline{x} + \underline{w} \qquad \underline{y} = \underline{C}\underline{x} \tag{25}$$

(where we may have $\underline{M} = \underline{B}\underline{L}$). In this case, for $\underline{w}(t) \equiv \underline{0}$ and $\underline{x}(0) = \underline{x}_2$, the output will still be identically zero because the feedback term is $-\underline{M}\underline{C}\underline{x} = -\underline{M}\underline{y}$, which must be zero by (24). Thus $\underline{x}_2$ must be an unobservable state of (25). The same argument establishes that any unobservable state of (25) is also an unobservable state of (23), and we conclude that feedback from the output has no effect on observability.

Finally, if the feedback is from output to input, then both of the arguments above apply, and the sets of controllable and observable states are identical for (1) and (4).

ΔΔΔ

The lemma above is concerned with the overall controllability and observability of a system. We shall now see that feedback *can* be used to alter controllability and/or observability with respect to particular components of the input and output vectors.

3. LEMMA. If a system

$$\dot{\underline{x}} = \underline{A}\underline{x} + \underline{B}\underline{u} \tag{26}$$

is controllable and $\underline{b}_i$ ($\neq \underline{0}$) is the i-th column of $\underline{B}$, then there exists a feedback control law

$$\underline{u}(t) = \underline{K}_i \underline{x}(t) + \underline{w}(t) \tag{27}$$

from the state to the input such that the closed-loop system is controllable using the i-th input alone. In other words, a matrix $\underline{K}_i$ may be found such that the single-input system

$$\dot{\underline{x}} = (\underline{A} + \underline{B}\underline{K}_i)\underline{x} + \underline{b}_i w_i \tag{28}$$

is controllable.

Proof: The derivation of $\underline{K}_i$ is somewhat involved, and may be found in Refs. 39 and 263.

ΔΔΔ

3'. LEMMA. If the system

$$\begin{aligned} \dot{\underline{x}} &= \underline{A}\underline{x} + \underline{v} \\ \underline{y} &= \underline{C}\underline{x} \end{aligned} \tag{29}$$

is observable and $\underline{c}_j'(\neq \underline{0}')$ is the j-th row of $\underline{C}$, then there exists a feedback control law

$$\underline{v}(t) = \underline{M}_j \underline{y}(t) + \underline{w}(t) \tag{30}$$

from the output to the state derivative such that the closed-loop system is observable using the j-th output alone. In other words, a matrix $\underline{M}_j$ may be found such that the single-output system

$$\begin{aligned} \dot{\underline{x}} &= (\underline{A} + \underline{M}_j \underline{C})\underline{x} + \underline{w} \\ y_j &= \underline{c}_j' \underline{x} \end{aligned} \tag{31}$$

is observable.

Proof: This is the dual of Lemma 3.

ΔΔΔ

Variations of Lemmas 3 and 3' exist in which the system becomes controllable using a single linear combination of the inputs, or observable using a single linear combination of the outputs: see Refs. 238c, 259 and 370. It is also interesting to note that these results carry over simultaneously to the case of output-to-input feedback:

4. *LEMMA*. Suppose the system

$$\dot{\underline{x}} = \underline{A}\underline{x} + \underline{B}\underline{u}$$
$$\underline{y} = \underline{C}\underline{x} \tag{32}$$

is controllable and observable, $\underline{b}_i \neq \underline{0}$ is the i-th column of $\underline{B}$, and $\underline{c}_j' \neq \underline{0}'$ is the j-th row of $\underline{C}$. Then there exists a feedback law

$$\underline{u}(t) = \underline{L}_{ij}\underline{y}(t) + \underline{w}(t) \tag{33}$$

from output to input such that the closed-loop system is controllable with the i-th input and observable from the j-th output. In other words, a matrix $\underline{L}_{ij}$ may be found such that the scalar system

$$\dot{\underline{x}} = (\underline{A} + \underline{B}\underline{L}_{ij}\underline{C})\underline{x} + \underline{b}_i w_i$$
$$y_j = \underline{c}_j'\underline{x} \tag{34}$$

is both controllable and observable.

Proof: See Ref. 251. ΔΔΔ

11.2 PLACEMENT OF CLOSED-LOOP POLES

We saw in Sec. 11.1 how various types of feedback affect the system matrix, and we observed in earlier chapters that closed-loop behavior is largely determined by the locations of the closed-loop poles. We shall now show that in the case of feedback from the state in a controllable system (or feedback to the state derivative in an observable system), *any* symmetric closed-loop pole configuration may be achieved. Recall that this result was anticipated in Example 9.3.1 and Problem 9.3-1.

1. *THEOREM (Pole placement)*. Consider the system

$$\dot{\underline{x}} = \underline{A}\underline{x} + \underline{B}\underline{u} \tag{1}$$

and any preassigned symmetric† configuration of n poles in the

†This simply means that any complex poles occur in conjugate pairs, which of course must be the case if $\underline{A}$, $\underline{B}$, and $\underline{K}$ are real valued. The theorem holds without restriction if $\underline{K}$ can contain complex numbers.

complex plane. There exists a feedback control law

$$\underline{u}(t) = \underline{K}\underline{x}(t) + \underline{w}(t) \tag{2}$$

from the state to the input such that the poles of the closed-loop system

$$\dot{\underline{x}} = (\underline{A} + \underline{B}\underline{K})\underline{x} + \underline{B}\underline{w} \tag{3}$$

lie in the specified locations if and only if (1) is controllable.

Proof: If the system is uncontrollable, it can be decomposed as indicated in Theorem 10.3.3, and the uncontrollable poles are clearly unaffected by any feedback law of the form (2). This proves the "only if" part of the theorem.

Next, consider the case where (1) is a controllable single-input system,

$$\dot{\underline{x}} = \underline{A}\underline{x} + \underline{b}u \tag{4}$$

with the scalar feedback law

$$u(t) = \underline{k}'\underline{x}(t) + \underline{w}(t) \tag{5}$$

The system has been assumed controllable, and so according to Theorem 10.4.6 there exists a similarity transformation $\hat{\underline{x}} = \underline{Q}\underline{x}$ which puts (4) into controllable canonical form

$$\dot{\hat{\underline{x}}} = \hat{\underline{A}}\hat{\underline{x}} + \hat{\underline{b}}u \tag{6}$$

where

$$\hat{\underline{A}} = \underline{Q}\underline{A}\underline{Q}^{-1} = \begin{bmatrix} 0 & 1 & 0 & \cdots & 0 \\ 0 & 0 & 1 & \cdots & 0 \\ \cdots & \cdots & \cdots & \cdots & \cdots \\ 0 & 0 & 0 & \cdots & 1 \\ -p_0 & -p_1 & -p_2 & \cdots & -p_{n-1} \end{bmatrix}$$

$$\hat{\underline{b}} = \underline{Q}\underline{b} = \begin{bmatrix} 0 \\ 0 \\ \vdots \\ 0 \\ 1 \end{bmatrix} \tag{7}$$

and $s^n + p_{n-1}s^{n-1} + \dots + p_1 s + p_0$ is the characteristic polynomial of both $\hat{\underline{A}}$ and $\underline{A}$.

If we now use a feedback law

$$u(t) = \hat{\underline{k}}'\hat{\underline{x}}(t) + w(t) \tag{8}$$

where $\hat{\underline{k}}' \triangleq [\hat{k}_0 \quad \hat{k}_1 \quad \dots \quad \hat{k}_{n-1}]$, we find that the closed-loop system is

$$\dot{\hat{\underline{x}}} = (\hat{\underline{A}} + \hat{\underline{b}}\hat{\underline{k}}')\hat{\underline{x}} + \hat{\underline{b}}w \tag{9}$$

Because of the form of $\hat{\underline{b}}$, all the rows of $\hat{\underline{b}}\hat{\underline{k}}'$ are zero except the last one, which is $\hat{\underline{k}}'$, and

$$\hat{\underline{A}} + \hat{\underline{b}}\hat{\underline{k}}' = \begin{bmatrix} 0 & 1 & 0 & \dots & 0 \\ 0 & 0 & 1 & \dots & 0 \\ \dots & \dots & \dots & \dots & \dots \\ 0 & 0 & 0 & \dots & 1 \\ \hat{k}_0 - p_0 & \hat{k}_1 - p_1 & \hat{k}_2 - p_2 & \dots & \hat{k}_{n-1} - p_{n-1} \end{bmatrix} \tag{10}$$

Thus the characteristic polynomial of $(\hat{\underline{A}} + \hat{\underline{b}}\hat{\underline{k}}')$ is

$$s^n + (p_{n-1} - k_{n-1})s^{n-1} + \dots + (p_1 - k_1)s + (p_0 - k_0) \tag{11}$$

and its coefficients may be set to any arbitrary values by choosing $\hat{k}_0, \hat{k}_1, \dots, \hat{k}_{n-1}$ appropriately. Since the system poles are the roots of this polynomial, we see that a gain vector $\hat{\underline{k}}'$ exists to achieve any symmetric closed-loop pole configuration.

Returning to the original system, we apply the inverse similarity transformation $\underline{x} = \underline{Q}^{-1}\hat{\underline{x}}$ to (9) to find that the feedback law (8) is

$$u(t) = \hat{\underline{k}}'\underline{Q}\underline{x}(t) + w(t) = \underline{k}'\underline{x}(t) + w(t) \tag{12}$$

and the closed-loop system (9) becomes

$$\begin{aligned} \dot{\underline{x}} &= \underline{Q}^{-1}(\hat{\underline{A}} + \hat{\underline{b}}\hat{\underline{k}}')\underline{Q}\underline{x} + \underline{Q}^{-1}\hat{\underline{b}}w \\ &= (\underline{A} + \underline{b}\underline{k}')\underline{x} + \underline{b}w \end{aligned} \tag{13}$$

where $\underline{k}' \triangleq \hat{\underline{k}}'\underline{Q}$. The similarity transformation leaves the characteristic polynomial (11) of the system unaltered, so that $\underline{k}'$ is the required feedback gain vector.

The argument above proves the theorem for single-input systems. The case of a multiple-input system is more difficult, and was first established in Ref. 317. A proof using a canonical form may be found in Ref. 210, and another proof using algebraic arguments is in Ref. 365.

An alternative approach to the multivariable case makes use of Lemma 11.1.3. We first apply feedback with a gain matrix $\underline{K}_1$ in order to make (1) controllable using only the first[†] input. This means that the single-input system

$$\dot{\underline{x}} = (\underline{A} + \underline{B}\underline{K}_1)\underline{x} + \underline{b}_1 w_1 \triangleq \underline{A}_1\underline{x} + \underline{b}_1 w_1 \tag{14}$$

is controllable, and the argument above establishes the existence of a control law

$$w_1(t) = \tilde{\underline{k}}_1'\underline{x}(t) + \tilde{w}_1(t) \tag{15}$$

to achieve any preassigned symmetric pole configuration in the closed-loop system

$$\dot{\underline{x}} = (\underline{A}_1 + \underline{b}_1\tilde{\underline{k}}_1')\underline{x} + \underline{b}_1\tilde{w}_1 \tag{16}$$

This is illustrated in Fig. 1, where the intermediate system (14) is enclosed by broken lines.

Now we rewrite (15) as

$$\underline{w}(t) = \begin{bmatrix} \tilde{\underline{k}}_1' \\ \underline{0}' \\ \vdots \\ \underline{0}' \end{bmatrix} \underline{x}(t) + \tilde{\underline{w}}(t) \triangleq \tilde{\underline{K}}\underline{x}(t) + \tilde{\underline{w}}(t) \tag{17}$$

and note that

$$\underline{A}_1 + \underline{b}_1\tilde{\underline{k}}_1' = \underline{A}_1 + \underline{B}\tilde{\underline{K}} = \underline{A} + \underline{B}\underline{K}_1 + \underline{B}\tilde{\underline{K}} \tag{18}$$

[†] Clearly the same argument may be pursued using one of the other inputs, or some linear combination of the inputs, leading to various control laws which achieve the same pole configuration.

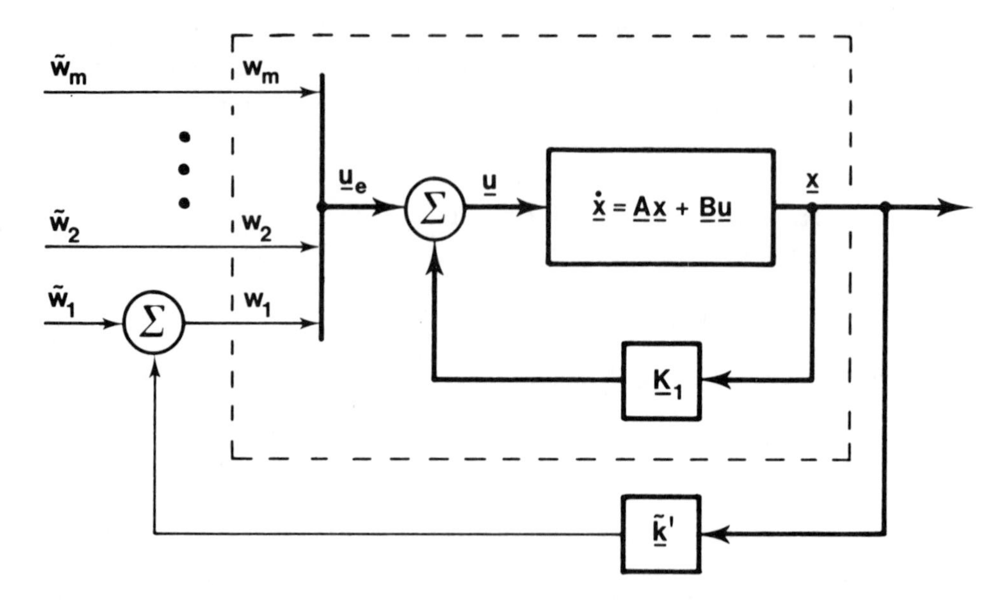

FIG. 1. Pole placement with dual-mode feedback (heavy lines represent vector quantities).

Thus the closed-loop system

$$\dot{\underline{x}} = [\underline{A} + \underline{B}(\underline{K}_1 + \tilde{\underline{K}})]\underline{x} + \underline{B}\tilde{\underline{w}} \tag{19}$$

has the required pole configuration, as a result of the composite feedback law

$$\underline{u}(t) = (\underline{K}_1 + \tilde{\underline{K}})\underline{x}(t) + \tilde{\underline{w}}(t) \tag{20}$$

ΔΔΔ

If a system is *not* controllable, then Theorem 1 says that its closed-loop poles cannot be arbitrarily specified. Nevertheless, the system may still be stabilizable:

2. *COROLLARY*. The closed-loop system (3) can be made stable if and only if the uncontrollable poles of (1) are all stable. In this case, the system is said to be *stabilizable* with state feedback, and the closed-loop poles of the controllable subsystem may be placed in any symmetric configuration.

Proof: The state transformation of Theorem 10.3.3 may be used to decompose the system into two subsystems. The reader should use Theorem 1 to verify that the closed-loop poles of the controllable subsystem (10.3-14) can be placed arbitrarily with feedback in the original (untransformed) system, and hence that the overall system can be stabilized.

ΔΔΔ

In the case of feedback from the output to the state derivative, we have the following dual results, the proofs of which are left to the reader (Problem 1).

1'. THEOREM (Pole placement). Consider the system

$$\begin{aligned} \dot{\underline{x}} &= \underline{A}\underline{x} + \underline{v} \\ \underline{y} &= \underline{C}\underline{x} \end{aligned} \tag{21}$$

and any preassigned symmetric configuration of n poles in the complex plane. There exists a feedback control law

$$\underline{v}(t) = \underline{M}\underline{y}(t) + \underline{w}(t) \tag{22}$$

from the output to the state derivative such that the poles of the closed-loop system

$$\begin{aligned} \dot{\underline{x}} &= (\underline{A} + \underline{M}\underline{C})\underline{x} + \underline{w} \\ \underline{y} &= \underline{C}\underline{x} \end{aligned} \tag{23}$$

lie in the specified locations if and only if (21) is observable.

ΔΔΔ

2'. COROLLARY. The closed-loop system (23) can be made stable if and only if the unobservable poles of (21) are all stable. In this case the system is said to be *stabilizable* with feedback from output to state derivative, and the closed-loop poles of the observable subsystem may be placed in any symmetric configuration.

ΔΔΔ

Theorem 1 indicates the great power of feedback from the state to the input (state-variable feedback) to influence the dynamic behavior of a controllable system. If the system is unstable, it can always be stabilized, and in principle any prespecified stability margin or damping ratio can be achieved. Although the closed-loop poles may be placed anywhere in the complex plane, some caution must be used in drawing conclusions about the resulting closed-loop transient behavior.

One reason for caution is that we have not taken into account the *zeros* of the system, and these can have a profound effect upon the shape of the transient response. We saw in Problem 5.3-3 and Example 9.1.2 that a zero can cause a pronounced overshoot even in the absence of any oscillatory (complex) poles. In Problem 5.3-4 and Example 9.3.4, cancellation or near-cancellation of a pole by a zero was shown to be very effective in removing the corresponding exponential mode from the transient response. Non-minimum phase zeros (i.e., those which lie in the right half-plane) can also be the source of some rather perverse behavior; this was discussed at the end of Sec. 6.2 and illustrated in Problem 6.2-10.

In Problem 2 the reader is asked to show that zeros are completely unaffected by feedback. They can, of course, be altered by compensation, i.e., by introducing additional dynamics into the system. Zeros of multivariable systems are the subject of a good deal of current research, although there is not even complete agreement on the basic definition, particularly when the input and output vectors have different dimensions [16, 105, 149, 244, 250, 308, 314, 360, 361].

Another reason for caution is that there are always practical limits to the speed of response that can be achieved in a system. A fast response requires large feedback gains, which can lead to unexpected transients or oscillations and magnify errors due to parameter uncertainty or roundoff. Moreover, the gains are always limited by physical constraints in the form of saturation, overheating, and various other nonlinear effects. While reading about various systematic and computer-oriented design procedures in this chapter and the next, it is very important to bear in mind the "rules of thumb" discussed in Chapters 8 and 9 in the context of single-loop feedback systems. Even with the cleverest of computer algorithms, there is no substitute for good, sound engineering judgment.

The state feedback control law which yields a particular pole configuration in a single-input system is unique. For a multiple-input system, however, there will be many control laws which achieve the same pole configuration, and the designer will in general have free parameters remaining even after specifying the pole positions. Although there are various algorithms available for determining specific control laws [210, 238, 259], effective usage of the remaining degrees of design freedom remains a topic of active research. One approach is to choose an *optimal* control law which minimizes a quadratic index of system performance. This will be the subject of Chapter 13.

3. EXAMPLE. The broom-balancing system of Example 2.3.2 is shown in Fig. 2(a). Recall that the state vector is

$$\underline{x}(t) = \begin{bmatrix} z(t) \\ \dot{z}(t) \\ \Theta(t) \\ \dot{\Theta}(t) \end{bmatrix} \tag{24}$$

and the linearized equations of the system (for $\Theta(t) \simeq 0$) are

$$\dot{\underline{x}} = \underline{\underline{A}}\underline{x} + \underline{b}u = \begin{bmatrix} 0 & 1 & 0 & 0 \\ 0 & 0 & -1 & 0 \\ 0 & 0 & 0 & 1 \\ 0 & 0 & 11 & 0 \end{bmatrix} \underline{x} + \begin{bmatrix} 0 \\ 1 \\ 0 \\ -1 \end{bmatrix} u \tag{25}$$

(a)

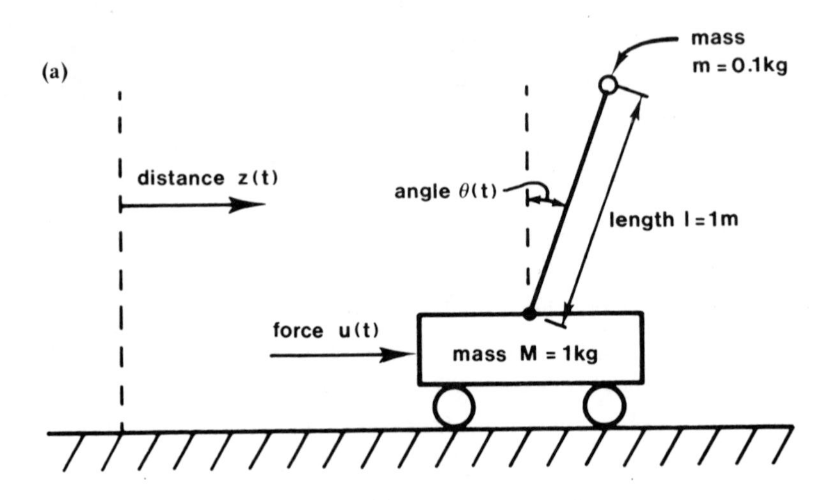

(b)

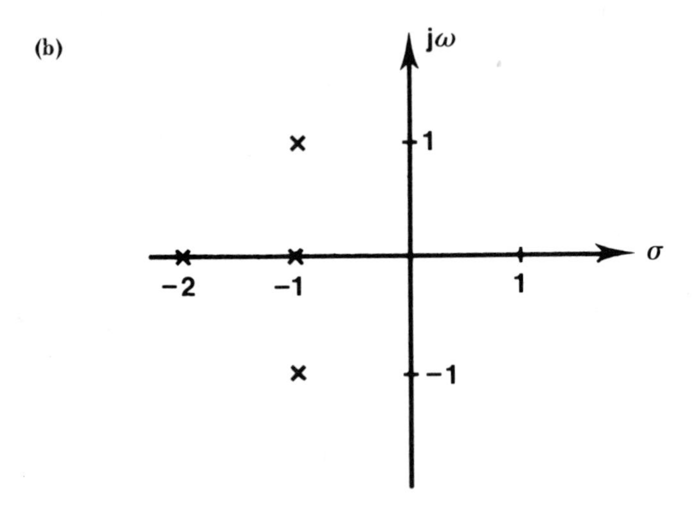

FIG. 2. System of Example 3: (a) broom-balancing system and (b) desired closed-loop pole configuration.

It is easy to verify that the characteristic polynomial is

$$p(s) = \det(s\underline{I} - \underline{A}) = s^4 - 11s^2$$
$$= s^2(s - \sqrt{11})(s + \sqrt{11}) \tag{26}$$

Since there are poles at 0, 0, $\sqrt{11}$, and $-\sqrt{11}$, the system is quite unstable, as one would expect from physical reasoning.

Suppose we require a feedback control law of the form

$$u(t) = \underline{k}'\underline{x}(t) + w(t) = \begin{bmatrix} k_1 & k_2 & k_3 & k_4 \end{bmatrix} \underline{x}(t) + w(t) \tag{27}$$

such that the closed-loop system has the stable pole configuration shown in Fig. 2(b). We verified in Example 10.1.9 that (25) is a controllable system, so Theorem 1 tells us that such a feedback gain vector $\underline{k}'$ does exist. We could determine the required $\underline{k}'$ by first transforming the system to controllable canonical form, as in the proof of Theorem 1, but in this case it is easier to proceed directly to the solution.

The desired characteristic polynomial is

$$\tilde{p}(s) = (s + 1)(s + 2)(s + 1 + j)(s + 1 - j)$$
$$= s^4 + 5s^3 + 10s^2 + 10s + 4 \tag{28}$$

Substituting the control law (27) into (25), we find that the closed-loop system is

$$\dot{\underline{x}} = (\underline{A} + \underline{b}\underline{k}')\underline{x} + \underline{b}w$$

$$= \begin{bmatrix} 0 & 1 & 0 & 0 \\ k_1 & k_2 & k_3 - 1 & k_4 \\ 0 & 0 & 0 & 1 \\ -k_1 & -k_2 & 11 - k_3 & -k_4 \end{bmatrix} \underline{x} + \begin{bmatrix} 0 \\ 1 \\ 0 \\ -1 \end{bmatrix} \underline{x} \triangleq \underline{\tilde{A}}\underline{x} + \underline{b}w \tag{29}$$

with the characteristic polynomial

$$\tilde{p}(s) = \det(s\underline{I} - \underline{\tilde{A}})$$
$$= s^4 + (k_4 - k_2)s^3 + (k_3 - k_1 - 11)s^2 + (10k_2)s + (10k_1) \tag{30}$$

The required feedback gain vector $\underline{k}'$ is that which makes the coefficients equal in the two expressions for $\tilde{p}(s)$, that is,

$$\begin{aligned} k_1 &= 0.4 \\ k_2 &= 1 \\ k_3 &= 10 + 11 + k_1 = 21.4 \\ k_4 &= 5 + k_2 = 6 \end{aligned} \tag{31}$$

This feedback control law yields a stable closed-loop system, so that the entire state vector $\underline{x}$ decays asymptotically to zero in the absence of disturbances. This means not only that the broom is balanced $(\theta \to 0)$, but that the cart returns to its origin as well $(z \to 0)$. It is interesting to note that if all four states are not available for feedback, then this system cannot be stabilized at all (see Problem 4).

ΔΔΔ

Determining a feedback gain matrix which achieves a particular pole configuration may involve a substantial computational effort, particularly for multivariable systems. If the exact pole positions are not crucial, however, the following theorem shows how a stabilizing control can be found more rapidly. This is also important because there are a number of iterative design algorithms which require a stable feedback law as a starting point.

4. *THEOREM*. If the system

$$\dot{\underline{x}} = \underline{A}\underline{x} + \underline{B}\underline{u} \tag{32}$$

is controllable, then the state feedback control law

$$\underline{u}(t) = -\underline{B}'\underline{W}_c^{-1}(0,T)\underline{x}(t) + \underline{w}(t) \tag{33}$$

results in an asymptotically stable closed-loop system

$$\dot{\underline{x}} = (\underline{A} - \underline{B}\underline{B}'\underline{W}_c^{-1})\underline{x} + \underline{B}\underline{w} \tag{34}$$

for any $T > 0$, where

$$W_c(0,T) = \int_0^T e^{-A\tau} BB' e^{-A'\tau} \, d\tau \tag{35}$$

Proof: See Ref. 281.

ΔΔΔ

The matrix $W_c(0,T)$ was defined in Theorem 10.1.4, and in principle it is nonsingular for any $T > 0$. However, choosing T so that $W_c(0,T)$ is numerically well conditioned for inversion is a nontrivial task. The case of an uncontrollable system (and hence a singular W_c) is considered in Ref. 328 and in Problem 6.

So far in this section we have assumed that feedback could be applied from the state to the input (or from the output to the state derivative). Often, however, not all of the states will be available and we must make do with feedback from output to input, of the form

$$u(t) = Ly(t) + w(t) \tag{36}$$

In this case the pole-placement theorems (1 and 1') no longer apply, and the precise relationship between L and the closed-loop pole configuration is only partially understood. Indeed, even the problem of determining whether an L exists to stabilize a given multivariable system remained one of the major unsolved problems of linear control theory until recently when a solution using decision methods was given in Ref. 209. Since this solution is very complex and computationally expensive, the problem will continue to motivate a good deal of research activity.

Another approach to the output feedback problem is given in Ref. 246. If we assume that the system is controllable and observable and that rank (B) + rank (C) > n, then it can be shown that for *almost all*[†] matrices B and C, there exists a feedback control law of the form (36) such that the closed-loop poles lie *arbitrarily close* to any preassigned symmetric configuration.

[†]This means that there are some B and C for which the result does not hold, but an arbitrarily small perturbation will produce $\hat{B}$ and $\hat{C}$ for which it does hold. Note, however, that perturbation of certain elements of B and C (e.g., zero elements) is often physically unreasonable.

No other general procedures exist for determining a stable multivariable feedback law from output to input, of the form (36). If we allow *dynamic feedback* (or *compensation*), then another linear system such as an *observer* can be used in the feedback loop to estimate the state vector, and the estimate can be used in a state feedback law as in Theorem 1. This will be the subject of Chapter 12.

For scalar systems, of course, the feedback gain matrix $\underline{L}$ reduces to a scalar ℓ, and answers to these questions are provided by the root-locus diagram (introduced in Sec. 8.3), which is just a plot of the closed-loop poles as functions of ℓ. Because scalar systems are easily analyzed using the techniques of Chapter 8, it might appear reasonable to "convert" a multiple-loop feedback system to a single-loop system using Lemma 11.1.4. Indeed, this lemma and the two preceding it are quite interesting, particularly since they can be used to prove the pole-placement theorems (1 and 1') in the multivariable case. However, this amounts to "discarding" all but one of the inputs (or outputs), and intuition suggests that a better control law may exist which makes use of all the available variables. Even where a satisfactory closed-loop pole configuration can be achieved with feedback to a single input (or from a single output), the use of additional inputs (or outputs) may improve the sensitivity of the closed-loop system to noise or parameter variations, decrease the control energy required, or provide a safety margin in the event of component failures.

More importantly, the use of Lemma 11.1.4 on a system which is stabilizable with multivariable feedback can lead to a scalar system which is more easily analyzed with root-locus or frequency response techniques, but cannot be stabilized *at all* with linear output feedback. This situation is illustrated by the following example.

5. *EXAMPLE*. Consider the double-input, single-output system described by

$$\dot{\underline{x}} = \underline{A}\underline{x} + \underline{B}\underline{u} = \begin{bmatrix} 0 & 0 & 5 \\ 1 & 0 & -1 \\ 0 & 1 & -3 \end{bmatrix} \underline{x} + \begin{bmatrix} -2 & 0 \\ 1 & -2 \\ 0 & 1 \end{bmatrix} \underline{u} \tag{37}$$

$$y = \underline{c}'\underline{x} = \begin{bmatrix} 0 & 0 & 1 \end{bmatrix} \underline{x}$$

Since this is in observable canonical form, it must be observable, and the characteristic polynomial is

$$\begin{aligned} p(s) &= s^3 + 3s^2 + s - 5 \\ &= [(s+2)^2 + 1^2](s-1) \end{aligned} \tag{38}$$

The pole at $s = 1$ indicates that the system is unstable, and we shall now investigate the possibility of stabilizing it by means of feedback from the output to the inputs.

The controllability matrices for each of the two inputs are

$$\underline{P}_{c1} = \begin{bmatrix} -2 & 0 & 5 \\ 1 & -2 & -1 \\ 0 & 1 & -5 \end{bmatrix} \qquad \underline{P}_{c2} = \begin{bmatrix} 0 & 5 & -25 \\ -2 & -1 & 10 \\ 1 & -5 & 14 \end{bmatrix} \tag{39}$$

both of which are nonsingular. Thus the system is already controllable from each of the inputs independently and observable from the only output, and there is no need to use Lemma 11.1.4 at all.

With feedback of the form

$$\underline{u}(t) = \underline{\ell} y(t) + \underline{w}(t) \tag{40}$$

from the one output to both inputs, the closed-loop system becomes

$$\begin{aligned} \dot{\underline{x}} &= \underline{\tilde{A}}\underline{x} + \underline{w} \\ y &= \underline{c}'\underline{x} \end{aligned} \tag{41}$$

where

$$\begin{aligned} \underline{\tilde{A}} &= \underline{A} + \underline{B}\underline{\ell}\underline{c}' \\ &= \begin{bmatrix} 0 & 0 & 5 \\ 1 & 0 & -1 \\ 0 & 1 & -3 \end{bmatrix} + \begin{bmatrix} -2 & 0 \\ 1 & -2 \\ 0 & 1 \end{bmatrix} \begin{bmatrix} \ell_1 \\ \ell_2 \end{bmatrix} \begin{bmatrix} 0 & 0 & 1 \end{bmatrix} \\ &= \begin{bmatrix} 0 & 0 & 5-2\ell_1 \\ 1 & 0 & \ell_1-2\ell_2-1 \\ 0 & 1 & \ell_2-3 \end{bmatrix} \end{aligned} \tag{42}$$

The closed-loop characteristic polynomial is thus

$$\tilde{p}(s) = s^3 + (3-\ell_2)s^2 + (1-\ell_1+2\ell_2)s + (2\ell_1 - 5) \tag{43}$$

If we choose the feedback gains to be

$$\ell_1 = 3 \qquad \ell_2 = 2 \tag{44}$$

then this becomes

$$\tilde{p}(s) = s^3 + s^2 + 2s + 1 = s^3 + \tilde{p}_2 s^2 + \tilde{p}_1 s + \tilde{p}_0 \tag{45}$$

Using the Routh-Hurwitz conditions (Corollary 5.2.6), we find that $\tilde{p}_0$, $\tilde{p}_1$, $\tilde{p}_2$, and $(\tilde{p}_2\tilde{p}_1 - \tilde{p}_0)$ are all positive, which implies that the closed-loop system is stable. In fact, the roots of $\tilde{p}$ are -0.57 and $-0.22 \pm j1.3$. Now suppose we had tried to treat the system as a scalar one, and to stabilize it using only the first input. In this case $\ell_2 = 0$, and (43) becomes

$$\tilde{p}(s) = s^3 + 3s^2 + (1 - \ell_1)s + (2\ell_1 - 5) \tag{46}$$

Since there is no choice of ℓ_1 for which $(1 - \ell_1)$ and $(2\ell_1 - 5)$ are both positive, we conclude that the system cannot be stabilized in this way. Similarly, if only the second input is used, we have

$$\tilde{p}(s) = s^3 + (3 - \ell_2)s^2 + (1 + 2\ell_2)s - 5 \tag{47}$$

which is unstable for all choices of ℓ_2 because the constant term is always negative.

The reader may wish to verify these conclusions by constructing root-locus diagrams.

ΔΔΔ

This example illustrates that reduction of a multivariable system to a scalar one, although it simplifies the analysis, may have disastrous consequences. In this case a stabilizable system is rendered unstabilizable, regardless of which input is used in the feedback loop.

11.3 STEADY-STATE PERFORMANCE

In the previous section, we saw how feedback from state to input could be used to position the closed-loop poles of a system and hence determine its transient performance. Now we shall turn to the complementary subject of steady-state behavior,[†] which was discussed for single-loop feedback systems in Sec. 8.2.

Consider a system of the form

$$\begin{aligned} \dot{\underline{x}}(t) &= \underline{A}\underline{x}(t) + \underline{B}\underline{u}(t) + \underline{d}(t) \\ \underline{y}(t) &= \underline{C}\underline{x}(t) \end{aligned} \tag{1}$$

where $\underline{u}(t)$ is the *control* vector, $\underline{d}(t)$ is a vector of unknown *disturbance* inputs, and the output $\underline{y}(t)$ is required to remain as close as possible to a *reference input* (or *desired output*) $\underline{y}_r(t)$. The vectors $\underline{x}$, $\underline{d}$, $\underline{u}$, $\underline{y}$, and $\underline{y}_r$ have dimensions n, n, m, p, and p, respectively. When $m = p = 1$, this reduces to the problem formulated in Sec. 8.1 (with multiple disturbances). We assume that (1) is controllable from the control input $\underline{u}$ (if $\underline{d} \equiv \underline{0}$).

In Sec. 8.2 we saw that perfect static accuracy could be achieved in a scalar system with constant reference and disturbance inputs by ensuring that the control input contained a term proportional to the integral of the error. Restricting our attention for the moment to $\underline{d}$ and $\underline{y}_r$ which are step functions, we may extend the reasoning of Sec. 8.2 to multiple inputs and outputs and conjecture that the control in this case should include a term proportional to

$$\underline{q}(t) = \int_0^t \underline{e}(\tau)\, d\tau = \int_0^t [\underline{y}(\tau) - \underline{y}_r(\tau)]\, d\tau \tag{2}$$

where $\underline{e} = \underline{y} - \underline{y}_r$ is the *error vector*. Since $\underline{q}(t)$ satisfies the differential equation

$$\dot{\underline{q}}(t) = \underline{e}(t) = \underline{C}\underline{x}(t) - \underline{y}_r(t) \tag{3}$$

with $\underline{q}(0) = \underline{0}$, it is easily included by augmenting the original system (1) as follows:

$$\begin{aligned} \begin{bmatrix} \dot{\underline{x}} \\ \dot{\underline{q}} \end{bmatrix} &= \begin{bmatrix} \underline{A} & \underline{0} \\ \underline{C} & \underline{0} \end{bmatrix} \begin{bmatrix} \underline{x} \\ \underline{q} \end{bmatrix} + \begin{bmatrix} \underline{B} \\ \underline{0} \end{bmatrix} \underline{u} + \begin{bmatrix} \underline{d} \\ -\underline{y}_r \end{bmatrix} \\ \underline{y} &= \begin{bmatrix} \underline{C} & \underline{0} \end{bmatrix} \begin{bmatrix} \underline{x} \\ \underline{q} \end{bmatrix} \end{aligned} \tag{4}$$

[†]This topic has been studied extensively in the recent literature. See, for example, Refs. 240-243, 256a, 256b, 273, 329, 332a, 339, 357b, 362, and 370.

where the new state vector has dimension $n+p$. This amounts to adding p integrators, each driven by one component of the error vector $\underline{e}$.

The first question which arises is whether the new system is controllable† from the input $\underline{u}$. This is answered as follows:

1. LEMMA. The composite system (4) is controllable if and only if the original system (1) is controllable *and*

$$\text{rank}\begin{bmatrix} \underline{A} & \underline{B} \\ \underline{C} & \underline{0} \end{bmatrix} = n+p \tag{5}$$

Note that (5) can hold *only if* $m \geq p$ (i.e., there are at least as many inputs as outputs) and rank $(\underline{C}) = p$ (i.e., the outputs are independent).

Proof: The system (4) clearly cannot be controllable unless (1) is controllable. Thus we may assume that the $n \times nm$ controllability matrix of (1),

$$\underline{P}_1 \triangleq [\underline{B} \quad \underline{AB} \quad \underline{A}^2\underline{B} \quad \ldots \quad \underline{A}^{n-1}\underline{B}] \tag{6}$$

has full rank (n). Additional columns cannot reduce the rank of (6), so the $n \times (n+p-1)m$ matrix

$$\underline{P}_2 = [\underline{B} \quad \underline{AB} \quad \underline{A}^2\underline{B} \quad \ldots \quad \underline{A}^{n+p-2}\underline{B}] \tag{7}$$

must also have full rank. Similarly, the system (4) will be controllable if and only if its $(n+p) \times (n+p)m$ controllability matrix

$$\underline{P} = \begin{bmatrix} \underline{B} & \underline{AB} & \underline{A}^2\underline{B} & \ldots & \underline{A}^{n+p-1}\underline{B} \\ \underline{0} & \underline{CB} & \underline{CAB} & \ldots & \underline{CA}^{n+p-2}\underline{B} \end{bmatrix} = \begin{bmatrix} \underline{B} & \underline{AP}_2 \\ \underline{0} & \underline{CP}_2 \end{bmatrix} \tag{8}$$

has full rank $(n+p)$. But this can be written as

$$\underline{P} = \begin{bmatrix} \underline{A} & \underline{B} \\ \underline{C} & \underline{0} \end{bmatrix} \begin{bmatrix} \underline{0} & \underline{P}_2 \\ \underline{I} & \underline{0} \end{bmatrix} \tag{9}$$

† Such a system is said to be controllable if it satisfies the usual controllability conditions when $\underline{d}$ and $\underline{y}_r$ are both zero.

where the $(n+m) \times (n+p)m$ matrix on the right clearly has full rank because $\underline{P}_2$ does. We conclude (using Problem A.5-11) that

$$\text{rank } \underline{P} = \text{rank} \begin{bmatrix} \underline{A} & \underline{B} \\ \underline{C} & \underline{0} \end{bmatrix} \tag{10}$$

Moreover, since the matrix in (10) is $(n+p) \times (n+m)$, its rank can be $n+p$ only if $m \geq p$ (Problem A.5-3), and only if rank $(\underline{C}) = p$ (otherwise, one of the last p rows in (10) would be dependent).

ΔΔΔ

Having determined a test for the controllability of the composite system (4), we now propose to apply a state feedback control law

$$\underline{u}(t) = \begin{bmatrix} \underline{K}_1 & \underline{K}_2 \end{bmatrix} \begin{bmatrix} \underline{x}(t) \\ \underline{q}(t) \end{bmatrix} = \underline{K}_1\underline{x}(t) + \underline{K}_2\underline{q}(t) \tag{11}$$

to obtain the closed-loop system

$$\begin{aligned} \begin{bmatrix} \dot{\underline{x}} \\ \dot{\underline{q}} \end{bmatrix} &= \begin{bmatrix} \underline{A}+\underline{B}\underline{K}_1 & \underline{B}\underline{K}_2 \\ \underline{C} & \underline{0} \end{bmatrix} \begin{bmatrix} \underline{x} \\ \underline{q} \end{bmatrix} + \begin{bmatrix} \underline{d} \\ -\underline{y}_r \end{bmatrix} \\ \underline{y} &= \begin{bmatrix} \underline{C} & \underline{0} \end{bmatrix} \begin{bmatrix} \underline{x} \\ \underline{q} \end{bmatrix} \end{aligned} \tag{12}$$

shown in Fig. 1. In (11), note that the first term $\underline{K}_1\underline{x}$ is ordinary state feedback for the original system (and hence includes both *proportional* and *derivative* control, as defined in Chapters 8 and 9) while the second term $\underline{K}_2\underline{q}$ provides *integral* control action to improve the static accuracy.

If (1) is controllable and (5) holds, then (4) is controllable and Theorem 11.2.1 establishes that the gain matrix $[\underline{K}_1 \;\; \underline{K}_2]$ may be chosen so as to place the closed-loop poles of (12) in any desired locations. It can also happen that (4) is not controllable, or the full plant state is not available for measurement, but the system is stabilizable anyway. For instance, if only $\underline{y}$ and $\underline{q}$ are available for feedback, then (11) must be replaced by

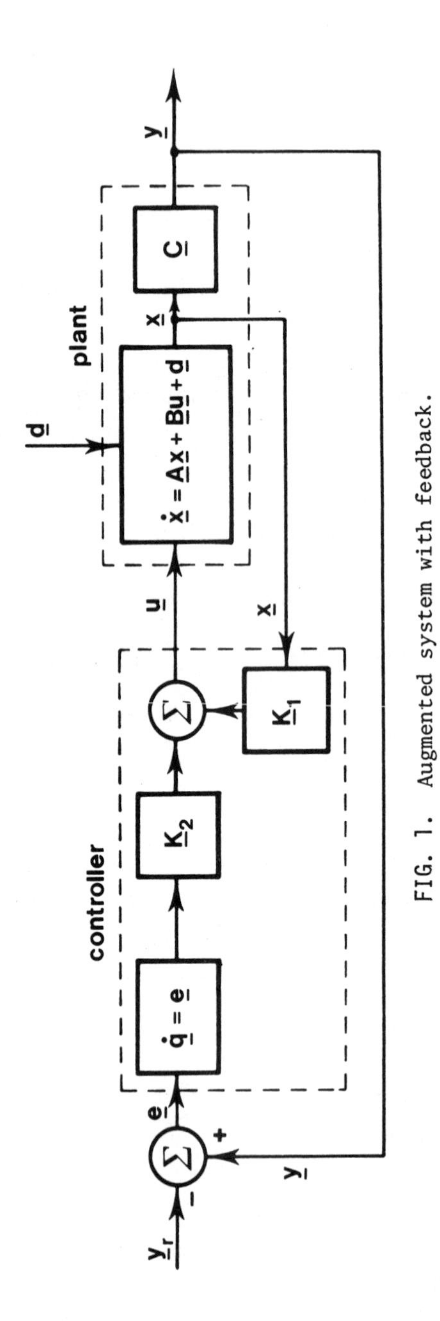

FIG. 1. Augmented system with feedback.

$$\underline{u}(t) = \underline{L}_1\underline{y}(t) + \underline{K}_2\underline{q}(t) = \underline{L}_1\underline{C}\underline{x}(t) + \underline{K}_2\underline{q}(t) \tag{13}$$

so that $[\underline{K}_1 \;\; \underline{K}_2]$ is constrained to have the form $[\underline{L}_1\underline{C} \;\; \underline{K}_2]$. In any case, if (12) can be stabilized, then it achieves perfect static accuracy for constant inputs.

2. *THEOREM.* Suppose that $[\underline{K}_1 \;\; \underline{K}_2]$ has been chosen so as to make the closed-loop system (12) stable, and that the disturbance and reference inputs are step functions,

$$\underline{d}(t) = \overline{\underline{d}}1(t) \qquad \underline{y}_r(t) = \overline{\underline{y}}_r 1(t) \tag{14}$$

Then $\underline{x}(t)$ and $\underline{q}(t)$ converge to constant steady-state values $\overline{\underline{x}}$ and $\overline{\underline{q}}$, and

$$\lim_{t\to\infty} \underline{y}(t) = \overline{\underline{y}}_r \tag{15}$$

Proof: Since the system is stable, Theorem 6.1.5(b) establishes the convergence to constant values. But this implies that $\dot{\underline{x}}(t)$ and $\dot{\underline{q}}(t)$ both approach zero, which in turn implies (15) because of (3). An alternative proof is requested in Problem 1.

ΔΔΔ

We have now seen how a multivariable system can be made to follow a constant reference input with perfect static accuracy, despite an unmeasurable constant disturbance input. The closed-loop system (12), shown in Fig. 1, resembles the scalar (or single-loop) feedback system of Fig. 8.2-3. As in Theorems 8.2.2 and 8.2.3, we see that perfect static accuracy is achieved by means of integrators (i.e., poles at $s = 0$) in the controller. In fact, Theorem 8.2.2 established that integrators in the *plant* would suffice if there were no disturbance input. These results may be extended using the multivariable definition of a *type ℓ system* [329, 357b].

The introduction of integrators, of course, tends to slow the system down. As in the scalar case, this can be corrected by increasing the feedback gains. These must be chosen so as to yield (via the closed-loop pole configuration) a transient response which is sufficiently fast for the job at hand but avoids excessive oscillations.

Inputs involving polynomials in t, such as $t1(t)$, $t^2 1(t)$, etc., can be handled by adding more integrators (Problem 2). A disturbance which is measurable directly can be dealt with by means of a *feed-forward controller* [240, 243, 250a, 339].

The condition (5) for controllability of the composite system has a number of interpretations and appears in the literature in various contexts. The following lemma should serve to further illuminate it.

3. LEMMA. The rank condition of Lemma 1 is unaffected by state feedback applied to the system (1), that is,

$$\text{rank}\begin{bmatrix} \underline{A} & \underline{B} \\ \underline{C} & \underline{0} \end{bmatrix} = \text{rank}\begin{bmatrix} \underline{A} + \underline{B}\underline{K}_1 & \underline{B} \\ \underline{C} & \underline{0} \end{bmatrix} \tag{16}$$

for all $m \times n$ feedback matrices $\underline{K}_1$. If $\underline{A}$ is nonsingular, then

$$\text{rank}\begin{bmatrix} \underline{A} & \underline{B} \\ \underline{C} & \underline{0} \end{bmatrix} = n + \text{rank}(\underline{C}\underline{A}^{-1}\underline{B}) = n + \text{rank}(\underline{H}(0)) \tag{17}$$

where $\underline{H}(s) = \underline{C}(s\underline{I} - \underline{A})^{-1}\underline{B}$ is the transfer function matrix of (1). If $\underline{A}$ is singular but the system (1) is stabilizable, then $\underline{A} + \underline{B}\underline{K}_1$ can be made nonsingular (no zero eigenvalues) by suitable choice of $\underline{K}_1$, and $\underline{A} + \underline{B}\underline{K}_1$ can be used in (17) in place of $\underline{A}$.

Proof: The product of a matrix of rank r and a nonsingular square matrix must also have rank r (Problem A.5-10). Thus (16) follows because

$$\begin{bmatrix} \underline{A} + \underline{B}\underline{K}_1 & \underline{B} \\ \underline{C} & \underline{0} \end{bmatrix} = \begin{bmatrix} \underline{A} & \underline{B} \\ \underline{C} & \underline{0} \end{bmatrix}\begin{bmatrix} \underline{I} & \underline{0} \\ \underline{K}_1 & \underline{I} \end{bmatrix} \tag{18}$$

where the matrix on the right is nonsingular by virtue of its block-triangular structure (Theorem A.2.7(f)). The same argument applies to the product

$$\begin{bmatrix} \underline{I} & \underline{0} \\ \underline{0} & \underline{C}\underline{A}^{-1}\underline{B} \end{bmatrix} = \begin{bmatrix} \underline{I} & \underline{0} \\ -\underline{C}\underline{A}^{-1} & \underline{I} \end{bmatrix}\begin{bmatrix} \underline{A} & \underline{B} \\ \underline{C} & \underline{0} \end{bmatrix}\begin{bmatrix} \underline{A}^{-1} & \underline{A}^{-1}\underline{B} \\ \underline{0} & -\underline{I} \end{bmatrix} \tag{19}$$

whose rank is clearly equal to

$$\operatorname{rank}(\underline{I}) + \operatorname{rank}(\underline{C}\underline{A}^{-1}\underline{B}) = n + \operatorname{rank}(\underline{C}\underline{A}^{-1}\underline{B}) \tag{20}$$

ΔΔΔ

This lemma is particularly interesting because it establishes (assuming $\underline{A}^{-1}$ exists) that the condition (5) is equivalent to

$$\operatorname{rank}(\underline{H}(0)) = \operatorname{rank}(\underline{C}\underline{A}^{-1}\underline{B}) = p \tag{21}$$

where $\underline{H}(0) = -\underline{C}\underline{A}^{-1}\underline{B}$ is the *static gain matrix* of the system (1), defined in Theorem 6.1.5(b). With the same number of inputs and outputs ($m = p$), this simply means that $\underline{H}(0)$ is nonsingular, a condition which has been called *nonsingular d.c. gain* [329]. The same condition is crucial to the concept of static decoupling, as we shall see in the next section.

4. EXAMPLE. In Example 11.2.3 a state feedback control law was found to stabilize the broom-balancing system of Fig. 11.2-2. Recall that the state vector is $\underline{x} = [z \quad \dot{z} \quad \Theta \quad \dot{\Theta}]'$. Suppose now that the wind exerts a horizontal force w(t) on the pendulum and 5w(t) on the cart. The reader can easily verify (Problem 2.3-2) that the dynamic equations in this case become

$$\dot{\underline{x}} = \underline{A}\underline{x} + \underline{b}u + \underline{d} = \begin{bmatrix} 0 & 1 & 0 & 0 \\ 0 & 0 & -1 & 0 \\ 0 & 0 & 0 & 1 \\ 0 & 0 & 11 & 0 \end{bmatrix} \underline{x} + \begin{bmatrix} 0 \\ 1 \\ 0 \\ -1 \end{bmatrix} u + \begin{bmatrix} 0 \\ 4 \\ 0 \\ 6 \end{bmatrix} w \tag{22}$$

Suppose further that we designate the position z as output,

$$z = \underline{c}'\underline{x} = \begin{bmatrix} 1 & 0 & 0 & 0 \end{bmatrix} \underline{x} \tag{23}$$

and require that it be maintained equal to the reference input (or set point) z_r. Note that we have *not* designated the angle Θ as an output since any control law which stabilizes the system must maintain the pendulum in balance regardless of z_r.

According to the arguments presented at the beginning of this section (and in Sec. 8.2), steady-state errors in z can be removed using the integral

$$q(t) = \int_0^t [z(\tau) - z_r(\tau)]\, d\tau \tag{24}$$

that is, by augmenting the system as in (4):

$$\begin{bmatrix} \dot{\underline{x}} \\ \dot{q} \end{bmatrix} = \begin{bmatrix} 0 & 1 & 0 & 0 & 0 \\ 0 & 0 & -1 & 0 & 0 \\ 0 & 0 & 0 & 1 & 0 \\ 0 & 0 & 11 & 0 & 0 \\ 1 & 0 & 0 & 0 & 0 \end{bmatrix} \begin{bmatrix} \underline{x} \\ q \end{bmatrix} + \begin{bmatrix} 0 \\ 1 \\ 0 \\ -1 \\ 0 \end{bmatrix} u + \begin{bmatrix} 0 \\ 4w \\ 0 \\ 6w \\ -z_r \end{bmatrix} \tag{25}$$

$$z = \begin{bmatrix} 1 & 0 & 0 & 0 & 0 \end{bmatrix} \begin{bmatrix} \underline{x} \\ q \end{bmatrix}$$

Using Lemma 1, this composite system is controllable if and only if the matrix

$$\begin{bmatrix} \underline{A} & \underline{B} \\ \underline{c}' & \underline{0} \end{bmatrix} = \begin{bmatrix} 0 & 1 & 0 & 0 & 0 \\ 0 & 0 & -1 & 0 & 1 \\ 0 & 0 & 0 & 1 & 0 \\ 0 & 0 & 11 & 0 & -1 \\ 1 & 0 & 0 & 0 & 0 \end{bmatrix} \tag{26}$$

has rank 5 (i.e., is nonsingular). Since the determinant of this matrix is -10, it is nonsingular and (25) must be controllable. Therefore, we know that it can be stabilized using a state feedback control law of the form[†]

$$\underline{u}(t) = \underline{k}' \begin{bmatrix} \underline{x}(t) \\ q(t) \end{bmatrix} = \begin{bmatrix} k_1 & k_2 & k_3 & k_4 & k_5 \end{bmatrix} \begin{bmatrix} \underline{x}(t) \\ q(t) \end{bmatrix} \tag{27}$$

[†]Note that the $\underline{w}$ term which appeared in control laws in Secs. 11.1 and 11.2 is replaced in this case by the vector of disturbance and reference inputs in (25).

Next, suppose that we wish to place the five poles of the closed-loop system at the locations shown in Fig. 11.2-2 with the additional one also at $s = -2$. From (11.2-28), the required closed-loop characteristic polynomial is

$$p(s) = (s+2)\tilde{p}(s) = s^5 + 7s^4 + 20s^3 + 30s^2 + 24s + 8 \tag{28}$$

Substitution of (27) into (25) yields the closed-loop system

$$\begin{bmatrix} \dot{\underline{x}} \\ \dot{q} \end{bmatrix} = \begin{bmatrix} 0 & 1 & 0 & 0 & 0 \\ k_1 & k_2 & k_3-1 & k_4 & k_5 \\ 0 & 0 & 0 & 1 & 0 \\ -k_1 & -k_2 & 11-k_3 & -k_4 & -k_5 \\ 1 & 0 & 0 & 0 & 0 \end{bmatrix} \begin{bmatrix} \underline{x} \\ q \end{bmatrix} + \begin{bmatrix} \underline{d} \\ -z_r \end{bmatrix} \tag{29}$$

$$z = \begin{bmatrix} 1 & 0 & 0 & 0 & 0 \end{bmatrix} \begin{bmatrix} \underline{x} \\ q \end{bmatrix}$$

with characteristic polynomial

$$\begin{aligned} p(s) &= \det(s\underline{I} - \underline{A}_1) \\ &= s^5 + (k_4 - k_2)s^4 + (k_3 - k_1 - 11)s^3 + (10k_2 - k_5)s^2 + 10k_1 s + 10k_5 \end{aligned} \tag{30}$$

Equating coefficients in (28) and (30), we find that the required feedback gains are

$$\begin{aligned} k_5 &= 0.8 \\ k_1 &= 2.4 \\ k_2 &= \frac{30 + k_5}{10} = 3.08 \\ k_3 &= 20 + k_1 + 11 = 33.4 \\ k_4 &= 7 + k_2 = 10.08 \end{aligned} \tag{31}$$

and the closed-loop system (29) becomes

$$\begin{bmatrix} \dot{\underline{x}} \\ \dot{q} \end{bmatrix} = \begin{bmatrix} 0 & 1 & 0 & 0 & 0 \\ 2.4 & 3.08 & 32.4 & 10.08 & 0.8 \\ 0 & 0 & 0 & 1 & 0 \\ -2.4 & -3.08 & -22.4 & -10.08 & -0.8 \\ 1 & 0 & 0 & 0 & 0 \end{bmatrix} \begin{bmatrix} \underline{x} \\ q \end{bmatrix} + \begin{bmatrix} \underline{d} \\ -z_r \end{bmatrix} \triangleq \underline{A}_1 \begin{bmatrix} \underline{x} \\ q \end{bmatrix} + \begin{bmatrix} \underline{d} \\ -z_r \end{bmatrix} \tag{32}$$

$$z = \begin{bmatrix} 1 & 0 & 0 & 0 & 0 \end{bmatrix} \begin{bmatrix} \underline{x} \\ q \end{bmatrix} \triangleq \underline{c}_1' \begin{bmatrix} \underline{x} \\ q \end{bmatrix}$$

Now let us verify that the steady-state behavior of (32) is satisfactory. If the disturbance $\underline{d}$ and reference input z_r are both constant (i.e., step functions), then Theorem 6.1.5(b) states that

$$\lim_{t\to\infty} z(t) = -\underline{c}_1'\underline{A}_1^{-1} \begin{bmatrix} \underline{d} \\ -z_r \end{bmatrix}$$

$$= -\begin{bmatrix} 1 & 0 & 0 & 0 & 0 \end{bmatrix} \begin{bmatrix} 0 & 0 & 0 & 0 & 1 \\ 1 & 0 & 0 & 0 & 0 \\ 0 & 0.1 & 0 & 0.1 & 0 \\ 0 & 0 & 1 & 0 & 0 \\ -3.85 & -2.8 & -12.6 & -4.05 & -3 \end{bmatrix} \begin{bmatrix} \underline{d} \\ -z_r \end{bmatrix} \tag{33}$$

$$= \begin{bmatrix} 0 & 0 & 0 & 0 & -1 \end{bmatrix} \begin{bmatrix} \underline{d} \\ -z_r \end{bmatrix} = z_r$$

Thus $z(t)$ approaches the steady-state value z_r, as required.

To check on Θ, observe that

$$\Theta = \begin{bmatrix} 0 & 0 & 1 & 0 & 0 \end{bmatrix} \begin{bmatrix} \underline{x} \\ q \end{bmatrix} = \underline{c}_2' \begin{bmatrix} \underline{x} \\ q \end{bmatrix} \tag{34}$$

so that

$$\lim_{t\to\infty} \Theta(t) = -\underline{c}_2'\underline{A}_1^{-1} \begin{bmatrix} \underline{d} \\ -z_r \end{bmatrix} = -\begin{bmatrix} 0 & 0.1 & 0 & 0.1 & 0 \end{bmatrix} \begin{bmatrix} \underline{d} \\ -z_r \end{bmatrix} \tag{35}$$

$$= 0.1(d_1 + d_4) = -w$$

A nonzero steady-state value of Θ is predictable, of course, since the pendulum must lean into the wind to remain in balance.

The reader should verify that the steady-state values of $\dot{\Theta}$ and $\dot{z}$ are both zero. Some other aspects of this example are explored in Problems 4 and 5.

ΔΔΔ

11.4 DECOUPLING

In this section we shall consider a system

$$\begin{aligned} \dot{\underline{x}} &= \underline{A}\underline{x} + \underline{B}\underline{u} \\ \underline{y} &= \underline{C}\underline{x} \end{aligned} \tag{1}$$

whose input and output vectors have the same dimension, m. If $\underline{x}(0) = \underline{0}$, these are related by the transfer function matrix,

$$\hat{\underline{y}}(s) = \underline{H}(s)\hat{\underline{u}}(s) = \underline{C}(s\underline{I} - \underline{A})^{-1}\underline{B} \tag{2}$$

which may be expanded into

$$\begin{aligned} \hat{y}_1(s) &= h_{11}(s)\hat{u}_1(s) + h_{12}(s)\hat{u}_2(s) + \ldots + h_{1m}(s)\hat{u}_m(s) \\ \hat{y}_2(s) &= h_{21}(s)\hat{u}_1(s) + h_{22}(s)\hat{u}_2(s) + \ldots + h_{2m}(s)\hat{u}_m(s) \\ &\cdots\cdots\cdots\cdots\cdots\cdots \\ \hat{y}_m(s) &= h_{m1}(s)\hat{u}_1(s) + h_{m2}(s)\hat{u}_2(s) + \ldots + h_{mm}(s)\hat{u}_m(s) \end{aligned} \tag{3}$$

These equations are said to be *coupled*, since each individual input influences all of the outputs. If it is necessary to adjust one of the outputs without affecting any of the others, determining appropriate inputs $\hat{u}_1, \hat{u}_2, \ldots, \hat{u}_m$ will be a difficult task in general. Consequently, there is considerable interest in devising control laws which remove this coupling, so that each input controls *only* the corresponding output.

1. DEFINITION. A system of the form (1) is said to be *decoupled* (or *noninteracting*) if its transfer function matrix $\underline{H}(s)$ is diagonal and nonsingular, that is,

$$\begin{aligned}
\hat{y}_1(s) &= h_{11}(s)\hat{u}_1(s) \\
\hat{y}_2(s) &= h_{22}(s)\hat{u}_2(s) \\
&\cdots\cdots\cdots\cdots \\
\hat{y}_m(s) &= h_{mm}(s)\hat{u}_m(s)
\end{aligned} \tag{4}$$

and none of the $h_{ii}(s)$ are identically zero. Such a system may be viewed as consisting of m independent subsystems, as shown in Fig. 1. Note that this definition depends upon the ordering of inputs and outputs, which is of course quite arbitrary.

ΔΔΔ

One example of a system requiring decoupling is a vertical take-off aircraft. The sixth-order linearized longitudinal equations of motion for such a vehicle in a hovering condition are given in Ref. 253. The outputs of interest are pitch angle, horizontal position, and altitude, and the control variables consist of three different fan inputs. Since these are all coupled, the pilot must acquire considerable skill in order to simultaneously manipulate the three inputs and successfully control the aircraft. The system may be decoupled using state-variable feedback, however, in order to provide the pilot with three independent and highly stable subsystems governing his pitch angle, horizontal position, and altitude. Details may be found in Ref. 253.

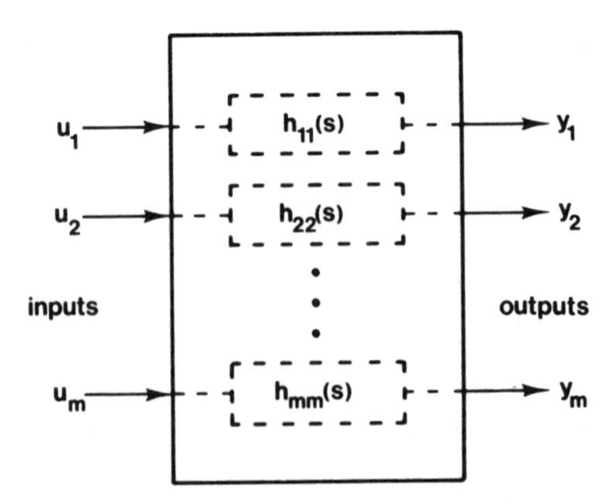

FIG. 1. Decoupled system.

We shall now examine some of the principal decoupling results which are currently available. Because the proofs are rather complex, they will be omitted, but the interested reader may wish to consult Refs. 39, 253 and 258. A more abstract geometric approach to these problems is surveyed in Refs. 184a, 310 and 369.

We shall assume for convenience that the system (1) is controllable[†] and consider a linear state-variable feedback control law of the form

$$\underline{u}(t) = \underline{K}\underline{x}(t) + \underline{G}\underline{w}(t) \tag{5}$$

where $\underline{G}$ is a nonsingular $m \times m$ matrix and $\underline{w}$ is the new input vector. Note that this is a slightly more general form than (11.1-16) since the inputs may be "rearranged" (e.g., reordered) by $\underline{G}$. The question of interest here is whether such a control law can be used to produce a closed-loop system which is both decoupled and stable.

For $i = 1, 2, \ldots, m$, we denote the rows of $\underline{C}$ by $\underline{c}_i'$ and define d_i to be the smallest integer between 0 and $n - 1$ for which

$$\underline{c}_i'\underline{A}^{d_i}\underline{B} \neq \underline{0}' \tag{6}$$

or if

$$\underline{c}_i'\underline{A}^j\underline{B} = \underline{0}' \qquad (j = 0, 1, \ldots, n-1) \tag{7}$$

then d_i is set equal to $n - 1$. We shall also make use of the matrices

$$\underline{D} \triangleq \begin{bmatrix} \underline{c}_1'\underline{A}^{d_1} \\ \underline{c}_2'\underline{A}^{d_2} \\ \vdots \\ \underline{c}_m'\underline{A}^{d_m} \end{bmatrix} \qquad \underline{E} \triangleq \underline{D}\underline{B} = \begin{bmatrix} \underline{c}_1'\underline{A}^{d_1}\underline{B} \\ \underline{c}_2'\underline{A}^{d_2}\underline{B} \\ \vdots \\ \underline{c}_m'\underline{A}^{d_m}\underline{B} \end{bmatrix} \qquad \underline{F} \triangleq \underline{D}\underline{A} = \begin{bmatrix} \underline{c}_1'\underline{A}^{d_1+1} \\ \underline{c}_2'\underline{A}^{d_2+1} \\ \vdots \\ \underline{c}_m'\underline{A}^{d_m+1} \end{bmatrix} \tag{8}$$

[†]No generality is lost with this assumption since the uncontrollable subsystems (Corollary 10.3.5) do not affect the transfer function matrix. If any of the uncontrollable poles are unstable, of course, there is no way to stabilize them.

With these definitions, the principal decoupling result may now be stated (alternative definitions of $\underline{E}$ and $d_1, d_2, \ldots, d_m$ in terms of the transfer function matrix are given in Problem 1).

2. *THEOREM.* The system (1) may be decoupled using a state-variable feedback control law of the form (5) if and only if the $m \times m$ matrix $\underline{E}$ is nonsingular. If so, the choice

$$\underline{K} = -\underline{E}^{-1}\underline{F} \qquad \underline{G} = \underline{E}^{-1} \tag{9}$$

results in a closed-loop system

$$\begin{aligned} \dot{\underline{x}} &= (\underline{A} - \underline{BE}^{-1}\underline{F})\underline{x} + \underline{BE}^{-1}\underline{w} \\ \underline{y} &= \underline{Cx} \end{aligned} \tag{10}$$

with transfer function matrix

$$\underline{H}(s) = \underline{C}(s\underline{I} - \underline{A} + \underline{BE}^{-1}\underline{F})^{-1}\,\underline{BE}^{-1} = \begin{bmatrix} \frac{1}{s^{d_1+1}} & 0 & \cdots & 0 \\ 0 & \frac{1}{s^{d_2+1}} & \cdots & 0 \\ \cdots & \cdots & \cdots & \cdots \\ 0 & 0 & \cdots & \frac{1}{s^{d_m+1}} \end{bmatrix} \tag{11}$$

Such a system is said to be in *integrator-decoupled* form.

ΔΔΔ

This theorem provides conditions for determining whether a system can be decoupled. If so, it specifies a particular feedback law which results in m independent subsystems, the i-th being equivalent to $d_i + 1$ levels of integration, that is,

$$\hat{y}_i(s) = h_{ii}(s)\hat{w}_i(s) = \frac{\hat{w}_i(s)}{s^{d_i+1}} \tag{12}$$

for $i = 1, 2, \ldots, m$. The question which now arises is whether additional feedback may be applied so as to retain decoupling while altering the pole locations (and hence the dynamic behavior) of the individual subsystems. To answer this, it is convenient to put the system into a canonical form.

3. *THEOREM.* There exists a state transformation

$$\underline{Q}\underline{x} = \tilde{\underline{x}} = \begin{bmatrix} \tilde{\underline{x}}_1 \\ \tilde{\underline{x}}_2 \\ \vdots \\ \tilde{\underline{x}}_m \\ \tilde{\underline{x}}_{m+1} \end{bmatrix} \tag{13}$$

such that the subvectors $\tilde{\underline{x}}_i$ have dimensions p_i ($\geq d_i + 1$ for $i = 1, 2, \ldots, m$) and the closed-loop system (10) takes on the *canonically decoupled* form,

$$\dot{\tilde{\underline{x}}} = \begin{bmatrix} \tilde{\underline{A}}_1 & \underline{0} & \cdots & \underline{0} & \underline{0} \\ \underline{0} & \tilde{\underline{A}}_2 & \cdots & \underline{0} & \underline{0} \\ \cdots & \cdots & \cdots & \cdots & \cdots \\ \underline{0} & \underline{0} & \cdots & \tilde{\underline{A}}_m & \underline{0} \\ \overline{\underline{A}}_1 & \overline{\underline{A}}_2 & \cdots & \overline{\underline{A}}_m & \tilde{\underline{A}}_{m+1} \end{bmatrix} \tilde{\underline{x}} + \begin{bmatrix} \tilde{\underline{b}}_1 & 0 & \cdots & \underline{0} \\ \underline{0} & \tilde{\underline{b}}_2 & \cdots & \underline{0} \\ \cdots & \cdots & \cdots & \cdots \\ \underline{0} & \underline{0} & \cdots & \tilde{\underline{b}}_m \\ \overline{\underline{b}}_1 & \overline{\underline{b}}_2 & \cdots & \overline{\underline{b}}_m \end{bmatrix} \underline{w}$$

$$\underline{y} = \begin{bmatrix} \tilde{\underline{c}}_1' & \underline{0}' & \cdots & \underline{0}' & \underline{0}' \\ \underline{0}' & \tilde{\underline{c}}_2' & \cdots & \underline{0}' & \underline{0}' \\ \cdots & \cdots & \cdots & \cdots & \cdots \\ \underline{0}' & \underline{0}' & \cdots & \tilde{\underline{c}}_m' & \underline{0}' \end{bmatrix} \tilde{\underline{x}} \tag{14}$$

where the various submatrices[†] are dimensioned appropriately. The independent scalar subsystems

---

[†]Their detailed structure is set out in Ref. 258.

$$\begin{aligned} \dot{\tilde{\underline{x}}}_i &= \tilde{\underline{A}}_i \tilde{\underline{x}}_i + \tilde{\underline{b}}_i w_i \\ y_i &= \tilde{\underline{c}}_i' \tilde{\underline{x}}_i \end{aligned} \qquad (i = 1, 2, \ldots, m) \tag{15}$$

are all controllable (but not necessarily observable), and the other subsystem

$$\dot{\tilde{\underline{x}}}_{m+1} = \tilde{\underline{A}}_{m+1} \tilde{\underline{x}}_{m+1} + \sum_{i=1}^{m} \overline{\underline{A}}_i \tilde{\underline{x}}_i + \sum_{i=1}^{m} \overline{\underline{b}}_i w_i \tag{16}$$

is completely unobservable if it is present at all (p_{m+1} may be 0).

ΔΔΔ

With this format, the class of *all* linear state-variable feedback control laws which decouple the system may be characterized:

4. *THEOREM*. The canonically decoupled system (14) remains decoupled under additional feedback

$$\underline{w}(t) = \tilde{\underline{K}} \tilde{\underline{x}}(t) + \overline{\underline{w}}(t) \tag{17}$$

if and only if the feedback gain matrix $\tilde{\underline{K}}$ has the form

$$\tilde{\underline{K}} = \begin{bmatrix} \tilde{\underline{k}}_1' & \underline{0}' & \cdots & \underline{0}' & \underline{0}' \\ \underline{0}' & \tilde{\underline{k}}_2' & \cdots & \underline{0}' & \underline{0}' \\ \cdots & \cdots & \cdots & \cdots & \cdots \\ \underline{0}' & \underline{0}' & \cdots & \tilde{\underline{k}}_m' & \underline{0} \end{bmatrix} \tag{18}$$

where $\tilde{\underline{k}}_i$ is a p_i-vector for $i = 1, 2, \ldots, m$. In other words, state-variable feedback

$$w_i(t) = \tilde{\underline{k}}_i' \tilde{\underline{x}}(t) + \overline{w}_i(t) \qquad (i = 1, 2, \ldots, m) \tag{19}$$

is allowed *within* each of the independent scalar subsystems (15), but any feedback within the unobservable subsystem (16) or linking different subsystems will result in a closed-loop system which is no longer decoupled.

ΔΔΔ

The implications of Theorem 4 are very important. From Theorem 11.2.1, each feedback law in (19) may be chosen so as to place the closed-loop poles of the corresponding subsystem in *any* desired locations. Thus if a system can be decoupled at all, the designer is free to determine its dynamic behavior by specifying the poles of each transfer function in Fig. 1. The catch is that the poles of the unobservable subsystem cannot be adjusted at all without recoupling the system. If (16) turns out to be stable, then this should be acceptable since $\tilde{\underline{x}}_{m+1}$ never affects any of the outputs, but an unstable eigenvalue of $\tilde{\underline{A}}_{m+1}$ means that no stable decoupling control law exists. In many cases, however, this subsystem will not be present at all ($p_{m+1} = 0$).

There are many systems, of course, which cannot be both decoupled and stabilized using linear feedback either because the conditions of Theorem 2 fail or because the unobservable subsystem of Theorem 3 is unstable. In such cases, stable decoupling can always be achieved by *compensation*, that is, by augmenting the system with additional dynamic elements. The requisite theory may be found in Refs. 184a, 311 and 369.

5. *EXAMPLE*. The linearized equations governing the equatorial motion of a satellite in a circular orbit were found in Example 2.4.3 to be

$$\dot{\underline{x}} = \underline{A}\underline{x} + \underline{B}\underline{u} = \begin{bmatrix} 0 & 1 & 0 & 0 \\ 3\omega^2 & 0 & 0 & 2\omega \\ 0 & 0 & 0 & 1 \\ 0 & -2\omega & 0 & 0 \end{bmatrix} \underline{x} + \begin{bmatrix} 0 & 0 \\ 1 & 0 \\ 0 & 0 \\ 0 & 1 \end{bmatrix} \underline{u}$$

$$\underline{y} = \underline{C}\underline{x} = \begin{bmatrix} 1 & 0 & 0 & 0 \\ 0 & 0 & 1 & 0 \end{bmatrix} \underline{x} \tag{20}$$

where the outputs and inputs are radial and tangential,

$$\underline{y} = \begin{bmatrix} r \\ \theta \end{bmatrix} \qquad \underline{u} = \begin{bmatrix} u_r \\ u_\theta \end{bmatrix} \tag{21}$$

These are coupled by the orbital dynamics, as indicated by the transfer function matrix

$$\underline{H}(s) = \begin{bmatrix} \dfrac{1}{s^2+\omega^2} & \dfrac{2\omega}{s(s^2+\omega^2)} \\ \dfrac{-2\omega}{s(s^2+\omega^2)} & \dfrac{s^2-3\omega^2}{s^2(s^2+\omega^2)} \end{bmatrix} \tag{22}$$

which was calculated in Problem 4.3-6.

To determine whether this system can be decoupled, we compute

$$\begin{aligned} \underline{c}_1'\underline{B} &= [\,0 \quad 0\,] \\ \underline{c}_1'\underline{A}\underline{B} &= [\,1 \quad 0\,] \\ \underline{c}_2'\underline{B} &= [\,0 \quad 0\,] \\ \underline{c}_2'\underline{A}\underline{B} &= [\,0 \quad 1\,] \end{aligned} \tag{23}$$

and find that $d_1 = d_2 = 1$ and

$$\underline{E} = \begin{bmatrix} \underline{c}_1'\underline{A}\underline{B} \\ \underline{c}_2'\underline{A}\underline{B} \end{bmatrix} = \begin{bmatrix} 1 & 0 \\ 0 & 1 \end{bmatrix} = \underline{I} \tag{24}$$

is nonsingular. Thus decoupling can be achieved with

$$\underline{u}(t) = -\underline{E}^{-1}\underline{F}\underline{x}(t) + \underline{E}^{-1}\underline{w}(t) = -\begin{bmatrix} 3\omega^2 & 0 & 0 & 2\omega \\ 0 & -2\omega & 0 & 0 \end{bmatrix}\underline{x}(t) + \underline{w}(t) \tag{25}$$

which results in the closed-loop system

$$\begin{aligned} \dot{\underline{x}} &= (\underline{A} - \underline{B}\underline{F})\underline{x} + \underline{B}\underline{w} = \begin{bmatrix} 0 & 1 & 0 & 0 \\ 0 & 0 & 0 & 0 \\ 0 & 0 & 0 & 1 \\ 0 & 0 & 0 & 0 \end{bmatrix}\underline{x} + \begin{bmatrix} 0 & 0 \\ 1 & 0 \\ 0 & 0 \\ 0 & 1 \end{bmatrix}\underline{w} \\ \underline{y} &= \underline{C}\underline{x} = \begin{bmatrix} 1 & 0 & 0 & 0 \\ 0 & 0 & 1 & 0 \end{bmatrix}\underline{x} \end{aligned} \tag{26}$$

Application of Theorem 3 is unnecessary, as (26) is already in canonically decoupled form, and there is no extra unobservable subsystem in this case.

Using Theorem 4, we may apply additional feedback

$$\underline{w} = \underline{\underline{K}}\underline{x} + \overline{\underline{w}} = \begin{bmatrix} k_1 & k_2 & 0 & 0 \\ 0 & 0 & k_3 & k_4 \end{bmatrix} \underline{x} + \begin{bmatrix} \overline{w}_r \\ \overline{w}_\Theta \end{bmatrix} \tag{27}$$

resulting in the closed-loop system

$$\dot{\underline{x}} = (\underline{A} - \underline{\underline{B}}\underline{\underline{F}} + \underline{\underline{B}}\underline{\underline{K}})\underline{x} + \overline{\underline{w}} = \begin{bmatrix} 0 & 1 & 0 & 0 \\ k_1 & k_2 & 0 & 0 \\ 0 & 0 & 0 & 1 \\ 0 & 0 & k_1 & k_2 \end{bmatrix} \underline{x} + \begin{bmatrix} 0 & 0 \\ 1 & 0 \\ 0 & 0 \\ 0 & 1 \end{bmatrix} \overline{\underline{w}}$$

$$\underline{y} = \underline{\underline{C}}\underline{x} = \begin{bmatrix} 1 & 0 & 0 & 0 \\ 0 & 0 & 1 & 0 \end{bmatrix} \underline{x} \tag{28}$$

If we choose the feedback gains $k_1 = k_3 = -1$ and $k_2 = k_4 = -2$, each of the two independent subsystems will be stable with a double pole at $s = -1$. The transfer function matrix may easily be shown to be

$$\underline{H}(s) = \begin{bmatrix} \dfrac{1}{(s+1)^2} & 0 \\ 0 & \dfrac{1}{(s+1)^2} \end{bmatrix} \tag{29}$$

as shown in Fig. 2.

This closed-loop system is not only stable (so that perturbations from the nominal trajectory will always decay to zero), but adjustments to r and Θ may be made independently via the external inputs $\overline{w}_r$ and $\overline{w}_\Theta$.

ΔΔΔ

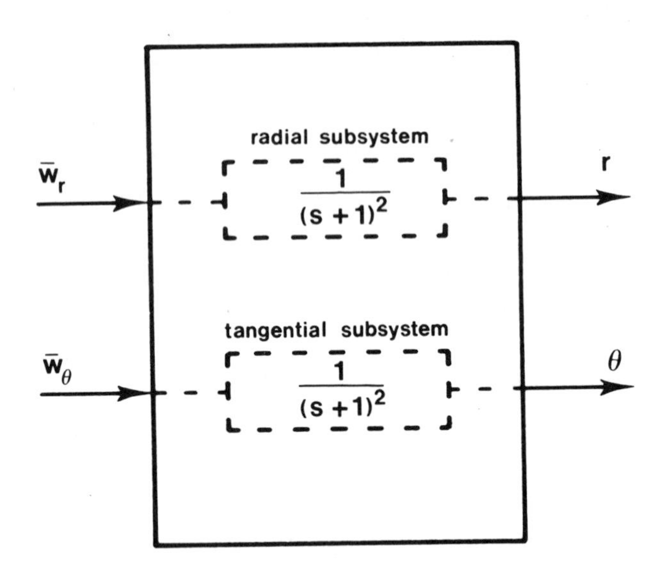

FIG. 2. Satellite system decoupled and stabilized with state-variable feedback.

A system which is decoupled in the sense of Definition 1 has very tightly constrained operating characteristics: A signal applied to input u_i must control output y_i and have *no* effect whatsoever on the other outputs. In many cases this requires a complex and highly sensitive control law, and in others it cannot be achieved at all (without additional compensation). It is useful, therefore, to consider a less stringent definition which involves only the steady-state behavior of the system [359].

6. *DEFINITION*. A system of the form (1) is said to be *statically decoupled* if it is stable and its static gain matrix $\underline{H}(0)$ is diagonal and nonsingular. This means that for a step function input $\underline{u}(t) = \underline{\alpha}1(t)$ the outputs satisfy

$$\begin{aligned} \lim_{t\to\infty} y_1(t) &= h_{11}(0)\alpha_1 \\ \lim_{t\to\infty} y_2(t) &= h_{22}(0)\alpha_2 \\ &\cdots\cdots\cdots \\ \lim_{t\to\infty} y_m(t) &= h_{mm}(0)\alpha_m \end{aligned} \tag{30}$$

where $h_{ii}(0) \neq 0$, $i = 1, 2, \ldots, m$. Again, note that the definition depends upon the ordering of inputs and outputs.

ΔΔΔ

In many ways, static decoupling has more practical significance than the dynamic decoupling of Definition 1, particularly if the latter is difficult or impossible to achieve. A step change in one input u_i will in general cause transients to appear at all of the outputs, but (30) ensures that y_j, $j \neq i$, will be unchanged in steady state.

As one might expect, the conditions for static decoupling are much simpler than those of Theorem 2. In fact, one of them turns out to be identical to the rank condition of Lemma 11.3.1, except that here we are assuming† $m = p$. As in (5), we consider a control law of the form

$$\underline{u}(t) = \underline{K}\underline{x}(t) + \underline{G}\underline{w}(t) \tag{31}$$

which yields the closed-loop system

$$\dot{\underline{x}} = (\underline{A} + \underline{B}\underline{K})\underline{x} + \underline{B}\underline{G}\underline{w} \triangleq \tilde{\underline{A}}\underline{x} + \tilde{\underline{B}}\underline{w} \tag{32}$$

with transfer function matrix

$$\tilde{\underline{H}}(s) = \underline{C}(s\underline{I} - \tilde{\underline{A}})^{-1}\tilde{\underline{B}} \tag{33}$$

7. *THEOREM*. The system (1) may be statically decoupled if and only if it can be stabilized by ordinary state feedback *and*

$$\det \begin{bmatrix} \underline{A} & \underline{B} \\ \underline{C} & \underline{0} \end{bmatrix} \neq 0 \tag{34}$$

This is accomplished by choosing a $\underline{K}$ which stabilizes the closed-loop system (32) and then choosing

$$\underline{G} = -(\underline{C}\tilde{\underline{A}}^{-1}\underline{B})^{-1}\underline{D} \tag{35}$$

---

†This is a convenient but noncritical assumption. If there are more inputs than outputs ($m > p$), then the system may be statically decoupled if and only if (11.3-5) holds.

where $\underline{D}$ is any nonsingular, diagonal matrix. If $\underline{A}$ is nonsingular, then (34) is equivalent to

$$\det(\underline{C}\underline{A}^{-1}\underline{B}) \neq 0 \tag{36}$$

Proof: By Definition 6, the system clearly must be stabilizable if it is to be statically decoupled. Thus an appropriate feedback gain matrix $\underline{K}$ can be found to stabilize the closed-loop system (32), and since stability implies that $\tilde{\underline{A}}$ is nonsingular, the static gain matrix is

$$\tilde{\underline{H}}(0) = -\underline{C}\tilde{\underline{A}}^{-1}\underline{B}\underline{G} \tag{37}$$

But by Lemma 11.3.3, (34) is equivalent to the nonsingularity of $\underline{C}\tilde{\underline{A}}^{-1}\underline{B}$, and the choice (35) for $\underline{G}$ makes $\tilde{\underline{H}}(0) = \underline{D}$, which is diagonal and nonsingular as required.

If $\underline{A}^{-1}$ exists, then Lemma 11.3.3 also establishes the equivalence of (34) and (36).

ΔΔΔ

11.5 SUMMARY

In this chapter we have investigated the use of linear feedback to alter the dynamic behavior of a multivariable system. Three types were identified: output feedback (from output to input), state-variable feedback (from state to input), and feedback from output to state derivative. With the latter two we saw in Sec. 11.2 that the poles of the closed-loop system could be given any desired values (assuming controllability and observability, respectively), but with output-to-input feedback the design problem is somewhat more difficult.

The use of integrators to improve steady-state performance was considered in Sec. 11.3, and the results were seen to closely parallel those of Sec. 8.2. In Sec. 11.4 we saw how to decouple a multivariable system with state-variable feedback, so that in the closed-loop system each input controls only its corresponding output, independently of the others.

In this chapter we have considered only linear feedback controls, using a constant gain matrix in the feedback loop. In the next chapter we will allow *dynamic feedback*, and install an additional dynamic system known as an *observer* in the feedback loop to provide an estimate of the state when only the output is available for measurement.

PROBLEMS

Section 11.1

1. Verify equation (11.1-15). *Hint:* Write h(s) as q(s)/p(s), substitute into (11.1-14), and let $\ell = 1$.

2. Prove Lemma 11.1.2 by using the controllability and observability matrices $\underline{P}_c$ and $\underline{P}_o$.

Section 11.2

1. (a) Prove Theorem 11.2.1' from Theorem 11.2.1 by duality.

   (b) Sketch a constructive proof which is analogous to the proof of Theorem 11.2.1.

   (c) Prove Corollary 11.2.2'.

2. (a) Show that the zeros of a scalar system are invariant under feedback to the input, i.e., that if feedback to the input of the system

   $$\dot{\underline{x}} = \underline{A}\underline{x} + \underline{b}u$$

   $$y = \underline{c}'\underline{x}$$

   is used to alter the poles of its transfer function, the zeros always remain unchanged. *Hint:* Show it under the assumptions that the system is controllable and is in controllable canonical form, and then explain how both of these assumptions may be removed.

   (b) What is the dual of this result?

(c) Consider an m-input, m-output system

$$\dot{\underline{x}} = \underline{A}\underline{x} + \underline{B}\underline{u}$$

$$\underline{y} = \underline{C}\underline{x}$$

with characteristic polynomial $p(s) = \det(s\underline{I} - \underline{A})$ and transfer function

$$\underline{H}(s) = \underline{C}(s\underline{I} - \underline{A})^{-1}\underline{B} = \frac{\underline{Q}(s)}{p(s)}$$

Define the zeros of this system to be the roots of $\det \underline{Q}(s)$, and generalize the results of (a) and (b) to this case. *Hint:* Use Theorem A.2.8 to show that

$$\det \underline{Q}(s) = \det \begin{bmatrix} s\underline{I} - \underline{A} & \underline{B} \\ -\underline{C} & \underline{0} \end{bmatrix}$$

and then prove that this determinant is unaffected by feedback to the input.

3. What is the dual of Theorem 11.2.4?

4. Show that the broom-balancing system of Example 11.2.3 cannot be stabilized without feedback from all four states.

5. (a) Consider the satellite problem of Example 10.2.10 with inputs u_r and u_Θ and outputs r and Θ. Is the system asymptotically stable? Can it be made asymptotically stable by applying feedback from the output to the input?

   (b) Suppose the outputs are Θ and $\dot{\Theta}$. Now can the system be stabilized with output-to-input feedback?

6. Show how to generalize Theorem 11.2.4 to the case where the system is stabilizable but not controllable. *Hint:* Use Theorem 10.3.3.

Section 11.3

1. Explain how you could solve for the steady-state values $\overline{\underline{x}}$ and $\overline{\underline{q}}$ in Theorem 11.3.2. Verify (11.3-15) by showing explicitly that $\underline{C}\overline{\underline{x}} = \underline{y}_r$. *Hint:* Use Theorems 6.1.5 and A.2.8.

2. How would you alter the system (11.3-4) to handle ramp inputs,

$$\underline{d}(t) = \bar{\underline{d}}\, t\, 1(t) \qquad \underline{y}_r(t) = \bar{\underline{y}}_r\, t\, 1(t)$$

Change Lemma 11.3.1 and Theorem 11.3.2 accordingly, and re-prove them for this case.

3. Are the eigenvalues of $\underline{A} + \underline{B}\underline{K}_1$ a subset of the closed-loop poles of (11.3-12)? Prove or disprove this conjecture.

4. (a) In Example 11.3.4, suppose that Θ (rather than z) is designated as the output of the system (22), and is required to track a reference input $\Theta_r \neq 0$ in steady-state. Augment the system with an integral similar to (24) and write out the equation corresponding to (25) for the composite system. Is this new system controllable? Can you find a state feedback control law to stabilize it?

   (b) Explain why your answers are consistent with physical intuition.

5. (a) In Example 11.3.4, suppose z is required to track the reference input $z_r(t)$ which is a step function, but there is *no* disturbance input. Apply Theorem 8.2.2 to determine whether perfect static accuracy can be achieved *without* using the additional state q.

   (b) Use Theorem 8.2.3 to find out what happens when a constant wind disturbance is included as in (11.3-22).

   (c) Verify your answers to (a) and (b) by applying a suitable stabilizing control law and then computing the static gain matrix of the closed-loop system (you may find the feedback gains computed in Example 11.2.3 useful here).

6. Extend all of the results of Sec. 11.3 to the system

$$\dot{\underline{x}} = \underline{A}\underline{x} + \underline{B}\underline{u} + \underline{d}$$

$$\underline{y} = \underline{C}\underline{x} + \underline{D}\underline{u}$$

Section 11.4

1. Show that d_i, $i = 1, 2, \ldots, m$ and $\underline{E}$ in (11.4-6) through (11.4-8 may be defined equivalently as follows:

   Denote the i-th row of $\underline{H}(s) = \underline{C}(s\underline{I} - \underline{A})^{-1}\underline{B}$ by $\underline{h}_i'(s)$ and define d_i to be the integer between 0 and n-1 such that

   $$\lim_{s\to\infty} s^{d_i+1}\,\underline{h}_i'(s)$$

   is finite and nonzero. Then

   $$\underline{E} = \underline{D}\underline{B} = \lim_{s\to\infty}\begin{bmatrix} s^{d_1+1}\underline{h}_1'(s) \\ s^{d_2+1}\underline{h}_2'(s) \\ \vdots \\ s^{d_m+1}\underline{h}_m'(s) \end{bmatrix}$$

   Hint: Express $(s\underline{I} - \underline{A})^{-1}$ as in Lemma 4.3.1, and use the resolvent algorithm (Theorem 4.3.2) to evaluate $\underline{Q}_0$, $\underline{Q}_1$, $\ldots$, $\underline{Q}_{n-1}$.

2. Show that a necessary (but not sufficient) condition for a system (11.4-1) to be decouplable is that $\underline{B}$ and $\underline{C}$ have full rank.

3. (a) Find a control law (if one exists) which decouples the system

   $$\dot{\underline{x}} = \begin{bmatrix} -1 & 0 & 0 \\ 0 & -2 & -3 \\ 1 & 0 & 1 \end{bmatrix}\underline{x} + \begin{bmatrix} 1 & 0 \\ 0 & 1 \\ 0 & -1 \end{bmatrix}\underline{u}$$

   $$\underline{y} = \begin{bmatrix} 1 & 0 & 0 \\ 0 & 1 & 1 \end{bmatrix}\underline{x}$$

   (b) Find a control law (if one exists) which statically decouples the system above.

Chapter 12

STATE ESTIMATION IN MULTIVARIABLE SYSTEMS

12.0 INTRODUCTION

The utility of state feedback for closed-loop pole placement and other purposes was established in the previous chapter. In many cases, however, the entire state vector is not available for feedback, and in this chapter we will see how it may be estimated using a *state estimator* or *observer* which operates on output and input measurements.

An observer may be used to realize a state feedback control law and achieve arbitrary closed-loop pole locations. By introducing additional dynamic elements into the system, it plays the role of a multivariable compensator, analogous to the single-loop compensators discussed in Chapter 9.

The subject of *noise* in linear systems will also be introduced, at a very rudimentary level, and state estimation using the celebrated *Kalman-Bucy filter* will be discussed.

12.1 STATE ESTIMATION USING OBSERVERS

In Corollary 10.2.5 we saw that if the state $\underline{x}$ of an observable linear system†

†The dimensions of $\underline{x}$, $\underline{u}$, and $\underline{y}$ are n, m, and p, respectively.

$$\dot{\underline{x}}(t) = \underline{A}\underline{x}(t) + \underline{B}\underline{u}(t)$$
$$\underline{y}(t) = \underline{C}\underline{x}(t) \tag{1}$$

is not directly available for measurement, then it can be determined indirectly from the output $\underline{y}$ (and the input $\underline{u}$). This involves multiplying $\underline{y}(t)$ by a weighting function and integrating over a finite time interval, and $\underline{x}$ is determined only at a single instant of time.

In this section we shall see how $\underline{x}(t)$ may be continually estimated from $\underline{y}$ and $\underline{u}$ with a dynamic system known as a *state observer* (or *estimator* or *reconstructor*). One obvious approach is to denote the estimate of $\underline{x}(t)$ by $\hat{\underline{x}}(t)$ and construct a *model* of the original system,

$$\dot{\hat{\underline{x}}}(t) = \underline{A}\hat{\underline{x}}(t) + \underline{B}\underline{u}(t) \tag{2}$$

as shown in Fig. 1. If the initial state $\hat{\underline{x}}(0) = \underline{x}(0)$, then this model should provide an exact estimate $\hat{\underline{x}}(t) = \underline{x}(t)$ for all t, but in practice there will always be errors due to noise or inaccuracies in the model and initial state.

The estimate $\hat{\underline{x}}$ provided by (2) is an open-loop one (as was that of Corollary 10.2.5) since the available information $\underline{y}$ from the system (1) is not being utilized for error correction. In order to make use of $\underline{y}$, we denote its estimate as $\hat{\underline{y}} = \underline{C}\hat{\underline{x}}$ and form the output

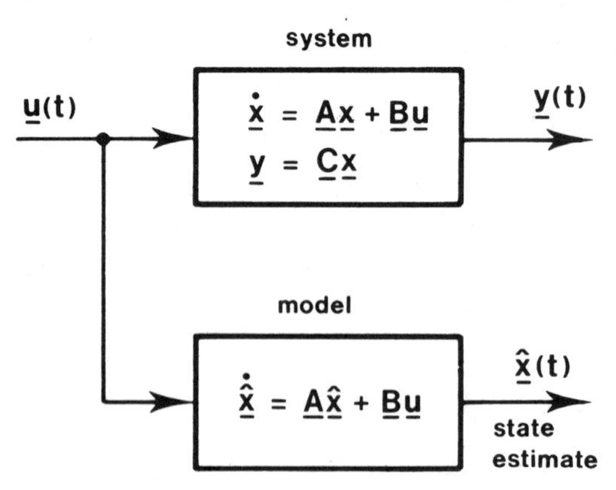

FIG. 1. Open-loop state estimation.

error vector

$$\tilde{\underline{y}}(t) \overset{\Delta}{=} \underline{y}(t) - \hat{\underline{y}}(t) = \underline{C}\underline{x}(t) - \underline{C}\hat{\underline{x}}(t) \tag{3}$$

It then seems reasonable to include in the model a correction term which depends upon the output error, replacing (2) by

$$\dot{\hat{\underline{x}}}(t) = \underline{A}\hat{\underline{x}}(t) + \underline{B}\underline{u}(t) - \underline{M}\tilde{\underline{y}}(t) \tag{4}$$

where the $n \times p$ matrix $\underline{M}$ is to be determined. This is illustrated in Fig. 2.

In order to select $\underline{M}$, we denote the state error vector as

$$\tilde{\underline{x}}(t) = \underline{x}(t) - \hat{\underline{x}}(t) \tag{5}$$

and derive the differential equation governing its behavior by subtracting (4) from (1):

$$\dot{\tilde{\underline{x}}} = \dot{\underline{x}} - \dot{\hat{\underline{x}}} = \underline{A}\underline{x} + \underline{B}\underline{u} - \underline{A}\hat{\underline{x}} - \underline{B}\underline{u} + \underline{M}\tilde{\underline{y}} \tag{6}$$

Substitution of (3) yields

$$\dot{\tilde{\underline{x}}} = (\underline{A} + \underline{M}\underline{C})(\underline{x} - \hat{\underline{x}}) = (\underline{A} + \underline{M}\underline{C})\tilde{\underline{x}} \tag{7}$$

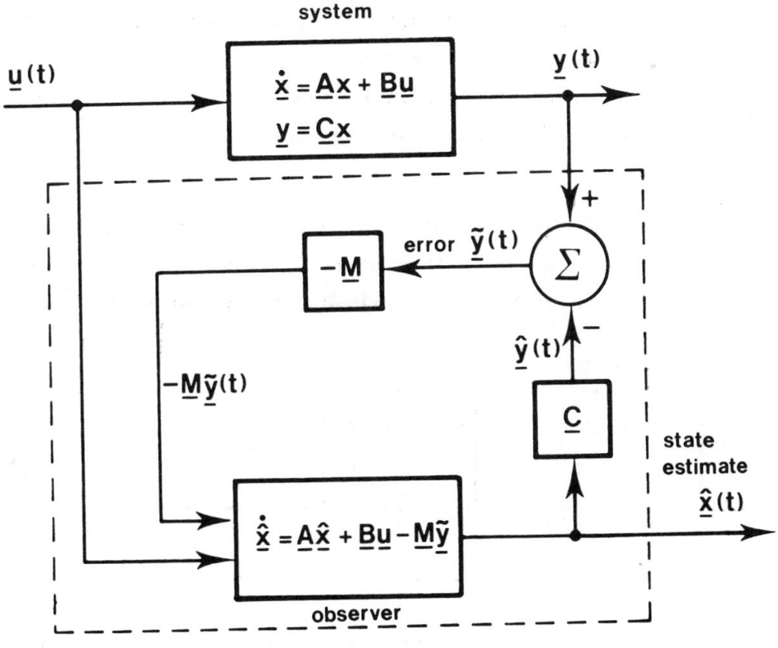

FIG. 2. State observer [Eq. (4)].

and we conclude that

$$\tilde{\underline{x}}(t) = e^{(\underline{A} + \underline{M}\underline{C})t}\tilde{\underline{x}}(0) \tag{8}$$

independently of the control $\underline{u}$ which is applied.

The error $\tilde{\underline{x}}$ will decay to $\underline{0}$ if $\underline{M}$ is chosen so that (7) is stable, that is, so that all the eigenvalues of $\underline{A} + \underline{M}\underline{C}$ lie in the left half-plane. Because the original system (1) was assumed observable, we may apply Theorem 11.2.1', which states that the eigenvalues of $\underline{A} + \underline{M}\underline{C}$ may be placed in any symmetric configuration by choosing an appropriate $\underline{M}$. This is all summarized in Theorem 1.

1. THEOREM. If the system (1) is observable, its state may be estimated with an n-dimensional observer of the form

$$\dot{\hat{\underline{x}}} = (\underline{A} + \underline{M}\underline{C})\hat{\underline{x}} + \underline{B}\underline{u} - \underline{M}\underline{y} \tag{9}$$

as shown in Fig. 3. The matrix $\underline{M}$ may be chosen so as to place the poles of (9) in any desired configuration, and these determine the rate at which the error (8) decays to zero.

Proof: The observer equation (9) is a rearrangement of (4) with (3) substituted for $\tilde{\underline{y}}$. The rest of the theorem follows from the discussion above. It may also be shown [105] that *every* n-dimensional observer has this same structure.

ΔΔΔ

The choice of observer poles is completely arbitrary in principle. A designer may select $\underline{M}$ (via Theorem 11.2.1') so as to make the error decay as fast as he wishes, constrained only by the bandwidth of the

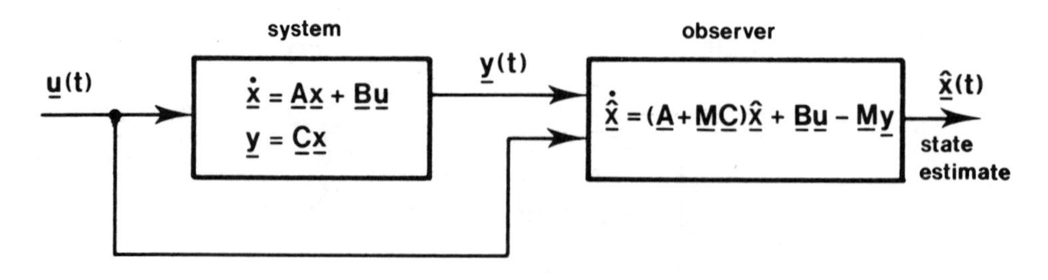

FIG. 3. Alternate representation of observer (Theorem 1).

available components. In practice, however, it is often difficult to make the observer substantially faster than the system itself. To do so requires large gains, which typically lead to a variety of problems. The discussion of this and other related points following Corollary 11.2.2' is equally relevant here.

Another practical consideration is that the output $\underline{y}$ will inevitably contain at least a small amount of measurement noise, and this will tend to be magnified if the observer is too fast. Put another way, a very fast observer approximates a set of differentiators, and these are known to be inherently noisy (see Problem 1). If the statistical structure of the noise is known, it can be utilized in the design of an observer; the Kalman-Bucy filter will be introduced in Sec. 12.4 as an observer of the form (9) with $\underline{M}$ chosen so as to minimize the mean-square error of the estimate.

We conclude this section by constructing an observer for a particular system:

2. *EXAMPLE.* Consider once again the broom-balancing system of Example 2.3.2, shown in Fig. 4(a), with state vector

$$\underline{x}(t) = [z(t) \quad \dot{z}(t) \quad \Theta(t) \quad \dot{\Theta}(t)] \tag{10}$$

and suppose that the only output available for measurement is z(t), the position of the cart. The linearized equations governing this system [for $\Theta(t) \simeq 0$] are

$$\dot{\underline{x}} = \underline{A}\underline{x} + \underline{b}u = \begin{bmatrix} 0 & 1 & 0 & 0 \\ 0 & 0 & -1 & 0 \\ 0 & 0 & 0 & 1 \\ 0 & 0 & 11 & 0 \end{bmatrix} \underline{x} + \begin{bmatrix} 0 \\ 1 \\ 0 \\ -1 \end{bmatrix} u \tag{11}$$

$$y = z = \underline{c}'\underline{x} = \begin{bmatrix} 1 & 0 & 0 & 0 \end{bmatrix} \underline{x}$$

and we established in Example 10.2.9 that this system is completely observable from the measurement of z.

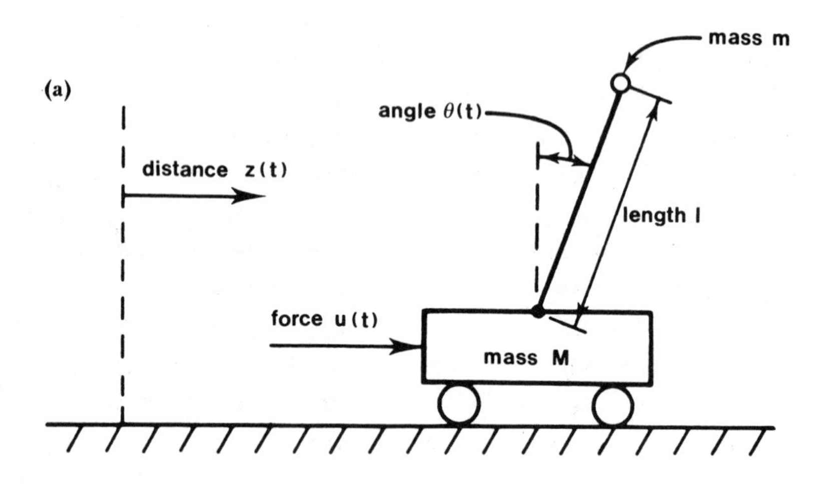

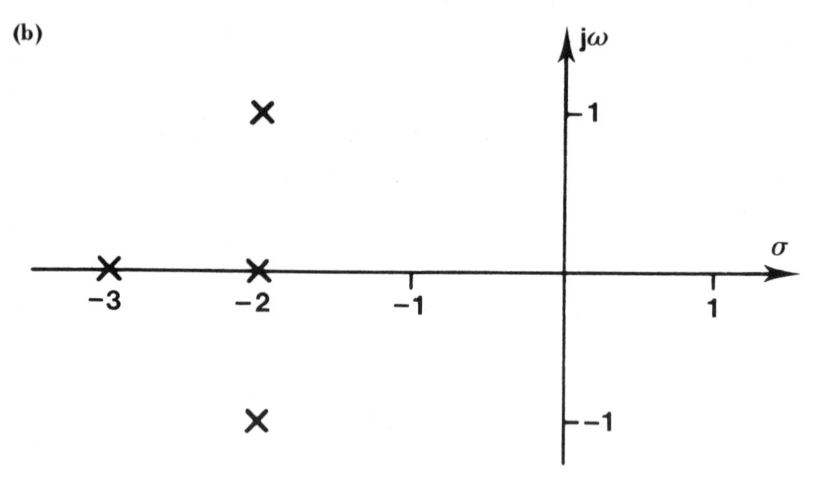

FIG. 4. System of Example 2: (a) broom-balancing system and (b) desired observer pole configuration.

Using Theorem 1, the entire state may be estimated from z with an observer of the form (9),

$$\dot{\hat{\underline{x}}} = (\underline{A} + \underline{m}\underline{c}')\hat{\underline{x}} + \underline{b}u - \underline{m}y \tag{12}$$

where $\underline{m} = [m_1 \quad m_2 \quad m_3 \quad m_4]'$, or

$$\dot{\hat{\underline{x}}} = \begin{bmatrix} m_1 & 1 & 0 & 0 \\ m_2 & 0 & -1 & 0 \\ m_3 & 0 & 0 & 1 \\ m_4 & 0 & 11 & 0 \end{bmatrix} \hat{\underline{x}} + \begin{bmatrix} 0 \\ 1 \\ 0 \\ -1 \end{bmatrix} u - \begin{bmatrix} m_1 \\ m_2 \\ m_3 \\ m_4 \end{bmatrix} y \tag{13}$$

In order to select $\underline{m}$, suppose the observer is required to have the pole configuration shown in Fig. 4(b), or a characteristic polynomial

$$\begin{aligned} r(s) &= (s+2)(s+3)(s+2+j)(s+2-j) \\ &= s^4 + 9s^3 + 31s^2 + 49s + 30 \end{aligned} \tag{14}$$

It is easy to derive the characteristic polynomial of $\underline{A} + \underline{m}\underline{c}'$ in terms of $\underline{m}$,

$$\begin{aligned} r(s) &= \det(s\underline{I} - \underline{A} - \underline{m}\underline{c}') \\ &= s^4 - m_1 s^3 - (m_2 + 11)s^2 + (11m_1 + m_3)s + (11m_2 + m_4) \end{aligned} \tag{15}$$

and if we choose the coefficients

$$\begin{aligned} m_1 &= -9 \\ m_2 &= -42 \\ m_3 &= 49 - 11m_1 = 148 \\ m_4 &= 30 - 11m_2 = 464 \end{aligned} \tag{16}$$

then the expressions in (14) and (15) coincide.

With this $\underline{m}$, the observer (13) will process the cart position z(t) and input u(t) to continuously provide an estimate $\hat{\underline{x}}(t)$ of the entire state vector, and any errors in the estimate will decay at least as fast as e^{-2t}.

The linearized equations are valid only for small Θ, of course, and the observer design assumes that a control will be applied which maintains this condition (i.e., keeps the broom balanced). ΔΔΔ

12.2 REDUCED-ORDER OBSERVERS

The state observer of Theorem 12.1.1 was derived by setting up a model of the system in question and feeding back a "correction" term proportional to the difference between the actual and estimated outputs. Such an observer contains some redundancy, however, since certain linear combinations of the states (specified by the rows of $\underline{C}$) are available directly as outputs and need not be estimated. Thus one is led to suspect that the dimension of the observer can be reduced.

To be more explicit, consider an (n-dimensional) observable system whose output vector consists of the last p states,[†]

$$\begin{bmatrix} \dot{\underline{x}}_1 \\ \hline \dot{\underline{x}}_2 \end{bmatrix} = \left[\begin{array}{c|c} \underline{A}_{11} & \underline{A}_{12} \\ \hline \underline{A}_{21} & \underline{A}_{22} \end{array}\right] \begin{bmatrix} \underline{x}_1 \\ \hline \underline{x}_2 \end{bmatrix} + \begin{bmatrix} \underline{B}_1 \\ \hline \underline{B}_2 \end{bmatrix} \underline{u}$$

$$\underline{y} = \left[\begin{array}{c|c} \underline{0} & \underline{I} \end{array}\right] \begin{bmatrix} \underline{x}_1 \\ \hline \underline{x}_2 \end{bmatrix} = \underline{x}_2 \tag{1}$$

as illustrated in Fig. 1. We lose no generality in choosing this form, as *any* system can be put into it with a simple state transformation, provided $\underline{C}$ has full rank (see Problem 1).

It would be an easy matter to construct an n-dimensional observer to estimate the state of this system as in Theorem 12.1.1. However, there is clearly no need to estimate the last p states, as these are available as outputs for direct measurement. Consequently, we may treat $\underline{x}_2(t) = \underline{y}(t)$ as a known time function and estimate only the n - p states which make up $\underline{x}_1$. This will result in an (n - p)-dimensional estimator, known as a *Luenberger observer* [290,291]; the derivation given here will be similar to those found in Refs. 259 and 293.

[†]Note that a system in observable canonical form (10.4-11) or (10.4-13) fits into this category.

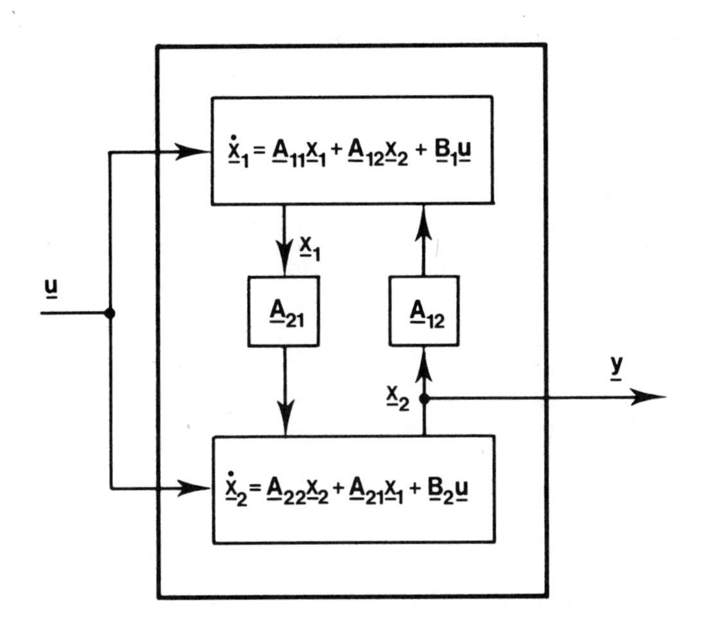

FIG. 1. Partitioned system.

By manipulating the equations in (1), x_1 may be viewed as the state of an (n - p)-dimensional subsystem,

$$\begin{aligned} \dot{x}_1 &= A_{11}x_1 + v \\ z &= A_{21}x_1 \end{aligned} \tag{2}$$

There is no difficulty in treating

$$v \triangleq A_{12}x_2 + B_1u = A_{12}y + B_1u \tag{3}$$

as a known input since u is known and y is directly measurable. Similarly, the "output" vector in (2) may be expressed as

$$\begin{aligned} z \triangleq A_{21}x_1 &= \dot{x}_2 - A_{22}x_2 - B_2u \\ &= \dot{y} - A_{22}y - B_2u \end{aligned} \tag{4}$$

where u is known and y is directly measurable (the need to differentiate $\dot{y}$ will be eliminated shortly).

We are now in a position to construct an observer for the subsystem (2), but first we must verify that it is observable.

1. LEMMA. The subsystem (2) is observable, provided the original system (1) is observable.

Proof: See Problem 2.

ΔΔΔ

Having verified that (2) is observable, it is an easy matter to apply Theorem 12.2.1 and estimate $\underline{x}_1$ with an observer

$$\dot{\hat{\underline{x}}}_1 = (\underline{A}_{11} + \underline{M}_1\underline{A}_{21})\hat{\underline{x}}_1 + \underline{v} - \underline{M}_1\underline{z} \tag{5}$$

where the $(n - p) \times p$ matrix $\underline{M}_1$ may be chosen so as to place the poles of (5) in any desired locations. Substituting (3) and (4) for $\underline{v}$ and $\underline{z}$, we have

$$\dot{\hat{\underline{x}}}_1 = (\underline{A}_{11} + \underline{M}_1\underline{A}_{21})\hat{\underline{x}}_1 + \underline{A}_{12}\underline{y} + \underline{B}_1\underline{u} - \underline{M}_1\dot{\underline{y}} + \underline{M}_1\underline{A}_{22}\underline{y} + \underline{M}_1\underline{B}_2\underline{u} \tag{6}$$

which is an $(n - p)$-dimensional observer for (1).

The only apparent difficulty in implementing the observer (6) is that differentiation of the output $\underline{y}$ is required. This can easily be avoided, however, by redefining the state of the observer to be

$$\underline{w}(t) \triangleq \hat{\underline{x}}_1(t) + \underline{M}_1\underline{y}(t) \tag{7}$$

Substitution in (6) yields

$$\begin{aligned}\dot{\underline{w}} &= \dot{\underline{x}}_1 + \underline{M}_1\dot{\underline{y}} \\ &= (\underline{A}_{11} + \underline{M}_1\underline{A}_{21})\underline{w} + (\underline{B}_1 + \underline{M}_1\underline{B}_2)\underline{u} + [\underline{A}_{12} + \underline{M}_1\underline{A}_{22} - (\underline{A}_{11} + \underline{M}_1\underline{A}_{21})\underline{M}_1]\underline{y}\end{aligned} \tag{8}$$

which is entirely equivalent[†] and can be implemented without differentiating $\underline{y}$.

The estimate of the full state vector $\underline{x}$ is given by

$$\hat{\underline{x}} = \begin{bmatrix} \hat{\underline{x}}_1 \\ \hline \underline{y} \end{bmatrix} = \begin{bmatrix} \underline{w} - \underline{M}_1\underline{y} \\ \hline \underline{y} \end{bmatrix} = \begin{bmatrix} \underline{I} \\ \hline \underline{0} \end{bmatrix} \underline{w} + \begin{bmatrix} -\underline{M}_1 \\ \hline \underline{I} \end{bmatrix} \underline{y} \tag{9}$$

[†] For a physical interpretation, see Problem 3.

There is no error in estimating the last p states, of course, since $\underline{x}_2 = \underline{y}$ is directly measurable. To verify that the error in estimating $\underline{x}_1$ behaves as predicted by Theorem 12.1.1, we may use (1) and (6) or (8) to determine the equations governing $\tilde{\underline{x}}_1 = \underline{x}_1 - \hat{\underline{x}}_1$:

$$\begin{aligned}\dot{\tilde{\underline{x}}}_1 &= \dot{\underline{x}}_1 - \dot{\hat{\underline{x}}}_1 \\ &= \underline{A}_{11}\underline{x}_1 + \underline{A}_{12}\underline{y} + \underline{B}_1\underline{u} - (\underline{A}_{11} + \underline{M}_1\underline{A}_{21})\hat{\underline{x}}_1 - \underline{A}_{12}\underline{y} - \underline{B}_1\underline{u} \\ &\quad + \underline{M}_1(\dot{\underline{y}} - \underline{A}_{22}\underline{y} - \underline{B}_2\underline{u})\end{aligned} \tag{10}$$

Cancelling terms and substituting (4), we have

$$\dot{\tilde{\underline{x}}}_1 = (\underline{A}_{11} + \underline{M}_1\underline{A}_{21})\tilde{\underline{x}}_1 \tag{11}$$

as expected. Since the poles of (11) can be arbitrarily placed by choosing $\underline{M}_1$ appropriately, the error

$$\tilde{\underline{x}}(t) = e^{(\underline{A}_{11} + \underline{M}_1\underline{A}_{21})t}\tilde{\underline{x}}(0) \tag{12}$$

can be made to decay to zero as fast as the designer wishes.

To summarize, we have the following theorem.

2. *THEOREM*. Any observable system with n states and p outputs may be put into the form (1) with a state transformation, and its state may be estimated with the (n - p)-dimensional (Luenberger) observer specified by (8) and (9), as shown in Fig. 2. The $(n - p) \times p$ matrix $\underline{M}_1$ may be chosen so as to place the poles of (8) in any desired configuration, and these determine the rate at which the error (12) decays to zero.

ΔΔΔ

The state transformation has been used here in order to keep the derivation on an intuitive level. The reader is asked in Problem 2 to transform the observer back to the original system. The Luenberger observer may also be derived directly by solving a set of algebraic equations [291]. We will now illustrate these ideas by deriving a reduced-order observer for the example of the last section.

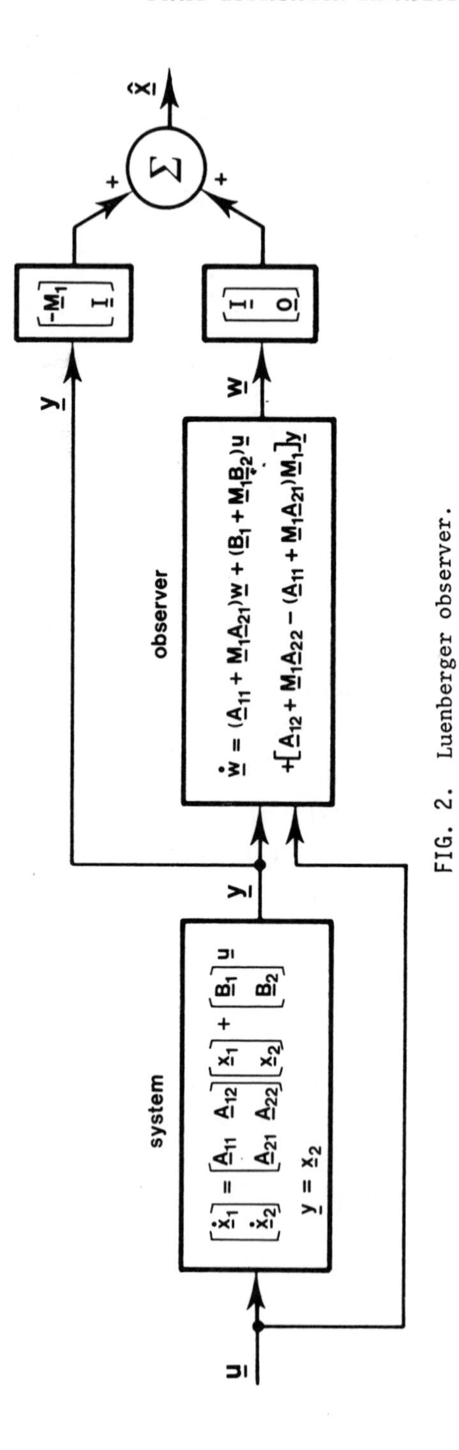

FIG. 2. Luenberger observer.

3. *EXAMPLE*. In Example 12.1.2, an observer of dimension 4 was constructed for a broom-balancing system with 4 states and 1 output. However, according to Theorem 2, this could just as well be done with an observer of dimension 4 - 1 = 3.

In order to put the broom-balancing system into the appropriate form† (1), we may simply reorder its states and write (12.1-11) as

$$\frac{d}{dt}\begin{bmatrix} \dot{z} \\ \theta \\ \dot{\theta} \\ \hline z \end{bmatrix} = \left[\begin{array}{ccc|c} 0 & -1 & 0 & 0 \\ 0 & 0 & 1 & 0 \\ 0 & 11 & 0 & 0 \\ \hline 1 & 0 & 0 & 0 \end{array}\right]\begin{bmatrix} \dot{z} \\ \theta \\ \dot{\theta} \\ \hline z \end{bmatrix} + \begin{bmatrix} 1 \\ 0 \\ -1 \\ \hline 0 \end{bmatrix} u$$

$$y = \left[\begin{array}{ccc|c} 0 & 0 & 0 & 1 \end{array}\right]\begin{bmatrix} \dot{z} \\ \theta \\ \dot{\theta} \\ \hline z \end{bmatrix} = z \tag{13}$$

where the dotted lines separate $\underline{A}_{11}$, $\underline{A}_{12}$, etc.

Having already established that this system is observable, we may now make use of Theorem 2 and determine a three-dimensional observer by substituting into (8) and (9). Since $\underline{A}_{12}$, $\underline{A}_{22}$, and $\underline{B}_2$ are all zero, this yields

$$\dot{\underline{w}} = (\underline{A}_{11} + \underline{m}\underline{A}_{21})\underline{w} + \underline{B}_1 u - (\underline{A}_{11} + \underline{m}\underline{A}_{21})\,\underline{m}y$$

$$\hat{\underline{x}} = \begin{bmatrix} \underline{I} \\ \hline \underline{0}' \end{bmatrix}\underline{w} + \begin{bmatrix} -\underline{m} \\ \hline 1 \end{bmatrix} y \tag{14}$$

where we have designated $\underline{M}_1$ as $\underline{m}$ for convenience.

Now we must select $\underline{m}$ in order to place the eigenvalues of

$$\underline{A}_{11} + \underline{m}\underline{A}_{21} = \begin{bmatrix} 0 & -1 & 0 \\ 0 & 0 & 1 \\ 0 & 11 & 0 \end{bmatrix} + \begin{bmatrix} m_1 \\ m_2 \\ m_3 \end{bmatrix}\begin{bmatrix} 1 & 0 & 0 \end{bmatrix} = \begin{bmatrix} m_1 & -1 & 0 \\ m_2 & 0 & 1 \\ m_3 & 11 & 0 \end{bmatrix} \tag{15}$$

†This is not really necessary: See Problem 4.

The characteristic polynomial of this matrix is

$$s^3 - m_1 s^2 + (m_2 - 11)s + (11m_1 + m_3) \tag{16}$$

and the choice

$$m_1 = -7 \qquad m_2 = 28 \qquad m_3 = 92 \tag{17}$$

yields the pole configuration of Fig. 12-1.4(b), omitting the pole at $s = -2$.

Substituting these values in (14), we see that the required observer has the form

$$\begin{aligned} \dot{\underline{w}} &= \begin{bmatrix} -7 & -1 & 0 \\ 28 & 0 & 1 \\ 92 & 11 & 0 \end{bmatrix} \underline{w} + \begin{bmatrix} 1 \\ 0 \\ -1 \end{bmatrix} u + \begin{bmatrix} -21 \\ 104 \\ 336 \end{bmatrix} y \\ \hat{\underline{x}} &= \begin{bmatrix} 1 & 0 & 0 \\ 0 & 1 & 0 \\ 0 & 0 & 1 \\ 0 & 0 & 0 \end{bmatrix} \underline{w} + \begin{bmatrix} 7 \\ -28 \\ -92 \\ 1 \end{bmatrix} y \end{aligned} \tag{18}$$

This will provide a continuous estimate $\hat{\underline{x}}(t)$ of the state vector, with any estimation errors decaying at least as fast as e^{-2t}.

Again, recall that the linearized equations are valid only for small Θ, so the observer design assumes that a control will be applied which maintains this condition (i.e., keeps the broom balanced).

ΔΔΔ

In this section we have examined a system with n states and p outputs and seen how to construct an observer with $n - p$ states and arbitrary pole positions which will provide a continuous estimate of the system state vector. Some caution is required in implementing such an observer because the state estimate $\hat{\underline{x}}$ in (9) contains a *feedforward* term proportional to the system output $\underline{y}$. If $\underline{y}$ should be corrupted by any noise, this will find its way directly into $\underline{x}$, bypassing the natural "filtering" action of the observer dynamics. If output noise appears to be a significant problem, then a full-order observer is probably needed, with its gain matrix chosen via the Kalman-Bucy theory of Sec. 12.4.

In some cases the dimension of the observer can be reduced still further. Suppose, for instance, that the system being observed has m inputs and that the object of producing a state estimate $\hat{\underline{x}}$ is to form the m-vector $\underline{K}\hat{\underline{x}}$ for use in a feedback control law.† Since m is usually a good deal smaller than n (in many cases $m = 1$), one suspects that it is possible to estimate the m-vector $\underline{K}\underline{x}$ directly, making use of a smaller observer than that required to estimate the entire n-vector $\underline{x}$. That this is indeed the case is established in Refs. 255 and 324, where procedures for constructing reduced-order observers are developed. The error dynamics are no longer arbitrary, however, and the smaller the dimension of the observer, the less freedom one has in selecting its poles.

Luenberger [291, 293] established a similar result in the case $m = 1$, where only an estimate of the scalar quantity $\underline{k}'\underline{x}$ is required. He showed that such an estimate could always be provided by an observer of dimension $\nu_o - 1$ (with arbitrary pole locations), where the *observability index* ν_o of the system is defined to be the smallest positive integer for which the matrix

$$\begin{bmatrix} \underline{C} \\ \underline{C}\underline{A} \\ \underline{C}\underline{A}^2 \\ \vdots \\ \underline{C}\underline{A}^{\nu_o - 1} \end{bmatrix} \tag{19}$$

has full rank (n). The dimension of this observer can also be reduced by the procedures mentioned above, at the cost of some freedom in selecting the poles.

---

†This subject is taken up in the next section.

12.3. APPLICATIONS AND EXTENSIONS

In the two preceding sections we saw how the state of an observable system can be estimated from output information using an observer of dimension n or n - p. In Chapter 11 we saw that a state feedback control law can be used on a system to place its closed-loop poles in any desired positions. If the entire state is not available for feedback, it seems reasonable to estimate it with an observer and use the *estimate* in the control law. Of course, there is no obvious guarantee that such an arrangement will prove satisfactory, or even be stable, for that matter; so now we shall analyze the behavior of such a feedback system.

Consider a controllable and observable system

$$\begin{aligned} \dot{\underline{x}} &= \underline{A}\underline{x} + \underline{B}\underline{u} \\ \underline{y} &= \underline{C}\underline{x} \end{aligned} \tag{1}$$

We know from Theorem 11.2.1 that if the state is available for feedback, a control law of the form

$$\underline{u}(t) = \underline{K}\underline{x}(t) + \overline{\underline{u}}(t) \tag{2}$$

can always be found to place the poles of the closed-loop system

$$\begin{aligned} \dot{\underline{x}} &= (\underline{A} + \underline{B}\underline{K})\underline{x} + \underline{B}\overline{\underline{u}} \\ \underline{y} &= \underline{C}\underline{x} \end{aligned} \tag{3}$$

in any desired configuration. When the state is not available, we propose to make use of Theorem 12.1.1 and construct an observer†

$$\dot{\hat{\underline{x}}} = (\underline{A} + \underline{M}\underline{C})\hat{\underline{x}} + \underline{B}\underline{u} - \underline{M}\underline{y} \tag{4}$$

to process $\underline{y}$ and $\underline{u}$ and estimate the state. The feedback control law will then be

$$\underline{u}(t) = \underline{K}\hat{\underline{x}}(t) + \overline{\underline{u}}(t) \tag{5}$$

as indicated in Fig. 1.

For the purpose of analysis, we may look upon the plant (1) and observer (4) as a composite system of dimension 2n,

†We could equally well use an (n - p)-dimensional observer, as in Theorem 12.2.2 (see Problem 1).

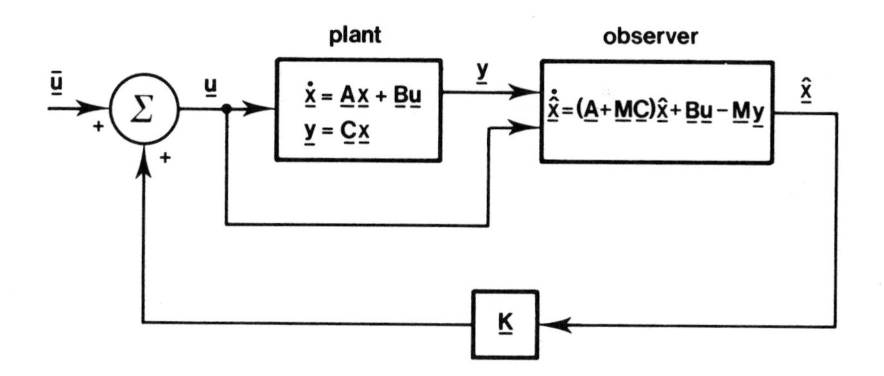

FIG. 1. Use of observer to implement control law.

$$
\begin{bmatrix} \dot{x} \\ \dot{\hat{x}} \end{bmatrix} = \begin{bmatrix} A & 0 \\ -MC & (A + MC) \end{bmatrix} \begin{bmatrix} x \\ \hat{x} \end{bmatrix} + \begin{bmatrix} B \\ B \end{bmatrix} u
$$

$$
y = \begin{bmatrix} C & 0 \end{bmatrix} \begin{bmatrix} x \\ \hat{x} \end{bmatrix} \tag{6}
$$

and introducing the feedback law (5), we have

$$
\begin{bmatrix} \dot{x} \\ \dot{\hat{x}} \end{bmatrix} = \begin{bmatrix} A & BK \\ -MC & (A + MC + BK) \end{bmatrix} \begin{bmatrix} x \\ \hat{x} \end{bmatrix} + \begin{bmatrix} B \\ B \end{bmatrix} \bar{u}
$$

$$
y = \begin{bmatrix} C & 0 \end{bmatrix} \begin{bmatrix} x \\ \hat{x} \end{bmatrix} \tag{7}
$$

It is more illustrative, however, to express the state of this composite system in terms of the error $\tilde{x}$ rather than the estimate $\hat{x}$. This is easily done via the state transformation

$$
\begin{bmatrix} x \\ \tilde{x} \end{bmatrix} = \begin{bmatrix} x \\ x - \hat{x} \end{bmatrix} = \begin{bmatrix} I & 0 \\ I & -I \end{bmatrix} \begin{bmatrix} x \\ \hat{x} \end{bmatrix} \tag{8}
$$

which puts (7) into the form

$$
\begin{bmatrix} \dot{x} \\ \dot{\tilde{x}} \end{bmatrix} = \begin{bmatrix} (A + BK) & -BK \\ 0 & (A + MC) \end{bmatrix} \begin{bmatrix} x \\ \tilde{x} \end{bmatrix} + \begin{bmatrix} B \\ 0 \end{bmatrix} \bar{u}
$$

$$
y = \begin{bmatrix} C & 0 \end{bmatrix} \begin{bmatrix} x \\ \tilde{x} \end{bmatrix} \tag{9}
$$

Note that another essential property of the observer becomes obvious in (9): the error states $\tilde{x}$ are all completely uncontrollable from the input u. This is as expected, of course, since the error decays to zero regardless of what input is applied.

The dynamic behavior of the closed-loop system is now clear since the eigenvalues of a block-triangular matrix are precisely the eigenvalues of the diagonal blocks (Problem A.3-11). The closed-loop system poles consist of two distinct sets: the poles of $A + BK$, which determine the behavior of the state x, and the poles of $A + MC$, which determine the behavior of the error $\tilde{x}$. Each of these sets of poles may be placed arbitrarily by selecting K and M, respectively.

This is a fortunate state of affairs, indeed, as the determination of a feedback control law and the construction of an observer to realize it can be considered as separate problems. We summarize as follows:

1. THEOREM (Separation property). If a feedback control law of the form (2) for the system (1) is realized by means of an observer,† the poles of the resulting composite system consist of

(a) The poles of $A + BK$, determined completely by the feedback matrix K, and

(b) The poles of the observer, determined completely by the observer parameters [in (9), these were the poles of $A + MC$, determined by M].

The behavior of the state vector is governed by the poles in (a), and that of the error vector by the poles in (b).

ΔΔΔ

The use of an observer in the feedback loop of Fig. 1 may be viewed as a type of *compensation* since it introduces additional dynamic elements into the control system. The following example illustrates the relation between observers and the compensators discussed in Chapter 9.

---

†The reader is asked in Problem 1 to verify that this theorem holds for an observer of any dimension.

2. *EXAMPLE.* The second-order system

$$\dot{\underline{x}} = \begin{bmatrix} 0 & 1 \\ -1 & 0 \end{bmatrix} \underline{x} + \begin{bmatrix} 1 \\ 0 \end{bmatrix} \underline{u} \tag{10}$$

$$y = \begin{bmatrix} 0 & 1 \end{bmatrix} \underline{x}$$

is known as a *harmonic oscillator* since it has a pair of undamped oscillatory poles on the imaginary axis (at $s = \pm j$). If we attempt to stabilize this system with output feedback $u = ky$, the closed-loop system matrix becomes

$$\begin{bmatrix} 0 & 1 \\ -1 & 0 \end{bmatrix} + \begin{bmatrix} 1 \\ 0 \end{bmatrix} k \begin{bmatrix} 0 & 1 \end{bmatrix} = \begin{bmatrix} 0 & 1+k \\ -1 & 0 \end{bmatrix} \tag{11}$$

with characteristic polynomial $p(s) = s^2 + k + 1$, which is not stable for any value of the gain k. The reader will find it instructive to sketch the root-locus diagram of (10) for both positive and negative values of k.

If the other state were available for feedback, we could use a state feedback law

$$u = \begin{bmatrix} k_1 & k_2 \end{bmatrix} \underline{x} \tag{12}$$

resulting in the closed-loop system matrix

$$\begin{bmatrix} 0 & 1 \\ -1 & 0 \end{bmatrix} + \begin{bmatrix} 1 \\ 0 \end{bmatrix} \begin{bmatrix} k_1 & k_2 \end{bmatrix} = \begin{bmatrix} k_1 & 1 + k_2 \\ -1 & 0 \end{bmatrix} \tag{13}$$

with characteristic polynomial $p(s) = s^2 - k_1 s + k_2 + 1$, and the closed-loop poles could be placed anywhere. To be more specific, the gain matrix $[k_1 \;\; k_2] = [-1 \;\; 1]$ would result in stable closed-loop poles at $s = -0.5 \pm j1.32$.

According to Theorem 1, this control law can be realized by estimating the state with an observer. Since (10) is already in the form (12.2-1), a first-order observer may be constructed by substituting into (12.2-8) and (12.2-9):

$$\begin{aligned} \dot{w} &= -mw + u + (1 + m^2)y \\ \hat{x}_1 &= w - my \\ \hat{x}_2 &= y = x_2 \end{aligned} \tag{14}$$

The observer gain matrix $\underline{M}_1$ reduces to a single scalar m in this case, and the choice $m = 2$ places the observer pole at $s = -2$.

Combining (12) and (14), where $k_1 = -1$, $k_2 = 1$, and $m = 2$, we have

$$\begin{aligned} u &= -\hat{x}_1 + y = -w + 3y \\ \dot{w} &= -2w + u + 5y = -3w + 8y \end{aligned} \tag{15}$$

The closed-loop system is thus

$$\begin{aligned} \begin{bmatrix} \dot{\underline{x}} \\ \dot{w} \end{bmatrix} &= \begin{bmatrix} 0 & 4 & -1 \\ -1 & 0 & 0 \\ 0 & 8 & -3 \end{bmatrix} \begin{bmatrix} \underline{x} \\ w \end{bmatrix} \\ y &= \begin{bmatrix} 0 & 1 & 0 \end{bmatrix} \begin{bmatrix} \underline{x} \\ w \end{bmatrix} \end{aligned} \tag{16}$$

and the reader should verify that its poles are at $s = -2$ and $s = -0.5 \pm j1.32$, as expected.

It is interesting to interpret the observer and feedback law as a compensator placed in the feedback loop from output to input. Indeed, from (15) it is easy to derive the transfer function

$$g(s) = \frac{3s + 1}{s + 3} = 3\,\frac{s + 1/3}{s + 3} \tag{17}$$

from y to u, which has a zero at $s = -1/3$ and a pole at $s = -3$. This is a *lead compensator* of the form (9.3-6), with $\alpha = 9$ and an additional gain factor of 3 (i.e., a gain of 1/3).

ΔΔΔ

Disturbance Inputs and Bias

In Sections 12.1 and 12.2 we assumed that any inputs to the system under consideration were also available as inputs to the observer. This is not always true, of course, since the inputs often include *unmeasurable disturbances* which must be taken into account if the observer is to produce an accurate estimate of the state. Consider the design of an observer for the system

$$\begin{aligned} \dot{\underline{x}} &= \underline{A}\underline{x} + \underline{B}\underline{w} \\ \underline{y} &= \underline{C}\underline{x} \end{aligned} \tag{18}$$

which has an m-dimensional input $\underline{w}(t)$ with an unknown but constant value.[†] We shall see that this is *dual* to the problem of tracking a constant desired output, considered in Sec. 11.3.

Since the input satisfies

$$\dot{\underline{w}}(t) = \underline{0} \tag{19}$$

a reasonable approach is to combine (19) with (18) and estimate $\underline{w}$ along with $\underline{x}$. This amounts to adding m integrators and results in the augmented system

$$\begin{aligned} \begin{bmatrix} \dot{\underline{x}} \\ \dot{\underline{w}} \end{bmatrix} &= \begin{bmatrix} \underline{A} & \underline{B} \\ \underline{0} & \underline{0} \end{bmatrix} \begin{bmatrix} \underline{x} \\ \underline{w} \end{bmatrix} \\ \underline{y} &= \begin{bmatrix} \underline{C} & \underline{0} \end{bmatrix} \begin{bmatrix} \underline{x} \\ \underline{w} \end{bmatrix} \end{aligned} \tag{20}$$

where the new state vector has dimension $n + m$. The feasibility of this approach will depend upon whether (20) is observable:

3. LEMMA. The composite system (20) is observable if and only if the original system (18) is observable *and*

$$\text{rank} \begin{bmatrix} \underline{A} & \underline{B} \\ \underline{C} & \underline{0} \end{bmatrix} = n + m \tag{21}$$

Note that this condition can hold *only if* $p \geq m$ (i.e., there are at least as many outputs as inputs) and $\text{rank}(\underline{B}) = m$ (i.e., the inputs are independent).

Proof: This is the dual of Lemma 11.3.1.

ΔΔΔ

Having determined a test for the observability of (20), it is an easy matter to construct a reduced-order (Sec. 12.2) or full-order (Sec. 12.1) observer which will estimate $\underline{x}$ and $\underline{w}$. In the latter case we substitute (20) into (12.1-9) to form the $(n + m)$-dimensional

---

[†] This problem is considered in Refs. 339 and 370. A combination of known and unknown inputs is easily accommodated, and the reader is asked to supply details in Problem 4. Ramps and other inputs are the subject of Problem 5.

observer

$$\begin{bmatrix} \dot{\hat{\underline{x}}} \\ \dot{\hat{\underline{w}}} \end{bmatrix} = \begin{bmatrix} \underline{A} + \underline{M}_1\underline{C} & \underline{B} \\ \underline{M}_2\underline{C} & \underline{0} \end{bmatrix} \begin{bmatrix} \hat{\underline{x}} \\ \hat{\underline{w}} \end{bmatrix} - \begin{bmatrix} \underline{M}_1 \\ \underline{M}_2 \end{bmatrix} \underline{y} \tag{22}$$

According to Theorem 12.1.1, if (20) is observable, then the gain matrix

$$\underline{M} = \begin{bmatrix} \underline{M}_1 \\ \underline{M}_2 \end{bmatrix} \tag{23}$$

may be chosen to place the observer poles in any desired (stable) locations, so that

$$\hat{\underline{x}}(t) \to \underline{x}(t) \qquad \hat{\underline{w}}(t) \to \underline{w} \tag{24}$$

as rapidly as necessary. The reader is asked to verify in Problem 4 that $\underline{x}$ can still be estimated if $\text{rank}(\underline{B}) < m$.

One often encounters a similar type of disturbance, known as *bias error*, which causes the measured output to be offset from its true value. The system

$$\begin{aligned} \dot{\underline{x}} &= \underline{A}\underline{x} + \underline{B}\underline{w} \\ \underline{y} &= \underline{C}\underline{x} + \underline{D}\underline{w} \end{aligned} \tag{25}$$

has an unknown bias vector $\underline{D}\underline{w}$ corrupting its output, as well as an input disturbance term $\underline{B}\underline{w}$. The same disturbance vector $\underline{w}$ is shown here in both equations for the sake of convenience, but separate input and output disturbances may easily be incorporated by partitioning:

$$\begin{aligned} \dot{\underline{x}} &= \underline{A}\underline{x} + \begin{bmatrix} \underline{B}_1 & \underline{0} \end{bmatrix} \begin{bmatrix} \underline{w}_1 \\ \underline{w}_2 \end{bmatrix} = \underline{A}\underline{x} + \underline{B}_1\underline{w}_1 \\ \underline{y} &= \underline{C}\underline{x} + \begin{bmatrix} \underline{0} & \underline{D}_2 \end{bmatrix} \begin{bmatrix} \underline{w}_1 \\ \underline{w}_2 \end{bmatrix} = \underline{C}\underline{x} + \underline{D}_2\underline{w}_2 \end{aligned} \tag{26}$$

In the case of a constant disturbance ($\dot{\underline{w}} = \underline{0}$), we may augment the system as in (20):

$$\begin{bmatrix} \dot{\underline{x}} \\ \dot{\underline{w}} \end{bmatrix} = \begin{bmatrix} \underline{A} & \underline{B} \\ \underline{0} & \underline{0} \end{bmatrix} \begin{bmatrix} \underline{x} \\ \underline{w} \end{bmatrix}$$
$$\underline{y} = \begin{bmatrix} \underline{C} & \underline{D} \end{bmatrix} \begin{bmatrix} \underline{x} \\ \underline{w} \end{bmatrix} \tag{27}$$

where the new state vector again has dimension $n + m$. Extending Lemma 3, we have the following:

4. LEMMA. The composite system (27) is observable if and only if the original system (25) is observable *and*

$$\text{rank} \begin{bmatrix} \underline{A} & \underline{B} \\ \underline{C} & \underline{D} \end{bmatrix} = n + m \tag{28}$$

Note that this condition can hold *only if* $p \geq m$ (i.e., there are at least as many outputs as disturbances).

Proof: This is a simple extension of Lemma 3 (Problem 6).

ΔΔΔ

Having determined a test for the observability of (27), it is straightforward to construct an observer as in (22). Substituting (27) into (12.1-9) yields

$$\begin{bmatrix} \dot{\hat{\underline{x}}} \\ \dot{\hat{\underline{w}}} \end{bmatrix} = \begin{bmatrix} \underline{A} + \underline{M}_1\underline{C} & \underline{B} + \underline{M}_1\underline{D} \\ \underline{M}_2\underline{C} & \underline{M}_2\underline{D} \end{bmatrix} \begin{bmatrix} \hat{\underline{x}} \\ \hat{\underline{w}} \end{bmatrix} - \begin{bmatrix} \underline{M}_1 \\ \underline{M}_2 \end{bmatrix} \underline{y} \tag{29}$$

and the gain matrices $\underline{M}_1$ and $\underline{M}_2$ may be chosen so as to place the observer poles in any desired locations (provided (28) holds). Finally, we note once again that the use of *observers*, particularly in feedback loops, may be regarded as a form of *compensation*. The entire subject of compensation, disturbance inputs, servo operation, bias, and stabilization has received a great deal of attention in the recent literature. Some alternative approaches treat a compensator simply as a group of additional dynamic elements (typically integrators) to be utilized in devising a feedback control law with specified characteristics; this is done without explicit recourse to the concept of state estimation.

The interested reader may wish to pursue the topic of compensation further by browsing in Refs. 229, 240-243, 273, 274, 293 and 329 and the references therein. The papers by Davison appear to be the most complete and general. An interesting optimal control application is explored in Ref. 303 and the use of observers for failure detection is discussed in Ref. 238a.

12.4. THE KALMAN-BUCY FILTER

In this section we shall touch very briefly (and superficially) on a new concept -- that of *noise* and *stochastic systems*. Up until now, we considered only systems which were *deterministic*, in the sense that all inputs could be specified exactly and all outputs could be measured with unlimited precision. These assumptions are mathematically convenient and have led to many powerful and useful theoretical developments. In practice, of course, they are never really satisfied. Input and output transducers are subject to unpredictable fluctuations and disturbances, and communication channels are corrupted by all manner of interference.

Such uncertainties are present in all physical systems and are usually referred to by the generic term *noise*. In some cases the noise is inconsequential, and a deterministic analysis will suffice. In others, the effect of the noise is too great to be ignored and it must be modeled explicitly. Probabilistic or statistical models are most common, although other conceptual frameworks have been found useful in certain situations.

We shall now describe very briefly and heuristically a few key elements of probability theory. The discussion is quite incomplete, particularly with respect to random processes and stochastic differential equations, and a great deal more rigorous detail is necessary to put the entire subject on a sound mathematical footing. Indeed, the reader should treat this section as simply a qualitative glimpse into a complex and powerful body of theory.

The first and most important concept is that of a *random variable*, whose value is determined by the outcome of a *trial* or *experiment*. Examples include the number of tails occurring in 10 tosses of a fair coin, the distance traveled by a golf ball, the amount of rainfall detected at a particular location in 24 hours, and the voltage read from a meter at the output of a turbogenerator. It is important to be aware of the distinction between a random variable (the abstract concept) and a *sample* of the same random variable (i.e., the outcome of a single trial). Most books on probability theory adopt a notation which distinguishes the two, but in this brief discussion we will not do so.

Although the value of a sample of a random variable can never be predicted with certainty, various parameters may be determined which summarize the aggregate behavior[†] of many samples. The first of these is the *mean*, or *average*, or *expected value*. If the random variable in question is denoted by x, its mean is denoted $\bar{x}$ or $E\{x\}$, where the linear operator E is known as the *expectation operator*. For example, if x is the number of tails occurring in 10 tosses of a fair coin, then $\bar{x} = 5$. In a particular trial, any number of tails from 0 to 10 may occur, but the average number of tails over many trials will be 5. Similarly, samples of the voltage v measured at the output of a generator will exhibit fluctuations due to disturbances in the generator, the load, and the meter, but their mean value should be fairly close to the specified line voltage, say $\bar{v} = 240$ volts.

Just as important as the mean value of a random variable x is its expected variation about the mean. The *variance* of x is the average value (over many trials) of the squared difference between x and its mean, denoted

$$\operatorname{var}(x) \triangleq E\{(x - \bar{x})^2\} \tag{1}$$

[†]A random variable is completely characterized by its associated *probability density function*, but we will not be making use of this concept.

The square root of the variance is called the *standard deviation*. This is a measure of the "randomness" of x, or the spread of the distribution of sample values. A nonrandom variable has zero variance.

The level of dependence between two random variables x and y is measured principally by their *covariance*,

$$\text{cov}(x,y) \triangleq E\{(x-\bar{x})(y-\bar{y})\} = E\{xy\} - \bar{x}\bar{y} \tag{2}$$

Note that $\text{cov}(x,x) = \text{var}(x)$. The covariance is bounded by[†]

$$[\text{cov}(x,y)]^2 \le \text{var}(x)\text{var}(y) \tag{3}$$

so that the *correlation coefficient* of x and y,

$$\rho(x,y) \triangleq \frac{\text{cov}(x,y)}{\sqrt{\text{var}(x)\text{var}(y)}} \tag{4}$$

may take on values between -1 and 1. If $|\rho(x,y)|$ is close to 1, then x and y are said to be highly *correlated*, or *dependent*. We would expect a large correlation, for instance, between simultaneous wind velocity measurements taken at two locations 10 m apart.

On the other hand, if $\rho(x,y) = \text{cov}(x,y) = 0$ (that is, $\overline{xy} = \bar{x}\,\bar{y}$), then x and y are *uncorrelated*. Random variables that are completely independent (such as the price of beans in Boston and the winning time in the Melbourne Cup) are always uncorrelated. The converse is not necessarily true, but is often at least a good approximation.

A *random process* (or *stochastic process*) is a collection of random variables $\{x(t) : t_0 \le t \le T\}$ defined at each point on a time interval $[t_0, T]$. Examples include the reading on a voltmeter at each point in time over a 5-sec period and the instantaneous power output of an electric guitar during a rock concert. A *sample* of a random process consists of a single time function $x(t)$, $t_0 \le t \le T$. The variable t may represent quantities other than time, of course.

The mean and variance of a random process x are defined as

$$\bar{x}(t) = E\{x(t)\} \tag{5}$$

[†]This is the *Schwartz inequality* in the space of random variables.

and

$$\text{var}(x(t)) = E\{[x(t) - \bar{x}(t)]^2\} \tag{6}$$

for $t_0 \leq t \leq T$. Also of interest is the *autocovariance function*

$$r(t,\tau) \triangleq \text{cov}(x(t),x(\tau)) \tag{7}$$

which measures the correlation of the process with itself at different instants of time.

A *(wide-sense) white noise process* x is one that is completely uncorrelated from one instant to the next, with autocovariance function

$$r(t,\tau) = \text{var}(x(t))\ \delta(t - \tau) \tag{8}$$

where $\delta(t - \tau)$ is a unit impulse[†] at $\tau = t$. If the mean and variance are already known, this implies that observing (or sampling) the value of x at one time provides no additional information about its value at other times. White noise, like the delta function, is a physically unrealizable but extremely useful mathematical artifice.

If $x_1, x_2, \ldots, x_n$ are all random variables (or random processes), it is often convenient to collect them into a *random vector*

$$\underline{x} = [x_1\ x_2\ \ldots\ x_n]' \tag{9}$$

with mean

$$\bar{\underline{x}} = E\{\underline{x}\} = [\bar{x}_1\ \bar{x}_2\ \ldots\ \bar{x}_n]' \tag{10}$$

We can define a *covariance matrix*

$$\underline{P} \triangleq E\{(\underline{x} - \bar{\underline{x}})(\underline{x} - \bar{\underline{x}})'\} \geq \underline{0} \tag{11}$$

whose ij-th element is $\text{cov}(x_i,x_j)$ and whose i-th diagonal element is $\text{var}(x_i)$. It is not hard to show (Problem 1) that such a matrix must be symmetric and nonnegative definite.

[†] Impulse functions are discussed in Secs. 3.4 and B.1. The Fourier transform (or spectrum) of an impulse is constant (at least in the frequency range of interest), and the term "white" is used because of the analogy with white light.

Now consider a stochastic linear system of the following form,[†]

$$\dot{\underline{x}}(t) = \underline{A}\underline{x}(t) + \underline{B}\underline{w}(t) + \underline{B}_1\underline{u}(t)$$
$$\underline{y}(t) = \underline{C}\underline{x}(t) + \underline{v}(t) \tag{12}$$

where $\underline{w}(t)$ is a vector random process called *driving noise, input noise,* or *process noise,* and $\underline{u}(t)$ is a known (deterministic) input vector. The latter may be interpreted as a control, as the mean of $\underline{w}$, or both (see Problem 2). Similarly, $\underline{w}$ may represent fluctuations of the applied control signal $\underline{u}$, unknown disturbances to the system, or both. The vector random process $\underline{v}(t)$ is known as *output noise, observation noise,* or *measurement noise* and represents uncertainties or deviations in the output measurements. A classic example of this is thermal noise in the amplifier used to make the measurements.

The system (12) is assumed to be both controllable and observable, and the processes $\underline{w}$ and $\underline{v}$ are assumed to have zero means and to be white,[††]

$$E\{\underline{w}(t)\} = \underline{0} \qquad E\{\underline{v}(t)\} = \underline{0} \qquad \text{(for all } t) \tag{13}$$

$$E\{\underline{w}(t)\underline{w}'(\tau)\} = \underline{Q}\,\delta(t-\tau) \qquad E\{\underline{v}(t)\underline{v}'(\tau)\} = \underline{R}\,\delta(t-\tau) \qquad \text{(for all } t,\tau) \tag{14}$$

Here the covariance matrix $\underline{Q}$ specifies relative levels of correlation among the various components of the vector $\underline{w}(t)$, while the delta function implies that all components of $\underline{w}(t)$ are uncorrelated with those of $\underline{w}(\tau)$ for $t \neq \tau$. An identical statement applies to $\underline{R}$ and $\underline{v}(t)$. We shall assume that $\underline{R}$ and $\underline{Q}$ are positive definite (i.e.,

[†]The differential equation in (12) is not well defined, because of the white noise input. It is nevertheless customary to use this form, and we do so for the sake of convenience. In discrete-time systems, white noise is well defined and the problem does not arise.

[††]Controllability and observability are defined as in Secs. 10.1 and 10.2, with the noise terms $\underline{w}$ and $\underline{v}$ set equal to $\underline{0}$. There is also an implicit but nonessential assumption here that $\underline{w}$ and $\underline{v}$ are *wide-sense stationary* processes since $\underline{Q}$ and $\underline{R}$ do not depend on the time t.

nonsingular) and that $\underline{w}$ and $\underline{v}$ are uncorrelated with one another:

$$E\{\underline{w}(t)\underline{v}'(\tau)\} = \underline{0} \qquad \text{(for all } t,\tau) \tag{15}$$

The initial state of the system (12) at time t_0 is usually assumed to be a random vector $\underline{x}(t_0) = \underline{x}_0$ with mean $\overline{\underline{x}}_0$ and covariance

$$\underline{P}_0 = E\{(\underline{x}_0 - \overline{\underline{x}}_0)(\underline{x}_0 - \overline{\underline{x}}_0)'\} \tag{16}$$

which is also assumed to be uncorrelated with the noise processes $\underline{w}$ and $\underline{v}$. Alternatively, we can assume that the initial time was $t_0 = -\infty$ and that the system (if it is stable) has reached a stochastic "steady state," in the sense that the effect of $\underline{x}_0$ has long since died away.

All of the assumptions above ensure that the state $\underline{x}(t)$ and output $\underline{y}(t), t_0 \leq t < \infty$, are themselves well-behaved vector random processes. We could, with a more rigorous probabilistic framework, make use of the formulas developed in Chapter 3 and investigate the properties of the processes $\underline{x}$ and $\underline{y}$. However, this section is intended only as a survey, so we shall resist such a temptation and turn instead to the topic of estimation.

State Estimation

Consider now the problem of estimating the state $\underline{x}(T)$ of (12) at time $T > t_0$, using the noisy measurement data $\{\underline{y}(t) : t_0 \leq t < T\}$. To determine an estimate $\hat{\underline{x}}(T)$ of $\underline{x}(T)$, it is common practice to form the state error vector $\tilde{\underline{x}} \triangleq \underline{x} - \hat{\underline{x}}$ and then minimize the *mean-square error*,[†]

$$e(T) \triangleq E\{\|\tilde{\underline{x}}(T)\|^2\} = E\{\sum_{i=1}^{n} \tilde{x}_i^2(T)\} = \sum_{i=1}^{n} E\{\tilde{x}_i^2(T)\} \tag{17}$$

In Sec. 12.1 we attacked the deterministic version of this problem by constructing a model of the system and then adding a correction term proportional to the output error. The result was the observer

[†]It turns out, in fact, that the solution of the problem is unchanged if we minimize $E\{\tilde{\underline{x}}'(T)\underline{W}\tilde{\underline{x}}(T)\}$, where $\underline{W}$ is any symmetric positive definite weighting matrix. The particular choice $\underline{W} = \underline{I}$ yields (17).

$$\dot{\hat{\underline{x}}}(t) = \underline{A}\hat{\underline{x}}(t) + \underline{B}_1\underline{u}(t) - \underline{M}[\underline{y}(t) - \hat{\underline{y}}(t)]$$
$$\hat{\underline{y}}(t) = \underline{C}\hat{\underline{x}}(t) \tag{18}$$

where the gain matrix $\underline{M}$ was chosen so as to place the observer poles (eigenvalues of $\underline{A} + \underline{MC}$) in some desired configuration. It can be shown that the linear estimator that minimizes e(T) has precisely this form, with $\hat{\underline{x}}(t_0) = \overline{\underline{x}}_0$, except that the gain matrix is time varying and given by

$$\underline{M}(t) = -\underline{P}(t)\underline{C}'\underline{R}^{-1} \qquad (t_0 \leq t \leq T) \tag{19}$$

where $\underline{P}(t)$ is the symmetric, nonnegative definite solution of a nonlinear, time-varying matrix differential equation known as a *Riccati equation:*

$$\dot{\underline{P}}(t) = \underline{AP}(t) + \underline{P}(t)\underline{A}' - \underline{P}(t)\underline{C}'\underline{R}^{-1}\underline{CP}(t) + \underline{BQB}'$$
$$\underline{P}(t_0) = \underline{P}_0 \tag{20}$$

The initial condition in (20) is $\underline{P}_0$, the covariance matrix of the initial state $\underline{x}_0$. Because $\hat{\underline{x}}(t_0) = \overline{\underline{x}}_0$ (a nonrandom vector), it follows that the initial error has expected value

$$E\{\tilde{\underline{x}}(t_0)\} = E\{\underline{x}(t_0)\} - \overline{\underline{x}}_0 = \underline{0} \tag{21}$$

and covariance matrix

$$E\{\tilde{\underline{x}}(t_0)\tilde{\underline{x}}'(t_0)\} = E\{[\underline{x}(t_0) - \overline{\underline{x}}_0][\underline{x}(t_0) - \overline{\underline{x}}_0]'\} = \underline{P}_0 \tag{22}$$

Moreover, it can be shown that

$$E\{\tilde{\underline{x}}(t)\} = \underline{0} \qquad (\text{for all } t \geq t_0) \tag{23}$$

and

$$E\{\tilde{\underline{x}}(t)\tilde{\underline{x}}'(t)\} = \underline{P}(t) \qquad (\text{for all } t \geq t_0) \tag{24}$$

The first of these conditions means that $\hat{\underline{x}}(t)$ is an *unbiased* estimate of $\underline{x}(t)$: the expected value of the error is zero for all time. The second condition says that the error covariance matrix is given by the solution of the Riccati equation and hence is independent of the measurement data. This covariance matrix specifies the accuracy of the estimate: its diagonal elements are the variances of the individual errors $\tilde{x}_i(t)$, and for $t_0 \leq t \leq T$,

$$e(t) = \sum_{i=1}^{n} E\{\tilde{x}_i^2(t)\} = \sum_{i=1}^{n} P_{ii}(t) = \operatorname{tr} \underline{P}(t) \tag{25}$$

Since T is quite arbitrary, the estimate is valid at any $t \geq t_0$, and we may collect the discussion above into the following theorem.

1. THEOREM. Under the assumptions stated earlier, the "best" linear estimate of the state $\underline{x}(t)$ of the system (12) that can be obtained from the noisy measurements $\{\underline{y}(\tau) : t_0 < \tau < t\}$ is given by the state $\hat{\underline{x}}(t)$ of the *Kalman-Bucy filter,*

$$\begin{aligned} \dot{\hat{\underline{x}}}(t) &= \underline{A}\hat{\underline{x}}(t) + \underline{B}_1\underline{u}(t) - \underline{M}(t)[\underline{y}(t) - \hat{\underline{y}}(t)] \\ \hat{\underline{y}}(t) &= \underline{C}\hat{\underline{x}}(t) \end{aligned} \tag{26}$$

with initial state $\hat{\underline{x}}(t_0) = \bar{\underline{x}}_0$, where the *gain matrix* is

$$\underline{M}(t) = -\underline{P}(t)\underline{C}'\underline{R}^{-1} \tag{27}$$

and the state *error covariance matrix* $\underline{P}(t)$ satisfies the *Riccati equation*

$$\begin{aligned} \dot{\underline{P}}(t) &= \underline{A}\underline{P}(t) + \underline{P}(t)\underline{A}' - \underline{P}(t)\underline{C}'\underline{R}^{-1}\underline{C}\underline{P}(t) + \underline{B}\underline{Q}\underline{B}' \\ \underline{P}(t_0) &= \underline{P}_0 \end{aligned} \tag{28}$$

Both the filter and the Riccati equation are stable.

The estimate $\hat{\underline{x}}(t)$ is *unbiased* since its average error (23) is zero, and it is *optimal* in the sense that at each time t its mean-square error (25) is smaller than that achieved by any other linear[†] estimator. Similarly, $\hat{\underline{y}}(t)$ is the best linear estimate of $\underline{y}(t)$ given the noisy data; the name "filter" is used because $\hat{\underline{y}}$ is a filtered version of $\underline{y}$. The output error vector $\tilde{\underline{y}} = \underline{y} - \hat{\underline{y}}$ is also known as the

---

[†]If we also make the fairly common assumption that the initial state and the two noise processes satisfy *Gaussian* (or *normal*) probability distributions, then the mean-square error is less than that achieved by any other estimator, linear or nonlinear.

innovations process (or the *residual*) and plays a central role in estimation theory. It is a zero-mean white noise process,

$$E\{\tilde{\underline{y}}(t)\tilde{\underline{y}}'(\tau)\} = \underline{R}\ \delta(t - \tau) \tag{29}$$

with the same covariance matrix $\underline{R}$ as the measurement noise, and $\tilde{\underline{y}}(t)$ is uncorrelated with all the past data $\{y(\tau) : t_0 \leq \tau < t\}$.

ΔΔΔ

A proof of the theorem is, of course, beyond the scope of this text, but the interested reader should consult any of the references cited at the end of this section.

The structure of the Kalman-Bucy filter, shown in Fig. 1, is

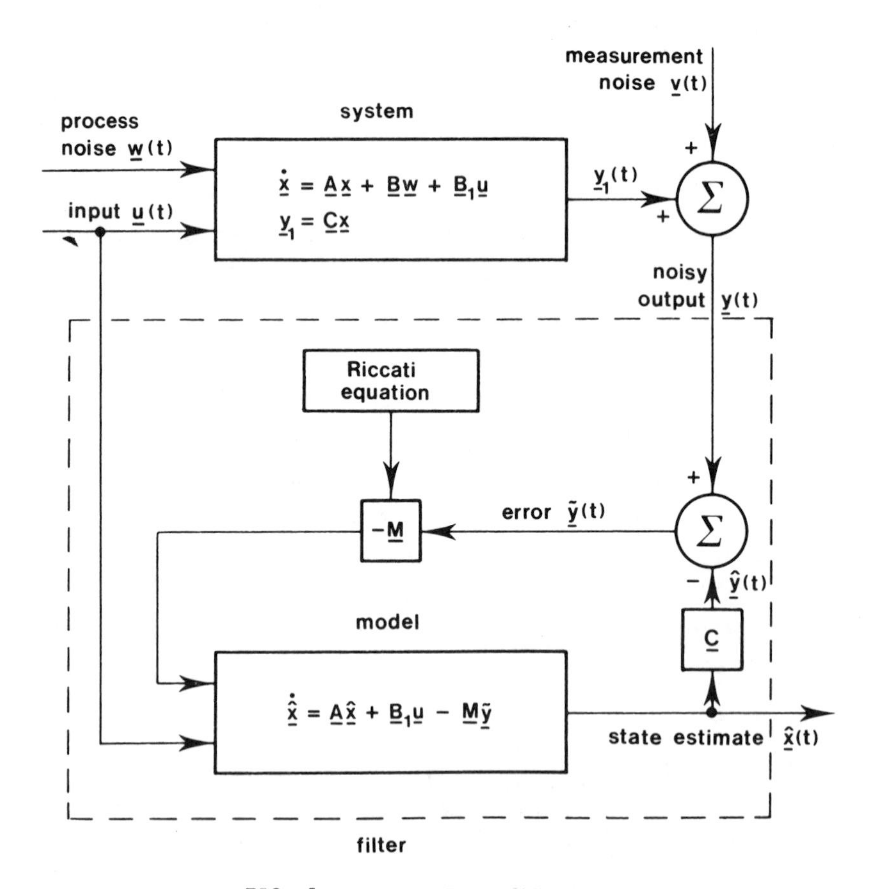

FIG. 1. Kalman-Bucy filter.

nearly identical to the full-order observer of Fig. 12.1-2. The only difference, in fact, is that in this case the gain matrix $\underline{M}(t)$ is time varying and must be computed by solving the Riccati equation (28). Standard numerical or analog techniques may be used to integrate (28) in parallel with the other filter operations, and it is often feasible to implement the whole operation in real time on a digital computer. Alternatively, the solution of the Riccati equation may be computed off line since it is completely independent of the measurement data, but this may require a large amount of storage.

The gain matrix $\underline{M}(t)$ determines the degree to which the measurement data are used to adjust the current state estimate $\hat{\underline{x}}(t)$, and the expression (27) is intuitively appealing. In particular, we see that a high degree of confidence in the current estimate [small $\underline{P}(t)$] and a high level of measurement noise (large $\underline{R}$) correspond to a small gain. Conversely, the gain will be large if the current state estimate is a poor one and the measurements are very accurate.

Because the gain matrix $\underline{M}(t)$ is time dependent, the Kalman-Bucy filter (26) lies outside the class of systems discussed in previous chapters (more will be said about time-varying systems in Chapter 14). However, it can be shown that as time goes on the solution of the Riccati equation (28) converges to a constant (steady-state) value which can also be found by setting $\dot{\underline{P}} = \underline{0}$ in (28) and solving the resulting algebraic equation† for $\underline{P}$. This means that $\underline{M} = -\underline{P}\underline{C}'\underline{R}^{-1}$ also approaches a constant value and the filter becomes time invariant in steady state.

The steady-state filter is a very useful concept, for in many situations (12) is stable (or stabilizable) and one may assume the initial time to be far in the past ($t_0 \to -\infty$). This means that the effect of the initial state $\underline{x}_0$ on the system (12) has long since disappeared, in which case $\underline{x}(t)$ and $\underline{y}(t)$ are said to be *(wide-sense) stationary* processes. More importantly, it means that the covariance

†Nevertheless, it is often convenient (but numerically inefficient) to find $\underline{P}$ by simply integrating (28) to convergence.

matrix $\underline{P}$ and gain matrix $\underline{M}$ can easily be precomputed off line with minimal storage requirements, thus simplifying the structure of the filter considerably. We summarize as follows:

2. *THEOREM.* The *steady-state Kalman-Bucy filter* for (12) is the stable system

$$\begin{aligned} \dot{\hat{\underline{x}}}(t) &= \underline{A}\hat{\underline{x}}(t) + \underline{B}_1\underline{u}(t) - \underline{M}[\underline{y}(t) - \hat{\underline{y}}(t)] \\ \hat{\underline{y}}(t) &= \underline{C}\hat{\underline{x}}(t) \end{aligned} \tag{30}$$

where the gain matrix is

$$\underline{M} = -\underline{P}\underline{C}'\underline{R}^{-1} \tag{31}$$

and the constant error covariance matrix $\underline{P}$ is the symmetric, positive definite solution of the *steady-state Riccati equation*

$$\underline{A}\underline{P} + \underline{P}\underline{A}' - \underline{P}\underline{C}'\underline{R}^{-1}\underline{C}\underline{P} + \underline{B}\underline{Q}\underline{B}' = \underline{0} \tag{32}$$

ΔΔΔ

The structure of the time-invariant filter is just as shown in Fig. 1, except that the box labeled "Riccati equation" now represents a single off-line calculation. As in the case of the observer, (30) can be rewritten in the form

$$\begin{aligned} \dot{\hat{\underline{x}}}(t) &= (\underline{A} + \underline{M}\underline{C})\hat{\underline{x}}(t) + \underline{B}_1\underline{u}(t) - \underline{M}\underline{y}(t) \\ \hat{\underline{y}}(t) &= \underline{C}\hat{\underline{x}}(t) \end{aligned} \tag{33}$$

which clearly shows that the filter's dynamic behavior is determined principally by the eigenvalues of $\underline{A} + \underline{M}\underline{C}$. These in turn are determined by the statistical properties of the noise (i.e., by $\underline{Q}$ and $\underline{R}$, via the Riccati equation). Thus the Kalman-Bucy filter is a state observer with the additional property that the mean-square error (25) is minimized.

The results of Sec. 12.3 apply to all observers, including the (steady-state) Kalman-Bucy filter. The separation property (Theorem 12.3.1) is especially significant since the filter is often used to implement a feedback control law. If $\underline{w}(t)$ has an unknown (but constant) mean value, this may be treated in the same way as a constant disturbance input. Similarly, a nonzero mean value of $\underline{v}(t)$ is equivalent to an output bias error. The results of Sec.12.2, however, are less easily applied. Except in special cases (such as a singular $\underline{R}$), any decrease in the dimension of the Kalman-Bucy filter leads to nonminimal mean-square error.

Many of the assumptions made above can be removed or relaxed without much difficulty. In the case of a nonlinear system, the *extended Kalman-Bucy filter* has been found to be extremely useful and versatile: it operates by continually relinearizing the system about the current state estimate.

The reader is encouraged to pursue these topics further [33,34, 69, 95, 105, 126, 155, 157, 171, 184, 236, 277, 279, 280, 340, 341]. Reference 202 contains several relevant articles, including an excellent tutorial paper [321].

12.5. SUMMARY

In this chapter we have examined the problem of state estimation. When the entire state of a system is not available for direct measurement, we saw how an *observer*, driven by the input and output of the original system, can be used to provide a continuous estimate of its state.

In Sec. 12.1 this was done by constructing a model, with the same dimension as the original system, and feeding back error information from the output to adjust the rate at which the state estimation error decays to zero. In Sec. 12.2 we observed that the output vector already provides p dimensions of direct measurement of the state vector, and showed how the dimension of the observer can be reduced by this amount.

In Sec. 12.3 the state estimate provided by an observer was used to realize a linear feedback control law. There we established the *separation property:* The poles of the resulting composite system consist of (a) the closed-loop poles that would result if the state were directly available for feedback, and (b) the poles of the observer.

We also saw how the observer's state vector can be augmented to account for disturbance inputs and output bias errors.

Then we digressed, in Sec. 12.4, to consider systems with random noise driving the inputs and corrupting the outputs. State estimation was achieved using the *Kalman-Bucy filter*, which is essentially an observer whose dynamic characteristics are determined by the statistical properties of the noise.

PROBLEMS

Section 12.1

1. (a) Construct a state observer for the system

$$\dot{\underline{x}} = \begin{bmatrix} 0 & 1 \\ 0 & 0 \end{bmatrix} \underline{x} + \begin{bmatrix} 0 \\ 1 \end{bmatrix} u$$

$$y = \begin{bmatrix} 1 & 0 \end{bmatrix} \underline{x}$$

such that the observer poles are $-\gamma$ and -2γ $(\gamma > 0)$.

(b) Consider the observer as a system with input vector $[u \quad y]'$ and output vector $\hat{\underline{x}}$ and determine the transfer matrix $\underline{H}_\gamma(s)$ which relates $\hat{\underline{x}}$ to $[u \quad y]'$.

(c) Determine the limiting observer transfer matrix

$$\underline{H}_\infty(s) = \lim_{\gamma \to \infty} \underline{H}_\gamma(s)$$

What operation is the observer performing in this case? Does this make sense in terms of the original system?

(d) Suppose the system output is corrupted by a small, high-frequency disturbance, so that the signal appearing at the input to the observer in (c) is

$$y(t) + 10^{-3} \sin 10^6 t$$

instead of $y(t)$, and determine the state estimate $\hat{\underline{x}}(t)$ which results. What sort of performance would you expect from this observer in the presence of wide-bandwidth noise? How does this compare with the observer in (a)?

2. Derive a state observer for a single-output system in the observable canonical form (10.4-11), and show *explicitly* how the vector $\underline{m}$ may be chosen to place the observer poles in any desired locations.

3. Suppose you have an n-th order scalar (single-input, single-output) system in observable canonical form. Draw a block-diagram using only *summers* and *differentiators* to show how the system state can be determined exactly from measurements of the output and input.

4. Suppose a system is unobservable, but stabilizable in the sense of Corollary 11.2.2' and Theorem 10.3.3'. Show how a stable observer can be constructed to estimate its state. In what sense does this observer "estimate" the unobservable states? Can its dimension be reduced below $n - p$?

Section 12.2

1. Consider the system

$$\begin{aligned} \dot{\underline{x}} &= \underline{A}\underline{x} + \underline{B}\underline{u} \\ \underline{y} &= \underline{C}\underline{x} \end{aligned} \qquad (*)$$

with n states and p outputs, where $\underline{C}$ is assumed to have full rank p. Define a state transformation

$$\bar{\underline{x}} = \underline{Q}\underline{x} = \begin{bmatrix} \underline{D} \\ \\ \underline{C} \end{bmatrix} \underline{x}$$

where $\underline{D}$ is any $(n - p) \times n$ matrix such that $\underline{Q}^{-1}$ exists. Show that the resulting system has the form

$$\dot{\bar{\underline{x}}} = \bar{\underline{A}}\bar{\underline{x}} + \bar{\underline{B}}\underline{u}$$

$$\underline{y} = \bar{\underline{C}}\bar{\underline{x}} = \begin{bmatrix} \underline{0} & \underline{I} \end{bmatrix} \bar{\underline{x}}$$

Now suppose an observer of the form (12.2-8) and (12.2-9) is designed for this system. Show how to obtain an equivalent observer for the original system (*).

2. Prove Lemma 12.2.1 as follows: Use Definition 10.2.2 and argue that if (12.2-2) has a nonzero unobservable state, then (12.2-1) must also have one, which is a contradiction.

3. Consider the two-dimensional scalar system

$$\dot{\underline{x}} = \begin{bmatrix} a_{11} & a_{12} \\ a_{21} & a_{22} \end{bmatrix} \underline{x} + \begin{bmatrix} b_1 \\ b_2 \end{bmatrix} u$$

$$y = \begin{bmatrix} 0 & 1 \end{bmatrix} \underline{x}$$

Derive an observer of the form (12.2-6) for this system and show how it can be implemented using an integrator and a differentiator. Now verify that the redefinition of the state in (12.2-7) is equivalent to moving one of the inputs to the output side of the integrator, and that this results in an observer of the form (12.2-8).

4. Show how to construct a Luenberger observer for

$$\begin{bmatrix} \dot{\underline{x}}_1 \\ \dot{\underline{x}}_2 \end{bmatrix} = \begin{bmatrix} \underline{A}_{11} & \underline{A}_{12} \\ \underline{A}_{12} & \underline{A}_{22} \end{bmatrix} \begin{bmatrix} \underline{x}_1 \\ \underline{x}_2 \end{bmatrix} + \begin{bmatrix} \underline{B}_1 \\ \underline{B}_2 \end{bmatrix} \underline{u}$$

$$\underline{y} = \begin{bmatrix} \underline{I} & \underline{0} \end{bmatrix} \begin{bmatrix} \underline{x}_1 \\ \underline{x}_2 \end{bmatrix} = \underline{x}_1$$

Section 12.3

1. Prove Theorem 12.3.1 for the case of an $(n - p)$-dimensional observer.

2. Show that the separation property (Theorem 12.3.1) does not hold for just any interconnection of systems, by considering the following two scalar systems:

$$\dot{x} = ax + bu$$

$$\dot{z} = cz + du + ex$$

Determine the poles of each of these systems separately, and then determine the poles of the composite system that results when the feedback law $u = kz$ is applied.

3. Verify that the state transformation (12.3-8) puts the system (12.3-7) into the form (12.3-9).

4. (a) Show how to extend Lemma 12.3.3 and the discussion following it to the case where some of the inputs are available to the observer and some are not.

   (b) Explain how to estimate $\underline{x}$ when rank$(\underline{B}) < m$.

5. (a) Show how to construct an observer for a system with an unmeasurable ramp input $\underline{u}(t) = \bar{\underline{u}}t1(t)$.

   (b) Do the same for the input

$$\underline{u}(t) = \begin{bmatrix} \bar{u}_1 1(t) \\ \bar{u}_2 t1(t) \end{bmatrix}$$

6. Show how to extend the proof of Lemma 11.3.1 in order to prove Lemma 12.3.4.

7. (a) Consider the design of an observer for the system

$$\dot{\underline{x}} = \underline{A}\underline{x} + \underline{B}_1\underline{w}_1$$

$$\underline{y} = \underline{C}\underline{x} + \underline{w}_2$$

   where $\underline{w}_1$ and $\underline{w}_2$ are unmeasurable constant disturbances. In the absence of disturbances, the system is observable. Augment the state vector with $\underline{w}_1$ and $\underline{w}_2$ and write out the expanded state equations. On intuitive grounds, do you expect the augmented system to be observable? In other words, do you think that $\underline{w}_1$ and $\underline{w}_2$ can be determined (independently) using only the measurements $\underline{y}$? Explain.

   (b) Verify your answer to (a) by forming the observability matrix of the augmented system and determining whether it has full rank. If not, determine which states are unobservable.

   (c) Repeat your analysis for the case $\underline{w}_1 = \underline{0}$. Show that the augmented system is observable if and only if $\underline{A}$ is nonsingular. Give an intuitive explanation of this result.

(d) Does your intuitive explanation in (c) apply to the case where $\underline{w}_2 = \underline{0}$ and $\underline{w}_1 \neq \underline{0}$? Test it on a system of dimension $n = 1$.

8. (a) Consider the broom-balancing system with horizontal wind forces, described in Example 11.3.4. Assume that the wind force is constant, set up an augmented system of the form (12.3-20), and verify that it is observable. Construct a reduced-order observer with the pole configuration shown in Fig. 12.1-4(b).

(b) Use this observer to realize the control law of Example 11.3.4. Assume that the extra state q is available for measurement as an output. Can you eliminate one of the two extra states (q or w) and still maintain zero steady-state error in z?

Section 12.4

1. Show that the covariance matrix $\underline{P}$ defined in (12.4-11) is symmetric and nonnegative definite. *Hint:* Use Theorem A.6.2.

2. Suppose the system

$$\dot{\underline{x}}(t) = \underline{A}\underline{x}(t) + \underline{B}\underline{w}(t)$$

$$\underline{y}(t) = \underline{C}\underline{x}(t) + \underline{v}(t)$$

has driving noise $\underline{w}(t)$ with a known nonzero mean vector $E\{\underline{w}(t)\} = \overline{\underline{w}}(t)$.

(a) Show how this system can be put into the form (12.4-12).

(b) Explain how you would handle measurement noise with a known nonzero mean.

(c) What can you do if $\underline{w}(t)$ or $\underline{v}(t)$ has an unknown mean?

3. Generalize (12.4-25) to show that

$$E\{\tilde{\underline{x}}'(t)\ \underline{W}\tilde{\underline{x}}(t)\} = \mathrm{tr}(\underline{W}\underline{P}(t))$$

Hint: Use Problem A.2-14.

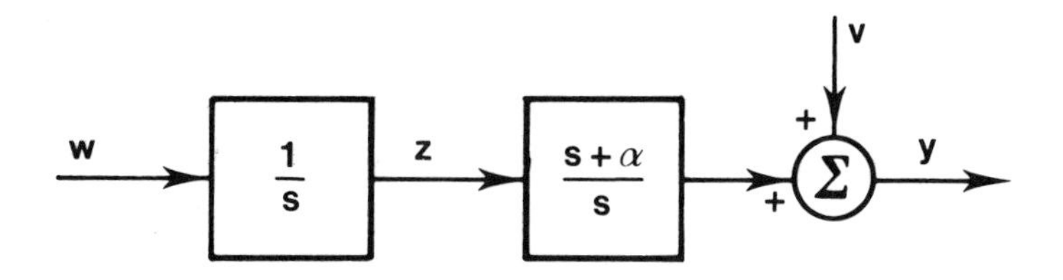

FIG. P1

4.[†] Consider the linear system shown in Fig. P1, where w and v are uncorrelated, zero mean, white noise processes with

$$E\{w(t)w(\tau)\} = \delta(t - \tau) \qquad E\{v(t)v(\tau)\} = r\delta(t - \tau)$$

(a) Suppose that $\alpha < 0$, and conjecture the transfer function of a filter which estimates z from y (in steady state) and becomes "optimal" as $r \to 0$. How does your answer change if $\alpha > 0$?

(b) Write down a state-variable representation and determine the steady-state Kalman-Bucy filter for this system. Find the transfer function h(s) relating the estimate $\hat{z}$ of z to y. Sketch $|h(j\omega)|$ in the limit as $r \to 0$. Find the limiting mean-square error $E\{(z - \hat{z})^2\}$ for $\alpha < 0$ and for $\alpha > 0$. Can you draw any conclusions about filtering of nonminimum phase systems?

5. (a) Suppose a signal z(t) is measured in the presence of additive noise, and the derivative $\dot{z}(t)$ is required. Discuss how the Kalman-Bucy theory could be used for this purpose and indicate what additional information and/or assumptions would be necessary.

(b) Apply your procedure to the system

$$\ddot{z}(t) = w(t)$$

$$y(t) = z(t) + v(t)$$

where w and v are white noise processes as in the previous problem.

[†]This problem and the next are due to Prof. Nils Sandell.

Part IV

RELATED TOPICS

Chapter 13

OPTIMAL CONTROL†

13.0. INTRODUCTION

In previous chapters we encountered various methods for designing feedback control laws, ranging from root-locus and compensation techniques in the case of scalar systems to pole-placement and state estimation for multivariable systems. In each case, the designer was left with decisions regarding the locations of closed-loop poles (and zeros).

Here a somewhat different approach will be taken: the performance of the system is measured with a single scalar quantity known as the *cost*, which in turn is minimized by the *optimal control law*. We shall see that such control laws have some computational advantages, as well as a number of useful properties. The task of the designer then shifts to one of specifying various parameters in the cost.

This chapter will take the form of a survey, describing (without proofs) basic results of linear optimal control theory. The interested reader who wishes to pursue the subject further is referred to Refs. 6, 11, 20, 32, 33, 105, and 202.

---

†This chapter assumes familiarity with Sec. A.6.

13.1. THE OPTIMAL LINEAR REGULATOR

Consider the linear system

$$\dot{\underline{x}}(t) = \underline{A}\underline{x}(t) + \underline{B}\underline{u}(t)$$
$$\underline{y}(t) = \underline{C}\underline{x}(t) \tag{1}$$

with $\underline{x}(0) = \underline{x}_0$, and suppose that our objective is to design a control law which will bring the output $\underline{y}(t)$ close to zero and maintain it there without an excessive expenditure of control "effort." In the terminology of Chapter 8, the system is to act as a *regulator*. If $\underline{C} = \underline{I}$, of course, then $\underline{y} = \underline{x}$ and it becomes a *state regulator*.

One measure of the magnitude of the output vector $\underline{y}(t)$ (or of its distance from the origin) is the norm $\|\underline{y}(t)\|$, defined by

$$\|\underline{y}(t)\|^2 = \underline{y}'(t)\underline{y}(t) \tag{2}$$

More generally, if $\underline{Q}$ is any symmetric, positive definite matrix ($\underline{Q} > \underline{0}$), then a *weighted* measure of the magnitude of $\underline{y}(t)$ is provided by the *quadratic form*†

$$\underline{y}'(t)\underline{Q}\underline{y}(t) \geq 0 \tag{3}$$

which is equal to zero if and only if $\underline{y}(t) = \underline{0}$. Similarly, if $\underline{R} > \underline{0}$, then

$$\underline{u}'(t)\underline{R}\underline{u}(t) \geq 0 \tag{4}$$

is a measure of the magnitude of the control vector, and hence of the instantaneous control effort. The matrices $\underline{Q}$ and $\underline{R}$ are often chosen to be diagonal, so that (3) and (4) become

$$\underline{y}'\underline{Q}\underline{y} = \sum_{i=1}^{p} q_{ii}y_i^2 \tag{5}$$

and

$$\underline{u}'\underline{R}\underline{u} = \sum_{i=1}^{m} r_{ii}u_i^2 \tag{6}$$

where the q_{ii} and r_{ii} are all positive.

---

†See Definition A.6.1.

If the system is to be controlled during the time interval $0 < t \leq T$, a measure of its performance is given by

$$J[\underline{x}_0;\underline{u}(\cdot)] = \int_0^T [\underline{y}'(t)\underline{Q}\underline{y}(t) + \underline{u}'(t)\underline{R}\underline{u}(t)]\, dt + \underline{y}'(T)\underline{S}\underline{y}(T) \tag{7}$$

where $\underline{S} \geq \underline{0}$ is another weighting matrix. The integral provides a weighted average of the output and the control over the interval, and the last term places special emphasis on the output at the terminal time T. The choice of $\underline{Q}$, $\underline{R}$, and $\underline{S}$ determines the relative weighting of the various terms. J is called a quadratic *cost functional* or *performance index*;[†] its value depends upon the initial state $\underline{x}_0$ and the entire control function $\underline{u}(t)$, $0 < t \leq T$.

Because this cost is quadratic in the components of $\underline{y}$ and $\underline{u}$, it tends to penalize large errors and large controls more heavily than small ones (the reader should plot y^2 as a function of y to verify this behavior in the scalar case). There are, of course, numerous other possible cost functionals, involving fourth powers, absolute values, maximum values, etc. Nevertheless, we shall restrict our attention to the quadratic case for two reasons: it has been found useful in a variety of applications, and the problem of minimizing J has a highly tractable solution. Accordingly, we now formulate the following optimization problem.

1. DEFINITION. The *linear regulator problem* is to determine an *optimal control* $\underline{u}^*(t)$, $0 < t \leq T$, for the system (1) which minimizes the value of (7). In other words, $\underline{u}^*$ must have the property that

$$J[\underline{x}_0;\underline{u}^*(\cdot)] \leq J[\underline{x}_0;\underline{u}(\cdot)] \tag{8}$$

for any other control $\underline{u}$.

ΔΔΔ

[†]This corresponds to the *objective function* in mathematical programming and operations research problems.

In view of the quadratic structure of J, a control law which minimizes it must in some sense satisfy our original objective of maintaining the output near zero with a moderate level of control effort. The assumption $\underline{R} > \underline{0}$ ensures that the control is bounded and the problem is well defined, and we have assumed $\underline{Q} > \underline{0}$ as well. (See Problem 1.) It turns out that the solution of the linear regulator problem can be expressed in a convenient feedback form.

2. *THEOREM.* The optimal control law which accomplishes the minimization (8) may be specified as

$$\underline{u}^*(t) = \underline{K}(t)\underline{x}(t) \qquad (0 < t \leq T) \tag{9}$$

where the time-varying *feedback gain matrix* is

$$\underline{K}(t) = -\underline{R}^{-1}\underline{B}'\underline{P}(t) \tag{10}$$

and $\underline{P}(t)$ is the symmetric, nonnegative definite solution of a nonlinear, time-varying matrix differential equation known as a *Riccati equation:*

$$\begin{aligned} \dot{\underline{P}}(t) &= -\underline{A}'\underline{P}(t) - \underline{P}(t)\underline{A} + \underline{P}(t)\underline{B}\underline{R}^{-1}\underline{B}'\underline{P}(t) - \underline{C}'\underline{Q}\underline{C} \\ \underline{P}(T) &= \underline{C}'\underline{S}\underline{C} \end{aligned} \tag{11}$$

The minimal value of the cost J for any initial state $\underline{x}_0$ is given by

$$J^*(\underline{x}_0) \triangleq J[\underline{x}_0;\underline{u}^*(\cdot)] = \underline{x}_0'\underline{P}(0)\underline{x}_0 \tag{12}$$

ΔΔΔ

Note that the boundary condition in (11) is specified at the terminal time T, so that the differential equation must be solved backwards in time and stored before the gain matrix $\underline{K}(t)$ can be computed. This makes on-line implementation of the optimal feedback law (9) somewhat difficult.

Because the gain matrix $\underline{K}(t)$ is time varying, the closed-loop system which results from using the control law (9) lies outside the class of systems discussed previously (the subject of time-varying systems will be taken up in the next chapter). However, if the system is controllable and observable, then as $T \to \infty$, the solution of the Riccati equation (11) for finite t converges to a constant (steady-state) value, which can be found by setting $\dot{\underline{P}} = \underline{0}$ in (11) and solving the resulting algebraic equation[†] for $\underline{P}$. This means that $\underline{K} = -\underline{R}^{-1}\underline{B}'\underline{P}$ also approaches a constant value and the control system becomes time invariant in steady state.

In other words, if the terminal time T is sufficiently far in the future, the effect of the last term in (7) becomes negligible, and Definition 1 and Theorem 2 become the following:

3. DEFINITION. The *steady-state linear regulator problem* is to determine an optimal control $\underline{u}^*(t)$, $0 < t < \infty$, for the system (1) which minimizes the value of

$$J[\underline{x}_0;\underline{u}(\cdot)] = \int_0^\infty [\underline{y}'(t)\underline{Q}\underline{y}(t) + \underline{u}'(t)\underline{R}\underline{u}(t)]\, dt$$

$$= \int_0^\infty [\underline{x}'(t)\underline{C}'\underline{Q}\underline{C}\underline{x}(t) + \underline{u}'(t)\underline{R}\underline{u}(t)]\, dt \tag{13}$$

ΔΔΔ

4. THEOREM. If (1) is controllable and observable, then the optimal control law which minimizes the cost (13) may be specified as

$$\underline{u}^*(t) = \underline{K}\underline{x}(t) \qquad (0 < t < \infty) \tag{14}$$

where the *feedback gain matrix* is

$$\underline{K} = -\underline{R}^{-1}\underline{B}'\underline{P} \tag{15}$$

and $\underline{P}$ is the symmetric, positive definite solution of the *steady-state Riccati equation*

$$\underline{A}'\underline{P} + \underline{P}\underline{A} - \underline{P}\underline{B}\underline{R}^{-1}\underline{B}'\underline{P} + \underline{C}'\underline{Q}\underline{C} = 0 \tag{16}$$

The minimal value of the cost is

$$\underline{J}^*(\underline{x}_0) = J[\underline{x}_0;\underline{u}^*(\cdot)] = \underline{x}_0'\,\underline{P}\underline{x}_0 \tag{17}$$

ΔΔΔ

[†]Nevertheless, it is often convenient (but numerically inefficient) to find $\underline{P}$ by simply integrating (11) to convergence.

Theorem 4 is a basic and well-known result in the theory of optimal control. Once the designer has specified $\underline{Q}$ and $\underline{R}$, representing his (often subjective) assessment of the relative importance of the various terms in (13), the solution of (16) specifies the optimal control law (14). This yields the optimal closed-loop system

$$\begin{aligned} \dot{\underline{x}} &= (\underline{A} + \underline{B}\underline{K})\underline{x} \\ \underline{y} &= \underline{C}\underline{x} \end{aligned} \tag{18}$$

whose poles are the eigenvalues of $\underline{A} + \underline{B}\underline{K}$. If the resulting transient response is unsatisfactory, the designer may alter the weighting matrices $\underline{Q}$ and $\underline{R}$ and try again. Thus the earlier trial-and-error selection of pole locations is replaced in this case by a similar selection of weightings.

The regulator problem formulated above involved the relatively simple task of keeping the output near zero. If there are disturbance inputs present or if a reference input is to be tracked, then a term of the form $\underline{B}\underline{w}$ will appear in (18) and some extensions are required. Various conceptual approaches have been proposed in the literature, but they generally amount to augmenting the state vector as in Sec. 11.3 and then applying the linear regulator theory. Since this involves a redefinition of states and inputs, some additional thought must be given to the appropriate form of the cost functional. Numerous other extensions of the basic problem may also be found in the literature, such as cross terms of the form $\underline{x}'\underline{M}\underline{u}$ in the cost functional, nonquadratic costs, and constraints on the permissible values of $\underline{u}$ and $\underline{x}$, but they will not be pursued here.

In essence, the linear regulator theory provides a systematic method for control system design which is easily implemented on a computer. Assuming that the weighting matrices have been suitably chosen, it specifies a closed-loop pole configuration which is appropriate to the control objectives. In addition, the optimal closed-loop system has a number of desirable properties:

5. *PROPERTY*. The optimal closed-loop system (18) is asymptotically stable.

ΔΔΔ

Our earlier assumptions that $\underline{Q} > \underline{0}$ and that the system is observable ensure that all state trajectories have some effect on the cost. Without this assumption, there can be unstable modes even when J is finite (see Problem 2). Similarly, controllability ensures that the system can be stabilized, and $\underline{R} > \underline{0}$ prevents the control from becoming unbounded. Most of these assumptions can be removed or relaxed in certain circumstances.

The remaining properties of interest are most conveniently presented in the context of a scalar system

$$\dot{\underline{x}} = \underline{A}\underline{x} + \underline{b}u \tag{19}$$

with an optimal control law

$$u(t) = \underline{k}'\underline{x}(t) + w(t) \triangleq z(t) + w(t) \tag{20}$$

which minimizes a cost functional of the form (13) and yields the optimal closed-loop system

$$\dot{\underline{x}} = (\underline{A} + \underline{b}\underline{k}')\underline{x} + \underline{b}w \tag{21}$$

The term w represents an external input. Since the transfer function matrix from input to state in (19) is $(s\underline{I} - \underline{A})^{-1}\underline{b}$, the closed-loop system has the form shown in Fig. 1(a). It is useful to rearrange this into the equivalent form shown in Fig. 1(b), so that

$$-\kappa g(s) \triangleq \underline{k}'(s\underline{I} - \underline{A})^{-1}\underline{b} \tag{22}$$

can be interpreted as the scalar *open-loop transfer function* in a unity-feedback system, and κ as the *loop gain* (see Definition 8.1.3). Recall from Lemma 8.1.2 that the closed-loop transfer function is

$$h(s) = \frac{\kappa g(s)}{1 + \kappa g(s)} \tag{23}$$

It turns out that for an optimal feedback control law, the *return difference* $1 + \kappa g(s)$, evaluated at $s = j\omega$, can be shown to satisfy

$$|1 + \kappa g(j\omega)| \geq 1 \tag{24}$$

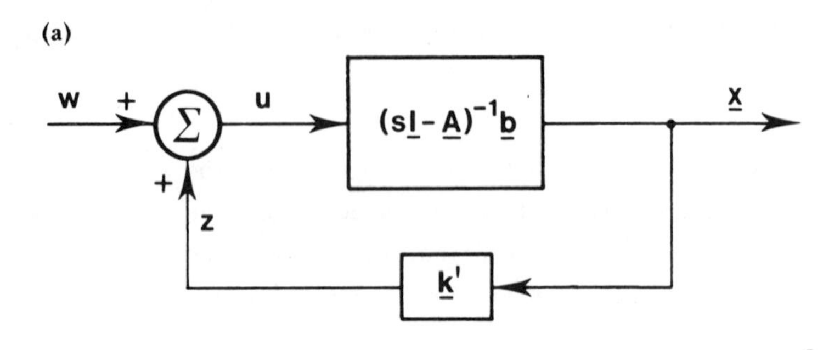

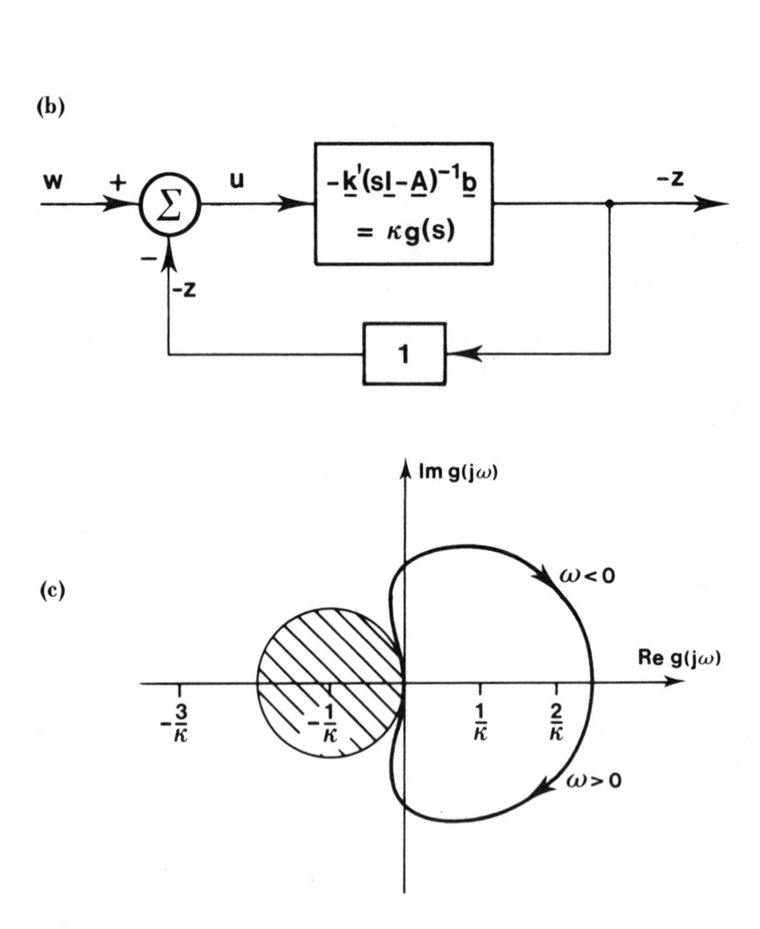

FIG. 1. (a) Optimal closed-loop system, (b) equivalent unity-feedback form, and (c) Nyquist plot of typical $\kappa g(j\omega)$.

or equivalently,

$$\left|\frac{1}{\kappa} + g(j\omega)\right| \geq \frac{1}{\kappa} \tag{25}$$

for $-\infty < \omega < \infty$. Interpreted graphically, this means that the Nyquist plot of the open-loop frequency response $g(j\omega)$ does not enter the shaded disc in Fig. 1(c). We shall now state four useful properties which are consequences of this inequality. The inequality and the properties can all be generalized to the case of a multivariable system [6,327a] , but we will confine our attention to the scalar case.

6. *PROPERTY*. The optimal closed-loop system (21) has *infinite gain margin*.

ΔΔΔ

The gain margin (Definition 8.4.5) is the factor by which the loop gain can be increased before instability results. Since the Nyquist plot cannot intersect the real axis between $-2/\kappa$ and 0, the loop gain κ can be multiplied by a factor f, $1/2 < f < \infty$, without altering the number of encirclements of $-1/\kappa$. In other words, the feedback gain of 1 in Figure 1(b) may be replaced by any value between 1/2 and ∞ without destroying the stability of the closed-loop system.

7. *PROPERTY*. The optimal closed-loop system (21) has a *phase margin of 60°*.

ΔΔΔ

Again referring to Definition 8.4.5, the phase margin is the amount of phase lag required to bring on instability. In Fig. 1(c), an arc through $-1/\kappa$ and centered at the origin intersects the edge of the shaded disc at ±60°. Thus any phase shift in the feedback loop of 60° or less cannot change the number of encirclements of $-1/\kappa$ and bring on instability. Note that this also implies that the system can tolerate a certain amount of *time delay*, which is equivalent to a frequency-dependent phase shift.

8. PROPERTY. The optimal closed-loop system (21) will tolerate a *memoryless nonlinearity* $\phi(\cdot)$ in the feedback loop without becoming unstable, provided that

$$\frac{1}{2} < \frac{1}{2} + \varepsilon \leq \frac{\phi(z)}{z} \leq M < \infty \tag{26}$$

for all z, where ε and M are constants.

ΔΔΔ

In other words, z in Fig. 1(b) can be replaced by any function $\phi(z)$ satisfying (26) and the closed-loop system remains stable. This is a generalization of Property 6, which states the same thing for the case of a linear function $\phi(z) = fz$, and it may be proved by constructing a Liapunov function (Sec. 5.4) or by applying the celebrated *circle criterion*. This property is covered in detail in Ref. 6, where the insertion of *relays* [which do not satisfy (26)] into the feedback loop is also discussed.

9. PROPERTY. The *sensitivity* of the optimal closed-loop transfer function (23) to plant parameter variations is less than or equal to that of the open-loop transfer function (22).

ΔΔΔ

This property is concerned with the situation where some of the parameter values in (22) are in error either because they cannot be precisely measured or because they undergo unpredictable variations. This means that the open-loop transfer function $\kappa g(s)$ must be replaced by $\kappa g(s) + \delta(s)$.† Expressed as a percentage, the open-loop perturbation is $\delta(s)/\kappa g(s)$.

The effect of this perturbation on the closed-loop transfer function h(s) in (23) is to change it to

$$\hat{h}(s) = \frac{\kappa g(s) + \delta(s)}{1 + \kappa g(s) + \delta(s)} \tag{27}$$

---

†In general, the perturbation will vary with time, but we assume that any such time dependence is sufficiently slow to be ignored.

Thus the percentage closed-loop change is easily seen to be

$$\frac{h(s) - \hat{h}(s)}{h(s)} = \frac{1 + \kappa g(s)}{\kappa g(s)} \left(\frac{\kappa g(s)}{1 + \kappa g(s)} - \frac{\kappa g(s) + \delta(s)}{1 + \kappa g(s) + \delta(s)} \right)$$

$$= \frac{-1}{1 + \kappa g(s) + \delta(s)} \frac{\delta(s)}{\kappa g(s)} \approx \frac{-1}{1 + \kappa g(s)} \frac{\delta(s)}{\kappa g(s)} \tag{28}$$

where the final approximation assumes that $\delta(s)$ is relatively small. Setting $s = j\omega$, we see that the closed-loop perturbation at each frequency ω is no greater than the open-loop perturbation $\delta(j\omega)/\kappa g(j\omega)$ provided

$$\left| \frac{-1}{1 + \kappa g(j\omega)} \right| = \frac{1}{|1 + \kappa g(j\omega)|} \leq 1 \tag{29}$$

The return difference condition (24) establishes that (29) holds for all frequencies ω. This means that the optimal feedback control law reduces (or at worst leaves unchanged) the sensitivity of the system to such perturbations.

Property 5 establishes the stability of an optimally designed system. It turns out that any desired *stability margin* (Sec. 5.3) can be guaranteed by altering the cost functional (13). With the assumptions made earlier, it is clear that both $\underline{x}(t)$ and $\underline{u}(t)$ must decay to zero as $t \to \infty$ in order for the integral in (13) to remain finite. If we replace (13) by

$$\begin{aligned} J[\underline{x}_0;\underline{u}(\cdot)] &= \int_0^\infty e^{2\sigma t} [\underline{y}'(t)\underline{Q}\underline{y}(t) + \underline{u}'(t)\underline{R}\underline{u}(t)]\, dt \\ &= \int_0^\infty e^{2\sigma t} [\underline{x}'(t)\underline{C}'\underline{Q}\underline{C}\underline{x}(t) + \underline{u}'(t)\underline{R}\underline{u}(t)]\, dt \end{aligned} \tag{30}$$

for some $\sigma > 0$, then $\underline{x}(t)$ and $\underline{u}(t)$ must decay at least as fast as $e^{-\sigma t}$ in order for the integral to remain finite. In other words, the closed-loop system must not only be stable but have stability margin σ : all of its poles must lie to the left of the line $\text{Re } s = -\sigma$.

A simple variable change allows Theorem 4 to be applied to the cost functional (30), and the optimal control law in this case is given by (14) and (15) with the Riccati equation (16) replaced by

$$(\underline{A}' + \sigma\underline{I})\underline{P} + \underline{P}(\underline{A} + \sigma\underline{I}) - \underline{PBR}^{-1}\underline{B}'\underline{P} + \underline{C}'\underline{QC} = \underline{0} \tag{31}$$

The reader is asked to supply details in Problem 3. Thus it is a simple matter for the designer to specify a stability margin σ in the closed-loop system: he need only replace $\underline{A}$ by $\underline{A} + \sigma\underline{I}$ in the Riccati equation. It is not difficult to show that this closed-loop system satisfies the inequalities (24) and (25), so that Properties 6-9 still hold [6].

It has probably not escaped the reader's attention that the Riccati equation governing the optimal control laws in Theorems 3 and 4 looks very much like the one encountered in our discussion of the Kalman-Bucy filter in Sec. 12.4. This is no coincidence, as these are *dual* optimization problems (in Sec. 12.4 the mean-square estimation error was minimized), and the state estimation problem can actually be formulated as a linear regulator problem with time running backwards [6]. For our purposes, however, it will suffice to observe that (15) and (16) become identical to (12.4-31) and (12.4-32) if we make the following replacements:

$$\underline{A} \leftarrow \underline{A}' \qquad \underline{B} \leftarrow \underline{C}' \qquad \underline{C} \leftarrow \underline{B}' \qquad \underline{K} \leftarrow \underline{M}' \tag{32}$$

Thus the optimal steady-state regulator and estimator have precisely the same structure.

A similar statement applies to the time-varying versions of these problems, but in this case the time reversal plays a crucial role. The finite-time regulator solution requires the Riccati equation to be solved backwards from the terminal time before the control law can be implemented. In contrast, the estimator Riccati equation is solved forward in time, a computation which can proceed in parallel with the other estimation operations.

The linear regulator theory discussed here all relates to the deterministic system (1). The use of an optimal observer to realize the optimal control law is discussed in Ref. 303. The problem becomes even more interesting if the input and output are corrupted by noise, as in (12.4-12). Since only a noisy measurement of the state $\underline{x}$ (or the output $\underline{y}$) is available for feedback in this case, intuition suggests the control law

$$\underline{u}(t) = \underline{K}\hat{\underline{x}}(t) \tag{33}$$

where $\underline{K}$ is the optimal gain matrix of Theorem 4 (or 2), and $\hat{\underline{x}}(t)$ is the optimal state estimate produced by the Kalman-Bucy filter of Theorem 12.4.2 (or 12.4.1).

Under certain assumptions, it may be shown that this separation of the estimation and control operations results in an optimal control (in the sense that the expected value of J is minimized). This result is known as the *separation theorem* or *certainty equivalence principle* (depending on the assumptions), and it generalizes the separation property of Theorem 12.3.1. A recent paper indicates, however, that the cost may be reduced still further by adding to (33) a term involving the error (or innovation) vector $\tilde{\underline{y}}$ from the Kalman-Bucy filter [316]. This suggests that the assumptions mentioned above may be overly restrictive.

We conclude by applying the linear regulator theory to the control of an orbiting satellite.

10. EXAMPLE. The linearized equations of motion governing the equatorial motion of a satellite in circular orbit were found in Example 2.4.3 to be

$$\begin{aligned}
\dot{\underline{x}} &= \begin{bmatrix} 0 & 1 & 0 & 0 \\ 3\omega^2 & 0 & 0 & 2\omega \\ 0 & 0 & 0 & 1 \\ 0 & -2\omega & 0 & 0 \end{bmatrix} \underline{x} + \begin{bmatrix} 0 & 0 \\ 1 & 0 \\ 0 & 0 \\ 0 & 1 \end{bmatrix} \begin{bmatrix} u_r \\ u_\Theta \end{bmatrix} \\
\underline{y} &= \begin{bmatrix} 1 & 0 & 0 & 0 \\ 0 & 0 & 1 & 0 \end{bmatrix} \underline{x} = \begin{bmatrix} r \\ \Theta \end{bmatrix}
\end{aligned} \tag{34}$$

where $\underline{x} = [r \quad \dot{r} \quad \Theta \quad \dot{\Theta}]'$. Suppose we define a cost functional of the form (13) for this system, with $\underline{R} = \underline{I}$ and $\underline{Q} = \alpha\underline{I}$, where α is a scalar weighting parameter:

$$J[\underline{x}_0;\underline{u}(\cdot)] = \int_0^\infty [\alpha\underline{y}'(t)\underline{y}(t) + \underline{u}'(t)\underline{u}(t)]\ dt$$

$$= \int_0^\infty [\alpha r^2(t) + \alpha\Theta^2(t) + u_r^2(t) + u_\Theta^2(t)]\ dt \qquad (35)$$

Recall that r and Θ represent linearized perturbations in range and angle about a nominal orbital trajectory: nonzero values are penalized by the terms αr^2 and $\alpha\Theta^2$ in the cost. The expenditure of control energy to reduce the perturbations is penalized by the terms u_r^2 and u_Θ^2, and the relative importance of these two conflicting objectives is determined by the choice of α.

According to Theorem 4, the control which minimizes J may be found by solving (16) for $\underline{P}$ and then substituting it into (15) to get the gain matrix $\underline{K}$. This is a substantial computational task

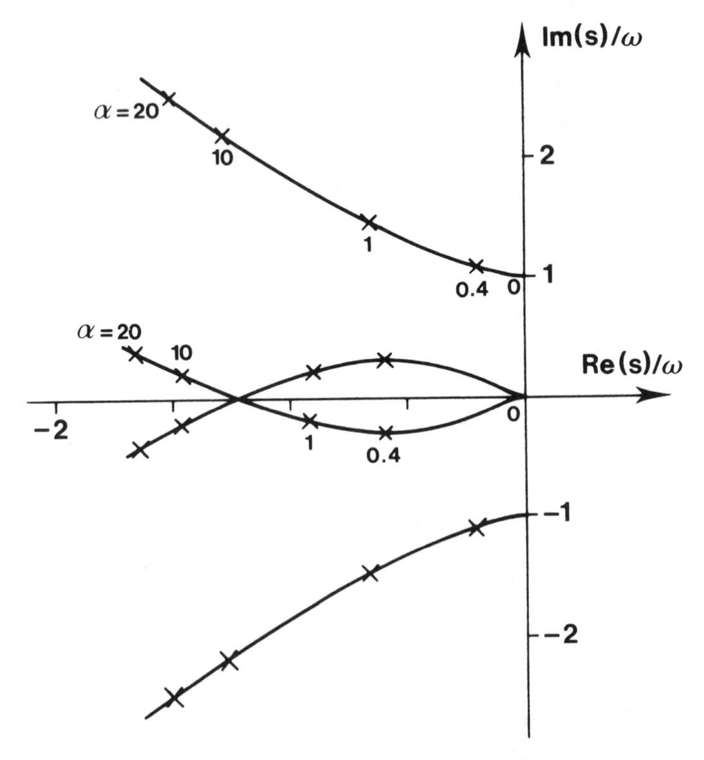

FIG. 2. Root locus of optimal closed-loop poles.

which will not be undertaken here. Rather, we shall make use of computations reported in Ref. 237, where $\underline{K}$ was determined for various values of the parameter α. Using this gain matrix, the closed-loop poles of the optimal system (eigenvalues of $\underline{A} + \underline{B}\underline{K}$) may be found, and these are plotted as functions of α in the root-locus diagram of Fig. 2. Note that the pole locations are normalized with respect to ω, the angular frequency of the nominal trajectory. The closed-loop system is stable for all values of $\alpha > 0$, as predicted by Property 5 (the controllability and observability of (34) were established in Examples 10.1.10 and 10.2.10).

The transient response of the optimal system to an initial state $\underline{x}_0 = [1 \quad 1 \quad 1 \quad 1]'$ is shown in Fig. 3 for $\alpha = 1$ (curve B)

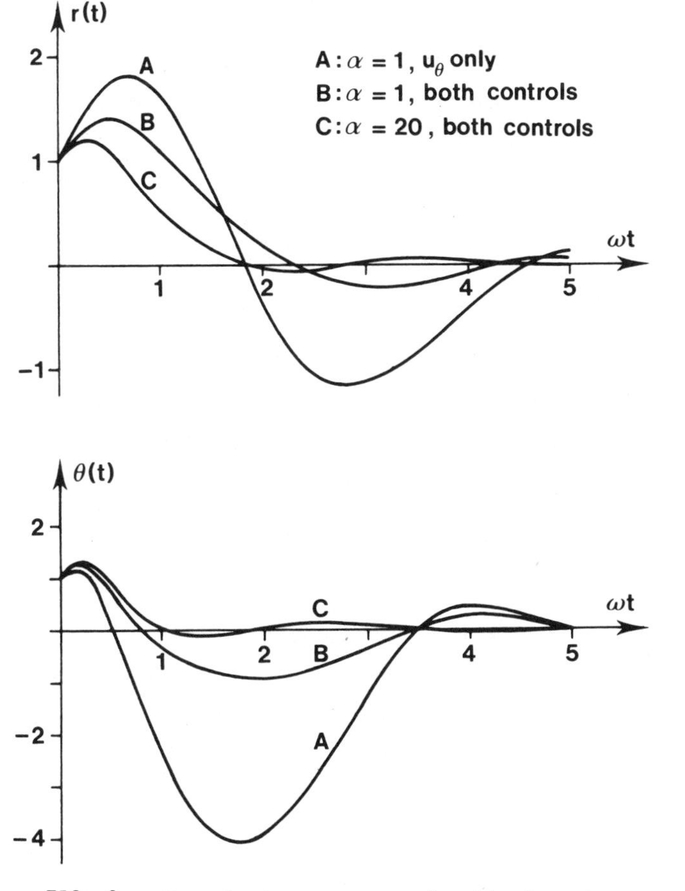

FIG. 3. Transient responses of optimal systems.

and $\alpha = 20$ (curve C). Note that the time scale is also normalized with respect to ω.

In Example 10.1.10 it was also found that the system (34) is controllable using only the single control u_Θ. Thus it is possible to do without the control u_r and minimize the cost functional

$$J_1[\underline{x}_0;u_\Theta(\cdot)] = \int_0^\infty [\alpha r^2(t) + \alpha\Theta^2(t) + u_\Theta^2(t)]\, dt \tag{36}$$

The optimal closed-loop transient response for this case, with $\alpha = 1$, is also shown in Fig. 3 for comparison (curve A).

The units in Fig. 3 are arbitrary, and the choice $\underline{x}_0 = [1\ \ 1\ \ 1\ \ 1]'$ is simply one of convenience. However, it is important to bear in mind that the validity of (34) depends upon the state $\underline{x}(t)$ remaining small. Before a linear control scheme of the sort described here could be implemented, it would be necessary to ensure that no expected excursions of $\underline{x}(t)$ would be large enough to invalidate the linearization carried out in Example 2.4.3.

ΔΔΔ

PROBLEMS

Section 13.1

1. (a) Determine the pole of the one-dimensional system

$$\dot{x} = x + u$$

$$y = x$$

and sketch its transient response. Find the control law which minimizes the cost functional

$$J[x_0;u] = \int_0^\infty [qy^2(t) + ru^2(t)]\, dt$$

Sketch the root-locus of the closed-loop system as q/r varies. What happens to the closed-loop transient response as $q \to 0$? What happens to the control $u(t)$ as $r \to 0$?

(b) Repeat (a) for the two-dimensional system

$$\dot{\underline{x}} = \begin{bmatrix} 1 & 1 \\ -1 & 1 \end{bmatrix} \underline{x} + \begin{bmatrix} 0 \\ 1 \end{bmatrix} u$$

$$y = \begin{bmatrix} 1 & 0 \end{bmatrix} \underline{x}$$

2. For the system

$$\dot{\underline{x}} = \begin{bmatrix} 1 & 0 \\ 0 & 2 \end{bmatrix} \underline{x} + \begin{bmatrix} 1 \\ 0 \end{bmatrix} u$$

$$y = \begin{bmatrix} 1 & 0 \end{bmatrix} \underline{x}$$

determine the control law which minimizes

$$J[\underline{x}_0;u] = \int_0^{\infty} [3y^2(t) + u^2(t)]\ dt$$

Compute the optimal cost $J^*(\underline{x}_0)$. Is the closed-loop system stable? Is this consistent with Property 13.1.5? Explain.

3. Determine the optimal control law for the cost functional (30) as follows:

(a) Make the variable changes

$$\hat{\underline{x}}(t) = e^{\sigma t}\underline{x}(t) \qquad \hat{\underline{u}}(t) = e^{\sigma t}\underline{u}(t)$$

and rewrite the system differential equation (13.1-1) in terms of $\hat{\underline{x}}$ and $\hat{\underline{u}}$.

(b) Verify that the system derived in (a) must be controllable and observable, provided the original system is.

(c) Apply Theorem 13.1.4.

(d) Change back to the original variables $\underline{x}$ and $\underline{u}$.

Chapter 14

TIME-VARYING SYSTEMS

14.0. INTRODUCTION

Up until now, we have considered almost exclusively systems whose parameters did not vary with time. The two exceptions to this were the finite-time versions of the Kalman-Bucy filter (Sec. 12.4) and the optimal linear regulator (Sec. 13.1), where time-varying feedback gain matrices appeared.

Fortunately, much of the theory we have discussed for systems in the state-variable format extends in a natural way to the time-varying case. In this chapter we will briefly indicate some of these extensions without proofs. For a complete discussion, the reader is referred to Ref. 32. Other useful sources are Refs. 39, 42, 98, 268, and 352.

Extension of transfer function and frequency response methods to the time-varying case is very difficult. The whole frequency domain concept essentially relies upon the property that a sinusoidal input evokes a sinusoidal output, and this does not hold in general for time-varying systems. Although certain ad hoc procedures have been proposed from time to time, there is really no satisfactory mathematical framework available for producing general results.

14.1. TIME-VARYING SYSTEMS

In Sec. 2.1 we first introduced the concept of *state* and formulated the differential equations describing a system's behavior in *state-variable format*. In the case of a *linear, time-varying system*, this was given by (2.1-16),

$$\begin{aligned} \dot{\underline{x}}(t) &= \underline{A}(t)\underline{x}(t) + \underline{B}(t)\underline{u}(t) \\ \underline{y}(t) &= \underline{C}(t)\underline{x}(t) + \underline{D}(t)\underline{u}(t) \end{aligned} \tag{1}$$

We then assumed a *time-invariant* system ($\underline{A}$, $\underline{B}$, $\underline{C}$, and $\underline{D}$ constant), and all subsequent results were based on this assumption. We shall now show how some of the results extend to the time-varying case. For convenience, we drop the final term in (1) and consider the system

$$\begin{aligned} \dot{\underline{x}}(t) &= \underline{A}(t)\underline{x}(t) + \underline{B}(t)\underline{u}(t) \\ \underline{y}(t) &= \underline{C}(t)\underline{x}(t) \end{aligned} \tag{2}$$

with $\underline{x}(t_0) = \underline{x}_0$ at the initial time t_0.

One situation where such systems commonly arise is the linearization of a nonlinear system (even a time-invariant one) about a nominal state trajectory $\bar{\underline{x}}(t)$, as described in Sec. 2.4. Unless the nominal trajectory is constant (as it was for the tunnel diode of Example 2.4.1), the linearized equations will change from one point to the next along the trajectory and have the form (2) or (1). The linearized equations of the satellite in Example 2.4.2 were time invariant only because the nominal orbit was circular. With a noncircular orbit, the linearized equations have the form (2) with $\underline{A}(t)$, $\underline{B}(t)$, and $\underline{C}(t)$ all periodic functions of time.

Some time-varying systems may also be described in the input-output format (2.2-1), where the coefficients p_i and q_i of the differential equation depend upon t. Unfortunately, there is no convenient way to find and characterize solutions in this format (the Laplace transform approach provides very little help), so we will omit any further discussion of it.

Solutions

If we assume that $\underline{A}(t)$ is a continuous function of time, then it can be shown that the unforced (homogeneous) differential equation

$$\dot{\underline{x}}(t) = \underline{A}(t)\underline{x}(t) \tag{3}$$

has a unique solution for every initial state $\underline{x}(t_0) = \underline{x}_0$. This solution can be written as

$$\underline{x}(t) = \underline{\Phi}(t,t_0)\underline{x}_0 \tag{4}$$

where $\underline{\Phi}(t,t_0)$ is called the *transition matrix* of (3). An infinite series expression for $\underline{\Phi}$ is given in Problem 1. The transition matrix is fundamental to the study of linear, time-varying differential equations, and itself satisfies

$$\begin{aligned} \dot{\underline{\Phi}}(t,t_0) &= \frac{\partial}{\partial t}\underline{\Phi}(t,t_0) = \underline{A}(t)\underline{\Phi}(t,t_0) \\ \underline{\Phi}(t_0,t_0) &= \underline{I} \end{aligned} \tag{5}$$

For a time-invariant system, $\underline{A}(t) = \underline{A}$, and $\underline{\Phi}$ is the matrix exponential

$$\underline{\Phi}(t,t_0) = e^{\underline{A}(t-t_0)} \tag{6}$$

which was first encountered in Sec. 3.1.

Various properties of the matrix exponential carry over to the transition matrix. The *composition rule* (Theorem 3.2.1) becomes

$$\underline{\Phi}(t_2,t_0) = \underline{\Phi}(t_2,t_1)\underline{\Phi}(t_1,t_0) \quad \text{(for all } t_2,t_1,t_0) \tag{7}$$

and $\underline{\Phi}(t,t_0)$ is nonsingular for all t and t_0. In fact,

$$\underline{\Phi}^{-1}(t,t_0) = \underline{\Phi}(t_0,t) \tag{8}$$

where for $t \geq t_0$, $\underline{\Phi}(t_0,t)$ specifies the solution of (3) backwards in time. The concept of a state trajectory and the discussion of Fig. 3.2-1 are equally relevant here. Theorem 3.2.3 becomes

$$\det \underline{\Phi}(t,t_0) = \exp \int_{t_0}^{t} \operatorname{tr} \underline{A}(\tau)\, d\tau \tag{9}$$

If we reinstate the driving term and output equation and assume $\underline{B}(t)$ and $\underline{u}(t)$ to be piecewise continuous, the complete solution looks much like Corollary 3.3.3:

1. *THEOREM.* The differential equation (2) has a unique solution for every initial state $\underline{x}_0$, given by

$$\begin{aligned}\underline{y}(t) &= \underline{C}(t)\underline{x}(t)\\ &= \underline{C}(t)\underline{\Phi}(t,t_0)\underline{x}_0 + \int_{t_0}^{t} \underline{C}(t)\underline{\Phi}(t,\tau)\underline{B}(\tau)\underline{u}(\tau)\, d\tau \end{aligned} \tag{10}$$

where $\underline{\Phi}(t,t_0)$ is the transition matrix described above, and the two terms on the right-hand side are the *natural* and *forced* responses, respectively.

The *impulse response matrix* (or *weighting pattern*) of (2) is

$$\underline{H}(t,\tau) \triangleq \underline{C}(t)\underline{\Phi}(t,\tau)\underline{B}(\tau) \tag{11}$$

and for $\underline{x}_0 = \underline{0}$ the system response is determined by the input-output relation

$$\underline{y}(t) = \int_{t_0}^{t} \underline{H}(t,\tau)\underline{u}(\tau)\, d\tau \tag{12}$$

This may be interpreted as a superposition integral, and the discussion in Sec. 3.4 also applies here.

ΔΔΔ

In Sec. 2.5 we saw how an alternate representation of a linear system could be obtained by means of a *state transformation*. In this case we may define

$$\hat{\underline{x}}(t) = \underline{Q}(t)\underline{x}(t) \tag{13}$$

where $\underline{Q}(t)$ is nonsingular and continuously differentiable for all t, and the effect on the system is summarized in Lemma 2.

2. *LEMMA.* In terms of the alternate state vector $\hat{\underline{x}}(t)$ defined in (13), the system (2) is described by

$$\begin{aligned}\dot{\hat{\underline{x}}}(t) &= \hat{\underline{A}}(t)\hat{\underline{x}}(t) + \hat{\underline{B}}(t)\underline{u}(t)\\ \underline{y}(t) &= \hat{\underline{C}}(t)\hat{\underline{x}}(t)\end{aligned} \tag{14}$$

where

$$\begin{aligned}\hat{\underline{A}}(t) &= \underline{Q}(t)\underline{A}(t)\underline{Q}^{-1}(t) + \dot{\underline{Q}}(t)\underline{Q}^{-1}(t)\\ \hat{\underline{B}}(t) &= \underline{Q}(t)\underline{B}(t)\\ \hat{\underline{C}}(t) &= \underline{C}(t)\underline{Q}^{-1}(t)\end{aligned} \tag{15}$$

The transition matrix of (14) is

$$\hat{\underline{\Phi}}(t,t_0) = \underline{Q}(t)\underline{\Phi}(t,t_0)\underline{Q}^{-1}(t_0) \tag{16}$$

where $\underline{\Phi}(t,t_0)$ is the transition matrix of (2). The impulse response matrix $\underline{H}(t,\tau)$ is unchanged.

ΔΔΔ

In the time-invariant case, we used a state transformation to put the matrix $\underline{A}$ in diagonal or block-diagonal form. This revealed the underlying structure of the transition matrix $e^{\underline{A}t}$: It is made up of exponentials and sinusoids involving the eigenvalues of $\underline{A}$. Unfortunately, the transition matrix $\underline{\Phi}(t,t_0)$ exhibits no such structure in general, and the usefulness of state transformations for time-varying systems is somewhat limited.

One notable exception is the case of a *periodic* system with period T,

$$\begin{aligned}\dot{\underline{x}}(t) &= \underline{A}(t)\underline{x}(t)\\ \underline{A}(t+T) &= \underline{A}(t) \qquad \text{(for all } t)\end{aligned} \tag{17}$$

In this case there exists a state transformation (see Problem 3) which yields a time-invariant version of the system,

$$\dot{\hat{\underline{x}}}(t) = \hat{\underline{A}}\hat{\underline{x}}(t) \tag{18}$$

with transition matrix

$$\hat{\underline{\Phi}}(t,t_0) = e^{\hat{\underline{A}}(t-t_0)} \tag{19}$$

Thus the transition matrix of the original system (17) is

$$\underline{\Phi}(t,t_0) = \underline{Q}^{-1}(t)e^{\hat{\underline{A}}(t-t_0)}\underline{Q}(t_0) \tag{20}$$

Stability

Determination of *stability* in the case of a time-varying system can be quite difficult. It is clear that the system is asymptotically stable in the sense of Definition 5.1.1 if and only if

$$\| \underline{\Phi}(t,t_0) \| \xrightarrow[t\to\infty]{} 0 \tag{21}$$

but there are no convenient tests (such as the Routh and Hurwitz criteria) for checking whether this condition holds. The problem is that the behavior of $\underline{\Phi}(t,t_0)$ cannot generally be characterized in terms of the eigenvalues of $\underline{A}(t)$. Even if these eigenvalues $\lambda_i(t)$ remain in the left half-plane for all time, the system can still be unstable (see Problem 5).

The Liapunov function approach outlined in Sec. 5.4 often proves useful for investigating the stability of time-varying systems. The quadratic function v in (5.4-3) can be allowed to vary with time,

$$v(\underline{x}(t),t) = \underline{x}'(t)\underline{Q}(t)\underline{x}(t) \tag{22}$$

and a number of useful theorems can be proved. Nevertheless, successful application of this technique depends principally on one's cleverness in guessing an appropriate Liapunov function.

In the case of the periodic system (17), stability can be determined using the state transformation to (18). It turns out that $\underline{Q}(t)$ is a *Liapunov transformation*, which means (among other things) that $\| \underline{Q}(t) \|$ and $\| \underline{Q}^{-1}(t) \|$ are bounded. Thus (17) is stable if and only if (18) is stable, which is true if and only if the eigenvalues of $\hat{\underline{A}}$ all lie in the left half-plane.

One other useful result has been established in the case of a *slowly varying* system [249,326]. This says that if the eigenvalues satisfy

$$\text{Re}\ \lambda_i(t) \le -k < 0 \qquad (i = 1,\ 2,\ \ldots,\ n) \tag{23}$$

for all t, <u>and</u> if the time variation of $\underline{A}(t)$ is sufficiently slow, then the system is stable.

Controllability and Observability

The concepts of controllability and observability discussed in Chapter 10 may be extended to the time-varying system (2) in a straightforward manner. This must be done with some care, particularly with regard to the time interval under consideration, as a number of special distinctions and subtleties arise which are not present in the time-invariant case. For example, the sets of controllable and reachable states are not always the same, nor are the sets of observable and reconstructible states. Rather than become immersed in a large collection of definitions, theorems, and special cases, we will simply state two results which are representative of the others available.

3. THEOREM. There exists a control function for every pair of states $\underline{x}_0, \underline{x}_1$ which drives the system (2) from $\underline{x}(t_0) = \underline{x}_0$ to $\underline{x}(t_1) = \underline{x}_1$ if and only if the controllability Gramian

$$\underline{W}_c(t_0,t_1) \triangleq \int_{t_0}^{t_1} \underline{\Phi}^{-1}(t,t_0)\underline{B}(t)\underline{B}'(t)\underline{\Phi}'^{-1}(t,t_0)\,dt \tag{24}$$

is nonsingular. In this case, the system is said to be both *controllable* and *reachable on the interval* $[t_0,t_1]$. For any given $\underline{x}_0, \underline{x}_1$ a control which makes the transfer is

$$\bar{\underline{u}}(t) = \underline{B}(t)\underline{\Phi}'^{-1}(t,t_0)\underline{W}_c^{-1}(t_0,t_1)[\underline{\Phi}^{-1}(t_1,t_0)\underline{x}_1 - \underline{x}_0] \tag{25}$$

and this control has the "minimum energy" property of Corollary 10.1.5 (also see Problem 10.1-4).

ΔΔΔ

4. THEOREM. Let the input to the system (2) be zero. If the state $\underline{x}(t)$ is unknown, it can be determined from the output $\underline{y}(\tau)$, $t_0 \le \tau \le t_1$ if and only if the observability Gramian

$$\underline{W}_o(t_0,t_1) \triangleq \int_{t_0}^{t_1} \underline{\Phi}'(t,t_0)\underline{C}'(t)\underline{C}(t)\underline{\Phi}(t,t_0)\,dt \tag{26}$$

is nonsingular. In this case, the system is said to be both *observable* and *reconstructible on the interval* $[t_0,t_1]$. The state may be determined by the linear operations

$$\underline{x}(t_0) = \int_{t_0}^{t_1} \underline{W}_o^{-1}(t_0,t_1)\underline{\Phi}'(t,t_0)\underline{C}'(t)\underline{y}(t)\,dt \tag{27}$$

and

$$\underline{x}(t) = \underline{\Phi}(t,t_0)\underline{x}(t_0) \qquad (\text{for } t \neq t_0) \tag{28}$$

ΔΔΔ

The controllability and observability conditions given above are rather difficult to check since the transition matrix $\underline{\Phi}(t,t_0)$ must be computed and the integrals (24) and (26) evaluated. If $\underline{A}(t)$, $\underline{B}(t)$, and $\underline{C}(t)$ are sufficiently differentiable, time-varying versions of the controllability and observability matrices in Theorems 10.1.6 and 10.2.6 may be defined via some matrix differential equations [39]. However, the resulting tests for controllability and observability are much more difficult to apply than they are in the time-invariant case.

The canonical decomposition Theorem (10.3.4) may be generalized as well, but it becomes somewhat awkward when the dimensions of the subspaces of controllable and observable states vary with time.

Since the transfer function concept does not extend in any useful way to time-varying systems, Definition 10.4.1 is not applicable. It is appropriate, however, to consider an impulse response matrix (or weighting pattern) $\underline{H}(t,\tau)$ which defines an input-output relation as in (12), and ask whether there exists a system of the form (2) for which (11) holds for all t and τ. Such a system is then said to be a *realization* of $\underline{H}(t,\tau)$. It can be shown that such a realization exists if and only if $\underline{H}(t,\tau)$ can be decomposed into

$$\underline{H}(t,\tau) = \underline{\Psi}(t)\underline{\Gamma}(\tau) \qquad (\text{for all } t,\tau) \tag{29}$$

where $\underline{\Psi}(t)$ and $\underline{\Gamma}(\tau)$ are finite-dimensional matrices. Moreover, it can also be shown that a realization is *minimal* if and only if it is both controllable and observable on some time interval (this generalizes Theorem 10.4.4).

Control and Estimation

Designing a control law for a time-varying system can be a difficult task indeed. Other than the minimum energy control (25) for a specific state transfer, the designer has little theory at his disposal. The pole-placement theorems (11.2.1 and 11.2.1') are not applicable, except in the case of a periodic system which has been transformed as in (18). The results of Sec. 11.3 on steady-state behavior are of some use.

Fortunately, the linear optimal control results of Chapter 13 extend easily to the time varying case. The designer may choose a cost functional of the form (13.1-7). Theorem 13.1.2 remains valid when the matrices $\underline{A}$, $\underline{B}$, $\underline{C}$, $\underline{R}$, and $\underline{Q}$ are time varying, and may be used to determine a stable feedback control law (assuming the appropriate controllability and observability conditions hold). Implementation of such a control law poses some difficulties, however, since the Riccati equation (13.1-11) must be solved backward in time (it is numerically unstable in the forward direction).

The situation with regard to state estimation is very similar. Although the theory of observers can be extended to time-varying systems, the results of Chapter 12 on pole placement are not applicable. The Kalman-Bucy filter, however, is very useful, and Theorem 12.4.1 is unchanged when the matrices $\underline{A}$, $\underline{B}$, $\underline{C}$, $\underline{R}$, and $\underline{Q}$ are time varying. Since the Riccati equation (12.4-28) propagates forward in time, it can be solved in parallel with the other filter operations. If reasonable values are chosen for the covariance matrices $\underline{R}$ and $\underline{Q}$, and if appropriate controllability and observability conditions hold, the resulting state estimator will be stable.

PROBLEMS

Section 14.1

1. The transition matrix of $\dot{\underline{x}} = \underline{A}(t)\underline{x}$ may be specified by the *Peano-Baker series:*

$$\underline{\Phi}(t,t_0) = \underline{I} + \int_{t_0}^{t} \underline{A}(\tau_1)\, d\tau_1 + \int_{t_0}^{t} \underline{A}(\tau_1) \int_{t_0}^{\tau_1} \underline{A}(\tau_2)\, d\tau_2\, d\tau_1 + \cdots$$

   (a) Verify that this expression satisfies (14.1-5).

   (b) Verify that when $\underline{A}(t) = \underline{A}$, this becomes the matrix exponential.

   (c) Compute $\Phi(t,t_0)$ explicitly for the scalar differential equation $\dot{x} = a(t)x$.

2. Prove Lemma 14.1.2.

3. (a) Show that the equation

$$\underline{\Phi}(T,0) = e^{\hat{\underline{A}}T}$$

   has a unique solution $\hat{\underline{A}}$, where $\underline{\Phi}$ is the transition matrix of the periodic system (14.1-17) and T is its period. *Hint:* Diagonalize $\underline{\Phi}(T,0)$.

   (b) Show that $\underline{Q}(t) = e^{\hat{\underline{A}}t}\underline{\Phi}^{-1}(t,0)$ defines the state transformation which puts (14.1-17) into the form (14.1-18). *Hint:* First show that

$$\frac{\partial}{\partial t}\underline{\Phi}^{-1}(t,0) = -\underline{\Phi}^{-1}(t,0)\underline{A}$$

   (c) Is $\underline{Q}(t)$ itself periodic?

4. Prove that a realization of $\underline{H}(t,\tau)$ exists if and only if it can be decomposed as in (14.1-28).

5. (a) Determine the eigenvalues $\lambda_i(t)$ of the time-varying system

$$\dot{\underline{x}}(t) = \begin{bmatrix} -1 + 1.5\cos^2 t & 1 - 1.5\cos t\,\sin t \\ -1 - 1.5\cos t\,\sin t & -1 + 1.5\sin^2 t \end{bmatrix} \underline{x}(t)$$

   Do you expect this system to be stable? Why?

(b) Show that the transition matrix of this system is

$$\underline{\Phi}(t,0) = \begin{bmatrix} e^{t/2}\cos t & e^{-t}\sin t \\ -e^{t/2}\sin t & e^{-t}\cos t \end{bmatrix}$$

by verifying that it satisfies (14.1-5). Is the system stable? Explain.

6. Consider the system

$$\dot{\underline{x}} = \begin{bmatrix} 0 & 1 \\ -2 & 2 \end{bmatrix}\underline{x} + \begin{bmatrix} 0 \\ 1 \end{bmatrix}u$$

$$y = \begin{bmatrix} -4 & 3 \end{bmatrix}\underline{x}$$

with the feedback control law $u(t) = ky(t)$ from output to input [233].

(a) Is the open-loop system controllable and observable? Is it stable?

(b) Does there exist a constant value of k which stabilizes the system?

(c) Show that the system is stabilized if k is replaced by

$$k(t) = \begin{cases} 0 & (iT \le t < iT + T_1) \\ 1 & [iT + T_1 \le t < (i+1)T] \end{cases}$$

for $i = 0,1,2,\ldots$, where $T_1 = \tan^{-1} 3$ and $T = T_1 + \pi$.

Chapter 15

DISCRETE-TIME SYSTEMS

15.0. INTRODUCTION

Throughout this book we have been concerned with dynamic systems which vary continuously in time and are described by differential equations. In practice, the use of computers and other digital devices to realize control systems has become very common, and in view of the rapid development, miniaturization, and decreasing costs of digital technology, this trend is likely to continue and perhaps accelerate.

Since the digital computer is by nature discrete, some accommodation with continuous systems is required, and two distinct issues are involved. First, variables must be *quantized* to a finite set of values in the computer; the accuracy of this approximation is limited by the number of available bits. Although this is a complex and important subject, we will not address it here. Instead, we will simply assume sufficient accuracy for all variables to be considered continuous.

The second issue is that variables in a computer can only be altered at certain discrete instants whose frequency is limited by the speed of the hardware and the complexity of the operations to be performed. If the computation is sufficiently rapid, this effect can also be ignored, but such is not often the case. Thus in this chapter we shall investigate a *discrete-time* framework for dynamic systems and see how the results of the previous chapters carry over to this case.

Many dynamic systems fall quite naturally into the discrete-time category, usually because their outputs can only be measured at certain discrete instants. Examples include economic systems [10], population models, digital filters, radar tracking systems, and industrial or chemical processes in which discrete batches of materials are processed. Nevertheless, if there is an underlying continuous system it can often provide valuable physical insights which are less obvious in the discrete-time formulation.

As in Chapters 13 and 14, we will only present a brief outline of the available results, and leave it to the reader to fill in the gaps. The literature on this subject is somewhat scattered, but the reader should find Refs. 36, 46, 52, 64, 102, 105, 139, 145, 173, 228a, 276, and 351 useful.

15.1. DISCRETE-TIME SYSTEMS

Suppose $s(t)$, $t \geq 0$, is a physical variable or measurement which varies continuously with time. For instance, $s(t)$ might be a communications signal received over a radio link, the angular velocity of a servomotor, or the voltage recorded at the output of a wind direction indicator. If $s(t)$ is to be processed by a computer or other digital apparatus, it must be turned into a finite collection of numbers in a form acceptable to the machine. This is usually done with a device called a *sampler* which accepts $s(t)$ and produces the sequence $s(i\Delta)$, $i = 0, 1, 2, \ldots$, where Δ is the time interval between samples. A digital voltmeter, for example, performs this function. A sampler is shown in Fig. 1(a), with the sequence of samples indicated schematically by an impulse train. This process is commonly known as *analog-to-digital conversion*, which emphasizes its quantization function, but for our purposes it is more appropriately called *continuous-to-discrete* conversion.

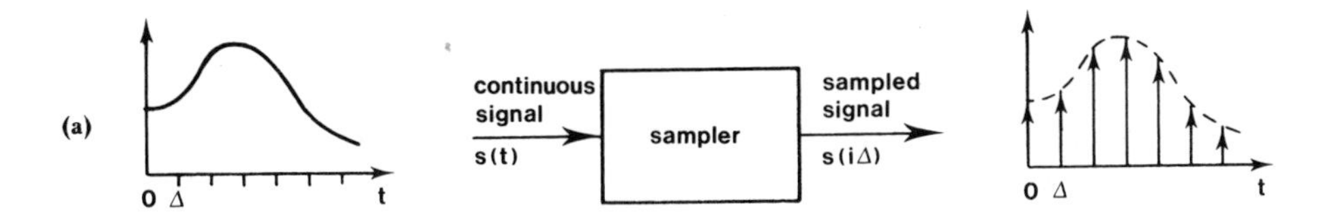

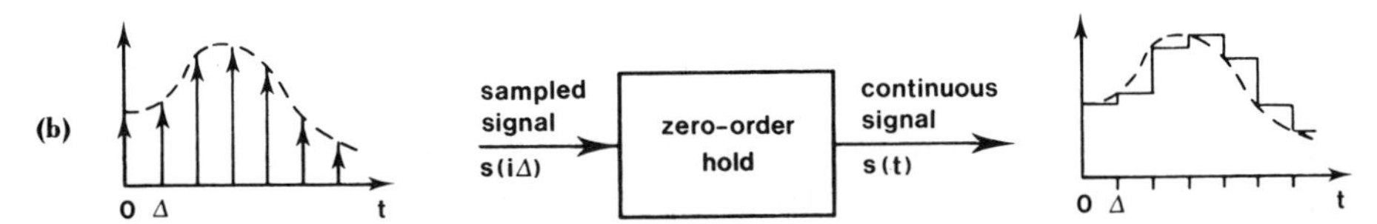

FIG. 1. Conversions between continuous and discrete signals: (a) continuous to discrete and (b) discrete to continuous.

There are many types of samplers, with a wide variety of characteristics. The sampling interval may be non-constant, for example, or the i-th sample might be the integral of s(t) over the sampling interval rather than $s(i\Delta)$. Indeed, a truly instantaneous sample is never obtained, since some inertia is always present in the sampling device. Nevertheless, this sampler will do for our purposes, and it is representative of most others which might be encountered in practice.

It might appear that a sampler discards part of the information present in a signal, and this is certainly the case if the sampling interval Δ is too long. However, the celebrated *sampling theorem* assures us that the continuous signal s(t) can be reconstructed perfectly from the samples if the sampling frequency $1/\Delta$ is more than twice the signal bandwidth. In other words, the transform $\hat{s}(j\omega)$ should be zero for all frequencies $f = \omega/2\pi \geq 1/2\Delta$. Although this condition can never be satisfied exactly, in practice a low-pass filter with cutoff frequency $1/2\Delta$ is usually installed at the input to the sampler to ensure that it holds approximately.

When a signal generated by the computer, (i.e., a sequence of numbers) is to be applied to a continuous system, the process is reversed, and an extrapolating device is often used to smooth the signal by filling in the gaps between data points. A *zero-order hold* accomplishes this by simply maintaining the previous value until the next one is available, and the resulting staircase function is shown at the output in Fig. 1(b). This is usually called *digital-to-analog conversion*, but for our purposes it is more appropriately termed *discrete-to-continuous conversion*.

The zero-order hold is chosen here as representative of the techniques available, but there are, of course, many others. First- and higher-order holds extrapolate the slope and higher-order derivatives as well in order to produce a smoother signal. Alternatively, the hold can be eliminated entirely and a string of narrow pulses with amplitudes proportional to the sample values can be used, allowing the inertia of the continuous system to smooth the signal. This is called *pulse amplitude modulation (PAM)* and related techniques make use of the pulse width (PWM) or frequency (PFM).

Now consider a time-invariant linear system of the form

$$\begin{aligned} \dot{\underline{x}}(t) &= \overline{\underline{A}}\,\underline{x}(t) + \overline{\underline{B}}\,\underline{u}(t) \\ \underline{y}(t) &= \overline{\underline{C}}\,\underline{x}(t) \end{aligned} \tag{1}$$

which is to be controlled using a computer or other digital device. This may be done by sampling $\underline{y}$ (or $\underline{x}$) and using a zero-order hold to smooth $\underline{u}$, as shown in Fig. 2. If the sampling period is Δ, the data sequence presented to the controller consists of the output vectors $\underline{y}(i\Delta)$ for $i = 0, 1, 2, \ldots$. The controller in turn puts out a sequence of control vectors $\underline{u}(i\Delta)$ for $i = 0, 1, 2, \ldots$, and $\underline{u}(i\Delta)$ is applied to the system during the interval $i\Delta < t \leq (i+1)\Delta$.

The forced response formula derived in Sec. 3.3 may be used to determine how the state propagates from one sampling instant to the next. Letting $t_0 = i\Delta$ and $t = (i+1)\Delta$, we have

$$\begin{aligned} \underline{x}(i\Delta + \Delta) &= e^{\overline{\underline{A}}\Delta}\underline{x}(i\Delta) + \int_{i\Delta}^{i\Delta+\Delta} e^{\overline{\underline{A}}(i\Delta+\Delta-\tau)}\overline{\underline{B}}\,\underline{u}(i\Delta)\, d\tau \\ &= e^{\overline{\underline{A}}\Delta}\underline{x}(i\Delta) + \int_0^{\Delta} e^{\overline{\underline{A}}\sigma}\, d\sigma\, \overline{\underline{B}}\,\underline{u}(i\Delta) \end{aligned} \tag{2}$$

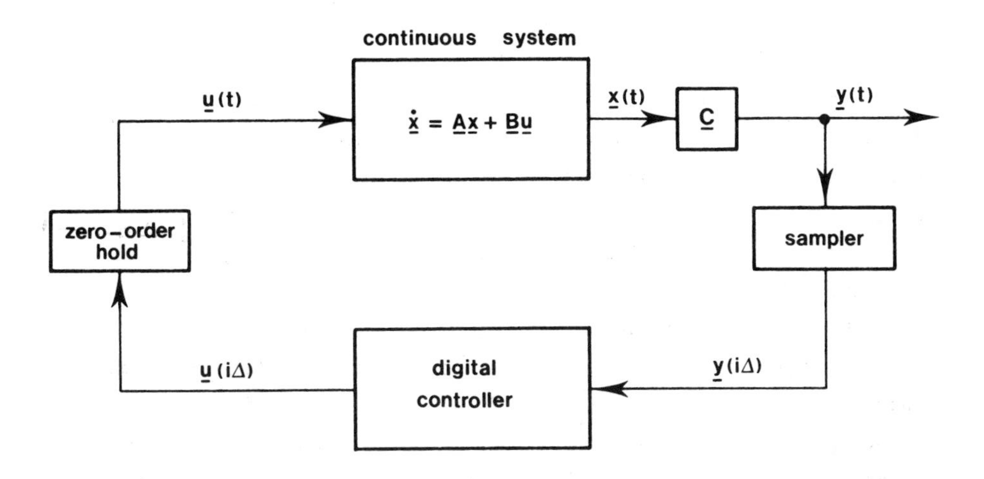

FIG. 2. Digital feedback control system.

where the variable of integration has been replaced by $\sigma = i\Delta + \Delta - \tau$. If we are interested only in the values of $\underline{x}$, $\underline{y}$, and $\underline{u}$ at the sampling instants, the behavior of the equivalent *discrete-time system* is described by

$$\begin{aligned} \underline{x}(i+1) &= \underline{A}\underline{x}(i) + \underline{B}\underline{u}(i) \\ \underline{y}(i) &= \underline{C}\underline{x}(i) \end{aligned} \tag{3}$$

with the initial condition $\underline{x}(0) = \underline{x}_0$, where

$$\underline{A} = e^{\overline{\underline{A}}\Delta} \qquad \underline{B} = \int_0^\Delta e^{\overline{\underline{A}}\sigma}\, d\sigma\, \overline{\underline{B}} \qquad \underline{C} = \overline{\underline{C}} \tag{4}$$

and the time variable $i\Delta$ has been replaced by i to simplify the notation. This is known as a linear, first-order, *vector difference equation*. As usual, the dimensions of $\underline{x}$, $\underline{u}$, and $\underline{y}$ are n, m, and p, respectively.

Alternatively, the zero-order hold may be removed and the control realized with a string of impulses $\Delta\underline{u}(i\Delta)\delta(t - i\Delta)$, $i = 0, 1, 2, \ldots$, where each impulse is assumed to occur just after $\underline{x}$ is sampled. An equation similar to (2) results, and the reader is asked in Problem 1 to verify that it reduces to the form (3), where now

$$\underline{A} = e^{\underline{\overline{A}}\Delta} \qquad \underline{B} = e^{\underline{\overline{A}}\Delta}\underline{\overline{B}}\Delta \qquad \underline{C} = \underline{\overline{C}} \tag{5}$$

Using other devices for the sample and hold operations causes further changes in the definitions of $\underline{B}$ and $\underline{C}$, and a term $\underline{D}\underline{u}(i)$ may appear in the output equation (see Problem 2). But $\underline{A}$ remains the same, and the discrete equivalents of all continuous linear systems have this same general form. Note that $\underline{A}$ is a matrix exponential and hence always nonsingular.

There are some dynamic systems, of course, which are inherently discrete and are conveniently formulated directly in terms of difference equations; some examples were given at the beginning of this chapter. If the system in question is linear, it can usually be described in the form (3). It is possible for $\underline{A}$ to be singular, but if the discrete system is derived from a continuous one, then (4) holds and $\underline{A}$ will be invertible. Nonlinear discrete systems can be linearized using techniques analogous to those described in Sec. 2.4.

A single-input, single-output discrete linear system may also be described by an n-th order scalar difference equation of the form

$$y(i+n) + p_{n-1}y(i+n-1) + \dots + p_1y(i+1) + p_0y(i)$$
$$= q_{n-1}u(i+n-1) + \dots + q_1u(i+1) + q_0u(i) \tag{6}$$

This is analogous to (2.2-1), and transformations between this form and (3) can be carried out as in Sec. 2.2.

Solution of Difference Equations

It is generally much easier to deal with difference equations than with differential equations. Questions of existence and uniqueness are more transparent, and solutions can be obtained recursively. In (3), for example, the solution is easily found by inspection:

$$\begin{aligned}
\underline{x}(0) &= \underline{x}_0 \\
\underline{x}(1) &= \underline{A}\underline{x}_0 + \underline{B}\underline{u}(0) \\
\underline{x}(2) &= \underline{A}\underline{x}(1) + \underline{B}\underline{u}(1) = \underline{A}^2\underline{x}_0 + \underline{A}\underline{B}\underline{u}(0) + \underline{B}\underline{u}(1) \\
\underline{x}(3) &= \underline{A}^3\underline{x}_0 + \underline{A}^2\underline{B}\underline{u}(0) + \underline{A}\underline{B}\underline{u}(1) + \underline{B}\underline{u}(2) \\
&\cdots\cdots\cdots\cdots\cdots\cdots \\
\underline{x}(i) &= \underline{A}^i\underline{x}_0 + \sum_{j=0}^{i-1} \underline{A}^{i-j-1}\underline{B}\underline{u}(j)
\end{aligned} \tag{7}$$

Thus the discrete-time analog of Corollary 3.3.3 is the following theorem.

1. THEOREM. The solution of the difference equation (3), for $\underline{x}(0) = \underline{x}_0$, is given by

$$\underline{y}(i) = \underline{C}\underline{x}(i) = \underline{C}\underline{A}^i\underline{x}_0 + \sum_{j=0}^{i-1} \underline{C}\underline{A}^{i-j-1}\underline{B}\underline{u}(j) \tag{8}$$

The first term on the right-hand side is due to the initial state $\underline{x}_0$ and is called the *natural response*, while the second term, due to the input $\underline{u}(\cdot)$, is called the *forced response*.

ΔΔΔ

The reader has probably noticed the similarity between this formula and (3.3-14). In this case $\underline{A}^i$ replaces the matrix exponential, and the integral involving $\underline{u}(\cdot)$ has become a summation (see Problem 3). To carry the analogy further, we may set $\underline{x}_0 = \underline{0}$ in (8) to get the input-output relation

$$\underline{y}(i) = \sum_{j=0}^{i-1} \underline{C}\underline{A}^{i-j-1}\underline{B}\underline{u}(j) = \sum_{j=0}^{i-1} \underline{H}(i - j - 1)\underline{u}(j) \tag{9}$$

where

$$\underline{H}(i) \triangleq \underline{C}\underline{A}^i\underline{B} \qquad (i = 0, 1, 2, \ldots) \tag{10}$$

is called the *pulse response matrix* of the system (see Problem 4). Equation (9) is a *convolution* or *superposition sum* of the input and the pulse response, just as in Sec. 3.4. For a time-varying discrete system, all of these equations take on forms analogous to those of Chapter 14 (see Problem 5).

As in the case of the matrix exponential $e^{\underline{A}t}$, much can be said about the structure of $\underline{A}^i$. Indeed, the reader is encouraged to go through Sec. 3.2 and show how the various results can be modified to apply to $\underline{A}^i$.

Z-Transforms

In Chapter 4 we found that the theory of Laplace transforms provided a very useful framework in which to study the behavior of linear differential equations. We shall now very briefly sketch a parallel and equally useful transform theory for discrete systems.

In this chapter we are concerned with functions of the discrete time variable i (or $i\Delta$), such as f(i), $i = 0, 1, 2, \ldots$, and we shall assume, for convenience, that $f(i) = 0$ for $i < 0$. This may be viewed, however, as a sequence of impulses with magnitudes f(0), f(1), f(2), ..., occurring at the times 0, Δ, 2Δ, ..., thus defining the impulse-modulated continuous time function

$$f(t) = f(0)\delta(t) + f(1)\delta(t-\Delta) + f(2)\delta(t-2\Delta) + \cdots = \sum_{i=0}^{\infty} f(i)\delta(t-i\Delta) \tag{11}$$

Since the Laplace transform of an impulse delayed by $i\Delta$ is just $e^{-si\Delta}$ (see Sec. B.4), the Laplace transform of f is

$$\hat{f}(s) = \sum_{i=0}^{\infty} f(i)e^{-si\Delta} = \sum_{i=0}^{\infty} f(i)(e^{s\Delta})^{-i} \tag{12}$$

If we now define a new complex variable $z \triangleq e^{s\Delta}$, this becomes

$$\hat{f}(z) = \sum_{i=0}^{\infty} f(i)\, z^{-i} \triangleq Z\{f(i)\} \tag{13}$$

which is known as the (one-sided) *Z-transform* of f.

The functions f(t) and $\hat{f}(s)$ are introduced here only to provide motivation; the Z-transform of the sequence f(0), f(1), ..., is defined by (13) at all values of the complex variable z for which the summation exists. This definition is analogous to (B.1-3), and in fact much of Appendix B could be repeated here by simply replacing s with z (see Problems 6 and 7).

An n-th order scalar difference equation of the form (6) can be solved using Z-transforms in the same way as an n-th order scalar differential equation is solved using Laplace transforms. If we transform (6) and use Problem 6(c), each side of the equation becomes a polynomial in z. Division by the left-hand polynomial yields the solution $\hat{y}(z)$, and the ratio of polynomials in z is a *transfer function* which specifies the input-output relationship between $\hat{u}(z)$ and $\hat{y}(z)$. Inversion of Z-transforms is conveniently carried out using partial fractions expansions, as described in Sec. B.2. The reader is asked to work out an example in Problem 19.

If we define Z-transforms of vector quantities in the obvious way, the vector difference equation (3) transforms to

$$\begin{aligned} z\hat{\underline{x}}(z) - z\underline{x}(0) &= \underline{A}\hat{\underline{x}}(z) + \underline{B}\hat{\underline{u}}(z) \\ \hat{\underline{y}}(z) &= \underline{C}\hat{\underline{x}}(z) \end{aligned} \tag{14}$$

Solving the first equation for $\hat{\underline{x}}$, we have

$$\hat{\underline{y}}(z) = \underline{C}\hat{\underline{x}}(z) = \underline{C}(z\underline{I} - \underline{A})^{-1}z\underline{x}_0 + \underline{C}(z\underline{I} - \underline{A})^{-1}\underline{B}\hat{\underline{u}}(z) \tag{15}$$

which is the discrete-time analog of (4.2-6) and is the Z-transform of (8). When $\underline{x}_0 = \underline{0}$, this yields the input-output relation

$$\hat{\underline{y}}(z) = \underline{C}(z\underline{I} - \underline{A})^{-1}\underline{B}\hat{\underline{u}}(z) = \hat{\underline{H}}(z)\hat{\underline{u}}(z) \tag{16}$$

The *transfer function matrix* of this system is

$$\hat{\underline{H}}(z) \triangleq \underline{C}(z\underline{I} - \underline{A})^{-1}\underline{B} = Z\{H(i-1)\} \tag{17}$$

where $\underline{H}(i)$ is the pulse response matrix defined in (10).

The *resolvent matrix* $(z\underline{I} - \underline{A})^{-1}$ plays a key role here, just as $(s\underline{I} - \underline{A})^{-1}$ did in Chapter 4. It is the Z-transform of the matrix sequence $\underline{0}$, $\underline{I}$, $\underline{A}$, $\underline{A}^2$, ..., and may be computed with the resolvent algorithm of Theorem 4.3.2.

We shall not pursue the Z-transform theory any further here, but the interested reader may consult the references cited earlier. Virtually all of the results of Chapter 4 (and Appendix B) carry over to difference equations and Z-transforms: in Problem 9 the reader is asked to make the appropriate modifications.

State Transformations and Stability

As with continuous systems (Sec. 2.5), a state transformation of the form

$$\hat{\underline{x}}(i) = \underline{Q}\underline{x}(i) \tag{18}$$

where $\underline{Q}$ is a nonsingular matrix, often proves to be very useful. The reader should verify that in terms of the new state vector, the system (3) is described by

$$\begin{aligned} \hat{\underline{x}}(i+1) &= \underline{Q}\underline{A}\underline{Q}^{-1}\hat{\underline{x}}(i) + \underline{Q}\underline{B}\underline{u}(i) \\ \underline{y}(i) &= \underline{C}\underline{Q}^{-1}\hat{\underline{x}}(i) \end{aligned} \tag{19}$$

It is easy to show that the pulse response and transfer function matrices are unaffected by the transformation, and that

$$(\underline{Q}\underline{A}\underline{Q}^{-1})^i = \underline{Q}\underline{A}^i\underline{Q}^{-1} \tag{20}$$

We shall now use such a transformation to investigate stability.

As in Sec. 5.1, we say that the discrete-time system (3) is *(asymptotically) stable* if whenever the input is zero,

$$\|\underline{x}(i)\| \xrightarrow[i\to\infty]{} 0 \tag{21}$$

for any initial state $\underline{x}_0$. Since $\underline{Q}$ is nonsingular, it should be clear that (3) is stable if and only if (19) is. To determine stability conditions, suppose that the matrix $\underline{A}$ has a linearly independent set of eigenvectors, and denote its eigenvalues by $\lambda_1, \lambda_2, \ldots, \lambda_n$ (these are also called the *poles* of the system). Then if we choose $\underline{Q}^{-1}$ to be the matrix of eigenvectors (as in Theorem A.3.3), and set $\underline{u}(i) \equiv \underline{0}$

(19) becomes

$$\hat{\underline{x}}(i+1) = \underline{QAQ}^{-1}\hat{\underline{x}}(i) = \begin{bmatrix} \lambda_1 & 0 & \cdots & 0 \\ 0 & \lambda_2 & \cdots & 0 \\ \cdots & \cdots & \cdots & \cdots \\ 0 & 0 & \cdots & \lambda_n \end{bmatrix} \hat{\underline{x}}(i) \tag{22}$$

From (7), the solution of (22) is

$$\hat{\underline{x}}(i) = \begin{bmatrix} \lambda_1 & 0 & \cdots & 0 \\ 0 & \lambda_2 & \cdots & 0 \\ \cdots & \cdots & \cdots & \cdots \\ 0 & 0 & \cdots & \lambda_n \end{bmatrix}^i \hat{\underline{x}}_0 = \begin{bmatrix} \lambda_1^i & 0 & \cdots & 0 \\ 0 & \lambda_2^i & \cdots & 0 \\ \cdots & \cdots & \cdots & \cdots \\ 0 & 0 & \cdots & \lambda_n^i \end{bmatrix} \hat{\underline{x}}_0 \tag{23}$$

and each component of $\underline{x}(i)$ is proportional to the i-th power of the the corresponding eigenvalue. Now each eigenvalue λ is a complex number with magnitude $|\lambda|$, and it is easy to see† that

$$|\lambda^i| = |\lambda|^i \underset{i\to\infty}{\longrightarrow} 0 \quad \Longleftrightarrow \quad |\lambda| < 1 \tag{24}$$

Thus we may conclude that $\hat{\underline{x}}(i) \to \underline{0}$ if and only if $|\lambda_1|, |\lambda_2|, \ldots, |\lambda_n|$ are all less than 1, and we leave it to the reader (Problem 10) to show how the argument extends to a system with repeated eigenvalues which cannot be diagonalized.

This conclusion carries over to the original state vector $\underline{x} = \underline{Q}^{-1}\hat{\underline{x}}$, and we summarize with the following discrete-time analog of Theorem 5.1.3.

2. *THEOREM.* The system (3) is (asymptotically) stable if and only if all of its poles lie strictly within the unit circle in the complex plane, that is,

---

†For example, express λ in polar form $\lambda = \rho e^{j\phi}$, where $\rho = |\lambda|$ is the magnitude and ϕ is the phase angle.

$$|\lambda_m| < 1 \qquad (m = 1, 2, \ldots, n) \tag{25}$$

where λ_m are the eigenvalues of $\underline{A}$.

If $|\lambda_m| \leq \rho < 1$ for all m, we may conclude that each mode of the transient response dies away at least as fast as ρ^i, and the system is said to have *stability margin* ρ (compare the analogous definition in Sec. 5.3).

ΔΔΔ

In Sec. 5.2 we saw that the Routh and Hurwitz tests were very useful for checking the stability of a continuous-time system. These tests operate on the coefficients of the characteristic polynomial and determine whether all of its roots lie in the left half-plane. Here we must determine whether the roots of the characteristic polynomial p(z) lie within the unit circle, and this can be done with a similar test involving *inners* [12, 97, 275]. These procedures can be rather tedious, even for a polynomial of moderate order, and in practice it is often easier to utilize a standard computer program which solves directly for the eigenvalues of $\underline{A}$.

Controllability and Observability

The concepts of controllability and observability discussed in Chapter 10 are equally relevant here. Some caution is required if $\underline{A}$ is singular, for then a controllable state may not be reachable (Problem 11) and a reconstructible state may not be observable. Rather than become involved in various definitions, theorems, and special cases, we will simply state and prove two results which are representative of the others available. The reader should also note that discrete-time controllability and observability Gramians $\underline{W}_c$ and $\underline{W}_o$ may be defined by sums analogous to (10.1-6) and (10.2-6), and results analogous to Theorems 10.1.4 and 10.2.4 and Corollaries 10.1.5 and 10.2.5 may be proved. However, we shall see that it is more convenient in the discrete-time case to deal directly with the controllability and observability matrices $\underline{P}_c$ and $\underline{P}_o$.

3. *THEOREM.* There exists a control sequence for every pair of states $\underline{x}_0$, $\underline{x}_1$ which drives the system (3) from $\underline{x}(0) = \underline{x}_0$ to $\underline{x}(n) = \underline{x}_1$ if and only if the $n \times nm$ controllability matrix

$$\underline{P}_c = [\underline{B} \quad \underline{AB} \quad \underline{A}^2\underline{B} \quad \dots \quad \underline{A}^{n-1}\underline{B}] \tag{26}$$

has full rank n. In this case the system is said to be both *controllable* and *reachable*. For any given $\underline{x}_0$, $\underline{x}_1$, a control sequence which makes the transfer is given by

$$\begin{bmatrix} \underline{u}(n-1) \\ \underline{u}(n-2) \\ \vdots \\ \underline{u}(0) \end{bmatrix} = \underline{P}_c' (\underline{P}_c' \underline{P}_c)^{-1} (\underline{x}_1 - \underline{A}^n \underline{x}_0) \tag{27}$$

and the "energy" expended by this control sequence,

$$E = \sum_{i=0}^{n-1} \| \underline{u}(i) \|^2 \tag{28}$$

is less than that expended by any other sequence which accomplishes the same transfer in n steps or less.

Here n is the dimension of $\underline{A}$, and the statement holds for n steps if and only if it holds for any number of steps greater than n. Note that if $m > 1$, the matrix $\underline{P}_c$ (and the number of steps) may be truncated at $n - m$, as in (10.1-42), provided $\underline{B}$ has rank m.

Proof: From (7) it is clear that the required control sequence must satisfy

$$\underline{x}_1 = \underline{x}(n) = \underline{A}^n \underline{x}_0 + \sum_{j=0}^{n-1} \underline{A}^{n-j-1} \underline{Bu}(j) \tag{29}$$

Rearranging this equation, we have

$$[\underline{B} \quad \underline{AB} \quad \underline{A}^2\underline{B} \quad \dots \quad \underline{A}^{n-1}\underline{B}] \begin{bmatrix} \underline{u}(n-1) \\ \underline{u}(n-2) \\ \vdots \\ \underline{u}(0) \end{bmatrix} = \underline{x}_1 - \underline{A}^n \underline{x}_0 \tag{30}$$

According to Theorem A.5.4(c), this equation has a solution for every value of the right-hand side if and only if the matrix on the left ($\underline{P}_c$) has full rank, and (27) is clearly one such solution. The case of $k > n$ steps follows from Problem 10.1-6. Of course, there will always be particular pairs of states (for example, $\underline{x}_0$ and $\underline{A}\underline{x}_0$) for which the transfer is possible in less than n steps.

ΔΔΔ

4. *THEOREM.* Let the input to the system (3) be zero. If the state is unknown, it can be determined from the output sequence $\underline{y}(0)$, $\underline{y}(1)$, ..., $\underline{y}(n-1)$ if and only if the $np \times n$ observability matrix

$$\underline{P}_o = \begin{bmatrix} \underline{C} \\ \underline{C}\underline{A} \\ \underline{C}\underline{A}^2 \\ \vdots \\ \underline{C}\underline{A}^{n-1} \end{bmatrix} \tag{31}$$

has full rank n. In this case the system is said to be both *observable* and *reconstructible*. The states may be determined by the linear operations

$$\underline{x}(0) = (\underline{P}_o' \underline{P}_o)^{-1} \underline{P}_o' \begin{bmatrix} \underline{y}(0) \\ \underline{y}(1) \\ \vdots \\ \underline{y}(n-1) \end{bmatrix} \tag{32}$$

and

$$\underline{x}(i) = \underline{A}^i \underline{x}(0) \tag{33}$$

for $i \neq 0$. Note that if $\underline{A}$ is singular, then (33) holds only for $i > 0$.

Here n is the dimension of $\underline{A}$, and the statement holds for n output measurements if and only if it holds for any number of measurements greater than n. Note that if $p > 1$, the matrix $\underline{P}_o$ may be truncated at n-p, as in (10.2-20), provided $\underline{C}$ has rank p.

Proof: The output is given by (8), which for $\underline{u} \equiv \underline{0}$ is

$$\underline{y}(i) = \underline{C}\underline{A}^i \underline{x}_0 \qquad (i = 0, 1, 2, \ldots) \tag{34}$$

or

$$\begin{bmatrix} \underline{y}(0) \\ \underline{y}(1) \\ \vdots \\ \underline{y}(n-1) \end{bmatrix} = \begin{bmatrix} \underline{C} \\ \underline{C}\underline{A} \\ \vdots \\ \underline{C}\underline{A}^{n-1} \end{bmatrix} \underline{x}(0) \tag{35}$$

According to Theorem A.5.4(b), this equation uniquely determines $\underline{x}(0)$ if and only if the matrix on the right ($\underline{P}_o$) has full rank, in which case (32) specifies the solution. The case of $k > n$ measurements follows from the dual of Problem 10.1-6.

ΔΔΔ

Various other results in Chapter 10 are applicable here. Controllability and observability are dual, of course, and a canonical decomposition theorem analogous to Theorem 10.3.4 applies to discrete-time systems as well. Virtually all of the discussion of canonical forms and realizations in Sec. 10.4 carries over with obvious modifications, one of which is that the integrators in Fig. 10.4-1 become delay elements. Conditions under which controllability and observability of a continuous-time system carry over to a sampled version of the same system are discussed in Ref. 39. These matters will not be pursued any further here.

Control and Estimation

Feedback control laws may be designed for discrete-time systems using all of the techniques that have been developed for the continuous-time case. Linear feedback from output to input, for example, has the form

$$\underline{u}(i) = \underline{L}\underline{y}(i) + \underline{w}(i) \tag{36}$$

where $\underline{L}$ is a constant $m \times p$ feedback gain matrix and $\underline{w}(i)$ is an external input. Applying this feedback law to (3) yields the closed-loop system

$$\underline{x}(i+1) = \tilde{\underline{A}}\underline{x}(i) + \underline{B}\underline{w}(i) = (\underline{A} + \underline{B}\underline{L}\underline{C})\underline{x}(i) + \underline{B}\underline{w}(i)$$
$$\underline{y}(i) = \underline{C}\underline{x}(i) \tag{37}$$

Just as in continuous time, the behavior of the closed-loop system depends upon the eigenvalues of $\tilde{\underline{A}} = \underline{A} + \underline{B}\underline{L}\underline{C}$ (i.e., the closed-loop poles). In this case, of course, the poles must lie within the unit circle for stability, rather than in the left half-plane.

As in Sec. 11.1, state feedback ($\underline{C} = \underline{I}$) results in the closed-loop system matrix $\tilde{\underline{A}} = \underline{A} + \underline{B}\underline{K}$, and feedback from the output to the state ($\underline{B} = \underline{I}$) yields $\tilde{\underline{A}} = \underline{A} + \underline{M}\underline{C}$. All of these feedback forms are as shown in Fig. 11.1-1, with the differential equations replaced by corresponding difference equations.

The key result on linear feedback is the pole-placement theorem (11.2.1), which applies here without alteration: *There exists a state feedback control law to place the closed-loop poles in any preassigned symmetric configuration if and only if the original system is controllable.* Corollary 11.2.2 applies as well, and says that a system is stabilizable with state feedback if all of its unstable poles are controllable. A discrete-time version of Theorem 11.2.4 is also available [282] for rapidly computing a stable control law without resorting to pole placement.

A particularly interesting closed-loop system results if a state feedback control law is chosen which places all of the closed-loop poles at the origin. In the absence of an external input, this means that the closed-loop system

$$\underline{x}(i+1) = (\underline{A} + \underline{B}\underline{K})\underline{x}(i) \tag{38}$$

has the characteristic equation $\lambda^n = 0$. But the Cayley-Hamilton theorem (A.3.9) then implies that $(\underline{A} + \underline{B}\underline{K})^n = \underline{0}$, and

$$\underline{x}(i) = (\underline{A} + \underline{B}\underline{K})^i \underline{x}_0 = \underline{0} \qquad (\text{for } i \geq n) \tag{39}$$

In other words, if the closed-loop poles are all zero, any initial state is driven to zero in (at most) n steps. This is known as a *deadbeat control system*.

If a deadbeat controller operates on a sampled version of a continuous system, then once the sampled state $\underline{x}(i)$ reaches zero, the state $\underline{x}(t)$ will remain at zero (unless disturbed) for all subsequent times t, even between sampling instants. The same conclusion does not follow, however, if the output $\underline{y}(i)$ is driven to zero. In this case the output $\underline{y}(t)$ may exhibit unacceptable behavior in between the sampling instants (see Problem 17). Thus the design of discrete-time control laws for sampled versions of continuous systems must be done with care. In some cases it is safer to design a continuous-time control law and implement a sampled approximation to it, but even this approach has its pitfalls.

In Sec. 11.3 we considered the steady-state performance of feedback systems and saw how steady-state errors due to reference and/or disturbance inputs could be eliminated by augmenting the state vector. In Problem 13, the reader is asked to make appropriate modifications and rederive these results for the discrete-time case.

State estimation, described in Chapter 12 for continuous-time systems, is equally useful in discrete time. The concept of a state observer carries over with minor modifications, both in its full-order (n) and reduced-order $(n - p)$ forms, as does the separation property (Theorem 12.3.1) for control laws realized with observers. The poles of a discrete-time observer may be placed in any symmetric configuration inside the unit circle for stability. If they are all placed at the origin, the error will satisfy an equation dual to (39) and go to zero in n or $n - p$ steps, in which case the term *deadbeat observer* is used. Once again, the details are left to the reader (Problem 14). This topic is also covered in Ref. 105.

Finally, we remark that all of the optimization results dealing with Kalman-Bucy filters (Sec. 12.4) and linear regulators (Sec. 13.1) have discrete-time analogs, and for these the reader is referred to the literature, particularly Ref. 252 and the references cited after Theorem 12.4.1.

We conclude this section by considering a sampled-data version of the satellite problem.

5. *EXAMPLE.* In Example 2.4.3 we considered an earth satellite in a circular, equatorial orbit and found that small perturbations about this orbit (in the equatorial plane) would be governed by the linearized equations

$$\dot{\underline{x}} = \overline{\underline{A}}\underline{x} + \overline{\underline{B}}\underline{u} = \begin{bmatrix} 0 & 1 & 0 & 0 \\ 3\omega^2 & 0 & 0 & 2\omega \\ 0 & 0 & 0 & 1 \\ 0 & -2\omega & 0 & 0 \end{bmatrix} \underline{x} + \begin{bmatrix} 0 & 0 \\ 1 & 0 \\ 0 & 0 \\ 0 & 1 \end{bmatrix} \underline{u} \tag{40}$$

$$\underline{y} = \underline{C}\underline{x} = \begin{bmatrix} 1 & 0 & 0 & 0 \\ 0 & 0 & 1 & 0 \end{bmatrix} \underline{x}$$

where

$$\underline{x} = \begin{bmatrix} r \\ \dot{r} \\ \Theta \\ \dot{\Theta} \end{bmatrix} \qquad \underline{u} = \begin{bmatrix} u_r \\ u_\Theta \end{bmatrix} \qquad \underline{y} = \begin{bmatrix} r \\ \Theta \end{bmatrix} \tag{41}$$

In order to simplify the computations, let the time be measured in days, and assume that the satellite is a synchronous one, with a period of one day. Thus $\omega = 2\pi$ and the matrix exponential computed in Problem 4.3-6 becomes

$$e^{\overline{\underline{A}}t} = \begin{bmatrix} 4 - 3\cos 2\pi t & \frac{1}{2\pi}\sin 2\pi t & 0 & (1 - \cos 2\pi t)/\pi \\ 6\pi \sin 2\pi t & \cos 2\pi t & 0 & 2\sin 2\pi t \\ 6(\sin 2\pi t - 2\pi t) & (\cos 2\pi t - 1)/\pi & 1 & \frac{2}{\pi}\sin 2\pi t - 3t \\ 12\pi(\cos 2\pi t - 1) & -2\sin 2\pi t & 0 & 4\cos 2\pi t - 3 \end{bmatrix} \tag{42}$$

Now suppose that measurement and control can only occur at certain specified times, separated by an interval Δ. Suppose further that the measurements are sampled just before the inputs are applied, and that each input consists of a single, brief thrust from a gas jet which may be approximated as an impulse. This leads us to the difference equation

$$\underline{x}(i+1) = \underline{A}\underline{x}(i) + \underline{B}\underline{u}(i)$$
$$\underline{y}(i) = \underline{C}\underline{x}(i) \tag{43}$$

where $\underline{A}$ and $\underline{B}$ are given by (5) and $\underline{C}$ is unchanged. If sampling and control take place *once per orbit* ($\Delta = 1$ day), these are

$$\underline{A} = e^{\underline{\bar{A}}\Delta} = \begin{bmatrix} 1 & 0 & 0 & 0 \\ 0 & 1 & 0 & 0 \\ -12\pi & 0 & 1 & -3 \\ 0 & 0 & 0 & 1 \end{bmatrix} \tag{44}$$

and

$$\underline{B} = e^{\underline{\bar{A}}\Delta}\underline{B}\Delta = \begin{bmatrix} 0 & 0 \\ 1 & 0 \\ 0 & -3 \\ 0 & 1 \end{bmatrix} \tag{45}$$

At this point it is relevant to ask whether the sampled system is controllable and observable, and we form the controllability matrix (using Problem 10.1-7)

$$\underline{P}_c = \begin{bmatrix} 0 & 0 & 0 & 0 & 0 & 0 \\ 1 & 0 & 1 & 0 & 1 & 0 \\ 0 & -3 & 0 & -6 & 0 & -9 \\ 0 & 1 & 0 & 1 & 0 & 1 \end{bmatrix} \tag{46}$$

This clearly does not have full rank, since it contains a row of zeros, and we conclude that the system is uncontrollable at this sampling rate (the reader should verify that it is the r component of the state vector which cannot be controlled). Similarly, the observability matrix

$$\underline{P}_o = \begin{bmatrix} 1 & 0 & 0 & 0 \\ 0 & 0 & 1 & 0 \\ 1 & 0 & 0 & 0 \\ -12\pi & 0 & 1 & -3 \\ 1 & 0 & 0 & 0 \\ -24\pi & 0 & 1 & -6 \end{bmatrix} \tag{47}$$

does not have full rank, and the system is not observable at this sampling rate (the reader should verify that $\dot{r}$ cannot be determined from output measurements).

Next, suppose the sampling rate is increased to *twice per orbit* ($\Delta = 1/2$ day), so that

$$\underline{A} = e^{\overline{\underline{A}}\Delta} = \begin{bmatrix} 7 & 0 & 0 & 2/\pi \\ 0 & -1 & 0 & 0 \\ -6\pi & -2/\pi & 1 & -3/2 \\ -24\pi & 0 & 0 & -7 \end{bmatrix} \tag{48}$$

$$\underline{B} = e^{\overline{\underline{A}}\Delta}\overline{\underline{B}}\Delta = \frac{1}{2}\begin{bmatrix} 0 & 2/\pi \\ -1 & 0 \\ -2/\pi & -3/2 \\ 0 & -7 \end{bmatrix} \tag{49}$$

In this case the controllability matrix

$$\underline{P}_c = \frac{1}{2}\begin{bmatrix} 0 & 2/\pi & 0 & 0 & 0 & 2/\pi \\ -1 & 0 & 1 & 0 & -1 & 0 \\ -2/\pi & -3/2 & 0 & -3 & -2/\pi & -9/2 \\ 0 & -7 & 0 & 1 & 0 & -7 \end{bmatrix} \tag{50}$$

has full rank, as the first four columns are independent. Recall from Example 10.1.10 that the continuous system is controllable with the tangential control u_Θ alone. In this case both controls are needed: the reader should verify that the sampled system is uncontrollable without u_r, and that $\dot{r}$ is the state which cannot be controlled.

The observability matrix in this case is

$$\underline{P}_o = \begin{bmatrix} 1 & 0 & 0 & 0 \\ 0 & 0 & 1 & 0 \\ 7 & 0 & 0 & 2/\pi \\ -6\pi & -2/\pi & 1 & -3/2 \\ 1 & 0 & 0 & 0 \\ -12\pi & 0 & 1 & -3 \end{bmatrix} \tag{51}$$

which has full rank, as the first four rows are independent. Recall from Example 10.2.10 that the continuous system is observable using measurements of Θ alone. In this case both measurements are needed: the reader should verify that the sampled system is unobservable without the r measurement.

If sampling and control occur *four times per orbit* (Δ = 1/4 day), then the system is controllable using u_Θ alone and observable using Θ alone (Problem 15).

Finally, we shall design a feedback control system for the case where sampling and control occur twice per orbit (Δ = 1/2 day). If a state feedback control law of the form

$$\underline{u}(i) = \underline{K}\underline{x}(i) \tag{52}$$

is applied to (43), where $\underline{A}$ and $\underline{B}$ are given by (48) and (49), the resulting closed-loop system is

$$\underline{x}(i+1) = (\underline{A} + \underline{B}\underline{K})\underline{x}(i) \tag{53}$$

where

$$\underline{A} + \underline{B}\underline{K} = \begin{bmatrix} 7 & 0 & 0 & 2/\pi \\ 0 & -1 & 0 & 0 \\ -6\pi-2/\pi & 1 & -3/2 \\ -24\pi & 0 & 0 & -7 \end{bmatrix} + \begin{bmatrix} 0 & 1/\pi \\ -1/2 & 0 \\ -1/\pi & -3/4 \\ 0 & -7/2 \end{bmatrix} \begin{bmatrix} k_{11} & k_{12} & k_{13} & k_{14} \\ k_{21} & k_{22} & k_{23} & k_{24} \end{bmatrix}$$

$$= \begin{bmatrix} 7 + \frac{k_{21}}{\pi} & \frac{k_{22}}{\pi} & \frac{k_{23}}{\pi} & \frac{2}{\pi} + \frac{k_{24}}{\pi} \\ -\frac{k_{11}}{2} & -1 - \frac{k_{12}}{2} & -\frac{k_{13}}{2} & -\frac{k_{14}}{2} \\ -6\pi - \frac{k_{11}}{\pi} - \frac{3k_{21}}{4} & -\frac{2}{\pi} - \frac{k_{12}}{\pi} - \frac{3k_{22}}{4} & 1 - \frac{k_{13}}{\pi} - \frac{3k_{23}}{4} & -\frac{3}{2} - \frac{k_{14}}{\pi} - \frac{3k_{24}}{4} \\ -24\pi - \frac{7k_{21}}{2} & -\frac{7k_{22}}{2} & -\frac{7k_{23}}{2} & -7 - \frac{7k_{24}}{2} \end{bmatrix} \tag{54}$$

Since the system is controllable, the pole-placement theorem says that the closed-loop poles may be put into any symmetric configuration by an appropriate choice of the gain matrix $\underline{K}$. Suppose the desired configuration has all the poles at the origin, in order to achieve a deadbeat response. We could now apply one of the techniques discussed in Chapter 11 to find a $\underline{K}$. In this case, however, a bit of judicious head scratching quickly yields

$$\underline{K} = \begin{bmatrix} 0 & -2 & \pi & 0 \\ -7\pi & 0 & 0 & -2 \end{bmatrix} \qquad \underline{A} + \underline{BK} = \begin{bmatrix} 0 & 0 & 0 & 0 \\ 0 & 0 & -\pi/2 & 0 \\ -3/4 & 0 & 0 & 0 \\ 1/2 & 0 & 0 & 0 \end{bmatrix} \tag{55}$$

The closed-loop system matrix is now lower block-triangular, and the lower 3×3 block is itself upper triangular, so that the eigenvalues are just the diagonal elements, which are all zero.

To verify the deadbeat property, we may compute

$$(\underline{A} + \underline{BK})^2 = \begin{bmatrix} 0 & 0 & 0 & 0 \\ 3\pi/8 & 0 & 0 & 0 \\ 0 & 0 & 0 & 0 \\ 0 & 0 & 0 & 0 \end{bmatrix} \qquad (\underline{A} + \underline{BK})^3 = \underline{0} \tag{56}$$

Since $\underline{x}(i) = (\underline{A} + \underline{BK})^i \underline{x}_0$, we see that any nonzero initial state is returned to zero in three steps (1.5 orbits) or less.

Since only the outputs r and Θ are available for measurement, an observer is required to estimate $\dot{r}$ and $\dot{\Theta}$. A reduced-order observer with dimension $4 - 2 = 2$ can be designed by a procedure analogous to that of Sec. 12.2 provided the states are reordered to put the system in the form (12.2-1). If both of the poles are placed at the origin, it will be a deadbeat observer and the error will go to zero in two steps (1 orbit) or less. Design of the observer is left as a task for the interested reader (Problem 16).

ΔΔΔ

PROBLEMS

Section 15.1

1. (a) If $\lambda_1, \lambda_2, \ldots, \lambda_n$ are the eigenvalues of $\bar{A}$, what are the eigenvalues of $A = e^{\bar{A}\Delta}$?

   (b) Show that if $\bar{A}^{-1}$ exists, the expression for B in (15.1-4) may be written as

   $$B = (e^{\bar{A}\Delta} - I)\,\bar{A}^{-1}\bar{B}$$

   (c) Show that the definitions (15.1-5) result when the input to (15.1-1) is a sequence of impulses. Why is each impulse multiplied by a factor of Δ?

   (d) Would you expect the B matrices in (15.1-4) and (15.1-5) to approach the same value as $\Delta \to 0$? Why? Do they?

2. (a) Suppose, in the system of Fig. 2, that the continuous output $y(t)$ is sampled at the times $t = i\Delta + \delta$, where $0 \leq \delta < \Delta$. Rederive the discrete-time system equations for this case. What happens to the output equation if $\delta = 0$?

   (b) Repeat your analysis for the case where the input is a sequence of impulses at the times $i\Delta$.

   (c) Repeat your analysis for the case where $\delta = 0$ but the continuous system is time varying, that is, A, B, and C all depend upon t.

3. Show that in the limit as the sampling interval $\Delta \to 0$, (15.1-8) becomes equivalent to (3.3-14).

4. Let $x_0 = 0$ in (15.1-3) and determine the response y to an input which is a single pulse at time $i = 0$.

5. (a) Find the solution of the time-varying homogeneous equation

   $$x(i+1) = A(i)x(i) \qquad x(0) = x_0$$

by determining a *transition matrix* $\underline{\Phi}(i,j)$ analogous to that used in Chapter 14. What difference equation does $\underline{\Phi}$ satisfy? State a *composition rule* for $\underline{\Phi}$.

(b) Use this transition matrix to find the solution of (15.1-3) when $\underline{A}$, $\underline{B}$, and $\underline{C}$ all depend upon i.

(c) Determine the pulse response matrix for this case.

6. Let $Z\{f(i)\} = \hat{f}(z)$ and $Z\{g(i)\} = \hat{g}(z)$, and prove the following theorems:

(a) Linearity:

$$Z\{\alpha f(i) + g(i)\} = \alpha\hat{f}(z) + \hat{g}(z)$$

(b) Delay:

$$Z\{f(i-1)\} = z^{-1}\hat{f}(z) \qquad \text{(analogous to integration)}$$

(c) Advance:

$$Z\{f(i+1)\} = z\hat{f}(z) - zf(0) \quad \text{(analogous to differentiation)}$$

(d) Convolution:

$$Z\{\sum_{j=0}^{i} f(j)g(i-j)\} = \hat{f}(z)\hat{g}(z)$$

7. Determine the Z-transforms of the following sequences [assuming $f(i) = 0$ for $i < 0$]:

(a) 1, 1, 1, 1, ...

(b) 1, 0, 0, 0, ...

(c) 0, 1, 0, 0, ...

(d) $1, \lambda, \lambda^2, \lambda^3, \ldots$

(e) $\lambda, \lambda^2, \lambda^3, \lambda^4, \ldots$

8. (a) Suppose the system

$$\dot{\underline{x}} = \begin{bmatrix} \sigma & \omega \\ -\omega & \sigma \end{bmatrix} \underline{x} + \begin{bmatrix} 1 \\ 0 \end{bmatrix} \underline{u} = \overline{\underline{A}}\underline{x} + \overline{\underline{b}}u$$

$$\underline{y} = \begin{bmatrix} 0 & 1 \end{bmatrix} \underline{x} = \underline{c}'\underline{x}$$

is discretized as in Problem 2(b), with $\delta = 0$, so that it becomes

$$\underline{x}(i+1) = e^{\overline{\underline{A}}\Delta}\underline{x}(i) + \overline{\underline{b}}\Delta u(i) = \underline{\underline{A}}\underline{x}(i) + \underline{b}u(i)$$
$$y(i) = \underline{c}'\underline{x}(i)$$

Compute $\underline{A}$ and find its eigenvalues. Find a simple expression for $\underline{A}^i$, $i = 0, 1, 2, \ldots$.

(b) Determine the transfer function and pulse response of the discrete-time system. How do these compare with the continuous-time transfer function and impulse response?

By making appropriate changes in parts (a) and (b), determine the Z-transforms of the following sequences:

(c) $e^{\sigma i\Delta} \sin \omega i\Delta$

(d) $e^{\sigma i\Delta} \cos \omega i\Delta$

(e) $\sin \omega i\Delta$

(f) $\cos \omega i\Delta$

$(i = 0, 1, 2, \ldots)$

9. Show how to alter Chapter 4 so that it applies to discrete-time systems and Z-transforms. The results of Problems 6, 7, and 8 will be useful for this task.

10. Show how the arguments leading up to Theorem 15.1.2 may be extended to the case of repeated eigenvalues and an $\underline{A}$ which is put into Jordan form.

11. Invent an example of a system having some states which are controllable (to $\underline{0}$) but not reachable (from $\underline{0}$). State and prove theorems characterizing the subspaces of controllable and reachable states, respectively.

12. Prove the minimum energy property stated in Theorem 3. *Hint:* Denote the nm-dimensional vector of controls in (29) by $\overline{\underline{\mu}}$ and let $\underline{\mu}$ be any other such vector which accomplishes the transfer. Show that $\overline{\underline{\mu}} \perp (\overline{\underline{\mu}} - \underline{\mu})$, and proceed as in the proof of Corollary 10.1.5.

13. Show how to modify the results of Sec. 11.3 for discrete-time systems. *Hint:* Start by augmenting the state vector to include

$$\underline{q}(i) = \sum_{j=0}^{i-1} [\underline{y}(j) - \underline{y}_r(j)]$$

14. Rederive the observers of Secs. 12.1 and 12.2 in discrete-time and prove the separation property (Theorem 12.3.1) for this case. Show how disturbance inputs and bias (Sec. 12.3) can be handled in a manner similar to that of Problem 13.

15. Check the controllability and observability of the satellite (Example 15.1.5) when sampling and control occur four times per orbit.

16. In Example 15.1.5, design a discrete-time, reduced-order deadbeat observer (using two samples per orbit) to implement the state feedback control law (15.1-52). How many time steps are required for the state of the closed-loop system to reach the origin?

17. (a) Show that if a deadbeat controller drives the sampled state to $\underline{x}(i\Delta) = \underline{0}$, then $\underline{x}(t) = \underline{0}$ for all $t \geq i\Delta$.

    (b) Devise an example of a single-output system for which $y(i\Delta) = 0, 1, 2, 3, \ldots$, but $y(t) \neq 0$ for $t \neq i\Delta$.

18. What is the analog of the deadbeat property for a continuous-time system? Can a continuous, linear feedback control ever drive the state of a continuous system to zero in finite time?

19. (a) Write out the Z-transform of (15.1-6) and express it in a form analogous to (4.1-4). Identify the transfer function.

    (b) Find an explicit (time-domain) solution of
    $$y(i+2) + 1.3y(i+1) + 0.4y(i) = u(i)$$
    where $y(1) = y(0) = 1$ and u consists of a single unit pulse at time $i = 1$. Is this system stable? What is its transfer function?

20. *Jacobi's method.* A classic iterative scheme for solving the linear equation $\underline{A}\underline{x} = \underline{b}$, where $\underline{A}$ is a square matrix, is to let $\underline{D}$ be the diagonal part of $\underline{A}$ and compute

$$\underline{x}_{i+1} = -\underline{D}^{-1}(\underline{A} - \underline{D})\underline{x}_i + \underline{D}^{-1}\underline{b} \qquad (i = 1,2,3,\ldots)$$

where $\underline{x}_0 = \underline{0}$. State conditions under which the iterations will converge to a constant $\overline{\underline{x}}$. If so, does $\overline{\underline{x}}$ solve the original equation? What can you say about the speed of convergence? What happens if $\underline{A}\underline{x} = \underline{b}$ has no solution, or if the solution is not unique? What can you do if $\underline{A}$ has a zero on its main diagonal?

Chapter 16

EPILOGUE

This book began with a discussion of the history of control systems. In subsequent chapters we developed the theory of deterministic, finite-dimensional, linear, time-invariant, differential systems to the point where the reader should now feel comfortable with the relevant literature and be able to proceed to more advanced work of either a theoretical or a practical nature. It therefore seems appropriate to close by briefly surveying various extensions of the basic theory and areas of current research activity. We do not attempt to provide a complete bibliography; rather, we cite textbooks, monographs, and survey papers wherever possible.

Certain extensions have already been discussed. In Chapter 14 we saw that time-varying systems can be treated in an analogous way to time-invariant systems, with two fundamental differences: the matrix exponential $e^{\underline{A}t}$ must be replaced by a less easily computable transition matrix $\underline{\Phi}(t,t_0)$ and the crucial characterization of system behavior and stability in terms of poles (eigenvalues) is absent. In Chapter 15 the theory of discrete-time systems was seen to be very similar to that of continuous-time systems; in this case $e^{\underline{A}t}$ is replaced by the more easily computable power matrix $\underline{A}^i$, and stable poles must lie within the unit circle rather than the left half-plane.

One topic of great theoretical interest has been the use of abstract algebra and category theory to study finite-state systems and other automata, as well as systems described by differential equations. Material in this area may be found in Refs. 5, 98, 119, 140, and 213-216.

Some interesting perspectives on trends and directions in the field of control theory may be found in Refs. 213a and 223.

Control System Design Techniques

Throughout this book we have been mainly concerned with providing a firm theoretical basis for control system design and, in the case of single-loop systems, with presenting such standard design tools as the root-locus and Nyquist methods. A great deal more can be said about the design of practical feedback control systems, and a wide variety of other techniques have been developed over the years, particularly for single-loop systems. The interested reader may consult such textbooks as Refs. 43, 48, 57, 59, 71, 72, 75, 89, 99, 137, 139, 158, 168, and many others.

The impact of large-scale integrated circuits and increasingly powerful microprocessors on control system design cannot be ignored. *Direct digital control* has been used in a variety of aerospace systems, and further reductions in size and price seem certain to spread this technology into other areas. Chapter 15 and the references therein relate to this topic.

Similarly, the availability of interactive computers with graphical displays has led to significant developments in *computer-aided design* [150, 296, 297]. The root-locus, Nyquist, and other graphical techniques have been extended and mechanized so that the designer can select the relevant display and change various parameters by simply typing commands on a terminal.

Some interesting issues regarding the role of the designer are raised in Ref. 327.

Linear Systems Theory

The underlying mathematical *structure* of linear multivariable systems has been investigated in great depth, far beyond the basic theory of controllability and observability presented in Chapter 10. Some topics of interest are *decoupling* (see Sec. 11.4), *inverse systems*, characterization of *input-output relationships*, and *model matching* (where closed-loop behavior is made to duplicate that of a given model). These are discussed in Refs. 307, 310, 325, 335, 337, 349, 358, and 368. A number of other structural issues have been explored as well [184a, 234, 250a, 251a, 252b, 256a, 256b, 260, 264, 264a, 308, 309, 318, 350a], including the properties and parameters which remain invariant under a variety of transformations. Many of these play critical roles in the context of system identification, which will be discussed below.

The *sensitivity* of a system to changes or uncertainties in some of its parameters has long been a topic of both theoretical and practical interest; material on this subject may be found in Refs. 43, 44, 89, 167, 201, 256b, and 332a.

Although state-space methods have dominated the literature for nearly two decades, there has also been a more recent resurgence of interest in frequency-domain analysis of multivariable systems, yielding a number of important results. These include frequency-domain characterizations of controllability and observability, multivariable extensions of the Nyquist results, and placement of closed-loop zeros as well as poles.

Most of these topics are surveyed in Refs. 207 and 296-298; more details may be found in various textbooks [39, 53a, 149, 183]. The reader may also find it interesting to review Ref. 267, where certain gaps and practical difficulties in the state-variable theory are pointed out. It is also argued there that conventional frequency-domain methods are vastly superior to the state-space approach, but in the opinion of the present authors, the case is greatly overstated. Another perspective may be found in Ref. 256.

The use of modules rather than vector spaces for the study of linear systems has also been explored. In some ways, this provides a more natural framework in which to work. Details may be found in Refs. 98 and 140. Still other frameworks are discussed below under nonlinear systems.

Nonlinear Systems and Stability

Dealing with nonlinear systems is both a challenging and a frustrating task. Stability is generally a local rather than a global property, and different control laws may be required for different initial states. Unfortunately, there is no unifying theoretical framework such as we have established for linear systems. Nevertheless, various techniques have been developed for dealing with nonlinearities in conventional control systems; books on this topic include Refs. 45, 73, 75, 78, 133, 158, 159, and 164. The method of *describing functions* extends frequency response concepts to the nonlinear case by ignoring higher-order harmonics, and has proved to be quite useful [70, 257, 300].

As the state-space formulation of problems has become more common, interest in stability theory has also increased. A great deal of literature has appeared on this topic, much of it based on the celebrated theorems of Liapunov and Popov [3, 22, 43, 77, 106a, 107, 136, 181, 230, 319]. A certain amount of work has also been carried out using an operator framework and the tools of functional analysis [53a, 88, 180, 204, 372]. Other approaches may be found in Ref. 21.

At the same time, optimal control theory was recognized as a powerful tool for the control of nonlinear systems. More will be said about this shortly.

More recently, some techniques from the theory of differential geometry have been found to provide powerful theoretical tools for the analysis of various classes of nonlinear systems. We have seen in this book that ordinary vector spaces provide a very natural setting for the study of linear systems, and the general aim of this research is to find mathematical settings which are equally natural for various classes of nonlinear systems [122, 232, 262b].

The class of *bilinear systems* has proved amenable to this approach, using such frameworks as Lie groups, Lie algebras, and Volterra series expansions [53, 134, 135, 231]. This class includes a variety of interesting physical systems, including those with multiplicative noise, and applications to the design of closed-form bilinear estimators have recently appeared [288].

Optimal Control Theory

The simplest linear optimal control results were presented in Chapter 13, but they barely scratched the surface of this vast field. Various optimization techniques have been around for a long time, and many optimal control problems can be formulated in the framework of the classical calculus of variations. However, it was not until Bellman described his method of *dynamic programming* and Pontryagin proved his celebrated *minimum principle* in the late 1950's that the field of optimal control theory began to flourish. This was also due to the availability at about the same time of large digital computers which could handle the resulting computational tasks.

Optimal control problems are typically formulated as in Chapter 13, but the system need not be linear and the cost functional need not be quadratic. Moreover, the terminal state may be specified and constraints may be placed on allowable values of the control and the state trajectory. This leads to a variety of theoretical issues and much more difficult computational problems, but it is often the *only* tractable approach for dealing with a complex nonlinear system.

The minimum principle and its diverse applications are presented in Ref. 11, which remains the standard textbook on this subject. An excellent capsule survey of theory, computational methods, and applications, together with a bibliography up to 1966, is given in Ref. 219, and the original proof of the minimum principle may be found in Ref. 144. The dynamic programming approach is described in Refs. 17, 22, 24, 27, 29, 55, 94, 106, and 313. Other works on optimal control include Refs. 6, 26, 33, 37, 50, 56, 84, 85, 101, 105, 108, 110-112, 154, 172, and 202.

Infinite-Dimensional Systems

All the systems discussed in this book have had n-dimensional state vectors and been described by ordinary differential equations. There are many systems, however, which fall outside this class. Consider, for example, a heated rod of finite length and negligible cross section whose temperature is a function of both time and distance along the rod. It is not hard to show that the temperature must satisfy a partial differential equation, and for initial conditions to be specified, we must know the initial temperature at every point along the rod. Designating temperature as the state, we see that it has not n but a whole continuum of components, corresponding to all the points along the rod. Thus the state in this case is *infinite-dimensional*, and we have what is known as a *distributed parameter system*. Many other such examples occur in the area of chemical process control and elsewhere.

It is not surprising that the control of distributed parameter systems, described by partial differential equations, is a rather difficult task. One technique, often called *modal analysis*, is to isolate a finite-dimensional subsystem which includes the most important modes (eigenvalues) and apply the conventional theory. Other techniques retain the distributed essence of the problem and make use of some fairly deep theoretical tools from functional analysis.

The literature in this area includes books on chemical process control [72] and on distributed systems [25a, 35, 50, 115, 116, 143, 147], survey papers [239, 304, 323, 348], and other works [22, 38, 53a, 65, 163, 175, 295].

Another interesting class of systems includes those which contain *time delays* and are described by differential-difference equations, for example,

$$\dot{\underline{x}}(t) = \underline{A}_1\underline{x}(t) + \underline{A}_2\underline{x}(t - \Delta) + \underline{B}\underline{u}(t)$$

The vector $\underline{x}(t_0)$ is an insufficient initial condition for such an equation at time t_0; a moment's reflection reveals that $\underline{x}(\tau)$, $t_0 - \Delta < \tau \leq t_0$, is needed to specify the solution from t_0 on. Systems of this type are often called *hereditary*, and they are infinite-dimensional in the same sense as the heated rod example mentioned above. References on this topic include Refs. 23, 25a, 55a, 226, and 247.

Stochastic Systems

Systems which involve random processes (noise) are currently the subject of a great deal of research effort. Problems of *state estimation* and *filtering* in the presence of noise, upon which we touched in Sec. 12.4, constitute a major research area by themselves. The linear theory is well developed, and much work has also been done on the nonlinear case. Literature on this subject includes the annual *Proceedings of the Symposium on Nonlinear Estimation Theory and its Applications*, survey papers [236, 239, 277, 321, 340, 341, 364], and numerous books [13, 33, 34, 69, 95, 105, 105a, 126, 155, 157, 160, 171, 184].

Parameter estimation is another important topic, both here and in the section below on system identification. The distinction between states and parameters can be rather nebulous, particularly in the case of a nonlinear system, but it is often true that parameters are either constant or slowly varying relative to the states. Also, parameters are usually uncontrollable.

When the task of controlling a system is added to that of estimation, the problems become even more interesting. This is the subject of *stochastic control theory*, on which a survey [366], and a number of textbooks [7, 9, 15, 27, 63, 103, 104, 127, 162a] have appeared. The limits of optimal control in the presence of parameter uncertainty are explored in Ref. 222a.

Some fundamental results in this area are concerned with the *separation* of estimation and control [357], which was mentioned briefly in Chapter 13. Under certain assumptions, such as a linear system, Gaussian statistics, a quadratic cost functional, and a single controller, the stochastic optimal control problem separates into two independent problems: state estimation (stochastic) and control (deterministic). The situation becomes much more complicated when there are multiple controllers, each having access to only part of the available data. This leads to the important concept of an *information pattern*, which will be discussed in the section on decentralized control.

Another important problem is that of selecting an appropriate input signal to "probe" the system and facilitate the estimation of states or parameters [301]. An even more challenging problem arises when the designer must trade off this role of the input with the often conflicting aim of controlling the system [227].

The stochastic processes described in Sec. 12.4 and considered in most of the references above are of the *diffusion* type, that is, they have continually changing variables. More recently, interest has begun to develop in systems that are more naturally modeled with *jump processes* in which discrete events (such as random price changes) occur at random times. Discussions of these problems and further references may be found in Refs. 161, 228b, 333, 334, 343, and 367.

Still another research topic concerns the modeling of uncertainty in a nonstochastic fashion. One approach to this problem is to replace the standard topology with one based on *fuzzy sets* [100, 186, 261, 280a, 371]. With an ordinary set, any given point or object is either a member of the set or it is not. In contrast, each object has a certain "degree" of membership in a fuzzy set, varying from none to full (examples are the set of beautiful women and the set of small integers). The framework of fuzzy sets can be used to model variables whose values are uncertain without any recourse to the concepts of randomness and probability.

A conceptually different approach is to consider variables whose values are confined to lie within certain boundaries but are otherwise completely unknown [157, 228]. No probability distributions are specified, and the variables are said to be *unknown but bounded*. If the vectors of interest (state, input noise, measurement noise, etc.) are all bounded by ellipsoids, a variety of useful results can be proved which specify how these ellipsoids propagate in time. The estimation procedures which have been developed in this framework closely parallel those discussed in Sec. 12.4 since the bounding ellipsoids are analogous to ellipsoids of constant probability which are defined by the covariance matrix of a random vector.

System Identification and Adaptive Control

Heretofore we have assumed that each system under consideration was completely known, in the sense that we could specify differential equations governing its behavior. Exceptions occurred in Sec. 6.2, where we mentioned the possibility of determining these equations from frequency response measurements, and in Sec. 10.4, where we discussed the problem of finding a *realization* (in state-variable format) of a given transfer function or impulse response matrix.

These are both examples of the *system identification problem*, which arises whenever the equations governing a system are partially or completely unknown. If the equations are known except for the values of certain parameters, then the problem is often said to be one of parameter identification or estimation, and it overlaps to a significant degree with the estimation problems which were discussed above. Questions of system structure (mentioned under linear systems theory), canonical forms, and parameter identifiability (analogous to observability) are especially important in this context.

When identification must be carried out using noisy data the nonstatistical methods discussed in Secs. 6.2 and 10.4 are usually inadequate. Fortunately, other approaches have been found to be more effective, and extensive literature has appeared in recent years, including books [4, 30, 60, 74, 127a, 128, 155], survey articles [218, 225, 262, 262a, 271, 312, 332], and a special journal issue [203].

In addition to various assumptions about linearity, time invariance, finite dimensionality, etc., most identification procedures require the user to guess the order (state dimension) of the system being identified. This is a critical step because a choice which is too low can result in an inadequate model, while one which is too high can render the system unidentifiable because of an excessive number of parameters. An attempt to resolve this dilemma by minimizing a weighted sum of the system order and the log of the identification error has recently been described [206].

Another important issue is that of selecting inputs to aid the identification/estimation process, which was mentioned under estimation [227, 301].

The problem of *adaptive control* [204a] combines identification (or parameter estimation) and control: The controller must apply inputs so as to adequately control the system, and at the same time it must adapt itself to unknown or changing parameters. Once again, the hazy distinction between states and parameters tends to cloud the issues, but it is not uncommon to select state variables so as to obtain a linear system and a linear control law, and then designate anything that enters nonlinearly as a parameter.

One approach to this problem is known as *model-reference adaptive control*, where a model system is constructed with the desired characteristics and the output of the actual system is compared with that of the model. The resulting error signal is then fed back and used to alter the parameters of the controller so as to reduce the error. Two surveys of this topic are available [284, 285]. It is interesting to note that the dual of this scheme is *model-reference identification*, where the parameters of the model are altered so as to reduce errors and force the model to become a copy of the system being identified.

Other adaptive control strategies are described in Refs. 19, 58, 129, 169, 217, 224, 243, 252a, 272, and 357a. *Failure detection* schemes are discussed in Refs. 306, 353, and 354.

The design of *adaptive observers* has also received some attention. The objective is to simultaneously estimate both the state of a linear system and its unknown parameters. Some recent papers are Refs. 208, 283a, and 289, which contain references to most of the other relevant literature.

Decentralized Control and Large-Scale Systems

The systems in this book have all had a single control vector which could be manipulated by a central controller operating on the system output or state. Some of the first systems to be studied with multiple controllers were *differential games*, where two "players" each manipulate certain inputs to a dynamic system and either cooperate or compete in controlling it. A linear example is

$$\dot{\underline{x}}(t) = \underline{A}\underline{x}(t) + \underline{B}_1\underline{u}_1(t) + \underline{B}_2\underline{u}_2(t)$$

where different players with possibly conflicting objectives control $\underline{u}_1$ and $\underline{u}_2$. A classic problem in this field is that of pursuit and evasion, where one player attempts to capture the other, who in turn tries to escape. References in this area include Refs. 28, 33, 66, 92, 152, and 269.

More recently, a great deal of interest has developed in *decentralized* and *hierarchical* control structures for large-scale systems, where various subsystems are controlled by separate local controllers, possibly coordinated by a supervisory controller at a higher level. A key issue which arises in this context is that of the *information pattern*, and it establishes a very strong link to the theory of stochastic control (discussed above). Memory of past inputs and outputs may be imperfect, communications may be limited or noisy, and different controllers may not have access to the same data base, all of which make for some very interesting and challenging problems which are only just beginning to be understood. Papers on this subject include Refs. 86, 212, 238b, 241, 243a, 330, 350, 355, and 356. Several survey papers [220, 266, 331, 338, 344, 357], books [130, 182], and a special journal issue [203a] are also available.

Applications

The applications of control systems theory are so diverse and numerous that it would take several more chapters to describe them in any detail. In addition to the ubiquitous servomechanism, the authors have seen modern and classical control applications described in all of the following areas:

Aerospace systems	Man-machine systems
Chemical processes	Ecology
Transportation systems	Air traffic control
Urban systems	Agriculture
Vehicular traffic	Electric power systems
Municipal services	Nuclear reactors
Industrial processes	Ocean surveillance
Economics	Digital systems
Physiology	Navigation
Medicine	Communications
Social systems	Biology

Some interesting applications are described in Refs. 7a, 76, 91, 110, 113, 131, 132, 148, 162, 166, 174, 203b, 211, 212a, 221-224, 228c, 228d, 238a, 248, 253a, 254, 258a, 262, 270, 280a, 283, 286, 294, 315, 320, 342, 343a, 346 and 347, and many more are listed in the bibliography of Ref. 302.

The reader is encouraged to browse further in the literature. Journals in this area include the following:

IEEE Transactions on Automatic Control

SIAM Journal on Control and Optimization

Automatica

International Journal of Control

Information and Control

ASME Journal on Dynamic Systems, Measurement, and Control

IEEE Transactions on Systems, Man, and Cybernetics

Relevant papers may also be found in the proceedings of the *Joint Automatic Control Conference* (annual), the *IEEE Conference on Decision and Control* (annual), the *IFAC Congress* (triennial), and numerous other conferences and symposia. As a rule, however, most conference papers of major significance are subsequently republished in one of the journals.

Material on specific topics may be located using one of the indexing services, such as the *Computer and Control Abstracts* and the *Scientific Citation Index*.

Appendix A

REVIEW OF LINEAR ALGEBRA

A.0. INTRODUCTION

This appendix is intended to serve as a brief review for the reader who is already familiar with vectors and matrices. It is neither rigorous nor complete, and we include only those concepts and results which are essential to the main text. The reader is also referred to Refs. 1, 16, 18, 67, 68, 79, 87, 118, 120, 121, 137a, 142, 151, and 170, or any other good textbook on linear algebra or matrix analysis. Numerical methods are discussed in Refs. 2, 80, 81, 83, 90, 93, 146, and 179, and an excellent introduction to more general vector space issues may be found in Ref. 117.

A.1. LINEAR VECTOR SPACES

1. *DEFINITIONS*. The *linear vector space* of interest in this book is the n-dimensional Euclidean space R^n whose elements are column *vectors* of real numbers, denoted by underlined lowercase letters,

$$\underline{x} = \begin{bmatrix} x_1 \\ x_2 \\ \vdots \\ x_n \end{bmatrix} \qquad \underline{y} = \begin{bmatrix} y_1 \\ y_2 \\ \vdots \\ y_n \end{bmatrix} \tag{1}$$

The *transpose* of a vector $\underline{x}$ is a row vector

$$\underline{x}' = [x_1 \quad x_2 \quad \dots \quad x_n] \tag{2}$$

Addition and *scalar multiplication* are defined on R^n in the obvious way,

$$\alpha\underline{x} + \underline{y} = \alpha \begin{bmatrix} x_1 \\ x_2 \\ \vdots \\ x_n \end{bmatrix} + \begin{bmatrix} y_1 \\ y_2 \\ \vdots \\ y_n \end{bmatrix} = \begin{bmatrix} \alpha x_1 + y_2 \\ \alpha x_2 + y_2 \\ \vdots \\ \alpha x_n + y_n \end{bmatrix} \tag{3}$$

and the *zero vector* is denoted

$$\underline{0} = \begin{bmatrix} 0 \\ 0 \\ \vdots \\ 0 \end{bmatrix} \tag{4}$$

The Euclidean *norm* (or *length*) of a vector $\underline{x}$ is defined as

$$\| \underline{x} \| \triangleq \sqrt{x_1^2 + x_2^2 + \dots + x_n^2} \tag{5}$$

and has the properties[†]

$$\| \underline{x} \| \geq 0, \qquad \text{and} \qquad \| \underline{x} \| = 0 \iff \underline{x} = \underline{0}$$

$$\| \alpha\underline{x} \| = |\alpha| \, \| \underline{x} \| \qquad \text{(for any scalar } \alpha) \tag{6}$$

$$\| \underline{x} + \underline{y} \| \leq \| \underline{x} \| + \| \underline{y} \| \qquad \text{(the } \textit{triangle inequality})$$

A *subspace* S of R^n is any set of vectors which is *closed* under addition and scalar multiplication in the sense that

$$\underline{x}, \underline{y} \in S \implies (\alpha\underline{x} + \underline{y}) \in S \qquad \text{(for any scalar } \alpha) \tag{7}$$

ΔΔΔ

2. *EXAMPLE*. In the plane R^2, the vectors

$$\underline{x} = \begin{bmatrix} 3 \\ 1 \end{bmatrix} \quad \underline{y} = \begin{bmatrix} 1 \\ 2 \end{bmatrix} \quad 2\underline{x} = \begin{bmatrix} 6 \\ 2 \end{bmatrix} \quad 2\underline{x} + \underline{y} = \begin{bmatrix} 7 \\ 4 \end{bmatrix} \tag{8}$$

---

† These three properties are often used to *define* a norm on a linear vector space.

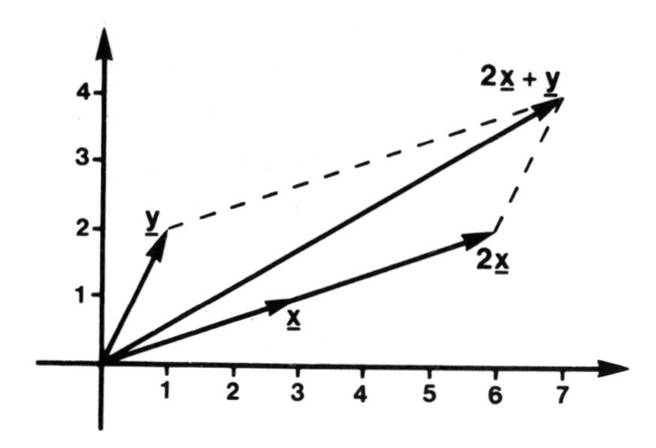

FIG.1. Vectors in R^2.

are as shown in Fig. 1, where the first component of each vector is plotted horizontally and the second component vertically. The norms (lengths) of these vectors are easily computed; for example,

$$\| \underline{x} \| = \sqrt{3^2 + 1^2} = \sqrt{10} = 3.162 \tag{9}$$

Any straight line through the origin constitutes a (one-dimensional) subspace of R^2.

ΔΔΔ

3. *DEFINITION.* A *linear combination* of m vectors in R^n is a sum of the form

$$\sum_{i=1}^{m} \alpha_i \underline{x}_i = \alpha_1 \underline{x}_1 + \alpha_2 \underline{x}_2 + \cdots + \alpha_m \underline{x}_m \tag{10}$$

and it is said to be *nontrivial* if the scalars $\alpha_1, \alpha_2, \ldots, \alpha_n$ are not all zero. A collection of vectors $\{\underline{x}_1, \underline{x}_2, \ldots, \underline{x}_m\}$ is said to be *linearly independent* if there exists no nontrivial linear combination of them with

$$\sum_{i=1}^{m} \alpha_i \underline{x}_i = 0 \tag{11}$$

Otherwise, they are said to be *linearly dependent* and at least one of them may be expressed as a linear combination of the others (see Problem 2).

ΔΔΔ

4. *DEFINITION*. The *subspace spanned by* a set of vectors $\{\underline{x}_1, \underline{x}_2, \ldots, \underline{x}_m\}$ in R^n consists of all linear combinations of the vectors and is denoted $\text{span}\{\underline{x}_1, \underline{x}_2, \ldots, \underline{x}_m\}$. It is a linear vector space in its own right, contained in R^n. If the set $\{\underline{x}_1, \underline{x}_2, \ldots, \underline{x}_m\}$ is linearly independent, then it is said to be a *basis* for the subspace and the number of elements in the basis, m, is the *dimension* of the subspace.

Any set of n linearly independent vectors is a basis for the space R^n, and the *standard basis* is

$$\underline{e}_1 = \begin{bmatrix} 1 \\ 0 \\ 0 \\ \vdots \\ 0 \end{bmatrix} \qquad \underline{e}_2 = \begin{bmatrix} 0 \\ 1 \\ 0 \\ \vdots \\ 0 \end{bmatrix} \qquad \cdots \qquad \underline{e}_n = \begin{bmatrix} 0 \\ 0 \\ 0 \\ \vdots \\ 1 \end{bmatrix} \tag{12}$$

ΔΔΔ

5. *DEFINITION*. A vector which is a function of time is denoted

$$\underline{x}(t) = \begin{bmatrix} x_1(t) \\ x_2(t) \\ \vdots \\ x_n(t) \end{bmatrix} \tag{13}$$

and its derivative is

$$\frac{d}{dt}\underline{x}(t) = \dot{\underline{x}}(t) = \begin{bmatrix} \dot{x}_1(t) \\ \dot{x}_2(t) \\ \vdots \\ \dot{x}_n(t) \end{bmatrix} \tag{14}$$

ΔΔΔ

A.2. LINEAR TRANSFORMATIONS AND MATRICES

1. *DEFINITIONS.* An $m \times n$ *matrix* is an array of mn real numbers denoted by an underlined capital letter,

$$\underline{A} = \begin{bmatrix} a_{11} & a_{12} & \cdots & a_{1n} \\ a_{21} & a_{22} & \cdots & a_{2n} \\ \cdots & \cdots & \cdots & \cdots \\ a_{m1} & a_{m2} & \cdots & a_{mn} \end{bmatrix} = [\, \underline{a}_1 \;\; \underline{a}_2 \;\; \cdots \;\; \underline{a}_n \,] \tag{1}$$

where the vectors $\underline{a}_1, \underline{a}_2, \ldots, \underline{a}_n$ are its *columns*, and its *transponse* is the $n \times m$ matrix

$$\underline{A}' = \begin{bmatrix} a_{11} & a_{21} & \cdots & a_{m1} \\ a_{12} & a_{22} & \cdots & a_{m2} \\ \cdots & \cdots & \cdots & \cdots \\ a_{1n} & a_{2n} & \cdots & a_{mn} \end{bmatrix} = \begin{bmatrix} \underline{a}_1' \\ \underline{a}_2' \\ \vdots \\ \underline{a}_n' \end{bmatrix} \tag{2}$$

with rows $\underline{a}_1', \underline{a}_2', \ldots, \underline{a}_n'$. Note that an $m \times 1$ matrix (such as $\underline{a}_1$) may also be considered a vector in R^m. A square matrix $\underline{A}$ is said to be *symmetric* if $\underline{A} = \underline{A}'$. Addition and scalar multiplication are defined as for vectors, and the *zero matrix* is denoted $\underline{0}$. The *subspace spanned* by the columns of $\underline{A}$ is denoted span($\underline{A}$).

A *partitioned matrix* is one written in the form

$$\underline{A} = \begin{bmatrix} \underline{A}_{11} & \underline{A}_{12} & \cdots & \underline{A}_{1p} \\ \underline{A}_{21} & \underline{A}_{22} & \cdots & \underline{A}_{2p} \\ \cdots & \cdots & \cdots & \cdots \\ \underline{A}_{m1} & \underline{A}_{m2} & \cdots & \underline{A}_{mp} \end{bmatrix} \tag{3}$$

where the *blocks* (or *submatrices*) $\underline{A}_{ij}$ have appropriate dimensions.

ΔΔΔ

2. *DEFINITIONS. Multiplication* of an $m \times n$ matrix $\underline{A}$ by an $n \times p$ matrix $\underline{B}$ results in an $m \times p$ product matrix $\underline{C}$ whose elements are

$$c_{ij} = \sum_{k=1}^{n} a_{ik}b_{kj} \qquad (i = 1, 2, \ldots, m \,;\; j = 1, 2, \ldots, p) \tag{4}$$

In the case of a row vector ($1 \times n$ matrix) $\underline{x}'$ and a column vector ($n \times 1$ matrix) $\underline{y}$ the product is

$$\underline{x}'\underline{y} = \begin{bmatrix} x_1 & x_2 & \cdots & x_n \end{bmatrix} \begin{bmatrix} y_1 \\ y_2 \\ \vdots \\ y_n \end{bmatrix} = x_1y_1 + x_2y_2 + \cdots + x_ny_n \tag{5}$$

and it is convenient to think of the matrix product (4) in terms of rows and columns, that is,

$$\underline{A}\underline{B} = \begin{bmatrix} \underline{\alpha}_1' \\ \underline{\alpha}_2' \\ \vdots \\ \underline{\alpha}_m' \end{bmatrix} \begin{bmatrix} \underline{b}_1 & \underline{b}_2 & \cdots & \underline{b}_p \end{bmatrix} = \begin{bmatrix} \underline{\alpha}_1'\underline{b}_1 & \underline{\alpha}_1'\underline{b}_2 & \cdots & \underline{\alpha}_1'\underline{b}_p \\ \underline{\alpha}_2'\underline{b}_1 & \underline{\alpha}_2'\underline{b}_2 & \cdots & \underline{\alpha}_2'\underline{b}_p \\ \cdots & \cdots & \cdots & \cdots \\ \underline{\alpha}_m'\underline{b}_1 & \underline{\alpha}_m'\underline{b}_2 & \cdots & \underline{\alpha}_m'\underline{b}_p \end{bmatrix} \tag{6}$$

Two matrices $\underline{A}$ and $\underline{B}$ are said to *commute* if $\underline{A}\underline{B} = \underline{B}\underline{A}$. Note also that $(\underline{A}\underline{B})' = \underline{B}'\underline{A}'$. *Powers* of a square matrix are defined as $\underline{A}^2 = \underline{A}\underline{A}$, $\underline{A}^3 = \underline{A}\underline{A}\underline{A}$, etc.

ΔΔΔ

3. *DEFINITION*. Multiplication of a vector $\underline{x}$ in R^n by an $m \times n$ matrix $\underline{A}$ produces a vector

$$\underline{y} = \underline{A}\underline{x} = \begin{bmatrix} \underline{\alpha}_1' \\ \underline{\alpha}_2' \\ \vdots \\ \underline{\alpha}_m' \end{bmatrix} \underline{x} = \begin{bmatrix} \underline{\alpha}_1'\underline{x} \\ \underline{\alpha}_2'\underline{x} \\ \vdots \\ \underline{\alpha}_n'\underline{x} \end{bmatrix} \tag{7}$$

in R^m, and this operation is known as a *linear transformation* from R^n into R^m, denoted $\underline{A} : R^n \to R^m$. It is also convenient to visualize $\underline{y}$ as a linear combination of the columns of $\underline{A}$,

$$\underline{y} = \underline{A}\underline{x} = \begin{bmatrix} \underline{a}_1 & \underline{a}_2 & \cdots & \underline{a}_n \end{bmatrix} \begin{bmatrix} x_1 \\ x_2 \\ \vdots \\ x_n \end{bmatrix} = x_1\underline{a}_1 + x_2\underline{a}_2 + \dots + x_n\underline{a}_n \tag{8}$$

The transpose of $\underline{A}$ is another transformation $\underline{A}' : R^m \to R^n$. If $\underline{B} : R^p \to R^n$ is an $n \times p$ matrix, then its product with $\underline{A}$ is $\underline{A}\underline{B}: R^p \to R^m$.

ΔΔΔ

The $n \times n$ *identity matrix* is

$$\underline{I} \triangleq \begin{bmatrix} 1 & 0 & \dots & 0 \\ 0 & 1 & \dots & 0 \\ \dots & \dots & \dots & \dots \\ 0 & 0 & \dots & 1 \end{bmatrix} \tag{9}$$

and it is clear that this transformation from R^n into R^n leaves $\underline{x}$ unchanged. The notation $\underline{I}_n$ will occasionally be used to explicity indicate its dimension.

Square $(n \times n)$ matrices of the form

$$\begin{bmatrix} a_{11} & 0 & 0 & \dots & 0 \\ 0 & a_{22} & 0 & \dots & 0 \\ 0 & 0 & a_{33} & \dots & 0 \\ \dots & \dots & \dots & \dots & \dots \\ 0 & 0 & 0 & \dots & a_{nn} \end{bmatrix} \qquad \begin{bmatrix} a_{11} & a_{12} & a_{13} & \cdots & a_{1n} \\ 0 & a_{22} & a_{23} & \cdots & a_{2n} \\ 0 & 0 & a_{33} & \dots & a_{3n} \\ \dots & \dots & \dots & \dots & \dots \\ 0 & 0 & 0 & \dots & a_{nn} \end{bmatrix} \tag{10}$$

are called *diagonal* and *(upper) triangular*, respectively. *Block-diagonal* and *block-triangular* matrices are defined analogously.

4. *DEFINITION.* An $n \times n$ matrix $\underline{A}$ is said to be *invertible* or *non-singular* if its $n \times n$ *inverse* $\underline{A}^{-1}$ exists such that

$$\underline{A}\underline{A}^{-1} = \underline{A}^{-1}\underline{A} = \underline{I} \tag{11}$$

Otherwise, it is said to be *singular.* Note that

$$(\underline{A}\underline{B})^{-1} = \underline{B}^{-1}\underline{A}^{-1} \tag{12}$$

whenever $\underline{A}$ and $\underline{B}$ are both invertible.

ΔΔΔ

The reader should recall the following standard results in linear algebra.

5. *THEOREM.* The $n \times n$ matrix $\underline{A}$ in (1) is invertible if and only if $\det \underline{A} \neq 0$, in which case

$$\underline{A}^{-1} = \frac{\underline{C}'}{\det \underline{A}} \tag{13}$$

where $\det \underline{A}$ is the *determinant* of $\underline{A}$ and the elements c_{ij} of the matrix† $\underline{C}$ are the *cofactors* of A. These may be computed (recursively) as follows:

$$\begin{aligned} \det \underline{A} &= a_{11}c_{11} + a_{12}c_{12} + \cdots + a_{1n}c_{1n} \\ &= a_{11}c_{11} + a_{21}c_{21} + \cdots + a_{n1}c_{n1} \end{aligned} \tag{14}$$

$$c_{ij} = (-1)^{i+j} \det \underline{A}^{ij} \tag{15}$$

where $\underline{A}^{ij}$ is the $(n-1) \times (n-1)$ matrix formed by deleting the i-th row and j-th column from $\underline{A}$. The determinant of a scalar (1×1 matrix) is defined as the scalar itself.

Proof: See any elementary linear algebra text.

ΔΔΔ

---

†$\underline{C}'$ is sometimes called the *adjoint* or *adjugate* matrix of $\underline{A}$.

6. *THEOREM.* The following statements are all equivalent for an $n \times n$ matrix $\underline{A}$:

(a) $\underline{A}$ is singular.

(b) $\underline{A}'$ is singular.

(c) The columns of $\underline{A}$ are linearly dependent.

(d) The rows of $\underline{A}$ are linearly dependent.

(e) There exists $\underline{x} \neq \underline{0}$ in R^n such that $\underline{A}\underline{x} = \underline{0}$.

Proof: Left to the reader (Problem 5).

ΔΔΔ

7. *THEOREM.* The determinant function obeys the following rules, where $\underline{A}$ and $\underline{B}$ are $n \times n$ matrices:

(a) $\det \underline{A}\underline{B} = \det \underline{A} \det \underline{B}$ (16)

(b) $\det \underline{A}^{-1} = 1/\det \underline{A}$ (17)

(c) $\det k\underline{A} = k^n \det \underline{A}$ (18)

(d) $\det \underline{A} = \det \underline{A}'$ (19)

(e) The determinant of a diagonal or triangular matrix is the product of the diagonal elements.

(f) The determinant of a block-diagonal or block-triangular matrix is the product of the determinants of the diagonal blocks.

Proof: Left to the reader (Problem 5).

ΔΔΔ

8. *THEOREM.* The inverse of a nonsingular partitioned matrix,

$$\begin{bmatrix} \underline{A} & \underline{B} \\ \underline{C} & \underline{D} \end{bmatrix}^{-1} = \begin{bmatrix} \underline{E} & \underline{F} \\ \underline{G} & \underline{H} \end{bmatrix} \tag{20}$$

is specified by

$$\underline{E} = \underline{A}^{-1} + \underline{A}^{-1}\underline{BHCA}^{-1} = (\underline{A} - \underline{BD}^{-1}\underline{C})^{-1} \tag{21}$$

$$\underline{F} = -\underline{A}^{-1}\underline{BH} = -\underline{EBD}^{-1} \tag{22}$$

$$\underline{G} = -\underline{HCA}^{-1} = -\underline{D}^{-1}\underline{CE} \tag{23}$$

$$\underline{H} = (\underline{D} - \underline{CA}^{-1}\underline{B})^{-1} = \underline{D}^{-1} + \underline{D}^{-1}\underline{CEBD}^{-1} \tag{24}$$

provided the necessary inverses exist† (which they may not). The formula (21) is a well-known matrix identity.

The determinant of a partitioned matrix is

$$\det \begin{bmatrix} \underline{A} & \underline{B} \\ \underline{C} & \underline{D} \end{bmatrix} = \det \underline{D} \det(\underline{A} - \underline{BD}^{-1}\underline{C}) = \det \underline{A} \det(\underline{D} - \underline{CA}^{-1}\underline{B}) \tag{25}$$

again, provided either $\underline{A}^{-1}$ or $\underline{D}^{-1}$ exists.

Proof: The reader may verify the two symmetric sets of equations in (21)-(24) by substituting one set or the other into (20) and then multiplying by the original partitioned matrix (see Problem 10). Similarly, the two forms of (25) follow by multiplying the partitioned matrix on the left by

$$\begin{bmatrix} \underline{I} & -\underline{BD}^{-1} \\ \underline{0} & \underline{I} \end{bmatrix} \quad \text{and} \quad \begin{bmatrix} \underline{I} & \underline{0} \\ -\underline{CA}^{-1} & \underline{I} \end{bmatrix} \tag{26}$$

respectively.

ΔΔΔ

9. *DEFINITION.* Suppose $\underline{A}$, $\underline{P}$, and $\underline{P}^{-1}$ are all $n \times n$ matrices. Then the matrix

$$\hat{\underline{A}} = \underline{P}^{-1}\underline{AP} \tag{27}$$

is said to be *similar* to $\underline{A}$, and the operation yielding $\hat{\underline{A}}$ is known as a *similarity transformation.*††

ΔΔΔ

†That is, either both $\underline{A}^{-1}$ and $(\underline{D} - \underline{CA}^{-1}\underline{B})^{-1}$ exist, or both $\underline{D}^{-1}$ and $(\underline{A} - \underline{BD}^{-1}\underline{C})^{-1}$ exist.

††$\underline{A}$ and $\hat{\underline{A}}$ may be viewed as representations of the same linear transformation in different bases. In the main text, this is used in the context of a *state transformation*.

The *derivative* of a time-varying $m \times n$ matrix $\underline{A}(t)$ is defined as in (A.1-14),

$$\frac{d}{dt}\underline{A}(t) = \dot{\underline{A}}(t) = \begin{bmatrix} \dot{a}_{11}(t) & \dot{a}_{12}(t) & \cdots & \dot{a}_{1n}(t) \\ \dot{a}_{21}(t) & \dot{a}_{22}(t) & \cdots & \dot{a}_{2n}(t) \\ \cdots & \cdots & \cdots & \cdots \\ \dot{a}_{m1}(t) & \dot{a}_{m2}(t) & \cdots & \dot{a}_{mn}(t) \end{bmatrix} \tag{28}$$

The usual rule for differentiating products is easily extended to the matrix (or vector) case as follows.

10. THEOREM. If $\underline{A}(t)$ and $\underline{B}(t)$ are time-varying $m \times n$ and $n \times p$ matrices, respectively, and

$$\underline{C}(t) = \underline{A}(t)\underline{B}(t) \tag{29}$$

is their $m \times p$ product, then

$$\dot{\underline{C}}(t) = \frac{d}{dt}\underline{A}(t)\underline{B}(t) = \dot{\underline{A}}(t)\underline{B}(t) + \underline{A}(t)\dot{\underline{B}}(t) \tag{30}$$

Proof: According to (4),

$$c_{ij}(t) = \sum_{k=1}^{n} a_{ik}(t)b_{kj}(t) \tag{31}$$

Using the product rule for scalar functions, we may differentiate this to yield

$$\begin{aligned} \dot{c}_{ij}(t) &= \sum_{k=1}^{n} [\dot{a}_{ik}(t)b_{kj}(t) + a_{ik}(t)\dot{b}_{kj}(t)] \\ &= \sum_{k=1}^{n} \dot{a}_{ik}(t)b_{kj}(t) + \sum_{k=1}^{n} a_{ik}(t)\dot{b}_{kj}(t) \end{aligned} \tag{32}$$

which establishes (30).

ΔΔΔ

One other useful definition is the following (see Problems A.2-14 and A.3-9).

11. DEFINITION. The *trace* of an $n \times n$ matrix $\underline{A}$ is the sum of its diagonal elements,

$$\mathrm{tr}\ \underline{A} \triangleq \sum_{i=1}^{n} a_{ii} \qquad (33)$$

ΔΔΔ

A.3. SPECTRAL THEORY

1. DEFINITION. If $\underline{A}$ is an $n \times n$ matrix, a complex number λ is said to be an *eigenvalue* of $\underline{A}$ if there exists a complex† *eigenvector* $\underline{u} \neq \underline{0}$ such that

$$\underline{A}\underline{u} = \lambda \underline{u} \qquad \text{or} \qquad (\lambda \underline{I} - \underline{A})\underline{u} = \underline{0} \qquad (1)$$

Since the eigenvalues of $\underline{A}$ are those complex numbers which make $(\lambda \underline{I} - \underline{A})$ singular, they must be the roots of the *characteristic equation*

$$\det(\lambda \underline{I} - \underline{A}) = 0 \qquad (2)$$

ΔΔΔ

It is easily verified (Problem 3) that the left-hand side of (2) is an n-th order *characteristic polynomial* in λ,

$$\begin{aligned}\det(\lambda \underline{I} - \underline{A}) &= \lambda^n + \alpha_{n-1}\lambda^{n-1} + \dots + \alpha_1 \lambda + \alpha_0 \\ &= (\lambda - \lambda_1)(\lambda - \lambda_2) \dots (\lambda - \lambda_n) \qquad (3)\end{aligned}$$

whose n complex roots†† are the eigenvalues $\lambda_1, \lambda_2, \dots \lambda_n$. It is also easy to show (Problem 2) that a matrix is singular if and only if it has one or more eigenvalues equal to zero.

---

†When dealing with eigenvalues, we tacitly assume that R^n has been replaced by the vector space C^n of complex-valued n-vectors.

††By a slight abuse of terminology, we refer interchangeably to the roots of a *polynomial* $a(\lambda)$ and the roots of the *polynomial equation* $a(\lambda) = 0$.

Next we shall see that the eigenvalues of a matrix are unchanged when the matrix undergoes a similarity transformation, and that the matrix may be *diagonalized* by such a transformation.

2. *THEOREM.* The eigenvalues of a matrix are *invariant* under a similarity transformation, that is, the matrices $\underline{A}$ and $\underline{QAQ}^{-1}$ have the same set of eigenvalues for any nonsingular $\underline{Q}$.

Proof: Left to the reader (Problem 5).

ΔΔΔ

3. *THEOREM.* Suppose $\underline{A}$ has eigenvalues $\lambda_1, \lambda_2, \ldots, \lambda_n$ and a linearly independent† set of eigenvectors $\underline{u}_1, \underline{u}_2, \ldots, \underline{u}_n$. Then the matrix of eigenvectors,

$$\underline{U} = [\underline{u}_1 \quad \underline{u}_2 \quad \cdots \quad \underline{u}_n] \tag{4}$$

is nonsingular and may be used to put $\underline{A}$ in *diagonal form,*

$$\underline{U}^{-1}\underline{AU} = \underline{\Lambda} = \begin{bmatrix} \lambda_1 & 0 & \cdots & 0 \\ 0 & \lambda_2 & \cdots & 0 \\ \cdots & \cdots & \cdots & \cdots \\ 0 & 0 & \cdots & \lambda_n \end{bmatrix} \tag{5}$$

Proof: By the definition (1) we have

$$\underline{AU} = [\lambda_1\underline{u}_1 \quad \lambda_2\underline{u}_2 \quad \cdots \quad \lambda_n\underline{u}_n] \tag{6}$$

and $\underline{U}$ is invertible because the eigenvectors are linearly independent. Thus $\underline{U}^{-1}\underline{U} = \underline{I}$, and (A.2-6) implies that

$$\underline{v}_i'\underline{u}_j = \delta_{ij} = \begin{cases} 1 & (i = j) \\ 0 & (i \neq j) \end{cases} \tag{7}$$

---

†If the eigenvalues are all *distinct*, the eigenvectors must be linearly independent (see Problem 4).

where $\underline{v}_i'$ are the rows of $\underline{U}^{-1}$. Using (A.2-6) again, we see that the elements of $\underline{U}^{-1}\underline{A}\underline{U}$ are $\lambda_j \delta_{ij}$.

ΔΔΔ

4. THEOREM. Suppose $\underline{A}$ has eigenvalues λ_i, i = 1, 2, ..., p, each of which is repeated μ_i times, where $\sum_{i=1}^{p} \mu_i = n$. Then there exists a matrix $\underline{U}$ which may be used to put $\underline{A}$ in *Jordan form*,

$$\underline{U}^{-1}\underline{A}\underline{U} = \underline{\Lambda} = \begin{bmatrix} \underline{J}_1 & \underline{0} & \cdots & \underline{0} \\ \underline{0} & \underline{J}_2 & \cdots & \underline{0} \\ \cdots & \cdots & \cdots & \cdots \\ \underline{0} & \underline{0} & \cdots & \underline{J}_p \end{bmatrix} \tag{8}$$

where the diagonal blocks are $\mu_i \times \mu_i$ and have the form†

$$\underline{J}_i = \left[\begin{array}{ccccc|ccccc} \lambda_i & 1 & 0 & \cdots & 0 & & & & & \\ 0 & \lambda_i & 1 & \cdots & 0 & & & & & \\ \cdots & \cdots & \cdots & \cdots & \cdots & & & \underline{0} & & \\ 0 & 0 & 0 & \cdots & 1 & & & & & \\ 0 & 0 & 0 & \cdots & \lambda_i & & & & & \\ \hline & & & & & \lambda_i & 0 & 0 & \cdots & 0 \\ & & & & & 0 & \lambda_i & 0 & \cdots & 0 \\ & & \underline{0} & & & \cdots & \cdots & \cdots & \cdots & \cdots \\ & & & & & 0 & 0 & 0 & \cdots & 0 \\ & & & & & 0 & 0 & 0 & \cdots & \lambda_i \end{array}\right] \tag{9}$$

(the left block columns span η_i; the right block columns span $\mu_i - \eta_i$)

for i = 1, 2, ..., p. When $\mu_i = 1$, $\underline{J}_i$ reduces to a scalar (1 × 1 matrix) λ_i.

Proof: The proof may be found in various texts [67, 68, 87, 138, 185]. In Ref. 39, it is shown that the columns of $\underline{U}$ form a set of *generalized eigenvectors*.

ΔΔΔ

---

†If the eigenvectors are linearly independent, then $\eta_i = 0$ in all cases, and (8) reduces to (5).

5. *EXAMPLE*. If a 4×4 matrix has eigenvalues 2, 2, 2, and 3, its Jordan form will be

$$\begin{bmatrix} 2 & 1 & 0 & 0 \\ 0 & 2 & 1 & 0 \\ 0 & 0 & 2 & 0 \\ 0 & 0 & 0 & 3 \end{bmatrix} \quad \text{or} \quad \begin{bmatrix} 2 & 1 & 0 & 0 \\ 0 & 2 & 0 & 0 \\ 0 & 0 & 2 & 0 \\ 0 & 0 & 0 & 3 \end{bmatrix} \quad \text{or} \quad \begin{bmatrix} 2 & 0 & 0 & 0 \\ 0 & 2 & 0 & 0 \\ 0 & 0 & 2 & 0 \\ 0 & 0 & 0 & 3 \end{bmatrix}$$

depending upon whether it has 2, 3, or 4 linearly dependent eigenvectors. ΔΔΔ

The reader will note that because a real matrix $\underline{A}$ may have complex eigenvalues, the matrices $\underline{U}$, $\underline{U}^{-1}$, and $\underline{\Lambda}$ defined above will have complex entries in general. It will be convenient to derive a diagonal form which avoids complex numbers entirely by separating their real and imaginary parts.

Because $\underline{A}$ is a real matrix, the characteristic polynomial (3) has real coefficients and any complex roots must occur in conjugate pairs. That is, if

$$\lambda_i = \sigma_i + j\omega_i \qquad (\omega_i \neq 0) \tag{10}$$

is an eigenvalue of $\underline{A}$, then its complex conjugate

$$\lambda_{i+1} = \overline{\lambda}_i = \sigma_i - j\omega_i \tag{11}$$

is also[†] an eigenvalue of $\underline{A}$. Moreover, it is easy to establish that if

$$\underline{u}_i = \underline{v}_i + j\underline{w}_i \tag{12}$$

is the eigenvector associated with λ_i, then

$$\underline{u}_{i+1} = \overline{\underline{u}}_i = \underline{v}_i - j\underline{w}_i \tag{13}$$

is the eigenvector associated with $\lambda_{i+1} = \overline{\lambda}_i$. With these facts at hand, we can state the following theorems.

[†] We assume consecutive numbering of conjugate pairs of eigenvalues for convenience.

6. *THEOREM*. Suppose $\underline{A}$ has eigenvalues

$$\left.\begin{aligned} \lambda_i &= \sigma_i + j\omega_i \\ \lambda_{i+1} &= \sigma_i - j\omega_i = \overline{\lambda}_i \end{aligned}\right\} \quad (i = 1, 3, 5, \ldots, m-1)$$
$$\lambda_i = \overline{\lambda}_i \qquad (i = m+1, m+2, \ldots, n) \tag{14}$$

and a linearly independent[†] set of eigenvectors

$$\left.\begin{aligned} \underline{u}_i &= \underline{v}_i + j\underline{w}_i \\ \underline{u}_{i+1} &= \underline{v}_i - j\underline{w}_i = \overline{\underline{u}}_i \end{aligned}\right\} \quad (i = 1, 3, 5, \ldots, m-1)$$
$$\underline{u}_i = \overline{\underline{u}}_i \qquad (i = m+1, m+2, \ldots, n) \tag{15}$$

Then the real-valued matrix

$$\underline{U} = \begin{bmatrix} \underline{v}_1 & \underline{w}_1 & \underline{v}_3 & \underline{w}_3 & \cdots & \underline{v}_{m-1} & \underline{w}_{m-1} & \underline{u}_{m+1} & \underline{u}_{m+2} & \cdots & \underline{u}_n \end{bmatrix} \tag{16}$$

is nonsingular and may be used to put $\underline{A}$ in *block-diagonal form*,

$$\underline{U}^{-1}\underline{A}\underline{U} = \underline{\Lambda} = \begin{bmatrix} \underline{\Lambda}_1 & \underline{0} & \cdots & \underline{0} & \underline{0} \\ \underline{0} & \underline{\Lambda}_3 & \cdots & \underline{0} & \underline{0} \\ \cdots & \cdots & \cdots & \cdots & \cdots \\ \underline{0} & \underline{0} & \cdots & \underline{\Lambda}_{m-1} & \underline{0} \\ \underline{0} & \underline{0} & \cdots & \underline{0} & \underline{\Lambda}_{m+1} \end{bmatrix} \tag{17}$$

where

$$\underline{\Lambda}_i = \begin{bmatrix} \sigma_i & \omega_i \\ -\omega_i & \sigma_i \end{bmatrix} \qquad (i = 1, 3, 5, \ldots, m-1) \tag{18}$$

and

$$\underline{\Lambda}_{m+1} = \begin{bmatrix} \lambda_{m+1} & 0 & \cdots & 0 \\ 0 & \lambda_{m+2} & \cdots & 0 \\ \cdots & \cdots & \cdots & \cdots \\ 0 & 0 & \cdots & \lambda_n \end{bmatrix} \tag{19}$$

[†]If the eigenvalues are all *distinct*, the eigenvectors must be linearly independent (see Problem 4).

This result also holds for nondistinct eigenvalues, provided the eigenvectors are linearly independent.

Proof: Left to the reader (Problem 7).

ΔΔΔ

7. *THEOREM.* Suppose $\underline{A}$ has eigenvalues

$$\begin{aligned} &\lambda_i = \sigma_i + j\omega_i && (i = 1, 3, 5, \ldots, m-1) \\ &\lambda_{i+1} = \sigma_i - j\omega_i = \overline{\lambda}_i && \\ &\lambda_i = \overline{\lambda}_i && (i = m+1,\ m+2,\ \ldots,\ p) \end{aligned} \tag{20}$$

each of which is repeated μ_i times, where $\sum_{i=1}^{p} \mu_i = n$. Then there exists a matrix $\underline{U}$ which may be used to put $\underline{A}$ in *block-Jordan form,*

$$\underline{U}^{-1}\underline{A}\underline{U} = \underline{\Lambda} = \begin{bmatrix} \underline{J}_1 & \underline{0} & \cdots & \underline{0} & \underline{0} & \cdots & \underline{0} \\ \underline{0} & \underline{J}_3 & \cdots & \underline{0} & \underline{0} & \cdots & \underline{0} \\ \cdots & \cdots & \cdots & \cdots & \cdots & \cdots & \cdots \\ \underline{0} & \underline{0} & \cdots & \underline{J}_{m-1} & \underline{0} & \cdots & \underline{0} \\ \underline{0} & \underline{0} & \cdots & \underline{0} & \underline{J}_{m+1} & \cdots & \underline{0} \\ \cdots & \cdots & \cdots & \cdots & \cdots & \cdots & \cdots \\ \underline{0} & \underline{0} & \cdots & \underline{0} & \underline{0} & \cdots & \underline{J}_p \end{bmatrix} \tag{21}$$

For $i = 1, 3, 5, \ldots, m-1$, the $\mu_i \times \mu_i$ diagonal blocks have the form[†]

$$\underline{J}_i = \left[\begin{array}{c|c} \overbrace{\begin{matrix} \underline{\Lambda}_i & \underline{I} & \underline{0} & \cdots & \underline{0} \\ \underline{0} & \underline{\Lambda}_i & \underline{I} & \cdots & \underline{0} \\ \cdots & \cdots & \cdots & \cdots & \cdots \\ \underline{0} & \underline{0} & \underline{0} & \cdots & \underline{I} \\ \underline{0} & \underline{0} & \underline{0} & \cdots & \underline{\Lambda}_i \end{matrix}}^{2\eta_i} & \overbrace{\quad \underline{0} \quad}^{2(\mu_i - \eta_i)} \\ \hline \underline{0} & \begin{matrix} \underline{\Lambda}_i & \underline{0} & \underline{0} & \cdots & \underline{0} \\ \underline{0} & \underline{\Lambda}_i & \underline{0} & \cdots & \underline{0} \\ \cdots & \cdots & \cdots & \cdots & \cdots \\ \underline{0} & \underline{0} & \underline{0} & \cdots & \underline{0} \\ \underline{0} & \underline{0} & \underline{0} & \cdots & \underline{\Lambda}_i \end{matrix} \end{array}\right] \tag{22}$$

[†] If the eigenvectors are linearly independent, then $\eta_i = 0$ in all cases, and (21) reduces to (17).

where $\mathbf{\Lambda}_i$ was defined in (18). When $\mu_i = 1$, $\mathbf{J}_i$ reduces to a 2×2 matrix $\mathbf{\Lambda}_i$. For $i = m+1, m+2, \ldots, p$, the diagonal blocks $\mathbf{J}_i$ are as in (9).

Proof: The proof is a straightforward but complicated extension of the proofs of Theorems 4 and 6. It is left as an exercise for the interested reader. An alternative form may be found in Ref. 39, p. 232.

ΔΔΔ

8. EXAMPLE. If a 5×5 matrix has eigenvalues $1 + j2$, $1 - j2$, $1 + j2$, $1 - j2$, and 3, its block-Jordan form will be

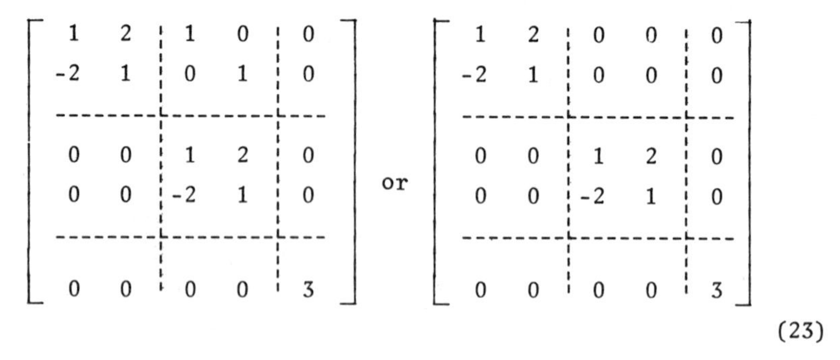

$$\left[\begin{array}{cc|cc|c} 1 & 2 & 1 & 0 & 0 \\ -2 & 1 & 0 & 1 & 0 \\ \hline 0 & 0 & 1 & 2 & 0 \\ 0 & 0 & -2 & 1 & 0 \\ \hline 0 & 0 & 0 & 0 & 3 \end{array}\right] \quad \text{or} \quad \left[\begin{array}{cc|cc|c} 1 & 2 & 0 & 0 & 0 \\ -2 & 1 & 0 & 0 & 0 \\ \hline 0 & 0 & 1 & 2 & 0 \\ 0 & 0 & -2 & 1 & 0 \\ \hline 0 & 0 & 0 & 0 & 3 \end{array}\right] \tag{23}$$

depending on whether it has 3 or 5 linearly independent eigenvectors.

ΔΔΔ

The *Cayley-Hamilton theorem* states that a matrix always satisfies its own characteristic equation:

9. THEOREM (Cayley-Hamilton). If $\mathbf{A}$ is an $n \times n$ matrix with characteristic equation

$$\lambda^n + \alpha_{n-1}\lambda^{n-1} + \cdots + \alpha_1\lambda + \alpha_0 = 0 \tag{24}$$

then

$$\mathbf{A}^n + \alpha_{n-1}\mathbf{A}^{n-1} + \cdots + \alpha_1\mathbf{A} + \alpha_0\mathbf{I} = \mathbf{0} \tag{25}$$

Proof: See Problem 4.3-1, or any linear algebra text.

ΔΔΔ

This theorem implies, in particular, that $\underline{A}^m$, $m \geq n$, may be expressed as a linear combination of the matrices $\underline{I} = \underline{A}^0$, $\underline{A}$, $\underline{A}^2$, ..., $\underline{A}^{n-1}$ (see Problem 10).

A.4 GEOMETRIC CONCEPTS

1. DEFINITION. The *inner product* of two vectors $\underline{x}$ and $\underline{y}$ in R^n will be taken here to be

$$\langle \underline{x},\underline{y} \rangle \overset{\Delta}{=} \underline{x}'\underline{y} = x_1y_1 + x_2y_2 + \cdots + x_ny_n \tag{1}$$

so that the norm of a vector $\underline{x}$ is

$$\| \underline{x} \| = \sqrt{\langle \underline{x},\underline{x} \rangle} \tag{2}$$

ΔΔΔ

Three important properties† of the inner product are that it is *symmetric*,

$$\langle \underline{x},\underline{y} \rangle = \langle \underline{y},\underline{x} \rangle \tag{3}$$

linear in each variable,

$$\langle \alpha \underline{x} + \underline{y},\underline{z} \rangle = \alpha\langle \underline{x},\underline{z} \rangle + \langle \underline{y},\underline{z} \rangle \tag{4}$$

and *positive definite,*

$$\langle \underline{x},\underline{x} \rangle = \| \underline{x} \|^2 > 0 \qquad \text{(for } \underline{x} \neq \underline{0}) \tag{5}$$

It also satisfies the *Schwartz Inequality* (Problem 1),

$$|\langle \underline{x},\underline{y} \rangle| \leq \| \underline{x} \| \; \| \underline{y} \| \tag{6}$$

2. DEFINITION. The inner product (1) provides R^n with a geometric structure. We say that two vectors $\underline{x}$ and $\underline{y}$ are *orthogonal* if their inner product is zero, that is,

$$\underline{x} \perp \underline{y} \iff \langle \underline{x},\underline{y} \rangle = 0 \tag{7}$$

---

†These three properties are often used to *define* an inner product on a linear vector space.

Similarly, if S and T are subspaces of $\underline{R}^n$, we say that

$$\underline{x} \perp S \Longleftrightarrow \langle \underline{x}, \underline{y} \rangle = 0 \text{ for all } \underline{y} \in S \tag{8}$$

$$T \perp S \Longleftrightarrow \langle \underline{x}, \underline{y} \rangle = 0 \text{ for all } \underline{y} \in S, \ \underline{x} \in T \tag{9}$$

Note that Pythagoras' Theorem generalizes to n dimensions (Problem 3) as

$$\| \underline{x} + \underline{y} \|^2 = \| \underline{x} \|^2 + \| \underline{y} \|^2 \quad (\text{if } \underline{x} \perp \underline{y}) \tag{10}$$

A set of vectors $\{\underline{u}_1, \underline{u}_2, \ldots, \underline{u}_m\}$ in R^n is said to be *orthonormal* if they are mutually orthogonal and have norm 1, that is,

$$\langle \underline{u}_i, \underline{u}_j \rangle = \delta_{ij} = \begin{cases} 1 & (i = j) \\ 0 & (i \neq j) \end{cases} \tag{11}$$

In particular, the standard basis (A.1-12) for R^n has this property.

ΔΔΔ

3. *DEFINITION.* The vector

$$\underline{P}_{\underline{y}}\underline{x} \triangleq \left(\frac{\underline{y}\underline{y}'}{\| \underline{y} \|^2} \right) \underline{x} = \frac{\langle \underline{y}, \underline{x} \rangle}{\| \underline{y} \|^2} \underline{y} \tag{12}$$

is the *orthogonal projection* of $\underline{x}$ on $\underline{y}$, in the sense that the difference vector is orthogonal to $\underline{y}$,

$$(\underline{x} - \underline{P}_{\underline{y}}\underline{x}) \perp \underline{y} \tag{13}$$

(see Problem 4). This is indicated schematically in Fig. 1, where

$$\alpha = \cos^{-1} \left(\frac{\langle \underline{x}, \underline{y} \rangle}{\| \underline{x} \| \, \| \underline{y} \|} \right) \tag{14}$$

is the generalized *angle* between $\underline{x}$ and $\underline{y}$.

Spaces and subspaces are conveniently defined in terms of orthonormal bases (see Problem 6). If S is the subspace of R^n spanned by an orthonormal set $\{\underline{u}_1, \underline{u}_2, \ldots, \underline{u}_m\}$, the *orthogonal projection* of $\underline{x}$ on S is given by

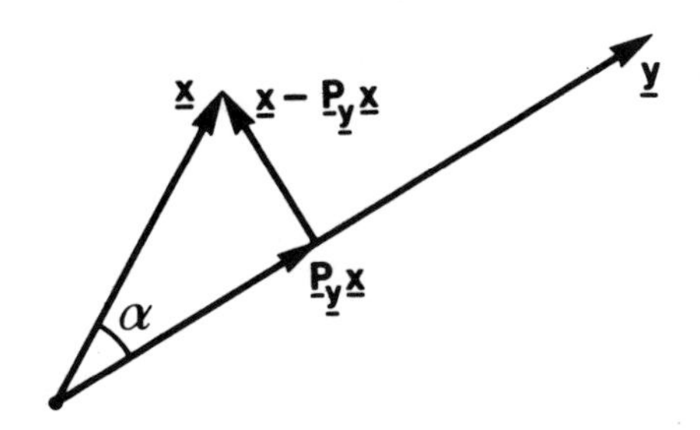

FIG. 1. Projection of $\underline{x}$ on $\underline{y}$.

$$\underline{P}_S \underline{x} = \sum_{i=1}^{m} \underline{P}_{\underline{u}_i} \underline{x} = \sum_{i=1}^{m} \langle \underline{u}_i, \underline{x} \rangle \, \underline{u}_i \tag{15}$$

ΔΔΔ

4. DEFINITION. The *orthogonal complement* of a subspace S of R^n is

$$S^{\perp} \triangleq \{\underline{x} \in R^n : \underline{x} \perp S\} \tag{16}$$

Note that $S^{\perp}$ is itself a subspace of R^n (Problem 7) and that $S \perp S^{\perp}$. The subspaces S and $S^{\perp}$ provide a *decomposition* of R^n in the sense that any $\underline{x} \in R^n$ can be uniquely expressed as

$$\underline{x} = \underline{P}_S \underline{x} + \underline{P}_{S^{\perp}} \underline{x} \triangleq \underline{y} + \underline{z} \tag{17}$$

where $\underline{y} \perp \underline{z}$. Equivalently, R^n is said to be a *direct sum* of S and $S^{\perp}$, denoted

$$R^n = S \oplus S^{\perp} \tag{18}$$

ΔΔΔ

The reader should verify (Problem 8) that the dimensions of S and $S^{\perp}$ sum to n and that only the vector $\underline{0}$ belongs to both S and $S^{\perp}$.

5. EXAMPLE. Consider R^3 with the standard orthonormal basis $\{\underline{e}_1, \underline{e}_2, \underline{e}_3\}$ defined in (A.1-12), as illustrated in Fig. 2. If the subspace S is defined to be all vectors along the vertical axis,

$$S = \{\underline{x} \in R^3 : \underline{x} = \gamma \underline{e}_3 \, , \ \gamma \text{ scalar}\} \tag{19}$$

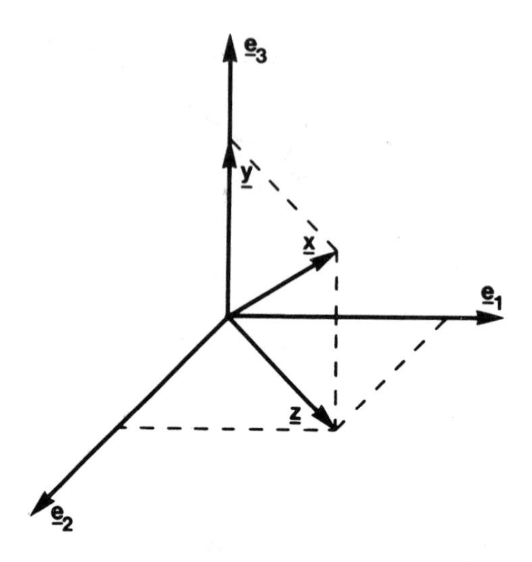

FIG. 2. Decomposition of $\underline{x}$ in R^3.

then its orthogonal complement must consist of all vectors in the horizontal plane,

$$S^{\perp} = \{\underline{x} \in R^3 : \underline{x} = \alpha\underline{e}_1 + \beta\underline{e}_2 \ , \ \alpha,\beta \text{ scalars}\} \tag{20}$$

Thus any vector $\underline{x} \in R^n$ may be decomposed into its projections on S and $S^{\perp}$,

$$\underline{x} = \begin{bmatrix} x_1 \\ x_2 \\ x_3 \end{bmatrix} = \underline{y} + \underline{z} = \begin{bmatrix} 0 \\ 0 \\ x_3 \end{bmatrix} + \begin{bmatrix} x_1 \\ x_2 \\ 0 \end{bmatrix} \tag{21}$$

ΔΔΔ

A.5. STRUCTURE OF LINEAR TRANSFORMATIONS

Recall from Definition A.2.3 that an $m \times n$ matrix $\underline{A}$ constitutes a linear transformation $\underline{A} : R^n \to R^m$, and that its transpose is another transformation $\underline{A}' : R^m \to R^n$. In this section we shall examine the structure of such transformations.

1. DEFINITION. The *range space* (or *image*) of $\underline{A}$ is the subspace of R^m defined by

$$R(\underline{A}) \triangleq \{\underline{y} \in R^m : \underline{y} = \underline{A}\underline{x} \quad \text{for some } \underline{x} \in R^n\} \tag{1}$$

ΔΔΔ

From (A.2-8), $\underline{A}\underline{x}$ is a linear combination of the columns of $\underline{A}$, i.e.,

$$R(\underline{A}) = \text{span}(\underline{A}) \tag{2}$$

The *rank* of $\underline{A}$ is the dimension of $R(\underline{A})$, i.e., the number of linearly independent columns of $\underline{A}$. A matrix is said to have *full rank* if its rank is equal to the smaller of its two dimensions (it can be no greater: see Problem 3). A square matrix has full rank if and only if it is invertible (see Theorem A.2.6).

2. DEFINITION. The *null space* (or *kernel*) of $\underline{A}$ is the subspace of R^n defined by

$$N(A) \triangleq \{\underline{x} \in R^n : \underline{A}\underline{x} = \underline{0}\} \tag{3}$$

ΔΔΔ

Since $N(\underline{A})$ consists of all those vectors which are orthogonal to the rows of $\underline{A}$ (or to the columns of $\underline{A}'$), we have

$$N(\underline{A}) = [\text{span}(\underline{A}')]^{\perp} = R(\underline{A}')^{\perp} \tag{4}$$

The *nullity* of $\underline{A}$ is the dimension of $N(\underline{A})$, and from (4) it follows that

$$\text{rank}(\underline{A}') + \text{nullity}(\underline{A}) = n \tag{5}$$

The reader should verify (Problem 4) that

$$\text{rank}(\underline{A}) + \text{nullity}(\underline{A}) = n \tag{6}$$

and hence that

$$\text{rank}(\underline{A}) = \text{rank}(\underline{A}') \tag{7}$$

Another useful relation is the following.

3. *LEMMA*. Suppose $\underline{A} : R^n \to R^m$, then

$$\mathcal{R}(\underline{A}) = \mathcal{R}(\underline{A}\underline{A}') \tag{8}$$

$$\mathcal{N}(\underline{A}) = \mathcal{N}(\underline{A}'\underline{A}) \tag{9}$$

where $\underline{A}\underline{A}' : \mathcal{R}^m \to \mathcal{R}^m$ and $\underline{A}'\underline{A} : \mathcal{R}^n \to \mathcal{R}^n$.

Proof: If $\underline{y} \in \mathcal{R}(\underline{A}\underline{A}')$, then by definition

$$\underline{y} = \underline{A}\underline{A}'\underline{z} \qquad (\text{for some } \underline{z} \in R^m) \tag{10}$$

Letting $\underline{A}'\underline{z} = \underline{x} \in R^n$, we have $\underline{y} = \underline{A}\underline{x}$ and conclude that $\underline{y} \in \mathcal{R}(\underline{A})$.

Conversely, if $\underline{y} \in \mathcal{R}(\underline{A})$, then

$$\underline{y} = \underline{A}\underline{x} \qquad (\text{for some } \underline{x} \in R^n) \tag{11}$$

From (A.4-15), recall that $\underline{x}$ may be decomposed into $\underline{x} = \underline{x}_1 + \underline{x}_2$, where

$$\underline{x}_1 \in \mathcal{R}(\underline{A}') \implies \underline{x}_1 = \underline{A}'\underline{z} \qquad (\text{for some } \underline{z} \in R^m) \tag{12}$$

$$\underline{x}_2 \in \mathcal{R}(\underline{A}')^{\perp} = \mathcal{N}(\underline{A}) \implies \underline{A}\underline{x}_2 = \underline{0} \tag{13}$$

Thus we have

$$\underline{y} = \underline{A}\underline{x} = \underline{A}\underline{x}_1 + \underline{A}\underline{x}_2 = \underline{A}\underline{A}'\underline{z} + \underline{0} \qquad (\underline{z} \in R^m) \tag{14}$$

and conclude that $\underline{y} \in \mathcal{R}(\underline{A}\underline{A}')$.

The argument above establishes (8), and (9) follows because

$$\mathcal{N}(\underline{A}) = \mathcal{R}(\underline{A}')^{\perp} = \mathcal{R}(\underline{A}'\underline{A})^{\perp} = \mathcal{N}(\underline{A}'\underline{A}) \tag{15}$$

ΔΔΔ

The relations

$$\text{rank}(\underline{A}) = \text{rank}(\underline{A}\underline{A}') = \text{rank}(\underline{A}'\underline{A}) = \text{rank}(\underline{A}') \tag{16}$$

and

$$\text{nullity}(\underline{A}) = \text{nullity}\,(\underline{A}'\underline{A}) \tag{17}$$

follow from this lemma and (7) and provide a test for determining if a matrix $\underline{A}$ has full rank: Compute $\underline{A}\underline{A}'$ or $\underline{A}'\underline{A}$ (whichever is smaller) and test for a nonzero determinant.

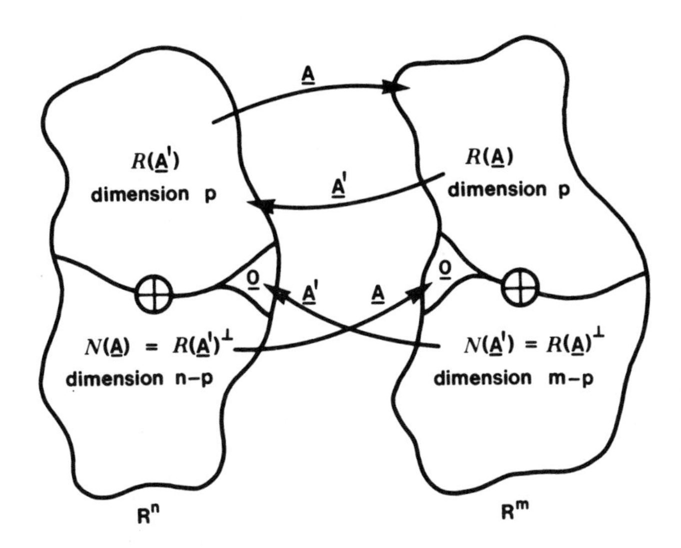

FIG. 1. Decomposition of the linear transformation $\underline{A}$.

The foregoing results are indicated schematically in Fig. 1 for an $m \times n$ matrix $\underline{A}$ with rank p. The spaces R^n and R^m are shown decomposed into

$$R^n = R(\underline{A}') \oplus N(\underline{A}) \tag{18}$$

$$R^m = R(\underline{A}) \oplus N(\underline{A}') \tag{19}$$

as indicated in (4), where the vector $\underline{0}$ belongs to *both* subspaces in each case. The arrows indicate that $\underline{A}$ maps $N(\underline{A})$ into $\underline{0}$ and its orthogonal complement $R(\underline{A}')$ into $R(\underline{A})$. An identical statement holds in the opposite direction for $\underline{A}'$.

We now consider the solution of a linear equation

$$\underline{A}\underline{x} = \underline{b} \tag{20}$$

for an unknown $\underline{x} \in R^n$, where $\underline{b} \in R^m$ and the $m \times n$ matrix $\underline{A}$ are given. In view of the decomposition above, questions of *existence* and *uniqueness* of solutions become transparent:

4. THEOREM.

(a) The linear equation (20) has a solution $\underline{x}$ if and only if

$$\underline{b} \in R(\underline{A}) = R(\underline{AA}') \tag{21}$$

or, equivalently,

$$\underline{b} \perp N(\underline{A}') = N(\underline{AA}') \tag{22}$$

or, equivalently,

$$\text{rank}(\underline{A}) = \text{rank}([\underline{A} \quad \underline{b}]) \tag{23}$$

(b) There is at most one solution if and only if

$$N(\underline{A}) = N(\underline{A}'\underline{A}) = \{\underline{0}\} \tag{24}$$

or, equivalently,

$$\text{rank}(\underline{A}) = n \tag{25}$$

Otherwise, if $\underline{x}$ is a solution, then so is $\underline{x} + \hat{\underline{x}}$ for any $\hat{\underline{x}} \in N(A)$.

(c) There is a solution for each $\underline{b} \in R^n$ if and only if

$$R(\underline{A}) = R(\underline{AA}') = R^m \tag{26}$$

or, equivalently,

$$\text{rank}(\underline{A}) = m \tag{27}$$

(d) There is one and only one solution for each $\underline{b} \in R^m$ if and only if $\underline{A}$ is square and nonsingular, in which case the solution is $\underline{x} = \underline{A}^{-1}\underline{b}$.

Proof:

(a) Condition (21) restates the definition (1); $\underline{Ax}$ must belong to $R(\underline{A})$ for any $\underline{x} \in R^n$. Condition (22) follows by applying (4) to $\underline{A}'$. The relations $R(\underline{A}) = R(\underline{AA}')$ and $N(\underline{A}') = N(\underline{AA}')$ simply reiterate Lemma 3. Condition (23) recognizes that $\underline{b} \in R(\underline{A})$ if and only if $\underline{b}$ is linearly dependent upon the columns of $\underline{A}$.

(b) Now suppose $N(\underline{A})$ contains only the zero vector and there are multiple solutions of (2), that is,

$$\underline{Ax}_1 = \underline{Ax}_2 = \underline{b} \tag{28}$$

Then

$$\underline{Ax}_1 - \underline{Ax}_2 = \underline{A}(\underline{x}_1 - \underline{x}_2) = \underline{0} \tag{29}$$

which means that $(\underline{x}_1 - \underline{x}_2) \in N(\underline{A})$, and we conclude that $\underline{x}_1 = \underline{x}_2$. Conversely, if $\underline{x}$ is a solution of (20) and $N(\underline{A})$ contains a vector $\hat{\underline{x}} \neq 0$, then

$$\underline{A}(\underline{x} + \hat{\underline{x}}) = \underline{A}\underline{x} + \underline{A}\hat{\underline{x}} = \underline{b} + \underline{0} \tag{30}$$

so that $(\underline{x} + \hat{\underline{x}})$ is also a solution of (20). Condition (25) follows from (6).

(c) This follows trivially from (a).

(d) In this case both (b) and (c) must hold, that is, $\text{rank}(\underline{A}) = n = m$, which implies that $\underline{A}$ is square and nonsingular.

ΔΔΔ

When $\underline{A}$ is not an invertible matrix, approximate and exact solutions of (20) may be found by means of a *pseudo-inverse* or *generalized inverse* matrix. This topic is explored in Problems 5-7.

A.6. QUADRATIC FORMS

1. DEFINITION. A *quadratic form* is a scalar function of $\underline{x} \in R^n$ given by

$$f(\underline{x}) = \underline{x}'\underline{Q}\underline{x} \tag{1}$$

where $\underline{Q}$ is a symmetric $n \times n$ matrix (see Problem 1). The matrix (or the quadratic form) is said to be *positive definite* if

$$f(\underline{x}) = \underline{x}'\underline{Q}\underline{x} > 0 \qquad \text{(for all } \underline{x} \neq \underline{0}) \tag{2}$$

and *nonnegative definite* (or *positive semi-definite*) if

$$f(\underline{x}) = \underline{x}'\underline{Q}\underline{x} \geq \underline{0} \qquad \text{(for all } \underline{x}) \tag{3}$$

These two conditions are denoted

$$\underline{Q} > \underline{0} \qquad \underline{Q} \geq \underline{0} \tag{4}$$

respectively.

ΔΔΔ

A quadratic form is generalization of the norm[†] defined in (A.1-5), in that it is a measure of the size of $\underline{x}$, "weighted" by the matrix $\underline{Q}$. If $\underline{Q} > \underline{0}$, then the measure is unambiguous.

[†]Indeed, $\underline{x}'\underline{Q}\underline{x}$ is the norm of the vector $\underline{Q}^{\frac{1}{2}}\underline{x}$ (see Problem 3).

2. THEOREM. The eigenvalues $\lambda_1, \lambda_2, \ldots, \lambda_n$ of a real symmetric matrix $\underline{Q}$ are all real, and there is an *orthogonal* matrix $\underline{P}$ (that is, $\underline{P}^{-1} = \underline{P}'$) which diagonalizes $\underline{Q}$,

$$\underline{PQP}' = \underline{\Lambda} = \begin{bmatrix} \lambda_1 & 0 & \cdots & 0 \\ 0 & \lambda_2 & \cdots & 0 \\ \cdots & \cdots & \cdots & \cdots \\ 0 & 0 & \cdots & \lambda_n \end{bmatrix} \tag{5}$$

Proof: This standard result may be found in Refs. 18, 67, 87, 137a, and 138.

ΔΔΔ

3. THEOREM. If $\underline{Q}$ is a symmetric $n \times n$ matrix, then

$$\underline{Q} > \underline{0} \iff \lambda_i > 0 \qquad (i = 1, 2, \ldots, n) \tag{6}$$

$$\underline{Q} \geq \underline{0} \iff \lambda_i \geq 0 \qquad (i = 1, 2, \ldots, n) \tag{7}$$

Proof: Using Theorem 2, the quadratic form (1) may be expressed as

$$f(\underline{x}) = \underline{x}'\underline{Qx} = \underline{x}'\underline{P}'\underline{\Lambda Px} \tag{8}$$

Defining the vector $\underline{y} = \underline{Px}$, we have

$$\underline{x}'\underline{Qx} = \underline{y}'\underline{\Lambda y} = \lambda_1 y_1{}^2 + \lambda_2 y_2{}^2 + \cdots + \lambda_n y_n{}^2 \tag{9}$$

From the definitions (2) and (3), it is clear that $\underline{x}'\underline{Qx}$ is positive [nonnegative] for all $\underline{x} \neq \underline{0}$ if and only if $\underline{y}'\underline{\Lambda y}$ is positive [nonnegative] for all $\underline{y} \neq \underline{0}$. It follows from (9) that this condition holds if and only if the λ_i are all positive [nonnegative].

ΔΔΔ

Since a matrix is invertible if and only if its eigenvalues are all nonzero, Theorem 3 establishes that a positive definite matrix is always invertible, while one which is only nonnegative definite must be singular.

Another useful formula which follows from (9) is

$$\lambda_{min}(Q)\ \|x\|^2 \leq x'Qx \leq \lambda_{max}(Q)\ \|x\|^2 \tag{10}$$

where $\lambda_{min}(Q)$ and $\lambda_{max}(Q)$ are the smallest and largest eigenvalues of Q, respectively (see Problem 2).

Finally, we state some well-known criteria for a matrix to be nonnegative or positive definite:

4. *DEFINITION.* The *minors* of a square matrix

$$Q = \begin{bmatrix} q_{11} & q_{12} & \cdots & q_{1n} \\ q_{21} & q_{22} & \cdots & q_{2n} \\ \cdots & \cdots & \cdots & \cdots \\ q_{n1} & q_{n2} & \cdots & q_{nn} \end{bmatrix} \tag{11}$$

are the determinants of submatrices formed by striking out an equal number of rows and columns. The *principal minors* are those minors which result from striking out pairs of identically numbered rows and columns. The *leading principal minors* p_k are principal minors formed by striking out the last $n - k$ pairs of rows and columns, that is,

$$p_k \triangleq \det \begin{bmatrix} q_{11} & q_{12} & \cdots & q_{1k} \\ q_{21} & q_{22} & \cdots & q_{2k} \\ \cdots & \cdots & \cdots & \cdots \\ q_{k1} & q_{k2} & \cdots & q_{kk} \end{bmatrix} \qquad (k = 1, 2, \ldots, n) \tag{12}$$

ΔΔΔ

5. *THEOREM (Sylvester's tests).* Let Q be a symmetric $n \times n$ matrix.

(a) Q is positive definite if and only if all the leading principal minors (12) are positive (> 0).

(b) Q is nonnegative definite if and only if all the principal minors are nonnegative (≥ 0).

Proof:

(a) See Refs. 18, 87, 137a and 138.

(b) See Ref. 170. Note that $p_k \geq 0$, $k = 1, 2, \ldots, n$ does *not* imply[†] that $\underline{Q} \geq \underline{0}$; *all* principal minors must be nonnegative.

ΔΔΔ

We have been making use of the Euclidian vector norm $\| \underline{x} \|$ defined in (A.1-5), and it is useful to define the matrix norm [93, 137a] *induced by* this vector norm as follows.

6. *DEFINITION*. The norm of a square matrix $\underline{A}$ is

$$\| \underline{A} \| \triangleq \max_{\underline{x} \neq \underline{0}} \frac{\| \underline{A}\underline{x} \|}{\| \underline{x} \|} = \max_{\| \underline{x} \| = 1} \| \underline{A}\underline{x} \| \tag{13}$$

and hence

$$\| \underline{A}\underline{x} \| \leq \| \underline{A} \| \; \| \underline{x} \| \tag{14}$$

ΔΔΔ

It is not difficult to show (Problem 7) that $\| \underline{A} \|$ has the properties

$$\begin{aligned} &\| \underline{A} \| \geq 0 \quad \text{and} \quad \| \underline{A} \| = 0 \Longleftrightarrow \underline{A} = \underline{0} \\ &\| \alpha \underline{A} \| = |\alpha| \, \| \underline{A} \| \qquad \text{(for any scalar } \alpha) \\ &\| \underline{A} + \underline{B} \| \leq \| \underline{A} \| + \| \underline{B} \| \qquad \text{(the } \textit{triangle inequality}) \\ &\| \underline{A}\underline{B} \| \leq \| \underline{A} \| \; \| \underline{B} \| \end{aligned} \tag{15}$$

The first three of these are identical to the properties (A.1-6) of the vector norm. From (9) and (10) we have

$$\| \underline{A}\underline{x} \|^2 = \underline{x}'\underline{A}'\underline{A}\underline{x} \leq \lambda_{max}(\underline{A}'\underline{A}) \| \underline{x} \|^2 \tag{16}$$

with equality when $\underline{x}$ is the eigenvector corresponding to $\lambda_{max}(\underline{A}'\underline{A})$, and it follows that

$$\| \underline{A} \| = \sqrt{\lambda_{max}(\underline{A}'\underline{A})} \tag{17}$$

[†] The statement in various textbooks [40, 138, 139] to this effect is incorrect, as demonstrated by the following counterexample: $\begin{bmatrix} 0 & 0 \\ 0 & -1 \end{bmatrix}$

In Problem 3(a) the reader is asked to verify that $\lambda_{max}(\underset{-}{A}'\underset{-}{A}) \geq 0$.

The norm defined above is known as the *spectral norm*. There are numerous other matrix norms satisfying (15), one of which is the *Euclidian matrix norm*

$$\| \underset{-}{A} \|_E = \sqrt{\sum_{i,j=1}^{n} a_{ij}^2} = \sqrt{tr\ (\underset{-}{A}'\underset{-}{A})} \tag{18}$$

This one is particularly useful because it is easier to calculate than (17) and it provides a bound,

$$\| \underset{-}{A} \| \leq \| \underset{-}{A} \|_E \tag{19}$$

PROBLEMS

Section A.1

1. Sketch the following vectors in R^3:

$$\underset{-}{x} = \begin{bmatrix} 1 \\ 2 \\ 3 \end{bmatrix} \qquad \underset{-}{y} = \begin{bmatrix} 2 \\ -1 \\ 1 \end{bmatrix} \qquad \underset{-}{x} + \underset{-}{y} \qquad \underset{-}{x} - 2\underset{-}{y}$$

Compute $\| \underset{-}{x} \|$ and $\| \underset{-}{y} \|$.

2. Show that if a set of vectors $\{\underset{-}{x}_1, \underset{-}{x}_2, \ldots, \underset{-}{x}_m\}$ in R^n is linearly dependent, at least one of the vectors $\underset{-}{x}_i$ may be expressed as a linear combination of the others.

3. Determine whether each set of vectors is linearly independent or linearly dependent. If the latter, express as many vectors as possible as linear combinations of the others.

(a)

$$\underset{-}{x}_1 = \begin{bmatrix} 1 \\ 2 \\ 3 \end{bmatrix} \qquad \underset{-}{x}_2 = \begin{bmatrix} -1 \\ 3 \\ -2 \end{bmatrix} \qquad \underset{-}{x}_3 = \begin{bmatrix} 0 \\ 5 \\ 1 \end{bmatrix}$$

(b)

$$\underset{-}{x}_1 = \begin{bmatrix} 1 \\ 2 \\ 0 \end{bmatrix} \qquad \underset{-}{x}_2 = \begin{bmatrix} -2 \\ -4 \\ 0 \end{bmatrix} \qquad \underset{-}{x}_3 = \begin{bmatrix} 2 \\ -1 \\ 3 \end{bmatrix}$$

(c)

$$\underline{x}_1 = \begin{bmatrix} 1 \\ 1 \end{bmatrix} \qquad \underline{x}_2 = \begin{bmatrix} 1 \\ -1 \end{bmatrix}$$

4. For each of the following sets of vectors, determine the subspace of R^3 spanned by $\{\underline{x}_1, \underline{x}_2, \ldots, \underline{x}_m\}$. Indicate the subspace on a sketch, and determine a basis for it.

(a)

$$\underline{x}_1 = \begin{bmatrix} 1 \\ 2 \\ 0 \end{bmatrix} \qquad \underline{x}_2 = \begin{bmatrix} -3 \\ 4 \\ 0 \end{bmatrix} \qquad \underline{x}_3 = \begin{bmatrix} 2 \\ -1 \\ 0 \end{bmatrix}$$

(b)

$$\underline{x}_1 = \begin{bmatrix} 1 \\ 0 \\ 0 \end{bmatrix} \qquad \underline{x}_2 = \begin{bmatrix} 1 \\ 1 \\ 0 \end{bmatrix} \qquad \underline{x}_3 = \begin{bmatrix} 1 \\ 1 \\ 1 \end{bmatrix}$$

(c)

$$\underline{x}_1 = \begin{bmatrix} 1 \\ 1 \\ 1 \end{bmatrix} \qquad \underline{x}_2 = \begin{bmatrix} -2 \\ -2 \\ -2 \end{bmatrix}$$

5. Prove the three properties stated in (A.1-6), and explain why the triangle inequality is so named.

6. Show that the concept of dimension is well defined, i.e., that any two bases of a linear vector space (or subspace) must contain the same number of elements.

7. Determine whether a set of m vectors can be a basis for R^n in each of the following cases and give reasons for your answers.

(a) $m < n$

(b) $m = n$

(c) $m > n$

Section A.2

1. Define the vectors and matrices,

$$\underline{A} = \begin{bmatrix} -1 & 2 \\ 3 & -4 \end{bmatrix} \qquad \underline{x} = \begin{bmatrix} 7 \\ 1 \end{bmatrix} \qquad \underline{z} = \begin{bmatrix} 3 \\ -5 \end{bmatrix}$$

$$\underset{-}{B} = \begin{bmatrix} -2 & 1 \\ 3 & -2 \\ 0 & 5 \end{bmatrix} \qquad \underset{-}{D} = \begin{bmatrix} 2 & -6 & 4 \\ 1 & 7 & -2 \\ 3 & -3 & 0 \end{bmatrix} \qquad \underset{-}{y} = \begin{bmatrix} 0 \\ 1 \\ -2 \end{bmatrix}$$

and compute the following products:

$\underline{BA}$, $\underline{Ax}$, $\underline{DB}$, $\underline{y}'\underline{B}$, $\underline{x}'\underline{z}$, $\underline{x}'\underline{A}^2\underline{x}$, $\underline{xz}'$, and $\underline{x}'\underline{AB}'\underline{y}$.

2. Show that $\underline{Ix} = \underline{x}$ for any vector $\underline{x}$.

3. (a) Show that a diagonal matrix is invertible if and only if its diagonal elements are all nonzero, by explicitly determining the inverse.

   (b) Show that a triangular matrix is invertible if and only if its diagonal elements are all nonzero, by computing the determinant. Determine the inverse explicitly for a 3×3 matrix.

4. Compute the determinant and inverse of each matrix:

$$\underset{-}{A} = \begin{bmatrix} a & b \\ c & d \end{bmatrix} \qquad \underset{-}{B} = \begin{bmatrix} 1 & 2 \\ 3 & 4 \end{bmatrix}$$

$$\underset{-}{C} = \begin{bmatrix} 0 & 1 & 2 \\ 3 & 4 & 5 \\ 6 & 7 & 8 \end{bmatrix} \qquad \underset{-}{D} = \begin{bmatrix} 1 & 2 & 3 & 4 \\ 0 & 5 & 6 & 7 \\ 0 & 0 & 8 & 9 \\ 0 & 0 & 0 & 0 \end{bmatrix}$$

5. Prove Theorems A.2.6 and A.2.7. You may wish to prove Theorem A.2.7(a) for the 2×2 case only.

6. Prove or disprove the following statements, where $\underline{A}$ and $\underline{B}$ are any $m \times n$ matrices, and $\underline{x}_1$ is a particular nonzero vector in R^n.

   (a) If $\underline{Ax}_1 = \underline{Bx}_1$, then $\underline{A} = \underline{B}$.

   (b) If $\underline{Ax} = \underline{Bx}$ *for all* $\underline{x} \in R^n$, then $\underline{A} = \underline{B}$.

   (c) $\det \underline{AB} = \det \underline{BA}$

7. Show that $(\underline{QAQ}^{-1})^k = \underline{QA}^k\underline{Q}^{-1}$.

8. (a) Prove that

$$(\underline{A} + \underline{B})^k = \underline{A}^k + k\underline{A}^{k-1}\underline{B} + \ldots + k\underline{A}\underline{B}^{k-1} + \underline{B}^k$$

if and only if $\underline{A}$ and $\underline{B}$ commute, that is,

$$\underline{A}\underline{B} = \underline{B}\underline{A}$$

Hint: Prove it for $k = 2$ and use an inductive argument.

(b) Derive expressions for $(\underline{A} + \underline{B})^2$ and $(\underline{A} + \underline{B})^3$ for the case where $\underline{A}\underline{B} \neq \underline{B}\underline{A}$.

9. Show that

$$\begin{bmatrix} \underline{A}_1 & \underline{0} & \ldots & \underline{0} \\ \underline{0} & \underline{A}_2 & \ldots & \underline{0} \\ \ldots & \ldots & \ldots & \ldots \\ \underline{0} & \underline{0} & \ldots & \underline{A}_m \end{bmatrix}^k = \begin{bmatrix} \underline{A}_1^k & \underline{0} & \ldots & \underline{0} \\ \underline{0} & \underline{A}_2^k & \ldots & \underline{0} \\ \ldots & \ldots & \ldots & \ldots \\ \underline{0} & \underline{0} & \ldots & \underline{A}_m^k \end{bmatrix}$$

10. Suppose the product

$$\underline{A}\underline{B} = \begin{bmatrix} \underline{A}_{11} & \underline{A}_{12} \\ \underline{A}_{21} & \underline{A}_{22} \end{bmatrix} \begin{bmatrix} \underline{B}_{11} & \underline{B}_{12} \\ \underline{B}_{21} & \underline{B}_{22} \end{bmatrix} = \begin{bmatrix} \underline{C}_{11} & \underline{C}_{12} \\ \underline{C}_{21} & \underline{C}_{22} \end{bmatrix} = \underline{C}$$

of partitioned matrices is well defined, where $\underline{A}_{11}$ and $\underline{B}_{11}$ are $n \times n$ and $\underline{A}_{22}$ and $\underline{B}_{22}$ are $m \times m$. Determine the dimensions of $\underline{A}$, $\underline{B}$, $\underline{C}$, $\underline{A}_{12}$, $\underline{A}_{21}$, $\underline{B}_{12}$, $\underline{B}_{21}$, $\underline{C}_{11}$, $\underline{C}_{12}$, $\underline{C}_{21}$, and $\underline{C}_{22}$. Calculate the blocks of $\underline{C}$ in terms of those of $\underline{A}$ and $\underline{B}$.

11. Suppose $\underline{A}_{11}$ and $\underline{A}_{22}$ are invertible $n \times n$ matrices. Compute the inverse of the matrix

$$\begin{bmatrix} \underline{A}_{11} & \underline{A}_{12} \\ \underline{0} & \underline{A}_{22} \end{bmatrix}$$

12. (a) Show that

$$\frac{d}{dt} \| \underline{x}(t) \|^2 = 2\underline{x}'(t)\dot{\underline{x}}(t) = 2\dot{\underline{x}}'(t)\underline{x}(t)$$

(b) Compute $\frac{d}{dt} \| \underline{x}(t) \|$.

13. (a) Assume that $\underline{A}(t)$ is nonsingular for all t and determine the derivative of $\underline{A}^{-1}(t)$. *Hint:* Differentiate $\underline{I} = \underline{A}(t)\underline{A}^{-1}(t)$.

    (b) Determine the second derivative of $\underline{A}^{-1}(t)$.

    (c) Determine the derivative of $[\underline{A}(t)\underline{B}(t)]^{-1}$.

14. If $\underline{A}$ and $\underline{B}$ are $n \times m$ and $m \times n$, respectively, show that $\text{tr}\ \underline{AB} = \text{tr}\ \underline{BA}$.

Section A.3

1. Determine the characteristic polynomials and eigenvalues of

$$\begin{bmatrix} 2 & 1 \\ 4 & 9 \end{bmatrix} \qquad \begin{bmatrix} 1 & -1 & 5 \\ 0 & -7 & 3 \\ -1 & -6 & -2 \end{bmatrix}$$

2. Prove that a matrix is invertible if and only if all of its eigenvalues are nonzero.

3. Verify that $\det(\lambda\underline{I} - \underline{A})$ is an n-th order polynomial in λ, where $\underline{A}$ is an $n \times n$ matrix.

4. Verify that a matrix with no repeated eigenvalues has a linearly independent set of eigenvectors. Show that the converse statement is untrue (i.e., determine a matrix with repeated eigenvalues and linearly independent eigenvectors). *Hint:* Suppose $\sum_{i=1}^{n} \alpha_i \underline{u}_i = \underline{0}$, where $\alpha_k \neq 0$. Multiply this relation on the left by the matrix

$$(\lambda_1\underline{I} - \underline{A})(\lambda_2\underline{I} - \underline{A}) \ \cdots \ (\lambda_{k-1}\underline{I} - \underline{A})(\lambda_{k+1}\underline{I} - \underline{A}) \ \cdots \ (\lambda_n\underline{I} - \underline{A})$$

   Note that all the factors of this matrix commute, and derive a contradiction.

5. Prove Theorem A.3.2.

6. Show that the eigenvalues of a diagonal or a triangular matrix are precisely the elements on the main diagonal.

7. (a) Show that if λ is an eigenvalue of $\underline{A}$ with eigenvector $\underline{u}$, then the eigenvector associated with $\overline{\lambda}$ is $\overline{\underline{u}}$.

(b) Show that if the eigenvectors of $\underline{A}$ are linearly independent, then so are the columns of $\underline{U}$ in (A.3-16).

(c) Prove Theorem A.3.6.

Hint: Show that

$$\left.\begin{aligned} \underline{A}\underline{v}_i &= \sigma_i \underline{v}_i - \omega_i \underline{w}_i \\ \underline{A}\underline{w}_i &= \omega_i \underline{v}_i + \sigma_i \underline{w}_i \end{aligned}\right\} \qquad (i = 1, 3, 5, \ldots, m-1)$$

and proceed as in Theorem A.3.3.

8. Find the eigenvalues and eigenvectors of the following matrices and use $\underline{U}$ put them into diagonal or block diagonal form, whichever is appropriate:

$$\begin{bmatrix} 1 & 2 \\ 5 & 4 \end{bmatrix} \qquad \begin{bmatrix} 1 & 2 \\ 2 & 4 \end{bmatrix} \qquad \begin{bmatrix} 1 & 2 & 0 \\ -1 & 3 & 0 \\ 2 & -4 & 3 \end{bmatrix}$$

9. If the characteristic polynomial of $\underline{A}$ is

$$\det(\lambda\underline{I} - \underline{A}) = \lambda^n + \alpha_{n-1}\lambda^{n-1} + \ldots + \alpha_1\lambda + \alpha_0$$

$$= (\lambda - \lambda_1)(\lambda - \lambda_2)\ldots(\lambda - \lambda_n)$$

where $\lambda_1, \lambda_2, \ldots, \lambda_n$ are its eigenvalues, show that

$$\lambda_1 + \lambda_2 + \ldots + \lambda_n = \text{tr } \underline{A} = -\alpha_{n-1}$$

and

$$\lambda_1\lambda_2\ldots\lambda_n = \det \underline{A} = (-1)^n\alpha_0$$

10. If $\underline{A}$ is an $n \times n$ matrix and $m \geq n$, show that $\underline{A}^m$ may be expressed as

$$\underline{A}^m = \lambda_0\underline{I} + \lambda_1\underline{A} + \lambda_2\underline{A}^2 + \ldots + \lambda_{n-1}\underline{A}^{n-1}$$

for some coefficients λ_i. *Hint:* Use the Cayley-Hamilton theorem (A.3.9) recursively.

11. (a) Show that the eigenvalues of a block-diagonal matrix consist of all the eigenvalues of its respective diagonal blocks.

(b) Do the same for a block-triangular matrix.

Section A.4

1. Prove the Schwartz inequality (A.4-6),

$$|\langle \underline{x},\underline{y}\rangle| \leq \| \underline{x} \| \; \| \underline{y} \|$$

and show that the equality sign holds if and only if $\underline{x} = \gamma \underline{y}$ for some scalar γ.

2. Compute $\langle \underline{x},\underline{y}\rangle$, $\langle \underline{y},\underline{z}\rangle$, $\langle \underline{z},\underline{x}\rangle$, and $\langle \underline{x},\underline{y}+\underline{z}\rangle$, where

$$\underline{x} = \begin{bmatrix} 1 \\ 2 \\ 3 \end{bmatrix} \qquad \underline{y} = \begin{bmatrix} -1 \\ 4 \\ 6 \end{bmatrix} \qquad \underline{z} = \begin{bmatrix} 6 \\ 3 \\ -4 \end{bmatrix}$$

Which of these vectors are orthogonal?

3. *Pythagoras' theorem.* Show that if $\underline{x} \perp \underline{y}$, then

$$\| \underline{x}+\underline{y} \|^2 = \| \underline{x} \|^2 + \| \underline{y} \|^2$$

4. Show that the difference vector of (A.4-13) is orthogonal to $\underline{y}$. Verify that (A.4-12) is a projection of $\underline{x}$ on $\underline{y}$ by constructing a two-dimensional example. Verify (A.4-14) in your example.

5. Show that an orthonormal set of vectors is always linearly independent.

6. *Gram-Schmidt orthogonalization process.* Verify that a linearly independent set of vectors $\{\underline{x}_1, \underline{x}_2, \ldots, \underline{x}_m\}$ can always be used to generate an orthonormal set, as follows:

$$\underline{u}_1 = \frac{\underline{x}}{\| \underline{x}_1 \|}$$

$$\underline{u}_2 = \frac{\underline{x}_2 - P_{\underline{u}_1}\underline{x}_2}{\| \underline{x}_2 - P_{\underline{u}_1}\underline{x}_2 \|} \perp \underline{u}_1$$

$$\underline{u}_3 = \frac{\underline{x}_3 - P_{\underline{u}_2}\underline{x}_3 - P_{\underline{u}_1}\underline{x}_3}{\| \underline{x}_3 - P_{\underline{u}_2}\underline{x}_3 - P_{\underline{u}_1}\underline{x}_3 \|} \perp \underline{u}_1,\underline{u}_2$$

7. Prove that $S^{\perp}$ is a subspace of R^n by verifying (A.1-7).

8. If S is a subspace of R^n with an orthonormal basis $\{\underline{u}_1, \underline{u}_2, \ldots, \underline{u}_m\}$, how can you find a basis for $S^\perp$? Show that the dimensions of S and $S^\perp$ must sum to n and that only the zero vector belongs to both S and $S^\perp$.

Section A.5

1. Determine the range space and null space of each matrix:

$$\underline{A} = \begin{bmatrix} 1 & 1 \\ 6 & 6 \end{bmatrix} \qquad \underline{B} = \begin{bmatrix} -1 & 2 \\ 3 & -5 \\ -6 & 10 \end{bmatrix} \qquad \underline{C} = \begin{bmatrix} 4 & -2 & 1 \\ 3 & 1 & 6 \end{bmatrix}$$

2. Verify (A.5-4) through (A.5-7) for each of the matrices in Problem A.5-1.

3. Show that $\text{rank}(\underline{A}) \le \min\{m, n\}$, where $\underline{A}$ is an $m \times n$ matrix.

4. Prove that $\text{rank}(\underline{A}) + \text{nullity}(\underline{A}) = n$ and $\text{rank}(\underline{A}) = \text{rank}(\underline{A}')$, where $\underline{A}$ is an $m \times n$ matrix. *Hint:* Let $\{\underline{u}_1, \underline{u}_2, \ldots, \underline{u}_p\}$ and $\{\underline{u}_{p+1}, \underline{u}_{p+2}, \ldots, \underline{u}_n\}$ be bases for $N(\underline{A})$ and $N(\underline{A})^\perp$, respectively, and show that $\{\underline{A}\underline{u}_{p+1}, \underline{A}\underline{u}_{p+2}, \ldots, \underline{A}\underline{u}_n\}$ is a basis for $R(\underline{A})$.

5. Suppose $\underline{A}$ is an $m \times n$ matrix of rank m, where $m < n$.

(a) Sketch the decomposition of Fig. A.5-1 for this case. Verify that the linear equation

$$\underline{A}\underline{x} = \underline{b} \tag{L}$$

has a solution for *any* $\underline{b} \in R^m$.

(b) Show that $\underline{A}\underline{A}'$ is nonsingular and explain why the matrix $\underline{A}'(\underline{A}\underline{A}')^{-1}$ is called a *pseudo-inverse* or *generalized inverse* of $\underline{A}$ [16, 117, 137a, 151, 185].

(c) Show that the vector

$$\underline{x}^* = \underline{A}'(\underline{A}\underline{A}')^{-1}\underline{b}$$

is the *minimum norm* solution of (L) in the sense that

$$\| \underline{x}^* \| \le \| \underline{x} \|$$

for all other solutions $\underline{x}$. *Hint:* Decompose any solution $\underline{x}$ into its projections on $R(\underline{A}')$ and $N(\underline{A})$,

$$\underline{x} = \underline{P}_R\underline{x} + \underline{P}_N\underline{x} = \underline{x}_R + \underline{x}_N$$

show that $\underline{x}_R = \underline{x}^*$ for *all* solutions, and then use (A.4-10).

6. Suppose $\underline{A}$ is an $m \times n$ matrix of rank n, where $n < m$.

   (a) Sketch the decomposition of Fig. A.5-1 for this case. Verify that the linear equation

   $$\underline{A}\underline{x} = \underline{b} \qquad \text{(L)}$$

   has *at most* one solution and may have none.

   (b) Show that $\underline{A}'\underline{A}$ is nonsingular and explain why the matrix $(\underline{A}'\underline{A})^{-1}\underline{A}'$ is called a *pseudo-inverse* or *generalized inverse* of $\underline{A}$ [16, 117, 137a, 151, 185].

   (c) Show that if (L) has a solution, it is given by

   $$\underline{x}^* = (\underline{A}'\underline{A})^{-1}\underline{A}'\underline{b}$$

   (d) In the case where (L) has no solution, show that $\underline{x}^*$ is the "best" approximation to a solution, in the sense that

   $$\| \underline{A}\underline{x}^* - \underline{b} \| \leq \| \underline{A}\underline{x} - \underline{b} \|$$

   for all other $\underline{x} \in R^n$. *Hint:* Show that $(\underline{A}\underline{x}^* - \underline{b}) \perp R(\underline{A})$, write

   $$\underline{A}\underline{x} - \underline{b} = (\underline{A}\underline{x}^* - \underline{b}) + \underline{A}(\underline{x} - \underline{x}^*)$$

   and then use (A.4-10).

7. What happens in the two problems above if $\underline{A}$ is square? Discuss generalizations of the pseudo-inverse concept to cases where $\underline{A}$ does not have full rank.

8. Suppose $\underline{A}$ is a square $(n \times n)$ matrix. Sketch Fig. A.5-1 for this case. Find a counterexample to show that $N(\underline{A}) \neq R(\underline{A})^{\perp}$ in general. In other words, find an $\underline{A}$ and an $\underline{x} \neq \underline{0}$ such that $\underline{x} \in R(\underline{A})$ *and* $\underline{x} \in N(\underline{A})$.

9. Show that nullity $(\underline{A}) \neq$ nullity $(\underline{A}')$ in general.

10. Show that the product of a nonsingular square matrix and a matrix of rank p must have rank p.

11. (a) Show that the product of a matrix of rank p and one of rank q must have rank $\leq \min\{p,q\}$. Give an example for which the strict inequality holds.

    (b) Show that equality holds in (a) if the matrices are $n \times p$ and $p \times r$, with $n > p$, or if they are $n \times q$ and $q \times r$, with $r > q$.

Section A.6

1. Verify that there is no loss of generality in assuming $\underline{Q}$ symmetric in Definition A.6.1 by showing that if $\underline{Q} \neq \underline{Q}'$, then

$$\underline{x}'\underline{Q}\underline{x} = \underline{x}'\underline{Q}'\underline{x} = \underline{x}'\underline{B}\underline{x}$$

   for some symmetric matrix $\underline{B}$.

2. Establish the inequalities in (A.6-10). For what vectors $\underline{x}_{min}$ and $\underline{x}_{max}$ do the inequalities become equalities? *Hint:* Show that $\|\underline{x}\| = \|\underline{y}\|$ and use (A.6-9).

3. (a) Show that a matrix of the form $\underline{Q} = \underline{A}'\underline{A}$ must be nonnegative definite and that $\text{rank}(\underline{Q}) = \text{rank}(\underline{A})$.

   (b) Show that if $\underline{A}$ is square and nonsingular, then $\underline{Q} = \underline{A}'\underline{A}$ is positive definite.

   (c) Note that the converses of both (a) and (b) are true. That is, *any* nonnegative definite $\underline{A}$ may be written as $\underline{Q} = \underline{A}'\underline{A}$ with $\text{rank}(\underline{A}) = \text{rank}(\underline{Q})$. If $\underline{Q} > \underline{0}$, then $\underline{A}$ can be square, nonsingular, and symmetric, in which case it is known as the *square root* of $\underline{Q}$. Attempt a proof if you wish.

4. Determine whether the following matrices are positive (or nonnegative) definite

$$\begin{bmatrix} 1 & 0 & 0 \\ 0 & 2 & 0 \\ 0 & 0 & 3 \end{bmatrix} \qquad \begin{bmatrix} 1 & -2 \\ -2 & 5 \end{bmatrix} \qquad \begin{bmatrix} 3 & 2 \\ 2 & 1 \end{bmatrix} \qquad \begin{bmatrix} 1 & 0 & 2 \\ 0 & 0 & 0 \\ 2 & 0 & 1 \end{bmatrix}$$

5. (a) Show that eigenvectors associated with *distinct* eigenvalues of a symmetric matrix must be orthogonal.

   (b) Show how to determine the diagonalizing matrix $\underline{P}$ in Theorem A.6.2.

6. Show that the two expressions in (A.6-13) are equivalent.

7. Verify the properties (A.6-15) of $\|\underline{A}\|$.

8. Prove (A.6-19).

Appendix B

REVIEW OF LAPLACE TRANSFORMS

B.0. INTRODUCTION

In this appendix we shall briefly review the theory of Laplace transforms, assuming that the reader has had a previous introduction to this topic. For a more complete and rigorous treatment of the subject, see Refs. 114, 177, and 185.

The basic idea of a Laplace transformation is straightforward. We perform an integral operation on a function of a time variable t which transforms it to a function of a complex-valued variable s. This operation is invertible, in the sense that each function in the *complex domain* corresponds to one and only one function in the *time domain*.

This means that we can transform a time function, perform various operations in the complex domain which are easier than the equivalent operations in the time domain, and then use the inverse transform to determine the resultant time function. This procedure is particularly useful in the case of differential equations, which transform to algebraic equations in the complex domain and are easily solved.

B.1. DEFINITION AND PROPERTIES

We shall restrict our attention here to *one-sided (unilateral) Laplace transforms* and assume throughout that all time functions of interest vanish for $t < t_0$. Moreover, we note that t can always be replaced by $t - t_0$, so there is no loss of generality in setting $t_0 = 0$.

1. *DEFINITION.* If f(t) is an integrable function of time with

$$f(t) = 0 \qquad \text{(for } t < 0\text{)} \tag{1}$$

then the *Laplace transform* of f is a function $\hat{f}$ of a complex variable

$$s = \sigma + j\omega \tag{2}$$

It is defined by the Laplace integral,[†]

$$L\{f(t)\} = \hat{f}(s) = \int_0^\infty f(t)e^{-st}\,dt \tag{3}$$

at all values of s for which the integral exists (see Problem 6). It can be extended to other values of s, as indicated in Example 2. In particular, if the integral exists for all s such that

$$\sigma > \sigma_c \tag{4}$$

and does not exist otherwise, then σ_c is called the *abscissa of convergence* for the function f(t), as shown in Fig. 1. The region to the right of the abscissa of convergence is known as the *region of convergence* of the Laplace integral.

ΔΔΔ

2. *EXAMPLE.* Consider the *exponential function*

$$f(t) = \begin{cases} 0 & (t \leq 0) \\ e^{at} & (t > 0) \end{cases} \tag{5}$$

[†]The use of $L\{f(t)\}$ instead of $L\{f\}$ or $L\{f(\cdot)\}$ is not strictly correct, but this abuse of notation is both common and convenient and we shall retain it.

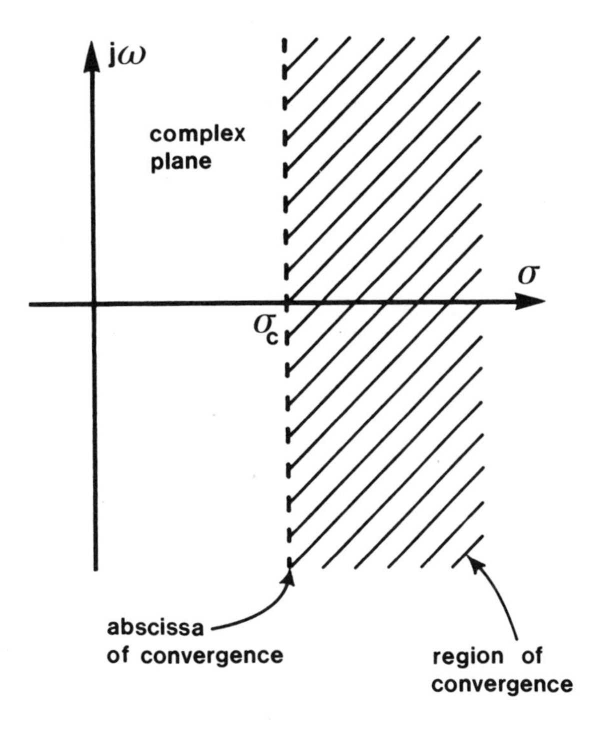

FIG. 1. Region of convergence of the Laplace integral.

where $a = \alpha + j\beta$ may be any real or complex number. The Laplace integral in this case is

$$L\{f(t)\} = \hat{f}(s) = \int_0^\infty f(t)e^{-st}\,dt \tag{6}$$

Since $|e^{-j\omega t}| = |e^{j\omega t}| = 1$, we have

$$\begin{aligned}|\hat{f}(s)| &= \left|\int_0^\infty e^{\alpha t}e^{j\beta t}e^{-\sigma t}e^{-j\omega t}\,dt\right| \\ &\leq \int_0^\infty e^{(\alpha-\sigma)t}\,dt \end{aligned} \tag{7}$$

It is clear that the integral is finite for $\alpha - \sigma < 0$ and infinite otherwise, so that $\sigma_c = \alpha$ is the abscissa of convergence for f and

$$\hat{f}(s) = \int_0^\infty e^{(a-s)t}\, dt$$

$$= \frac{1}{a-s} e^{(a-s)t} \Big|_0^\infty = \frac{1}{s-a} \tag{8}$$

for all s such that Re $s = \sigma > a$. In fact, we may use the Analytic Continuation theorem [114] and let $\hat{f}(s) = 1/(s-a)$ for all s. This means that $\hat{f}(s) = \infty$ at the *singular point* $s = a$.

ΔΔΔ

Thus we have established that[†]

$$L\{e^{at}\} = \frac{1}{s-a} \tag{9}$$

Note that the *step function*

$$1(t) \triangleq \begin{cases} 0 & (t \leq 0) \\ 1 & (t > 0) \end{cases} \tag{10}$$

is a special case of (5) with $a = 0$, so that its transform is

$$L\{1(t)\} = \frac{1}{s} \tag{11}$$

We shall now investigate several properties of the Laplace transform, the most important of which is linearity.

3. *THEOREM (Linearity).* If

$$L\{f_1(t)\} = \hat{f}_1(s) \qquad L\{f_2(t)\} = \hat{f}_2(s) \tag{12}$$

then[††]

$$L\{\alpha f_1(t) + f_2(t)\} = \alpha\hat{f}_1(s) + \hat{f}_2(s) \tag{13}$$

[†]Recall that we assumed all time functions to be zero for $t < 0$, so $L\{e^{at}\}$ really means $L\{f(t)\}$ or $L\{1(t)e^{at}\}$.

[††]If σ_1 and σ_2 are the abscissae of convergence for f_1 and f_2, respectively, then $\sigma_c = \max\{\sigma_1, \sigma_2\}$ will be the abscissa of convergence for $\alpha f_1 + f_2$.

where α is any real or complex number.

Proof: This follows from the linearity of the Laplace integral,

$$\begin{aligned} L\{\alpha f_1(t) + f_2(t)\} &= \int_0^\infty [\alpha f_1(t) + f_2(t)]e^{-st}\, dt \\ &= \alpha \int_0^\infty f_1(t)e^{-st}\, dt + \int_0^\infty f_2(t)e^{-st}\, dt \\ &= \alpha \hat{f}_1(s) + \hat{f}_2(s) \end{aligned} \tag{14}$$

ΔΔΔ

4. THEOREM (Delay). If

$$L\{f(t)\} = \hat{f}(s) \tag{15}$$

then

$$L\{f(t - \Delta)\} = \hat{f}(s)e^{-s\Delta} \tag{16}$$

where $f(t - \Delta)$ is the original function delayed† by a time $\Delta > 0$, as illustrated in Fig. 2.

Proof: By definition

$$L\{f(t - \Delta)\} = \int_0^\infty f(t - \Delta)e^{-st}\, dt = \int_\Delta^\infty f(t - \Delta)e^{-st}\, dt \tag{17}$$

If we let $\tau = t - \Delta$, this becomes

†Actually, Δ may be negative as long as $f(t - \Delta) = 0$ for $t < 0$.

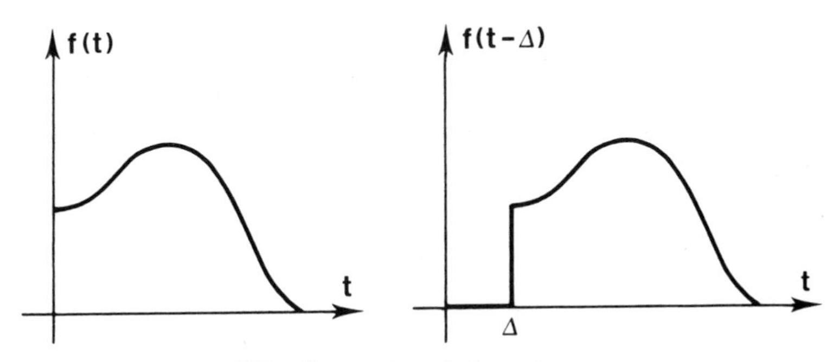

FIG. 2. Delayed function.

$$\int_0^\infty f(\tau)e^{-s(\tau+\Delta)}\,d\tau = e^{-s\Delta}\int_0^\infty f(\tau)e^{-s\tau}\,d\tau = e^{-s\Delta}\hat{f}(s) \tag{18}$$

ΔΔΔ

5. *THEOREM (Exponential factor).* If

$$L\{f(t)\} = \hat{f}(s) \tag{19}$$

then

$$L\{f(t)e^{\alpha t}\} = \hat{f}(s - \alpha) \tag{20}$$

where α is any complex number.

Proof: Left to the reader (Problem 1).

ΔΔΔ

6. *THEOREM (Change of time scale).* If

$$L\{f(t)\} = \hat{f}(s) \tag{21}$$

then

$$L\{f(t/T)\} = T\hat{f}(Ts) \tag{22}$$

where T is any positive real number.

Proof: Left to the reader (Problem 1).

ΔΔΔ

Using the results above, we can derive the Laplace transforms of various functions.

7. *EXAMPLE.* If the exponential function of Example 1 is multiplied by t^n, for any $n = 0, 1, 2, \ldots$, then

$$L\{t^n e^{at}\} = \int_0^\infty t^n e^{(a-s)t}\,dt = \frac{n!}{(s-a)^{n+1}} \tag{23}$$

When $n = 0$, this reduces to (9). The transform of a power of t may be found by setting $a = 0$,

$$L\{t^n\} = \frac{n!}{s^{n+1}} \tag{24}$$

ΔΔΔ

8. EXAMPLE. The product of a real exponential and a sinusoid may be written as

$$e^{\sigma t} \cos \omega t = \frac{e^{\sigma t}(e^{j\omega t} + e^{-j\omega t})}{2} = \frac{e^{(\sigma+j\omega)t} + e^{(\sigma-j\omega)t}}{2} \tag{25}$$

and

$$e^{\sigma t} \sin \omega t = \frac{e^{\sigma t}(e^{j\omega t} - e^{-j\omega t})}{2j} = \frac{e^{(\sigma+j\omega)t} - e^{(\sigma-j\omega)t}}{2j} \tag{26}$$

The transforms of these are found by using (9), with $a = \sigma \pm j\omega$, and Theorem 3:

$$\begin{aligned} L\{e^{\sigma t} \cos \omega t\} &= \frac{1/(s - \sigma - j\omega) + 1/(s - \sigma + j\omega)}{2} \\ &= \frac{s - \sigma}{(s - \sigma)^2 + \omega^2} \end{aligned} \tag{27}$$

$$\begin{aligned} L\{e^{\sigma t} \sin \omega t\} &= \frac{1/(s - \sigma - j\omega) - 1/(s - \sigma + j\omega)}{2j} \\ &= \frac{\omega}{(s - \sigma)^2 + \omega^2} \end{aligned} \tag{28}$$

In the case where $\sigma = 0$, these become

$$L\{\cos \omega t\} = \frac{s}{s^2 + \omega^2} \tag{29}$$

$$L\{\sin \omega t\} = \frac{\omega}{s^2 + \omega^2} \tag{30}$$

ΔΔΔ

9. EXAMPLE. Consider the *unit pulse* function shown in Fig. 3 and defined by

$$p(t) = \begin{cases} \varepsilon^{-1} & (0 < t \leq \varepsilon) \\ 0 & (\text{otherwise}) \end{cases} \tag{31}$$

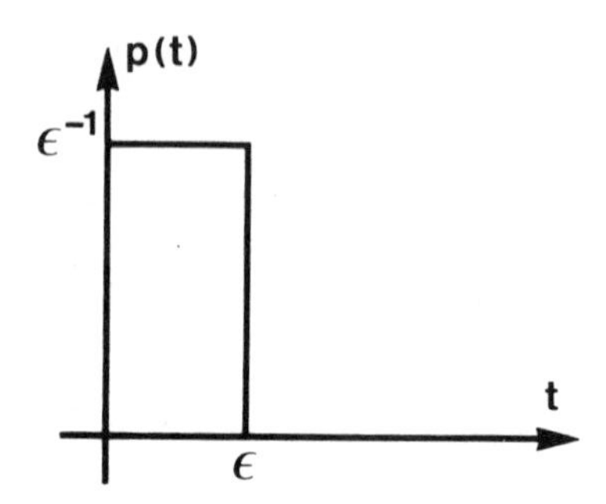

FIG. 3. Pulse function of Example 9.

This may also be expressed as the difference of two step functions, one delayed by ε,

$$p(t) = \varepsilon^{-1}1(t) - \varepsilon^{-1}1(t-\varepsilon) \tag{32}$$

and the transform may be found by using (11) and Theorems 3 and 4:

$$\mathcal{L}\{p(t)\} = \frac{\varepsilon^{-1}}{s} - \frac{\varepsilon^{-1}}{s} e^{-s\varepsilon} = \frac{\varepsilon^{-1}}{s}(1 - e^{-s\varepsilon}) \tag{33}$$

If we let $\varepsilon \to 0$, the pulse becomes very narrow but still has unit area, and we denote it as

$$\lim_{\varepsilon \to 0} p(t) = \delta(t) \tag{34}$$

which is called a *unit impulse*, or *Dirac delta function* at $t = 0$. In fact, this limit does not really exist in the ordinary sense (see Definition 3.4.1 and the discussion following it), but (33) is well defined in the limit,

$$\lim_{\varepsilon \to 0} \left[\frac{\varepsilon^{-1}}{s} (1 - e^{-s\varepsilon})\right]$$

$$= \lim_{\varepsilon \to 0} \left[\frac{\varepsilon^{-1}}{s} \left(1 - 1 + s\varepsilon - \frac{s^2\varepsilon^2}{2!} + \ldots\right)\right] = 1 \tag{35}$$

Thus we may use the transform

$$\mathcal{L}\{\delta(t)\} = 1 \tag{36}$$

and overlook the fact that $\delta(t)$ is ill defined.

ΔΔΔ

The transforms derived in Examples 2, 7, 8, and 9, along with various others, are collected into a table in Sec. B.4. We shall now derive a few more useful properties of the Laplace transform.

10. THEOREM (Differentiation). If

$$L\{f(t)\} = \hat{f}(s) \tag{37}$$

then

$$L\{\dot{f}(t)\} = s\hat{f}(s) - f(0) \tag{38}$$

and

$$L\{f^{(n)}(t)\} = s^n\hat{f}(s) - s^{n-1}f(0) - \ldots - sf^{(n-2)}(0) - f^{(n-1)}(0) \tag{39}$$

Proof: Integrate the Laplace integral by parts,

$$L\{\dot{f}(t)\} = \int_0^\infty \dot{f}(t)e^{-st}\,dt = f(t)e^{-st}\Big|_0^\infty - \int_0^\infty f(t)(-se^{-st})\,dt$$

$$= s\hat{f}(s) - f(0) \tag{40}$$

We obtain (39) by using (38) on itself $n - 1$ times.

ΔΔΔ

11. THEOREM (Integration). If

$$L\{f(t)\} = \hat{f}(s) \tag{41}$$

and any impulses occur after $t = 0$ (for example at $t = 0^+$), then

$$L\{\int_0^t f(\tau)\,d\tau\} = \frac{\hat{f}(s)}{s} \tag{42}$$

and

$$L\{\int_0^t \int_0^{\tau_1} \ldots \int_0^{\tau_{n-1}} f(\tau_n)\,d\tau_n\,d\tau_{n-1} \ldots d\tau_1\} = \frac{\hat{f}(s)}{s^n} \tag{43}$$

Proof: Left to the reader (Problem 4).

ΔΔΔ

12. THEOREM (Final value). Suppose

$$L\{f(t)\} = \hat{f}(s) \tag{44}$$

and $\dot{f}(t)$ is an integrable function. Then if $f(t)$ approaches a limit as $t \to \infty$, it is given by

$$\lim_{t \to \infty} f(t) = \lim_{s \to 0} s\hat{f}(s) \tag{45}$$

Note that the limit on the left *may not exist*, even though the one on the right does.

Proof: Using (38), we have

$$L\{\dot{f}(t)\} = \int_0^\infty \dot{f}(t) e^{-st}\, dt = s\hat{f}(s) - f(0) \tag{46}$$

Taking the limit as $s \to 0$, and interchanging limit with integral on the left-hand side, we have

$$\int_0^\infty \dot{f}(t) \lim_{s \to 0} e^{-st}\, dt = \int_0^\infty \dot{f}(t)\, dt$$

$$= f(\infty) - f(0) = \lim_{s \to 0} s\hat{f}(s) - f(0) \tag{47}$$

provided $f(\infty)$ exists. Addition of $f(0)$ yields (45).

ΔΔΔ

13. THEOREM (Initial value). Suppose

$$L\{f(t)\} = \hat{f}(s) \tag{48}$$

and $f(t)$ is an integrable function. Then if $f(t)$ approaches a limit as $t \to 0$, it is given by

$$\lim_{t \to 0} f(t) = \lim_{s \to \infty} s\hat{f}(s) \tag{49}$$

Note that the limit on the left *may not exist* and may not be equal to $f(0)$.

Proof: As in the previous proof, take the limit of (38) as $s \to \infty$ and interchange the limit with the integral to yield

$$\int_0^\infty \dot{f}(t) \lim_{s\to\infty} e^{-st}\, dt = 0 = \lim_{s\to\infty} s\hat{f}(s) - f(0) \tag{50}$$

ΔΔΔ

14. THEOREM (Convolution). If

$$L\{f_1(t)\} = \hat{f}_1(s) \qquad L\{f_2(t)\} = \hat{f}_2(s) \tag{51}$$

and the convolution $g = f_1 * f_2$ is defined by

$$g(t) \triangleq \int_0^t f_1(t-\tau)f_2(\tau)\, d\tau = \int_0^t f_1(\tau)f_2(t-\tau)\, d\tau \tag{52}$$

then

$$L\{g(t)\} = \hat{g}(s) = \hat{f}_1(s)\hat{f}_2(s) \tag{53}$$

Proof: Since $f_1(t) = f_2(t) = 0$ for $t < 0$, we may replace the upper limit on the convolution integral (52) by ∞ and write

$$\hat{g}(s) = \int_0^\infty \left\{ \int_0^\infty f_1(\tau)f_2(t-\tau)\, d\tau \right\} e^{-st}\, dt$$

Interchanging integrals and using Theorem 4, we have

$$\hat{g}(s) = \int_0^\infty f_1(\tau) \left\{ \int_0^\infty f_2(t-\tau)e^{-st}\, dt \right\} d\tau$$

$$= \int_0^\infty f_1(\tau)\, \{\hat{f}_2(s)e^{-s\tau}\}\, d\tau$$

$$= \left\{ \int_0^\infty f_1(\tau)e^{-s\tau}\, d\tau \right\} \hat{f}_2(s) = \hat{f}_1(s)\hat{f}_2(s) \tag{54}$$

ΔΔΔ

All of the above transforms and transform properties, along with some additional ones are collected into a table in Sec. B.4. We now turn to the problem of inverting a Laplace transform and determining its corresponding time function.

B.2. INVERSION OF LAPLACE TRANSFORMS

The value of the Laplace transformation (B.1-3) lies in the fact that it is *invertible*, so that each transform in the complex domain corresponds to one and only one function in the time domain. The appropriate time function may be determined as follows:

1. THEOREM. If $\hat{f}(s)$ is the Laplace transform of f(t), then f(t) may be found with the *inversion formula*,

$$f(t) = \frac{1}{2\pi j} \int_{\sigma - j\infty}^{\sigma + j\infty} \hat{f}(s) e^{st} \, ds \qquad (t > 0) \tag{1}$$

where $\sigma > \sigma_c$ and σ_c is the abscissa of convergence for f(t). Thus the path of integration may be any vertical line in the complex plane which lies in the region of convergence for f.

Proof: See Ref. 114.

ΔΔΔ

Inversion of Laplace transforms directly using (1) is not an easy task since the integral will generally be difficult to evaluate. Fortunately, it is possible to use the theory of *contour integration* [114] and evaluate the integral indirectly.

Rather than present this theory, we shall proceed directly to the standard inversion method which follows from it. We have already generated a table in Sec. B.4 which may be used to invert simple transforms. The next result insures that a large class of transforms (those which are rational) may be reduced to linear combinations of the simple transforms in the table for easy inversion.

2. THEOREM. Suppose $\hat{f}(s)$ is a strictly proper, rational function† of s with real coefficients,

---

†The function $\hat{f}(s)$ is *rational* if it is a ratio of polynomials, with a numerator of degree m and a denominator of degree n. It is *proper* if $m \leq n$ and *strictly proper* if $m < n$, an assumption which will be relaxed in Example 8.

$$\hat{f}(s) = \frac{q(s)}{p(s)} = \frac{q_{n-1}s^{n-1} + \ldots + q_1 s + q_0}{s^n + p_{n-1}s^{n-1} + \ldots + p_1 s + p_0}$$

$$= \frac{q(s)}{(s - \lambda_1)(s - \lambda_2) \ldots (s - \lambda_n)} \tag{2}$$

where the complex numbers

$$\lambda_1, \lambda_2, \ldots \lambda_n \tag{3}$$

are the roots[†] of $p(s)$ and are all distinct. Then $\hat{f}(s)$ may be written as a *partial fractions expansion*

$$\hat{f}(s) = \frac{R_1}{s - \lambda_1} + \frac{R_2}{s - \lambda_2} + \ldots + \frac{R_n}{s - \lambda_n} \tag{4}$$

where the *residues* $R_1, R_2, \ldots, R_n$ are complex numbers given by

$$R_i = [(s - \lambda_i)\hat{f}(s)]_{s=\lambda_i} \qquad (i = 1, 2, \ldots, n) \tag{5}$$

The inverse Laplace transform of $\hat{f}(s)$ may thus be written as

$$f(t) = R_1 e^{\lambda_1 t} + R_2 e^{\lambda_2 t} + \ldots + R_n e^{\lambda_n t} \tag{6}$$

Proof: The fact that any strictly proper rational function has an expansion of the form (4) may be found in Ref. 114. To evaluate one of the residues R_i, multiply (4) by $(s - \lambda_i)$ and write it as

$$(s - \lambda_i)\hat{f}(s) = R_i + (s - \lambda_i) \text{ [other terms]} \tag{7}$$

Since the terms in square brackets contain no factors of $s - \lambda_i$ in their denominators, all terms on the right-hand side except the first vanish for $s = \lambda_i$, leaving (5).

[†]By a slight abuse of terminology, we refer interchangeably to the roots of the *polynomial* $p(s)$ and the roots of the *polynomial equation* $p(s) = 0$. The abscissa of convergence is $\max_i\{\text{Re } \lambda_i\}$.

Finally, (6) follows by using the Laplace transform pair (B.1-9) and the linearity property (Theorem B.1.3).

ΔΔΔ

Assuming that the coefficients of q(s) and p(s) in (2) are real numbers, each real root λ_k has a residue R_k which is also real. Moreover, any complex roots occur in conjugate pairs,

$$\begin{aligned} \lambda_i &= \sigma_i + j\omega_i \\ \lambda_{i+1} &= \sigma_i - j\omega_i = \overline{\lambda}_i \end{aligned} \tag{8}$$

and have conjugate residues (see Problem 1),

$$\begin{aligned} R_i &= \frac{1}{2}C_i - \frac{1}{2}jS_i \\ R_{i+1} &= \frac{1}{2}C_i + \frac{1}{2}jS_i = \overline{R}_i \end{aligned} \tag{9}$$

Using these facts, the complex terms in the partial fractions expansion may be put in a more convenient form:

3. *COROLLARY*. If λ_i and λ_{i+1} in (8) are a pair of complex roots of p(s) with residues (9), then the corresponding pair of complex terms in the partial fractions expansion may be written as

$$\frac{R_i}{s - \lambda_i} + \frac{\overline{R}_i}{s - \overline{\lambda}_i} = \frac{C_i(s - \sigma_i) + S_i\omega_i}{(s - \sigma_i)^2 + \omega_i^2} \tag{10}$$

where $C_i = 2\,\mathrm{Re}\,R_i$ and $S_i = -2\,\mathrm{Im}\,R_i$ may also be found from

$$2jR_i = S_i + jC_i = \frac{1}{\omega_i}\,[(s - \sigma_i)^2 + \omega_i^2]\hat{f}(s)\Big|_{s=\sigma_i+j\omega_i} \tag{11}$$

The inverse Laplace transform of (10) is

$$C_i e^{\sigma_i t}\cos\omega_i t + S_i e^{\sigma_i t}\sin\omega_i t \tag{12}$$

which may also be written as

$$\rho_i e^{\sigma_i t}\sin(\omega_i t + \phi_i) \tag{13}$$

where ρ_i and ϕ_i are defined by $S_i + jC_i = \rho_i e^{j\phi_i}$, that is

$$\rho_i = \sqrt{S_i^2 + C_i^2} = 2|R_i|$$
$$\phi_i = \tan^{-1} \frac{C_i}{S_i} = 90° + \angle R_i \tag{14}$$

Proof: Left to the reader (Problem 3).

ΔΔΔ

4. *EXAMPLE*. Consider the function

$$\begin{aligned}\hat{f}(s) &= \frac{2s^3 - 3s^2 + 49s - 52}{s^4 - 2s^3 - 11s^2 + 52s} \\ &= \frac{2s^3 - 3s^2 + 49s - 52}{(s^2 - 6s + 13)(s + 4)s} \\ &= \frac{2s^3 - 3s^2 + 49s - 52}{[(s - 3)^2 + 2^2](s + 4)s}\end{aligned} \tag{15}$$

whose denominator has distinct roots

$$\begin{aligned}\lambda_1 &= 3 + j2 \\ \lambda_2 &= 3 - j2 = \bar{\lambda}_1 \\ \lambda_3 &= -4 \\ \lambda_4 &= 0\end{aligned} \tag{16}$$

According to Theorem 2 and Corollary 3, $\hat{f}(s)$ may be expanded in partial fractions as

$$\hat{f}(s) = \frac{C_1(s - 3) + S_1(2)}{(s - 3)^2 + 2^2} + \frac{R_3}{s + 4} + \frac{R_4}{s - 0} \tag{17}$$

where

$$\begin{aligned}S_1 + jC_1 &= \frac{1}{2}\left[\frac{2s^3 - 3s^2 + 49s - 52}{(s + 4)s}\right]_{s=3+j2} = 3 + j1 \\ R_3 &= \left[\frac{2s^3 - 3s^2 + 49s - 52}{(s^2 - 6s + 13)s}\right]_{s=-4} = 2 \\ R_4 &= \left[\frac{2s^3 - 3s^2 + 49s - 52}{(s^2 - 6s + 13)(s + 4)}\right]_{s=0} = -1\end{aligned} \tag{18}$$

Thus we have

$$\hat{f}(s) = \frac{(s-3) + 3(2)}{(s-3)^2 + 2^2} + \frac{2}{s+4} - \frac{1}{s} \tag{19}$$

and the inverse Laplace transform is

$$\begin{aligned} f(t) &= (\cos 2t + 3 \sin 2t)e^{3t} + 2e^{-4t} - 1 \\ &= \sqrt{10}\, e^{3t} \sin(2t + 18°) + 2e^{-4t} - 1 \end{aligned} \tag{20}$$

ΔΔΔ

If p(s) has repeated roots, the partial fractions expansion takes on a slightly different form and Theorem 2 generalizes to the following:

5. *THEOREM.* Suppose $\hat{f}(s)$ is a strictly proper, rational function of s with real coefficients,

$$\hat{f}(s) = \frac{q(s)}{p(s)} = \frac{q(s)}{(s-\lambda_1)^{\mu_1}(s-\lambda_2)^{\mu_2} \cdots (s-\lambda_m)^{\mu_m}} \tag{21}$$

where the complex numbers

$$\lambda_i \text{ (repeated } \mu_i \text{ times)} \qquad (i = 1, 2, 3, \ldots, m)$$

are the roots of p(s). Then the partial fractions expansion is

$$\hat{f}(s) = \sum_{i=1}^{m} \left[\frac{R_{i1}}{s-\lambda_i} + \frac{R_{i2}}{(s-\lambda_i)^2} + \cdots + \frac{R_{i,\mu_i}}{(s-\lambda_i)^{\mu_i}} \right] \tag{22}$$

where the residues may be evaluated by

$$R_{i,\mu_i} = [(s-\lambda_i)^{\mu_i} \hat{f}(s)]_{s=\lambda_i} \tag{23}$$

$$R_{i,\mu_i - 1} = \frac{1}{1!} \left[\frac{d}{ds} (s-\lambda_i)^{\mu_i} \hat{f}(s) \right]_{s=\lambda_i} \tag{24}$$

....

$$R_{i1} = \frac{1}{(\mu_i - 1)!} \left[\frac{d^{\mu_i - 1}}{ds^{\mu_i - 1}} (s-\lambda_i)^{\mu_i} \hat{f}(s) \right]_{s=\lambda_i} \tag{25}$$

for $i = 1, 2, \ldots, m$. The inverse Laplace transform of $\hat{f}(s)$ is

$$f(t) = \sum_{i=1}^{m} e^{\lambda_i t} \left(R_{i1} + \frac{R_{i2} t}{1!} + \ldots + \frac{R_{i,\mu_i} t^{\mu_i - 1}}{(\mu_i - 1)!} \right) \tag{26}$$

Proof: Again, the existence of an expansion of the form (22) is established in Ref. 114. The rest of the proof is straightforward, but tedious.

ΔΔΔ

As above, all the residues associated with a repeated real root are real. In the case of a pair of complex conjugate roots,

$$\begin{aligned} \lambda_i &= \sigma_i + j\omega_i \\ \lambda_{i+1} &= \sigma_i - j\omega_i = \overline{\lambda}_i \end{aligned} \tag{27}$$

which are repeated μ_i times, the residues also occur in conjugate pairs,

$$\begin{aligned} R_{ik} &= \frac{1}{2} C_{ik} - \frac{1}{2} jS_{ik} \\ & \qquad (k = 1, 2, \ldots, \mu_i) \\ R_{i+1,k} &= \frac{1}{2} C_{ik} + \frac{1}{2} jS_{ik} = \overline{R}_{ik} \end{aligned} \tag{28}$$

Thus the complex terms in (26) may be combined as follows:

6. *COROLLARY*. If λ_i and λ_{i+1} in (27) are a pair of complex roots of $p(s)$ with residues (28), then the inverse Laplace transform of the terms

$$\frac{R_{ik}}{(s - \lambda_i)^k} + \frac{R_{i+1,k}}{(s - \lambda_{i+1})^k} \tag{29}$$

is given by

$$\frac{e^{\sigma_i t} t^{k-1}}{(k-1)!} (C_{ik} \cos \omega_i t + S_{ik} \sin \omega_i t) \tag{30}$$

or alternatively by

$$\frac{e^{\sigma_i t} t^{k-1}}{(k-1)!} \rho_{ik} \sin(\omega_i t + \phi_{ik}) \tag{31}$$

where ρ_{ik} and ϕ_{ik} are defined as in (14) by

$$S_{ik} + jC_{ik} = \rho_{ik} e^{j\phi_{ik}} \tag{32}$$

Proof: Left to the reader.

ΔΔΔ

7. *EXAMPLE*. Consider the function

$$\hat{f}(s) = \frac{11s^2 + 72s + 109}{s^4 + 8s^3 + 18s^2 - 27} = \frac{11s^2 + 72s + 109}{(s+3)^3(s-1)} \tag{33}$$

whose denominator has roots

$$\begin{aligned} \lambda_1 &= -3 \qquad (\mu_1 = 3) \\ \lambda_2 &= +1 \qquad (\mu_2 = 1) \end{aligned} \tag{34}$$

According to Theorem 5, $\hat{f}(s)$ may be expanded in partial fractions as

$$\hat{f}(s) = \frac{R_{11}}{s+3} + \frac{R_{12}}{(s+3)^2} + \frac{R_{13}}{(s+3)^3} + \frac{R_2}{s-1} \tag{35}$$

where

$$\begin{aligned} R_{13} &= \left[\frac{11s^2 + 72s + 109}{s-1}\right]_{s=-3} = 2 \\ R_{12} &= \frac{1}{1!}\left[\frac{11s^2 - 22s - 181}{(s-1)^2}\right]_{s=-3} = -1 \\ R_{11} &= \frac{1}{2!}\left[\frac{384}{(s-1)^3}\right]_{s=-3} = -3 \\ R_2 &= \left[\frac{11s^2 + 72s + 109}{(s+3)^3}\right]_{s=1} = 3 \end{aligned} \tag{36}$$

Thus we have

$$\hat{f}(s) = \frac{-3}{s+3} + \frac{-1}{(s+3)^2} + \frac{2}{(s+3)^3} + \frac{3}{s-1} \tag{37}$$

and from the Laplace transform table,

$$f(t) = -3e^{-3t} - te^{-3t} + t^2e^{-3t} + 3e^{t} \tag{38}$$

ΔΔΔ

In Theorems 2 and 5 we assumed the rational transform to be strictly proper, but this was done only for convenience. If $\hat{f}(s)$ is only proper, it can easily be put in the form of a constant plus a strictly proper function, as illustrated in Example 8.

8. EXAMPLE. Consider the function

$$\hat{f}(s) = \frac{s^2 + 4s + 1}{s^2 + 3s + 2} \tag{39}$$

Using long division, we find

$$\begin{array}{r|l} & \quad 1 \\ \hline s^2 + 3s + 2 & s^2 + 4s + 1 \\ & \underline{s^2 + 3s + 2} \\ & 0 \;+\; s - 1 \end{array}$$

so that $\hat{f}(s)$ may be written as

$$\hat{f}(s) = 1 + \frac{s - 1}{s^2 + 3s + 2} \tag{40}$$

Expanding the second term in partial fractions, we have

$$\hat{f}(s) = 1 + \frac{3}{s + 2} - \frac{2}{s + 1} \tag{41}$$

and the inverse transform is

$$f(t) = \delta(t) + 3e^{2t} - 2e^{t} \tag{42}$$

ΔΔΔ

Thus we see that a transform which is not strictly proper is associated with a time function containing an impulse. A numerator of degree greater than that of the denominator yields terms proportional to s, s^2, ..., and these correspond to "derivatives" of the impulse $\delta(t)$ (see Problem 3).

B.3. SOLUTION OF DIFFERENTIAL EQUATIONS

Laplace transforms are used primarily to solve linear, time-invariant differential equations, as we shall illustrate here with a simple example. The general case is taken up in Sec. 4.1.

1. EXAMPLE. Consider the differential equation

$$\dddot{y}(t) + 4\ddot{y}(t) + 29\dot{y}(t) = 29u(t) \tag{1}$$

where the initial conditions are

$$y(0) = 0 \qquad \dot{y}(0) = 17 \qquad \ddot{y}(0) = -122 \tag{2}$$

and the input function is a unit step $u(t) = 1(t)$. Using the linearity and differentiation theorems (B.1.3 and B.1.10), this differential equation transforms to the algebraic equation

$$\begin{aligned} &[s^3\hat{y}(s) - s^2 y(0) - s\dot{y}(0) - \ddot{y}(0)] \\ &\quad + 4[s^2\hat{y}(s) - sy(0) - \dot{y}(0)] \\ &\qquad + 29[s\hat{y}(s) - y(0)] = 29\hat{u}(s) \end{aligned} \tag{3}$$

Rearranging terms, substituting (2), and using (B.1-11), we have

$$\begin{aligned} (s^3 + 4s^2 + 29s)\hat{y}(s) &= s^2 y(0) + s[\dot{y}(0) + 4y(0)] \\ &\quad + [\ddot{y}(0) + 4\dot{y}(0) + 29y(0)] + 29\hat{u}(s) \\ &= (17s - 54) + \frac{29}{s} \end{aligned} \tag{4}$$

which can be solved for $\hat{y}(s)$,

$$\hat{y}(s) = \frac{17s - 54 + 29/s}{s^3 + 4s^2 + 29s} = \frac{17s^2 - 54s + 29}{[(s+2)^2 + 5^2]s^2} \tag{5}$$

The denominator polynomial in (5) has two real roots at $s = 0$ and a pair of complex roots at $s = -2 \pm j5$. Using Corollary B.2.3 and Theorem B.2.5, we may expand this in partial fractions,

$$\frac{C_1(s+2) + S_1(5)}{(s+2)^2 + 5^2} + \frac{R_{21}}{s} + \frac{R_{22}}{s^2} \tag{6}$$

where

$$S_1 + jC_1 = \frac{1}{5}\left[\frac{17s^2 - 54s + 29}{s^2}\right]_{s=-2+j5} = 4 + j2$$

$$R_{22} = \left[\frac{17s^2 - 54s + 29}{(s+2)^2 + 5^2}\right]_{s=0} = 1 \tag{7}$$

$$R_{21} = \frac{1}{1!}\left[\frac{122s^2 + 928s - 1682}{[(s+2)^2 + 5^2]^2}\right]_{s=0} = -2$$

Thus we have

$$\hat{y}(s) = \frac{2(s+2) + 4(5)}{(s+2)^2 + 5^2} - \frac{2}{s} + \frac{1}{s^2} \tag{8}$$

and the solution of the differential equation (1) is the inverse Laplace transform

$$y(t) = 2e^{-2t}\cos 5t + 4e^{-2t}\sin 5t - 2 + t$$

$$= \sqrt{20}\, e^{-2t}\sin(5t + 0.464) - 2 + t \tag{9}$$

ΔΔΔ

It will also be useful to consider first-order vector differential equations of the form

$$\begin{aligned} \dot{\underline{x}}(t) &= \underline{A}\,\underline{x}(t) + \underline{v}(t) \\ \underline{x}(0) &= \underline{x}_0 \end{aligned} \tag{10}$$

where

$$\underline{x}(t) = \begin{bmatrix} x_1(t) \\ x_2(t) \\ \vdots \\ x_n(t) \end{bmatrix} \qquad \underline{v}(t) = \begin{bmatrix} v_1(t) \\ v_2(t) \\ \vdots \\ v_n(t) \end{bmatrix} \tag{11}$$

are n-vectors and $\underline{A}$ is an $n \times n$ matrix. Defining the Laplace transform of a vector quantity to be

$$L\{\underline{x}(t)\} = \hat{\underline{x}}(s) = \begin{bmatrix} \hat{x}_1(s) \\ \hat{x}_2(s) \\ \vdots \\ \hat{x}_n(s) \end{bmatrix} = \begin{bmatrix} L\{x_1(t)\} \\ L\{x_2(t)\} \\ \vdots \\ L\{x_n(t)\} \end{bmatrix} \tag{12}$$

we note that the transform of $\dot{\underline{x}}(t)$ is as follows.

2. *LEMMA (Vector differentiation).* If

$$L\{\underline{x}(t)\} = \hat{\underline{x}}(s) \tag{13}$$

then

$$L\{\dot{\underline{x}}(t)\} = s\hat{\underline{x}}(s) - \underline{x}(0) \tag{14}$$

Proof: This is simply an application of Theorem B.1.10 to each component of $\dot{\underline{x}}(t)$,

$$L\{\dot{\underline{x}}(t)\} = \begin{bmatrix} L\{\dot{x}_1(t)\} \\ L\{\dot{x}_2(t)\} \\ \vdots \\ L\{\dot{x}_n(t)\} \end{bmatrix} = \begin{bmatrix} s\hat{x}_1(s) - x_1(0) \\ s\hat{x}_2(s) - x_2(0) \\ \vdots \\ s\hat{x}_n(s) - x_n(0) \end{bmatrix} = s\hat{\underline{x}}(s) - \underline{x}(0) \tag{15}$$

ΔΔΔ

With this result, (10) may be transformed to

$$s\hat{\underline{x}}(s) - \underline{x}(0) = \underline{\underline{A}}\hat{\underline{x}}(s) + \hat{\underline{v}}(s) \tag{16}$$

or

$$(s\underline{I} - \underline{A})\hat{\underline{x}}(s) = \underline{x}_0 + \hat{\underline{v}}(s) \tag{17}$$

Since $s\underline{I} - \underline{A}$ is nonsingular except where s is an eigenvalue of $\underline{A}$, this may be solved for

$$\hat{\underline{x}}(s) = (s\underline{I} - \underline{A})^{-1}\underline{x}_0 + (s\underline{I} - \underline{A})^{-1}\hat{\underline{v}}(s) \tag{18}$$

where $(s\underline{I} - \underline{A})^{-1}$ is known as a *resolvent matrix*.† The solution of

†The abscissa of convergence associated with $(s\underline{I} - \underline{A})^{-1}$ is $\max_i\{\text{Re }\lambda_i\}$, where λ_i are the eigenvalues of $\underline{A}$.

(10) is obtained by taking the inverse Laplace transform of (18), which is the subject of Sec. 4.2.

B.4. TABLE OF LAPLACE TRANSFORMS

All the Laplace transform properties and pairs derived in Sec. B.1 are collected in the following table. A few other useful pairs are also included.

Laplace Transform Properties

A. *Linearity:* $L\{\alpha f_1(t) + f_2(t)\} = \alpha\hat{f}_1(s) + \hat{f}_2(s)$

B. *Differentiation:* $L\{\dot{f}(t)\} = s\hat{f}(s) - f(0)$

$$L\{f^{(n)}(t)\} = s^n\hat{f}(s) - s^{n-1}f(0) - \dots - sf^{(n-2)}(0) - f^{(n-1)}(0)$$

C. *Integration:*† $L\left\{\int_0^t f(\tau)\, d\tau\right\} = \hat{f}(s)/s$

$$L\left\{\int_0^t \int_0^{\tau_1} \dots \int_0^{\tau_{n-1}} f(\tau_n)\, d\tau_n\, d\tau_{n-1} \dots d\tau_1\right\} = \hat{f}(s)/s^n$$

D. *Delay:* $L\{f(t - \Delta)\} = \hat{f}(s)e^{-s\Delta}$ $\quad (\Delta > 0)$

E. *Exponential:* $L\{f(t)e^{\alpha t}\} = \hat{f}(s - \alpha)$

F. *Time scaling:* $L\{f(t/T)\} = T\hat{f}(Ts)$ $\quad (T > 0)$

G. *Time factor:* $L\{tf(t)\} = -\dfrac{d\hat{f}(s)}{ds}$

H. *Time division:* $L\{f(t)/t\} = \int_s^\infty f(s_1)\, ds_1$

I. *Convolution:* $L\left\{\int_0^t f_1(\tau)f_2(t - \tau)\, d\tau\right\} = \hat{f}_1(s)\hat{f}_2(s)$

J. *Final Value:** $\lim_{t\to\infty} f(t) = \lim_{s\to 0} s\hat{f}(s)$

K. *Initial Value:** $\lim_{t\to 0} f(t) = \lim_{s\to\infty} s\hat{f}(s)$

---

†Assuming any impulses occur at $t = 0^+$ or later.

*Assuming all limits exist and f(t) is integrable.

Laplace Transform Pairs (Each time function is zero for $t < 0$)

	Time function	Laplace transform	
1.	e^{at}	$\frac{1}{s-a}$	
2.	$1(t)$ (unit step)	$1/s$	
3.	$e^{\sigma t}\sin \omega t$	$\frac{\omega}{(s-\sigma)^2+\omega^2}$	
3a.†	$e^{-\zeta\omega_n t}\sin(\omega_n\sqrt{1-\zeta^2}\,t)$	$\frac{\omega_n\sqrt{1-\zeta^2}}{s^2+2\zeta\omega_n s+\omega_n^2}$	$(\zeta < 1)$
4.	$e^{\sigma t}\cos \omega t$	$\frac{s-\sigma}{(s-\sigma)^2+\omega^2}$	
4a.†	$e^{-\zeta\omega_n t}\cos(\omega_n\sqrt{1-\zeta^2}\,t)$	$\frac{s+\zeta\omega_n}{s^2+2\zeta\omega_n s+\omega_n^2}$	$(\zeta < 1)$
5.*	$Ce^{\sigma t}\cos\omega t + Se^{\sigma t}\sin\omega t = \rho e^{\sigma t}\sin(\omega t+\phi)$	$\frac{C(s-\sigma)+S\omega}{(s-\sigma)^2+\omega^2}$	
6.	$\sin\omega t$	$\frac{\omega}{s^2+\omega^2}$	
7.	$\cos\omega t$	$\frac{s}{s^2+\omega^2}$	
8.*	$C\cos\omega t + S\sin\omega t = \rho\sin(\omega t+\phi)$	$\frac{Cs+S\omega}{s^2+\omega^2}$	
9.	$t^n e^{at}$ $(n = 0,1,2,\ldots)$	$\frac{n!}{(s-a)^{n+1}}$	
10.	t^n $(n = 0,1,2,\ldots)$	$\frac{n!}{s^{n+1}}$	
11.	$\delta(t)$ (unit impulse)	1	
12.	$\delta(t-\Delta)$	$e^{-s\Delta}$	

† $\omega_n = \sqrt{\omega^2+\sigma^2}$ and $\zeta = -\sigma/\omega_n$; $\omega = \omega_n\sqrt{1-\zeta^2}$ and $\sigma = -\zeta\omega_n$.

* $S + jC = \rho e^{j\phi}$; $\rho = \sqrt{S^2+C^2}$ and $\phi = \tan^{-1}(C/S)$.

PROBLEMS

Section B.1

1. Prove Theorems B.1.5 and B.1.6.

2. (a) Derive (B.1-30) by using Theorem B.1.10 on (B.1-29).
   (b) Do the same using Theorem B.1.4 on (B.1-29).

3. (a) Sketch the function

$$p^1(t) = \begin{cases} -\varepsilon^{-2} & (0 < t \leq \varepsilon/2) \\ \varepsilon^{-2} & (\varepsilon/2 < t \leq \varepsilon) \\ 0 & (\text{otherwise}) \end{cases}$$

   and determine its Laplace transform $\hat{p}^1(s)$. As ε becomes very small, this function is denoted

$$\lim_{\varepsilon \to 0} p^1(t) = \delta^1(t)$$

   and called a *unit doublet*. Determine $L\{\delta^1(t)\}$ by finding the limit of $\hat{p}^1(s)$ as $\varepsilon \to 0$.

   (b) The unit doublet is the "derivative" of the unit impulse (B.1-34). Is this terminology consistent with Theorem B.1.10?

   (c) Determine $\int_{-\infty}^{\infty} x(t)\delta(t)\,dt$ and $\int_{-\infty}^{\infty} x(t)\delta^1(t)\,dt$, where $x(t)$ is any continuously differentiable function. This property is often used to define $\delta(t)$ and $\delta^1(t)$.

4. Prove Theorem B.1.11 using integration by parts. Why is it necessary to be careful with delta functions at $t = 0$?

5. Prove Properties G and H in the table of Sec. B.4. *Hint:* Note the relationships with Theorems B.1.10 and B.1.11, respectively.

6. Show that a *sufficient* condition for the existence of a Laplace transform of $f(t)$ is that $f(t)$ be of *exponential class*, i.e., that there exist constants k and α such that

$$|f(t)| \leq ke^{\alpha t} \qquad (t \geq 0)$$

$$|f(t)| = 0 \qquad (t < 0)$$

Give an example of a function whose Laplace transform does not exist.

7. Give an example of a function for which the final value theorem (B.1.12) does not hold. Do the same for the initial value theorem (B.1.13).

Section B.2

1. In Theorem B.2.2 show that the residues of real roots must be real and that the residues of a pair of complex conjugate roots must be conjugate (assuming p(s) and q(s) have real coefficients).

2. In Theorem B.2.2, under what circumstances can a real residue or a pair of complex residues be equal to zero? *Hint:* What relation exists in this case between q(s) and p(s)?

3. Prove Corollary B.2.3 as follows:

   (a) Show that (B.2-10) holds.
   (b) Establish (B.2-11) with an argument similar to the one used to prove (B.2-5).
   (c) Explain how (B.2-12) follows from (B.2-10).
   (d) Derive (B.2-13) from (B.2-12).

4. In Example B.2.4, multiply out the partial fractions expansion (19) of $\hat{f}(s)$ and verify that it is equal to (15).

5. Expand the following transforms in partial fractions and determine the corresponding time functions:

   (a) $\dfrac{2s + 1/2}{s^2 + (5/3)s - 2/3}$

   (b) $\dfrac{s^2 + 3s - 19/4}{s^3 - 2s^2 - (7/4)s - 15/4}$ (One root is at $s = 3$.)

   (c) $\dfrac{s^2 + 2s + 1}{(s^2 - 6s + 19)(s + 4)^2}$

Section B.3

1. Find the solutions of the following differential equations (assume that u(t) and all of its derivatives are zero for $t \leq 0$):

(a) $\dddot{y} - \ddot{y} - (3/4)\dot{y} + (5/4)y = \dot{u} + u$

$y(0) = 4 \qquad \dot{y}(0) = -11 \qquad \ddot{y}(0) = 17 \qquad u(t) = 1(t)$

(One root is at $s = -1$.)

(b) $\dddot{y} + 2\ddot{y} = \ddot{u} + \dot{u} - 2u$

$y(0) = \dot{y}(0) = \ddot{y}(0) = 0 \qquad u(t) = \delta(t)$

BIBLIOGRAPHY

This bibliography is divided into books [1-186] and papers [201-372]. Book reviews in the *IEEE Transactions on Automatic Control (T-AC)* are indicated in parentheses at the end of the book citation.

Books

1. R. Abraham, *Linear and Multilinear Algebra*, Benjamin, 1966.
2. F. S. Acton, *Numerical Methods That Work*, Harper and Row, 1970.
3. J. K. Aggarwal, *Notes on Nonlinear Systems*, Van Nostrand Reinhold, 1972 (*T-AC*, December 1973, p. 693).
4. A. E. Albert and L. A. Gardner, Jr., *Stochastic Approximation and Nonlinear Regression*, M. I. T. Press, 1967.
5. B. D. O. Anderson, M. A. Arbib, and E. G. Manes, *Foundations of System Theory--Finitary and Infinitary Conditions*, Springer-Verlag, 1976.
6. B. D. O. Anderson, *Linear Optimal Control*, Prentice-Hall, 1971.
7. M. Aoki, *Optimization of Stochastic Systems*, Academic Press, 1967.

7a. M. Aoki, *Optimal Control and System Theory in Dynamic Economic Analysis*, North-Holland, 1975.

8. T. Apostol, *Mathematical Analysis*, Addison-Wesley, 1957.
9. K. J. Åström, *Introduction to Stochastic Control Theory*, Academic Press, 1970 (*T-AC*, June 1972, p. 422).

10. M. Athans, M. L. Dertouzos, R. N. Spann, and S. J. Mason, *Systems, Networks, and Computation: Multivariable Methods*, McGraw-Hill, 1974.

11. M. Athans and P. L. Falb, *Optimal Control*, McGraw-Hill, 1966 (*T-AC*, June 1967, p. 345).

12. D. M. Auslander, Y. Takahashi, and M. J. Rabins, *Introducing Systems and Control*, McGraw-Hill, 1974.

13. A. B. Baggeroer, *State Variables, the Fredholm Theory and Optimal Communications*, M. I. T. Press, 1970.

14. N. Balabanian and T. A. Bickart, *Electrical Network Theory*, Wiley, 1969.

15. A. V. Balakrishnan, *Stochastic Differential Systems I: Filtering and Control--A Function Space Approach*, Springer-Verlag, 1973.

16. S. Barnett, *Matrices in Control Theory*, Van Nostrand Reinhold, 1971.

17. R. Bellman, *Dynamic Programming*, Princeton University Press, 1957.

18. R. Bellman, *Introduction to Matrix Analysis*, McGraw-Hill, 1960.

19. R. Bellman, *Adaptive Control Processes: A Guided Tour*, Princeton University Press, 1961.

20. R. Bellman, *Introduction to the Mathematical Theory of Control Processes, Volume I: Linear Equations and Quadratic Criteria*, Academic Press, 1967 (*T-AC*, June 1970, p. 406).

21. R. Bellman, *Methods of Nonlinear Analysis, Volume I*, Academic Press, 1970 (*T-AC*, December 1972, p. 848).

22. R. Bellman, *Introduction to the Mathematical Theory of Control Processes, Volume II: Nonlinear Processes*, Academic Press, 1971 (*T-AC*, April 1973, p. 195).

23. R. Bellman and K. L. Cooke, *Differential-Difference Equations*, Academic Press, 1963.

24. R. Bellman and S. E. Dreyfus, *Applied Dynamic Programming*, Princeton University Press, 1962.

25. R. Bellman and R. Kalaba (eds.), *Selected Papers on Mathematical Trends in Control Theory*, Dover, 1964.

25a. A. Bensoussan, M. C. Delfour, and S. K. Mitter, *Representation and Control of Infinite Dimensional Linear Systems*, to appear.

26. L. D. Berkovitz, *Optimal Control Theory*, Springer-Verlag, 1974.

27. D. P. Bertsekas, *Dynamic Programming and Stochastic Control*, Academic Press, 1976.

28. A. Blaquière, F. Gerard, and G. Leitmann, *Quantitative and Qualitative Games*, Academic Press, 1969.

29. R. Boudarel, J. Delmas, and P. Guichet, *Dynamic Programming and its Application to Optimal Control*, Academic Press, 1971 (*T-AC*, June 1973, p. 327).

30. G. E. P. Box and G. M. Jenkins, *Time Series Analysis, Forecasting, and Control,* Holden-Day, 1970 (*T-AC*, April 1972, p. 281).

31. R. M. Bracewell, *The Fourier Transform and its Applications*, McGraw-Hill, 1965.

32. R. W. Brockett, *Finite-Dimensional Linear Systems*, Wiley, 1970 (*T-AC*, October 1972, p. 753).

33. A. E. Bryson and Y.-C. Ho, *Applied Optimal Control*, Blaisdell, 1969 (*T-AC*, February 1972, p. 186).

34. R. S. Bucy and P. D. Joseph, *Filtering for Stochastic Processes with Applications to Guidance*, Wiley, 1968 (*T-AC*, February 1972, p. 184).

35. A. G. Butkovskiy, *Distributed Control Systems*, American Elsevier, 1969 (*T-AC*, June 1971, p. 286).

36. J. A. Cadzow, *Discrete-Time Systems*, Prentice-Hall, 1973 (*T-AC*, October 1974, p. 633).

37. M. D. Canon, C. D. Cullum, and E. Polak, *Theory of Optimal Control and Mathematical Programming*, McGraw-Hill, 1970.

38. R. W. Carroll, *Abstract Methods in Partial Differential Equations*, Harper and Row, 1969.

39. C.-T. Chen, *Introduction to Linear System Theory*, Holt, Rinehart, and Winston, 1970 (*T-AC*, October 1972, p. 748).

40. C.-F. Chen and I. J. Haas, *Elements of Control Systems Analysis*, Prentice-Hall, 1968.

41. H. Chestnut and R. W. Mayer, *Servomechanisms and Regulating System Design*, Wiley, 1951 and 1955.

42. E. Coddington and N. Levinson, *Theory of Ordinary Differential Equations*, McGraw-Hill, 1955.

43. J. B. Cruz, Jr. (ed.), *Feedback Systems*, McGraw-Hill, 1972 (*T-AC*, October 1972, p. 751).

44. J. B. Cruz, Jr. (ed.), *System Sensitivity Analysis*, Dowden, Hutchinson and Ross, 1973.

45. W. J. Cunningham, *Introduction to Nonlinear Analysis*, McGraw-Hill, 1958.

46. R. E. Curry, *Estimation and Control with Quantized Measurements*, M. I. T. Press, 1970 (*T-AC*, October 1971, p. 523).

47. H. D'Angelo, *Linear Time-Varying Systems: Analysis and Synthesis*, Allyn and Bacon, 1970.

48. J. D'Azzo and C. H. Houpis, *Linear Control Systems: Conventional and Modern*, McGraw-Hill, 1975.

49. No reference.

50. M. Denn, *Optimization by Variational Methods*, McGraw-Hill, 1969.

51. M. L. Dertouzos, M. Athans, R. N. Spann and S. J. Mason, *Systems, Networks, and Computation: Basic Concepts*, McGraw-Hill, 1972.

52. P. DeRusso, R. Roy and C. Close, *State Variables for Engineers*, Wiley, 1965 (*T-AC*, February 1967, p. 124).

53. C. A. Desoer (ed.), *Applications of Lie Group Theory to Nonlinear Network Problems* (supplement to IEEE International Symposium on Circuit Theory), Western Periodicals Co., April 1974.

53a. C. A. Desoer and M. Vidyasagar, *Feedback Systems: Input-Output Properties*, Academic Press, 1975 (*T-AC*, April 1976, p. 300).

54. J. Dieudonné, *Foundations of Modern Analysis*, Academic Press, 1960.

55. S. E. Dreyfus, *Dynamic Programming and the Calculus of Variations*, Academic Press, 1965.

55a. R. D. Driver, *Ordinary and Delay Differential Equations*, Springer-Verlag, 1977.

56. P. Dyer and S. R. McReynolds, *The Computation and Theory of Optimal Control*, Academic Press, 1970 (*T-AC*, December 1972, p.850).

57. O. Elgerd, *Control Systems Theory*, McGraw-Hill, 1967.

58. V. W. Eveleigh, *Adaptive Control and Optimization Techniques*, McGraw-Hill, 1967 (*T-AC*, October 1968, p. 602).

59. V. W. Eveleigh, *Introduction to Control Systems Design*, McGraw-Hill, 1972 (*T-AC*, October 1973, p. 565).

60. P. Eykhoff, *System Identification: Parameter and State Estimation*, Wiley, 1974.

61. D. K. Faddeev and V. N. Faddeeva, *Computational Methods in Linear Algebra*, Freeman, 1963.

62. V. N. Faddeeva, *Computational Methods in Linear Algebra*, Dover, 1959.

63. A. A. Fel'dbaum, *Optimal Control Systems*, Academic Press, 1965 (*T-AC*, April 1968, p. 224).

64. H. Freeman, *Discrete-Time Systems*, Wiley, 1965 (*T-AC*, April 1966, p. 336).

65. A. Friedman, *Partial Differential Equations*, Holt, Rinehart, and Winston, 1969.

66. A. Friedman, *Differential Games*, Wiley-Interscience, 1971 (*T-AC*, October 1974, p. 630).

67. F. R. Gantmacher, *The Theory of Matrices, Volume I*, Chelsea, 1959.

68. F. R. Gantmacher, *The Theory of Matrices, Volume II*, Chelsea, 1959.

69. A. Gelb (ed.), *Applied Optimal Estimation*, M. I. T. Press, 1974.

70. A. Gelb and W. E. VanderVelde, *Multiple-Input Describing Functions and Nonlinear System Design*, McGraw-Hill, 1968.

71. J. C. Gille, M. J. Pelegrin and P. Decaulne, *Feedback Control Systems*, McGraw-Hill, 1959.

72. L. A. Gould, *Chemical Process Control: Theory and Applications*, Addison-Wesley, 1969 (*T-AC*, April 1973, p. 198).

73. D. Graham and D. McRuer, *Analysis of Nonlinear Control Systems*, Dover, 1961.

74. D. Graupe, *Identification of Systems*, Van Nostrand Reinhold, 1972.

75. A. L. Greensite, *Elements of Modern Control Theory*, Spartan, 1970 (*T-AC*, December 1973, p. 693).

76. A. L. Greensite, *Analysis and Design of Space Vehicle Flight Control Systems*, Spartan, 1970 (*T-AC*, February 1974, p. 96).

77. W. Hahn, *Theory and Application of Lyapunov's Direct Method*, Prentice-Hall, 1963.

78. W. Hahn, *Stability of Motion*, Springer-Verlag, 1967 (*T-AC*, February 1969, p. 118).

79. P. Halmos, *Finite Dimensional Vector Spaces*, Van Nostrand, 1958.

80. R. W. Hamming, *Numerical Methods for Scientists and Engineers*, McGraw-Hill, 1962.

81. R. W. Hamming, *Introduction to Applied Numerical Analysis*, McGraw-Hill, 1971.

82. P. Henrici, *Discrete Variable Methods in Ordinary Differential Equations*, Wiley, 1962.

83. P. Henrici, *Elements of Numerical Analysis*, Wiley, 1964.

84. H. Hermes and J. P. LaSalle, *Functional Analysis and Time-Optimal Control*, Academic Press, 1969 (*T-AC*, February 1972, p. 189).

85. M. R. Hestenes, *Calculus of Variations and Optimal Control Theory*, Wiley, 1966.

86. Y.-C. Ho and S. J. Mitter (eds.), *Post-IFAC Symposium on Large-Scale System Theory*, Plenum Press, 1976.

87. K. Hoffman and R. Kunze, *Linear Algebra*, Prentice-Hall, 1960.

88. J. M. Holtzman, *Nonlinear System Theory: a Functional Analysis Approach*, Prentice-Hall, 1970 (*T-AC*, August 1972, p. 584).

89. I. Horowitz, *Synthesis of Feedback Systems*, Academic Press, 1963.

90. A. S. Householder, *The Theory of Matrices in Numerical Analysis*, Blaisdell, 1964.

91. M. D. Intriligater, *Mathematical Optimization and Economic Theory*, 1971 (*T-AC*, December 1972, p. 846).

92. R. Isaacs, *Differential Games*, Wiley, 1965 (*T-AC*, October 1965, p. 501).

93. E. Isaacson and H. B. Keller, *Analysis of Numerical Methods*, Wiley, 1966.

94. D. H. Jacobson and D. Q. Mayne, *Differential Dynamic Programming*, American Elsevier, 1970 (*T-AC*, August 1971, p. 389).

95. A. H. Jazwinski, *Stochastic Processes and Filtering Theory*, Academic Press, 1970 (*T-AC*, October 1972, p. 752).

96. G. M. Jenkins and D. G. Watts, *Spectral Analysis and its Applications*, Holden-Day, 1968.

97. E. I. Jury, *Inners and Stability of Dynamic Systems*, Wiley, 1974 (*T-AC*, October 1976, p. 809).

98. R. E. Kalman, P. L. Falb, and M. A. Arbib, *Topics in Mathematical System Theory*, McGraw-Hill, 1969 (*T-AC*, February 1972, p. 181).

99. W. Kaplan, *Operational Methods for Linear Systems*, Addison-Wesley, 1962.

100. A. Kaufmann, *Introduction to the Theory of Fuzzy Subsets*, Academic Press, 1975.

101. D. E. Kirk, *Optimal Control Theory: an Introduction*, Prentice-Hall, 1970 (*T-AC*, June 1972, p. 423).

102. B. C. Kuo, *Discrete-Data Control Systems*, Prentice-Hall, 1970 (*T-AC*, June 1972, p. 418).

103. H. J. Kushner, *Stochastic Stability and Control*, Academic Press, 1967.

104. H. J. Kushner, *Introduction to Stochastic Control*, Holt, Rinehart, and Winston, 1971 (*T-AC*, December 1972, p. 851).

105. H. Kwakernaak and R. Sivan, *Linear Optimal Control Theory*, Wiley, 1972 (*T-AC*, October 1974, p. 631).

105a. D. G. Lainiotis (ed.), *Estimation Theory*, Elsevier, 1974 (*T-AC*, February 1977, p. 152).

106. R. E. Larson, *State Increment Dynamic Programming*, Elsevier, 1968 (*T-AC*, February 1970, p. 162).

106a. J. P. LaSalle, *The Stability of Dynamical Systems*, SIAM Publications, 1976.

107. J. P. LaSalle and S. Lefschetz, *Stability by Liapunov's Direct Method with Applications*, Academic Press, 1961.

108. E. B. Lee and L. Markus, *Foundations of Optimal Control Theory*, Wiley, 1967 (*T-AC*, April 1968, p. 222).

109. S. Lefschetz, *Differential Equations: Geometric Theory*, Interscience, 1957.

110. G. Leitmann, editor, *Optimization Techniques with Applications to Aerospace Systems*, Academic Press, 1962.

111. G. Leitmann, *An Introduction to Optimal Control*, McGraw-Hill, 1966 (*T-AC*, December 1967, p. 802).

112. G. Leitmann (ed.), *Topics in Optimization*, Academic Press, 1967 (*T-AC*, August 1969, p. 437).

113. C. T. Leondes (ed.), *Advances in Control Systems: Theory and Applications, Volumes 1-12*, Academic Press, 1964-1976.

114. W. R. LePage, *Complex Variables and the Laplace Transform for Engineers*, McGraw-Hill, 1961.

115. J. L. Lions, *Optimal Control of Systems Governed by Partial Differential Equations*, translated by S. K. Mitter, Springer-Verlag, 1971 (*T-AC*, August 1972, p. 586).

116. J. L. Lions, *Some Aspects of the Optimal Control of Distributed Parameter Systems*, SIAM Publications, 1971.

117. D. G. Luenberger, *Optimization by Vector Space Methods*, Wiley, 1969 (*T-AC*, February 1970, p. 160).

118. S. MacLane and G. Birkhoff, *Algebra*, MacMillan, 1967.

119. E. G. Manes (ed.), *Category Theory Applied to Computation and Control*, Springer-Verlag, 1975.

120. M. Marcus and H. Minc, *A Survey of Matrix Theory and Matrix Inequalities*, Allyn and Bacon, 1964.

121. M. Marcus and H. Minc, *Introduction to Linear Algebra*, MacMillan, 1965.

122. D. Q. Mayne and R. W. Brockett (eds.), *Geometric Methods in System Theory*, Reidel, 1973.

123. O. Mayr, *The Origins of Feedback Control*, M. I. T. Press, 1970 (*T-AC*, April 1972, p. 283). (Also see article in *Scientific American, vol. 223*, 1970, pp. 110-118.)

124. L. McColl, *Fundamental Theory of Servomechanisms*, Van Nostrand, 1945.

125. D. D. McCracken and W. S. Dorn, *Numerical Methods and Fortran Programming*, Wiley, 1964.

126. T. McGarty, *Stochastic Systems and State Estimation*, Wiley, 1974.

127. J. S. Meditch, *Stochastic Optimal Linear Estimation and Control*, McGraw-Hill, 1969 (*T-AC*, June 1972, p. 423).

127a. R. K. Mehra and D. G. Lainiotis (eds.), *System Identification: Advances and Case Studies*, Academic Press, 1976.

128. J. M. Mendel, *Discrete Techniques of Parameter Estimation*, Dekker, 1973.

129. J. M. Mendel and K. S. Fu (eds.), *Adaptive, Learning, and Pattern Recognition Systems: Theory and Applications*, Academic Press, 1970 (*T-AC*, April 1972, p. 278).

130. M. D. Mesarovic, D. Macko and Y. Takahara, *Theory of Hierarchical Multilevel Systems*, Academic Press, 1970 (*T-AC*, April 1972, p. 280).

131. H. T. Milhorn, *The Application of Control Theory to Physiological Systems*, Saunders, 1966.

132. J. H. Milsum, *Biological Control Systems Analysis*, McGraw-Hill, 1966 (*T-AC*, June 1967, p. 347).

133. N. Minorsky, *Theory of Nonlinear Control Systems*, McGraw-Hill, 1969 (*T-AC*, October 1971, p. 524).

134. R. R. Mohler, *Bilinear Control Processes, with Applications to Engineering, Ecology, and Medicine*, Academic Press, 1973.

135. R. R. Mohler and A. Ruberti (eds.), *Theory and Applications of Variable Structure Systems*, Academic Press, 1972.

136. K. S. Narendra and J. H. Taylor, *Frequency Domain Criteria for Absolute Stability*, Academic Press, 1973 (*T-AC*, August 1975, p. 583).

137. G. C. Newton, Jr., L. A. Gould, and J. Kaiser, *Analytic Design of Linear Feedback Controls*, Wiley, 1957.

137a. B. Noble, *Applied Linear Algebra*, Prentice-Hall, 1969.

138. K. Ogata, *State-Space Analysis of Control Systems*, Prentice-Hall, 1967 (*T-AC*, December 1967, p. 805).

139. K. Ogata, *Modern Control Engineering*, Prentice-Hall, 1970 (*T-AC*, June 1972, p. 419).

140. L. Padulo and M. A. Arbib, *System Theory: A Unified State-Space Approach to Discrete and Continuous Systems*, Saunders, 1974.

141. A. Papoulis, *The Fourier Integral and its Applications*, McGraw-Hill, 1962.

142. M. C. Pease, *Methods of Matrix Algebra*, Academic Press, 1965.

143. G. A. Phillipson, *Identification of Distributed Systems*, American Elsevier, 1971 (*T-AC*, August 1972, p. 586).

144. L. S. Pontryagin, et.al., *The Mathematical Theory of Optimal Processes*, Interscience, 1962.

145. J. R. Ragazzini and G. F. Franklin, *Sampled-Data Control Systems*, McGraw-Hill, 1958.

146. A. Ralston, *A First Course in Numerical Analysis*, McGraw-Hill 1965.

147. W. H. Ray and J. Szekely, *Process Optimization*, Wiley, 1973.

148. D. S. Riggs, *Control Theory and Physiological Feedback Mechanisms*, Williams and Wilkins, 1970.

149. H. H. Rosenbrock, *State-Space and Multivariable Theory*, Nelson, 1970 (*T-AC*, August 1972, p. 583).

150. H. H. Rosenbrock, *Computer-Aided Control System Design*, Academic Press, 1974.

151. H. H. Rosenbrock and C. Storey, *Mathematics of Dynamical Systems*, Nelson, 1970.

152. E. O. Roxin, et.al. (eds.), *Differential Games and Control Theory*, Dekker, 1975.

153. W. Rudin, *Principles of Mathematical Analysis*, McGraw-Hill, 1964.

154. A. P. Sage, *Optimum System Controls*, Prentice-Hall, 1968 (*T-AC*, August 1969, p. 434).

155. A. P. Sage and J. L. Melsa, *Estimation Theory with Applications to Communications and Control*, McGraw-Hill, 1971 (*T-AC*, August 1972, p. 585).

156. A. P. Sage and J. L. Melsa, *System Identification*, Academic Press, 1971 (*T-AC*, February 1973, p. 85).

157. F. C. Schweppe, *Uncertain Dynamic Systems*, Prentice-Hall, 1973.

158. S. M. Shinners, *Modern Control System Theory and Application*, Addison-Wesley, 1972 (*T-AC*, June 1973, p. 328).

159. D. D. Siljak, *Nonlinear Systems*, Wiley, 1969.

160. D. L. Snyder, *The State-Variable Approach to Continuous Estimation*, M. I. T. Press, 1969.

161. D. L. Snyder, *Random Point Processes*, Wiley, 1975 (*T-AC*, October 1976, p. 808).

162. L. Stark, *Neurological Control Systems: Studies in Bioengineering* Plenum Press, 1968.

162a. D. Sworder, *Optimal Adaptive Control Systems*, Academic Press, New York, 1966.

163. Y. Takahashi, M. J. Rabins, and D. M. Auslander, *Control and Dynamic Systems*, Addison-Wesley, 1970.

164. G. J. Thaler and M. H. Pastel, *Analysis and Design of Nonlinear Feedback Control Systems*, McGraw-Hill, 1962.

165. L. K. Timothy and B. E. Bona, *State-Space Analysis*, McGraw-Hill 1968 (*T-AC*, August 1969, p. 436).

166. T. Tintner and J. K. Sengupta, *Stochastic Economics*, Academic Press, 1972 (*T-AC*, August 1975, p. 583).

167. R. Tomovic and M. Vukobratovic, *General Sensitivity Theory*, Elsevier, 1972 (*T-AC*, April 1973, p. 199).

168. J. G. Truxal, *Control System Synthesis*, McGraw-Hill, 1955.

169. Ya. Z. Tsypkin, *Adaptation and Learning in Automatic Systems*, Academic Press, 1971 (*T-AC*, April 1974, p.173).

170. H. W. Turnbull and A. C. Aitken, *An Introduction to the Theory of Canonical Matrices*, Dover, 1961.

171. H. L. Van Trees, *Detection, Estimation and Modulation*, Wiley, 1968.

172. P. P. Varaiya, *Notes on Optimization*, Van Nostrand Reinhold, 1972 (*T-AC*, October 1973, p. 565).

173. P. Vidal, *Nonlinear Sampled-Data Systems*, Gordon and Breach, 1969 (*T-AC*, February 1971, p. 114).

174. K. P. Vishwakarma, *Macro-Economic Regulation*, Rotterdam University Press, 1974 (*T-AC*, October 1975, p. 721).

175. J. Warga, *Optimal Control of Differential and Functional Equations*, Academic Press, 1972 (*T-AC*, April 1974, p. 174).

176. L. A. Weinberg, *Network Analysis and Synthesis*, McGraw-Hill, 1962.

177. D. V. Widder, *An Introduction to Transform Theory*, Academic Press, 1971.

178. N. Wiener, *Cybernetics, or Control and Communication in the Animal and the Machine*, M. I. T. Press, 1965 (originally published in 1948).

179. J. H. Wilkinson, *The Algebraic Eigenvalue Problem*, Clarendon Press, 1965.

180. J. C. Willems, *The Analysis of Feedback Systems*, M. I. T. Press, 1971 (*T-AC*, October 1972, p. 745).

181. J. L. Willems, *Stability Theory of Dynamical Systems*, Wiley, 1970 (*T-AC*, February 1973, p. 83).

182. D. A. Wismer (ed.), *Optimization Methods for Large-Scale Systems*, McGraw-Hill, 1971.

183. W. A. Wolovich, *Linear Multivariable Systems*, Springer-Verlag, 1974.

184. E. Wong, *Stochastic Processes in Information and Dynamical Systems*, McGraw-Hill, 1971 (*T-AC*, December 1972, p. 847).

184a. W. M. Wonham, *Linear Multivariable Control: A Geometric Approach*, Springer, 1974. (*T-AC*, December 1977, p. 1000)

185. L. A. Zadeh and C. A. Desoer, *Linear System Theory*, McGraw-Hill, 1963.

186. L. A. Zadeh, et.al. (eds.), *Fuzzy Sets and Their Applications to Cognitive and Decision Processes*, Academic Press, 1975.

Papers

201. *Automatica* (Collection of papers on sensitivity), *5*, May 1969, pp. 249-317.

202. *IEEE Trans. Auto. Control* (Special issue on the linear-quadratic-gaussian problem) *AC-16*, December 1971.

203. *IEEE Trans. Auto. Control* (Special issue on system identification and time-series analysis) *AC-19*, December 1974.

203a. *IEEE Trans. Auto. Control* (Special issue on decentralized control and large-scale systems) *AC-23*, February 1978, to appear.

203b. *IEEE Trans. Auto. Control* (Mini-issue on the F-8 aircraft) *AC-22*, October 1977, pp. 752-806.

204. *Proc. IEEE* and *J. Franklin Institute* (Joint special issue on recent trends in system theory), January 1976.

204a. *Proc. IEEE* (Special issue on adaptive systems), *64*, August 1976.

205. J. E. Ackermann and R. S. Bucy, Canonical minimal realization of a matrix of impulse response sequences, *Info. and Control*, *19*, 1971, pp. 224-231.

206. H. Akaike, A new look at the statistical model identification, *IEEE Trans. Auto. Control*, *AC-19*, December 1974, pp. 716-723.

207. B. D. O. Anderson, Linear multivariable control systems: a survey, *Proc. IFAC Fifth World Congress*, 1972.

208. B. D. O. Anderson, An approach to multivariable system identification, *Automatica*, *13*, July 1977, to appear.

209. B. D. O. Anderson, N. K. Bose, and E. I. Jury, Output feedback stabilization and related problems--solution via decision methods, *IEEE Trans. Auto. Control*, *AC-20*, February 1975, pp. 53-66.

210. B. D. O. Anderson and D. G. Luenberger, Design of multivariable feedback systems, *Proc. IEE, 114*, March 1967, pp. 395-399.

211. Anonymous, Optimal control: a mathematical supertool, *Business Week*, May 19, 1973, pp. 74-75.

212. M. Aoki, On feedback stabilizability of decentralized dynamic systems, *Automatica, 8*, March 1972, pp. 163-173.

212a. M. Aoki, Stochastic control in economic theory and economic systems, *IEEE Trans. Auto. Control, AC-21*, April 1976, pp. 213-220.

213. M. A. Arbib, Decomposition theory for automata and biological systems, *System Structure*, (A. S. Morse, ed.), IEEE Control Systems Society, 1971.

213a. M. A. Arbib, Trends and focus in control, *Automatica, 12*, November 1976, pp. 629-632.

214. M. A. Arbib and E. G. Manes, Machines in a category: an expository introduction, *SIAM Review, 16*, April 1974, pp. 163-192.

215. M. A. Arbib and E. G. Manes, Foundations of system theory: decomposable systems, *Automatica, 10*, May 1974, pp. 285-302.

216. M. A. Arbib and H. P. Zeiger, On the relevance of abstract algebra to control theory, *Automatica, 5*, September 1969, pp. 589-606.

217. K. J. Åström, et al., Theory and applications of adaptive regulators based on recursive parameter estimation, *Proc. IFAC Sixth World Congress*, 1975, *Automatica*, Sept. 1977.

218. K. J. Åström and P. Eykhoff, System identification--a survey, *Automatica, 7*, March 1971, pp. 123-162.

219. M. Athans, The status of optimal control theory and applications for deterministic systems, *IEEE Trans. Auto. Control, AC-11*, July 1966, pp. 580-596.

220. M. Athans, Survey of decentralized control methods, *Ann. Economic and Social Measurement, 4*, April 1975, pp. 345-355.

221. M. Athans and D. Kendrick, Control theory and economics: a survey, forecast and speculations, *IEEE Trans. Auto. Control, AC-19*, October 1974, pp. 518-524.

222. M. Athans, et al., The stochastic control of the F-8C aircraft using the multiple model adaptive control (MMAC) method, *Proc. IEEE Conference on Decision and Control*, December 1975, pp. 217-228.

222a. M. Athans, R. Ku, and S. B. Gershwin, The uncertainty threshold principle: some fundamental limitations of optimal decision making under dynamic uncertainty, *IEEE Trans. Auto. Control, AC-22*, June 1977, pp. 491-495.

223. G. S. Axelby, Round-table discussion on the relevance of control theory, *Automatica, 9*, March 1973, pp. 279-281.

224. A. V. Balakrishnan, Identification and adaptive control: an application to flight control systems, *J. Opt. Theory and Appl., 9*, March 1972, pp. 187-213.

225. A. V. Balakrishnan and V. Peterka, Identification in automatic control systems, *Automatica, 5*, November 1969, pp. 817-829.

226. H. T. Banks and A. Manitius, Application of abstract variational theory to hereditary systems--a survey, *IEEE Trans. Auto. Control, AC-19*, October 1974, pp. 524-533.

227. Y. Bar-Shalom and E. Tse, Dual effect, certainty equivalence, and separation in stochastic control, *IEEE Trans. Auto. Control, AC-19*, October 1974, pp. 494-500.

227a. S. Bennett, The emergence of a discipline: automatic control 1940-1960, *Automatica, 12*, March 1976, pp. 113-121.

228. D. P. Bertsekas and I. B. Rhodes, Recursive state estimation for a set-membership description of uncertainty, *IEEE Trans. Auto. Control, AC-16*, April 1971, pp. 117-128.

228a. F. J. Beutler, Recovery of randomly sampled signals by simple interpolators, *Info. and Control, 26*, December 1974, pp. 312-340.

228b. R. Boel and P. Varaiya, Optimal control of jump processes, *SIAM J. Control and Opt., 15*, January 1977, pp. 92-119.

228c. U. Borisson and R. Syding, Self-tuning control of an ore crusher, *Automatica, 12*, January 1976, pp. 1-7.

228d. L. W. Botsford, H. E. Rauch and R. A. Shleser, Optimal temperature control of a lobster plant, *IEEE Trans. Auto. Control, AC-19*, October 1974, pp. 541-543.

229. F. M. Brasch and J. B. Pearson, Pole placement using dynamic compensators, *IEEE Trans. Auto. Control, AC-15*, February 1970, pp. 34-43.

230. R. W. Brockett, The status of stability theory for deterministic systems, *IEEE Trans. Auto. Control, AC-11*, July 1966, pp. 596-606.

231. R. W. Brockett, Algebraic decomposition methods for nonlinear systems, *System Structure*, (A. S. Morse, ed.), IEEE Control Systems Society, 1971.

232. R. W. Brockett, Nonlinear systems and differential geometry, *Proc. IEEE, 64*, January 1976, pp. 61-72.

233. R. W. Brockett and H. B. Lee, Frequency-domain instability criteria for time-varying and nonlinear systems, *Proc. IEEE, 55*, May 1967, pp. 604-619.

234. P. Brunovsky, A classification of linear controllable systems, *Kybernetika, 3*, 1970, pp. 173-188.

235. R. S. Bucy, Canonical forms for multivariable systems, *IEEE Trans. Auto. Control, AC-13*, October 1968, pp. 567-569.

236. R. S. Bucy, Linear and nonlinear filtering, *Proc. IEEE, 58*, June 1970, pp. 854-864.

236a. P. C. Byrne, Comments on "On the Routh-Hurwitz criterion", *IEEE Trans. Auto. Control, AC-20*, February 1975, pp. 178-179.

237. I. B. Chammas, Least-squares design of orbit controller, M.S. thesis, Massachusetts Institute of Technology, May 1968.

237a. T.-S. Chang and C.-T. Chen, On the Routh-Hurwitz criterion, *IEEE Trans. Auto. Control, AC-19*, June 1974, pp. 250-251.

238. M. R. Chidambara, R. B. Broen, and J. Zaborszky, A simple algorithm for pole assignment in a multiple-input linear time-invariant dynamic system, *Trans. ASME: J. Dynamic Systems, Measurement and Control, 96*, March 1974, pp. 13-18.

238a. R. N. Clark, D. C. Fosth, and V. M. Walton, Detecting instrument malfunctions in control systems, *IEEE Trans. Aerospace and Electronic Systems, AES-11*, July 1975, pp. 465-473.

238b. J. P. Corfmat and A. S. Morse, Decentralized control of linear multivariable systems, *Automatica, 12*, September 1976, pp. 479-495.

238c. J. P. Corfmat and A. S. Morse, Control of linear systems through specified input channels, *SIAM J. Control and Opt., 14*, January 1976, pp. 163-175.

239. R. F. Curtain, A survey of infinite-dimensional filtering, *SIAM Review, 17*, July 1975, pp. 395-411.

240. E. J. Davison, The systematic design of control systems for large multivariable linear time-invariant systems, *Automatica, 9*, July 1973, pp. 441-452.

241. E. J. Davison, The robust decentralized control of a general servomechanism problem, *IEEE Trans. Auto. Control, AC-21*, February 1976, pp. 14-24.

242. E. J. Davison, The robust control of a servomechanism problem for linear time-invariant multivariable systems, *IEEE Trans. Auto. Control, AC-21*, February 1976, pp. 25-34.

243. E. J. Davison, Multivariable tuning regulators: the feedforward and robust control of a general servomechanism problem, *IEEE Trans. Auto. Control, AC-21*, February 1976, pp. 35-47.

243a. E. J. Davison, Connectability and structural controllability of composite systems, *Automatica, 13*, March 1977, pp. 109-123.

244. E. J. Davison and S.-H. Wang, Properties and calculation of transmission zeros of linear multivariable systems, *Automatica, 10*, December 1974, pp. 643-658.

245. E. J. Davison and S.-H. Wang, New results on the controllability and observability of general composite systems, *IEEE Trans. Auto. Control, AC-20*, February 1975, pp. 123-128.

246. E. J. Davison and S.-H. Wang, On pole assignment in linear multivariable systems using output feedback, *IEEE Trans. Auto. Control, AC-20*, August 1975, pp. 516-518.

247. M. C. Delfour, C. McCalla and S. K. Mitter, Stability and the infinite-time quadratic cost problem for linear hereditary differential systems, *SIAM J. Control, 13*, January 1975, pp. 48-88.

248. M. M. Denn, Chemical reaction engineering: optimization, control, and stability, *Ann. Reviews of Industrial and Engineering Chemistry, 1970*, (V. W. Weekman, Jr., ed.), Amer. Chem. Soc., 1972, pp. 251-262.

249. C. A. Desoer, Slowly varying system $\dot{x} = A(t)x$, *IEEE Trans. Auto. Control, AC-14*, December 1969, pp. 780-781.

250. C. A. Desoer and J. D. Schulman, Zeros and poles of matrix transfer functions and their dynamical interpretation, *IEEE Trans. Circuits and Systems, CAS-21*, January 1974, pp. 3-8.

250a. B. W. Dickinson, The classification of linear state variable control laws, *SIAM J. Control and Opt., 14*, May 1976, pp. 467-477.

251. C. Y. Ding, F. M. Brasch, Jr., and J. B. Pearson, On multivariable linear systems, *IEEE Trans. Auto. Control, AC-15*, February 1970, pp. 96-97.

251a. S. Dolecki and D. L. Russell, A general theory of observation and control, *SIAM J. Control and Opt., 15*, February 1977, pp. 185-220.

252. P. Dorato and A. H. Levis, Optimal linear regulators: the discrete-time case, *IEEE Trans. Auto. Control, AC-16*, December 1971, pp. 613-620.

252a. R. F. Drenick, Feedback control of partly known systems, *SIAM J. Control and Opt., 15*, May 1977, pp. 506-519.

252b. E. Emre, L. M. Silverman, and K. Glover, Generalized dynamic covers for linear systems with applications to deterministic identification and realization problems, *IEEE Trans. Auto. Control, AC-22*, February 1977, pp. 26-35.

253. P. L. Falb and W. A. Wolovich, Decoupling in the design and synthesis of multivariable systems, *IEEE Trans. Auto. Control, AC-12*, December 1967, pp. 651-659.

253a. R. E. Fenton, G. C. Melocik, and K. W. Olson, On the steering of automated vehicles: theory and experiment, *IEEE Trans. Auto. Control, AC-21*, June 1976, pp. 306-315.

254. J. W. Forrester, Dynamics of socioeconomic systems, *Proc. IFAC Sixth World Congress*, 1975.

255. T. E. Fortmann and D. Williamson, Design of minimal-order observers for linear feedback control laws, *IEEE Trans. Auto. Control, AC-17*, June 1972, pp. 301-308.

256. A. S. Foss, Critique of chemical process control theory, *IEEE Trans. Auto. Control, AC-18*, December 1973, pp. 646-652.

256a. B. A. Francis, The linear multivariable regulator problem, *SIAM J. Control and Opt., 15*, May 1977, pp. 486-505.

256b. B. A. Francis and W. M. Wonham, The internal model principle of control theory, *Automatica, 12*, September 1976, pp. 457-465.

257. A. Gelb and R. S. Warren, Direct statistical analysis of non-linear systems: CADET, *AIAA Journal, 11*, May 1973, pp. 689-694.

258. E. Gilbert, The decoupling of multivariable systems by state feedback, *SIAM J. Control, 7*, February 1969, pp. 50-63.

258a. C. L. Golliday, Jr. and H. Hemami, Postural stability of the two-degree-of-freedom biped by general linear feedback, *IEEE Trans. Auto. Control, AC-21*, February 1976, pp. 74-79.

259. B. Gopinath, On the control of linear multiple input-output systems, *Bell System Tech. J., 50*, March 1971, pp. 1063-1081.

260. R. Guidorzi, Canonical structures in the identification of multivariable systems, *Automatica, 11*, July 1975, pp. 361-374.

261. M. M. Gupta and E. H. Mamdani, Second IFAC round table on fuzzy automata and decision processes, *Automatica, 12*, May 1976, pp. 291-296.

262. I. Gustavsson, Survey of application of identification in chemical and physical processes, *Automatica, 11*, January 1975, pp. 3-24.

262a. I. Gustavsson, L. Ljung, and T. Soderstrom, Identification of processes in closed loop: identifiability and accuracy aspects, *Automatica, 13*, January 1977, pp. 59-75.

262b. R. Hermann and C. F. Martin, Applications of algebraic geometry to systems theory - Part 1, *IEEE Trans. Auto. Control, AC-22*, February 1977, pp. 19-25.

263. M. Heymann, Comments on "Pole assignment in multi-input controllable linear systems", *IEEE Trans. Auto. Control, AC-13*, December 1968, pp. 748-749.

264. M. Heymann, The prime structure of linear dynamical systems, *SIAM J. Control, 10*, August 1972, pp. 460-469.

264a. M. Heymann, Controllability subspaces and feedback simulation, *SIAM J. Control and Opt., 14*, July 1976, pp. 769-789.

265. B. L. Ho and R. E. Kalman, Effective contruction of linear state-variable models from input/output functions, *Proc. Third Allerton Conf.*, 1965, pp. 449-459, also in *Regelungstechnik, 14*, 1966, pp. 545-548.

266. Y.-C. Ho and K. C. Chu, Information structure in dynamic multi-person control problems, *Automatica, 10*, July 1974, pp. 341-351.

267. I. M. Horowitz and U. Shaked, Superiority of transfer function over state-variable methods in linear time-invariant feedback system design, *IEEE Trans. Auto. Control, AC-20*, February 1975, pp. 84-97.

268. M. Ikeda, H. Maeda, and S. Kodama, Estimation and feedback in linear time-varying systems: a deterministic theory, *SIAM J. Control, 13*, February 1975, pp. 304-326.

269. R. Isaacs, Differential games: their scope, nature, and future, *J. Opt. Theory and Appl., 3*, May 1969, pp. 283-295.

270. L. Isaksen and H. J. Payne, Suboptimal control of linear systems by augmentation with application to freeway traffic regulation, *IEEE Trans. Auto. Control, AC-18*, June 1973, pp. 210-219.

271. R. Isermann, et al., Comparison of six on-line identification and parameter estimation methods, *Automatica, 10*, January 1974, pp. 81-103.

272. R. A. Jarvis, Optimization strategies in adaptive control: a selective survey, *IEEE Trans. Systems, Man, and Cybernetics, SMC-5*, January 1975, pp. 83-94.

273. C. D. Johnson, Accommodation of external disturbances in linear regulator and servomechanism problems, *IEEE Trans. Auto. Control, AC-16*, December 1971, pp. 635-644.

274. T. L. Johnson and M. Athans, On the design of optimal constrained dynamic compensators for linear constant systems, *IEEE Trans. Auto. Control, AC-15*, December 1970, pp. 658-660.

275. E. I. Jury, The theory and applications of the inners, *Proc. IEEE, 63*, July 1975, pp. 1044-1068.

276. E. I. Jury and Ya. Z. Tsypkin, On the theory of discrete systems, *Automatica, 7*, 1971, pp. 89-107.

277. T. Kailath, A view of three decades of linear filtering theory, *IEEE Trans. Info. Theory, IT-20*, March 1974, pp. 145-181.

278. R. E. Kalman, Mathematical description of linear dynamical systems, *SIAM J. Control, 1*, 2, 1963, pp. 152-192.

279. R. E. Kalman, A new approach to linear filtering and prediction problems, *Trans. ASME: J. Basic Eng., 82*, March 1960, pp. 35-45.

280. R. E. Kalman and R. S. Bucy, New results in filtering and prediction theory, *Trans. ASME: J. Basic Eng., 83*, March 1961, pp. 95-108.

280a. W. J. M. Kickert and H. R. Van Nauta Lemke, Application of a fuzzy controller in a warm water plant, *Automatica, 12*, July 1976, pp. 301-308.

281. D. L. Kleinman, An easy way to stabilize a linear constant system, *IEEE Trans. Auto. Control, AC-15*, December 1970, p. 692.

282. D. L. Kleinman, Stabilizing a discrete, constant, linear system with application to iterative methods for solving the Riccati equation, *IEEE Trans. Auto. Control, AC-19*, June 1974, pp. 252-254.

283. D. L. Kleinman, S. Baron, and W. Levison, A control-theoretic approach to manned-vehicle systems analysis, *IEEE Trans. Auto. Control, AC-16*, December 1971, pp. 824-832.

283a. G. Kreisselmeier, Adaptive observers with exponential rate of convergence, *IEEE Trans. Auto. Control, AC-22*, February 1977, pp. 2-8.

284. I. D. Landau, Model-reference adaptive systems--a survey, *Trans. ASME: J. Dynamic Systems, Measurement, and Control, 94*, June 1972, pp. 119-132.

285. I. D. Landau, A survey of model-reference adaptive techniques--theory and applications, *Automatica, 10*, July 1974, pp. 353-379.

286. A. H. Levis and M. Athans, On the optimal sampled data control of strings of vehicles, *Transportation Science, 2*, November 1968, pp. 362-382.

287. C.-T. Lin, Structural controllability, *IEEE Trans. Auto. Control, AC-19*, June 1974, pp. 201-208.

288. J. T.-H. Lo and A. S. Willsky, Estimation for rotational processes with one degree of freedom--Parts I, II, and III, *IEEE Trans. Auto. Control, AC-20*, February 1975, pp. 10-33.

289. G. Lüders and K. S. Narendra, Stable adaptive schemes for state estimation and identification of linear systems, *IEEE Trans. Auto. Control, AC-19*, December 1974, pp. 841-847.

290. D. G. Luenberger, Observing the state of a linear system, *IEEE Trans. Military Electronics, MIL-8*, April 1964, pp. 74-80.

291. D. G. Luenberger, Observers for multivariable systems, *IEEE Trans. Auto. Control, AC-11*, April 1966, pp. 190-197.

292. D. G. Luenberger, Canonical forms for linear multivariable systems, *IEEE Trans. Auto. Control, AC-12*, June 1967, pp. 290-293.

293. D. G. Luenberger, An introduction to observers, *IEEE Trans. Auto. Control, AC-16*, December 1971, pp. 596-602.

294. D. G. Luenberger, A nonlinear economic control problem with a linear feedback solution, *IEEE Trans. Auto. Control, AC-20*, April 1975, pp. 184-191.

295. D. L. Lukes and D. L. Russell, The quadratic criterion for distributed systems, *SIAM J. Control, 7*, February 1969, pp. 101-121.

296. A. G. J. MacFarlane, Multivariable-control-system design techniques: a guided tour, *Proc. IEE, 117*, May 1970, pp. 1039-1047.

297. A. G. J. MacFarlane, A survey of some recent results in linear multivariable feedback theory, *Automatica, 8*, July 1972, pp. 455-492.

298. A. G. J. MacFarlane, Report on roundtable discussion: trends in linear multivariable theory, *Automatica, 9*, March 1973, pp. 273-277.

299. D. Q. Mayne, Computational procedure for the minimal realization of transfer function matrices, *Proc. IEE, 115*, September 1968, pp. 1363-1368.

300. A. I. Mees and A. R. Bergen, Describing functions revisited, *IEEE Trans. Auto. Control, AC-20*, August 1975, pp. 473-478.

301. R. K. Mehra, Optimal input signals for parameter estimation in dynamic systems--survey and new results, *IEEE Trans. Auto. Control, AC-19*, December 1974, pp. 753-768.

302. J. M. Mendel and D. L. Gieseking, Bibliography on the linear-quadratic-gaussian problem, *IEEE Trans. Auto. Control, AC-16*, December 1971, pp. 847-869.

303. R. A. Miller, Specific optimal control of the linear regulator using a minimal order observer, *Int. J. Control, 18*, July 1973, pp. 139-159.

304. S. K. Mitter, Optimal control of distributed parameter systems, *Proc. Joint Automatic Control Conference*, 1969, pp. 13-48.

305. B. C. Moore and L. M. Silverman, Model-matching by state feedback and dynamic compensation, *IEEE Trans. Auto. Control, AC-17*, August 1972, pp. 491-497.

306. R. C. Montgomery and D. B. Price, Management of optimal redundancy in digital flight control systems for aircraft, *Proc. AIAA Mechanics and Control of Flight Conference*, August 1974.

307. A. S. Morse, Structure and design of linear model-following systems, *IEEE Trans. Auto. Control, AC-18*, August 1973, pp. 346-354.

308. A. S. Morse, Structural invariants of linear multivariable systems, *SIAM J. Control, 11*, August 1973, pp. 446-465.

309. A. S. Morse and L. M. Silverman, Structure of index-invariant systems, *SIAM J. Control, 11*, May 1973, pp. 215-225.

310. A. S. Morse and W. M. Wonham, Status of noninteracting control, *IEEE Trans. Auto. Control, AC-16*, December 1971, pp. 568-581.

311. A. S. Morse and W. M. Wonham, Decoupling and pole assignment by dynamic compensation, *SIAM J. Control, 8*, August 1970, pp. 317-337.

312. R. E. Nieman, D. G. Fisher, and D. E. Seborg, A review of process identification and parameter estimation techniques, *Int. J. Control, 13*, 1971, pp. 209-264.

313. U. Passy and L. Silman, One-at-a-time dynamic programming over polytope, *Automatica, 11*, September 1975, pp. 499-508.

314. R. V. Patel, On zeros of multivariable systems, *Int. J. Control, 21*, April 1975, pp. 599-608.

315. H. J. Payne, W. A. Thompson and L. Isaksen, Design of a traffic-responsive control system for a Los Angeles freeway, *IEEE Trans. Systems, Man, and Cybernetics, SMC-3*, May 1973, pp. 213-224.

316. L. K. Platzman and T. L. Johnson, A linear-quadratic-gaussian control problem with innovations feedthrough solution, *IEEE Trans. Auto. Control, AC-21*, October 1976, pp. 721-725.

317. V. M. Popov, Hyperstability and optimality of automatic systems with several control functions, *Rev. Roum. Sci.-Electrotechn. et Energ., 9*, 1964, pp. 629-690.

318. V. M. Popov, Invariant description of linear, time-invariant controllable systems, *SIAM J. Control, 10*, May 1972, pp. 252-264.

319. V. M. Popov, Dichotomy and stability by frequency-domain methods, *Proc. IEEE, 62*, May 1974, pp. 548-562.

320. B. V. Raja Rao, A survey of automatic control in the glass industry: 1961-1973, *Automatica, 11*, January 1975, pp. 37-52.

321. I. B. Rhodes, A tutorial introduction to estimation and filtering, *IEEE Trans. Auto. Control, AC-16*, December 1971, pp. 688-706.

322. J. Rissanen, Recursive identification of linear systems, *SIAM J. Control, 9*, August 1971, pp. 420-430.

323. A. C. Robinson, A survey of optimal control of distributed-parameter systems, *Automatica, 7*, May 1971, pp. 371-387.

324. J. R. Roman and T. E. Bullock, Design of minimal order stable observers for linear functions of the state via realization theory, *IEEE Trans. Auto. Control, AC-20*, October 1975, pp. 613-622.

325. W. L. Root, On the modeling of systems for identification, Parts I and II, *SIAM J. Control, 4*, July 1975, pp. 927-974.

326. H. H. Rosenbrock, The stability of linear time-dependent control systems, *J. Electronics and Control, 15*, July 1963, pp. 73-80.

327. H. H. Rosenbrock, The future of control, *Proc. IFAC Sixth World Congress*, 1975, *Automatica*, July 1977.

327a. M. G. Safonov and M. Athans, Gain and phase margin for multiloop LQG regulators, *IEEE Trans. Auto. Control, AC-22*, April 1977, pp. 173-179.

328. N. R. Sandell, Jr., On Newton's method for Riccati equation solution, *IEEE Trans. Auto. Control, AC-19*, June 1974, pp. 254-255.

329. N. R. Sandell, Jr. and M. Athans, On "type ℓ" multivariable linear systems, *Automatica, 9*, January 1973, pp. 131-136.

330. N. R. Sandell, Jr. and M. Athans, A finite-state, finite-memory minimum principle, *J. Opt. Theory and Appl.*, to appear.

331. N. R. Sandell, Jr., P. P. Varaiya, and M. Athans, A survey of decentralized control methods for large-scale systems, *Proc. Engineering Foundation Conference on Systems Engineering for Power*, August 1975.

332. G. N. Saridis, Comparison of six on-line identification algorithms, *Automatica, 10*, January 1974, pp. 69-79.

332a. O. A. Sebakhy and W. M. Wonham, A design procedure for multivariable regulators, *Automatica, 12*, September 1976, pp. 467-478.

333. A. Segall, M. H. A. Davis, and T. Kailath, Nonlinear filtering with counting observations, *IEEE Trans. Info. Theory, IT-21*, March 1975, pp. 143-149.

334. A. Segall and T. Kailath, The modeling of randomly modulated jump processes, *IEEE Trans. Info. Theory, IT-21*, March 1975, pp. 135-143.

334a. R. W. Shields and J. B. Pearson, Structural controllability of multiinput linear systems, *IEEE Trans. Auto. Control, AC-21*, April 1976, pp. 203-212.

335. L. M. Silverman, Inversion of multivariable linear systems, *IEEE Trans. Auto. Control, AC-14*, June 1969, pp. 270-276.

336. L. M. Silverman, Realization of linear dynamical systems, *IEEE Trans. Auto. Control, AC-16*, December 1971, pp. 554-567.

337. L. M. Silverman and H. J. Payne, Input-output structure of linear systems with application to the decoupling problem, *SIAM J. Control, 9*, May 1971, pp. 199-233.

338. M. G. Singh, S. A. W. Drew and J. F. Coales, Comparisons of practical hierarchical control methods for interconnected dynamical systems, *Automatica, 11*, July 1975, pp. 331-350.

339. H. W. Smith and E. J. Davison, Design of industrial regulators: integral feedback and feedforward control, *Proc. IEE, 119*, August 1972, pp. 1210-1216.

340. H. W. Sorenson, Least-squares estimation: from Gauss to Kalman, *IEEE Spectrum, 7*, July 1970, pp. 63-68.

341. H. W. Sorenson, Estimation for dynamic systems: a perspective, *Proc. Fourth Symp. Nonlinear Estimation Theory and Applns.*, 1973, pp. 291-318.

342. J. Van Amerongen and A. J. Udink Ten Cate, Model reference adaptive autopilots for ships, *Automatica, 11*, September 1975, pp. 441-449.

343. P. P. Varaiya, The martingale theory of jump processes, *IEEE Trans. Auto. Control, AC-20*, February 1975, pp. 34-42.

343a. P. P. Varaiya, On the design of rent control, *IEEE Trans. Auto. Control, AC-21*, June 1976, pp. 316-319.

344. P. P. Varaiya, Trends in the theory of decision-making in large systems, *Ann. Economic and Social Measurement, 1*, October 1972, pp. 493-500.

345. M. Vidyasagar, A characterization of e^{At} and a constructive proof of the controllability criterion, *IEEE Trans. Auto. Control, AC-16*, August 1971, pp. 370-371.

346. T. L. Vincent, E. M. Cliff, and B.-S. Goh, Optimal direct control of prey-predator systems, *Trans. ASME: J. Dynamic Systems, Measurement and Control, 96*, March 1974, pp. 71-76.

347. M. K. Vukobratovic and D. E. Okhocimskii, Control of legged locomotion robots, *Proc. IFAC Sixth World Congress*, 1975.

348. P. K. C. Wang, Control of distributed parameter systems, *Advances in Control Systems, 1*, (C. T. Leondes, ed.) Academic Press, 1964.

349. S.-H. Wang and E. J. Davison, A minimization algorithm for the design of linear multivariable systems, *IEEE Trans. Auto. Control, AC-18*, June 1973, pp. 220-225.

350. S.-H. Wang and E. J. Davison, The stabilization of decentralized control systems, *IEEE Trans. Auto. Control, AC-18*, October 1973, pp. 473-478.

350a. S.-H. Wang and E. J. Davison, Canonical forms of linear multivariable systems, *SIAM J. Control, 14*, February 1976, pp. 236-250.

351. L. Weiss, Controllability, realization, and stability of discrete-time systems, *SIAM J. Control, 10*, May 1972, pp. 230-251.

352. J. C. Willems and S. K. Mitter, Controllability, observability, pole allocation, and state reconstruction, *IEEE Trans. Auto. Control, AC-16*, Dec. 1971, pp. 582-595.

353. A. S. Willsky, A survey of design methods for failure detection in dynamic systems, *Automatica, 12*, November 1976, pp. 601-611.

354. A. S. Willsky and H. L. Jones, A generalized likelihood ratio approach to state estimation in linear systems subject to abrupt changes, *IEEE Trans. Auto. Control, AC-21*, February 1976, pp. 108-112.

355. H. S. Witsenhausen, A counterexample in stochastic optimum control, *SIAM J. Control, 6*, February 1968, pp. 131-147.

356. H. S. Witsenhausen, On information structures, feedback, and causality, *SIAM J. Control, 9*, May 1971, pp. 149-160.

357. H. S. Witsenhausen, Separation of estimation and control for discrete-time systems, *Proc. IEEE, 59*, November 1971, pp. 1557-1566.

357a. B. Wittenmark, Stochastic adaptive control methods: a survey, *Int. J. of Control, 21*, May 1975, pp. 705-730.

357b. C. A. Wolfe and J. S. Meditch, Theory of system type for linear multivariable servomechanisms, *IEEE Trans. Auto. Control, AC-22*, February 1977, pp. 36-46.

358. W. A. Wolovich, The use of state feedback for exact model matching, *SIAM J. Control, 10*, August 1972, pp. 512-523.

359. W. A. Wolovich, Static decoupling, *IEEE Trans. Auto. Control, AC-18*, October 1973, pp. 536-537.

360. W. A. Wolovich, On determining the zeros of state-space systems, *IEEE Trans. Auto. Control, AC-18*, October 1973, pp. 542-544.

361. W. A. Wolovich, On the numerators and zeros of rational transfer matrices, *IEEE Trans. Auto. Control, AC-18*, October 1973, pp. 544-546.

362. W. A. Wolovich, Multivariable system synthesis with step disturbance rejection, *IEEE Trans. Auto. Control, AC-19*, April 1974, pp. 127-130.

363. W. A. Wolovich and P. L. Falb, On the structure of multivariable systems, *SIAM J. Control, 7*, August 1969, pp. 437-451.

364. E. Wong, Recent progress in stochastic processes--a survey, *IEEE Trans. Info. Theory, IT-19*, May 1973, pp. 262-275.

365. W. M. Wonham, On pole assignment in multi-input controllable linear systems, *IEEE Trans. Auto. Control, AC-12*, December 1967, pp. 660-665.

366. W. M. Wonham, Optimal stochastic control, *Automatica, 5*, January 1969, pp. 113-117.

367. W. M. Wonham, Random differential equations in control theory, *Probabilistic Methods in Applied Mathematics, Vol. 2* (A. T. Bharucha-Reid, ed.), Academic Press, 1970, pp. 131-212.

368. W. M. Wonham, Algebraic methods in linear multivariable control, *System Structure*, (A. S. Morse, ed.), IEEE Control Systems Society, 1971.

369. W. M. Wonham and A. S. Morse, Decoupling and pole assignment in linear multivariable systems: a geometric approach, *SIAM J. Control, 8*, February 1970, pp. 1-18.

370. P. C. Young and J. C. Willems, An approach to the linear multivariable servomechanism problem, *Int. J. Control, 15*, 1972, pp. 961-979.

371. L. A. Zadeh, Outline of a new approach to the analysis of complex systems and decision processes, *IEEE Trans. Systems, Man, and Cybernetics, SMC-3*, January 1973, pp. 28-44.

372. G. Zames, On the input-output stability of time-varying nonlinear feedback systems, Parts I and II, *IEEE Trans. Auto. Control, AC-11*, April and July 1966, pp. 228-238 and 465-476.

INDEX

The letters n and p following page numbers indicate footnotes and problems, respectively.